W9-AAV-518

DISCARD

Nature Revealed

Edward O. Wilson

Nature Revealed

Selected Writings,
1949–2006

The Johns Hopkins University Press
Baltimore

Printed in the United States of America on acid-free paper
9 8 7 6 5 4 3 2 1

The Johns Hopkins University Press
2715 North Charles Street
Baltimore, Maryland 21218-4363

www.press.jhu.edu

Library of Congress Cataloging-in-Publication Data

Wilson, Edward O.
Nature revealed : selected writings, 1949–2006 / Edward O. Wilson.
p. cm.
Includes index.
ISBN 0-8018-8329-6 (hardcover : alk. paper)
1. Ants. 2. Insect societies. 3. Biological diversity. 4. Biological diversity conservation.
I. Title.
QL568.F7W63 2006
570—dc22 2005023389

A catalog record for this book is available from the British Library.

Contents

Preface

It was my great good fortune to participate in biology throughout the second half of the twentieth century, thereby witnessing close at hand its emergence as a premier science equal to physics and chemistry. I began as a boy who turned a fascination with insects into a determination to become a professional entomologist. My dream was not visionary. What I truly hoped for was to stay outdoors as long as possible to luxuriate in the pleasures of natural history.

During my teens, spent in Alabama and northern Florida, I waffled a bit. I had a butterfly and ant period, then a snake period, and next an excursion into birds and freshwater fishes. Finally, at the age of sixteen, I decided that to succeed in life I had better pick a group of insects on which to be a world authority. Such are the pleasures of growing up in an intellectual vacuum, with no competitors and no sense of limitation on my own potential. I chose flies. Flies are important! There are myriads of species, including the tiny jewel-like dolichopodids that dance everywhere on sunlit leaves. I soon discovered, however, that to make collections of flies it is necessary to use the specialized insect pins that can be purchased from scientific supply houses. But the pins were not available because of shortages caused by World War II. So I turned back to ants, which can be preserved and studied in local pharmacy prescription bottles filled with rubbing (isopropyl) alcohol.

In 1949, while a senior at the University of Alabama, I published my first article on ants, a brief report on the spread of the red imported fire ant. Thereafter, through graduate studies and a teaching career at Harvard University in the decades that followed, ants retained their fascination for me. Like so many naturalists before me, I found that ants are a magical well; as Karl von Frisch said of his beloved honeybees: the more you draw from them, the more there is to draw.

From the study of ants, and all their diversity, I was able to venture away from entomology into other subdisciplines of biology, including systematics, biogeography, animal communication, sociobiology, and environmental science. I followed instinctively one of the guiding principles for the conduct of biological research: for every organism there exist scientific problems the solutions to which that organism is ideally suited. Colon bacteria serve well for molecular genetics, fruitflies for the genetics of developmental programs, and nematode worms for the mapping of nervous systems. Ants have proved optimal, as one would expect, for the biology of social behavior, but also for pheromones and many aspects of ecology.

The papers collected here are those subjects to which ants and my boyhood passions led me. Together they reflect, I hope faithfully, some of the broader events that have occurred in the disciplines they represent and the times in which they were written.

I am grateful to Vincent Burke, Senior Editor at the Johns Hopkins Press, for first suggesting this collection of my scientific articles and book chapters and then shepherding them through to publication; and to Kathleen M. Horton, who not only assisted in preparation of the volume but also played a key role over four decades in editing and launching most of the original papers themselves.

PART I

Ants and Sociobiology

1949

"*Richteri*, the fire ant," *Alabama Conservation* 20(12) (June): 8–9

This article was written when I was a nineteen-year-old senior at the University of Alabama, following a three-month's stint as an entomologist for the Alabama Department of Conservation in collaboration with James H. Eads, a fellow student. My first published contribution to science, it is a report on the notorious red imported fire ant, depicting this species in its earliest expansion across the southern United States. (A decade later, I would assist Rachel Carson as she prepared to make the fire ant a star player in *Silent Spring.*) The article was not, however, my first publishable observation. In 1942, while exploring the ants of my neighborhood in Mobile, I discovered a nest of the invader near the port docks and subsequently reported it as one of the first two observations of the species in the United States.

In 1949, we at the Alabama Department of Conservation called the invader the Argentine fire ant, in reference to its supposed native country. Soon, however, not wishing to offend our neighbors to the south, we changed the common name to the imported fire ant. The U.S. Department of Agriculture then officially labeled it the red imported fire ant. When the first studies were made, the accepted scientific name was *Solenopsis saevissima* var. *richteri.* Later researchers realized that the introduced U.S. population comprises two species, the dark imported fire ant *Solenopsis richteri* and the red imported fire ant, aptly named by William F. Buren *Solenopsis invicta* (literally, the "unconquered" *Solenopsis*).

From Any Angle, It's A Nuisance

RICHTERI, THE FIRE ANT

By Edward O. Wilson, Jr.
Department of Conservation Entomologist

A survey of the damage being done by this imported fire ant is being conducted by Entomologists Wilson and James Eads under sponsorship of the Department of Conservation with the University of Alabama cooperating in making its laboratory facilities available to the two field men. This article tells about the ant and its habits.

THE stinging, mound-building fire ants of southern Alabama and Mississippi have received a good deal of publicity lately, and it is well deserved. They were first called to the attention of the public in April of last year and since that time have been accused of causing enormous amounts of damage to crops, wildlife, and forests. Considerable effort has been made to obtain federal aid in controlling them.

While much attention has been focused on the damage, little has been said about the ants themselves, because actually very little is known about them. What they are, what their distribution is, exactly what sort of damage they are doing and how much, what their life cycle is, are some of the things that must be known before a full-scale program of research is initiated. The surveys made so far have been preliminary in scope, but with them we have been able to find out at least enough to estimate the size of the problem.

First, the imported fire ants, which entomologists call **Solenopsis saevissima** var. **richteri**, have been well named. Their stings feel very much like a touch of fire. They were introduced about 30 years ago into Mobile from Argentina. In the Gulf States they apparently found an ideal home. Without serious competition and nautral predators they are getting out of control. Today, their range extends in a semi-circle of between eight and ten thousand square miles around the city of Mobile, and they are found in at least eight additional isolated areas in Alabama and Mississippi.

Although the individual worker ants are small, never exceeding a quarter of an inch in length, they occur in prodigious numbers. A single nest may contain hundreds of thousands of them. They are exceedingly aggressive and attack on the slightest provocation. As any person living in the infested area can probably attest, their sting is quite painful. It results in an itching papule which may later develop into a little sore, or pustule. The sexual forms, the males and queens, are larger than the workers but do not ordinarily sting. They periodically emerge from the colony to mate and start new colonies. All the innumerable workers in a colony are produced by a single mother queen. She is the vital spot and the one individual which must be killed to destroy the colony.

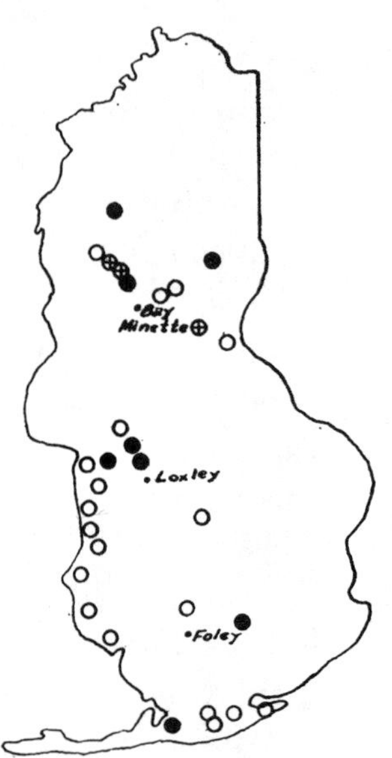

Baldwin County Forest Damage Survey

○ Uninfested
⊕ Pines infested by aphids.
● Pines infested by scale insects.

The worker ants, all sterile females, do the damage. They are omnivorous, preferring other insects for food but also feeding readily on seeds, young plants and animals, and the secretions of aphids, scale insects, and mealybugs. Because of their varied diet they are potentially very important insect pests. It is obvious that enough of them in any area could prove a definite hazard to nearly every form of life, and this is apparently what they are doing in the Gulf States. Thickly populating every region they infest, they are about as versatile and harmful as an insect pest can be. Because of their versatility, it has been impossible to determine the exact amount of damage the ants are doing, but the type of damage is fairly well known. An abbreviated tabulation would run something like this:

Crops (nearly every kind may be included): (1) Destroying whole plants in the seedling stage. (2) Eating the stems, leaves, and fruits of adult plants. (3) Tending the "ant cows" (aphids, scale insects, and mealybugs) that are crop pests. (4) Because of their vicious stings, making it difficult to gather crops.

Forests: (1) Tending the "ant cows" of pines. (2) Eating the seeds as they fall to the ground.

Small game and barnyard animals: (1) Killing young animals and birds at birth. (2) Driving the mothers from the nests.

As a general nuisance: (1) Entering houses, raiding food, and making life thoroughly uncomfortable for the inhabitants. (2) Killing and stunting garden plants, and disfiguring lawns. (3) Destroying bee hives.

That the Argentine fire ants are serious crop pests is easily demonstrated, but while there seems to be a general concurrence of opinion among the people of the infested areas that they are rapidly decimating wildlife and stunting considerably the growth of forests, this point needs careful checking. If it is true, they are without doubt among the most destructive insect pests ever to be introduced into this part of the country.

In the area of infestation, the imported fire ants are easily found and recognized. The colonies are housed in large, dome-shaped mounds, which are often over two feet in height. These are frequently built against stumps, logs, and the bases of shrubs. This one species can be separated from the native fire ants in that it builds such large mounds and that the largest workers have relatively small heads with toothed jaws. The native forms have never been overly abundant in southern Alabama, and now they are so completely subdued by the new pest as to be relatively rare. The imported fire ant should not be confused with the little "sugar" or "Argentine" ant, **Iridomyrme humilis.** The only thing the two have in common is that both are pests, and as far as I can determine they live amicably together where the latter is found.

The imported fire ants apparently nest in any kind of soil under all conditions of humidity, their mounds being constructed in everything from marsh muck to pure sand. Heavy rainfall and drought do not seem to affect them; they can also withstand the contrast of intense summer heat and freezing winter temperatures. Only very arid conditions, prolonged freezing weather, and, strangely enough, heavy carpets of leaf litter seem to stop them. In general, they thrive in grasslands, open pine woods, and cultivated areas, but avoid pine woods with thick leaf litters and nearly all types of deciduous woods. Judging from their present distribution in the Gulf States and in South America, I think it reasonable to assume that they are capable of spreading over much of the coastal plain and probably throughout the Gulf and South Atlantic states, from Texas to North Carolina. In support of this estimate, it is significant to note that one native species of fire ant already occupies most of this range and that the species **Solonopsis saevissima** ranges over the whole of South America in a variety of climates.

Solenopsis saevissima is what biologists call an "explosive" species. It is very variable in its appearance and over its large range geographic races have evolved, each one best suited for the region in which it lives. This apparently is what has. happened, or is happening, in the Gulf States.

Two color varieties are to be found in the main area of infestation around Mobile. For convenience they are both called "var. **richteri**", but only one of them resembles the original Argentine race. This variety is very deep blackish-brown in color and seems to be limited in its distribution to grassy lowlands. The other is reddish and smaller in size; it is the predominant form and the more adaptable of the two. It is the familiar fire ant of southwestern Alabama, the one that is rapidly spreading to the north and northeast. However, Dr. W. S. Creighton, who has done a great deal of work on this group of ants, stated to me in a recent letter that this has not always been the case.

Twenty years ago, when the Argentine fire ant was still limited mostly to the city of Mobile, the dark, typical "var. **richteri**" was the predominant, if not the only, form. This is still true in the isolated Mississippi populations, but in southwestern Alabama and around Selma the reddish form seems to be absorbing and replacing it. Because the reddish form is the more widespread and because its appearance and proportional increase is coincidental with the rapid spread of the ant in the last few years, it is possible that this species has undergone a permanent adaptive change. This very probably means that the Argentine fire ant is quickly adjusting itself to its new home and can be expected to continue its rapid spread.

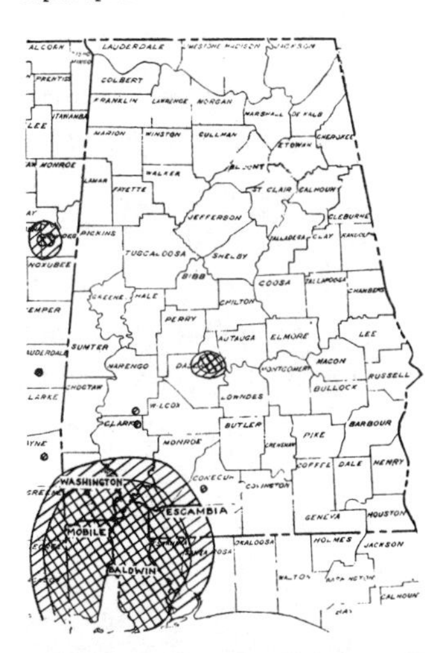

RANGE OF THE ARGENTINE FIRE ANTS. HEAVY INFESTATION IS INDICATED BY CROSS-HATCHING; LIGHT INFESTATION BY SINGLE HATCHING. THE EXTENT OF THE ISOLATED POPULATIONS IS STILL IMPERFECTLY KNOWN.

Occurring in phenomenal abundance, with sometimes as many as 50 to 100 mounds to the acre in many areas, the fire ants have more than doubled their range in the last ten years. In this period they have spread outward at an average rate of three miles a year, and this seems to be just an initial acceleration. The most rapid spread has been to the northeast, a convenient explanation for this being the prevailing southwestern winds of the warm season. Although the ants are spread mainly by the nuptial flight of the winged queens, most of the isolated populations were probably started by queens directly transported by freight or highway traffic. The ants evidently reached Dauphin Island from Cedar Point as stowaways on the ferry or fishing boats.

The job of controlling these pests is going to be enormous, because they are not ordinary, crop-specific insects with a highly predictable life cycle and limited habitat. Even for ants, they are unusually adaptable with a very long breeding season. They cannot be destroyed by applying poisons in certain places at certain times of the year, and the killing of a large number of workers means little to the colony; they are quickly replaced by the prolific queen.

Control, if it is attempted, will necessitate the destruction of the sexual forms and broods of every colony in all the numerous habitats where they occur. How this can best be accomplished we do not yet know. Preliminary experimentation with insecticides by Mississippi State College showed that chlordane, a newly developed chemical, is among the cheapest and most effective, and it was used in limited control work around Artesia. Bill Ziebach, working independently in Mobile, corroborated this, but the work of both the college and Mr. Ziebach was indecisive as to efficiency in the field. However, they did reach the conclusion that if the ants must be killed by direct mechanical application of poisons, the technique will have to be flawless and the cost will run high. The other possibilities, the use of poison baits and the introduction of natural enemies, have not been investigated yet. It is quite possible that they will prove fruitless. Ants have few enemies capable of destroying whole colonies, and the use of poison baits over large areas would have to be accompanied with extreme caution.

It is apparent, however, that the ants are going to have to be controlled soon. We are not certain yet of the danger they constitute to our wildlife and forests, but we are sure, at least, that they are farm pests of the first magnitude and that they have just begun to spread. Although the ants have proven unpredictable in the past and we might nourish the hope that they will begin to decline as suddenly as they have increased, we would be foolish to do so! The odds are greatly against it. I feel sure that Alabama and Mississippi have made a wise move in sponsoring research and pushing for federal aid. Those individuals who have figured prominently in initiating the campaign, particularly Bill Ziebach and Bert Thomas in Alabama and Clay Lyle in Mississippi, should be given credit for a real public service.

The world's largest rodent is the Capybara rat. It often grows to a length of four feet and sometimes weighs as much as 70 pounds. It is web-footed.

1951

"Variation and adaptation in the imported fire ant," *Evolution* 5(1): 68–79

In the course of my studies at the University of Alabama in the late 1940s, I broadened my interests to include evolutionary biology and applied what I had learned to the imported fire ants. This early article reports that two genetically distinct forms exist, now called the red imported fire ant *(Solenopsis invicta)* and the dark imported fire ant *(S. richteri)*. I proved that the difference is hereditary by switching queens from one color-type colony to the other; the offspring acquired their mother's color even when raised by workers of the other color. Field studies further revealed that the two species, both of South American origin, were hybridizing extensively in their new United States home. Also, the red *Solenopsis invicta* was replacing the blackish brown *S. richteri* by the late 1940s, a process that now, fifty years later, is almost complete.

Reprinted from EVOLUTION, Vol. V, No. 1, March, 1951
Printed in U. S. A.

VARIATION AND ADAPTATION IN THE IMPORTED FIRE ANT

EDWARD O. WILSON
University of Alabama

Received August 28, 1950

INTRODUCTION

The South American fire ant *Solenopsis saevissima richteri* Forel is one of the insect pests which has become recently established in the Southeastern United States. In 1949 it was reported from Florida, Alabama, and Mississippi (Wilson and Eads, 1949). Its known distribution in these three states at that time is shown in Figure 1. Many other populations have been found during 1950 in Louisiana, Mississippi, Alabama, and Georgia by members of the Federal Bureau of Entomology and Plant Quarantine stationed at Mobile, Alabama (*in litt.*); these are mostly very young and limited to nurseries which receive shipments of plants from the Mobile area.

The first published record of this ant in the United States was made by W. S. Creighton (1930), who found it initially in urban Mobile in 1925. Creighton was told by H. P. Loding, a local and reliable amateur naturalist, that the ant had first appeared in the bayfront area of Mobile around 1918, had been pushed north of the city by the subsequently invading Argentine ant, *Iridomyrmex humilis,* and had later re-entered its original range. As late as 1928 Creighton found *richteri* still "confined to a comparatively small area extending from the northwestern part of the town to Spring Hill" (*in litt.*). In 1932 L. C. Murphree, scouting Argentine ants in Alabama, recorded *richteri* from several localities in Mobile and Baldwin Counties, including Whistler, St. Elmo, and Fairhope (Murphree, 1947).

The first comprehensive study of this ant was undertaken during the period of March to July, 1949, by Wilson and Eads (1949) under the auspices of the Alabama State Department of Conservation. The ant was shown to be a versatile but erratic pest, doing extensive damage to seeds and young seedlings of a variety of crops. Part of the data used in the following study was presented in that report. The U. S. Department of Agriculture initiated its own study in July, 1949, and has continued it up to the time of this writing (August, 1950).

During the spring survey of 1949 there was observed an unusual amount of variability in color from nest to nest. This variation included one extreme blackish phase referable to the typical subspecies *richteri,* one extreme reddish phase referable to no described form, and intermediates between the two. Furthermore, the phase referable to *richteri* appeared to be mostly limited to part of the periphery of the range (see fig. 4). Since Creighton in 1930 described the forms he found in Mobile as typical *richteri* and has later stated (*in litt.*) that this form was the predominant one during that early period, the possibility of the present predominance of the extreme reddish phase representing an important change in the population must be considered. The following report is an attempt to analyze and explain the variation in this light from the point of view of the entire population.

Most of the field data presented in this report was obtained during the 1949 survey. Experiments with living colonies and morphological studies were conducted during the following year at the University of Alabama.

Grateful acknowledgment is made to Mr. J. H. Eads for his invaluable collaboration in the first survey of the ant, to Dr. R. L. Chermock for continuous

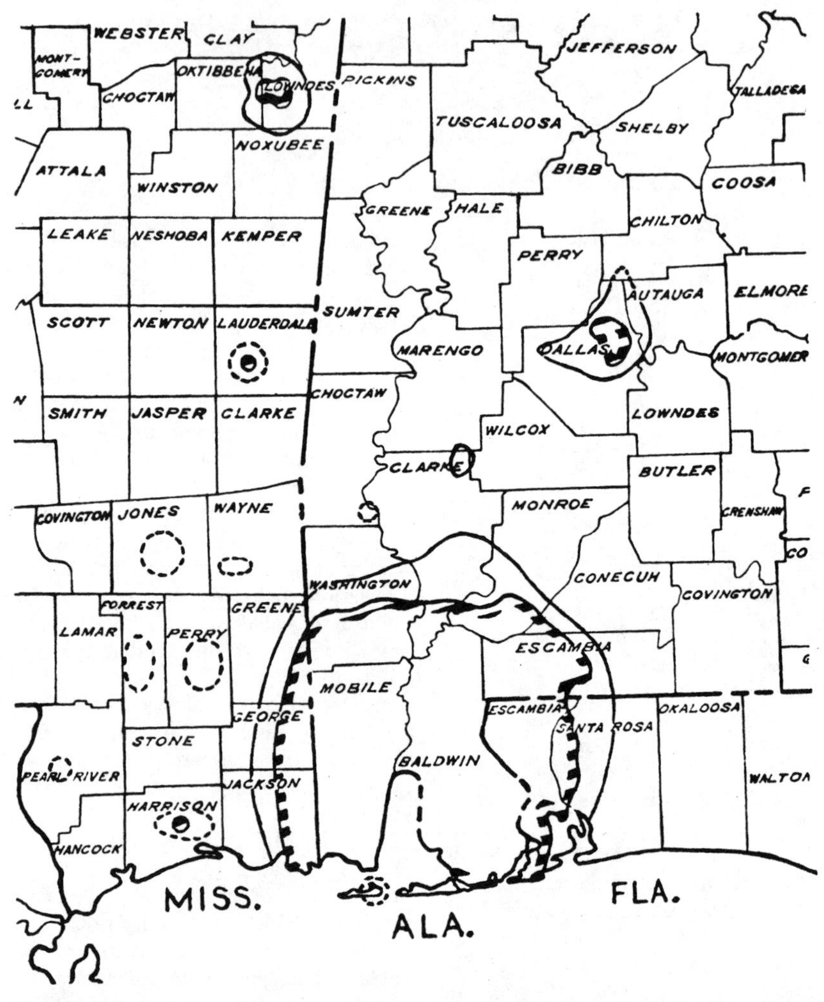

FIG. 1. Known range of the imported fire ant in the Gulf States in 1949. Inner barred lines enclose areas of heavy infestation; outer unbarred lines enclose the extreme range. Dotted circles indicate small populations the ranges of which were not accurately determined. The Choctaw County population and the populations in southern Mississippi were reported by the Federal survey crew (*in litt.*, 1949).

assistance and advice, and to Dr. W. S. Creighton for the hitherto unpublished information concerning the early infestation.

STUDY OF THE VARIATION

Solenopsis saevissima richteri Forel, as defined by W. S. Creighton in 1930, is apparently the southernmost race of a highly variable South American ant. Creighton described its range as extending from Uruguay south to the state of Rio Negro in Argentina, west to the Andes, and northwest almost to Bolivia. Another form, *S. saevissima* var. *quinquecuspis* Forel, is found over part of the range of *richteri* in eastern Argentina, but it has been distinguished from *rich-*

teri only on the basis of a highly variable color character and its validity is very questionable. *Solenopsis s. saevissima* (F. Smith), the typical subspecies, ranges with no great variability from the Guianas to the southern part of the state of Minas Geraes, Brazil. A vast zone of intergradation, containing a complexity of described forms, extends from Minas Geraes west through Bolivia to the Andes, south to Uruguay in the east, and south to the oases area of Argentina in the west. The taxonomic picture here is a very confused one; further study will probably reveal some of the forms to be intergrades of *saevissima* and *richteri,* while some may be shown to represent distinct species. A thorough discussion of the diagnostic characters and the known ranges of these forms has been presented by Creighton (1930). The extreme reddish phase, which plays so important a role in the Gulf States population, has not been formally described from South America, but it is possible that it exists there as a submerged element (see Discussion). The whole Gulf States population has been referred to in this report as *Solenopsis saevissima richteri* strictly as a matter of convenience. This has been its popular designation in reports and correspondence up to the present time. For reasons discussed later it is believed that the new form should not be given immediate formal taxonomic recognition.

The population found in the Gulf States exhibits a great deal of variation. In order to facilitate a more exact study, the extreme variants and their intermediates were divided into six arbitrary phases according to the color of the workers. These can be described as follows:

1. *Extreme dark.* Ground color of alitrunk and head piceous brown; a light brownish orange to light brownish fulvous stripe covering approximately the anterior three-fifths to four-fifths of the dorsum of the first gastric segment in all but the minimas and smallest medias; a similar stripe on the venter of the first gastric segment but the rest of the gaster piceous brown, the condition seen in all the phases. This is the form best referable to Forel's description of *richteri.*

2. *Dark intermediate.* Ground color of alitrunk and head dark brown but not piceous, and with no reddish tinge evident; stripe on dorsum of gaster similar to above.

3. *Intermediate.* Ground color dark reddish brown, lighter than dark intermediate phase and approximately intermediate between the two extreme phases; stripe on dorsum of gaster similar to two darker phases.

4. *Light intermediate.* Ground color medium reddish brown; stripe on gaster covering approximately one-half to two-thirds of the anterior surface, darker than in above phases, posterior border frequently indistinct.

5. *Dark red.* Ground color light reddish brown; stripe on dorsum of gaster covering approximately the same area as in light intermediate phase but darker, being very close to the color of the head and alitrunk and distinguishable only in contrast to the piceous brown of the rest of the gaster, its posterior border indistinct; stripe present only in larger medias and soldiers.

6. *Light red.* Ground color light reddish brown; stripe on gaster absent or present only in largest soldiers, very similar when present to that of dark red phase.

During the survey three colonies were found which could not correctly be assigned to any of the above phases. One of these, found near Gulf Shores, Baldwin Co., Ala., had workers with the ground color of the dark intermediate phase but with little or no gastric stripe. These are referable to Creighton's definition of *Solenopsis saevissima* var. *quinquecuspis* Forel (1930). The other two, found near Gulf Shores and Fairhope, Baldwin Co., Ala., had workers all of which possessed very nearly the same color as callows of the light red phase.

The queens of all the phases are very similar in coloration to the largest workers and exhibit the same variation. The males are all uniformly black.

Morphological studies of the six phases revealed significant differences only in size. It was found that in workers, queens, and males, the darker phases are larger on the average than the lighter phases. As demonstrated in figures 2 and 3, the successive intermediates tend to show successive differences in size, this being particularly pronounced in the worker caste. All the measurements shown were made of the alitrunk as seen in profile, from the dorsal base of the apronotal collar to the dorsum of the junction of the propodeum and petiole. The alitrunk was used because of its rigidity and the ease with which it is measured. To correlate the variability of the head with that of the alitrunk, the heads in addition to the alitrunks of fifty light red workers and fifty extreme dark and dark intermediate workers were measured in profile from the base of the clypeal spines to the extreme occipital border. No significant divergence was noted in the two sets of measurements nor could the two groups of ants be separated on the basis of head-alitrunk proportion.

In nearly all the areas where many of the phases were observed together it was noted that the darker forms tend to build larger and proportionally taller mounds. Ten mounds each of light red and extreme dark-dark intermediate phases were measured in an open field several miles south of Theodore, Mobile Co., Ala., in June, 1949, and the following differences recorded:

Light red

The smallest mound had a base diameter of 13″ × 13″ and a height of 5″; the largest had a base diameter of 24″ × 22″ and a height of 11″. Height varied from 3″ to 13″. The overall average of the base diameters was approximately 24″ × 22″ and of the heights, 7″.

Extreme dark and dark intermediate

The smallest mound had a base diameter of 22″ × 20″ and a height of 10″; the largest had a base diameter of 39″ × 24″ and a height of 14″. Height varied from 10″ to 18″. The overall average of the base diameters was approximately 27″ × 24″ and of the heights, 14″.

All of the phases tend to accumulate bits of vegetable detritus, small pieces of charcoal, and small pebbles on the surfaces of their mounds, but this is especially noticeable in the darkest three or

<table>
<tr><th></th><th>WORKERS</th><th>QUEENS</th><th>MALES</th></tr>
<tr><td>EXTREME DARK</td><td>1.264</td><td>2.682</td><td rowspan="3">2.712</td></tr>
<tr><td>DARK INTERMEDIATE</td><td>1.236</td><td>2.635</td></tr>
<tr><td>INTERMEDIATE</td><td rowspan="2">1.130</td><td rowspan="2">2.584</td></tr>
<tr><td>LIGHT INTERMEDIATE</td><td rowspan="3">2.604</td></tr>
<tr><td>DARK RED</td><td>1.095</td><td rowspan="2">2.499</td></tr>
<tr><td>LIGHT RED</td><td>1.038</td></tr>
</table>

Fig. 2. Means of the lengths in millimeters of the alitrunks of the color phases. For each phase a minimum of one hundred workers, twenty queens, and twenty males collected at two or more localities during at least two seasons were measured. Only statistically significant differences are shown.

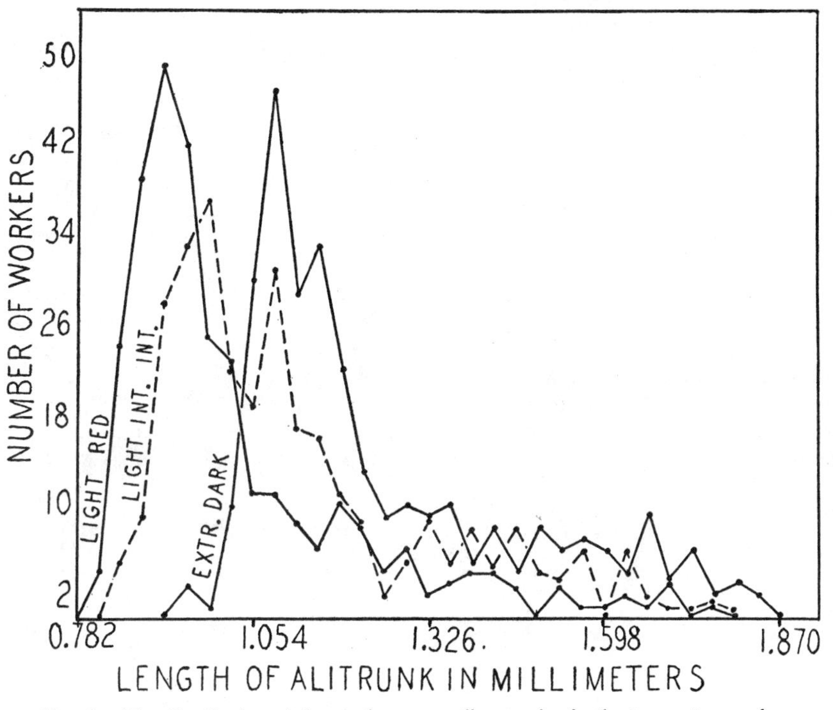

Fig. 3. The distribution of the workers according to size in the two extreme phases and two of the intermediate phases. Three hundred workers from four localities were measured in each group.

four phases. Occasionally mounds of the darkest phases are found nearly covered by a thin layer of this debris. In areas where they occur together it is often possible to tell with reasonable accuracy from a distance which are the darker and which are the lighter phase mounds, judging from the size, shape, and outer surface.

In order to plot the distribution of the color variants 10 colonies each in 84 relatively random localities over the main infested area were classified according to the six color phases. To supplement this a total of 193 colonies were classified in 18 other localities where ten nests could not be found together. Figure 4 shows the approximate distribution of the phases as determined by this survey. It will be noted at once that the darker phases are limited to the southern and eastern portions of the periphery of the range, to a small part of the western portion of the periphery, and to a number of small, isolated areas throughout the range. It is interesting to add here that the isolated areas are in nearly every case centered around marshy fields or grassland, but not all such situations contain darker phases. The Artesia and Meridian population in Mississippi are apparently homogeneously composed of the darkest one or two phases. In the Selma population the darker phases are scarce and irregularly distributed, and none at all have been found in the Thomasville population. The populations in Louisiana, Georgia, and southern Mississippi have not been carefully classified according to the color phases by the writer. However, Mr. G. H. Culpepper, of the Federal survey crew, has very kindly studied the color variation in these populations and has reported that with the exception of the one at Meridian all contain a predominance of light phase colonies (*in litt.*). He has expressed the opinion that all but those at Meridian and New Iberia are apparently small and recent in origin.

It is believed that the variation studied in the Gulf States has a genetic basis. This conclusion is based on the following observations:

1. In many areas where both extreme phases and their intermediates occur, colonies of nearly all the phases may be found in the same immediate area, sometimes within a few feet of one another, apparently under nearly identical conditions.

2. Variation within individual colonies is very slight, and in none of the nests examined were there found workers which covered more than two adjacent phases.

3. It would be very difficult to rationalize the distribution of the variants according to environment or colony age. In the main area of infestation the darker forms are mostly limited to the southern and eastern portions of the periphery, while the population at Artesia, in northern Mississippi, is apparently composed entirely of darker forms. The Artesia and Selma populations are both located in the clearly defined "Black Belt," but the Selma population contains very predominantly the two lightest phases.

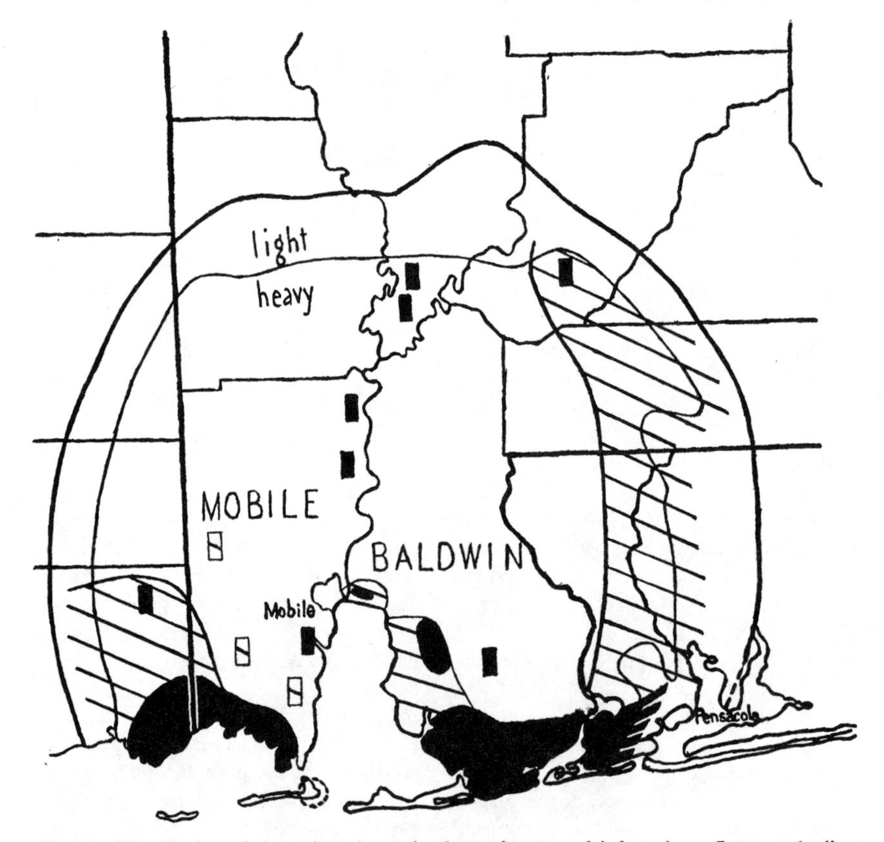

FIG. 4. Distribution of the color phases in the main area of infestation. Intense shading represents areas with incidence of four darkest phases greater than 20 per cent; hatching represents areas with incidence of these phases 5–20 per cent. Small rectangles represent small isolated dark phase populations.

4. Colonies of light red and dark intermediate phases have been maintained in the laboratory under a variety of conditions without appreciable change in the color of the original workers or those reared in artificial nests. Workers have been reared at temperatures above 30° C and below 20° C; others have been heated excessively and chilled. Still others have been reared variously at substarvation and near-optimum conditions. Minima workers of the dark intermediate phase produced under substarvation conditions tend to be lighter in color but are still distinguishable from those of the lightest phases.

5. The brood of two light phase queens adopted by dark intermediate phase workers in the laboratory and one adopted by *Solenopsis geminata* workers developed into workers with the color of their mothers. In each case young, recently fecundated queens were introduced into groups of twenty to thirty workers. These were maintained in Fielde nests modified by the addition of plaster-of-paris chambers and were fed honey, dogfood, and miscellaneous insects. It has been found that occasionally groups of workers, especially those from depauperate colonies, accepted alien queens readily, but in the majority of cases could be induced to do so only after being chilled to immobility from several hours to several days. Many remained hostile regardless of treatment.

Discussion

It is apparent that the Gulf States population of *Solenopsis saevissima richteri* has undergone a marked change during the period of 1929 through 1949. The original population was at least mostly composed of the darkest phases, as stated by Creighton. As late as 1929 this remained true (Creighton, 1930 and *in litt.*). In 1932 L. C. Murphree collected the ant from five localities in Mobile and Baldwin Counties, Ala., and judging from his description (1947), the material he collected must be assigned to one or both of the two darkest phases. In 1941 the writer observed a large number of colonies along the bayfront of urban Mobile and in several areas in the western part of the city, all of which belonged to the lightest two or three phases. By 1949 the darkest phases could be found in some abundance only along part of the periphery in the main area of infestation and even there were outnumbered by the lightest phases. The Selma and Thomasville populations were predominantly light phase, the Meridian and Artesia populations at least predominantly dark.

Two approaches may be used to explain this peculiar recent distribution, one considering the history of the population, the other the possible climatic preferences of the extreme forms. Figure 5 shows the approximate rate and direction of spread of the ant since its introduction, as based on the observations of W. S. Creighton, the records of L. C. Murphree, and the estimates of 65 residents of the main infested area (Wilson and Eads, 1949). It appears that by 1934, while the dark phases may still have been dominant, the ant spread south over the southern portions of Mobile and Baldwin Counties, a range now co-occupied by another introduced ant, *Brachymyrmex heeri obscurior* Forel. Estimates place the origin of the Artesia population around 1935 and the Meridian population around 1940. The Selma population was not in any case estimated to have originated before 1944 nor the Thomasville population before 1948. The light phases were apparently responsible for the ant's explosive spread to the north in the main area of infestation. The partly peripheral distribution of the darkest, presumably oldest phases seems to indicate that they were pushed outward by the expanding light forms, in a pattern somewhat similar to that first demonstrated by Matthew (1939) for the primitive members of some groups of mammals, that is, with the most primitive forms at the periphery and the most recently evolved toward the center. The relatively homogeneous ecologi-

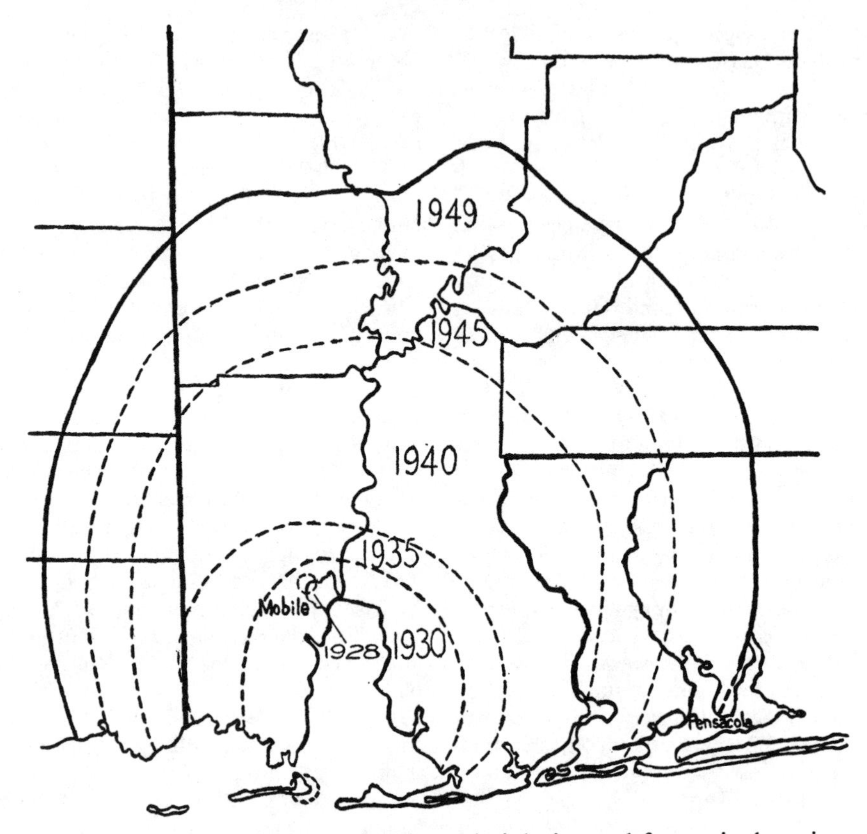

FIG. 5. Estimated rate and direction of spread of the imported fire ant in the main area of infestation.

cal conditions in and around the area of infestation, coupled with the absence of any significant geographical barriers, has allowed an even and steady spread of the ant since its introduction (see fig. 5). It has also resulted in the relatively clear preservation of the concentric pattern produced by the spread of the light forms. The darkest phases were best able to survive along the coast, along the eastern periphery, and in isolated spots through the infested area. Why marshy areas should suit these forms is not known. Large numbers of light phase nests often occur in the same places, with as many as seventy nests to an acre, and competition must be intense. Over much of the southern, continuous part of the range of the darkest phases, as around Gulf Shores, these forms occur in as wide a variety of situations as the lighter phases. The absence of light phases at Artesia and Meridian might be explained on the basis of early origin, that is, derivation from the early, nearly homogeneous stock. The lack of conspicuous success on the part of these populations, as compared with that of the younger Selma and Thomasville populations, and the southward expansion of the early population in the main area of infestation seem to indicate that the darker phases are poorly adapted to the climate of the Gulf States.

The present predominance of the light phases suggests an adaptive change within the population. The manner by which these phases have largely replaced the dark phases has been, in the opinion of the writer, partly rapid expansion in range and population size and subsequent dilution of the dark phase genes, resulting in a diversity of intermediate forms; it has also been partly through considerable populational pressure and elimination in direct competition. An indication of this latter process can be seen in the scarcity of dark and intermediate phases over much of the range formally occupied by the dark phases. Another indication can be seen in the scarcity of the native fire ants, *Solenopsis xyloni* and *S. geminata,* and of the Florida harvesting ant, *Pogonomyrmex badius,* in the infested areas. Colonies of the imported fire ant are extremely antagonistic to these ants as well as to alien colonies of their own species. During the survey no colonies of *Solenopsis xyloni* and very few of *S. geminata* were found in the heavily infested areas. *Pogonomyrmex badius* was more common but quite sporadic in distribution. These three species reach relative abundance around the periphery of the imported fire ant's range. Another important consideration in this light is the incidence of the phases. Although exactly random samples in quadrats were not taken during the survey, approximately twice as many nests of the two darker phases were found as those of the two intermediate phases in the counts taken. Assuming that the phases represent nearly equal arbitrary divisions of the successive genotypes, it is possible that this difference in incidence is a result of replacement by the light forms through direct competition as opposed to genetic dilution. The relative importance of these two modes of replacement can only be estimated on the basis of available data, but it seems safe to say that both play an important role.

When considering the significance of the light phases as an adaptive replacement, the possible role of the Argentine ant, *Iridomyrmex humilis* (Mayr), must be studied. This ant reached its peak in the latter part of the 1920's and during that time succeeded in eliminating nearly all of the native ants in the Mobile area (Creighton, 1950). By 1932 it had been partly controlled and the native ants had begun to infiltrate the infested area. It was this time also that *Solenopsis saevissima richteri* began its initial spread to the south. It would seem quite possible that the decline of the Argentine ant, and not the replacement of the dark phases by the light phases, was responsible for the fire ant's rapid spread to pest proportions. However, several observations have been made which seem to indicate that the Argentine ant had a minor influence on the phenomenon under consideration. One is that the Argentine ant appears to offer the imported fire ant little serious competition at the present time, even in areas where the former are abundant enough to affect the native ants (Wilson and Eads, 1949). The two species thrive together in some areas of Mobile. Another is that the Argentine ant is primarily an urban dweller; it never was able to blanket the wide variety of rural situations in which the imported fire ant thrives (*Ibid.*). There is no reason why the fire ant could not have spread by way of these situations, even assuming that the Argentine ant was able to hamper it during that ant's peak. Finally, the predominantly dark phase Artesia population and the predominantly light phase Selma population can be critically compared, since both are under very similar conditions. The latter is less than half as old as the former and yet much larger and denser. Both are situated mostly in rural areas, and neither could have been greatly affected by the Argentine ant since their inception.

In considering the possible genetic basis of the variation, there are several peculiarities in the expression of the characters which deserve mention. One of these is the small amount of variability observed in intergrade colonies. This, coupled with the great variability existing from nest to nest, cannot be explained on the basis of a single mutation. Admitting that the phases are totally arbitrary, that they overlap, and that occasional individuals can be assigned to an adjacent phase, at least four or five distinct genetic groups clearly exist; the workers of the six phases can be divided on the basis of size into five statistically significant units (fig. 2). Part of the difficulties are removed by explaining the variation on the basis of multiple factors or multiple alleles, which could control successional variation of the type observed. Of these two possibilities, multiple alleles seems to be the more tenable. From only three alleles a total of six combinations, or genotypes, is possible. Also, the greatest number of genotypes which could be produced by a single cross is three; the small amount of variability in single colonies studied in the field may well indicate such a limited number of genotypes. On the other hand, a dihybrid cross involving multiple factors showing no dominance would make possible nine genotypes and phenotypes, all of which, including the extremes, could be produced by a single heterozygous cross.

The close correlation between color and size variation and the present selective advantage the light phases hold seems to indicate a pleiotropic expression of whatever alleles are involved, producing a relatively clear-cut combination of characters for both extreme color forms. It would seem unlikely that these combinations of characters could maintain their identity in the extensive intergradation that has occurred unless they were very closely linked.

Even assuming multiple factors or multiple alleles, the small amount of variability within individual colonies is difficult to explain by the familiar laws of heredity. Provided that the queen is heterozygous and diploid, it would seem that the phenotypes of workers produced by an intermediate queen would be mixed according to random assortment. It appears very likely that the reason why this fails to occur in the imported fire ant (and in many other species of ants which exhibit the same peculiarity in their color varieties) entails an aberrant hereditary mechanism. One such possible mechanism is maternal determinism, a condition demonstrated in the snail *Limnaea peregra* by Diver, Boycott, and Garstang (1925). In this organism the direction of spiraling in the shell is determined by the genotype of the mother, regardless of the genotype of the offspring. On the other hand, in *Solenopsis* a mechanism involving haploidy in the queen and parthenogenesis is apparently rendered untenable, because unfertilized eggs are capable of developing at the most only to the late larval or pupal stage. Mature larvae were noted in fourteen out of fourteen nests containing only young alate queens; one of these nests produced several worker pupae, but since there was no absolute proof that all the queens were not fertilized, general conclusions concerning thelytoky should be avoided. One nest containing alate queens raised from larvae in the absence of males produced only larvae over a long period of time. This was noted in fourteen artificial nests out of fourteen which contained only virgin queens; mother queens in other nests kept up a prolific production of workers. Unfortunately, the interesting and significant problem of the genetic basis of the variation will prove to be a very difficult one to solve. For all practical purposes the colony is the individual in the life cycle of the ant, and with present knowledge of caste determination each generation of virgin queens will have to be preceded by hundreds or thousands of workers.

The origin of the light phases is another problem which can only be conjectured on the basis of available data. Three possibilities are treated below.

1. The light phases could have been a submerged recessive element in the genotype of the original invaders. However, this view is considerably weakened by the apparent homogeneity of the Artesia and Meridian populations, which were probably established while the dark phases

were still dominant. Also, the original invaders were probably few in number; it seems unlikely that they could have contained all the variability present today, unless this were to appear conspicuously in the early population.

2. The light phases could have been derived through mutations in the early, dark phase population. This possibility must always be considered, because the population mutation rate has been potentially tremendously high. Even in the early population, each of the many thousands of colonies were producing hundreds to thousands of sexual forms the year round. Today the number of colonies in the main infested area easily numbers into the millions (Wilson and Eads, 1949).

3. The light phases could have been derived from one or more later introductions. Despite the fact that nothing exactly comparable to the light red phase has ever been described, examination of representative material of most of the South American forms of *Solenopsis saevissima* has convinced the writer that the light red phase could have originated anywhere in the great zone of intergradation from southern Brazil to Argentina. Variation in this area is too great and as yet too poorly defined to dismiss the idea that the light phases were derived anywhere but in the Gulf States.

In final analysis the phenomenon can be described as an adaptive replacement, a shift in population dominance from one genetically distinct form (darkest phases, or *richteri* s. str.) to another (lightest phases). At present the population is quite unstable; it cannot be said to have reached an adaptive peak. Although it appears that a complete replacement is under way and a homogeneous population is in the making, new forms such as the two anomalous phases (see discussion of variation) may yet rise to dominance. The origin of the new form is not known. Because of this instability and uncertain origin, it does not seem wise to accord the light phases immediate formal taxonomic recognition, although by strict definition it might appear to some authorities to constitute a new subspecies. Otherwise, the question of the origin is an academic one only. Whether indigenous or introduced the light phases have functioned as a favorable mutation appearing in the population; they seem to have progressed very much as such a mutation would be expected to progress. As an example of this type of phenomenon the adaptive replacement in the Gulf States population is especially noteworthy. This is due to three significant conditions:

1. The new forms are easily identified and their history and present distribution can be carefully studied because of a distinct color character.

2. The adaptive change within the population has occurred so recently as to aid greatly the study of its history. The new forms constitute an almost vertical evolutionary change which has taken place in less than twenty years. Their history illustrates the extremely rapid rate with which such a change, which can be interpreted as an initial step in subspeciation, can occur in a population of insects. Similar changes have been observed in populations of the scale insects *Aonidiella aurantii, Coccus pseudomagnoliarum,* and *Sassieta oleae* (Quayle, 1938), and of the codling moth, *Carpocapsa pomonella* (Hough, 1934, and Boyce, 1935).

3. The very uniform terrain of the Gulf Coast area has allowed an identifiable preservation of what is apparently a concentric distribution of the old and new forms. That the dark forms have been pushed out along the periphery is still evident in their distribution today.

This populational change is indicative of the way that evolution can proceed in any population of ants. However, the rapidity with which this has occurred is probably rarely equaled by that in endemic species. Too little is known about North American ants at the present time to determine precisely the rates of evolution, but it appears that most holarctic genera have been evolving at a considerably lower rate than such better known groups of insects as the Rhophalocera. For instance, postglacial relics are rare or absent, and subspeciation initiated by glacial isolation seems to have progressed to a relatively slight degree. The accelerated evolution in the imported fire ant probably has the same origin as the explosive spreads of imported insects as a whole: the removal of the biotic pressures (parasites, competition with other insects, etc.) which controlled it in its native environment. If this is true, then the populational change in the imported fire ant can be regarded only as a swiftly enacted replica of the normal evolutionary processes and not as typical subspeciation.

Summary

1. The imported fire ant, *Solenopsis saevissima richteri* Forel, is the southernmost race of a widespread and highly variable South American ant. It was introduced into the port of Mobile, Alabama, sometime around 1918 and by 1949 had spread to parts of Florida, Mississippi, and Louisiana.

2. A great deal of color variation from nest to nest has been noted in the Gulf States population. This includes an extreme blackish phase referable to the original description of *richteri,* an extreme reddish phase referable to no described form, and intermediates between the two. This color variation is correlated with differences in size of the ants and in appearance and proportion of their nests.

3. The variation has a genetic basis. It is suggested in this study that the variation can be explained most readily on the basis of multiple pleiotropic alleles.

4. The history of the variation has been determined as follows:

The darkest forms, or *richteri* s. str., were the ones originally dominant from the time of the ant's introduction until at least 1929 and probably sometime after 1932. The origin of the new form is not known, although it is believed that it originated either through mutation within the population or through a second introduction. In 1949 it was by far the dominant form. It had apparently replaced the typical *richteri* partly by rapid expansion and subsequent genetic dilution and partly through natural selection by direct competition. Its predominance in the main population and in at least two smaller isolated populations has evidently been responsible for a much greater success of the species. In the main population in 1949 the typical *richteri* was mostly limited in distribution to portions of the periphery of the range, forming with the new form roughly the concentric pattern of Matthew's modified Age-and-Area hypothesis.

5. The new form has been interpreted as functioning, regardless of its origin, as a favorable mutation introduced into the population. Its rise to dominance has constituted an extremely rapid, nearly vertical evolutionary change.

Literature Cited

Boyce, A. M. 1935. The codling moth in Persian walnuts. J. Econ. Ent., **28**: 864–873.

Creighton, W. S. 1930. The new world species of the genus *Solenopsis.* Proc. Amer. Acad. Arts and Sci., **66**: 39–151.

——. 1950. The ants of North America. Bulletin of the Museum of Comparative Zoology at Harvard College, 104.

Diver, C., A. E. Boycott, and S. Garstang. 1925. The inheritance of inverse symmetry in *Limnaea peregra.* J. Genet., **15**: 113–200.

Hough, W. 1934. Colorado and Virginia strains of codling moth in relation to their ability to enter sprayed and unsprayed apples. J. Agric. Res., **48**: 533–553.

Matthew, W. D. 1939. Climate and evolution. Special Publications of the New York Academy of Science, **1**: 1–223.

Murphree, L. C. 1947. Alabama ants, description, distribution, and biology, with notes on the control of the most important household species. Unpublished Master's thesis, Mississippi State College.

Quayle, H. J. 1938. The development of resistance in certain scale insects to hydrocyanic acid. Hilgardia, **11**: 183–225.

Wilson, E. O., and J. H. Eads. 1949. A report on the imported fire ant *Solenopsis saevissima* var. *richteri* Forel in Alabama. Special mimeographed report to the Director of the Alabama State Department of Conservation.

1953

"The origin and evolution of polymorphism in ants," *Quarterly Review of Biology* 28(2): 136–156

Charles Darwin, in *Origin of Species,* considered the existence of sterile castes in ants the "one special difficulty, which at first appeared to me insuperable, and actually fatal to my whole theory." He solved it by conceiving the unit of natural selection to be the family, as opposed to the individual. The female that produces sterile daughters as helpers serves her own reproductive interests and is able to prevail, at least under some circumstances, over solitary females.

As a graduate student, I was enchanted by this idea, and even more by Julian Huxley's proposition, outlined in *Problems of Relative Growth* (1932), that ant castes arise in evolution by genetically programmed allometry. Thus differences among castes can yield proportionately larger or smaller heads or more or less fully developed ovaries, simply as a result of faster or slower growth in a given body part relative to growth of other body parts. A larger individual, given more food as a larva, reaches maturity with a bigger head, for example, or more fully developed ovaries.

Seizing Huxley's proposition, I ran with it. It seemed a wonderful way to study the evolution of social behavior. By comparing the anatomy of castes in hundreds of species of ants for which large colony samples were available, I showed that allometry is indeed a key process and that much of caste evolution can be tracked by changes in the magnitude of allometry. Improving on Huxley, I then added the second essential component: the size-frequency distribution of the worker caste as a more complete means of reconstructing the origin and evolution of subcastes—such as minor workers and soldiers. Comparisons among species further revealed that the changes in size-frequency distributions can be adjusted genetically by shifts in decision points along the pathway of larval development and the magnitude of diverging developmental rates following each decision point.

This overall model of the origin of castes has held up reasonably well across the ensuing five decades, although considerably improved upon and rendered more complex by later research. The evolution of ant castes provides an excellent example of a general principle in evolution—that, even when tweaked in very small ways, the genotype can produce large phenotypic changes.

THE ORIGIN AND EVOLUTION OF POLYMORPHISM IN ANTS

By EDWARD O. WILSON
Biological Laboratories, Harvard University

INTRODUCTION

IN THE biological literature of the past two decades the term *polymorphism* has been applied usually to one or the other of two almost completely separate phenomena. In genetics it is defined as the condition of two or more distinctive and discontinuous genetic types existing in a population (Ford, 1940). In the study of social insects it is defined as the existence within an individual colony of two or more phases or castes belonging to the same sex, without particular regard to their genetic or environmental origin. On the basis of usage by specialists on all groups of social insects a *caste* is properly defined as a differentiated morphological form with a specialized function, or at least the infrequent relict of such a form. It has been the practice of some entomologists to consider all abnormal, recurrent variants as castes and part and parcel of the species' colonial polymorphism. Actually, these cannot be considered apart from similar variants in non-social insects, and their inclusion obviously breaks down the original meaning of polymorphism by neglecting its relation to social organization.

The following paper is an attempt to summarize our existing knowledge of polymorphism in ants and to reinterpret it in terms of the hitherto unappreciated but tremendously important underlying features of adult allometry and intracolonial size frequency distribution. [In its broad sense allometry is defined as the regular disproportionate ontogenetic growth or adult variation of two organs or linear dimensions related as $y = bx^k$, where k is the *equilibrium constant* and the slope of the log-log plot of the curve. A fuller explanation is offered in a subsequent section.] The study of the latter phenomena has of late been shedding unexpected new light upon the nature and phylogenetic development of polymorphism. It is felt that further work in this direction will continue to yield information significant not only for the special fields of ant taxonomy and ant sociology but also for the general province of growth physiology.

CASTE DETERMINATION

In ants it is typically the female which is polymorphic, with an average colony consisting of two basic female castes, the queen (reproductive female) and the worker (sterile female). The queen and worker may in turn be connected by intermediate stages which are often stable enough to be recognized as distinct castes, while the workers may become subdivided into several additional castes. It is probable that sex is determined, as in most other Hymenoptera, by the genetic mechanism of haplo-diploidy. Although no direct cytological evidence has been adduced in support of this, it is indicated by a great body of indirect evidence which includes the existence of a number of undisputed gynandromorphs and the tendency for all unfertilized eggs to produce males. Probable cases of thelytoky have been reported in several species of ants (Gösswald, 1933; Haskins and Enzmann, 1945; Ledoux, 1950) and need additional investigation, but they are still explainable on the basis of genome duplication or the failure of meiosis, and may not form exceptions to haplo-diploidy.

The preponderance of evidence at the present time indicates that queens and workers are genetically similar and that the caste of an individual female is determined by the food given it as a larva. A miscellany of experiments over the past sixty years has always seemed to show that starved larvae tend to produce ordinary workers (minors), while well-fed larvae produce majors (soldiers) or queens (Light, 1942–1943). However, nearly all early work of this nature included no controls and was based on very small samples, so that most of the critical evidence is carried in the independent studies of three recent workers, R. E. Gregg, M. V. Brian, and A. Ledoux. Gregg (1942) followed the lead set in experimental work on termite castes by studying the effect of the complete removal of one caste on the major-minor ratio. He was able to show that colonies containing only majors produced a significantly smaller proportion of new majors than those containing only minors. From this he postulated the existence of a hormone secreted by the majors which tends to inhibit the production of additional members of their own caste. Certain discrepancies hinder the acceptance of this specific explanation, as Light (*op. cit.*) has pointed out. The efficiency of the majors as nurses was not determined, the adult mortality of most of the colonies was such to suggest abnormal conditions in the artificial nests, and the highly critical matter of larval mortality was not discussed at all. But while Gregg's data are not of a nature to allow interpretation of the exact trophic factors involved, they do strongly suggest that the destiny of an individual *Pheidole* as a minor or a major is decided during the larval period.

M. V. Brian's recent preliminary report on caste determination in *Myrmica rubra* L. (1951) is the most illuminating to appear on the subject to date. *Myrmica rubra* larvae hibernate in the third stadium, and in the spring those which are able to reach a critical size of about 6.5 mg. by a certain time (unspecified by Brian) are able to attain queenness, while those below this size become workers. The queen-potential larvae must then reach another critical size of about 7.0 mg. before pupation in order to become normal queens. Brian states that "it is necessary to suppose that an extra impetus to growth is given to queen determined larvae and withheld or actually withdrawn from those which fail to secure induction." This account of caste determination sounds especially promising, should Brian's as yet unpublished data warrant his conclusions, since it seems to be fully consonant with many of the details of trophic behavior in ants. For instance, the extra impetus assumed by Brian may originate from the proportionately larger fat body found in queen and soldier larvae of many ants. Such larvae have more "bargaining power" in trophallactic exchange with the adult workers and might thus accelerate their own growth. Weber (in Wheeler, 1937) observed in an *Atta cephalotes* colony that two of the larger larvae were actually segregated from the smaller larvae and fed more frequently by the adults, which were evidently attracted by their abundant fatty secretion and vigorous supplicatory "pouting" movements

In his studies on *Oecophylla longinoda* (Latreille), Ledoux (1950) has added significant information to the concept of critical developmental times and subsequent regulation of growth in larval development. *Oecophylla longinoda* possesses dimorphic workers and therefore three distinct female castes, the queen, major worker, and minor worker. There is good evidence to show that the critical time of queen-worker divergence is in very early larval life, while that of the major-minor divergence is at some time during the second larval stadium. Past the second ecdysis, collections of larvae of any given age can be separated into three size groups which correspond to the three female castes.

It would appear from the foregoing evidence that in the case of the determination of two segregated castes, growth trends are fixed at a certain age and proceed without interruption to form the adult ant. It should be emphasized, however, that the larval growth is merely a preliminary to the stage of growth which is actually destined to give rise to the final adult shape. As will be demonstrated in more detail later, the main features of adult caste formation are dependent solely on differential growth in the imaginal discs and probably only at the onset of pupal development, after the larval tissues have ceased growing. The important determination which occurs during larval development appears to be in the specific growth-rate potentials of the imaginal discs which reside in the larva. The growth of the larval tissues themselves, apart from that of the imaginal discs, is approximately regulated to allow the final expression of the disc potentials.

Considerable indirect evidence also strengthens the impression that trophogenesis of castes is universal in the Formicidae. In several species of ants, including members of the Ponerinae (*Pachycondyla*, *Neoponera*), Odontomachinae (*Odontomachus*), Myrmicinae (*Pheidole*), and Formicinae (*Camponotus*), infection of the worker by parasitic mermithid nematodes produces a peculiar feminization which is closely akin to parasitic castration. A thorough study of the external morphology of these modified individuals in *Pheidole pallidula* (Nylander) has been made by Vandel (1930), who has observed that in this species only the major is affected (producing "mermithostratiotes"). An examination of his graph and figures shows that the modified majors are in respect to the structure of their heads little more than major-minor or major-queen intercastes. Their thoraces are typically major-like in form. The mermithids are presumed to enter their hosts during late larval or early pupal life; in the case of the mermithostratiotes it is probably the former, since the adult form of the ants is assumed immediately upon the initiation of pupal development.

It is possible to relate Vandel's mermithization effect to the modifications in form described by Wheeler (1937) in his *Acromyrmex octospinosus* Reich "caste mosaics." The latter aberrations are of two principal types, "gynandromorphs" and "gynergates." The former possess normal male bodies with heads seemingly divided into male and female components, while the latter have normal major bodies with heads seemingly divided into major and queen components. Both are essentially the same type of phenomenon, but it is the gynergates upon which most of the theoretical consideration has been concentrated. Wheeler believed that his gynergates represent genetic mosaics no different in type from true gynandromorphs, and that female castes must therefore be genetically determined. In the year following, Whiting (1938), offered an alternative explanation which is the one generally accepted today. He pointed out that Wheeler's mosaics are quite different from true gynandromorphs as recognized in other Hymenoptera. The ant mosaic patterns are all limited to the head and are more or less regularly disposed, while in true gynandromorphs the male and female components vary in extent and are distributed at random over the entire body. Whiting proposed that Wheeler's mosaics are likely to be intersexes of the type described in the moth *Lymantria dispar*, and that the head alone was affected because this was the only part not past the final critical threshold when the transition from femaleness to maleness began. We might assume from Whiting's reasoning that the intercaste condition of the head could result secondarily from the shift toward maleness; in other words, the shift pulls the head across the threshold toward queenness.

Thus in both Vandel's mermithostratiotes and Wheeler's mosaics there appears to have been a late and forced trespassing of the queenness threshold by the head under exceptional physiological conditions. That the head should prove the most labile of any of the major parts of the body of a larva which has passed the critical time of the queen-worker divergence is fully in accord with present knowledge concerning the allometric basis of polymorphism. The data to be presented shortly show that cephalic allometry is nearly always involved in some way with the differentiation of female castes, and that in the case of the initial segregation of queen and worker it continues to be expressed long after other types of allometry are no longer evident.

One additional recent hypothesis concerning the trophogenic basis of caste determination in ants is of sufficient interest to mention in passing. Flanders (1945, 1952) has suggested that the amount of yolk in the egg could determine caste and that the caste might therefore be determined by the extent of ovisorption taking place in the oviducts of the queen. Differential ovisorption has been demonstrated to be the basis of some discontinuous variation in parasitic (terebrant) Hymenoptera, and there is little doubt that the same situation could hold in the aculeates also. According to Flanders, the hypothesis gains strength when it is recalled that ovisorption can be an inverse function of the rate of oviposition, and that the larger workers and sexual forms of ants tend to appear at times when the rate of oviposition is greatest in the colony history. Unfortunately, Flanders has passed by a great many published accounts of experiments which seem to indicate that the course of development in a larva from any given egg can be affected by nutritional changes regardless of the original condition of the egg. Also, he has overlooked previous attempts to distinguish ovarian caste differences which have yielded negative or contrary results (Bhattacharya, 1943; Ledoux, 1950). But the possibility still remains that the amount of yolk in an egg might affect the *chances* of a larva reaching the threshold values before the critical times. This may prove to be a valuable path of research to follow.

Polymorphism, as now understood, can be interpreted as a function of two variable characters of the adult females of any ant species: allometric growth series and intracolonial size variation. The most useful and applicable definition of polymorphism which also approximates the subjective concepts of previous workers is believed to be as follows: *allometry occurring over a sufficient range of size variation within a normal mature colony to produce individuals of detectably different form at the extremes of the size range.* In all ants, with the exception of a few degenerate parasitic forms and some ponerines, there are well differentiated queen and worker castes, so that in this sense polymorphism is an elementary condition. As the worker caste becomes divided into secondary castes through allometry, it is itself called polymorphic, and it must be remembered that in most taxonomic work application of the term *polymorphic* is usually restricted to the worker.

The principal allometric growth-center (point of greatest allometry; see Huxley, 1932) can occur in any part of the body. In the queen-worker differentiation it is in the mesothorax, with lesser ones following in the gaster and head. In major-minor differentiation, it is typically in the head, while lesser ones may occur in the gaster and various parts of the alitrunk. The allometric change may affect any or all features of the allometric structure, including total size, shape, spination, ommatidium number, details of sculpturing, and deposition of pigments, with the result that the extreme size forms are often strikingly different from one another. Dimorphism, or total bimodality with the elimination of intermediates, is a secondary condition and has produced in the evolution of the Formicidae first the segregation of the queen from the worker caste, and second, in many taxonomic groups independently, the segregation of the minor from the major worker caste. The condition of "partial dimorphism" refers only to the frequency and may be defined as clear-cut bimodality with overlap in the ranges of the two constituent curves.

It must be emphasized at this point that all degrees of intranidal allometry within the worker caste can be demonstrated, grading almost insensibly into absolute monomorphism. Each of the most prolific workers in systematic myrmecology, including Emery, Forel, and Wheeler, had his own idea of approximately where monomorphism stopped and polymorphism started, with the result that all rarely agreed on the status of feebly polymorphic species, and each tended to judge the matter according to how carefully he had examined the material before him. Within the gradient of increasingly allometric species it is impossible to draw a precise line as the lower limit of polymorphism so as both to adhere to previous usage and remain consistent throughout. It is necessary to set as an absolute standard the concept of detectable intranidal allometry. The objection may be raised that such a criterion will include many borderline cases that can be checked only upon careful measurements. Yet in many forms which are universally considered to be polymorphic, such as *Lasius fuliginosus* (Latreille) and a great many of the species of *Dorylus*, the allometry can be detected only by microscopic examination.

Some recent workers, including Falconer Smith (1942), Cole and Jones (1948), and Haskins (1950), have heavily emphasized the role of multimodality and intimated that worker polymorphism should somehow be defined in terms of frequency. Very possibly they were influenced by Wheeler's suggestion (1937), based on his predilection for blastogenesis and an erroneous interpretation of frequency curves of the sort published by Buckingham (1911), that the formation of castes is associated with a tendency toward frequency grouping. My own study has shown that bimodality, emerging from a skewed unimodal condition, closely follows upon the development of intranidal allometry and is correlated with later changes in the allometric regression line. However, trends toward the development of more than two modes are rare (see Fig. 5), and many polymorphic species are apparently unimodal.

TRENDS IN ALLOMETRY AND FREQUENCY

The queen and worker castes are usually separated from one another by a considerable gap with no intermediates, so that only in rare cases can their precise relationships be determined. The worker caste, on the other hand, shows within itself all transitions from monomorphism through elaborate allometry to total subdivision into major and minor castes. There is a good amount of evidence to indicate that this gradual development of dimorphism as seen from species to species represents the trend prevalent in the evolution of any one dimorphic species. The same steps have been encountered in nearly every group where a transition is present, and there can be no question that in most of these groups extreme polymorphism or dimorphism has been independently evolved. The principal stages as recognized at the present time are discussed in some detail below.

1. Monomorphism

The normal mature colony is either isometric or with limited size variability, or both, and its frequency distribution is unimodal. Secondary monomorphism often succeeds phylogenetically some development of polymorphism and is produced by a dropping out of a large segment of the size variation. This typically results in a caste which is even more uniform than the workers of a related, primarily monomorphic species. Well known examples of both types of monomorphism include the species of *Ectatomma*, *Myrmica*, and *Carebara*. (See Fig. 2A.)

2. *Monophasic allometry*

The allometric regression line has a single slope, the condition usually called "simple" allometry by students of growth. A common condition encountered in many genera, and exemplified by such forms as *Eciton nigrescens* (Cresson), *Formica obscuripes* Forel, and *Lasius fuliginosus* (Latreille), is that of feeble monophasic allometry expressed over a short range of size variation, which in turn shows a unimodal frequency curve. There is limited evidence to indicate that the extremes of this variation may serve as functional castes. Alpatov (1924), for instance, has observed some division of labor in the European *Formica rufa* Linné. An apparently later development in monophasic allometry involves an increased dispersion of the frequency curve for size, a condition combined with a marked tendency toward bimodality. This is the typical condition in many genera which are often cited as showing elementary polymorphism or partial dimorphism, including *Megaponera*, *Orectognathus*, *Azteca*, and *Camponotus* (Figs. 1, 2). Within *Camponotus*, polymorphism of this type has been demonstrated to be correlated with intracolonial division of labor (Buckingham, 1911; Lee, 1938).

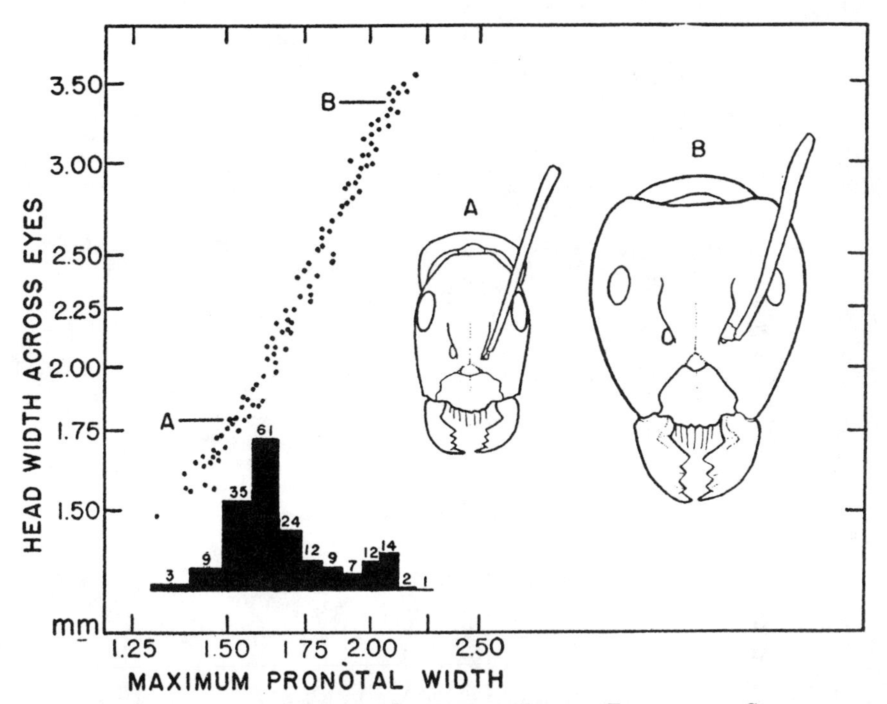

FIG. 1. INTRANIDAL MONOPHASIC ALLOMETRY AND INDISTINCTLY BIMODAL FREQUENCY IN CAMPONOTUS CASTANEUS (LATREILLE) (SUBFAMILY FORMICINAE)

Frontal views of heads and pronota of minor (A) and major (B) workers are shown, along with the placement of these individuals on the allometric regression line. Equilibrium constant (k) = 1.9. Calculated maximum error = ± 0.04 mm. Based on a single nest series, Knoxville, Tenn., W. J. Cloyd leg. In this and the following graphs, individual measurements pertaining to allometry are given, and both the allometric dimensions and the abscissa of the frequency curve are plotted *logarithmically*. All measurements have been taken with an ocular micrometer at magnifications of 24× to 90×. The calculated maximum error is defined as plus or minus the unit or unit-fraction to which the measurements are made; repeated checks have shown that the actual maximum error is in practice usually smaller than the calculated one.

3. *Diphasic allometry*

The log-log allometric regression line "breaks" and consists of two segments of different slopes meeting at a critical point. In the several instances where this condition has been demonstrated, the frequency curve is bimodal, and the trough between the overlapping constituent curves tends to

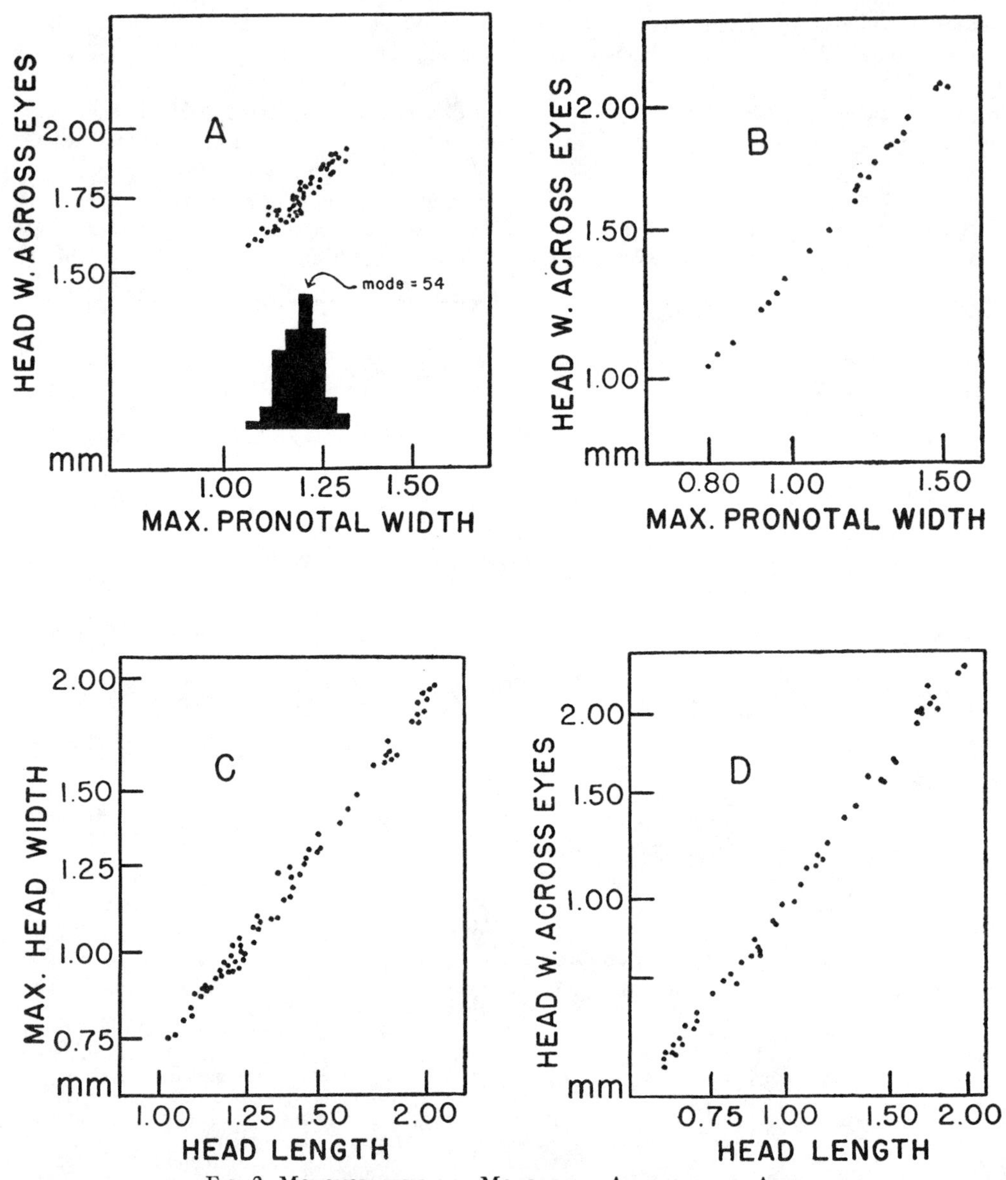

FIG. 2. MONOMORPHISM AND MONOPHASIC ALLOMETRY IN ANTS

A, monomorphism in *Formica exsectoides* Forel, characterized by isometry, limited size variation, and unimodal frequency. B–D, monophasic allometry over considerable size variation in B, *Megaponera foetans* (Fabricius) (subfamily Ponerinae), $k = 1.2$; in C, *Orectognathus versicolor* Donisthorpe (subfamily Myrmicinae), $k = 1.2$; in D, *Solenopsis geminata* (Fabricius) (subfamily Myrimicinae), $k = 1.2$. All three of these species show size frequency which is bimodal to some degree. Allometry in all four species shown as a log-log plot. Maximum calculated error in A, ± 0.010 mm; in B, ± 0.014 mm; in C, ± 0.020 mm; in D, ± 0.017 mm. A, single colony from Sudbury, Mass., E. O. Wilson leg.; B, material from several African localities, in Wheeler Coll.; C, material from five colonies several localities, Australia, W. L. Brown leg., data partly from Brown (unpub.); D, single colony from Mobile, Ala., E. O. Wilson leg.

fall just above the critical point of the break. Diphasic allometry appears to be a mechanism allowing the stabilization of a small caste, while at the same time providing for the production of a markedly different major caste over a slight size increase. The lower segment of the allometric regression line is usually nearly isometric, so that individuals falling within a large portion of the size range tend to be proportionately uniform in structure, but the upper segment is strongly allometric, with the result that a small increase in size yields a distinct new morphological type. In the fungus-growing ant *Atta texana* (Buckley) the cephalic allometry is actually reversed from negative to positive at the critical point (Fig. 3). The upper segment leads to the large-headed soldiers which in most species of *Atta* function primarily in defense of the nest (C. R. Gonçalves, pers. commun.). If this higher equilibrium constant were maintained past the critical point without a break, forms somewhat below the critical point would become microcephalic and probably inviable. A shift to a negative constant allows the production of very small normal-headed workers, which are very important or even essential for tending the fungus gardens on which the ants live.

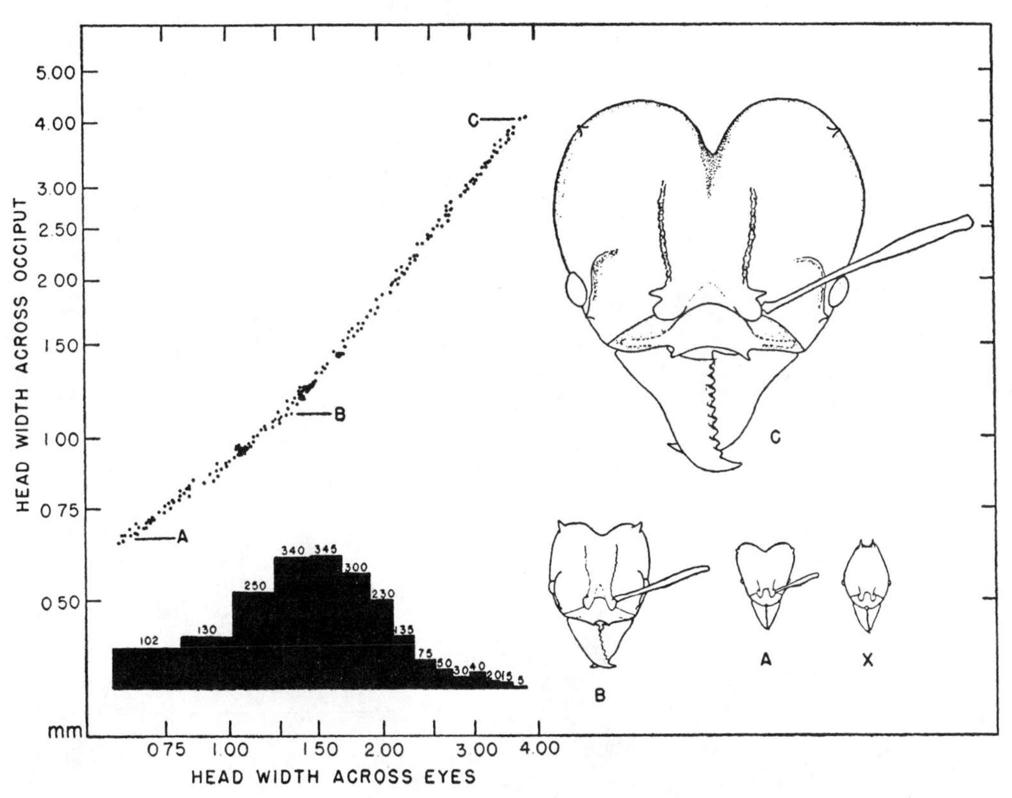

FIG. 3. DIPHASIC ALLOMETRY IN ATTA TEXANA (BUCKLEY) (SUBFAMILY MYRMICINAE)
Log-log plot. Note the great range of size variation and probably bimodal frequency curve. A, head of minima worker; B, head of small media at the point of the break in the allometric regression line, where the occiput width is relatively narrowest; C, head of major worker; X, head of hypothetical worker which would presumably result if the upper segment of the allometric regression line were to be extended to the minima size class. k of lower segment, 0.8; of upper segment, 1.3. Calculated maximum error, $\pm$ 0.013 mm. Based on a single colony collected by U.S.D.A. workers at Austin, Texas. The frequency curve was modified from measurements made on total length by W. J. Cloyd (1950, unpub. Master's thesis, Univ. of Tennessee).

4. *Triphasic allometry*

The log-log allometric regression line "breaks" at two critical points and consists of three segments; the terminal segments have varying equilibrium constants, but the middle segment is very highly allometric. The regression line greatly resembles a logistic curve and could be interpreted in terms of curvilinearity, but the concept of straight segments is retained here for consistency. The effect of triphasic allometry is the stabilization of both the major and minor castes. This condition may succeed phylogenetically diphasic allometry, which first stabilizes the minor caste, or it may emerge from monophasic allometry, as seems to be the case in the partially dimorphic *Camponotus* (see below). Triphasic allometry has been analyzed in two forms, *Oecophylla smaragdina*

(Fabricius) and *Pheidole rhea* Wheeler (Figs. 4, 5). Both approach complete dimorphism, with very few intermediates in a nest series connecting the constituent curves. In both cases the trough between the curves corresponds to the middle segment of the allometric regression line. Significantly, the middle segment comprises an apparently unstable size class, for its regression coefficient possesses an unusually high standard error of estimate. *Pheidole rhea* shows the very interesting and rare condition of trimodality. This species is one of several polymorphic members in a genus which is predominantly completely dimorphic, and its unusual polymorphism may actually have been derived from a dimorphic condition. Triphasic allometry probably occurs also in *Cataglyphis bombycina* (Roger) and species of *Paracryptocerus*, but insufficient material has been available to allow this to be determined with certainty. An interesting case which may represent triphasic allometry in its incipiency is seen in *Camponotus abdominalis floridanus* (Buckley) (Fig. 6).

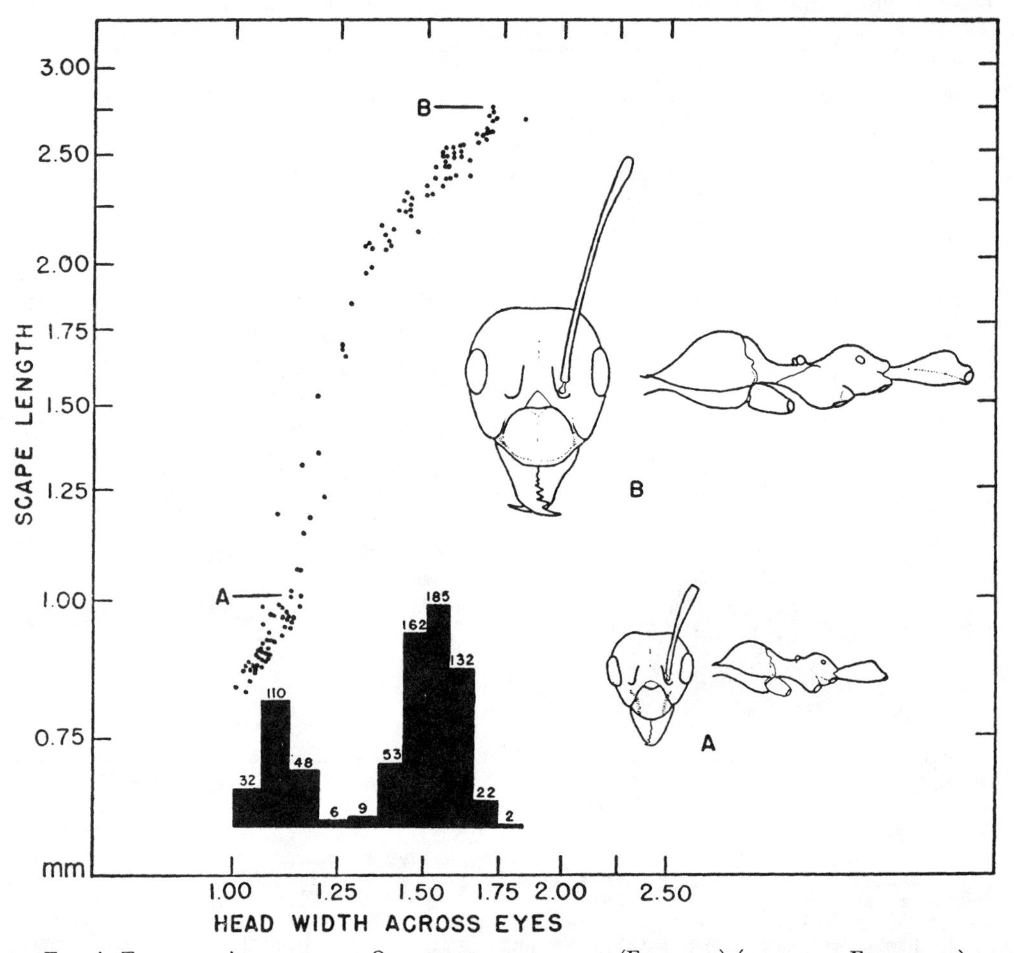

FIG. 4. TRIPHASIC ALLOMETRY IN OECOPHYLLA SMARAGDINA (FABRICIUS) (SUBFAMILY FORMICINAE)
Log-log plot. A, head and alitrunk-petiole of minor worker; B, same of major worker. Calculated maximum error ± 0.010 mm. Frequency modified from measurements by Cole and Jones (1948) on alitrunk length. Both curves based on two colonies from upper Assam; A. C. Cole leg. Figures based on Philippine material in the Wheeler Collection.

5. Complete dimorphism

Two separate size classes exist, separated by a gap in which no intermediates occur. Each class possesses an equilibrium constant approaching unity, but the allometric regression curves are not aligned, a fact suggesting that this condition may arise directly from triphasic allometry. Examples include most queen-worker segregations and a great many major-minor divisions in such genera as *Pheidole*, *Ischnomyrmex*, *Zatapinoma*, *Pseudolasius*, and *Camponotus*. Once complete dimorphism is established, the two resultant castes are capable of diverging further in some parts of the body, with no evident intergradient allometry. Profound changes may occur in the appearance or

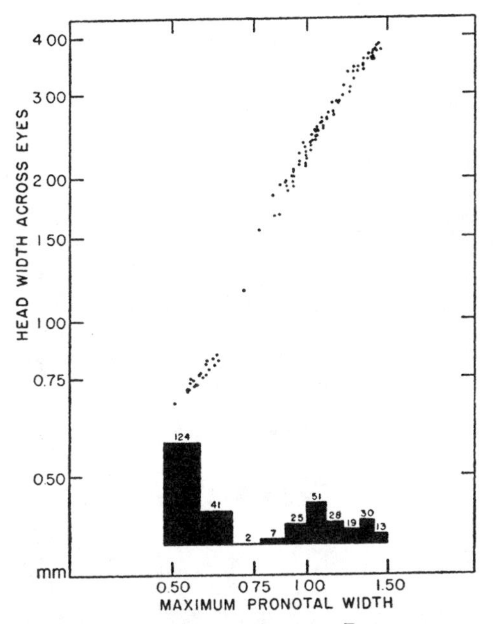

FIG. 5. TRIPHASIC ALLOMETRY IN PHEIDOLE RHEA WHEELER (SUBFAMILY MYRMICINAE)

Log-log plot. Note the exceptional trimodal frequency. The trough between the modes of the two larger size groups may correspond to an additional, fourth break in the allometric regression line; this is weakly evident in the above plot. Calculated maximum error, ± 0.010 mm. Based on a single colony, Nogales, Arizona, R. E. Gregg leg.

size of one or both of the castes, so that their original relationship might never be suspected if earlier phylogenetic stages were not exhibited by other species. In some groups the head of the largest workers comes to surpass in size that of the queens (some *Pheidole*, *Messor*, *Camponotus*, etc.), while in others the queen regresses in total size to become smaller than the average worker (*Formica microgyna* group). In at least two genera (*Melissotarsus*, *Acanthomyrmex*) there are species in which the non-allometric parts of the major may be equal in size to those of the minor, or smaller. It seems to be a rule among ants that once the secondary divergence of completely segregated castes becomes profound, the physiological thresholds separating them are no longer capable of being broken down to produce intercastes. At least one forceful exception exists, however; ergatogynes appear to be relatively common in the dimorphic genus *Oligomyrmex* (see, e.g., Kusnezov, 1951). Another feature of complete dimorphism is that it may be expressed in one character but not in another. In the queen-worker divergence, for instance, the alitrunk becomes dimorphic at a phylogenetically early stage, but recurrent allometric trends in the head may continue to link the two casts by a monophasic regression line.

DEFINITIONS OF CASTES

The nomenclature which has been built up in the past has always been excessively complicated and vacillating, which might be taken to be an implicit gesture of despair. The last attempt to treat comprehensively all of the recognized castes and to apply names to them was that of Wheeler in 1937. At this time Wheeler felt doubly justified in distinguishing each non-parasitogenic variant known to him, because he believed that each has arisen independently as a mutation. A reevaluation of Wheeler's classification reveals that many of the names proposed by him are superfluous, some are for practical purposes synonymous, and some are but stages in an allometric progression. Part of the reason for Wheeler's excess terminology was that he saw no difference between normal functional castes and true anomalies; to him all except the queen and typical male were basically anomalous forms, and he alternately referred to them as castes, phases, and anomalies. The necessity of distinguishing between normal castes and pathological forms has been previously emphasized. In the classification presented below, obviously pathological forms have been eliminated, and an attempt has been made to utilize names which will be useful in future descriptive analyses and most consistent with the allometric character of the female castes.

1. *Male*. Ordinarily possessing a generalized hymenopterous thorax and fully developed, non-deciduous wings. Apterous in some parasitic species such as the aberrant *Crematogaster* (*Apterocrema*) *atitlanica* Wheeler.

2. *Ergatomorphic male*. With normal male genitalia and a worker-like body. So close is the resemblance in some species of at least the anterior part of the body to that of the worker that it may

be eventually shown, as Wheeler has suggested, that these forms are actually persistent ergatandromorphs. Occurs in species of *Ponera, Cardiocondyla*, and *Formicoxenus*.

3. *Queen*. As recognized by Wheeler and others, the fully developed reproductive female, possessing a generalized hymenopterous thorax and fully developed, deciduous wings. This caste includes the teratogyne and microgyne of Wheeler.

4. *Worker*. The abortive, ordinarily sterile female, possessing reduced ovarioles and a greatly simplified thorax, the nota of which are usually represented by no more than a single sclerite each (Tulloch, 1935). Includes the macrergate, gynaecoid, and cryptogyne of Wheeler. Subcastes in a polymorphic worker series are designated approximately according to the relative size of the allometric organ as the *major* (soldier), *media*, and *minor*. In some cases of triphasic allometry these correspond to the three segments of the allometric regression line, and in complete dimorphism the media drops out.

5. *Ergatogyne*. Individuals falling along the allometric progression connecting the queen and worker castes, ranging from subapterous forms with queen-like alitrunks to slightly gynecoid workers. The normal major workers of genera such as *Pheidologeton* have ergatogynic thoraces, so that it is difficult to draw a precise line between these two castes. Perhaps it will prove best in the future to consider only those individuals in the upper segment of the diphasic allometric regression line as ergatogynes. Stable ergatogynes are the normal reproductives of some species of *Leptogenys* and *Monomorium*, and they are the only female caste known in the parasitic genus *Epixenus*. This caste probably includes the aberrant workers of the genus *Diacamma*, designated by Wheeler as "diacammatogynes." Pseudogynes, once assumed to be pathological ergatogynes, have recently been shown to be the result of an unrelated type of abnormal worker growth probably initiated by specific hormonal secretions (Novak, 1948).

6. *Physogastric ergatogyne*. Many ergatogynes show a divergent trend away from the normal queen-worker series in that the development of the gaster and postpetiole outpaces that of the thorax and head. The forms which are thus produced possess a gaster which approaches in size that of the typical female, while the thorax and head are only slightly ergatogynic. These intercastes may serve as complementals or may replace the normal queen. Includes the ergatoid and physergate of Wheeler.

7. *Dichthadiiform ergatogyne*. This caste appears to be the extreme stage of the physogastric trend, of which the reproductive female of *Acanthostichus quadratus* Emery might be said to be an intermediate stage. The total size is greatly increased, the gaster is huge, and the postpetiole is expanded

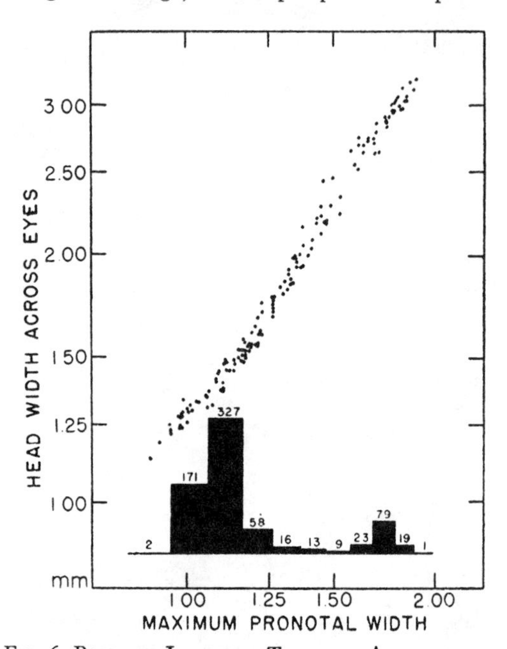

FIG. 6. POSSIBLE INCIPIENT TRIPHASIC ALLOMETRY IN CAMPONOTUS ABDOMINALIS FLORIDANUS BUCKLEY (SUBFAMILY FORMICINAE)

Log-log plot. Calculated maximum error, ± 0.013 mm. Frequency based on a single colony, Welaka Co., Fla., A. Van Pelt leg. Allometry based on this colony and on two others from Orange Beach, Baldwin Co., and Chattahoochee St. Pk., Houston Co., Ala., E. O. Wilson leg.

to the extent that it appears to be the first gastric segment. One gets the impression of a greatly increased positive allometry in the posterior part of the body, with a growth center located in the gaster and a growth gradient proceeding anteriorly to enlarge also the petiole and propodeum. In addition, the head is broadened and rounded and the mandibles are often falcate. Occurs typically in most groups with a legionary mode of life, and includes all known Dorylinae and Leptanillinae and the ponerine genera *Simopelta* and *Onychomyrmex*.

THE THEORETICAL BASIS OF ALLOMETRY

Huxley (1927, 1932) was the first to show that polymorphism in ants can be related to allometric growth in other animals. At the same time he made the excellent suggestion, surprisingly neglected by subsequent writers on caste determination, that the presence of a gradient allometric series connecting the major and minor castes of the same species can be explained most simply on the basis of variable nutrition. Since Huxley's inauguration of the allometry concept in the 1920's, a great deal of significant experimental and descriptive research has been added to it, with the result that much more is now known concerning its physiological basis and general applicability (see, e.g., Brody, 1945; Le Gros Clark et al., 1945).

In the present study all of the growth-ratio curves have been analyzed according to the allometric equation, but not all are known to conform exactly to allometry. The more strikingly polymorphic forms with considerable size variation, such as *Solenopsis*, *Atta*, and *Camponotus*, have growth curves which have been demonstrated in the present study to fit the power equation far more precisely than any other form of general equation. In many other cases, especially where the curve proves to be irregular or polyphasic, the logarithmic plotting is by far the most amenable to descriptive analysis. For pragmatic reasons alone, then, the power equation has been used consistently throughout.

Simple allometry is expressed as $y = bx^k$ where y is the size of the allometric organ, x is the size of the rest of the body or of an organ against which y is compared, b is the "initial growth-index," and k is the "equilibrium constant." This can be converted into logarithmic form as $\log y = \log b + k \log x$, so that no matter what the value of k, the regression line should be rectilinear on a double logarithmic plot. This equation can be directly derived from the differential multiplicative growth of two organs:

$$y = ce^{k_1 t_1} \quad \text{and} \quad x = de^{k_2 t_2}$$

where t is time. Taking the natural logarithms and differentiating, the two specific growth rates are obtained:

$$k_1 = \frac{dy}{ydt} \quad \text{and} \quad k_2 = \frac{dx}{xdt}$$

The equilibrium constant k is the ratio of the specific growth rates k_1/k_2 when $t_1 = t_2$. The times of development can be considered equal in the derivation of this equation if one of two conditions is assumed: either that the developments of the two organs started at the same time, or that the relationship is considered only over a segment of the developmental time. In considering allometric series in holometabolous insects, the full course of development is never discerned and only the latter condition of time can be assumed, so that the initial growth index b has no significance. For practical purposes, in the present study only the equilibrium constant k has been considered, and this has been obtained geometrically by measuring the slope of the log-log regression line. The value thus obtained has been found by repetitive tests to be precise to the first decimal place, which is sufficient for descriptive purposes. Future work, involving exact comparisons, may demand that the equilibrium constant be carried to a further decimal place. It should be remembered, however, that a constant based on linear measurements carries no physiological implications by itself, since it is only an approximate function of the absolute allometric value, which appertains to mass or volume.

It can be seen that allometry might have a very sound theoretical basis if growth could be shown to be always a uniformly multiplicative process. Unfortunately, this is rarely the case. The physiological mechanisms of differential growth are only partly understood, and in certain features, such as competition between parts and specific organ competence in relation to age, the data appear to be partly conflicting (Reeve and Huxley, 1945). There is also an outstanding paradox involved, in that certain complex histogenetic phenomena, such as the development of membrane bones, are totally unlike simple multiplicative growth and yet still conform to the power equation. However, despite these exceptions and theoretical difficulties which have come to light, allometry remains the most general and meaningful expression of relative growth.

A number of cases of deviations from typical allometry have been recorded from various groups of animals. Two types have been discovered in ants: polyphasic regression lines, which have already been shown to represent a fundamental trend in the development of castes; and lines which remain curvilinear on a log-log plot. Neither seem amenable to a physiological explanation at the present time. Polyphasic regression lines have been reported before in other arthropods, where their critical points are usually associated with successive molts (Teissier, 1937; Yasumatsu, 1946). Curvilinear regression on log-log plots has been obtained in scape length against head width in *Atta texana* (Fig. 7) and in mandible length against head length in *Orectognathus versicolor* Donisthorpe. It may be of significance that similar curvilinear trends have been found in the rostrum and mandible lengths of several genera of fish (Reeve and Huxley, 1945). The case of *Atta*, where two closely approximated structures show different types of allometry, brings to emphasis one of the weaker aspects of any statistical analysis of the sort conducted here on ants. Where complex allometry is involved, we speak of the allometric trend of a given species, but we are actually referring to the most prominent allometric character which has been chosen for study. Only by an exhaustive treatment of all parts of the body and the localization of growth centers and growth gradients can the real patterns of growth be described.

Still one more complication remains in the physiological interpretation of ant polymorphism. As in similar studies on other holometabolous insects, measurements are always taken from permanently formed imagines and therefore do not involve ontogenetic growth directly. Huxley (1932, 1945) is of the opinion that allometry in this case is derived through differential growth of adult tissues within the pupa, but a careful examination of the details of metamorphosis shows that the process is actually far more limited than this. The gross adult form, complete with the major allometric relationships of the external organs, is revealed suddenly in the early pupa at the last larval ecdysis, while the internal adult tissues which fill this mold do not begin development until a later period. It is evident that the allometry is first expressed in the configuration of the pupal hypodermis, so that the differential growth involves the multiplication and deployment of the epidermal cells. How such a process may occur has been determined recently by C. M. Williams (pers. commun.). In the moth *Platysamia cecropia* (Linné) the imaginal discs remain isometric in relation to the rest of the body and to one another through the larval period. During the prepupal stage, certain imaginal discs begin a precipitous growth while others, such as those destined to form the gonads and external genitalia, remain dormant and begin to grow rapidly only after the onset of adult development within the pupa. Also during the prepupal stage the pupal hypodermis is laid down in the course of a spurt of growth and forms molds in which the adult wings, antennae, and legs develop. Since the proliferation of the discs which form the various areas of the hypodermis proceeds at various rates, it is conceivable that at least the major features of allometry expressed between adult insects originates through simple multiplicative growth at this level. Lesser details of allometry in segmentation, spination, pigmentation, etc., may not appear until the onset of adult development. This particular aspect of allometric growth should prove a rich field for future research.

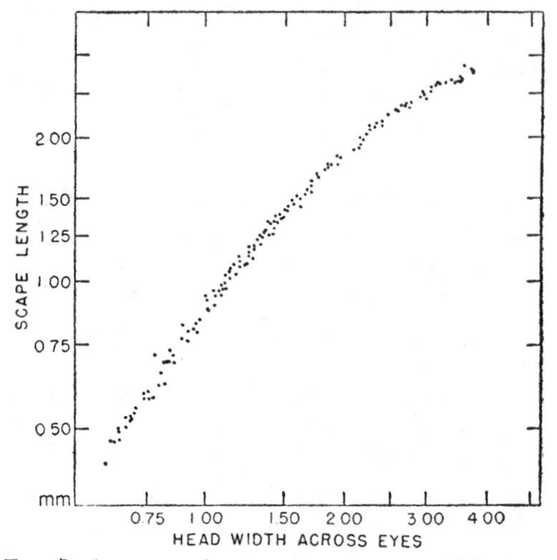

FIG. 7. ATYPICAL ALLOMORPHOSIS IN THE MANDIBLE LENGTH OF ATTA TEXANA, SHOWING EVEN CURVILINEARITY IN A LOG-LOG PLOT

Same colony as in Fig. 3. Calculated maximum error ± 0.013 mm.

AN INDEX OF POLYMORPHISM

It would be very desirable, if possible, to obtain a quantitative measure of polymorphism which could be used in comparing species or higher categories, because subjective evaluation has proven patently inadequate. It is especially difficult to decide which of two species is the more polymorphic when their allometry is complex and concerns different dimensions, and it is usually impossible to apply a simple statistical test in such a case. But where two species show the same type of allometry, as for instance in the broadening of the head, there is no reason why they cannot be compared directly. The degree of polymorphism of any species is a function of allometry and the range of the intranidal size variation. One index which is compounded of these two characteristics has been devised as follows:

$$\text{polymorphism index} = 100 \times \frac{v_2 - v_1}{m} \times |(k - 1)|$$

where $(v_2 - v_1)/m$ is the difference between the extreme measurements of the non-allometric dimension divided by their average and $|(k - 1)|$ is the absolute value of the equilibrium constant minus one. Such an index will have meaning only if a single organ of the body has marked allometry and the allometry is monophasic. It should prove expecially useful in comparing examples of feeble polymorphism across taxonomic groups and in descriptions of polymorphism within circumscribed taxonomic units. The indices of some representative polymorphic species are given in Table 1. In each case y = width of head across eyes, and x = maximum pronotal width. Those species with equilibrium constants between 1.0 and 1.1 have been assigned the constant 1.1.

TABLE 1

A comparison of the index of polymorphism in various species of ants

SPECIES	LOCALITY	k	POLYMORPHISM INDEX
Eciton nigrescens (Cresson)	Tuscaloosa, Alabama	1.2	9.4
Solenopsis saevissima richteri Forel	Iguazu, Misiones, Argentina	1.1	4.6
Solenopsis interrupta Santschi	Tucuman, Argentina	1.1	6.9
Solenopsis geminata (Fabricius)	Mobile, Alabama	1.5	52.0
Camponotus castaneus (Latreille)	Knoxville, Tennessee	1.9	96.9
Lasius fuliginosus spathepus Wheeler	Odawara, Japan	1.1	1.5

THE ORIGIN OF THE WORKER CASTE

The queen and worker castes are differentiated in all taxonomic groups by complete dimorphism in the structure of the alitrunk (thorax-propodeum) and the abdomen. In a few species of higher ants, including members of the myrmicine genera *Leptothorax*, *Monomorium*, and *Formicoxenus*, the two castes are connected by a nearly complete, gradient series of intermediates. These upon analysis show diphasic allometry in the alitrunk (see Fig. 9) and abdomen (gaster) and monophasic allometry in the head. Other species, including members of the ponerine genus *Leptogenys* and the myrmeciine genus *Myrmecia*, show occasional intermediates which maintain the same pattern. If our assumption is correct that this secondary breakdown of the queen-worker segregation follows the original allometric path of connection, it is probable that the worker caste has been derived phylogenetically from the queen by the same mechanism which later divided it into subcastes. The lower segment of the diphasic thoracic and abdominal line yields a progression of forms which differ little in proportion from the worker castes of other ants, while the upper segment gives rise over a short size increase to typical queens or queen-like ergatogynes. This modification of allometry appears to be directed toward the stabilization of the smaller worker caste, which was at first derived with little size change from the fully developed queen.

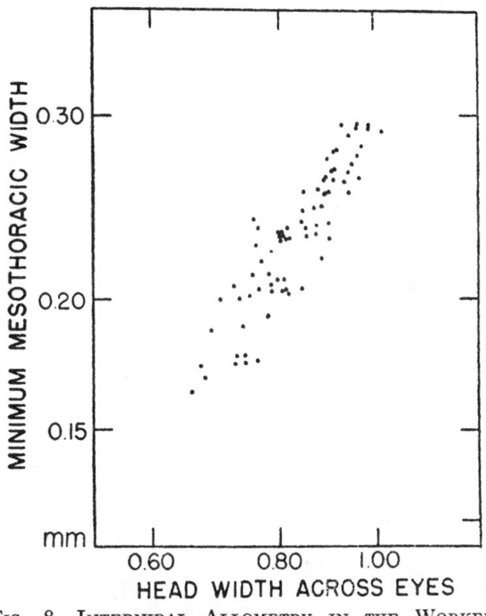

FIG. 8. INTERNIDAL ALLOMETRY IN THE WORKER CASTE OF A MONOMORPHIC FORMICINE ANT, PRENOLEPIS IMPARIS (SAY)

Log-log plot. Based on material from 45 localities over North America, exclusive of California and Oregon. Calculated maximum error, ± 0.006 mm.

The probable immediate result of the original queen-worker segregation can be seen in *Myrmecia*, considered with *Nothomyrmecia* to contain the most primitive of living ants. Here the queens are not much larger than the largest workers and are morphologically nearly identical to them except for the great development of the meso- and metathorax and consequent architectural effects on the rest of the alitrunk. Intercastes are common in this genus and with respect to the alitrunk fall along the steep incline of the upper segment of the allometric curve connecting the two principal castes. The condition seen in the several myrmicine genera mentioned above can be considered to be a secondary adaptive modification, while intercastes in the genus *Myrmecia* appear to be exceptional and superimposed upon a strongly marked dimorphism.

An important aspect of the usual queen-worker differentiation which has thus far escaped proper attention is that the head does not participate in

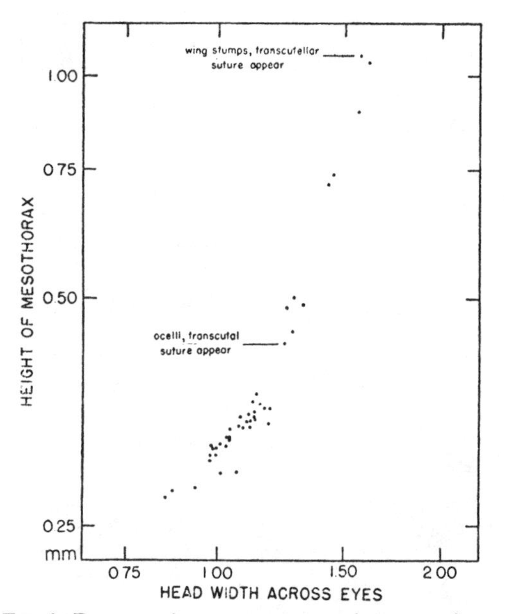

FIG. 9. DIPHASIC ALLOMETRY IN THE ALITRUNK SEEN IN QUEEN-WORKER INTERGRADATION IN MONOMORIUM CINCTUM WHEELER

Log-log plot. The lower segment of the regression line comprises the worker caste, while the upper segment comprises ergatogynes which range from worker-like forms to subapterous forms very close to the typical queen of related species. Based on the type series of *M. cinctum* and one series each from Ferntree Gulley and Lake Purrumbete, Victoria, W. L. Brown leg. Calculated maximum error, ± 0.006 mm.

the rigorous dimorphic segregation shown by the rest of the body, but tends to conform to worker allometry. This means that the shape of the head of the queen is such that various dimensions measured one against the other will fit an allometric regression line derived from those same measurements in the worker. Where the majors of a polymorphic worker series surpass in head-size the queen, the head shape of the queen is similar or identical to that of the media or major class with comparable head size. This is clearly shown in some very size-variable species such as *Orectognathus versicolor* Donisthorpe. Here the head of the queen approximates in size and shape that of a large media, and variability in size within the queen caste produces an allometric progression identical to that seen in the worker caste. A similar precise relationship is encountered in *Myrmecia*, which has basically polymorphic workers. It is also exhibited by various groups of the primitive subfamily Ponerinae (Amblyoponini—*Amblyopone*; Ponerini — *Paltothyreus*, *Odontoponera*, *Termitopone*, *Neoponera*, *Pachycondyla*, etc.) and in certain primitive Myrmicinae (Myrmicini—*Pogonomyrmex*). The relationship is only approximate in the higher Myrmicinae and in other subfamilies with polymorphic species. The queen shows obvious modifications surrounding the retention of fully developed eyes, her cephalic spination is often reduced, and her mandibles tend to retain the elementary worker dentition, which is aborted in the majors of many granivorous species. In some ants, especially those which show complete worker dimorphism, the head shape of the queen cannot be related in any way to that of the worker.

This peculiar cephalic bond occurring between the queen and worker defies physiological explanation at the present time. Allometry ordinarily is a relationship which can be drawn between any two parts or dimensions of the body, and the equilibrium constant is in theory most simply derived as the ratio of two specific growth rates. The queen-worker cephalic allometry clearly shows that relative growth of a different nature operates in ants. The alitrunk displays allometric variation which presumably has become diphasic and then dimorphic, although retaining the diphasic potentiality. The head has taken a different course, retaining monophasic allometry which is correlated with thoracic allometry only as this appears secondarily in the worker. This means that the various dimensions of the head which make up its shape and structure are actually related to the rest of the body only as this determines total head size and not as its constituent parts vary allometrically.

Yet even more extraordinarily, there remains one set of cephalic structures, the ocelli, which often varies directly with the body and independently of the rest of the head. All three ocelli are present and fully developed in all size classes of workers in *Myrmecia* and in a few other, scattered groups of ants, such as some Pseudomyrmicinae, the cerapachyine genus *Neophyracaces*, and many formicine genera. In most ants, however, the queens possess the three ocelli and the workers possess only one (the median) or none. That this is not an allometric character related to the head is seen in certain polymorphic species which have majors surpassing in head size the queen. In these, including many *Pogonomyrmex*, *Messor*, *Orectognathus*, *Pheidole*, *Pheidologeton*, and *Camponotus*, the queen head shape fits approximately the worker regression lines, but the media and major forms around and above the placement of the queen in these lines possess at the most only one ocellus. The most logical conclusion which presents itself is that the two posterior ocelli develop because of an influence, perhaps hormonal, which somehow originates from the massive bulk of the queen's body. The occasional appearance of the median ocellus in the upper segment of the worker allometric series indicates that either the total size of the body or some part of the body behind the head is the determining factor. Furthermore, the appearance of all three ocelli in some physogastric ergatogynes ("ergatoids" of Wheeler) which have worker-like alitrunks and heads and distended ovarian abdomens, suggests that the ovaries may be the most important specific loci influencing the ocelli.

THE EVOLUTION OF WORKER POLYMORPHISM

Since queen-worker cephalic continuity is seen to be widespread in polymorphic ants, the possibility is now presented that the diffrences in proportion between the worker and queen may be a measure of the potential allometry of the worker caste, even in essentially monomorphic species. Three important pieces of evidence support this. First, in those groups of ants which are totally monomorphic and have no evident immediate polymorphic ancestors, such as the Ectatommini, Typhlomyrmicini, and Proceratiini among the ponerines, the differences in head shape between the queen and worker castes tend to be very slight. Second, in some species which show internidal, but not intranidal, cephalic allometry, the queen conforms to the worker regression lines exactly as do the queens of polymorphic species; examples include *Paltothyreus tarsatus* (Fabricius), *Neoponera villosa* (Fabricius), *Pachycondyla crassinoda* (Latreille), *Pogonomyrmex badius* (F. Smith), *Vollenhovia pedestris* (F. Smith), and *Crematogaster chasei* Forel. Third, in a few small groups such as the myrmicine genus *Wasmannia*, the queen-worker cephalic differences seen from species to species seem to be a positive function of the difference in head size.

The queen-worker cephalic difference, involving typically an expansion of the head and particularly of the occipital lobes, is a character found throughout nearly every major taxonomic group of ants. If the queen head shape is truly consistent with the worker allometry, as the evidence above seems to indicate, this means that the great majority of genera possess latent cephalic allometry to some degree. It follows that *internidal* allometric variation would result from the acquisition of genetic internidal size differences, and *intranidal* allometric variation (worker polymorphism) would result from the increase of the normal phenotypic size variability which is based on the limited genotype possessed by a single average colony. This is an extremely significant concept, because it could explain why and how worker polymorphism appears in so many groups of ants independently, and nearly always follows the same trends in each.

From species to species through the Formicidae the worker caste shows nearly every conceivable step in a transition from complete monomorphism to complete dimorphism. The major stages into which this transition can be divided have been defined previously. The succession of these stages and their potentiality for reversal as understood at the present time are represented in Fig. 10. The evidence for such a phylogenetic series has been drawn from analyses of a number of taxonomic groups which show variable polymorphism and different associations of the several stages. For purposes of illustration some of the more significant groups are briefly described below.

1. The Ponerini. This tribe exhibits all nuances from monomorphism to strong monophasic allometry. *Ponera*, *Mesoponera*, *Euponera*, *Ectomomyrmex*, *Bothroponera*, and other genera show absolute monomorphism and little queen-worker cephalic difference. *Plectroctena mandibularis* F. Smith, *Pachycondyla crassinoda* (Latreille), *Neoponera villosa* (Fabricius), and other genera and species show internidal, monophasic cephalic allometry but not true worker polymorphism. The cephalic allometry is expressed within individual colonies to produce polymorphism in *Termitopone laevigata* (F. Smith) and *Megaponera foetans* (Fabricius). In the latter species strong bimodality is developed, with the intermediates becoming relatively scarce.

2. *Pogonomyrmex*. This genus shows within its

species very concisely the origin of simple polymorphism. The majority of species are completely monomorphic, and some, including members of the subgenus *Ephebomyrmex*, show little or no queen-worker cephalic difference. Most of the members of the subgenus *Pogonomyrmex*, however, show in addition considerable queen-worker cephalic difference, and in a few species, such as *P. barbatus* (F. Smith), these differences can be demonstrated to be linked exactly with internidal, monophasic worker allometry. In at least two species, *P. badius* (Latreille) and *P. wheeleri* Olsen, the allometry is expressed within normal nest series, producing worker polymorphism.

3. *Pheidole*. This genus shows the intimate connection between triphasic allometry and complete dimorphism. Most species are completely dimorphic, with each of the two subcastes limited to separate allometric regression lines, which in turn possess different equilibrium constants. In *P. rhea* Wheeler (Fig. 5) and *P. vasliti arizonica* Santschi the majors and minors appear on superficial examination to be connected by a complete series of intermediates. Analysis shows that actually the intermediates represent an extension of the major regression line to a point near the minor regression line. The minors, however, undergo no change and remain a segregated and size-limited group. If the major line were to be extended further down, it would by-pass the terminal point of the minor line; several intermediates found in the case of *P. rhea* indicate that the allometry is actually triphasic.

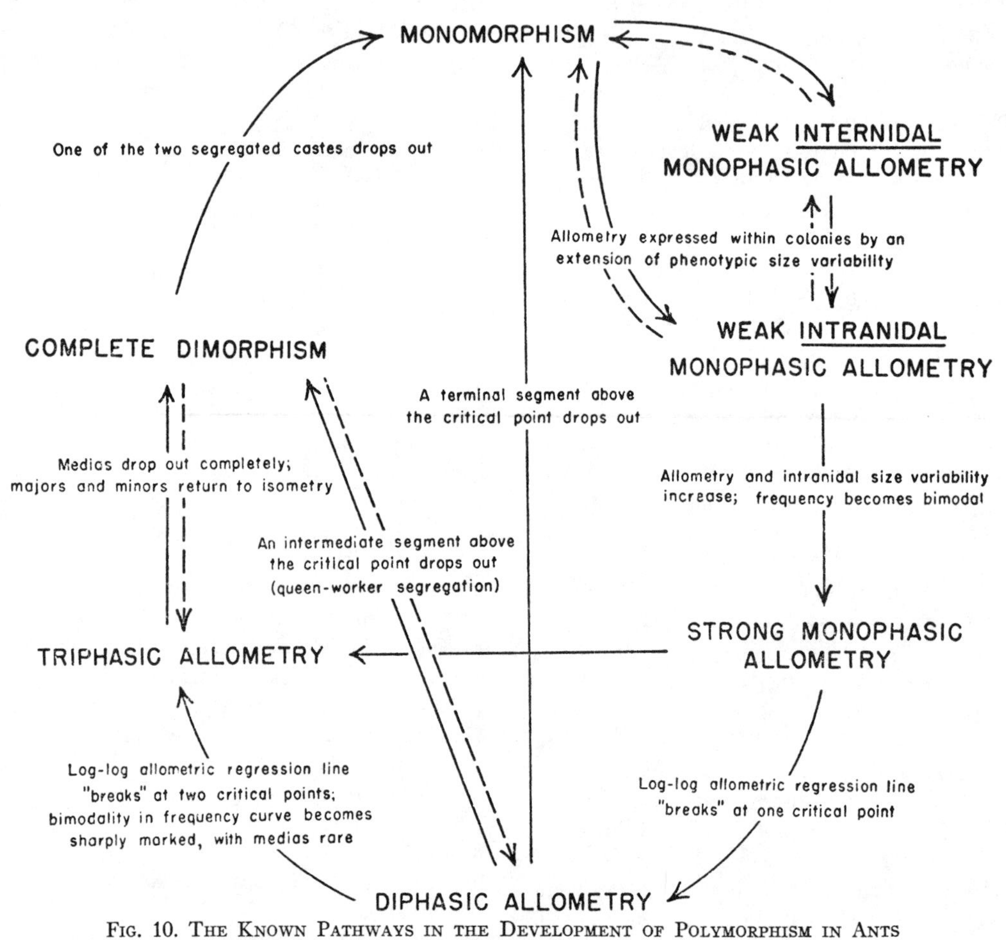

FIG. 10. THE KNOWN PATHWAYS IN THE DEVELOPMENT OF POLYMORPHISM IN ANTS
Further explanation in the text.

4. Miscellaneous genera of the pheidoline-myrmecinine-leptothoracine-etc. tribal complex. The genera composing this group, at present only imperfectly divided into tribes, run nearly the entire

gamut of polymorphism. Species of *Megalomyrmex* and *Huberia* are completely monomorphic, and only in the latter genus is there much queen-worker cephalic difference. *Vollenhovia pedestris* (F. Smith) shows internidal cephalic allometry in the worker but not true worker polymorphism. *Heteromyrmex rufiventris* (Forel) shows weak monophasic cephalic allometry expressed over a very great size range in a single colony. *Tranopelta subterranea* Mann and species of *Solenopsis* s. str. show relatively strong monophasic cephalic allometry over a fair size range and therefore are markedly polymorphic. The workers of *Monomorium* (*Holcomyrmex*) *criniceps* Mayr and *Solenopsis* (*Diplorhoptrum*) *robustior* Santschi, in the material which I have been able to examine, exhibit weak diphasic allometry over considerable size ranges. The worker castes of most of the species of *Solenopsis* (*Diplorhoptrum*), and possibly many of *Monomorium* also, are secondarily monomorphic, comprised of small individuals showing exceedingly little size variation and a form quite different from that of the queen. It is very possible that this condition has been derived from diphasic allometry of the type exhibited by *S. robustior* and *M. criniceps*, since it is only necessary to take away the upper segment of the diphasic log-log regression line to produce a worker caste similar to that of the monomorphic species. Extreme monomorphism, associated with small size and in some cases unquestionably derived secondarily from polymorphism, has been developed by several genera of the tribal complex and ordinarily for one or the other of two outstanding adaptive reasons. *Carebara*, *Carabarella*, *Liomyrmex*, and most *Solenopsis* (*Diplorhoptrum*) are thief ants, living in tiny galleries adjacent to the nests of larger species of ants or termites from which they steal food. *Allomerus*, probably *Xenomyrmex*, and some *Solenopsis* (*Diplorhoptrum*) live in the cavities of plants. Another, quite different trend which some genera have followed is the total loss of the worker caste through social parasitism on other ants. Examples include *Wheeleriella*, *Anergates*, *Epoecus*, and probably *Hagioxenus*. Finally, the parasitic genus *Epixenus* has apparently lost all female castes except for a worker-like ergatogyne.

5. The Dacetini. This tribe, which has recently been intensively studied by Brown (1953), is especially significant in that changes in polymorphism can now be correlated with generic phylogeny and partly with ecological evolution. According to Brown, dacetines can be divided into four basic stocks, the Dacetiti, the Orectognathiti, the Epopostrumiti, and the Strumigeniti. In each of these the primitive genera are polymorphic and usually epigaeic, and there is evidence to suggest that these features were derived from a common ancestor. Derivative genera have arisen in each which possess workers tending to be smaller, monomorphic, and hypogaeic (or at least cryptobiotic). In the genus *Strumigenys*, one species (*S. loriae* Emery) shows polymorphism of such extent that the major workers surpass in head size the queen. But the occurrence of this polymorphism is erratic, and it is often not expressed at all in smaller colonies. Furthermore, other *Strumigenys* and members of *Strumigenys*-derivative genera show monomorphism which gives the appearance of having been evolved from the kind of polymorphism seen in *S. loriae* by a suppression of intranidal size variability and a retention of only large medias with head size comparable to that of the queens. The association of monomorphism with a cryptobiotic mode of life in the Dacetini is illustrative of a trend common in ants.

6. *Camponotus*. This genus shows in gradual steps the transition from strong monophasic allometry to complete dimorphism. The majority of species show marked polymorphism based on strong allometry combined with considerable intranidal size variation; those which have been analyzed thus far also show some degree of bimodality. The bimodality increases, with intermediates becoming scarce, in a few species such as *C.* (*Myrmothrix*) *abdominalis floridanus* Buckley and *C.* (*Myrmobrachys*) *planatus* Roger. These upon analysis have shown what appears to be feeble triphasic allometry. Complete dimorphism has been attained several times independently within the genus, occurring in *C.* (*Tanaemyrmex*) *inaequalis* Roger (frequency data in Falconer Smith, 1943) and in species of the subgenera *Colobopsis*, *Pseudocolobopsis*, *Hypercolobopsis*, and *Paracolobopsis*.

DIVISION OF LABOR

The adaptive result of the development of female polymorphism is the division of labor within the colony. In many ants the queen and worker castes have assumed functions so different as to be almost mutually exclusive. In mature colonies the queens are ordinarily devoted entirely to oviposition and the occasional tending of the brood; as recognized by Wheeler and others, they are functionally the female "germ cells" of the social organism. The workers, on the other hand, are concerned almost entirely with the labor of the colony, tending the brood, foraging, and excavating the nest. On occasion they may act as supplementary reproductives or may even usurp the ovipository function of the queen.

The division of labor which occurs within the worker caste is less distinct. Thorough studies by Buckingham (1911), Chen (1937), and Lee (1938) on *Camponotus* and by Doflein (1920) on *Messor* show that pronounced division of labor occurs in species with strong monophasic allometry, but the division is by preponderance only, and any worker is capable of performing any given normal task. It is possible for some degree of specialization to follow upon the slightest degree of intranidal allometry. It has long been known, for instance, that even in monomorphic species workers often differ in temperament and inclination to forage, and that this is especially pronounced between nanitic and normal-sized individuals. Chen (1937) has shown that within a given size class of *Camponotus* a leader-follower relationship is maintained which has some of the features of the "peck order" established in gatherings of many semisocial animals. This relationship shifts from day to day, but so slowly that the temperaments of the extreme individuals remain opposed to one another for long periods of time. It is not difficult to imagine, therefore, that slight morphological variability of the sort produced by cephalic allometry would cause behavioral variability. An actual division of labor in a weakly polymorphic species, *Formica rufa* L., has been recorded by Alpatov (1924), but otherwise borderline cases have received little attention from students of ant behavior.

Worker polymorphism in ants is a special adaptive condition which has been assumed by many different phyletic lines with various results. Although it is a signal feature of advanced insect societies, it does not have the general effect within the ants of increasing the complexity of social organization, and it is not an advanced phylogenetic character, since it is possessed to a marked degree by *Myrmecia* and occurs commonly in the primitive subfamilies Ponerinae and Cerapachyinae. Similarly, it does not necessarily confer selective superiority over competing monomorphic species. It should be emphasized that *Iridomyrmex*, the most abundant ant of the Baltic Amber, and *Crematogaster*, probably the most rapidly expanding genus of modern times, are both eminently monomorphic. Also, primarily monomorphic genera such as *Myrmica, Formica, Monomorium, Leptothorax*, and *Tetramorium* are today among the outstanding dominants among ants in various parts of the world. In general, polymorphism clearly operates to apportion among morphological specialists the labor which must be performed by any ant colony, polymorphic or monomorphic, which occupies the same general ecological situation. In intermediate stages of polymorphism, a common situation is one where the majors are primarily concerned with defense and nest excavation, the medias with the majors are primarily concerned with foraging, and the minors are primarily concerned with care of the brood. The principal result of extreme polymorphism, involving triphasic allometry or complete dimorphism, is typically the production among the castes of a major form especially and elaborately constructed for one or a very limited number of tasks. For instance, the majors of *Pheidole* have enormously enlarged heads and serve in many species as meat trenchers or seed crackers for the colony. The major of *Cataglyphis bombycina* Roger is most active in nest excavation; it has greatly elongated mandibles which work in conjunction with the maxillary palps in carrying large pellets of sand (Bernard, 1951). Perhaps the most extraordinary specialization of this type is seen in the dimorphic species of *Paracryptocerus* and *Camponotus* (*Colobopsis*, etc.), the majors of which have shield- or plug-shaped heads used in blocking the nest entrances against invaders (see, e.g., Kempf, 1952).

It would be appropriate at this point to stress that feature of social polymorphism which makes it truly unique in organic evolution. This is its function in differentiating forms with regard to their contribution to the selective value of the colony as a whole but without regard to their selective values as individuals. Social polymorphism is thus strongly distinguished from other types of polymorphism, including the genetic, which are maintained in populations because of the selective advantage they confer on individuals. In terms of evolution, the following analogy can properly and instructively be drawn: the ant colony behaves as a superorganism, the basic unit upon which natural selection operates, and polymorphism, originating from allometric growth series, acts as the "morphogenetic" process underlying the somatic differentiation of this superorganism. Worker polymorphism is an advanced stage of the somatic differentiation which can be readily assumed or reduced according to the immediate needs imposed by the environment.

A PHYLOGENETIC CONSPECTUS

Worker polymorphism has arisen many times in the phylogenetic history of the Formicidae, possibly originating through the simple mechanism of the extension of intranidal size ranges of internidally allometric species. During the present survey the condition has been encountered in every subfamily with the exception of the Leptanillinae and Pseudomyrmicinae. It occurs in some species of at least a fourth of all ant genera recognized at the present time. Despite the uncertainty and confusion in our present knowledge of ant phylogeny, it still can be safely said that worker polymorphism has appeared independently in a minimum of eight major taxonomic groups. It occurs as a primitive character in *Myrmecia*, in the Dorylinae, and possibly in the ponerine tribe Amblyoponini, and it is found sporadically through the Cerapachyinae and Odontomachinae. It appears again in the ponerine tribe Ponerini. Most members of the subfamily Ponerinae are absolutely monomorphic, with little queen-worker cephalic difference, and this character is carried through the tribe Ectatommini to their probable derivatives, the myrmicine tribe Myrmicini. Within the Myrmicini, at least two species of *Pogonomyrmex* have developed worker polymorphism again. In the Myrmicinae past this group, the Myrmicariini, the Crematogastrini, and the extensive pheidoline-myrmecinine-leptothoracine tribal complex all show polymorphism of variable degree. Many of the genera in these groups contain both monomorphic and polymorphic species, and the internidal-intranidal allometry transition, or its reverse, is clearly operative. Worker polymorphism has appeared at least twice, and possibly three times, in the subfamily Dolichoderinae, occurring in *Tapinoma*, *Azteca*, and *Zatapinoma*. *Aneuretus* and the tribe Dolichoderini, considered the most primitive dolichoderines, are both monomorphic and show little queen-worker cephalic difference. Polymorphism appears as an elementary character in the section Alloformicinae of Emery, comprising the most primitive genera of the Formicinae. Many of the members of the more advanced section Euformicinae are also primarily polymorphic, but the character has very likely arisen again secondarily in the genus *Formica*.

Complete dimorphism is a distinct and easily recognized condition which has evolved, without question independently, in the following nine genera: Myrmicinae — *Ischnomyrmex*, *Pheidole*, *Pheidologeton* (?), *Oligomyrmex*, *Acanthomyrmex*, *Paracryptocerus;* Dolichoderinae — *Zatapinoma;* Formicinae—*Camponotus* (*Tanaemyrmex*, *Colobopsis*, etc.), *Pseudolasius*.

SUMMARY

The preponderance of evidence at the present time indicates that in ants sex is determined by the genetic mechanism of haplo-diploidy and the female castes are determined by larval nutrition. In the case of complete dimorphism of the female castes, as in queen-worker and major-minor segregations, there probably exists a critical developmental time at which the imaginal discs assume one or the other of two alternative specific growth-rate potentials, depending on whether or not the larva has reached a certain threshold size. Past the critical time, the larva starts along a more or less regulated trend of growth and at the prepupal stage metamorphoses into the essential form of the adult caste predetermined by the imaginal disc potentials. In incomplete dimorphism, whether in the more primitive types of worker subcaste differentiation or in the rare cases of queen-worker intergradation, the females form a gradient series which exhibits simple or modified allometry. For this reason polymorphism is here defined in terms of allometry, and its lower limits are set as follows: allometry occurring over a sufficient range of size variation within a normal mature colony to produce individuals of detectably different form at the extremes of the size range.

The trend of the evolution of worker polymorphism in most phyletic lines is first toward allometric differentiation along a gradient size series and second toward dimorphism, or the segregation of two extreme size groups with the elimination of intermediates. Several successive phylogenetic stages between monomorphism and dimorphism, characterized by correlated changes in the frequency and log-log allometric curves, have been encountered repeatedly. Dimorphism is preceded by a "breaking" of the allometric curve at one or two points; in the latter case the resultant middle segment corresponds to the deepening trough of the bimodal frequency curve. Dimorphism is finally attained by an elimination of the middle segment (the media subcaste) and a return of the terminal segments (the major and minor subcastes) to isometry.

The worker caste probably arose from the queen by the same mechanism. Queen-worker intergradation shows diphasic allometry in the alitrunk (thorax-propodeum) and abdomen, and monophasic allometry in the head. There is a tendency for the head to conform to secondarily acquired worker allometry, even after the rest of the parts of the body have become completely dimorphic.

With the restriction of the concept of the caste to normal, functional forms and an acceptance of the allometric background of polymorphism, a considerable simplification of the terminology of polymorphism can be introduced. Only ten morphological forms are recognized here as being distinct castes: the male, the ergatomorphic male, the queen, the worker (consisting in worker-polymorphic species of major, media, and minor subcastes), the typical ergatogyne, the physogastric ergatogyne, and the dichthadiiform ergatogyne.

The differential growth underlying at least the gross features of polymorphism in adult ants must occur during the brief period of the proliferation and deployment of the pupal hypodermis before the last larval ecdysis. Lesser details of allometry may become apparent only at the onset of adult development within the pupa.

The adaptive significance of polymorphism lies in its apportionment among morphological specialists of the labor which must be performed by any ant colony occupying a similar ecological situation. However, worker polymorphism is not an advanced phylogenetic character, it does not have the general effect of increasing the complexity of social organization, and it does not necessarily confer selective superiority over competing monomorphic species. It is a special adaptive character which has arisen independently in a minimum of eight major taxonomic groups and has resulted in various types and degrees of division of worker labor. In many cases it has probably started from the simple mechanism of an internidal-intranidal allometry shift, which is effected by an increase in the size variability of individual colonies. Worker dimorphism has evolved independently a minimum of nine times. It usually has the adaptive result of producing among the castes a major form especially and elaborately constructed to perform one or a very limited number of tasks.

ACKNOWLEDGMENTS

The writer wishes to express his appreciation to several individuals for aid given during the course of this study. Drs. W. L. Brown, W. J. Crozier, and C. M. Williams reviewed parts of the manuscript and original data and offered excellent criticism and additional information. Mr. W. J. Cloyd, and Drs. A. C. Cole, A. Van Pelt, and R. E. Gregg supplied from their own collections some of the most important nest series used in the study.

LIST OF LITERATURE

ALPATOV, W. W. 1924. Die Definition der untersten systematischen Kategorien vom Standpunkte des Studiums der Variabilitat der Ameisen und der Crustaceen. *Zool. Anz.*, 60: 161–168.

BERNARD, F. 1951. Adaptations au milieu chez les fourmis sahariennes. *Bull. Soc. Hist. nat. Toulouse,* 86: 161–168.

BHATTACHARYA, G. C. 1943. Reproduction and caste determination in aggressive red-ants *Oecophylla smaragdina* Fabr. Bose-Research Institute, Calcutta.

BRIAN, M. V. 1951. Caste determination in a myrmicine ant. *Experientia,* 7: 182–186.

BRODY, S. 1945. *Bioenergetics and Growth.* 1023 pp. Reinhold Pub. Co., New York.

BROWN, W. L. 1953. Revisionary studies in the ant tribe Dacetini. *Amer. Midl. Nat.,* in press.

BUCKINGHAM, E. N. 1911. Division of labor among ants. *Proc. Amer. Ass. Adv. Sci.,* 46: 425–507.

CHEN, S. C. 1937. The leaders and the followers among the ants in nest building. *Physiol. Zool.,* 10: 437–455.

COLE, A. C., and J. W. JONES. 1948. A study of the weaver ant, *Oecophylla smaragdina* (Fab.). *Amer. Midl. Nat.,* 39: 641–651.

DOFLEIN, F. 1920. *Mazedonischen Ameisen. Beobachtungen über ihre Lebensweise.* 74 pp. Gustav Fischer, Jena.

FLANDERS, S. 1945. Is caste differentiation in ants a function of the rate of egg deposition? *Science,* 101: 245–246.

——. 1952. Ovisorption as the mechanism causing worker development in ants. *J. econ. Ent.,* 45: 37–39.

FORD, E. B. 1940. Polymorphism and taxonomy. In *The New Systematics* (J. Huxley, ed.), pp. 493–513. Oxford University Press. London and New York.

GÖSSWALD, K. 1933. Weitere Untersuchungen über die Biologie von *Epimyrma gösswaldi* Men. und Bemerkungen über andere parasitische Ameisen. *Z. wiss. Zool.*, 144: 262–288.

GREGG, R. E. 1942. The origin of castes in ants with special reference to *Pheidole morrisi* Forel. *Ecology*, 23: 295–308.

HASKINS, C. P., and E. V. ENZMANN. 1945. On the occurrence of impaternate females in the Formicidae. *J. N. Y. ent. Soc.*, 53: 263–277.

——, and E. F. HASKINS. 1950. Notes on the biology and social behavior of the archaic ponerine ants of the genera *Myrmecia* and *Promyrmecia*. *Ann. ent. Soc. Amer.*, 43: 461–491.

HUXLEY, J. 1927. Further work on heterogonic growth. *Biol. Zbl.*, 47: 151–163.

——. 1932. *Problems of Relative Growth.* 276 pp. Dial Press, New York.

KEMPF, W. W. 1952. A synopsis of the pinelii-complex in the genus *Paracryptocerus* (Hym. Formicidae). *Stud. Entomol.* (Petropolis, Brazil), No. 1. 30 pp.

KUSNEZOV, N. 1951. "Dinergatogina" in *Oligomyrmex bruchi* Santschi (Hymenoptera Formicidae). *Rev. Soc. ent. argent.*, 15: 177–181.

LEDOUX, A. 1950. Recherche sur la biologie de la fourmi fileuse (*Oecophylla longinoda* Latr.). *Ann. Sci. nat., Zool.*, 12: 313–461.

LEE, J. 1938. Division of labor among the workers of the Asiatic carpenter ants (*Camponotus japonicus* var. *aterrimus*). *Peking nat. Hist. Bull.*, 13: 137–145.

LIGHT, S. F. 1942–1943. The determination of the castes of social insects. *Quart. Rev. Biol.*, 17: 312–326; 18: 46–63.

NOVAK, V. 1948. On the question of the origin of pathological creatures (pseudogynes) in ants of the genus *Formica*. *Věstn. čsl. zool. Společ.*, 12: 97–129.

REEVE, E. C. R., and J. HUXLEY. 1945. Some problems in the study of allometric growth. In *Essays on Growth and Form* (W. E. Le Gros Clark and P. B. Medawar, eds.), pp. 121–156. Clarendon Press, Oxford.

SMITH, FALCONER. 1942. Polymorphism in *Camponotus* (Hymenoptera-Formicidae). *J. Tenn. Acad. Sci.*, 17: 367–373.

TEISSIER, G. 1934. *Dysharmonies et Discontinuities dans la Croissance.* 38 pp. Hermann & Co., Paris.

TULLOCH, G. S. 1935. Morphological studies of the thorax of the ant. *Ent. amer.*, 15: 93–131.

VANDEL, A. 1930. La production d'intercastes chez la fourmi *Pheidole pallidula* sous l'action de parasites du genre *Mermis*. *Bull. biol.*, 64: 475–494.

WHEELER, W. M. 1937. *Mosaics and Other Anomalies among Ants.* 95 pp. Harvard Univ. Press, Cambridge, Mass.

WHITING, P. W. 1938. Anomalies and caste determination in ants. (Review of Wheeler, ref. above). *J. Hered.*, 29: 189–193.

YASUMATSU, K. 1946. Some analyses on the growth of insects, with special reference to a phasmid, *Phraortes kumamotoensis* Shiraki (Orthoptera). *J. Agric. Kyushu imp. Univ.*, 8: 1–579.

1957

"Quantitative studies of liquid food transmission in ants," *Insectes Sociaux* 4(2): 157–166 (with Tom Eisner, second author)

It had long been known that members of the same colony in some species of ants exchange liquid food among themselves. Following pioneering experiments by H. L. Nixon and G. R. Ribbands with honeybees, Thomas Eisner and I used radioactive sodium iodide to trace the dispersal by repeated regurgitation of honey consumed by a single ant worker through the remainder of the colony. We found that the amount and rate of flow varied enormously among ant species. This form of food sharing has since been demonstrated to be important in the transmission of pheromones, as well as the mixing of hydrocarbons that go into the colony odor. For those species in which sharing approaches uniformity, the process also allows the crop, a chitinized segment of the foregut, to serve as a "social stomach" for the colony as a whole. This arrangement allows each worker to use its own nutritional state to continuously monitor that of all her nestmates.

As a footnote, Tom Eisner, who has been a close and lifetime friend, went on to a distinguished career in entomology and chemical ecology, as described in part by his superb book, *For the Love of Insects.*

QUANTITATIVE STUDIES OF LIQUID FOOD TRANSMISSION IN ANTS

by

E. O. WILSON and T. EISNER (1)

(The Biological Laboratories, Harvard University.)

INTRODUCTION

It is well known that food transmission[2] is a process of central importance in the social life of ants. The majority of phyletic groups possess an elaborate type of proventriculus the principle function of which appears to be to facilitate the storage of liquid food in the crop so that it can be regurgitated later to other members of the colony (Eisner, 1957). As LeMasne (1953) has emphasized, food transmission is an integral part of the complex adult-larva relationship which is essential to social cohesion in the ant colony. There is a further possibility, as Ribbands (1953) has suggested for the honeybee, that food transmission may be an important means of communication among the workers. Moreover, it may facilitate the production and standardization of the colony "nest-odor".

In the present paper are presented the results of a preliminary quantitative study of liquid food transmission utilizing a radioactive tracer technique. In our approach we have followed the lead of Nixon and Ribbands (1952), who used radioactive phosphorus in syrup to trace food exchange in a honeybee community. Employing radioactive iodide in honey, we have made a similar study of five species of ants representing several widely divergent groups within the Formicidae. We are now in a position to describe to a limited extent some comparative aspects of transmission rates.

MATERIALS AND METHODS

The following experimental procedure was followed during this study. Colonies of ants, complete with dealate queens and brood, were collected in the field in the early spring of 1956, when most of the species involved were commencing their spring foraging. These were placed in artificial (Fielde) nests, with an open foraging arena, and left from one to three weeks to settle in their new surroundings. During this time little or no food was supplied the colonies. At the commencement of the experiments, a small group of workers were taken out of the nests, isolated in Petri dishes, and

(1) The authors wish to express their appreciation to Mr. E. W. Samuel, of the Harvard Biological Laboratories, whose expert aid and advice in the use of radioactive tracer technique made this study possible. Research was supported in part by a U. S. Public Health Service grant held by Eisner.

(2) In this paper we will use the expression "food transmission" in the restricted sense employed by Ribbands (1953) with respect to bees, i.e., meaning the direct exchange of food between individuals as opposed to separate feeding from a common food store. We wish to avoid the term "trophallaxis", since there is at the present time so much doubt concerning its exact meaning and significance. The reader is referred to the thought-provoking review of Brian and Brian (1952) on the evolution and current difficulties of the trophallaxis concept.

allowed to feed overnight on a dilute mixture of honey and radioactive iodide (NaI^{131}). After the group had fed, the workers were checked individually in a radiation counter, and the one containing the most radioactive material was reintroduced into the main nest. Sample groups of workers were taken from the colony thereafter at intervals to follow the progress of transmission of the radioactive material.

Data in each experiment were converted from counts/minute of gamma radiation to "volume-equivalents" based on the original dilution factor of the honey-iodide mixture. There are several inherent deficiencies in our present use of volume-equivalent, but these should have no effect on the final conclusions we have drawn if they are carefully borne in mind throughout. First, the volume-equivalent was usually based on one to several measurements and was derived from the radiation detected (in counts/minute) above the background radiation count plus three standard deviations of the background count. In other words, only the radiation which was "safely" (at a 99 % confidence level) above the background count, and therefore undoubtedly due to emission from the iodide, was recorded. This measurement was taken in lieu of the more accurate, but far more time-consuming method of determining the mean of a series of radiation counts for each individual and subtracting the background count. Since the background count is ordinarily very low, and the radiation emitted by the iodide proportionately very high, the calculation of volume by the short cut method will be distorted only when the iodide is present in vanishingly small amounts. This latter situation actually held in some of the borderline cases in *Pogonomyrmex*, *Crematogaster*, and *Solenopsis*. But in the great majority of measurements, the volume estimate could not have been significantly altered.

The second, and more important shortcoming of our measuring technique is that the "volume-equivalent" is reliable as a measure of *absolute* volume of honey ingested only in the ants allowed to feed directly on the honey and then only for a short period of time after the feeding. This is because the iodide probably was not chemically coupled with any of the honey ingredients and could have been absorbed differentially by the ant gut. Therefore, although the presence of iodide in workers not allowed access to honey is a sure indication of food transmission, the quantity of radiation is not a safe measure of the exact volume passed. Moreover, since the iodide could be excreted differentially, the amount present in an original forager is a safe measure of the absolute volume of honey contained by that individual only for a short time after it has fed on the honey.

With these limitations in mind, it is clear that the "volume-equivalent" we have employed here is only a rough index of volume. It cannot be accepted as a measure of absolute volume. It is of use only in studying transmission rates and in deriving approximate estimates of volumes for comparative studies.

Except for borderline measurements, as in the lower volume-equivalent classes of *Crematogaster lineolata*, the bulk of iodide detected in individuals can be safely considered to have been ingested and not to have been acquired through external contamination. This was established by dissections and detailed measurements of separate body parts in all of the species involved. Where sufficient iodide was present to compare body parts of individuals most of it was found to be concentrated in the gut lumen and wall, for at least two weeks after the initial feeding. A more detailed account of passage of tagged food material through the body is planned in a later report.

The data, in the form of frequency histograms of individuals classified according to volume-equivalents, are given in figures 1-6.

CONCLUSIONS

1. Perhaps the outstanding conclusion to be drawn from this preliminary study is that the rate of transmission of sugar solutions varies within the Formicidae, from near zero in *Pogonomyrmex badius* to virtually

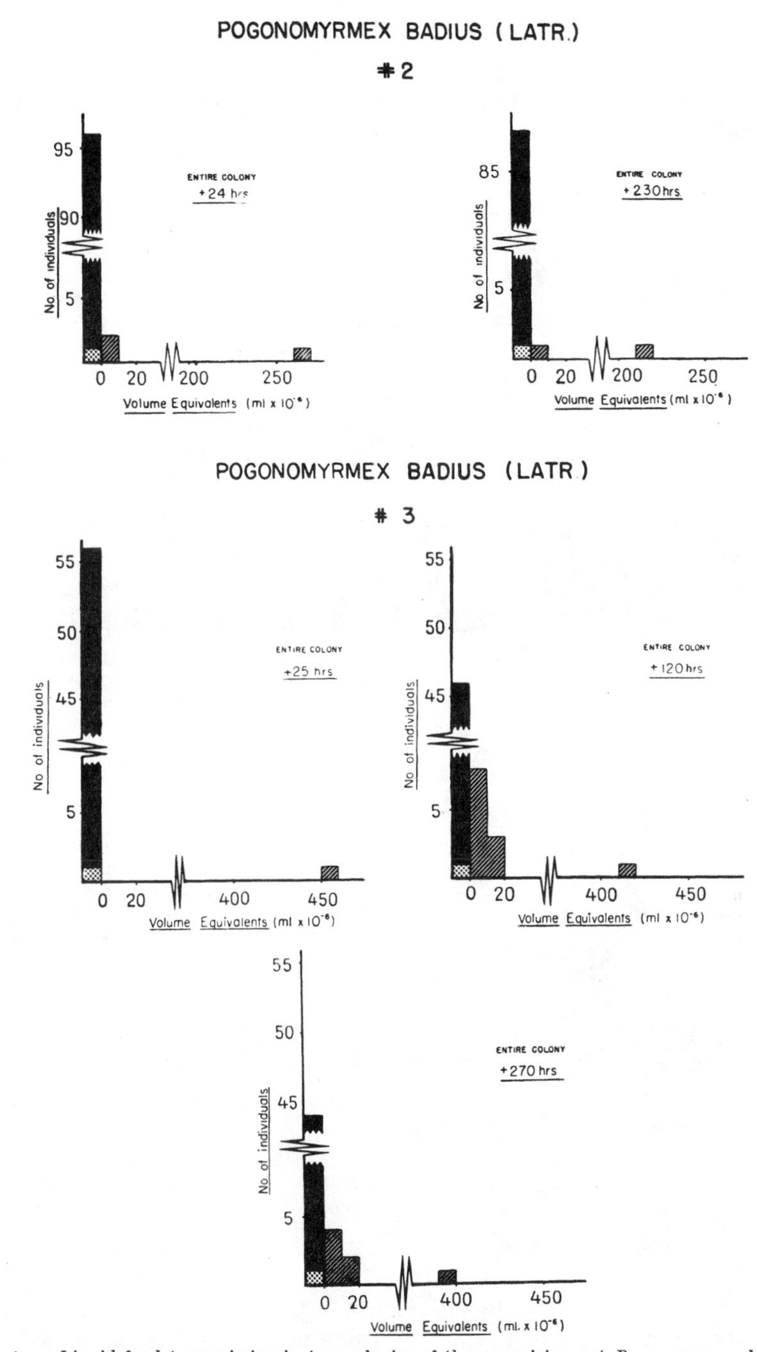

Fig. 1. — Liquid food transmission in two colonies of the myrmicine ant *Pogonomyrmex badius* (Latr.), presented in terms of frequency histograms of individuals classified by volume-equivalent of radioactive iodide. In colony no. 2 the original forager introduced into the nest contained 265 × 10^{-6} ml. of honey-iodide mixture ; in colony no. 3 the original forager contained 464 × 10^{-6} ml. of the mixture. *Dotting* indicates the nest queens, *black* indicates workers containing no iodide, and *hatching* indicates workers containing positive amounts of iodide.

complete colony saturation within thirty hours in the two species of *Formica*. Moreover, much of this variation can occur within a single subfamily, as evidenced by a comparison of the two myrmicine species *Pogonomyrmex badius* and *Crematogaster lineolata*.

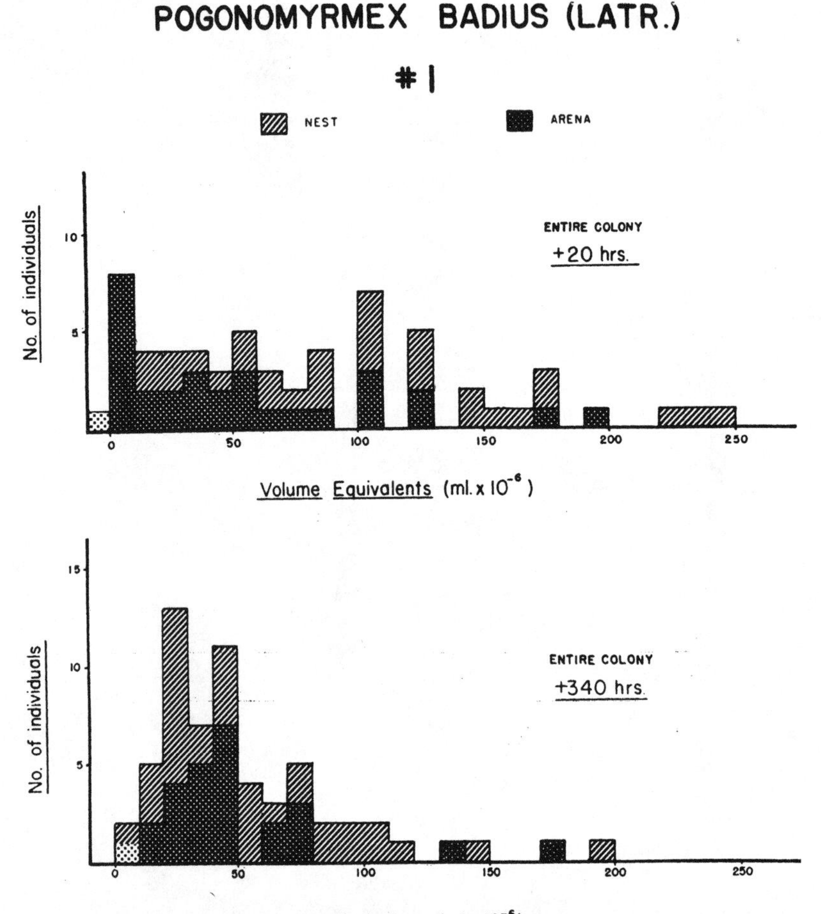

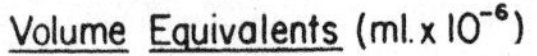

Fig. 2. — Frequency histograms by volume-equivalent of a colony of *Pogonomyrmex badius* allowed free access to a honey-iodide mixture for a period of 24 hours. Time intervals indicated are those following removal of the honey-iodide source. *Dotting* indicates the nest queen, *single hatching* the workers collected from within the nest, and *double hatching* the workers from the foraging arena outside the nest.

2. Although liquid transmission is much restricted in the *Pogonomyrmex*, all of the adult colony members may share in the same food source if it is readily accessible to the nest, as shown in figure 2. This is due to a high percentage of the workers visiting the food source and feeding

independently. The same results might be expected in the cases where solid food, such as insect prey or seeds, is brought into the nest itself. It thus appears that even in species where food transmission is limited, other aspects of colony behavior may insure that food is distributed

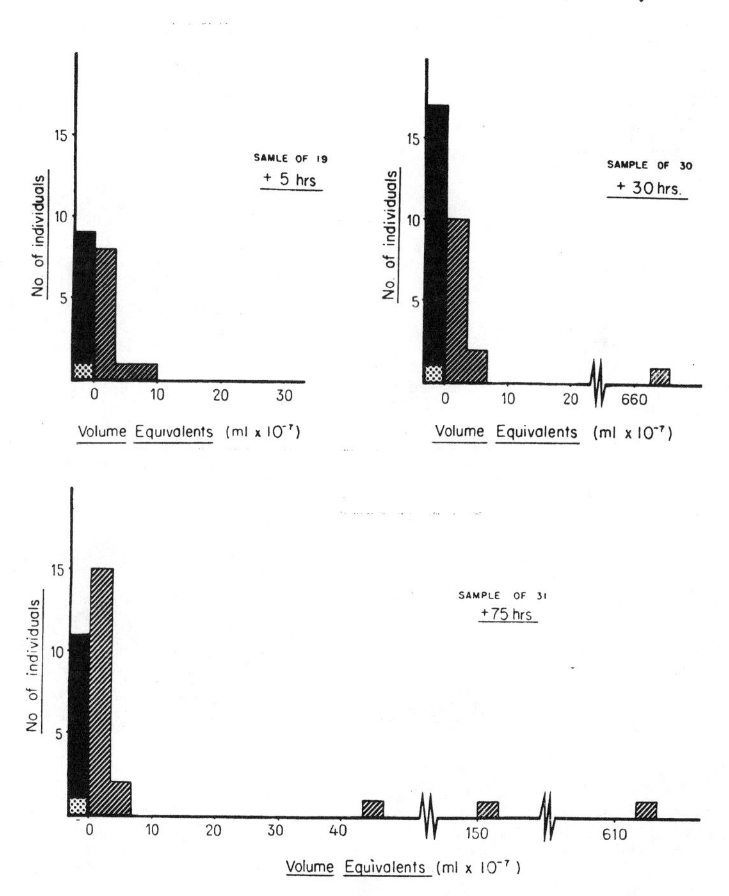

Fig. 3. — Liquid food transmission in a small colony fragment of 42 adults of the myrmicine ant *Solenopsis saevissima* (Fr. Smith). The original forager contained 1220×10^{-7} ml. of honey-iodide mixture. Conventions as in figure 1.

throughout the worker population, with the result that the colony diet remains uniform.

3. In the case of *Pogonomyrmex badius* presence of iodide in the gut of workers other than the original forager does not necessarily indicate oral transmission. No direct observations have yet been made to prove that oral transmission of liquid food occurs between adults of this species.

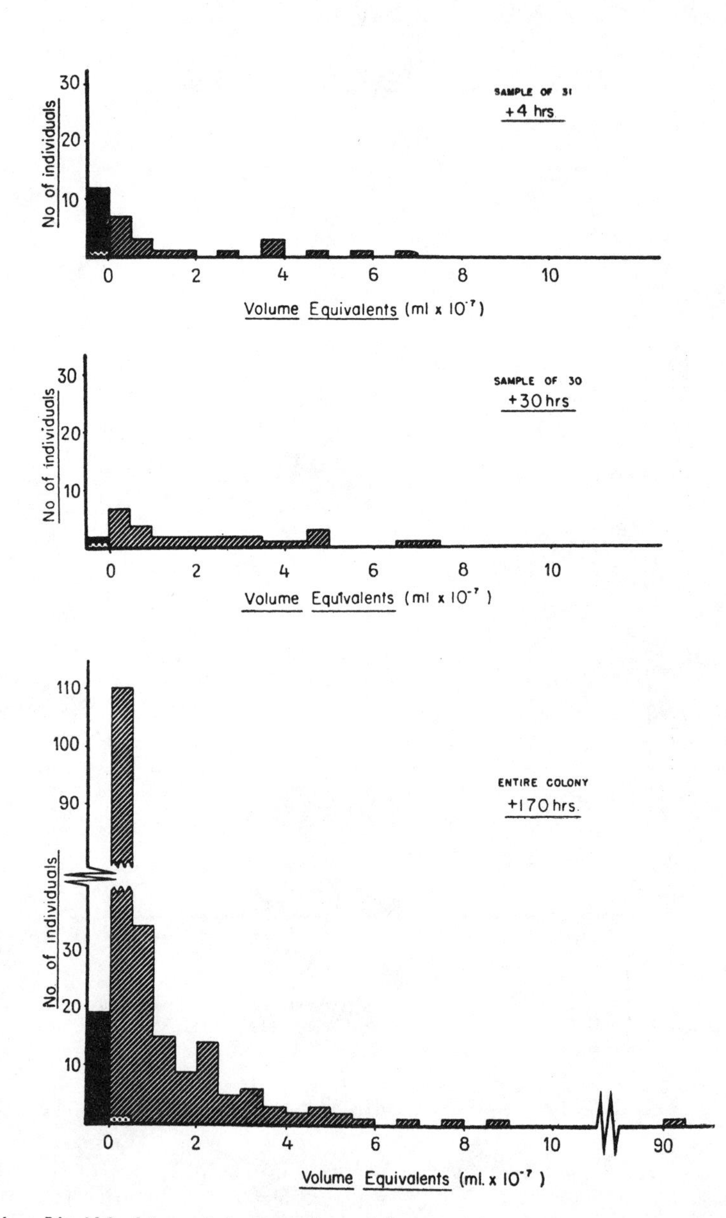

Fig. 4. — Liquid food transmission in a colony of the myrmicine ant *Crematogaster lineolata* (Say). The original forager contained 748×10^{-7} ml. of honey-iodide mixture. Conventions as in figure 1.

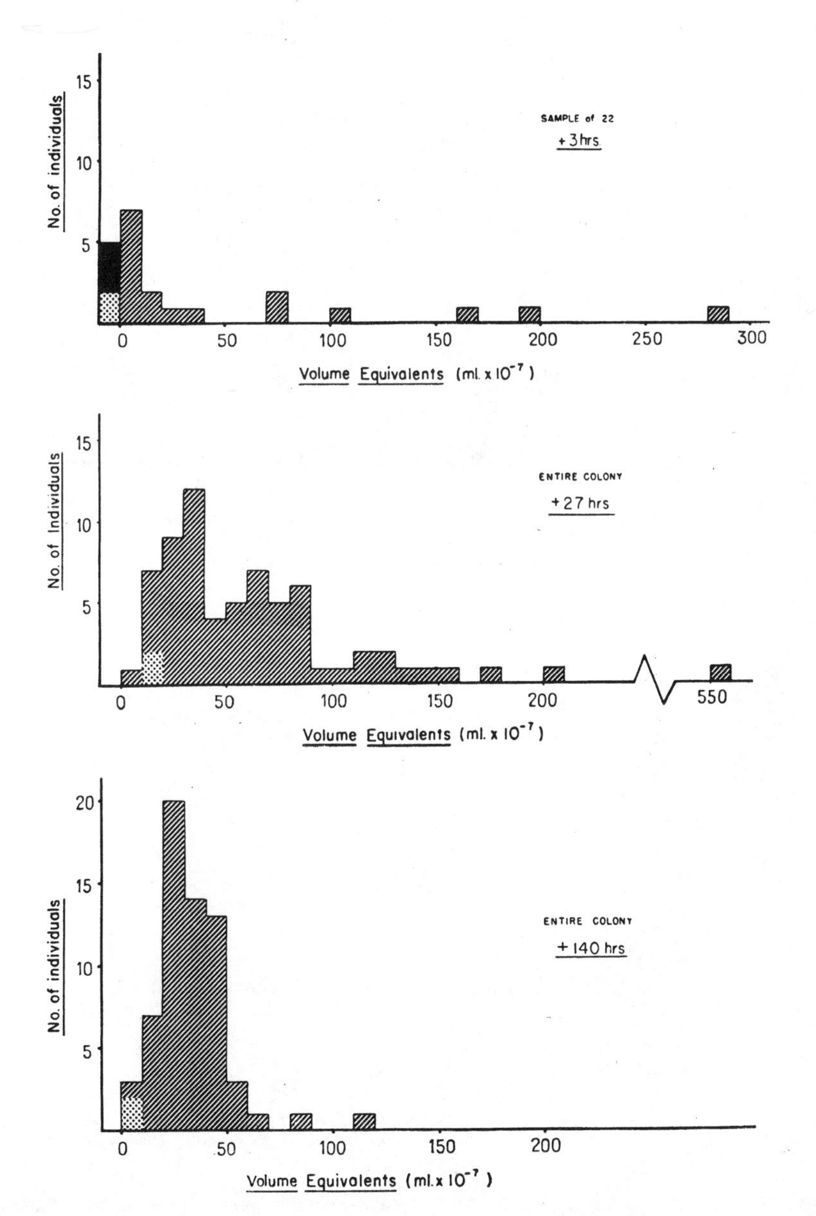

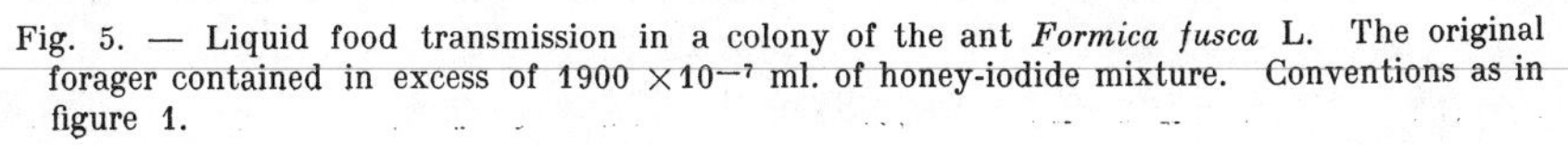
Fig. 5. — Liquid food transmission in a colony of the ant *Formica fusca* L. The original forager contained in excess of 1900 $\times 10^{-7}$ ml. of honey-iodide mixture. Conventions as in figure 1.

It is possible, although far from proven, that at least some of the iodide present in secondary workers could have been picked up by them from

FORMICA PALLIDEFULVA LATR.

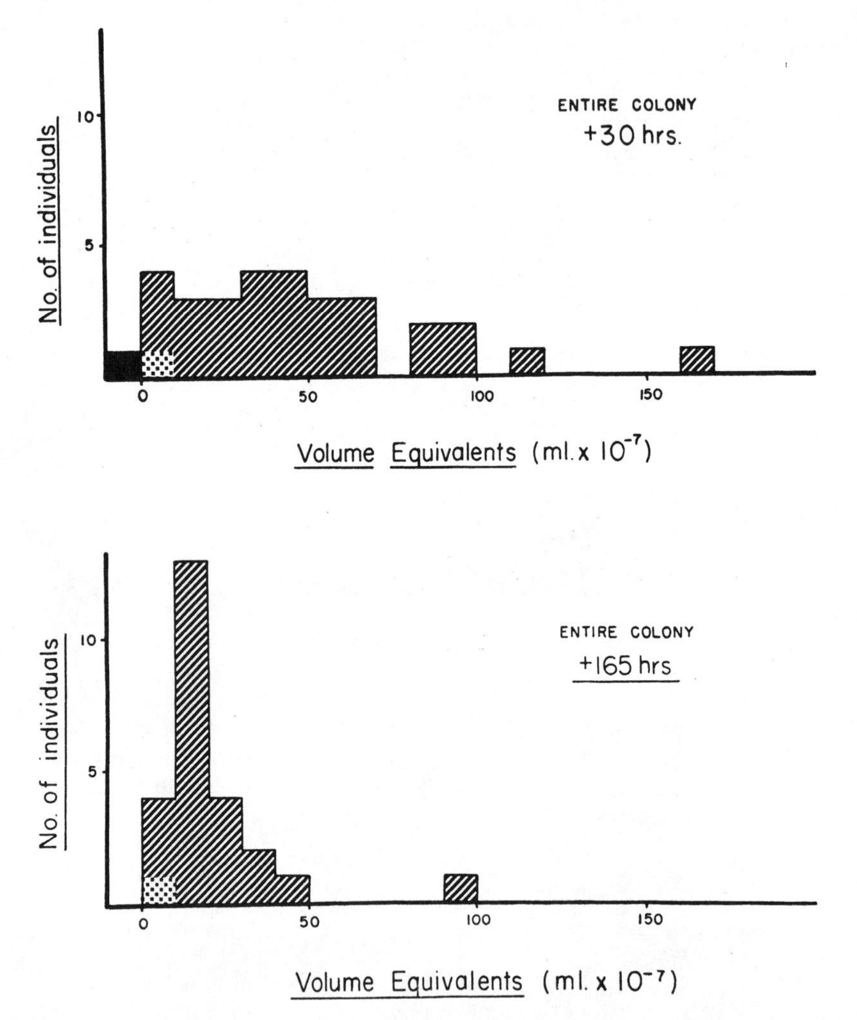

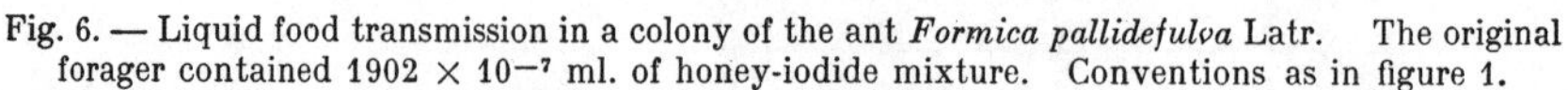
Fig. 6. — Liquid food transmission in a colony of the ant *Formica pallidefulva* Latr. The original forager contained 1902 × 10^{-7} ml. of honey-iodide mixture. Conventions as in figure 1.

material evacuated by the primary forager onto the floor of the nest. Separate observations on isolated workers fed with an honey-iodide mixture showed that these workers did evacuate in this fashion a half or more of the iodide within the two days. Whether honey or a derivative excretory product of the honey is passed with the iodide has not been determined.

4. The frequency curves obtained in the species with high transmission rates are of such a nature as to suggest that transmission occurred along sequences of individuals, following the primary donations given by the original forager. Had transmission involved only primary donations from the original forager, evenly proportioned frequency curves might have been expected. Instead, they are strongly skewed to the right. In our opinion, the simplest (but by no means the only) explanation for the skewness of the curves is that the original forager passed on fairly large donations to a limited number of workers, which in turn tended to pass on smaller donations to a larger total of workers and so on in a branching-chain pattern, with the largest number of workers receiving the smallest individual donations in the end.

5. The frequency curves tended to contract with time. Whether this leveling effect was due to continued chain transmission or to differential excretion is unknown. However, its result is clear enough: the increase in uniformity of gut content within the colony, a process which may have important sociological implications.

6. It is noteworthy that the queen and larvae received very little of the tagged honey. When these individuals showed traces of the iodide at all, it was usually in negligible amounts. It is very possible that differential feeding is practiced, and that when proteinaceous foods are used in similar experiments to those described here, very different results will be obtained.

Summary.

Tranmission of honey in several species of ants was studied using radioactive iodide as a tracer. Great variation in transmission rates between species was noted, ranging from negligible transmission over a ten-day period (in *Pogonomyrmex badius*) to complete colony saturation within thirty hours (in *Formica* spp.). The honey was passed mostly among workers, very little being given to the queens or larvae. Indirect evidence is cited which suggests the occurrence of chain transmission beyond the primary donations given by the original foragers.

Résumé.

On a étudié la transmission du miel chez plusieurs espèces de Fourmis en se servant d'iodure radio-actif comme traceur. On observe des différences considérables dans le taux de la transmission d'une espèce à l'autre, allant d'une transmission négligeable pour une période de dix jours (chez *Pogonomyrmex badius*) à une saturation intégrale de la colonie en trente heures (chez certaines espèces de *Formica*). Le plus souvent, le miel est passé d'une ouvrière à une autre, les reines et les larves recevant très peu de miel. Une preuve indirecte suggère une transmission en chaîne au-delà des dons primaires par les ouvrières fourragères nourries à la source.

Zusammenfassung.

Die Uebertragung von Honig bei verschiedenen Arten von Ameisen wurde mit Hilfe von Zugabe radioaktiven Iodids untersucht. Eine mit der Mischung gefütterte Arbeiterin wurde in das Nest gesetzt, und die Verteilung des von ihr abgegebenen Honigs gemessen. Die Verteilungsgeschwindigkeit war sehr verschieden je nach der untersuchten Art : praktisch zu vernachlässigende Werte, sogar nach 10 Tagen, bei *Pogonomyrmex badius* ; vollständige Sättigung des Staates, nach 30 Stunden, bei *Formica* spp. Der Honig wurde hauptsächlich unter Arbeiterinnen verteilt ; die Königinnen und Larven bekamen, wenn überhaubt, nur minimale Mengen. Die Resultate schienen darauf hinzuweisen, daß zur Verteilung im Staate nicht nur die Abgaben der ursprünglich gefütterten Arbeiterin verantwortlich waren, sondern auch Uebertragung der Empfänger untereinander.

LITERATURE.

1952. BRIAN (M. V.), BRIAN (A. D.). — The wasp, *Vespula sylvestris* Scopoli: feeding foraging and colony development (*Trans. Roy. Ent. Soc. Lond.*, vol. **103**, p. 1-26, 4 fig).

1957. EISNER (T.). — A comparative morphological study of the proventriculus of ants (*Bull. Mus. Comp. Zool. Harv.*, in press).

1953. LE MASNE (G.). — Observations sur les relations entre le couvain et les adultes chez les fourmis (*Ann. Sci. Nat. Zool. Biol. Animal.*, sér. 11, vol. **53**, p. 1-56).

1952. NIXON (H. L.), RIBBANDS (C. R.). — Food transmission within the honeybee community (*Proc. Roy. Soc.*, sér. B, vol. **140**, p. 43-50).

1953. RIBBANDS (C. R.). — The behaviour and social life of the honeybees (*Bee Research Association*).

1958

"The beginnings of nomadic and group-predatory behavior in the ponerine ants," *Evolution* 12(1): 24–31

This article contains one of the first attempts to reconstruct the origin and evolution of a complex form of social organization in the insects. Advanced army-ant behavior, characteristic of the ant subfamily Dorylinae, is one of the most absorbing, albeit miniature, spectacles of the natural world. T. C. Schneirla's superb field studies of New World ants in the genus *Eciton* in the 1940s and 1950s had revealed the existence of synchronized cyclic mass foraging and brood development in army ant colonies. In the report reproduced here, I introduced the evolutionary dimension to the phenomenon. By comparing similar behavior in the ants of the subfamily Ponerinae, some of which I had personally observed during fieldwork in New Guinea, I reconstructed the evolutionary steps that led from the ordinary colonial and foraging behavior of ants to the advanced patterns of the Dorylinae. This interpretation is still generally accepted.

Reprinted from EVOLUTION, Vol. XII, No. 1, March, 1958
Printed in U. S. A.

THE BEGINNINGS OF NOMADIC AND GROUP-PREDATORY BEHAVIOR IN THE PONERINE ANTS

E. O. WILSON
Biological Laboratories, Harvard University

Received May 22, 1957

It is a seldom recognized fact that the behavior patterns supposedly characteristic of the doryline "army ants" also occur, at least to a limited extent, in some groups of the primitive subfamily Ponerinae. In the past, entomologists have tended to speak of true army ants as occurring only in the Dorylinae or Leptanillinae, or the Dorylinae exclusively, but it must be admitted that this simple taxonomic qualification is not wholly adequate, particularly in view of the fact that there is now evidence to suggest that the Dorylinae may be polyphyletic (Brown, 1954). The best definition of the expression "army ant" may well be an operational one, of the sort offered by Webster's Unabridged International Dictionary (Second Edition): "Any species of ant that goes out in search of food in companies, esp. the driver and legionary ants."[1]

Actually, the above definition from Webster's Dictionary is incomplete. Upon closer examination one finds that there are really *two* discrete features that may be considered fundamental in army-ant behavior. These diagnostic features can perhaps be most conveniently distinguished under the concepts of "nomadism" and "group-predation." It is my purpose in the present paper to show that both nomadism and group-predation are developed to a variable degree in some genera of the Ponerinae, and that certain species of the genus *Leptogenys* combine these two features to achieve an adaptive "army-ant" type quite similar to that seen in the dorylines. Finally, I wish to take some account of the possible concurrence of nomadism and group-predation in still other groups of ponerines.

NOMADISM

Nomadism is a term well entrenched in the myrmecological literature and is probably best defined as *relatively* frequent colony emigration. Most, if not all, ant species shift their nest site if the environment of the nest area becomes unfavorable, and some, e.g., the famous *Iridomyrmex humilis,* are exceptionally restless and normally emigrate one or more times during the course of a single season. But hitherto none have been known in which emigration is undertaken so frequently or accomplished in such an orderly fashion, to cover so much new territory, as in the better known dorylines.

Of supplemental interest is the apparent existence of a limited and aberrant form of nomadism in some members of the ponerine tribe Amblyoponini. W. L. Brown (pers. commun.) has recently called my attention to a peculiar feature of the behavior of the Nearctic species *Amblyopone pallipes* (Haldeman) which certainly points in this direction. Small isolated groups of workers and larvae are often found clustered around recently

[1] There has been a tendency in some recent literature to use the terms "army ant" and "legionary ant" interchangeably. According to Webster's International Dictionary, the term "legionary ant" refers first to the New World army ants of the genus *Eciton,* and second to the "driver ants" of the genus *Anomma.* Most English-speaking authors of the past fifty years have used this term to mean specifically *Eciton* and its relatives, while employing the term "army ant" to cover all of the dorylines, although such application has been far from rigid. In the present paper I have used the two terms synonymously.

killed centipedes, the usual prey of the species. Since the larvae feed directly on the whole centipedes, without previous dismemberment, and since these insects seem too large and clumsy to transport for any distance intact, the suggestion offers itself that the workers transport the larvae to the prey rather than the more normal reverse. It is quite conceivable that colonies emigrate in part or in whole in this fashion to follow their food source.

I have observed similar behavior in the big amblyoponine *Myopopone castanea* (Fr. Smith) in New Guinea. On several occasions small groups of workers and larvae were found clustered about large wood-boring beetle larvae that could not possibly have been carried for any significant distance from the spot where they had been found and presumably killed by the workers. In one rotting log thoroughly dissected by myself, workers and larvae of *Myopopone* were found around two such prey at widely separated spots. The conclusion was inescapable in this case that the colony had distributed itself at least in part to be located near the prey. What constitutes "emigration" in *Myopopone* is open to question, however, since it can be argued that the entire subcortical surface of the log constitutes the "nest" of the ants.

Group-Predation

Group-predation as defined here includes both group-raiding and group-retrieving in the process of hunting living prey. These two processes must be carefully distinguished from each other, since they involve quite different innate behavior patterns and are not invariably linked. Many ant species, particularly those in the higher subfamilies, engage in group-retrieving of prey, but relatively few non-dorylines also group-raid. Various termitophagous ponerines, such as *Leptogenys* of the *processionalis* group and species of *Termitopone, Megaponera, Paltothyreus,* and *Ophthalmopone* (see Wheeler, 1936), do group-raid and are in this sense truly group-predatory. The Cerapachyinae, which appear to be primarily or exclusively robber ants (myrmecophagous), also include truly group-predatory species, a point I have developed at length in another paper on the biology of this group (Wilson, 1958). True group-predation is not to be confused with the specialized form of group-raiding conducted by the slave-making ants, including *Polyergus, Rossomyrmex, Strongylognathus,* and members of the *Formica sanguinea* group, which are specialized parasites hunting live captives rather than prey.

The Ponerine Army Ants

Now the very important question must be raised: To what extent are nomadism and group-predation associated in the non-doryline ants? Unfortunately, only a very incomplete answer can be supplied to this question, since we are handicapped by a scarcity of information on the extent of nomadism, or for that matter, emigration of any sort, in most ant groups. Perhaps it is sufficient at this point to establish the fact that the association of nomadism and group-predation, constituting at least the most general characteristics of true "army-ant" behavior as defined here, does exist in the Ponerinae. Below are given two examples from the genus *Leptogenys* [2] recently studied by present author in New Guinea, followed by a selection of probable and likely cases from other ponerine genera.

Leptogenys diminuta (Fr. Smith)

This species, belonging to the *L. kitteli* group and ranging from Ceylon and Botel Tobago to the Solomons and Queensland, was collected by the author in the following localities in eastern New Guinea: Karema, Brown River (accession no. 539); Bisianumu, 500 m. (acc. nos. 608,

[2] Group-predation *sans* nomadism is already a well-known phenomenon in the *processionalis* and *kitteli* groups of *Leptogenys*. The reader is referred to summaries of information on this subject by Wheeler (1910, 1936).

644, 659); Nadzab, Markham Valley (acc. nos. 1087, 1106, 1107); Bubia (acc. no. 1059); lower Busu River (acc. nos. 934, 940); Mongi River near Sambeang. On several occasions estimates of colony size and composition were made, and are given below:

1. *Acc. no. 1087.* May 20–22, 1955. Seventy to eighty workers, 10 males, and an undetermined amount of brood, consisting of larvae, one-quarter to full grown, and cocoons, the latter predominating.

2. *Acc. no. 1106.* May 20–22, 1955. One ergatogyne (highly ergatoid), about 150 workers, 10 males, and an undetermined amount of brood, consisting of larvae, one-half to full grown, and cocoons, the latter predominating.

3. *Acc. no. 539.* March 8, 1955. Somewhat in excess of 300 workers.

4. *Acc. no. 608.* March 15–20, 1955. About 200 workers and several males.

L. diminuta is one of the most widely ranging of New Guinea ponerines, occurring in dry evergreen forest as well as in rainforest at various elevations. It seems to favor forest borders and partial forest clearings. The nests are very temporary in aspect, consisting of nothing more than preformed cavities in rotting logs or simple bivouacs in open leaf litter. There is no evidence of excavation or nest building of any kind. Mann (1919) made the same observation with respect to this species ("var. *santschii*")[3] in the Solomons: "On several occasions I found masses of workers, accompanied by males, swarming on the ground, always in the forest, and numerous larvae and pupae beneath pieces of bark on the ground. Probably these were temporary nesting places." Neither Mann nor I observed actual colony emigration in this species, but Mann noted what was undoubtedly this process in a closely related, undescribed species from Malaita (Mann, 1919; cited erroneously as "var. *laeviceps*").

Group-predatory behavior, involving the essential process of group-raiding, as well as group-retrieving, is markedly developed in *L. diminuta.* At a lumbering camp on the edge of the Busu River, near Lae, New Guinea, where I stayed during May, 1955, I watched the workers of a *diminuta* colony forage in groups on two successive days. On both occasions approximately forty workers, forming a short, compact column of three to six individuals in width, emerged during late afternoon from the dense grass of a forest clearing out onto the bare ground surrounding an open firewood hut. As in dorylines, group cohesion was strong, each worker appearing unwilling to leave the file to forage on its own. Leadership of the group changed constantly from one worker to another. After a few minutes of exploration in this open space, the current of movement turned, and the group began to press back toward the grass border. One afternoon a secondary file was found ascending the supporting poles of the hut rafters. Here group cohesion was less marked than on the ground, and some of the workers ventured out singly for short periods of time. Nearby, in the Busu rainforest, a group of workers from another colony were found during midmorning foraging together around the base of a tree.

Food-retrieving in this species also entails extensive group activity, at least where large prey are involved. At the rainforest border near Bubia, also near Lae, about nine o'clock in the morning, six to ten workers were found carrying a rather large beetle larva together; they were all moving fairly easily in the same direction. No other workers were in evidence in the immediate vicinity. In the primary lowland rainforest near Karema, Papua, several hundred workers were found group-retrieving a very large millipede. They were first encountered at eight o'clock in the morning, when the first rays of sunlight were beginning to

[3] The formal synonymy of Mann's var. *santschii,* along with other proposed taxonomic changes in *Leptogenys,* is planned as part of the author's forthcoming revision of the Melanesian Ponerinae.

reach the forest floor. Large numbers of the workers were streaming back and forth between the millipede and the bivouac site, a distance of about five feet. The millipede was very feeble, but still alive and able to move slowly under its own power, despite the hindering action of the masses of ants which covered and pulled at it. By 9:00 it had moved to a point six feet from the nest. By 9:30, however, it was completely immobilized, and the ants had dragged it back to the five foot mark. By 10:30 they had dragged it all the way back to the nest site and covered it with leaf litter.

Leptogenys purpurea Emery

This is a second species of the *L. kitteli* group found on New Guinea. In the eastern part of the island it has been collected from sea level to 1,500 meters, and is apparently most abundant in the upper part of this elevational range, from about 800 to 1,500 meters. I encountered it on several occasions during a trip through the mountainous region around the headwaters of the Mongi and Mape Rivers, in the eastern part of the Huon Peninsula. The observations recorded below form the basis for my conclusion that this species shows nomadic and group-predatory behavior.

Near Tumnang (1,500 m.) a colony emigration of *L. purpurea* (accession no. 803) was discovered in progress at 12:30 P.M., just as the early afternoon mists were closing in on this mountainous area. What seemed to be the advance workers had made their appearance, forming a loose file which emerged from the impenetrably dense grass of a forest clearing onto a native trail, followed the open ground of the trail for about eight feet, and then turned off into the grass again. During the next thirty minutes a large number of workers, roughly estimated to be 2,000, emerged and passed along this route. At full strength the column they formed was tightly packed and measured about six individuals across. In its swift movement and compact structure this column strongly reminded me of the raiding columns of the larger ecitonines I have observed in tropical Mexico. Single *Leptogenys* workers periodically left the column and moved for several inches to a foot to the side behore returning to the main stream. These individuals were very aggressive and readily attacked objects offered them; in effect they were functioning as scouts or guards for the emigrating colony. Large numbers of workers were constantly returning against the main current and may have been on their way back to the original nest site to pick up additional brood. As many as 1,000 cocoons, along with a few mature to nearly mature larvae, were passed along in the course of the emigration. Unfortunately, only two of the cocoons were preserved; upon later dissection one was found to contain a partially pigmented worker pupa and the other an early prepupa of indeterminate caste. At least ten fully colored males also passed in transit; in each case they were being carried along (with their appendages folded up in "pupal" posture) by workers. A few callow workers were also present; these were moving on their own power and made their appearance toward the end of the emigration. The only food being transported was a fragment of a grasshopper, carried by a single worker. Despite careful scrutiny of the column, no female reproductive was sighted, although it is admitted that a worker-size ergatogyne might easily have been overlooked.

At Gemeheng, a colony of *purpurea,* containing at least a thousand workers (acc. no. 781), with a large quantity of brood, was found in a bivouac in the lower strata of a pile of recently cut grass and brush in a clearing in mid-mountain rainforest. At 10 A.M., during a clear, sunny morning, two independent columns of workers were proceeding out in different directions, and could be followed through the grass for distances of about six feet. One column was bringing in a large carabid larva, while the other

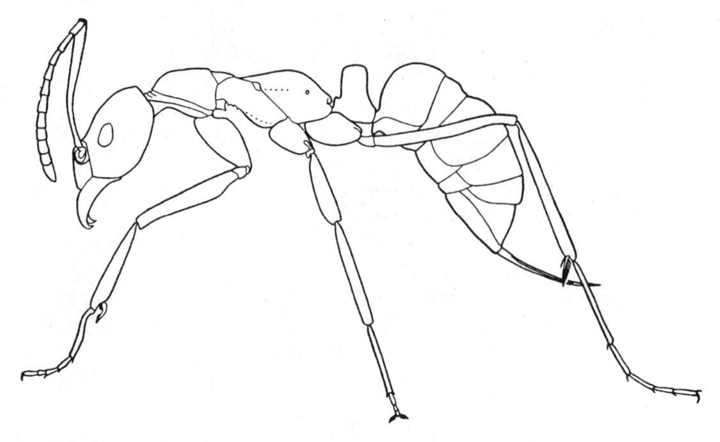

Worker of the ponerine ant *Leptogenys purpurea* Emery; Gemeheng, N.-E. New Guinea (acc. no. 781).

was apparently pushing out in the initial stage of foraging. Movement to and from the bivouac site was at this time about balanced. In the bright sunshine these tightly organized, fast-moving lines of metallescent ants made a striking sight. Their brood consisted of mature to nearly mature larvae and cocoons. Two of the cocoons were preserved and later dissected; both contained male pupae.

Near Wamuki, on the Mongi headwaters, about 500 workers were found nesting in a large preformed cavity in a rotting log on the floor of foot-hills rainforest. At midmorning, when the colony was discovered, a column of workers, apparently in the act of foraging, extended out from the log for a distance of about six feet.

Genus *Onychomyrmex*

Members of the aberrant amblyoponine genus *Onychomyrmex* resemble dorylines in many ways, leading W. M. Wheeler (1916) to suggest that "it is not improbable that the colonies move from place to place in search of their prey, like the colonies of the subterranean Dorylinae (*Eciton coecum* and *Dorylus*), which they very closely resemble in behavior, color, sculpture, and pilosity." Information on the habits of *Onychomyrmex*, a genus confined to the rain-forest areas of northern Queensland, is still very limited. Wheeler found workers of *O. hedleyi* group-foraging on two occasions, and of *O. mjobergi* says, "The ants moved rather slowly in long files through cracks in the wood [of rotting logs], evidently endeavoring to keep in close touch with one another by means of their antennae, after the manner of the Dorylinae."

These observations have been confirmed by W. L. Brown (pers. commun.), who studied *O. hedleyi* and *O. doddi* in the field in 1950. He found large numbers of workers moving in compact files without brood through rotting logs, evidently in the process of group-foraging. He was particularly impressed by the rapid, doryline-like vibration of the antennae by which the workers kept in touch with one another. Unfortunately, neither Wheeler nor Brown were able to make definitive observations on the frequency of colony emigration, and the existence of nomadism in this genus must be considered purely suppositional.

Megaponera foetans (Fabr.)

The group-predatory behavior of this giant termitophagous species has been described by several authors independently (see in Wheeler, 1922, 1936). In 1914 Arnold discovered a colony in emigration and was able to extract the reproductive female, which turned out to be an ergatogyne. The elaborateness of the emigration witnessed by Arnold, involving as it did the utilization of several "way stations," together with the circumstance that *M. foetans* evidently lives exclusively on a diet of termites, caused Wheeler (1936) to suggest that this species leads a nomadic life, i.e., constantly shifts its nest site to locate near fresh termite colonies. Much the same considerations apply to the other large, group-foraging, obligatorily termitophagous ponerines, such as *Termitopone* and *Ophthalmopone,* although unhappily we know even less about the biology of these forms than in the case of *Megaponera* (Wheeler, 1910, 1936).

Genus *Simopelta*

Simopelta is another ponerine genus whose members may eventually be found to qualify as true army ants.[4] The reproductive female of *S. pergandei,* recently described by Borgmeier (1950), is a dichthadiiform ergatogyne extraordinarily convergent to that found in the dorylines. Although this and other species of the genus occur widely through the New World tropics and have been collected repeatedly, virtually nothing is known of their habits. We have only the following suggestive statement by Mann (1916) concerning *S. jeckylli:* "The type colony was discovered quite accidentally, by scratching away some of the leaves and debris with which the ground in the forest is everywhere covered. As far as I could ascertain the ants were travelling in a definite direction."

[4] It is noteworthy that in at least one species of the closely related genus *Belonopelta* (*B. deletrix*) the workers forage solitarily and the reproductive female is a normal, alate queen (Wilson, 1955). No information is available on emigration frequency.

The Adaptive Value of Legionary Behavior

The combination of nomadic and group-predatory behavior has been achieved on at least two separate occasions in the evolutionary history of the ants: in the Dorylinae and in the ponerines of the *Leptogenys kitteli* group. It has been commonly assumed (but never proven) that the Leptanillinae are legionary, and if this is true these aberrant little ants undoubtedly represent a third independent army ant group. Finally, if the Dorylinae are diphyletic or triphyletic, as Brown (1954) has suggested, and if the members of *Onychomyrmex, Megaponera,* and *Simopelta* are both nomadic and group-predatory, as suggested here, then army-ant behavior must be considered to have arisen separately a minimum of seven times. This circumstance, coupled with the fact that the Dorylinae, which represent the ultimate in the army ant trend, are so eminently successful in tropical regions all over the world, indicates that legionary behavior confers a considerable selective advantage.

The adaptive value of nomadism seems abundantly clear. By continually moving into fresh hunting grounds, exclusively predatory ants are able to build up relatively large colonies (Wheeler, 1910). Schneirla (1956) has described how the great columns and swarms of *Eciton* strip the rainforest floor of much of its arthropod fauna during mass forays. These ants could not long survive if confined to the narrow trophophoric fields occupied by sedentary ants.

The selective advantage of group-predation is a matter of somewhat greater complexity. It has been noted repeatedly that compact armies of ants are more efficient at flushing and capturing prey than assemblages of lone foragers, and this is certainly a correct observation, but it is not the whole story. In my opinion there is another, primary function of group-predation that becomes clear only when the food habits (prey preference) of the group-predatory ponerines, cerapachyines, and dorylines are compared with those of the ponerines that forage in solitary fashion. The great bulk of non-legionary ponerine species of which the food habits are known take living prey of approximately the same size as their worker caste or smaller. As a rule they depend on proportionately small animals which can be captured and retrieved by lone foraging workers. Group-predatory ants, on the other hand, normally include large arthropods or the brood of other social insects in their diet, prey not normally accessible to solitary foraging ants. Thus *Leptogenys diminuta* and *L. purpurea* seem to specialize on large arthropods, *Eciton* and *Anomma* species prey on a wide variety of arthropods, including social wasps and other ants, *Megaponera foetans* specializes on termites, cerapachyines specialize on other ants, and so forth. It seems quite probable that group-predation, and with it the broader pattern of army-ant behavior, has arisen generally as a coadaptation to predation on large arthropods and social insects.

The Early Evolutionary Stages of Legionary Behavior

Thanks to the persistently conducted and exhaustive researches on *Eciton* by Schneirla and his associates (1956, 1957, and contained references), we now know perhaps more about the biology of this genus than about that of any other group of ants. A promising start has also been made on the African doryline genus *Anomma* by Raignier and van Boven (1955). But as enlightening as this work has proven to be throughout its many biological and ethological ramifications, there is still one important subject on which it has been unable to throw much light, and indeed cannot be expected to, and that is the early evolutionary history of army-ant behavior. To observe the beginnings of nomadism and group-predation, and the association of these two features to form true legionary behavior in a primitive state, we must leave behind the highly specialized dorylines and go to the Ponerinae. There are admittedly great gaps in our present knowledge of ponerine behavior, particularly with reference to the causal mechanisms behind foraging and emigratory activity, and it is impossible at this stage to speak about evolutionary trends with any degree of precision. Nevertheless, the point has been reached where it is at least possible to formulate a working hypothesis of the phyletic development of legionary behavior in terms of major adaptive steps. In the author's opinion, four such steps can be conceived, as follows:

1. *Group-predation is developed to allow specialized feeding on large arthropods and other social insects.* Group-predation without nomadism may occur in some of the cerapachyine ants, e.g., *Cerapachys* (*s. str.*) and *Phyracaces* (cf. Wilson, 1957), but this is probably a short-lived step, soon giving way to the stage described below.

2. *Nomadism is either developed concurrently with group-predatory behavior, or is added shortly afterwards.* The reason for this new adaptation is that large arthropods and social insects are more widely dispersed than other types of prey, and the group-predatory ant must constantly shift its trophophoric field to exploit these new food sources. With the acquisition of both group-predatory and nomadic behavior, the species is now truly "legionary" in the operational sense adopted here. Most of the group-predatory ponerines are believed to have reached this adaptive level. Colony size in these species averages larger than in related, non-legionary species, but does not approach that attained by *Eciton* and *Anomma*.

3. *As group-predation becomes more efficient, prey preference is expanded to include smaller, non-social insects and other arthropods.* The species tends to become a general predator showing little

prey discrimination. Colony size remains intermediate. No examples of this hypothetical intermediate stage are known at present, but some of the more poorly studied group-predatory ponerines may well prove to belong here.

4. *As the group-predatory and nomadic faculties are further improved, large colony size becomes possible.* This is the stage attained by most or all of the Dorylinae.

In the present paper, I have made an attempt to focus attention on what I have considered to be affirmed and probable examples of army ants in the Ponerinae. Whether or not my interpretations, theoretical and otherwise, prove correct in particular cases, I hope that I have at least succeeded in establishing the importance of extending studies of ponerine biology with this point of view included.

Summary

Army-ant (legionary) behavior is characterized by the combination of two discrete features, nomadism and group-predation, both of which are maximally developed in the subfamily Dorylinae. A growing body of evidence indicates that similar behavior, manifested at least in rudimentary form, has been evolved independently on several occasions in the more primitive subfamily Ponerinae.

An examination of the fragmentary information available on the biology of the various legionary and pre-legionary ponerines has led to the formulation of the theory presented herein, that group-predation has been evolved initially as a coadaptation to specialized feeding on large arthropods and social insects, while nomadism has been evolved secondarily to allow more efficient exploitation of this relatively widely dispersed food source. Later evolutionary steps, exhibited by the higher Dorylinae, have included the wide extension of prey preference and the increase of normal colony size.

Acknowledgments

The author wishes to express his gratitude to Dr. W. L. Brown and Dr. T. C. Schneirla for critically reading the manuscript and offering many helpful suggestions; and to Dr. J. J. Szent-Ivany, Mr. G. A. V. Stanley, Mr. Arnold Himson, and Mr. Robert Curtis for material aid received during the course of field work in New Guinea.

Literature Cited

Arnold, G. 1914. Nest-changing migrations of two species of ants. Proc. Rhod. Sci. Assoc., **13**: 25–32.

Borgmeier, T. 1950. A fêmea dichthadiiforme e os estádios evolutivos de Simopelta pergandei (Forel), e a descrição de S. bicolor, n. sp. (Hym. Formicidae). Revista Ent., **21**: 369–380.

Brown, W. L. 1954. Remarks on the internal phylogeny and subfamily classification of the family Formicidae. Insectes Sociaux, **1**: 21–31.

Mann, W. M. 1916. The Stanford Expedition to Brazil, 1911, John C. Branner, Director. The ants of Brazil. Bull. Mus. Comp. Zool. Harv., **60**: 399–490, 7 pls.

——. 1919. The ants of the British Solomon Islands. Bull. Mus. Comp. Zool. Harv., **63**: 273–391, 2 pls.

Raignier, A., and J. van Boven. 1955. Étude taxonomique, biologique et biométrique des Dorylus du sous-genre Anomma (Hymenoptera Formicidae). Ann. Mus. Roy. Congo Belg. (Zool.), **2**: 1–359.

Schneirla, T. C. 1956. The army ants. Smithsonian Report for 1955, pp. 379–406.

——. 1957. Theoretical consideration of cyclic processes in doryline ants. Proc. Amer. Phil. Soc., **101**: 106–133.

Wheeler, W. M. 1910. Ants. Columbia University Press.

——. 1916. The Australian ants of the genus Onychomyrmex. Bull. Mus. Comp. Zool. Harv., **60**: 45–54, 2 pls.

——. 1922. Ants of the American Museum Congo Expedition. Bull. Amer. Mus. Nat. Hist., **45**: 1–269, pls. 1–23.

——. 1936. Ecological relations of ponerine and other ants to termites. Proc. Amer. Acad. Arts Sci., **71**: 159–243.

Wilson, E. O. 1955. Ecology and behavior of the ant *Belonopelta deletrix* Mann (Hymenoptera: Formicidae). Psyche, **62**: 82–87.

——. 1958. Observations on the behavior of the cerapachyine ants. Insectes Sociaux, **5**: 129–140.

1959

"Source and possible nature of the odor trail of fire ants," *Science* 129: 643–644

The discovery reported here was exciting enough to keep me awake the whole night of the first successful experiment, thinking happily about the implications of what I had just learned. I had located for the first time the glandular origin of an ant pheromone. It was Dufour's gland, then called the accessory gland, and the species used was my old friend, the red imported fire ant, *Solenopsis invicta,* then called *Solenopsis saevissima.* More importantly, the trail substance turned out to be more than just a guide for ants stimulated by other signals. It also forms a large part of the remainder of the communication system, exciting workers, drawing them up the gradient to the source of emission, and finally inducing them to follow the trail away from the nest.

In the following year, working with two natural products chemists, I tried to identify the pheromone. We first gathered masses of fire ants in Florida and (after many painful ant stings) prepared active extracts. Using the new technique of coupled gas chromatography-mass spectrometry, we tentatively identified the substance as a farnesene, an odorous, low-molecular-weight compound ($C_{15}H_{24}$) that is widespread in other insects and plants. However, as the purification was refined, the ants' response weakened and the exact structure could not be determined. The mystery was solved later when, in a classic study of pheromone chemistry, Robert K. Vander Meer and his coworkers discovered that the active material is actually a mixture. The principal component for recruitment along the trail turned out to be Z,E-α-farnesene. This substance is fully active only when combined with two homofarnesene synergists and a still unidentified primer. By purifying the compound originally, we had unknowingly disposed of the synergists.

My research and that of many others after 1959 "broke" the ant pheromone code as a whole. We established that by far the largest part of communication among these insects is through taste and smell. The number of pheromone-producing exocrine glands known today from all ant species taken together is 39, with 10 or more the rule in each of all but the most anatomically primitive species. The most thoroughly studied ant species, including the red imported fire ant, employ at least ten kinds of signals. The great majority of these are pheromone mixtures. Ants also communicate by touch and sound but use those signals more frequently to supplement pheromones than to activate behavior.

Source and Possible Nature of the Odor Trail of Fire Ants

Abstract. Experimental evidence shows that the odor trail of the fire ant *Solenopsis saevissima* (Fr. Smith) is produced as a secretion of the accessory gland of the poison apparatus and released through the extruded sting. Preliminary studies suggest that this substance may be chemically allied to or even identical with the toxic principle of the venom.

Chemical trails laid down by worker ants are essential mechanisms in the organization of foraging and colony migration in many ant species. Yet only recently has much careful attention been paid to the topographic form of these trails, to their anatomical source, and to their chemical nature (*1, 2*). Chemical analyses conducted by Carthy show that in the formicine species *Lasius fuliginosus* (Latreille) the trail substance is a water-soluble anal emission containing uric acid, polysaccharides, and proteins (*2*). These data suggest that the bulk of the material is normal excretory and fecal matter rather than a special glandular secretion. But they do not exclude the possibility that special secretory products, serving as releasers of trail-following behavior, may be present in small amounts.

Now it is possible to show that in the myrmicine species *Solenopsis saevissima* (Fr. Smith) the essential trail substance is produced as a glandular secretion and is released through the sting. Workers of this species lay trails by dragging the tips of their abdomens over the ground with the stings fully extruded. To determine whether venom passed from the sting can induce trail following, a series of experiments was performed in which artificial trails of freshly extracted venom were drawn in the vicinity of foraging workers from a captive colony. In separate trials, venom was either collected directly on sharpened cork tips or (to reduce the possibility of contamination) collected by "milking" the ants with fine capillary tubes and then transferring the fluid to cork tips.

In preliminary tests, foraging workers could almost always be diverted to the artificial venom trails, although their response was generally weaker than it was to true trails laid in the same area by living minor workers. Under the experimental conditions described below, three such trails produced from venom collected directly on cork tips brought forth 6, 8, and 18 workers, respectively, the duration of the effect extending between 1 and 2.5 minutes. Numerous artificial trails made under the same conditions from the wall and contents of the principal gut divisions, as well as from crushed tissue and hemolymph of head, alitrunk, and abdomen (gaster), evoked various intensities of alarm, circling, antennal palpation, and even feeding (in the case of the crop contents), but no distinct trail following.

An attempt was next made to localize the source of the critical trail substance. Three organs are known to empty material through the sting or in its immediate vicinity: the hind-gut, the "true" poison glands (which are paired and empty their contents into the poison vesicle), and the accessory gland of the sting. In a series of experiments these organs were dissected out of freshly killed major and media workers, separated, doubly washed in insect Ringer's solution, and then crushed on separate cork tips to make artificial trails. All three organs were taken from each ant killed, and these were presented in varying sequences to eliminate possible bias in the results due to special sequential effects. As much uniformity as possible was obtained with respect to the number of foraging workers and their trophic "mood" by first allowing masses of workers to accumulate around a freshly killed meal worm (larva of *Tenebrio molitor*) pinned at the edge of the glass plate on which the trails were drawn. Under the particular conditions prevailing at the time of the experiments, a relatively stable concentration of 150 to 200 workers was reached within 10 minutes after the meal worm had been found by the first foraging worker. At this time a fringe of more or less idle workers milled in a tight group around the edge of the meal worm, and it was to these that the artificial trails were drawn.

The results, presented in Table 1, show clearly that the trail substance is concentrated in the accessory gland of the sting. The fact that both the hind-gut and the true poison glands (with vesicle) frequently gave quite negative results suggests that these structures do not normally contain any of the releaser substance at all, only picking it up by contamination during dissection. The relatively high variation in response to the accessory gland preparation may be explained, at least in part, by two irregularities difficult to control: variation in leakage of gland contents during dissection and variation in responsiveness of worker groups. It was noted that when several trials were made on the same day, responsiveness tended to decline progressively.

Table 1. Response of fire ant workers to artificial trails made from various abdominal organs of ten freshly killed workers. The positive responses recorded are those in which workers ran at least half the length of the artificial trails, or approximately 8 cm. The duration is the time interval from the first positive response observed to the last and is given to the nearest half minute.

No. of workers responding		Duration of group effect (min)	
Range	Mean ± standard error	Range	Mean
Hind-gut			
0– 18	2.3 ± 1.7	0 –1.5	0.4
Poison glands plus poison vesicle			
0– 26	8.2 ± 2.8	0 –4	1.4
Accessory gland			
31–164	107.5 ± 14.4	3.5–7	5.9

The foregoing results lead to the question: Does the accessory gland substance serve as both releaser and orientator, or is it only a releaser, with venom from the true poison glands functioning as the orienting agent in the trails? To solve this problem, the following experiment was devised. After the experimental conditions described previously had been arranged, an *accessory* gland preparation was drawn in a short sidewise stroke next to the peripheral group of ants, while simultaneously an artificial trail made from the *poison* glands (plus vesicle) was drawn outward from them. In each of five such trials, the ants showed intense excitement, spreading outward in random looping movements, and many new workers were attracted to the scene. But in only one case was the poison-gland trail followed, and then by the relatively small force of 20 workers, representing less than 30 percent of the outward-moving swarm of foragers.

It thus appears that the accessory gland secretion functions as both a releaser and an orientator of trail following. On the other hand, it was noted that occasionally when workers were following accessory-gland trails in large numbers they would also follow nearby old poison-gland trails that had been ignored previously. The implication seems to be that workers will follow other odor leads if there is some "knowledge" that a true (accessory-gland) trail exists. It also follows that only a small amount of the accessory-gland secretion need be in a trail to induce trail following. In fact, the venom from the true poison glands may be serving as a diluent for the accessory-gland secretion, although there is at present no direct evidence to support such a hypothesis (3).

The artificial trails made from accessory-gland preparations provide supernormal stimuli that attract far more workers than normal trails laid under similar circumstances by single living workers. The chemical nature of the releaser substance has not yet been precisely determined. However, the following data may be considered suggestive. A petroleum ether extract of steam distillate of whole ants prepared by M. S. Blum and his associates (4) produced trail-following responses of nearly comparable magnitude to those produced by accessory-gland preparations when it was tested under the experimental conditions described above. The number of workers drawn out by contact with the distillate was at least equal to the number attracted by the accessory-gland preparations, but orientation along the trails was somewhat less consistent. Blum *et al.* have shown that the infrared spectra of the distillate and of whole venom contain the same carbonyl band. On the basis of preliminary investigations, these authors have suggested that the carbonyl band is exhibited by the toxic principle itself, and that this constituent is manufactured by the accessory gland (5). It remains to be proved that the toxic principle and the trail-following releaser are one and the same (6).

EDWARD O. WILSON
Biological Laboratories, Harvard University, Cambridge, Massachusetts

References and Notes

1. E. C. Macgregor, *Behaviour* 1, 267 (1948).
2. J. D. Carthy, *ibid.* 3, 304 (1951).
3. P. S. Callahan (as he notes in correspondence with M. S. Blum) has found that the poison glands can be closed off at the base of the poison vesicle by a pair of highly developed muscle bundles; hence it is possible for the accessory gland to release its products independently of the poison glands. The possibility that such an operation occurs during trail laying should be considered in future studies.
4. I am indebted to Dr. Blum for supplying me with the fire ant extract used in this study and for granting permission to use unpublished data pertaining to it.
5. M. S. Blum (personal communication). For a report on the nature of whole venom, see M. S. Blum, J. R. Walker, P. S. Callahan, A. F. Novak, *Science* 128, 306 (1958).
6. It is interesting to note the significant observation by G. W. K. Cavill and D. L. Ford [*Chem. & Ind.* (*London*) 1953, 351 (1953)] that workers of the dolichoderine species *Iridomyrmex detectus* (Fr. Smith) follow artificial odor trails made from the steam distillate of other *detectus* workers. These authors have identified the distillate as 2-methylhept-2-en-6-one.

7 October 1958

Redox Absorption Spectra from Single Pigment Cells of Squid

Abstract. Single pigment cells from the squid *Loligo forbesi* have been studied by microspectrophotometry. The absorption spectra obtained show characteristic changes on reduction and oxidation which are compatible with those found in ommochromes. The presence of melanoid substances, however, cannot be excluded.

In several cephalopods, such as *Sepia officinalis, Octopus vulgaris*, and *Eledone moschata*, and also in arthropods such as Crustacea and Arachnoidea, a peculiar group of pigments, the ommochromes, has been found (1, 2). One of the significant properties of most ommochromes is that there is a characteristic change in the absorption spectrum on oxidation and reduction, although a few ommochromes do not behave in this manner (3). Pigments closely related are the ommatins (3, 4) and insectorubin (5), the latter being found in locusts and other insects.

In contrast to other investigations reported in the literature, the studies presented in this report were carried out on single pigment cells in the cutis of a cephalopod, *Loligo forbesi*, caught in the North Sea. The tissue was fixed in 4-percent Formalin, and sections were rinsed for 2 hours and immersed for 24 hours in a (reducing) $0.05M$ solution of $Na_2S_2O_5$. Microspectrophotometric measurements were made by comparing substrate and blank at each wavelength. The single pigment cells were magnified about 150 times. The absorption spectrum obtained after reduction is shown in Fig. 1 (curve 1). A maximum is found between 525 and 540 mμ, representing, when compared with measurements by Schwinck (2), a slight shift toward the longer wavelengths. This shift may be due in part to light scattering (6) or fixation. After oxidation for 24 hours in 7-percent H_2O_2, the maximum at 525 to 540 mμ essentially disappears (Fig. 1, curve 2).

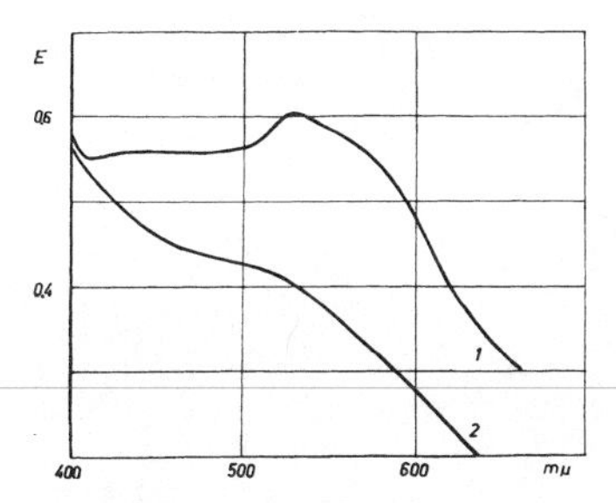

Fig. 1. Absorption spectra of a single pigment cell of *Loligo forbesi* after reduction (curve 1) and oxidation (curve 2) for 24 hours.

These results are in general agreement with bulk analyses on ommochromes reported by Becker (1) and Schwinck (2). They do not exclude, however, the presence of melanin or melanoid substances which show a gradually increasing absorption to the shorter-wavelength range (7); nor should it be postulated that the pigment found is identical with others already known. This study may merely show that, with suitable technique, redox absorption spectra can be obtained even from a single pigment cell and, thus, compared with analyses on extracted material.

MANFRED BAYER
Reinbek, Hamburg, Germany
JURGEN MEYER-ARENDT
Department of Pathology, Ohio State University, Columbus

References and Notes

1. E. Becker, *Naturwissenschaften* 29, 237 (1941).
2. I. Schwinck, *ibid.* 40, 365 (1953).
3. A. Butenandt and R. Beckmann, *Z. physiol. Chem., Hoppe-Seyler's* 301, 115 (1955).
4. A. Butenandt and G. Neubert, *ibid.* 301, 109 (1955).
5. T. W. Goodwin and S. Srisukh, *Biochem. J.* 47, 549 (1950).
6. P. Latimer, *Science* 127, 29 (1958).
7. E. Santamarina, *Can. J. Biochem. and Physiol.* 36, 227 (1958).

17 October 1958

Blood Groupings in Marshallese

Abstract. The absence of the Diego blood factor, the extremely low incidence of the M gene, and the unusually high R^1 gene frequency of the Marshallese more nearly resemble the blood groupings of the people of the western islands of Indonesia than the blood groupings of the Amerindians.

During March 1958, the annual medical survey of the Marshallese people of Rongelap Island was carried out, 4 years after they were accidentally exposed to radioactive fallout (March 1954) (1). These annual surveys are carried out by Brookhaven National Laboratory under the direction of R. A. Conard and are sponsored by the Atomic Energy Commission with the collaboration of the Department of Defense. During the course of these studies it became of interest to determine the blood groupings in the Marshallese people as an index of their origin and homogeneity. Blood samples were obtained by the survey team for this purpose.

The frequent movement of the Marshallese people among the various islands of Micronesia and, to a lesser extent, of Melanesia and other adjacent areas precludes any such concept as "pure" Marshallese. However, these people have lived for an estimated 2000 years on these islands with fewer outside contacts, perhaps, than most other groups. The findings presented consist of the

1962

"Chemical communication among workers of the fire ant *Solenopsis saevissima* (Fr. Smith), 1. The organization of mass-foraging," *Animal Behaviour* 10(1–2): 134–147

After the elements of trail communication in the red imported fire ant (*Solenopsis invicta,* but called *Solenopsis saevissima* in the following paper) had been worked out, it was possible to address the more complex problem of the higher-level control of foraging. In the study reported here, one of the first to analyze the self-organization of insect colonies, I introduced and documented the concept of "mass communication," the transfer of information that can be accomplished only from one group of individuals to another group of individuals. Thus from a relatively simple choice made by the individual workers, a more powerful and sometimes initially unpredictable higher-order response emerges. This is a principle, already apparent in the early 1960s, that has now been well documented throughout the social insects and remains a major focus of research.

REPRINT No. 247 from *Animal Behaviour*, 10, 1-2, January-April, 1962

CHEMICAL COMMUNICATION AMONG WORKERS OF THE FIRE ANT *Solenopsis saevissima* (Fr. Smith)

1. THE ORGANIZATION OF MASS-FORAGING*

By EDWARD O. WILSON
Biological Laboratories, Harvard University, Cambridge, Mass., U.S.A.

Introduction

In preliminary studies of the social behaviour of the fire ant, nine categories of communication among workers have been recognized. These are listed in Table I, in the approximate order of their frequency of occurrence. It is the purpose of the articles in this series to take up a preliminary analysis of several of the message categories, with the aim of reconstructing in part the complex events of mass behaviour and social organization in the normal life of the fire ant colony.

It is commonplace supposition that communication in ants, especially in small-eyed species such as *Solenopsis saevissima*, must be partly chemical. Yet the idea has received little critical examination, and detailed studies of chemosensory behaviour are relatively scarce in the literature. Two classes of chemical releasers, or "olfactorily acting pheromones" (Karlson & Butenandt, 1959), have been the object of physiological research in recent years: trail substances (MacGregor, 1948; Carthy, 1951; Wilson, 1959) and alarm substances (Sudd, 1957; Wilson, 1958; Butenandt, Linzen & Lindauer, 1959; Brown, 1959; Wilson & Pavan, 1959). The specific releasers of necrophoric behaviour have been preliminarily characterized (Wilson, Durlach & Roth, 1959). On the basis of this new information, chemosensory physiology deserves

Table 1. The Known Categories of Communication Among Fire Ant Workers.

Stimulus	Transmission	Response
Nest odour	Chemical	Nil, if odour is undisturbed.
Casual antennal or bodily contact	Tactual	Turning-toward movement or increased undirected movement
Body surface attractants	Chemical	Oral grooming; clustering
Oral grooming	Tactual	Decreased general movement, occasionally receptive posture
Ingluvial food solicitation	?	Regurgitation
Regurgitation	At least partly chemical	Feeding
Emission of Dufour's gland secretion as trail	Chemical	Attraction, followed by movement along trail
Emission of Dufour's gland secretion during attack	Chemical	Attraction to disturbed worker
Emission of cephalic substance	Chemical	Alarm behaviour

*Several persons provided essential assistance during the course of this study. Prof. Murray S. Blum generously contributed chemical fractions of fire ant workers used in his own biochemical studies of exocrine secretions. The technique of behavioural assays in the analysis of this material has been an unexpected and fruitful by-product of our collaboration. Prof. George A. Miller and Dr. Rudolf Jander followed the progress of the work, read the manuscript, and provided both new ideas and chastening criticisms. The work would be much less complete without the support of these colleagues, but the author alone must accept responsibility for any technical errors or misinterpretations that may remain lodged in the final report.

The research programme has been supported by a grant from the United States National Science Foundation.

to become an increasingly important aspect of social behaviour studies.

For the present at least, it is useful to explore the hypothesis that most of the social life of ants can be explained as the product of the interplay of stereotyped patterns, each of which is induced by one of a limited series of chemical releasers. If this assumption is correct, it should be possible to separate the releasers and employ them in the experimental analysis of individual patterns. It is further useful for purposes of investigation to suppose that at least some of the releasers are produced as glandular secretions and tend to be accumulated in gland reservoirs. This first paper is concerned with several aspects of behaviour underlying mass-foraging.

Materials and Methods

Colonies of *Solenopsis saevissima* were housed in circular artificial nests of the design shown in Fig. 4. A note concerning the construction of the nests may be of interest here. They are made entirely of plexiglas, the walls and ceiling being transparent and the base white opaque. Following a technique introduced by Brian (1951), a water channel leads into the base from the side to a central well stuffed with cotton wool, and water is allowed to pass in a continuous column from an outside reservoir. Four entrance spouts are set symmetrically in the wall, and are used to connect the nest, by means of clear plastic tubing, to additional nest units. The living area can thus be expanded indefinitely to accommodate the growing colony. The nests are placed on glass platforms supported on glass vials or bottles which in turn are set in dishes filled with mineral oil. The platforms serve as foraging fields for the ants. The nest interiors are constantly illuminated, so that within a year most of the living worker population has been reared in the light. As a result the entire colony can be examined conveniently at any moment without further disturbance.

The colonies used in the present study were reared from single recently fecundated light-phase queens collected at Baton Rouge, Lousiana. *Solenopsis saevissima* is an unusually fast-breeding species; under laboratory conditions healthy colonies raised from single queens reach "maturity", i.e. grow to include several thousands of workers, males, and virgin queens, in less than a year.

The Organization of Mass Foraging

In an earlier paper (Wilson, 1959) it was shown that the odour trail of *Solenopsis saevissima* is secreted by Dufour's gland* and released through the extruded sting. The pheromone component can be extracted in petroleum ether and concentrated by steam distillation and chromatographic separation. It has proven to be a powerful attractment, serving not only as the trail orientator but also as a true "releaser" of hunting behaviour on the part of worker ants. In this and in subsequent reports special attention will be devoted to the behavioural mechanisms by which fire ants use the Dufour's gland secretion in multiple roles: to communicate the location of new food finds, to organize efficiently the whole of mass food-retrieving, and to alarm sister workers.

Hunting by Solitary Workers

Foraging away from the nest is conducted preponderantly by workers in the small- and middle-sized classes. Only mature, fully-coloured individuals participate. Foraging is directed from the nest or way-stations by solitary workers, who move in irregular, looping paths.

When the periods of time spent away from the nest perimeter by 250 workers from a colony of about 3000 individuals were measured, these data formed the frequency curve shown in Fig. 1. When the measurements are transformed into number of workers left in the foraging field as a function of successive minute intervals following the departure from the nest of all workers simultaneously, a reasonably straight line is obtained in a double logarithmic plot. This shows that the function is moderately complex. In particular the probability of workers returning to the nest does not, as might be intuitively supposed, remain steady with time. Instead, the number remaining on the foraging field is best described in the double logarithmic form,

$$\log N = \log b - {}^{1}/_{a} \log t,$$

or

$$N = bt^{-1/}$$

Where N is the number of workers still foraging at any given time, *b* is the number of workers foraging at the end of the first second interval, and *t* is the number of expired second intervals. The data conform specifically to the equation,

*Dufour's gland has also been referred to by past authors as the "accessory" or "alkaline" gland of the sting. Its morphology has recently been described in detail in the admirable study of the fire ant poison apparatus by Callahan, Blum & Walker (1959).

ANIMAL BEHAVIOUR, X, 1-2 136

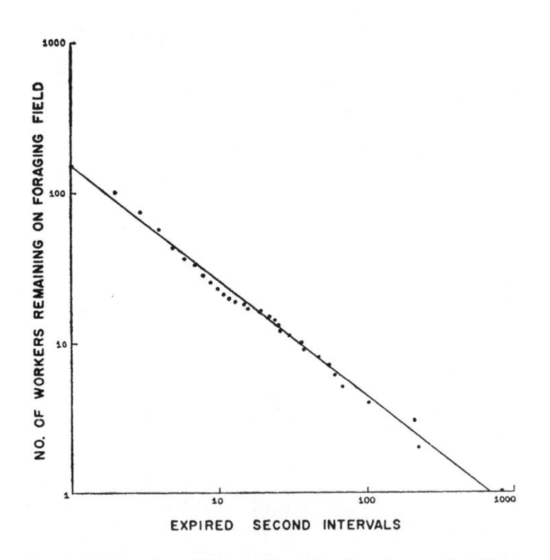

Fig. 1. Dispersion of foraging trip durations of solitary workers expressed as the decline of the number remaining on the foraging field with time. Colony size: about 3,000. Air temperature: 28 ± 1°C.

$$N = 100\ t^{-0.76};$$

which differentiated yields

$$\frac{dN}{dt} = -76t^{-1.76}$$

Thus the number of workers remaining foraging are decreasing at a steadily lessening rate with the passage of time. The longer a worker remains afield, the less becomes the probability that it will return home in the next second interval. This function can be at least partly explained by the common sense deduction that the longer a worker remains in the field, the farther it wanders from the nest, and hence the longer will be the journey home when it turns back. Nevertheless, the foraging time cannot be explained simply as the result of random movements out from the nest perimeter. The workers make frequent long straight runs and ordinarily follow a pattern of variable looping movements away from the nest perimeter. Evidence will be given later to show that in the laboratory hunting workers orient visually.

Rapid Exploration of New Territory

When larger laboratory colonies are given sudden access to a new foraging field, such as a clean glass platform or blank piece of poster card, the workers respond in a manner which is not predictable from observation of hunting behaviour on the old foraging field. During the first several hours the workers move onto the new area in large numbers. New foragers are recruited by the laying of odour trails. The emission of the trail substance under these circumstances is remarkable, since no reward in the usual sense is available; the only stimulation that can be inferred is the unfamiliarity of the new field. Workers do not concentrate at the end of the trails, and no specific attraction points can be observed. The majority of workers entering the new field follow the odour trails to the interior, then diverge individually to perform exploratory looping movements. As a result, new fields only a few square metres or less are thoroughly explored and patrolled during the first hour or two. After the initial "land-rush", the number of workers on the field gradually declines and finally stabilizes, in most instances at less than half the peak number. This unusual behaviour seems indicative of an "exploratory" or "manipulative" drive analogous to that described in some mammals (see Thorpe, 1956: 10).

Behaviour at the Food Find

When a solitary foraging worker is presented with a rich new food source, such as a freshly killed mealworm (*Tenebrio*) or roach (*Naupheta*), it abruptly mounts the object, and excitedly palpates it with its antennae while remaining motionless or creeping slowly and deliberately over the surface. If the food source is a single solid object and small enough, the ant soon picks it up and carries it forward or drags it backward in a homeward direction. If the source is an immovable solid object, the worker typically inspects it for a period ranging between 10 and 30 seconds without attempting to feed, and then commences laying a trail homeward. If the food source is a sugary fluid, the ant feeds to partial or whole repletion and returns homeward, laying a trail. Evidently there is nothing peculiar in the nature of liquid food that requires feeding before trail-laying, for if the liquid bait is taken away from the worker before it is able to feed, it nevertheless lays a trail.

Trail-laying Behaviour

The trail-laying worker is easy to distinguish in the throngs of foraging workers through which it is passing. It moves at a slower pace, crouching closer to the ground, and the forward part of its body periodically swings to the right or left of the main line of motion but usually back again to follow the same approximate direction as before.

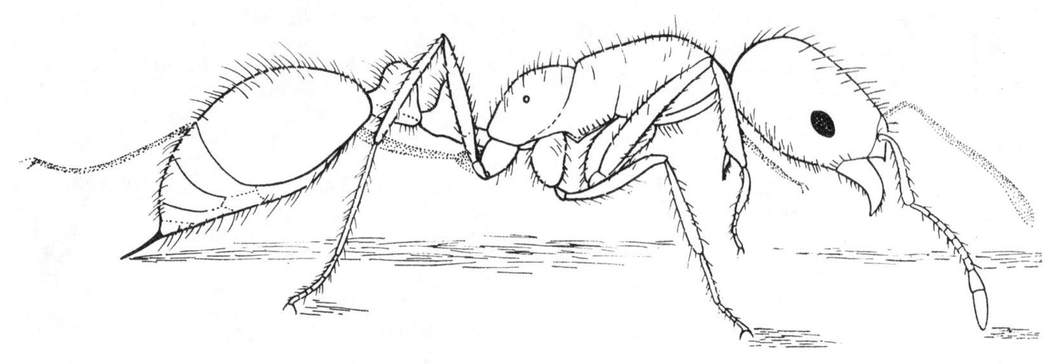

Fig. 2. Worker fire ant laying an odour trail, from left to right. Drawn from a photograph taken at 1/1000 second.

The gaster (abdomen) is bent downward slightly and the sting extruded. Examination of a series of photographs, taken with an electronic flash at 1/1000 sec. and showing lateral views of homing workers, reveal that only the tip of the sting touches the ground (Fig. 2). Furthermore, it is periodically lifted clear for short distances, or withdrawn altogether, breaking the trail into a series of streaks.

The chemical trail, proceeding from Dufour's gland and passing along the valves of the sting, is applied to the ground in a manner similar to a thin line being inked with a pen. It is nevertheless invisible to the naked eye and cannot be detected visually even with a dissecting microscope at magnifications of 50X. All efforts to render the trail visible, including dusting with fine powder (see Carthy, 1951), examination with near-ultraviolet light, and tracing with radioactive isotopes applied externally, have failed. Even strong trails, laid by hundreds of workers over a period of hours, are not visible.

While heading in a homeward direction, the trail-laying worker sometimes loops back in the direction of the bait, but only for short distances, before turning nestward again. During the backtracking movements, trail-laying continues. When another worker is contacted, the homing worker turns toward it. It may do no more than rush against the encountered worker for a fraction of a second before moving on again, but sometimes the reaction is stronger: it climbs partly on top of the worker and, in some instances, shakes its body lightly but vigorously, chiefly in a vertical plane. The precipitousness of the movement toward the encountered worker and the tendency to mount it recall strongly the behaviour shown when foragers first encounter food-finds (q.v.). The vibrating movement, however, is unique to these individual encounters. The behaviour at the encounter appears to function solely to bring the trail substance to the attention of the sister worker. It does not appear to communicate any important information about the food-find, because contacted workers do not seem to exhibit trail-following behaviour different from those not contacted. Moreover, as will be shown in a later section, the chemical trail is by itself sufficient to induce full and immediate trail-following behaviour when laid by artificial means. Trail-laying workers encountering other workers pause with them for only a fraction of a second, even when partially mounting them. Following contact they show a strong tendency to swing about and return in the direction of the food-find. At its weakest the fast-encounter movement involves at the least a partial deflection away from the homeward direction. Frequently the trail-layer turns a full 180°, heads for a short distance in the direction of the food-find, and then turns homeward again. As the ant approaches the nest perimeter, and contacts become more frequent, the looping movements become more pronounced. After several encounters (rarely, after only one) the worker now heads persistently in the direction of the food source. As it runs back, it continues laying a trail, so that a double trail is formed. It follows its own trail outward with no more precision than other workers are following it, wide deviations often occur, and hence an alternate trail is provided. If the target is sufficiently close for the trail-layer successfully to reach it, the ant usually turns homeward again, laying a third trail. Few workers get this far; in a single extreme case one was seen to lay a total of four trails, the final one ending in the vicinity of the target. When the target is so far away from the nest (on glass, about 50 cm. or more) that the outer stretch of the homeward-laid trail evaporates before the trail-layer can return along it, the trail-layer becomes confused. It continues for a while in an outward direction away from the nest, presumably either through kinesthetic momentum or menotactic orientation, but in the great majority of cases it is too far off to make an early second contact with the target.

Recruitment of New Workers

Most workers encountering a freshly laid trail respond by at once following it in an outward direction. They are able to detect it olfactorily over distances as great as 10 millimetres. Workers coming upon the trail by accident within the first half-minute respond with the same alacrity as those encountering the trail-layer directly. As noted before, there is no evidence that direct contact with the trail-layer confers any information beyond the attractive effect of the trail substance. No direct experimental test has yet been devised to determine whether or not the trail has an intrinsic orientation. However, several lines of indirect information virtually prove that it does not: (1) the trail is seemingly deposited in such a manner as to preclude the possibility of polarization by tapering or coding through varying intervals of segmentation; (2) the trail-layer commonly doubles back in looping motions while laying the trail, but followers continue along these inverted segments in an outbound direction without confusion; (3) in following trails workers are in direct contact with them only for short distances, often deflecting to one side and crossing them in an irregular zig-zag fashion; (4) workers that encounter nestward-bound trail-laying workers well away from the nest commonly follow closely behind them, sometimes most of the way back to the nest, rather than move toward the bait.*

Direct experimental evidence has been obtained to show that trails laid by multiple workers to a single food find contain no intrinsic orientation. In a series of tests, workers were forced to cross a bridge of white poster card 1·5 cm. wide and 30 cm. long in order to reach a bait. Approximately 10 minutes following maximum buildup, the bridge was removed, tapped free of ants, and placed sidewise on the near (nestward) side of the chasm in a position such that its centre lay athwart the oncoming stream of ants. Thus the ants approaching the bridge were all outbound and in order to proceed along the odour trail were required to choose between making right-angle turns to the right or left on it. If they were to choose predominantly the right-angle turn and thus follow the former outer half of the bridge, it could be assumed that they were orienting by some topographic or olfactory feature of the trail itself. If the stream divided evenly in both directions, it could be concluded that such features, if they exist, were not used. Counts were made of the numbers of ants on both outermost 10-cm. segment of the bridge 20 seconds after the bridge was turned. The direction of the bridge with respect to the lighting was alternated in successive trials to eliminate any special visual bias. In five replications a total of 59 workers travelled to the end of the bridge that previously would have led to the bait, and 62 travelled to the end that would have led back to the nest. In both the individual test scores, and the summed scores just cited, no significant difference was found in the direction taken by the ant stream. That the ants were truly following odour trails and not just randomly exploring was proven by control experiments in which newly constructed bridges were offered them in a sidewise position. The ants did not travel deliberately down the length of these as they had the test bridges but instead milled about in confusion near the centre. It was concluded that under the conditions of the experiment the odour trail offers no intrinsic features that allow polarity choices.

This negative conclusion is supported by the results of an additional experiment. The procedure just described was modified by simply removing the bridge, tapping it lightly to knock the ants from it, turning it around 180°, and replacing it across the gap. Under these conditions, both the outward- and inward-bound ant streams remounted the bridge and crossed it without evidence of confusion. Their transit was as swift and orderly as in other tests in which the bridge was replaced without reversal but notably faster than across fresh bridges lacking odour trails.

When the odour trail leading from the bait to the nest (or way-station) is short enough to allow workers to run its length before its outer segment evaporates, the build-up of workers at the bait is rapid. The build-up, in fact, is sigmoidal, accelerating at first and then stabilizing at an

*Macgregor (1948) and Carthy (1951) have demonstrated the lack of polarization communication in the odour trails of *Myrmica ruginodis* (=*rubra div. auct.*) and *Lasius fuliginosus*.

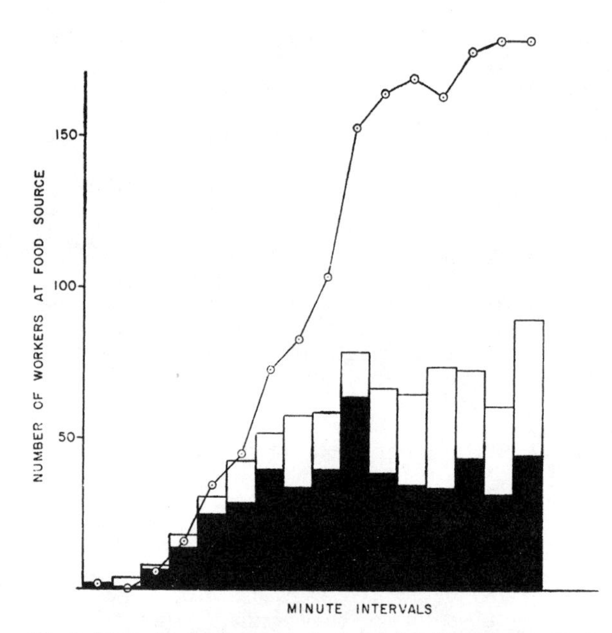

Fig. 3. The build-up of workers at a bait (adult *Naupheta* roach) placed near the edge of an artificial nest containing a large colony. Further explanation in the text.

asymptotic upper level (Fig. 3). The first newcomers to arrive at the bait repeat the behaviour of the discoverer worker: after a brief investigation they begin to lay trails homeward. The larger the number of trails, i.e. the greater quantity of Dufour's gland secretion, being laid at a given instant to the nest perimeter, the larger the number of workers that respond by going out to the bait. Since multiple workers are attracted by a single trail, and each of these in turn lay trails, the build-up at the bait is exponential.

The direct relation between the quantity of Dufour's gland deposit and the number of responding workers was demonstrated experimentally in the following manner. A highly concentrated preparation of Dufour's gland secretion was used.* To simulate the natural variation that obtains from the addition of multiple trials, the quantity of pheromone was varied by presenting differing areas of concentrate on the tips of glass rods. Rods of several diameters were dipped to measured distances into the concentrate and the adhering droplets removed to leave a smooth, flat liquid layer. By this means the quantity of pheromone released could be roughly measured in units of area of surface film exposed. Following treatment the glass rods were immediately placed at an entrance hole of a nest containing a small resting colony, as shown in Fig. 4. The evaporated pheromone was drawn over the colony by a steady air current being pulled by a vacuum apparatus through the nest. The number of workers attracted outside of the nest was then counted at the end of a ten-minute period. The results are shown in Fig. 5.

The data suggest that within the limits employed, the response increases approximately linearly with an increase in area of pheromone. The wide disparity in the slope of the curves obtained on two series of tests shows the influence of additional parameters of colony response not yet understood. Variation in trophic

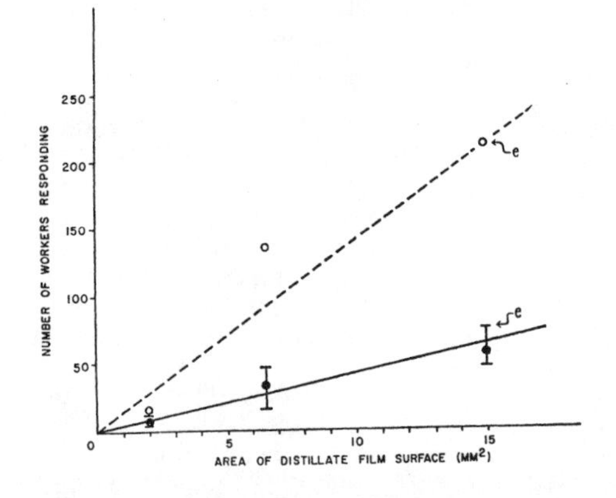

Fig. 5. Varying mass response of workers to different areas of Dufour's gland concentrate presented in the manner illustrated in Fig. 4. The results of two series of experiments on the same colony, containing 300-500 workers, are shown. In the lower series (joined by a solid line), three replications of each quantity of distillate were used. The range and mean are shown. In the upper series (joined by a broken line, only a single experiment was performed. The letter *e* designates the replications at which incipient emigration was induced. Further explanation in text.

*The concentrate used in this experiment was supplied by Dr. M. S. Blum, who obtained the pheromone by successively extracting wholeants in petroleum ether, steam-distilling the extract, and finally fractionating the distillate by column chromatography. The ethopotent chromatographic fraction was identified by the use of an artificial-trail assay (Wilson, 1959).

WILSON: ORGANIZATION AMONG FIRE ANTS

PLATE XII

Fig. 4. Response of workers in an artificial nest to evaporated trail substance. *Above:* before the start of the experiment, air is being drawn into the nest (by suction tubing inserted to the left) from the direction of the untreated glass rod; only a few workers are foraging outside the nest. *Below:* within a short time after the glass rod has been dipped into Dufour's gland concentrate and replaced, a large fraction of the worker force leaves the nest and moves in the direction of the rod.

condition was probably not sufficient to cause the difference, since the colony was given a uniform diet throughout the tests, and population structure did not change significantly. Unfortunately, not enough concentrate was available to extend these tests beyond the few measurements shown.

Under natural conditions the number of workers at the bait approaches an asymptotic limit, evidently for three reasons: (1) as workers begin to crowd on the bait, the number reaching the bait during a given period of time reaches a limit; (2) only workers that have contacted and examined the bait lay trails when they return homeward; (3) the odour trail is impermanent, individual and compound trails evaporating to below threshold level within two minutes on a clean glass surface and within ten to twenty minutes on highly absorbent paper.

To measure the duration of single odour trails, the following experiment was performed. A single worker was allowed to discover a bait (freshly killed *Naupheta*) mounted on a glass slide. As the worker investigated the bait, a fresh glass plate was inserted between it and the nest, so that it was forced to lay its trail over this fresh surface. As soon as it reached the nest, the worker was diverted so it could not lay a second trail back over the glass. Further, the bait was removed, so that workers reaching the end of the trail were not reinforced and did not lay trails of their own. The period during which other workers followed the trail, or segments of it, was timed. In five replications, the trail remained active between 85 and 125 seconds, with an average period of 104 seconds. The air temperature during the tests was 28°C. No attempt was made to determine whether the passage of the workers in some way reinforced the trail and extended its life; it is at least certain that these workers did not lay trails of their own by the conventional means of sting extrusion. In any case, the measurements given were collected under the more significant condition, since in natural circumstances trails die out while being followed by unrewarded workers.

Glass of course has an unnaturally smooth surface, from which substances can be expected to evaporate rapidly. The duration of three trails laid under similar conditions to the above but on highly absorbent ink-blotter paper lasted 450 seconds, 750 seconds, and approximately 20 minutes respectively. The build-up at baits on blotter-paper is correspondingly faster. In a series of tests using a 1M sucrose bait at 20 cm., the interval was measured between the departure of the first trail-layer and the moment the build-up of trail-followers at the bait reached ten. On glass this interval ranged between 253 and 409 seconds, with an average of 329·6 seconds, in five tests; on blotter paper the interval ranged between 78 and 202 seconds and averaged only 133·8 seconds in five tests. It is scarcely necessary to add that the mean difference is highly significant.

Finally, artificial trails made by smearing accessory glands, containing far more of the pheromone than single natural trails and drawing many times more workers, remained active on a glass surface between 3·5 and 7 minutes (mean: 5·9 minutes) in ten trials. The densest natural compound trails laid by columns of workers simultaneously on glass usually declined below threshold value within twenty minutes.

In summary, it has been concluded that the failure of trail emission in frustrated workers and the ephemeral nature of the trail itself combine to provide an effective negative-feedback device that prevents the excessive build-up of workers. As the bait becomes crowded, the number of workers laying trails (hence, rate of emission of trail substance) stabilizes. Since a single worker's contribution vanishes within a few minutes, the number of new workers coming to the bait also stabilizes. In order to increase the number of workers, it is only necessary to increase the area of the bait (see Fig. 6).

The Control of Build-up Rate and Limit

Starved colonies respond more swiftly to baits, and achieve a higher maximum density of attendance, than do well-fed ones. Similarly, the same colony responds more strongly to rich food sources than to poor ones; for example, the build-up at a 1M sucrose source is many times faster than at an adjacent pure water source, and the maximum attendance (build-up limit) is always greater at the former.

The basis of this variation in social response has been investigated by examining worker behaviour in detail at adjacent water and sucrose baits. Several key measurements are described in the sections to follow and summarized in Table II.

a-b. Workers approaching the water drop usually inspected it or fed from it in a brief and desultory manner, while those contacting the sugar solution invariably stopped at the point of

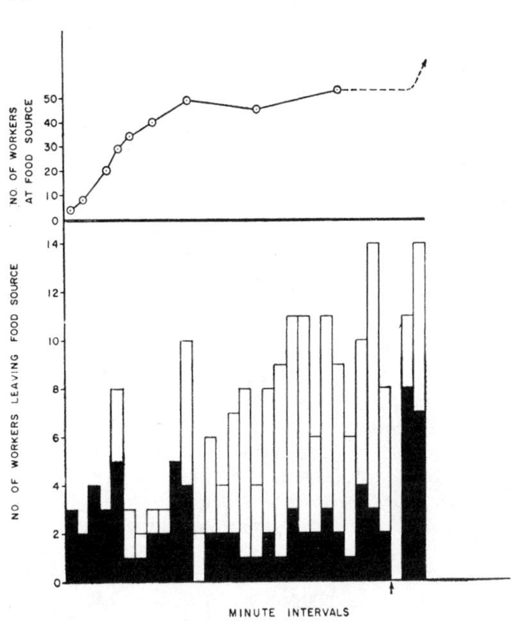

Fig. 6. A typical experiment demonstrating the relation between the area of the food source and the number of responding workers. *Lower:* number of laying workers (solid columns) and non-laying workers (open columns) leaving a bait at successive time intervals. At one point, designated by the arrow, the bait area was doubled. Following the disturbance, during which the total number of returning workers dropped off momentarily, the number of trail-laying workers increased significantly. *Upper:* the number of workers at the bait during the same experiment.

Table II. Responses of Foraging Workers to Water and Sucrose Baits (Explanation in text).

	a. Time spent at bait (range in secs.)	b. Time spent at bait (mean in secs.)	c. % extruding sting	d. Sting-extrusion period (range in secs.)	e. Sting-extrusion period (M±s.e.)
Pure water	2—196	31·8	30	18—29	23·8±2·1
1M sucrose	50—339	144·1	90	6—26	20·4±1·2

contact and fed lengthily. The first two columns give the periods of time spent in inspections and feeding by ten workers selected randomly as they first approached the water and by ten others selected as they approached the sugar solution. Only two of the workers drank enough water to distend their abdomens noticeably; one of these remained for 196 seconds and reached a replete condition, while the other remained for 42 seconds and became semi-replete. In contrast, the abdomens of all ten of the workers at the sugar solution distended noticeably, 8 becoming replete and 2 semi-replete. If the ants in each sample are classified as feeders vs. non-feeders, the difference in proportion of feeding between the two samples is statistically highly significant (probability of randomness less than 0·1 per cent.). Moreover, in other trials using sucrose and water baits simultaneously, the wide disparity in response was invariably noted.

c. In other random samples of 20 workers at each bait, a significantly lower percentage among those leaving water laid trails (30 per cent.) than among those leaving sugar (90 per cent.). The probability that this difference in proportion is due to chance is less than 0·1 per cent. Of those laying trails from the water, all had fed to partial or whole repletion. But 6 others (30 per cent. of total sample) fed to some degree of noticeable repletion without laying trails; this fraction is of course separate from the 30 per cent. that laid trails. All of the 20 workers at the sugar solution fed to some degree of repletion, and of these 18 (90 per cent.) laid trails. The 2 that failed to lay trails were nevertheless fully replete; it is noteworthy that these and other observations suggest that after the ant has fed no correlation exists between the degree of repletion and the tendency to lay trails.

d-e. As the forager (from sample *c*) left the bait it was followed for 30 seconds and the total amount of time spent in trail-laying measured with a stop-watch. Only data pertaining to workers laying trails are given. Actually, it is impossible under ordinary conditions to determine at each instant whether trail substance is being emitted, and it is furthermore very difficult to see whether at all times the tip of the sting touches the ground. Hence the measurement taken was the summed period during which the sting was extruded, on the assumption that this period is closely correlated with the duration of actual trail-laying and, similarly, quantity of trail substance emitted. The data in *d-e* show that trail laying is essentially an all-or-none response, since the average sting-extrusion periods differs very little. In fact the probability that the mean difference cited (3·4 seconds) is due to chance alone is 7 per cent., a magnitude greater than that conventionally accepted as indicating a significant difference. Workers that have fed on the bait either lay no trail at all or else engage in trail-laying (sting extrusion) more than half of the first thirty seconds. Other studies have shown that the all-or-none nature of the response holds until the ant returns to the nest. Note that a single worker leaving the sugar bait extruded its sting during only 6 seconds in the first 30-second period. The behaviour of this individual was exceptional in another, significant way. It was evidently disorientated for a short while after leaving the bait, wandering in irregular loops away from the homeward odour trail being followed by other foragers. During this time it did not extrude its sting. Only toward the end of the first 30-second period did it find the trail, start homeward, and commence extruding its sting. Thereafter it engaged in normal trail-laying, conforming to the general pattern. If this single individual is excluded, the periods of sting-extrusion by the remaining 17 workers ranged 14-26 seconds, and averaged 21·2 seconds with a standard error of 0·8 seconds. If these latter estimates are accepted, the probability that the new mean difference (2·6 seconds) is due to chance alone is 23 per cent.

Trail-laying by individual workers can therefore fairly be called an all-or-none response which is elicited, at least in the case of liquid foods, in a certain percentage of workers that successfully feed. The build-up at a poor source, such as pure water, is less than that at a rich source, chiefly because more foragers turn away without feeding at all. It may also be true that the poor source releases the trail-laying response in a smaller percentage of those workers that feed, but this control, if it exists, is clearly of secondary magnitude. The causes of failure of the response in a minority of feeding workers have not been further considered.

Properties of Overcompensation in the Mass Response

The sum of the experimental evidence points clearly to the trail pheromone being the paramount, and probably sole, mode of communication in the mass response.* Further, and of equal importance, the individual emission of the pheromone is an all-or-none response, remaining relatively constant when released regardless of the quantity or quality of the food find.

Together these two properties of individual behaviour make it unlikely that the single worker is able to communicate information about the quality or quantity of the food find. In fact, it appears able only to command sister workers to proceed outward, orienting them by means of the trail, without varying the number of recruits in any directed manner. By means of individual choice at the food site, newly arriving workers then "vote", i.e. independently choose whether to lay trails of their own. As shown in the preceding section, control of build-up is mass-communicated; i.e. the number of workers responding is controlled by the sum of workers participating in trail-laying.

Assuming that these simple properties of the mass response are not significantly complicated by other behavioural phenomena, it should be possible to make predictions of the outcome of experimental modifications of the response without undue difficulty. One prediction convenient to test is the following: if the bait is removed in the midst of the build-up, the ants should not be able to communicate this fact at once but should "overshoot", i.e. fresh workers should continue to pour into the now-empty bait area for at least several minutes. Furthermore, the "maximum overshoot", defined as the largest number of workers present in the bait area following bait removal, should appear after an interval which is proportional to the distance of the bait area from the nest. Finally, and most importantly, the maximum overshoot should be independent of (1) quantity of the bait, (2) quality of the bait, (3) stage of the build-up at which the bait is removed, and (4) distance of the bait area from the nest.

Most of these properties were demonstrated in actual experiments (Table III, Fig. 7, and below).

The following procedure was used to measure maximum overshoot. The bait was placed in the centre of a 4-cm.^{2} cover slip, which was in turn placed in the centre of a rectangular area 3 × 5 inches (7·6 × 12·7 cm.) marked on the lower

*That is, the mass response leading to the normal build-up at the food source. Later, extensive regurgitation occurs (Wilson & Eisner, 1957), with probable long-range social consequences that have not been adequately investigated.

Table III. Maximum Overshoot Obtained Under Various Experimental Conditions.

Colony size	Bait	Bait (target) distance	Initial build-up	Maximum overshoot (range)	Maximum overshoot (mean ± s.e.)	Time of maximum overshoot (mins.)
300±10	0·5 ml 1 M sucrose	10 cm.	10	1—8	5·0±1·2	1/2
13,000 ±2,000	,,	,,	,,	17—29	20·0±2·6	1 1/2
26,000 ±4,000	,,	,,	,,	7—33	19·0±3·6	1 1/2
,,	,,	20 cm.	,,	8—22	17·8±2·0	1 1/2
,,	,,	50 cm.	,,	10—33	19·0±3·6	3 1/2
,,	,,	80 cm.	,,	5—25	13·5±2·7	7
,,	,,	100 cm.	,,	10—17	13·0±1·2	7
,,	,,	150 cm.	,,	3—14	8·1±1·8	4·2
,,	,,	10 cm.	1	3—7	5·4±0·7	2
,,	,,	,,	50± 5	14—28	18·8±3·0	1 1/2
,,	,,	,,	100±10	10—37	20·9±3·5	2
,,	distilled water	,,	10	13,21*	17	2
,,	*Naupheta*	,,	,,	11—34	19·9±3·2	1 1/2

*Based on two trials only; see text.

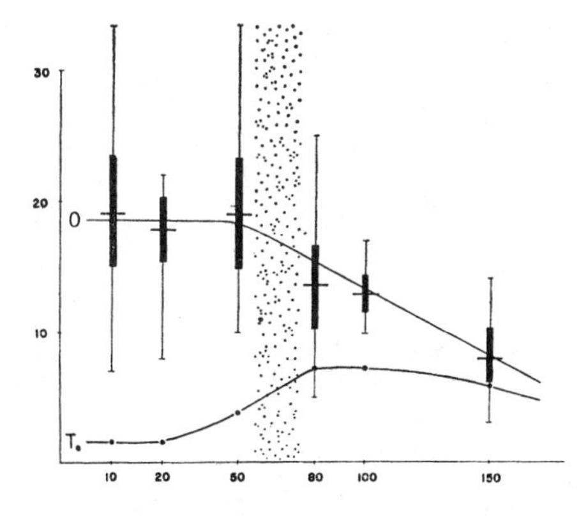

Fig. 7. Variation in maximum overshoot (O) and time of maximum shoot (To) as a function of distance of bait from the nest based in 8 replications for each distance. For the former, extreme limits, mean, and standard deviation are shown; for the latter only the mean is shown. The stippled area represents the target distance range in which the efficiency of the trail system of a larger colony begins a significant decline (see Fig. 8). Further explanation in the text.

surface of the glass platform. Workers were allowed to lay trails to the bait, and build up their numbers to a previously decided "initial build-up" level. Then the cover slip was lifted away, along with all the ants adhering to it. Thus at the outset, the ants feeding at the bait and those immediately arriving or leaving were removed. The number of ants remaining in the rectangle at this moment was noted. The increase in number of workers over the residue was then recorded at 30-second intervals. The excess is referred to as the "overshoot" (in the mass response) and the highest excess recorded at any half-minute interval the "maximum overshoot." The experiment was repeated eight times for each row given in Table III. One exception was made in the case of distilled water. Response to water was so slow that during repeated trials, an initial build-up of ten was attained only twice. The air temperature during the measurements was 28±1°C.

It will be seen from the data given in Table III and Fig. 7 that the maximum overshoot remains remarkably constant under a very wide variety of conditions. It is approximately the same, averaging between 17 and 21, for colonies of different sizes, for baits of different quality, for baits placed at varying distances from the nest, and for widely different initial build-ups allowed in the experiments. Specifically, the probability that the mean differences between the overshoot values are due to chance alone in most cases exceeds 95 per cent. The several statistically significant deviations from the constant maximum-overshoot value are not unexpected and do not detract from the significance of the general constancy. The very small colony (300 workers) overshot less because of the much smaller worker force at the nest periphery. The overshoot following an initial build-up of only one worker was significantly less because only a single odour trail was laid. Overshoot quickly reaches the expected level (by an initial build-up of 10) and does not exceed it in much higher build-ups (at least to 100). The reason for this can be deduced from the measurements given in Fig. 6. The upper limit of trail-laying activity, and hence the quantity of trail pheromone at any given instant, is quickly reached when the bait is relatively small. Finally, as seen in Fig. 7, the maximum overshoot significantly declines when the bait is greater than 50 cm. from the nest because by the time of an initial build-up of 10 the trail is still incomplete and recruitment less accurate.

Trail Formation Over Great Distances

Beyond the distance of approximately 50 cm. on a glass surface, the trail-laying worker is not able to bring back other workers with it before the outermost (first-laid) part of the trail evaporates to below threshold level. Under these conditions the ants are unable to proceed directly to the bait. Under laboratory conditions, nevertheless, baits placed at this and greater distances are soon rediscovered and a solid system of multiple trails laid down to connect them with the nest. This is accomplished in the following manner. Ants reaching the shrinking trail terminus continue searching in its vicinity; a large proportion continue moving outward in the general direction that the trail led them. The result is that the area around the bait, providing it is not excessively distant, is soon being patrolled by an increased number of hunting workers. The bait will probably be encountered by one of these at the close of shorter interval than was required for the initial discovery. When this occurs another trail is laid, more workers enter the area, and the intervals between discoveries and the laying of individual trails grow ever shorter. Soon multiple trails are being laid simultaneously at various distances between the bait. These form links that ultimately sum into a continuous trail.

Although continuous compound trails are manufactured with ease for some distance beyond the maximum continuous distance of a single trail, this distance does have a measureable upper limit. The limit is affected by at least two parameters, physical structure of the foraging surface and colony size. The former has not been well analysed, glass being consistently used as the standard surface for comparative experiments. The influence of colony size is illustrated in Fig. 8.

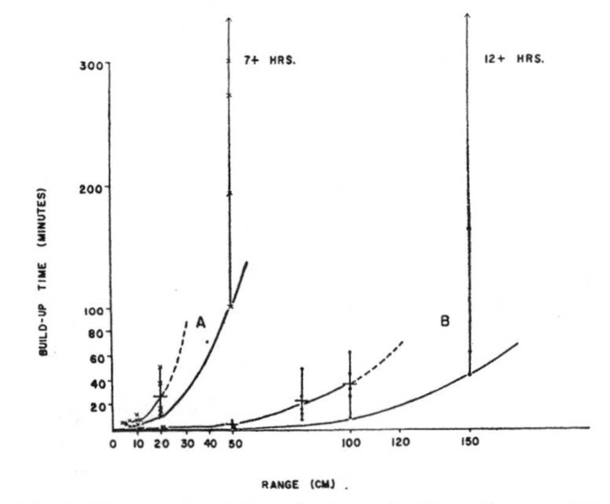

Fig. 8. Time required for colonies to build up forces of 10 workers at sucrose baits at various distances from the nest. All data are shown, with mean values for responses at each distance. A small colony (A), containing approximately 300 workers, is compared with a medium-sized colony (B), containing 26,000±4,000 workers. Two curves have been fitted by eye for the data from each colony, pertaining respectively to the lowermost and the mean values.

In nature, fire ant colonies commonly maintain odour trails leading to old food finds for long periods of time. Excavation often occurs along the trails, resulting in irregular roofs over the trail and gallery networks in the soil that serve as way-stations for foraging workers. These major trails, reminiscent of the "trunk routes" described in *Monomorium pharaonis* (L.) by Sudd (1960) and others, greatly enlarge the foraging fields of the fire ant colonies.

Specificity of the Trail Substance

Colony fragments of the fire ant species *Solenopsis geminata* (Fabricius), collected at

Table IV. Intra- and Inter-specific Responses to Dufour's Gland Secretion in the Artificial Trail Test.

Source species ⟶ test species	No. of replications	No. of workers responding (range)	Duration of response (range in mins.)
Geminata ⟶ *geminata*	4	34—40	4—7
Geminata ⟶ *saevissima*	4	0	—
Geminata ⟶ *xyloni*	2	5	5
Saevissima ⟶ *saevissima*	4	50—200+	5—15+
Saevissima ⟶ *geminata*	4	0—3	1/2—3
Saevissima ⟶ *xyloni*	0	—	—
Xyloni ⟶ *xyloni*	2	16—51	12—19
Xyloni ⟶ *geminata*	1	52	20
Xyloni ⟶ *saevissima*	3	100—200+	3—5

Palmar, Costa Rica, and *S. xyloni* (McCook), collected at Loma Linda, California, were used in tests with *S. saevissima* to determine specificity of the Dufour's gland secretions among these closely related species. Bioassays were based on the artificial trail test described elsewhere (Wilson, 1959). The results, presented in Table IV, show conclusively that the secretion of *geminata* is chemically different from those of *saevissima* and *xyloni*. Furthermore, the secretions of *saevissima* and *xyloni* differ in their effect on *geminata* in the limited trials made.

The Dufour's gland secretions of other ant species have also been tested in *Solenopsis saevissima* and vice versa. The secretion of *saevissima* caused no reaction in the distantly related "thief ant", *Solenopsis* (*Diplorhoptrum*) *molesta* (Say), and similar negative responses were obtained in the myrmicines *Myrmica rubra* (L.), *Pogonomyrmex badius* (Latr.), *Crematogaster lineolata* (Say), and *Xenomymex floridanus* Emery, and dolichoderines *Iridomyrmex humilis* Mayr, *Monacis bispinosa* (Olivier), and *Tapinoma sessile* (Say). The Dufour's gland secretions of the myrmicine species listed above were tested on *Solenopsis saevissima*, also with negative effect. However, a most surprising discovery was made when the secretion of *Monacis bispinosa* was tested on *S. saevissima*. In repeated tests it was found that a single *Monacis* Dufour's gland smeared out into an artificial trail attracted approximately as many *S. saevissima* workers as a single *S. saevissima* gland, over the same period of time. However, the *saevissima* gland does not cause a similar response in *Monacis* workers. The latter are known to emit the trail substance from another organ, the ventral scent gland (Wilson & Pavan, 1961). The specificity experiments have not been extended beyond the species listed here.

The Artificial Induction of Mass-Foraging Behaviour

If recruitment communication in *Solenopsis saevissima* is exclusively by means of a chemical releaser deposited in the trail, as the accumulated evidence seems to suggest, then it should be possible to evoke a full mass-foraging response from a laboratory colony solely by means of artificial trails made from Dufour's glands. That this is indeed the case is shown in the following experiment.

A freshly killed *Naupheta* roach was placed at a distance of a metre from the nest in a spot not being patrolled at the moment by workers. A thin artificial trail was then laid from the edge of the roach to the nest entrance, without allowing the applicator to touch workers during the inscription. In five replications of this procedure, the response was always the same. The great majority of workers within several millimetres distance of the freshly laid trail turned toward it and followed it at once in a direction away from the nest. Their initial response and movement along the trail differed in no detectable way from normal trail-following. When they reached the roach, they responded in the typical manner shown following the discovery of new food finds (q.v.), and when they departed they returned along the artificial trail, reinforcing it with their own secretions. In other tests workers led along artificial trails lacking rewards milled about in confusion at the end of the trail or continued on past the end a short distance before turning homeward, in movement patterns typical for natural dead-end trails. Workers could be led along tortuous paths in this manner, and even directed to form complete circles.

Summary

1. Recruitment in mass foraging is organized almost exclusively by the use of the trail substance, which is secreted by Dufour's gland and released through the extruded sting. The secretion is a powerful attractant, which excites workers and draws them out of the nest in the direction indicated by the trail. No other stimuli are required to induce this behaviour. The initial build-up of workers at food masses is exponential. The build-up decelerates towards a limit as workers become crowded on the food mass because (1) workers unable to reach the mass turn back without laying trails, and (2) the trail deposits of single workers evaporate within a few minutes. As a result, the number of workers at food masses tends to stabilize at a level which is a function of the area of the food mass. But the nature of the communication requires that the feed-back system have a lag of up to several minutes. The lag is measurable in the "overshoot" of numbers of approaching workers and is a function of the distance of the food mass from the nest.

2. Trail-laying, as measured by the extrusion of the sting, is an all-or-none response of individual workers; hence individual trails do not communicate quantity or quality of the food find. Quantity of food is communicated in the mass response by means of the negative feed-back control of build-up just described. Quality is communicated by means of an "electorate" mass response, in which individuals choose whether to lay trails after inspecting the food find. The more desirable the food find, the higher the percentage of positive responses, the more the trail substance presented to the colony, and hence the more the newcomer ants that emerge from the nest.

3. The trail substance is highly species-specific. In transposition experiments utilizing the fire ant species *Solenopsis saevissima*, *S. geminata*, and *S. xyloni*, it was found that the *geminata* secretion is certainly different from that of the other two species. The secretions of *saevissima* and *xyloni* are more interchangeable but appear to differ somewhat. The thief-ant *S. molesta* and members of other myrmicine genera tested also do not share the *saevissima* trail substances. Curiously, the Dufour's gland secretion of the dolichoderine species *Monacis bispinosa* causes a full response in *saevissima* workers, but *M. bispinosa* uses another substance for its trails.

4. The foraging and homing behaviour of individual workers is partially analysed. An unexpected response, in which workers are attracted and lay trails to unexplored areas near the nest, is described.

REFERENCES

Brian, M. V. (1951). Ant culture for laboratory experiment. *Entomologist's Monthly Mag.*, **87**, 134-136.

Brown, W. L. (1959). Releasers of alarm behavior in army ants. *Psyche*, **66**, 25-27.

Butenandt, A., Linzen, B. & Lindauer, M. (1959). Über einen Duftstoff aus der Mandibeldrüse der Blattschneiderameise *Atta sexdens rubropilosa* Forel. *Arch. anat. microscop. morph. Exp.*, **48**, 13-19.

Callahan, P. S., Blum, M. S. & Walker, J. R. (1959). Morphology and histology of the poison glands and sting of the imported fire ant (*Solenopsis saevissima* v. *richteri* Forel). *Ann. ent. Soc. Amer.*, **52**, 573-590.

Carthy, J. D. (1951). The orientation of two allied species of British ants. II. Odour trail laying and following in *Acanthomyops* (*Lasius*) *fuliginosus*. *Behaviour*, **3**, 304-318.

Karlson, P. & Butenandt, A. (1959). Pheromones (ectohormones) in insects. *Ann. Rev. Ent.*, **4**, 39-58.

Macgregor, E. C. (1948). Odour as a basis for oriented movement in ants. *Behaviour*, **1**, 267-296.

Sudd. J. H. (1957). A response of worker ants to dead ants of their own species. *Nature*, **179**, 431-432.

Sudd. J. H. (1960). The foraging method of Pharaoh's ant, *Monomorium pharaonis* (L.). *Animal Behaviour*, **8**, 67-76.

Thorpe, W. H. (1956). *Instinct and learning in animals.* Methuen.

Wilson, E. O. (1958). A chemical releaser of alarm and digging behavior in the ant *Pogonomyrmex badius* (Latreille). *Psyche*, **65**, 41-51.

Wilson, E. O. (1959). Source and possible nature of the odor trail of the fire ant *Solenopsis saevissima* (Fr. Smith). *Science*, **129**, 643-644.

Wilson, E. O. & Eisner, T. (1957). Quantitative studies of liquid food transmission in ants. *Insectes Sociaux*, **4**, 157-166.

Wilson, E. O., Durlach, N. & Roth L. M. (1959). Chemical releasers of necrophoric behavior in ants. *Psyche*, **65**, 108-114.

Wilson, E. O. & Pavan, M. (1959). Source and specificity of chemical releasers of social behavior in the dolichoderine ants. *Psyche*, **66**, 70-76.

(*Accepted for publication* 1*st January*, 1961).

1963

"Pheromones," *Scientific American* 208(5) (May): 100–114

The subject of pheromonal communication in animals had grown so swiftly that by 1963 the first general principles of chemical and physical properties could be formulated. Augmented over the ensuing four decades by further research and new ideas, these principles have for the most part stood the test of time. In contributing to this subject, I was blessed with the collaboration of natural products chemists and of William H. Bossert, who was then a graduate student at Harvard (and now a senior faculty member) and already a brilliant applied mathematician and biological theorist. The results of this first synthesis, presented here in popular form, included the functional distinction between primer and releaser pheromones; the first physical models of transmission, with the concept of the active space; and a theory of optimal molecular size according to the meaning of the signal. Coincidentally, and to the best of my knowledge, the article also contains the first mention of the possible existence of human pheromones.

PHEROMONES

A pheromone is a substance secreted by an animal that influences the behavior of other animals of the same species. Recent studies indicate that such chemical communication is surprisingly common

by Edward O. Wilson

It is conceivable that somewhere on other worlds civilizations exist that communicate entirely by the exchange of chemical substances that are smelled or tasted. Unlikely as this may seem, the theoretical possibility cannot be ruled out. It is not difficult to design, on paper at least, a chemical communication system that can transmit a large amount of information with rather good efficiency. The notion of such a communication system is of course strange because our outlook is shaped so strongly by our own peculiar auditory and visual conventions. This limitation of outlook is found even among students of animal behavior; they have favored species whose communication methods are similar to our own and therefore more accessible to analysis. It is becoming increasingly clear, however, that chemical systems provide the dominant means of communication in many animal species, perhaps even in most. In the past several years animal behaviorists and organic chemists, working together, have made a start at deciphering some of these systems and have discovered a number of surprising new biological phenomena.

In earlier literature on the subject, chemicals used in communication were usually referred to as "ectohormones." Since 1959 the less awkward and etymologically more accurate term "pheromones" has been widely adopted. It is used to describe substances exchanged among members of the same animal species. Unlike true hormones, which are secreted internally to regulate the organism's own physiology, or internal environment, pheromones are secreted externally and help to regulate the organism's external environment by influencing other animals. The mode of influence can take either of two general forms. If the pheromone produces a more or less immediate and reversible change in the behavior of the recipient, it is said to have a "releaser" effect. In this case the chemical substance seems to act directly on the recipient's central nervous system. If the principal function of the pheromone is to trigger a chain of physiological events in the recipient, it has what we have recently labeled a "primer" effect. The physiological changes, in turn, equip the organism with a new behavioral repertory, the components of which are thenceforth evoked by appropriate stimuli. In termites, for example, the reproductive and soldier castes prevent other termites from developing into their own castes by secreting substances that are ingested and act through the *corpus allatum,* an endocrine gland controlling differentiation [see "The Termite and the Cell," by Martin Lüscher; SCIENTIFIC AMERICAN, May, 1953].

These indirect primer pheromones do not always act by physiological inhibition. They can have the opposite effect. Adult males of the migratory locust *Schistocerca gregaria* secrete a volatile substance from their skin surface that accelerates the growth of young locusts. When the nymphs detect this substance with their antennae, their hind legs,

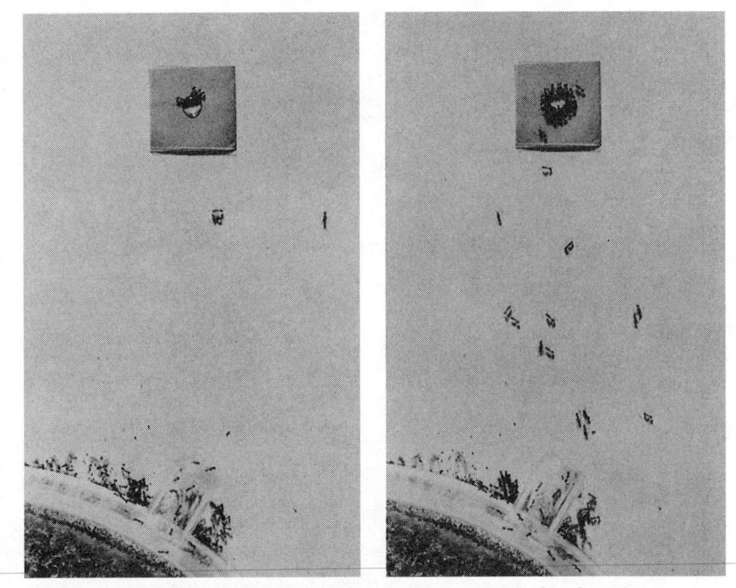

INVISIBLE ODOR TRAILS guide fire ant workers to a source of food: a drop of sugar solution. The trails consist of a pheromone laid down by workers returning to their nest after finding a source of food. Sometimes the chemical message is reinforced by the touching of antennae if a returning worker meets a wandering fellow along the way. This is hap-

some of their mouth parts and the antennae themselves vibrate. The secretion, in conjunction with tactile and visual signals, plays an important role in the formation of migratory locust swarms.

A striking feature of some primer pheromones is that they cause important physiological change without an immediate accompanying behavioral response, at least none that can be said to be peculiar to the pheromone. Beginning in 1955 with the work of S. van der Lee and L. M. Boot in the Netherlands, mammalian endocrinologists have discovered several unexpected effects on the female mouse that are produced by odors of other members of the same species. These changes are not marked by any immediate distinctive behavioral patterns. In the "Lee-Boot effect" females placed in groups of four show an increase in the percentage of pseudopregnancies. A completely normal reproductive pattern can be restored by removing the olfactory bulbs of the mice or by housing the mice separately. When more and more female mice are forced to live together, their oestrous cycles become highly irregular and in most of the mice the cycle stops completely for long periods. Recently W. K. Whitten of the Australian National University has discovered that the odor of a male mouse can initiate and synchronize the oestrous cycles of female mice. The male odor also reduces the frequency of reproductive abnormalities arising when female mice are forced to live under crowded conditions.

A still more surprising primer effect has been found by Helen Bruce of the National Institute for Medical Research in London. She observed that the odor of a strange male mouse will block the pregnancy of a newly impregnated female mouse. The odor of the original stud male, of course, leaves pregnancy undisturbed. The mouse reproductive pheromones have not yet been identified chemically, and their mode of action is only partly understood. There is evidence that the odor of the strange male suppresses the secretion of the hormone prolactin, with the result that the *corpus luteum* (a ductless ovarian gland) fails to develop and normal oestrus is restored. The pheromones are probably part of the complex set of control mechanisms that regulate the population density of animals [see "Population Density and Social Pathology," by John B. Calhoun; SCIENTIFIC AMERICAN, February, 1962].

Pheromones that produce a simple releaser effect—a single specific response mediated directly by the central nervous system—are widespread in the animal kingdom and serve a great many functions. Sex attractants constitute a large and important category. The chemical structures of six attractants are shown on page 9. Although two of the six—the mammalian scents muskone and civetone—have been known for some 40 years and are generally assumed to serve a sexual function, their exact role has never been rigorously established by experiments with living animals. In fact, mammals seem to employ musklike compounds, alone or in combination with other substances, to serve several functions: to mark home ranges, to assist in territorial defense and to identify the sexes.

The nature and role of the four insect sex attractants are much better understood. The identification of each represents a technical feat of considerable magnitude. To obtain 12 milligrams of esters of bombykol, the sex attractant of the female silkworm moth, Adolf F. J. Butenandt and his associates at the Max Planck Institute of Biochemistry in Munich had to extract material from 250,000 moths. Martin Jacobson, Morton Beroza and William Jones of the U.S. Department of Agriculture processed 500,000 female gypsy moths to get 20 milligrams of the gypsy-moth attractant gyplure. Each moth yielded only about .01 microgram (millionth of a gram) of

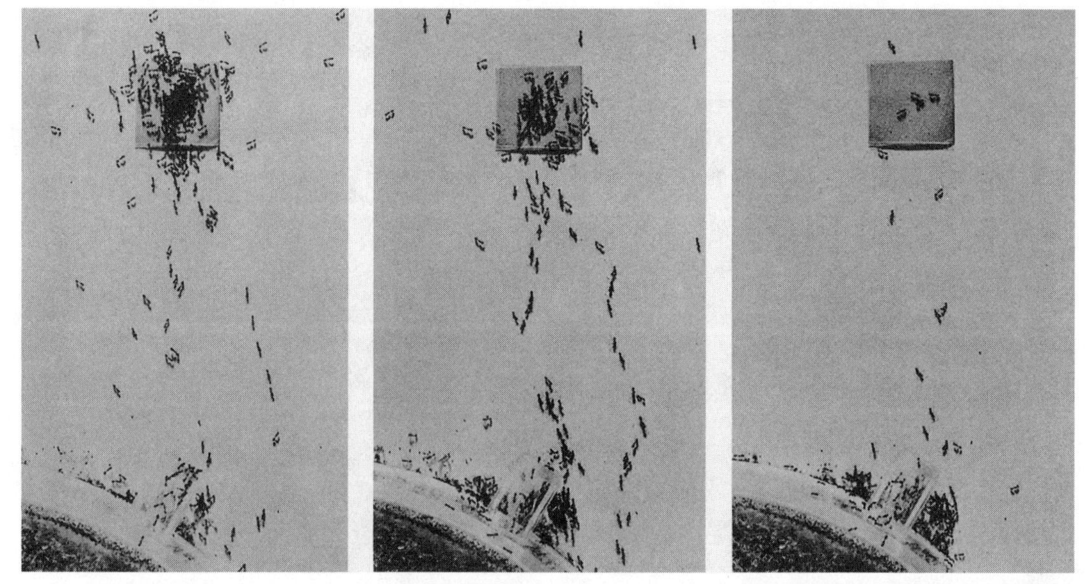

pening in the photograph at the far left. A few foraging workers have just found the sugar drop and a returning trail-layer is communicating the news to another ant. In the next two pictures the trail has been completed and workers stream from the nest in increasing numbers. In the fourth picture unrewarded workers return to the nest without laying trails and outward-bound traffic wanes. In the last picture most of the trails have evaporated completely and only a few stragglers remain at the site, eating the last bits of food.

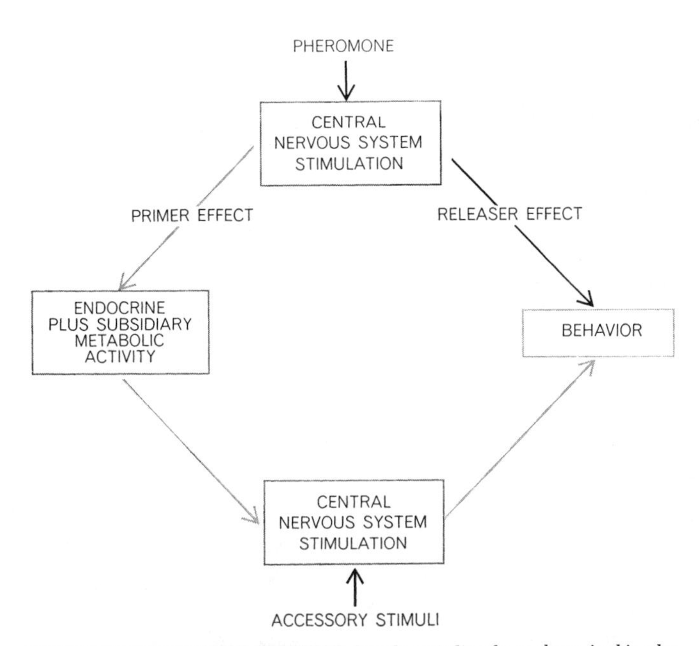

PHEROMONES INFLUENCE BEHAVIOR directly or indirectly, as shown in this schematic diagram. If a pheromone stimulates the recipient's central nervous system into producing an immediate change in behavior, it is said to have a "releaser" effect. If it alters a set of long-term physiological conditions so that the recipient's behavior can subsequently be influenced by specific accessory stimuli, the pheromone is said to have a "primer" effect.

gyplure, or less than a millionth of its body weight. Bombykol and gyplure were obtained by killing the insects and subjecting crude extracts of material to chromatography, the separation technique in which compounds move at different rates through a column packed with a suitable adsorbent substance. Another technique has been more recently developed by Robert T. Yamamoto of the U.S. Department of Agriculture, in collaboration with Jacobson and Beroza, to harvest the equally elusive sex attractant of the American cockroach. Virgin females were housed in metal cans and air was continuously drawn through the cans and passed through chilled containers to condense any vaporized materials. In this manner the equivalent of 10,000 females were "milked" over a nine-month period to yield 12.2 milligrams of what was considered to be the pure attractant.

The power of the insect attractants is almost unbelievable. If some 10,000 molecules of the most active form of bombykol are allowed to diffuse from a source one centimeter from the antennae of a male silkworm moth, a characteristic sexual response is obtained in most cases. If volatility and diffusion rate are taken into account, it can be estimated that the threshold concentration is no more than a few hundred molecules per cubic centimeter, and the actual number required to stimulate the male is probably even smaller. From this one can calculate that .01 microgram of gyplure, the minimum average content of a single female moth, would be theoretically adequate, if distributed with maximum efficiency, to excite more than a billion male moths.

In nature the female uses her powerful pheromone to advertise her presence over a large area with a minimum expenditure of energy. With the aid of published data from field experiments and newly contrived mathematical models of the diffusion process, William H. Bossert, one of my associates in the Biological Laboratories at Harvard University, and I have deduced the shape and size of the ellipsoidal space within which male moths can be attracted under natural conditions [*see bottom illustration on opposite page*]. When a moderate wind is blowing, the active space has a long axis of thousands of meters and a transverse axis parallel to the ground of more than 200 meters at the widest point. The 19th-century French naturalist Jean Henri Fabre, speculating on sex attraction in insects, could not bring himself to believe that the female moth could communicate over such great distances by odor alone, since "one might as well expect to tint a lake with a drop of carmine." We now know that Fabre's conclusion was wrong but that his analogy was exact: to the male moth's powerful chemoreceptors the lake is indeed tinted.

One must now ask how the male moth, smelling the faintly tinted air, knows which way to fly to find the source of the tinting. He cannot simply fly in the direction of increasing scent; it can be shown mathematically that the attractant is distributed almost uniformly after it has drifted more than a few meters from the female. Recent experiments by Ilse Schwinck of the University of Munich have revealed what is probably the alternative procedure used. When male moths are activated by the pheromone, they simply fly upwind and thus inevitably move toward the female. If by accident they pass out of the active zone, they either abandon the search or fly about at random until they pick up the scent again. Eventually, as they approach the female, there is a slight increase in the concentration of the chemical attractant and this can serve as a guide for the remaining distance.

If one is looking for the most highly developed chemical communication systems in nature, it is reasonable to study the behavior of the social insects, particularly the social wasps, bees, termites and ants, all of which communicate mostly in the dark interiors of their nests and are known to have advanced chemoreceptive powers. In recent years experimental techniques have been developed to separate and identify the pheromones of these insects, and rapid progress has been made in deciphering the hitherto intractable codes, particularly those of the ants. The most successful procedure has been to dissect out single glandular reservoirs and see what effect their contents have on the behavior of the worker caste, which is the most numerous and presumably the most in need of continuing guidance. Other pheromones, not present in distinct reservoirs, are identified in chromatographic fractions of crude extracts.

Ants of all castes are constructed with an exceptionally well-developed exocrine glandular system. Many of the most prominent of these glands, whose function has long been a mystery to entomologists, have now been identified as the source of pheromones [*see illustra-*

tion on page 7]. The analysis of the gland-pheromone complex has led to the beginnings of a new and deeper understanding of how ant societies are organized.

Consider the chemical trail. According to the traditional view, trail secretions served as only a limited guide for worker ants and had to be augmented by other kinds of signals exchanged inside the nest. Now it is known that the trail substance is extraordinarily versatile. In the fire ant (*Solenopsis saevissima*), for instance, it functions both to activate and to guide foraging workers in search of food and new nest sites. It also contributes as one of the alarm signals emitted by workers in distress. The trail of the fire ant consists of a substance secreted in minute amounts by Dufour's gland; the substance leaves the ant's body by way of the extruded sting, which is touched intermittently to the ground much like a moving pen dispensing ink. The trail pheromone, which has not yet been chemically identified, acts primarily to attract the fire ant workers. Upon encountering the attractant the workers move automatically up the gradient to the source of emission. When the substance is drawn out in a line, the workers run along the direction of the line away from the nest. This simple response brings them to the food source or new nest site from which the trail is laid. In our laboratory we have extracted the pheromone from the Dufour's glands of freshly killed workers and have used it to create artificial trails. Groups of workers will follow these trails away from the nest and along arbitrary routes (including circles leading back to the nest) for considerable periods of time. When the pheromone is presented to whole colonies in massive doses, a large portion of the colony, including the queen, can be drawn out in a close simulation of the emigration process.

The trail substance is rather volatile, and a natural trail laid by one worker diffuses to below the threshold concentration within two minutes. Consequently outward-bound workers are able to follow it only for the distance they can travel in this time, which is about 40 centimeters. Although this strictly limits the distance over which the ants can communicate, it provides at least two important compensatory advantages. The more obvious advantage is that old, useless trails do not linger to confuse the hunting workers. In addition, the intensity of the trail laid by many workers provides a sensitive index of the amount of food at a given site and the rate of its depletion. As workers move to and from

ANTENNAE OF GYPSY MOTHS differ radically in structure according to their function. In the male (*left*) they are broad and finely divided to detect minute quantities of sex attractant released by the female (*right*). The antennae of the female are much less developed.

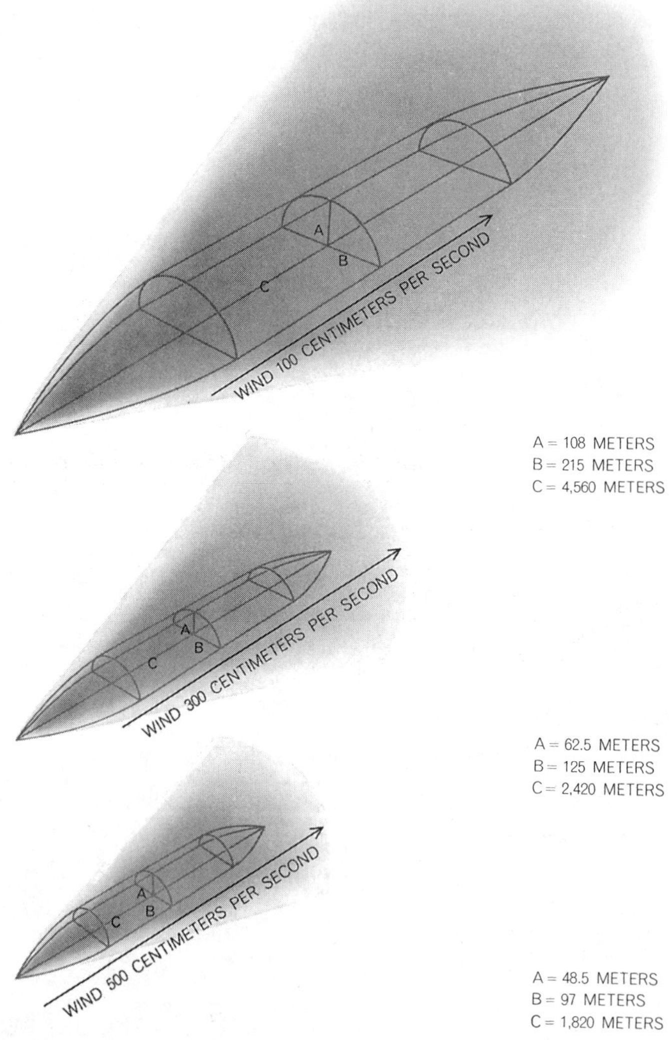

ACTIVE SPACE of gyplure, the gypsy moth sex attractant, is the space within which this pheromone is sufficiently dense to attract males to a single, continuously emitting female. The actual dimensions, deduced from linear measurements and general gas-diffusion models, are given at right. Height (*A*) and width (*B*) are exaggerated in the drawing. As wind shifts from moderate to strong, increased turbulence contracts the active space.

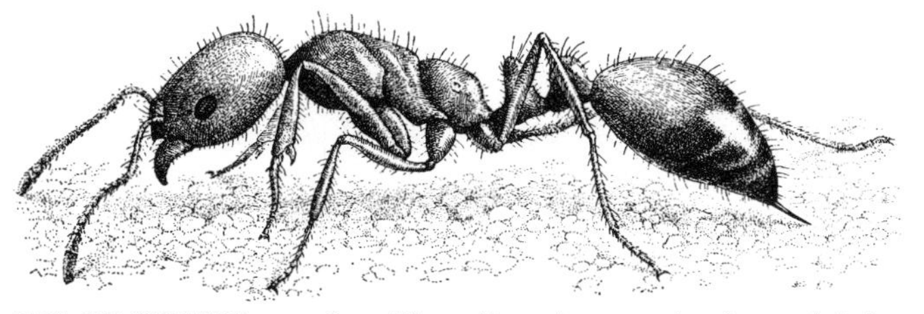

FIRE ANT WORKER lays an odor trail by exuding a pheromone along its extended sting. The sting is touched to the ground periodically, breaking the trail into a series of streaks.

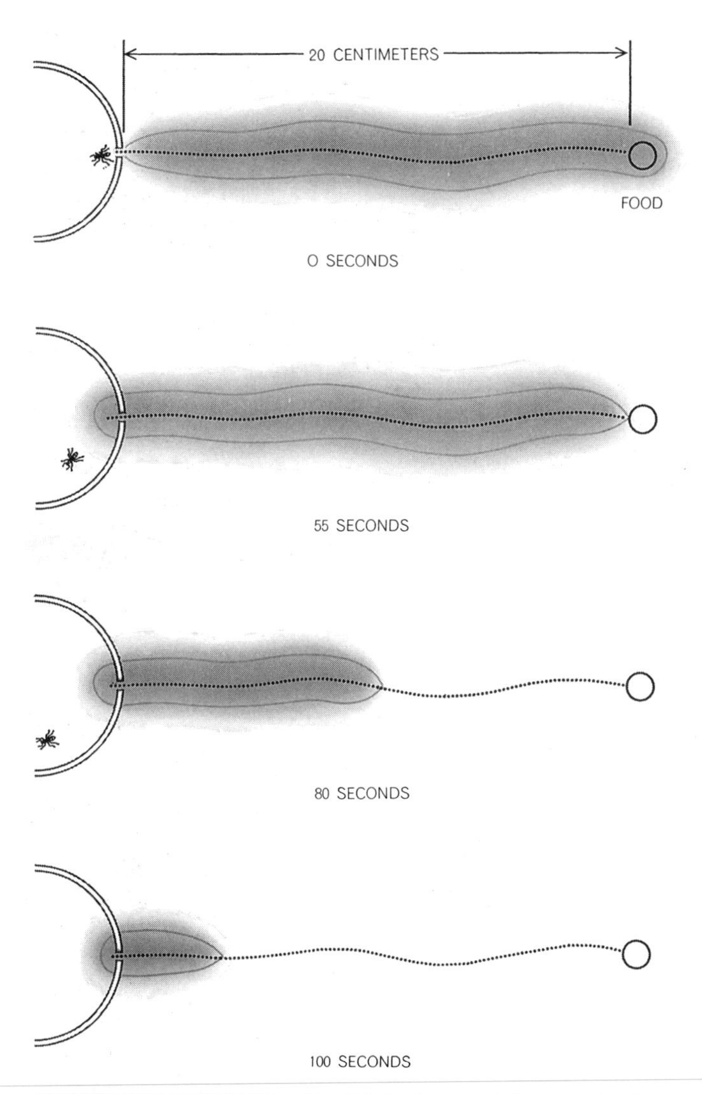

ACTIVE SPACE OF ANT TRAIL, within which the pheromone is dense enough to be perceived by other workers, is narrow and nearly constant in shape with the maximum gradient situated near its outer surface. The rapidity with which the trail evaporates is indicated.

the food finds (consisting mostly of dead insects and sugar sources) they continuously add their own secretions to the trail produced by the original discoverers of the food. Only if an ant is rewarded by food does it lay a trail on its trip back to the nest; therefore the more food encountered at the end of the trail, the more workers that can be rewarded and the heavier the trail. The heavier the trail, the more workers that are drawn from the nest and arrive at the end of the trail. As the food is consumed, the number of workers laying trail substance drops, and the old trail fades by evaporation and diffusion, gradually constricting the outward flow of workers.

The fire ant odor trail shows other evidences of being efficiently designed. The active space within which the pheromone is dense enough to be perceived by workers remains narrow and nearly constant in shape over most of the length of the trail. It has been further deduced from diffusion models that the maximum gradient must be situated near the outer surface of the active space. Thus workers are informed of the space boundary in a highly efficient way. Together these features ensure that the following workers keep in close formation with a minimum chance of losing the trail.

The fire ant trail is one of the few animal communication systems whose information content can be measured with fair precision. Unlike many communicating animals, the ants have a distinct goal in space—the food find or nest site—the direction and distance of which must both be communicated. It is possible by a simple technique to measure how close trail-followers come to the trail end, and, by making use of a standard equation from information theory, one can translate the accuracy of their response into the "bits" of information received. A similar procedure can be applied (as first suggested by the British biologist J. B. S. Haldane) to the "waggle dance" of the honeybee, a radically different form of communication system from the ant trail [see "Dialects in the Language of the Bees," by Karl von Frisch; Scientific American, August, 1962]. Surprisingly, it turns out that the two systems, although of wholly different evolutionary origin, transmit about the same amount of information with reference to distance (two bits) and direction (four bits in the honeybee, and four or possibly five in the ant). Four bits of information will direct an ant or a bee into one of 16 equally probable sectors of a circle and two bits will identify one of four equally probable distances. It is conceivable that these information values represent the maximum that can be achieved with the insect brain and sensory apparatus.

Not all kinds of ants lay chemical trails. Among those that do, however, the pheromones are highly species-specific in their action. In experiments in which artificial trails extracted from one species were directed to living colonies of other species, the results have almost always been negative, even among related species. It is as if each species had its own private language. As a result there is little or no confusion when the trails of two or more species cross.

Another important class of ant pheromone is composed of alarm substances. A simple backyard experiment will show that if a worker ant is disturbed by a clean instrument, it will, for a short time, excite other workers with whom it comes in contact. Until recently most students of ant behavior thought that the alarm was spread by touch, that one worker simply jostled another in its excitement or drummed on its neighbor with its antennae in some peculiar way. Now it is known that disturbed workers discharge chemicals, stored in special glandular reservoirs, that can produce all the characteristic alarm responses solely by themselves. The chemical structure of four alarm substances is shown on page 11. Nothing could illustrate more clearly the wide differences between the human perceptual world and that of chemically communicating animals. To the human nose the alarm substances are mild or even pleasant, but to the ant they represent an urgent tocsin that can propel a colony into violent and instant action.

As in the case of the trail substances, the employment of the alarm substances appears to be ideally designed for the purpose it serves. When the contents of the mandibular glands of a worker of the harvesting ant (*Pogonomyrmex badius*) are discharged into still air, the volatile material forms a rapidly expanding sphere, which attains a radius of about six centimeters in 13 seconds. Then it contracts until the signal fades out completely some 35 seconds after the moment of discharge. The outer shell of the active space contains a low concentration of pheromone, which is actually attractive to harvester workers. This serves to draw them toward the point of disturbance. The central region of the active space, however, contains a concentration high enough to evoke the characteristic frenzy of alarm. The "alarm sphere" expands to a radius of about three centimeters in eight seconds and, as might be expected, fades out more quickly than the "attraction sphere."

The advantage to the ants of an alarm signal that is both local and short-lived becomes obvious when a *Pogonomyrmex* colony is observed under natural conditions. The ant nest is subject to almost innumerable minor disturbances. If the

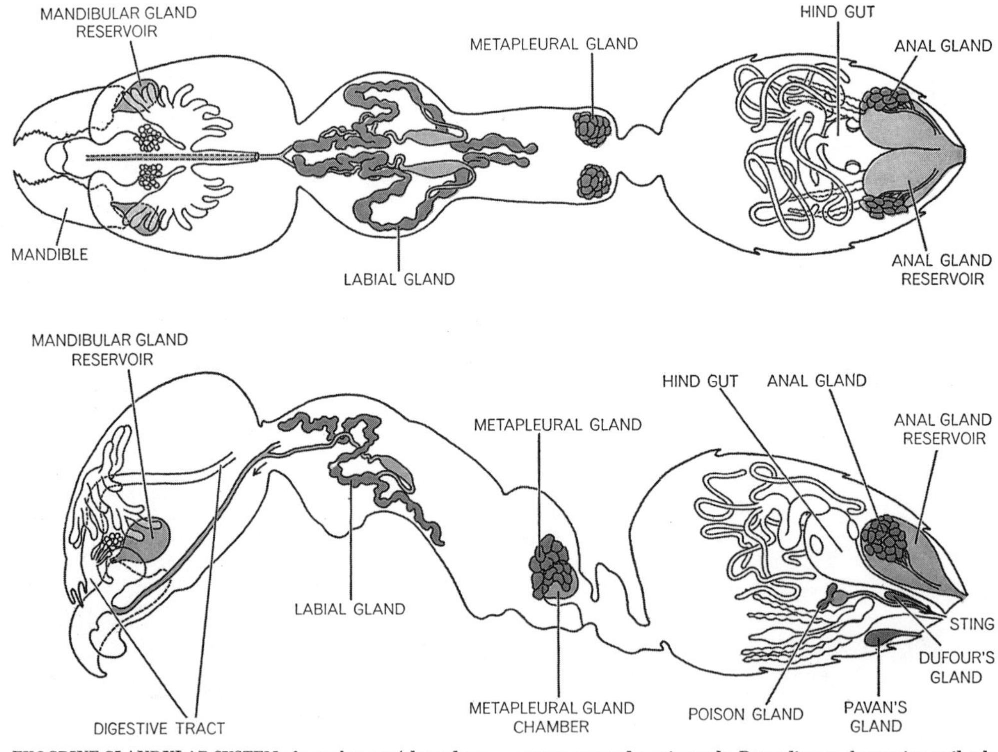

EXOCRINE GLANDULAR SYSTEM of a worker ant (*shown here in top and side cutaway views*) is specially adapted for the production of chemical communication substances. Some pheromones are stored in reservoirs and released in bursts only when needed; others are secreted continuously. Depending on the species, trail substances are produced by Dufour's gland, Pavan's gland or the poison glands; alarm substances are produced by the anal and mandibular glands. The glandular sources of other pheromones are unknown.

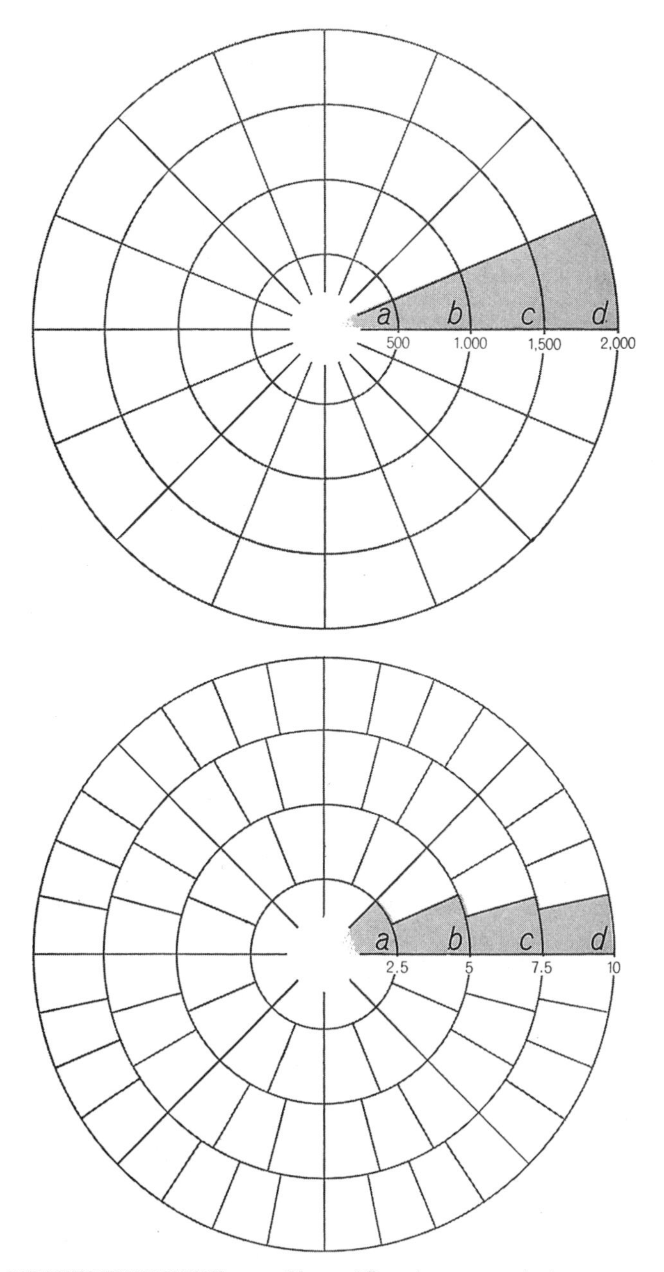

FORAGING INFORMATION conveyed by two different insect communication systems can be represented on two similar "compass" diagrams. The honeybee "waggle dance" (*top*) transmits about four bits of information with respect to direction, enabling a honeybee worker to pinpoint a target within one of 16 equally probable angular sectors. The number of "bits" in this case remains independent of distance, given in meters. The pheromone system used by trail-laying fire ants (*bottom*) is superior in that the amount of directional information increases with distance, given in centimeters. At distances *c* and *d*, the probable sector in which the target lies is smaller for ants than for bees. (For ants, directional information actually increases gradually and not by jumps.) Both insects transmit two bits of distance information, specifying one of four equally probable distance ranges.

alarm spheres generated by individual ant workers were much wider and more durable, the colony would be kept in ceaseless and futile turmoil. As it is, local disturbances such as intrusions by foreign insects are dealt with quickly and efficiently by small groups of workers, and the excitement soon dies away.

The trail and alarm substances are only part of the ants' chemical vocabulary. There is evidence for the existence of other secretions that induce gathering and settling of workers, acts of grooming, food exchange, and other operations fundamental to the care of the queen and immature ants. Even dead ants produce a pheromone of sorts. An ant that has just died will be groomed by other workers as if it were still alive. Its complete immobility and crumpled posture by themselves cause no new response. But in a day or two chemical decomposition products accumulate and stimulate the workers to bear the corpse to the refuse pile outside the nest. Only a few decomposition products trigger this funereal response; they include certain long-chain fatty acids and their esters. When other objects, including living workers, are experimentally daubed with these substances, they are dutifully carried to the refuse pile. After being dumped on the refuse the "living dead" scramble to their feet and promptly return to the nest, only to be carried out again. The hapless creatures are thrown back on the refuse pile time and again until most of the scent of death has been worn off their bodies by the ritual.

Our observation of ant colonies over long periods has led us to believe that as few as 10 pheromones, transmitted singly or in simple combinations, might suffice for the total organization of ant society. The task of separating and characterizing these substances, as well as judging the roles of other kinds of stimuli such as sound, is a job largely for the future.

Even in animal species where other kinds of communication devices are prominently developed, deeper investigation usually reveals the existence of pheromonal communication as well. I have mentioned the auxiliary roles of primer pheromones in the lives of mice and migratory locusts. A more striking example is the communication system of the honeybee. The insect is celebrated for its employment of the "round" and "waggle" dances (augmented, perhaps, by auditory signals) to designate the location of food and new nest sites. It is not so widely known that chemical signals play equally important roles in other aspects of honeybee life. The mother queen regulates the reproductive cycle of the colony by secreting from her mandibular glands a substance recently identified as 9-ketodecanoic acid. When this pheromone is ingested by the worker bees, it inhibits development of their ovaries and also their ability to manufacture the royal cells in which new queens are reared. The same pheromone serves as a sex attractant in the queen's nuptial flights.

Under certain conditions, including the discovery of new food sources, worker bees release geraniol, a pleasant-smelling alcohol, from the abdominal Nassanoff glands. As the geraniol diffuses through the air it attracts other workers and so supplements information contained in the waggle dance. When a worker stings an intruder, it discharges, in addition to the venom, tiny amounts of a secretion from clusters of unicellular glands located next to the basal plates of the sting. This secretion is responsible for the tendency, well known to beekeepers, of angry swarms of workers to sting at the same spot. One component, which acts as a simple attractant, has been identified as isoamyl acetate, a compound that has a banana-like odor. It is possible that the stinging response is evoked by at least one unidentified alarm substance secreted along with the attractant.

Knowledge of pheromones has advanced to the point where one can make some tentative generalizations about their chemistry. In the first place, there appear to be good reasons why sex attractants should be compounds that contain between 10 and 17 carbon atoms and that have molecular weights between about 180 and 300—the range actually observed in attractants so far identified. (For comparison, the weight of a single carbon atom is 12.) Only compounds of roughly this size or greater can meet the two known requirements of a sex attractant: narrow specificity, so that only members of one species will respond to it, and high potency. Compounds that contain fewer than five or so carbon atoms and that have a molecular weight of less than about 100 cannot be assembled in enough different ways to provide a distinctive molecule for all the insects that want to advertise their presence.

It also seems to be a rule, at least with insects, that attraction potency increases with molecular weight. In one series of esters tested on flies, for instance, a doubling of molecular weight resulted in as much as a thousandfold increase in efficiency. On the other hand, the molecule cannot be too large and complex or it will be prohibitively difficult for the insect to synthesize. An equally important limitation on size is

BOMBYKOL (SILKWORM MOTH)

GYPLURE (GYPSY MOTH)

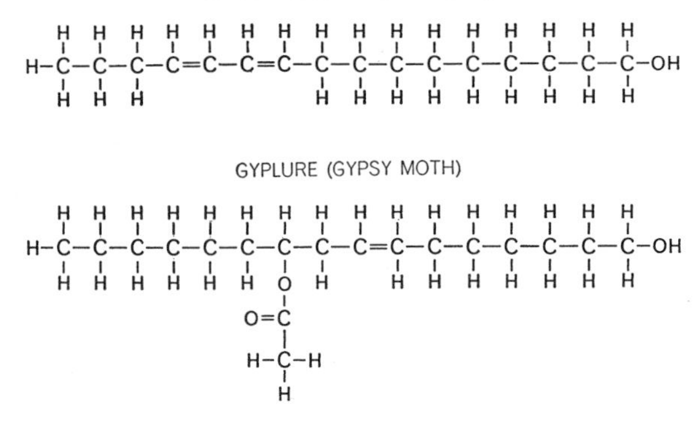

CIVETONE (CIVET)

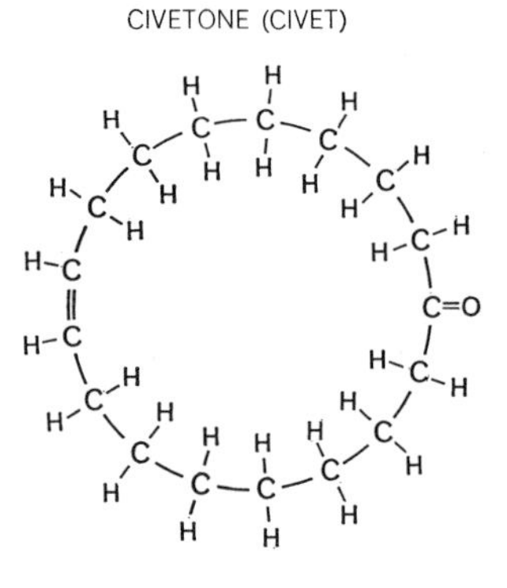

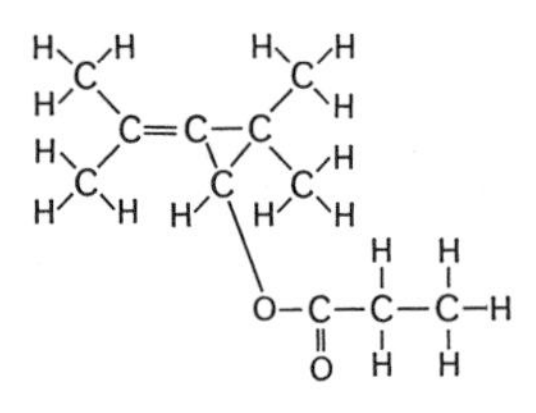

MUSKONE (MUSK DEER)

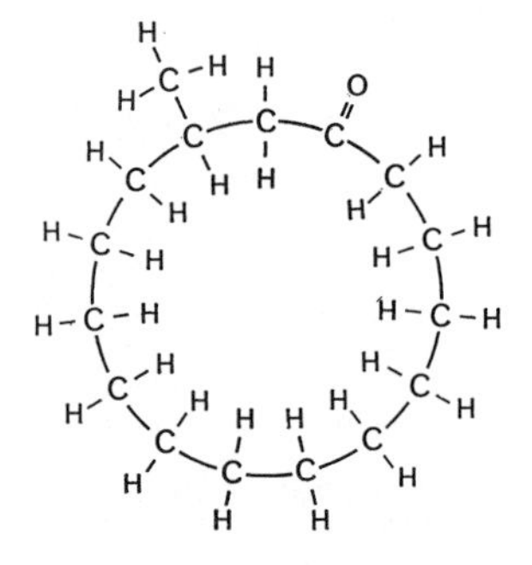

HONEYBEE QUEEN SUBSTANCE

SIX SEX PHEROMONES include the identified sex attractants of four insect species as well as two mammalian musks generally believed to be sex attractants. The high molecular weight of most sex pheromones accounts for their narrow specificity and high potency.

the fact that volatility—and, as a result, diffusibility—declines with increasing molecular weight.

One can also predict from first principles that the molecular weight of alarm substances will tend to be less than those of the sex attractants. Among the ants there is little specificity; each species responds strongly to the alarm substances of other species. Furthermore, an alarm substance, which is used primarily within the confines of the nest, does not need the stimulative potency of a sex attractant, which must carry its message for long distances. For these reasons small molecules will suffice for alarm purposes. Of seven alarm substances known in the social insects, six have 10 or fewer carbon atoms and one (dendrolasin) has 15. It will be interesting to see if future discoveries bear out these early generalizations.

Do human pheromones exist? Primer pheromones might be difficult to detect, since they can affect the endocrine system without producing overt specific behavioral responses. About all that can be said at present is that striking sexual differences have been observed in the ability of humans to smell certain

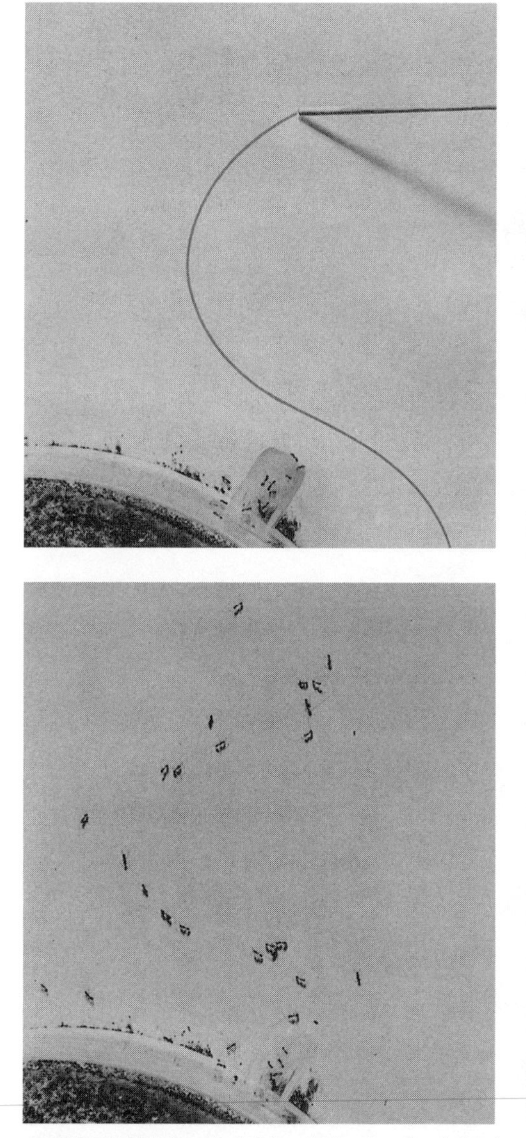

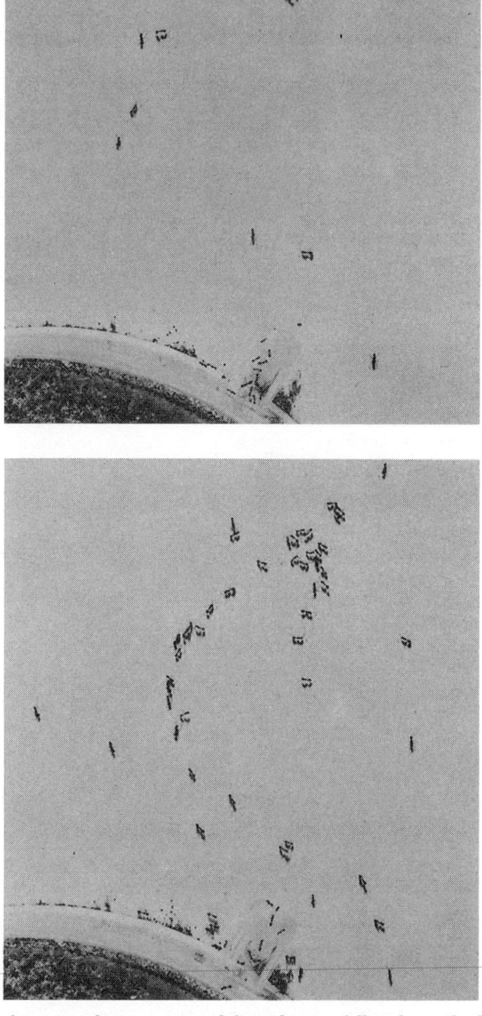

ARTIFICIAL TRAIL can be laid down by drawing a line (*colored curve in frame at top left*) with a stick that has been treated with the contents of a single Dufour's gland. In the remaining three frames, workers are attracted from the nest, follow the artificial route in close formation and mill about in confusion at its arbitrary terminus. Such a trail is not renewed by the unrewarded workers.

DENDROLASIN (*LASIUS FULIGINOSUS*)

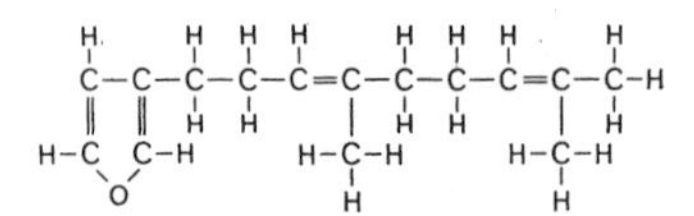

CITRAL (*ATTA SEXDENS*)

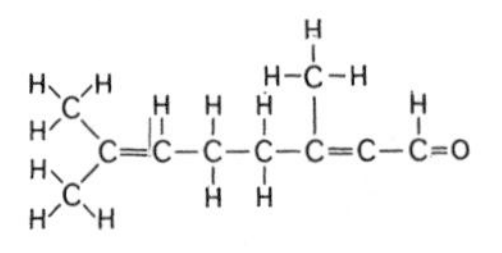

CITRONELLAL (*ACANTHOMYOPS CLAVIGER*)

2-HEPTANONE (*IRIDOMYRMEX PRUINOSUS*)

FOUR ALARM PHEROMONES, given off by the workers of the ant species indicated, have so far been identified. Disturbing stimuli trigger the release of these substances from various glandular reservoirs.

substances. The French biologist J. LeMagnen has reported that the odor of Exaltolide, the synthetic lactone of 14-hydroxytetradecanoic acid, is perceived clearly only by sexually mature females and is perceived most sharply at about the time of ovulation. Males and young girls were found to be relatively insensitive, but a male subject became more sensitive following an injection of estrogen. Exaltolide is used commercially as a perfume fixative. LeMagnen also reported that the ability of his subjects to detect the odor of certain steroids paralleled that of their ability to smell Exaltolide. These observations hardly represent a case for the existence of human pheromones, but they do suggest that the relation of odors to human physiology can bear further examination.

It is apparent that knowledge of chemical communication is still at an early stage. Students of the subject are in the position of linguists who have learned the meaning of a few words of a nearly indecipherable language. There is almost certainly a large chemical vocabulary still to be discovered. Conceivably some pheromone "languages" will be found to have a syntax. It may be found, in other words, that pheromones can be combined in mixtures to form new meanings for the animals employing them. One would also like to know if some animals can modulate the intensity or pulse frequency of pheromone emission to create new messages. The solution of these and other interesting problems will require new techniques in analytical organic chemistry combined with ever more perceptive studies of animal behavior.

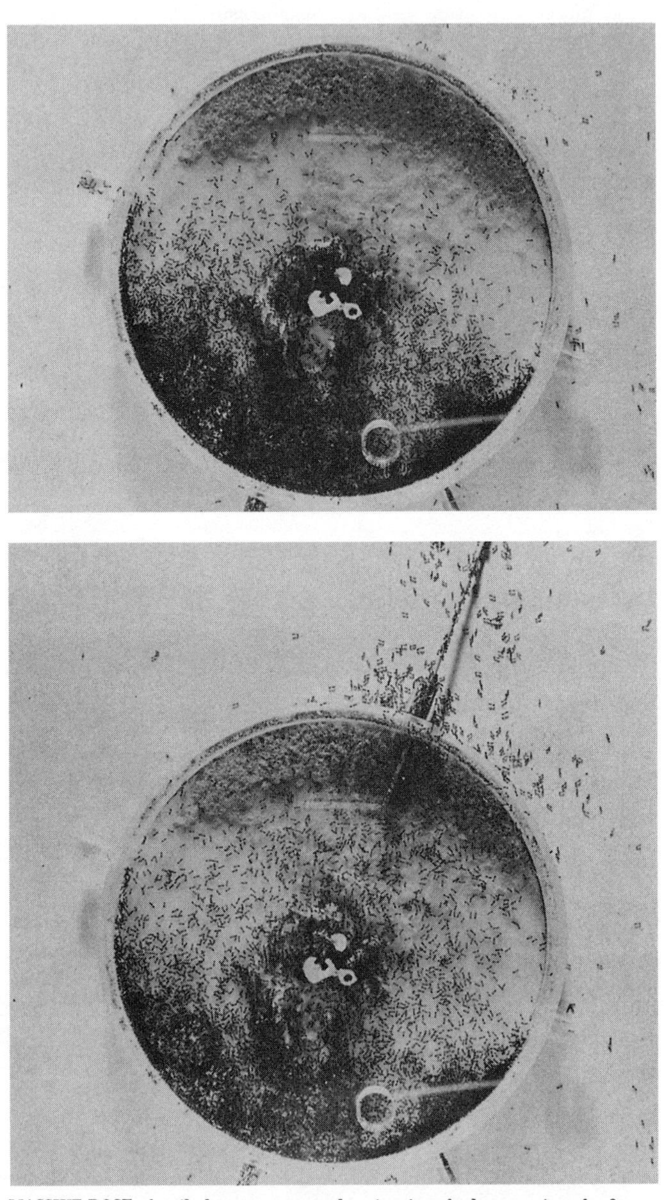

MASSIVE DOSE of trail pheromone causes the migration of a large portion of a fire ant colony from one side of a nest to another. The pheromone is administered on a stick that has been dipped in a solution extracted from the Dufour's glands of freshly killed workers.

The Author

EDWARD O. WILSON is associate professor of zoology at Harvard University. As a native of Alabama, Wilson fairly early in life became acquainted with the Southern agricultural pest known as the fire ant, which he discussed in an article for SCIENTIFIC AMERICAN ("The Fire Ant," March, 1958). Wilson received B.S. and M.S. degrees from the University of Alabama in 1949 and 1950. He took a Ph.D. in biology at Harvard, where he held a National Science Foundation fellowship and a junior fellowship in the Society of Fellows. He joined the Harvard faculty in 1956.

Bibliography

OLFACTORY STIMULI IN MAMMALIAN REPRODUCTION. A. S. Parkes and H. M. Bruce in *Science*, Vol. 134, No. 3485, pages 1049–1054; October, 1961.

PHEROMONES (ECTOHORMONES) IN INSECTS. Peter Karlson and Adolf Butenandt in *Annual Review of Entomology*, Vol. 4, pages 39–58; 1959.

THE SOCIAL BIOLOGY OF ANTS. Edward O. Wilson in *Annual Review of Entomology*, Vol. 8, pages 345–368; 1963.

1967

"The first Mesozoic ants," *Science* 157: 1038–1040 (with F. M. Carpenter and W. L. Brown, Jr., second and third authors)

Until the 1960s the oldest known ants in the fossil record were from Middle Eocene deposits, some 50 million years in age. The species represented by these remains are modern in aspect, with no evident ancestral ant present among them. By reconstructing phylogenetic trends within both fossil and contemporary faunas, William L. Brown and I had tried to deduce what the Ur-ants might have looked like. They must have existed well back in time, we believed, most likely in the middle or late Mesozoic. In 1967 the first two such specimens, both beautifully preserved, were discovered in the Late Cretaceous amber of New Jersey. They were nearly twice as old (actually 90 million years, rather than 100 million years as estimated then) as the oldest fossils previously known. Brown and I were gratified to find that our predictions of the anatomy of this species, which we named *Sphecomyrma freyi,* was not far off the reconstruction we had attempted earlier. Because *Sphecomyrma* (which means "wasp-ant," bespeaking its resemblance to the ancestral wasps) is so primitive, we were able to redraw the phylogeny of the ants, based for the first time on direct fossil evidence concerning their origin. The total picture has shifted considerably since then, but *Sphecomyrma* remains firmly in place near the base of the ant family tree.

ARTICLE 9

The First Mesozoic Ants

Abstract. *Two worker ants preserved in amber of Upper Cretaceous age have been found in New Jersey. They are the first undisputed remains of social insects of Mesozoic age, extending the existence of social life in insects back to approximately 100 million years. They are also the earliest known fossils that can be assigned with certainty to aculeate Hymenoptera. The species,* Sphecomyrma freyi, *is considered to represent a new subfamily (Sphecomyrminae), more primitive than any previously known ant group. It forms a near-perfect link between certain nonsocial tiphiid wasps and the most primitive myrmecioid ants.*

Until now the earliest known fossils of ants, and of social insects generally, have been Eocene in age (*1*). Large assemblages of ant species, most belonging to living tribes and even genera, occur in the Baltic Amber (Oligocene), the Sicilian and Chiapan ambers (Miocene), and the Florissant and Ruby Basin shales (Miocene) (*2*). The diversity of these faunas and the advanced phylogenetic position of many of their elements have long prompted entomologists to look to the Cretaceous for fossils that might link the ants to some ancestral nonsocial wasp group, but until now, with one doubtful exception, no relevant fossils have turned up.

The exception is the hymenopterous forewing described by Sharov (*3*) as *Cretavus sibiricus*, from the Upper Cretaceous of Siberia. This wing is rather similar to that of the wasp family Plumariidae, and also approaches a reasonable possible precursor pattern for the venations of known primitive ants. However, we have no guarantee that venational characters evolved concordantly with other, more truly diagnostic body characters, so we cannot even regard it as certain that *Cretavus* is an aculeate.

Cretaceous amber from Canada and Alaska contains a moderate number of insects (*4*), but no ants or aculeate Hymenoptera of any kind are present among them (a fact now suggesting that the Canadian amber, which has never been precisely dated within the Cretaceous, may have been formed in an earlier part of the period). Amber securely dated to the lower part of the Upper Cretaceous is fairly common from Maryland to New Jersey in deposits of the Magothy Formation, but until recently almost no insect inclusions had been reported. In 1965, Mr. and Mrs. Edmund Frey (*5*), mineral collectors of Mountainside, New Jersey, found a lump of amber in clay of the same formation at the base of seaside bluffs at Cliffwood, New Jersey. The fragile lump broke into pieces, and two of these bear insects, including two well-preserved worker ants.

The two specimens appear to belong to the same species; one is shown in the cover photograph. We judge this species, *Sphecomyrma freyi*, to be by far the most primitive member of the Formicidae (ants) yet discovered. It is sufficiently removed from all other ants to be received into a distinct subfamily, the Sphecomyrminae. The most distinctive morphological features, and our assessment of their phylogenetic significance, can be summarized as follows.

1) The head capsule resembles that of a generalized aculeate wasp or ant. The clypeus and frontal carinae are antlike, but are of such simple conformation as not to depart significantly from these structures in some aculeate wasp groups. We regard the large, convex form of the compound eyes and their placement near the center of the sides of the head as primitive characters for aculeates generally. The presence of three large ocelli is certainly primitive.

2) The mandibles are short, curvilinear, and bidentate, and closely resemble those of certain species of several existing aculeate wasp families.

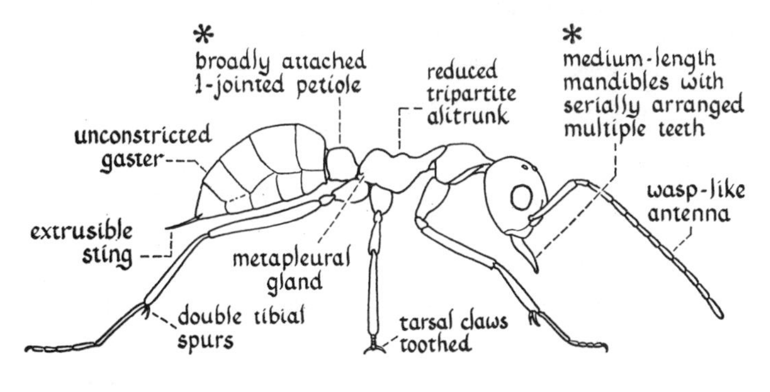

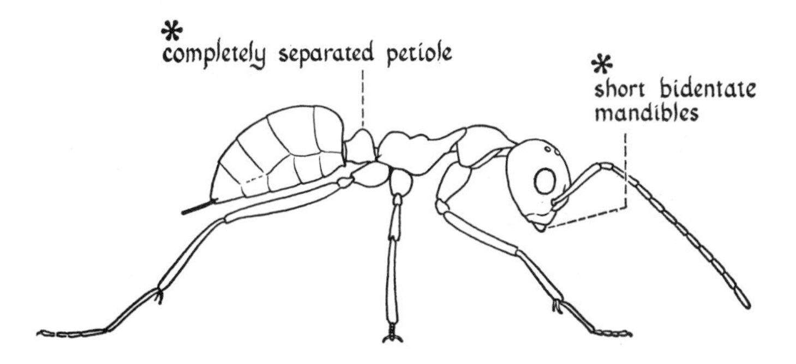

Fig. 1. A comparison of the main features previously hypothesized by the authors to characterize the external morphology of the ancestral ant, and *Sphecomyrma* itself. The minor details of body form are arbitrarily made the same. In the drawing of *Sphecomyrma*, the starred character states indicate where our phylogenetic hypothesis proved in error.

3) The antennal funiculi are long and filiform, a trait more wasplike than antlike. The antennal scapes (basal segments) are elongate, a characteristic of ants generally but exceptional among other aculeates; still, the scapes are shorter than is usual for worker ants.

4) The alitrunk (thorax + propodeum) is more completely sutured, and therefore more primitive, than that of any other worker ant, and is almost identical with that of the wingless females of the tiphiid genus *Methocha*. Prothorax, mesothorax, and metanotopropodeum are separated each from the next by two complete and possibly flexible sutures; and the mesonotum is composed of well-defined, convex scutum and scutellum, separated by a narrow sunken area. In fact, the only major alitruncal difference from *Methocha* is the presence in *Sphecomyrma* of apparently well-developed metapleural glands, which are peculiar to the Formicidae.

5) The single-segmented petiole, narrowly constricted behind, is an ant character state; the absence of a constriction in the gaster and the presence of a well-developed, extrusible sting are states shared by most wasps and primitive myrmecioid ants.

6) The legs show two character states that we have long regarded as primitive for ants: two spurs on each tibial apex of the middle and posterior legs, and toothed tarsal claws.

In summary, *Sphecomyrma* presents a mosaic of wasplike and antlike character states. There are nevertheless enough truly antlike traits to place *Sphecomyrma* within the Formicidae, where the most similar (but still quite different) forms are the living myrmeciine *Nothomyrmecia macrops* of Australia and the primitive aneuretine Dolichoderinae, such as *Paraneuretus* and *Protaneuretus*, of Oligocene age, described by Wheeler (*2*). These are primitive forms in the myrmecioid complex (*6*).

It is interesting to compare our earlier conception of the archetypal ant with the actuality presented by *Sphecomyrma*. This is done in pictorial form in Fig. 1. It can be seen that our vision of what was yet to be revealed differs from *Sphecomyrma* in only one essential respect: we guessed that antlike mandibles evolved before the antlike "waist" (petiole), but the reverse actually proved to be the case.

Compared with living wasp genera, *Sphecomyrma* appears to come closest

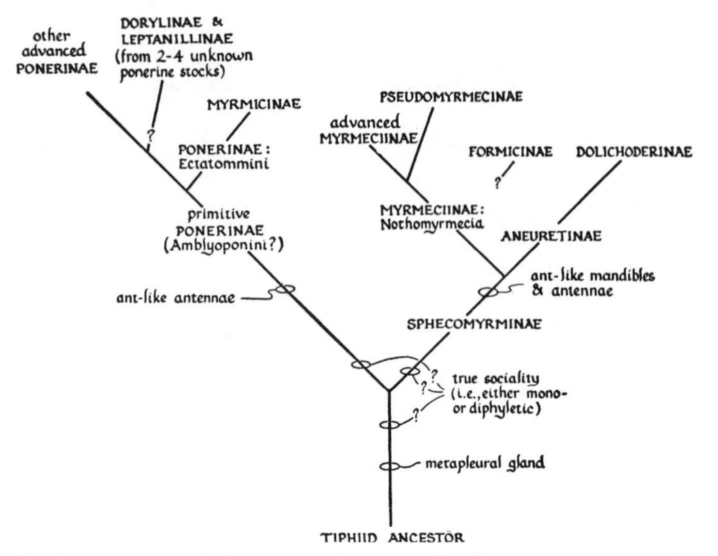

Fig. 2. A new hypothetical cladogram of the ant subfamilies taking into account the morphology of *Sphecomyrma*.

to the tiphiid genera *Methocha* (Methochinae) and *Rhagigaster* (Thynninae) (*7*). One interesting aspect of the morphology of *Sphecomyrma* is that in "ant characters" it does fall so close to the myrmecioid complex of genera, yet bears so little resemblance to *Amblyopone* and other genera of the Ponerinae previously regarded as nearly as primitive as the myrmecioids. The possibility is thus raised that divergence between myrmecioid and poneroid lines may already have taken place by the time *Sphecomyrma* lived. However, the presence of the complex metapleural gland in *Sphecomyrma* and all other primitive ants speaks for a monophyletic origin of the Formicidae from tiphiid ancestors. The function of the metapleural gland is still unknown, but if it turns out to mediate some phase of social behavior, then monophyletic origin of social life would be strongly implied for the ants as we know them. These new considerations are incorporated into a cladogram of the ant subfamilies (Fig. 2).

Finally, the origin of social life in the insects has now been put back from the Eocene, about 60 million years ago, to the middle or lower part of the Upper Cretaceous, about 100 million years ago. It may be true that social life in insects is not much older that that. *Sphecomyrma* is evidently only a little changed from tiphiid wasps, and it is possible that this relatively slight transformation indicates a correspondingly short period of social evolution. Perhaps as more hymenopteran fossils become available from the New Jersey and similar ambers, new light will be shed on the origin of the ants.

A fuller account of *Sphecomyrma* and its phylogenetic implications, together with a formal taxonomic description, is published elsewhere (*8*).

EDWARD O. WILSON
FRANK M. CARPENTER
Museum of Comparative Zoology, Harvard University, Cambridge, Massachusetts 02138
WILLIAM L. BROWN, JR.
Department of Entomology, Cornell University, Ithaca, New York; and *Museum of Comparative Zoology, Harvard University, Cambridge, Massachusetts 02138*

References and Notes

1. The oldest Eocene ant fossil is *Eoponera berryi*, based on a forewing from the Wilcox Clay of Tennessee; F. M. Carpenter, *J. Wash. Acad. Sci.* **19**, 300 (1929).
2. The Baltic Amber ants were monographed by W. M. Wheeler [*Schrift. Phys.-ökon. Ges. Königsberg* **55**, 1 (1914)]; and the Florissant ants by F. M. Carpenter [*Bull. Mus. Comp. Zool. Harvard* **70**, 1 (1930)]. W. L. Brown, Jr. (unpublished) has examined the few available ant fossils from the Ruby Basin (Montana) shales and found them to match the dominant Florissant species; he has also cursorily examined the ants of the Chiapas Amber and found them related to those of the Florissant and the existing tropical Mexican faunas.
3. A. G. Sharov, *Dokl. Akad. Nauk* **112**, 943

(1957). For a comparison with wings of ants and Plumariidae, see W. L. Brown, Jr., and W. L. Nutting, *Trans. Amer. Entomol. Soc.* **75**, 113 (1950).

4. F. M. Carpenter, J. W. Folsom, E. O. Essig, A. C. Kinsey, C. T. Brues, M. W. Boesel, H. E. Ewing, *Univ. Toronto Stud. Geol. Ser.* **40**, 7 (1934).
5. We gratefully acknowledge the splendid cooperation of Mr. and Mrs. Frey, as well as the intermediary aid of Dr. Donald Baird of Princeton University and Mr. David Stager of the Newark Museum.
6. W. L. Brown, Jr., *Insectes Sociaux* **1**, 21 (1954).
7. We acknowledge the aid of H. E. Evans, who gave us the benefit of extensive comparisons of *Sphecomyrma* characters with those of various wasp genera. In classifying tiphiids, we have arbitrarily followed the system of V. S. L. Pate, *J. N.Y. Entomol. Soc.* **55**, 115 (1947).
8. E. O. Wilson, F. M. Carpenter, W. L. Brown, Jr., *Psyche*, in press.

26 June 1967

1968

"The ergonomics of caste in the social insects," *The American Naturalist* 102: 41–66

Consider the stage at which sterile castes are established in an insect colony and the success or failure of the caste system that depends on the efficiency of their division of labor. What then in evolution determines the ratios of investments in the castes made by the colony and of the tasks assumed by the specialists? This might seem to be a problem appropriate to the theory of business management—and so it is. But the same principles apply to insect societies. Here I employed models of linear programming and came up with several interesting predictions. One was that when colony-level natural selection results in the improved performance of a task, the percentage of investment in workers specialized to perform the task declines—the opposite of what happens when natural selection acts solely on individuals and their offspring. Another prediction is that the greater the specialization of physical castes (examples of which are minor workers and soldiers), the more the species is "locked in" to that particular social organization. In other words, such species are less likely to lose the castes and change to another system. This effect is consistent with the principle of the "point of no return," supported by the rarity of known cases of social insect species, which after acquiring physical castes, then evolved back to the solitary condition.

Reprinted for private circulation from THE AMERICAN NATURALIST
Vol. 102, No. 923, Jan.-Feb. 1968

Vol. 102, No. 923 The American Naturalist January–February, 1968

THE ERGONOMICS OF CASTE IN THE SOCIAL INSECTS

EDWARD O. WILSON

The Biological Laboratories, Harvard University, Cambridge, Massachusetts 02138

Past studies of caste systems in the social insects have focused on the obvious genetic and physiological problems of the mechanisms that control caste determination in the individual insect. Several independent experimental investigations have now provided a body of definitive information on the subject. In the great majority of bees, and in most—and perhaps all—ants, wasps, and termites, caste is apparently environmentally determined. The environmental controls are diverse in nature and differ from group to group. They include the biasing influences of yolk nutrients, of various quantitative and qualitative factors in larval feeding, of temperature changes, and of the caste-specific pheromones (see reviews by Brian, 1965; Lüscher, 1961; Weaver, 1966).

There is a second major problem connected with caste which is evolutionary in nature and can be phrased as follows: Why do the *ratios* of the castes (in a whole colony population) vary among species of social insects? This question is much less obvious than the physiological one and can be considered only in the context of the ecology of the individual species. A large amount of empirical information on ratios exists but only a small amount of theory. Only recently have students of social insects begun to handle the subject in a systematic fashion, as, for example, in the work of Richards and Richards (1951), Lindauer (1961), Hamilton (1964), Brian (1966), and Wilson (1966). Still, very little theory on the subject has been formulated.

This matter of the presence or absence of a given caste, together with its relative abundance when present, should be susceptible to some form of optimization theory, provided we are able to assume selection at the colony level. In fact, colony selection in the advanced social insects does appear to be the one example of group selection that can be accepted unequivocally. It therefore seems a sound procedure to accept colony selection as a mechanism and to press on in search of an optimization theory based on the axiom that the mechanism operates generally. For if selection is mostly at the colony level, workers can be altruistic with respect to the remainder of the colony; and their numbers and behavior can be regulated to achieve maximum colony fitness. What has been lacking so far is an entree to the theory of group behavior, a way of abstracting our empirical knowledge of caste and colony ergonomics[1] into a form that can be used to analyze optimiza-

[1] In an earlier article (Wilson, 1963) I suggested the term "ergonomics," borrowed from human sociology, to identify the quantitative study of the distribution of work, performance, and efficiency in insect societies.

tion. The purpose of this article is to show the feasibility of a first formulation by means of the techniques of linear programming and to report on some interesting but still imperfect and mostly theoretical results that have been obtained.

A LINEAR PROGRAMMING MODEL

The Concept of Cost

As colonies grow, their caste ratios change. Very young colonies founded by single queens typically consist solely of the queens and minor workers. As they approach maturity, these same colonies may add medias and major workers (also known as soldiers). Finally, they produce males and new, virgin queens. Here we will consider the ergonomics of the mature colony alone. A mature colony is defined as a colony large enough to produce new, virgin queens. Also, for convenience, we will include under the term "caste" both *physical castes,* such as minor workers and soldiers, and *temporal "castes."* The latter are classes of individuals in those periods of labor specialization which most individual social insects pass through in the course of their lives. What determines the efficiency of the mature colony is the number of workers in each temporal caste at any given moment. This conception is spelled out in the examples given in Figure 1.

Consider the mature colony. Depending on the species, the adult force may contain anywhere from a few tens of workers to several millions. The number is a species characteristic. It has been evolved as an adaptation to ultimate limiting factors in the environment. An ultimate limit may be imposed by a peculiar nest site to which the species is adapted, or a restricted productivity of some prey species on which the species specializes, or conversely, a prey species or competitor so physically formidable as to require a large worker force as a minimum for survival. These and other ultimate, limiting factors have already been documented and discussed in the literature (Brian, 1966). The mature colony, on reaching its predetermined size, can be expected to contain caste ratios which approximate the *optimal mix.* This is simply the ratio of castes which can produce the maximum rate of production of virgin queens and males while the colony is at or near its maximum size.

It is helpful to think of a colony of social insects as operating somewhat like a factory inside a fortress. Entrenched in the nest site, harrassed by enemies and uncertain changes in the physical environment, the colony must send foragers out to gather food while converting the secured food inside the nest into virgin queens and males as rapidly and as efficiently as possible. The rate of production of the sexual forms is an important component of colony fitness. Suppose we are comparing two genotypes belonging to the same species. If we could but measure survival rates of queens and males belonging to the two genotypes from the moment they leave the nests on

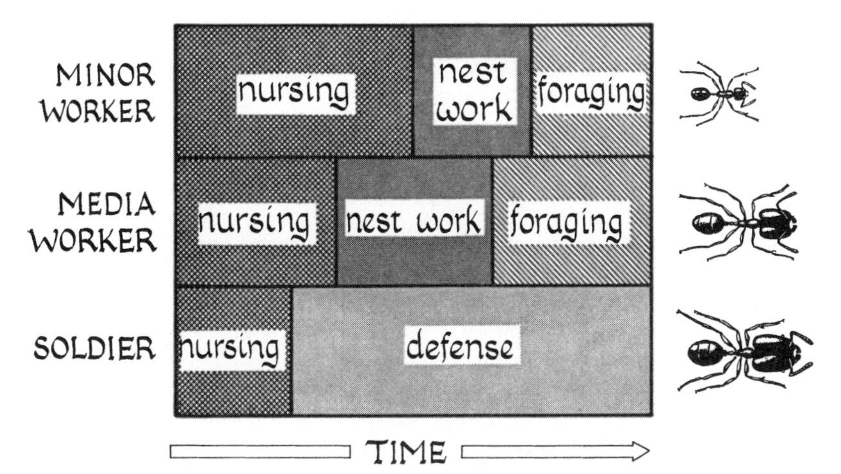

FIG. 1.—This diagram visualizes the principal work periods traversed in the life spans of three worker subcastes of a generalized polymorphic ant species. The work periods are those periods in which the indicated task is the most frequent one performed; other tasks may be performed at lesser frequencies. The form of the castes and the sequences of work periods within each caste are based on real species, but the precise durations of the periods are imaginary. In this case each of the eight periods, the (arbitrary) total encountered in the three castes, is treated as a separate "caste." The optimal mix can be evolved both by varying the relative numbers in each subcaste and the relative time spent in each work period. (The ants represented in this and subsequent figures belong to the myrmicine genus *Pheidole* and are shown here as an intuitive aid for the reader not very familiar with social insects.)

their nuptial flights, then record their mating success and the survival rate of the fecundated queens thereafter, together with the growth rates of the colonies the survivors produce, then we could calculate the relative fitness of the two genotypes. However, in order to develop an ergonomic theory in a stepwise fashion, we need now to restrict our comparison to the mature colonies. In order to do this and retain precision, we would have to do the following: take the difference in survivorship between the two genotypes outside the period of colony maturity and reduce it to a single weighting factor. But we can sacrifice precision without losing the potential for general qualitative results by taking the difference as zero. Now we are concerned only with the mature colony, and the production of sexual forms is (keeping in mind the artificiality of our convention) the exact measure of colony fitness. If colonies belonging to one genotype contain on the average 1,000 sterile workers and produce 10 new, virgin queens in their mature life span, and colonies belonging to the second genotype contain on the average only 100 workers but produce 20 new, virgin queens in their mature life span, the second genotype has twice the fitness of the first, despite its smaller colony size. As a result, selection would eliminate larger colony size.

The lower fitness of the first could be due to lower survival rate of mature colonies, or to a smaller average production of sexual forms per surviving mature colony, or to both. The important point is that the rate of production of sexual forms is the measure of fitness, and evolution can be expected to shape mature colony size and organization to maximize this rate.

The production of sexual forms is determined by the number of mistakes made by the mature colony as a whole in the course of its fortress-and-factory operations. A mistake is made when some potentially harmful contingency is not met—a predator successfully invades the nest interior, a breach in the nest wall is tolerated long enough to dessicate a brood chamber, a hungry larva is left unattended, and so forth. The *cost* of the mistakes for a given category of contingencies is the product of the number of times a mistake was made times the reduction in queen production per mistake. The total cost of all mistakes in a given period of time is visualized as the principal component in the reduction of colony fitness. In symbols,

$$\Delta N_f = f(W - W_m) - \sum_i F_i, \text{ and}$$

$$F_i = k_i x_i, \text{ where}$$

ΔN_f is the number of queens produced in a unit of time, say in one full year;
W is the weight of workers present during the rearing of the brood;
W_m is the minimum weight of workers required to rear virgin queens;
F_i is the cost (in number of virgin queens not produced) due to all errors in the ith error category;
x_i is the cost (in number of virgin queens) per error in the ith error category;
k_i is the frequency of errors in the ith error category.

The Lowest Tolerable Cost as a Uniform Value

At this early point, an interesting deduction can already be made concerning the behavioral responses of colonies to contingencies and the cost level tolerated by the colony. In brief, it can be argued that there is some threshold cost F at or above which the species evolves a behavioral response to hold the average cost (per colony per unit time) to F. Also, in cases where behavioral responses have been evolved, the cost tolerated per error category by the colony is in each case the same, namely, F. In other words, $F = F_1 = F_2 = \ldots$ *et seq.* for each error category to which a discrete behavioral response has been evolved. This convergence effect is inferred as follows. If $F_i > F_j$ for any two error categories i and j, the colony is more likely to increase fitness with the same amount of behavioral evolution by decreasing F_i than by decreasing F_j. As F_i is decreased, F_j will decrease more slowly; or, more likely, it will remain about the same or increase, providing the alteration of behavior to respond to contingency i adversely affected its ability to respond to j. As a result, F_i and F_j will tend to converge.

Not all contingencies will be frequent enough, or costly enough, to prompt the evolution of a specific behavioral response; in other words, their total cost will not exceed F. For example, it is natural to expect alarm communication and attack behavior specifically adapted to arthropodan intruders, but no particular behavioral adaptation to avoid having meteorites fall on the nest.

Suppose a new contingency a arose through a change in the environment, and $F_a > F$. The behavior of the colony would tend to evolve to bring F_a equal with the other $F_i (\doteq F)$. This is not to say that F is an absolute threshold. The adjustment to a or any increase in the cost of another category (F_i) due to a change in the environment would cause readjustment in F_i values generally and a new convergence to some value F. But the new F might be different from the old.

The Two-Contingency Case

The average output of queens is viewed as the difference between the ideal number made possible by the productivity of the foraging area of the colony and the number lost by failure to meet some of the contingencies. The evolutionary problem that we are postulating to have been faced by social insects can be solved as follows: the colony produces the mixture of castes which maximizes this output. In order to describe the solution by means of a simple linear programming model, it is necessary to restate the solution in terms of the dual of the first statement: the colony evolves the mixture of castes that allows it to produce a given number of queens with a minimum of workers. The objective, as I have formulated it here, is to minimize the energy cost.

The main purpose of the section to follow is to show that under a wide variety of conceivable conditions the proportions of castes can be related to minimum energy cost in a linear form.

Let us start with the case of two contingencies whose costs would exceed the threshold value if left unattended, together with two castes whose efficiencies at dealing with the two contingencies differ. The inferences to be made from this simplest situation can then be extended to any number of contingencies and castes.

The most important step is to relate the total weights, W_1 and W_2, of two castes at a given instant to the frequency and importance of two contingencies and the relative efficiencies of the castes at performing the necessary tasks. By stating the problem as the minimization of energy cost, the relation can be given in linear form as follows:

$$W_1 = \text{const (1)} - \frac{\alpha_{12} \ln (1 - q_{12})}{\alpha_{11} \ln (1 - q_{11})} W_2, \qquad q_{11} > 0; \tag{1a}$$

$$W_1 = \text{const (2)} - \frac{\alpha_{22} \ln (1 - q_{22})}{\alpha_{21} \ln (1 - q_{21})} W_2, \qquad q_{21} > 0; \tag{1b}$$

where const (1), const (2), and the coefficients of W_2 are constants determined by the frequency of occurrence and the importance of the contingencies, and by the relative efficiencies of the castes at meeting the contingencies. These two expressions give the relative numbers of caste 1 and caste 2 needed to hold costs (due to the two contingencies, respectively) to the tolerable level. It will now be shown how they are derived and why it is reasonable to employ linear relationships in the exploratory phase of ergonomic theory. The following conventions will be used:

W_1 is the weight of all members in an average colony belonging to caste 1;
W_2 is the weight of all members in an average colony belonging to caste 2;
F_1 and F_2 are the highest tolerable costs due to contingencies 1 and 2 (and by an earlier argument, $F_1 \approx F_2$);
α_{11} is a constant such that $\alpha_{11}W_1$ gives the average number of individual contacts with a contingency of type 1 by members of caste 1 during the existence of the contingency;
α_{12} is a constant such that $\alpha_{12}W_2$ gives the average number of individual contacts with a contingency of type 1 by members of caste 2 during the existence of the contingency;
α_{21} and α_{22} are constants similar to the above two but with reference to contingencies of type 2;
q_{11} is the probability that, on encountering contingency 1, a worker of caste 1 responds successfully;
q_{12} is the probability that, on encountering contingency 1, a worker of caste 2 responds successfully;
q_{21} and q_{22} are the probabilities similar to the above two but with reference to contingency 2;
x_1 and x_2 are the average costs (in nonproduction of virgin queens) per failure to meet contingencies 1 and 2, respectively;
k_1 and k_2 are the frequencies of contingencies 1 and 2, respectively, for a given period of time.

When a worker of, say, caste 1 encounters a contingency of type 1, the probability that it will respond incorrectly is $1-q_{11}$. This value can be put to use only if we know the number of encounters per caste. Now if we label as p_{10} the probability that *no* worker of caste 1 encounters a given contingency of type 1, p_{11} as the probability that exactly one worker encounters it, p_{12} as the probability that exactly two workers encounter it, and so forth, then the probability that *no* correct response will be made to a given contingency of type 1 is

$$p_{10}(1 - q_{11})^0 + p_{11}(1 - q_{11}) + p_{12}(1 - q_{11})^2 + \cdots ,$$

and the cost in a given interval of time is this sum, multiplied by the frequency the contingency occurs, multiplied in turn by the cost of each contingency that is not successfully met:

$$k_1x_1 \sum_i p_{1i}(1 - q_{11})^i.$$

The distribution of p_{1i} should be obtainable by empirical measurements; and very likely it can then be approximated by some general form, such as the Poisson. For our purposes, we can approximate the expression

$$\sum_i p_{1i}(1 - q_{11})^i$$

by a single number, which in turn is a function of W_1, the weight of caste 1 in the colony. In the simplest case, the single most appropriate number is the mean value of i, that is, the mean number of contacts per contingency, which value can be labeled $\bar{\imath}$. Keeping in mind that $\bar{\imath}$ is only an approximation of

$$\sum_i p_{1i}(1 - q_{11})^i,$$

we next recognize that it must be a function of the number of workers of caste 1 present, or, translated into terms of the present model, of the total weight (W_1) of workers of caste 1. If the nest volume were kept constant and W_1 increased, the number of contacts per contingency by members of caste 1 should increase linearly; in other words, $\bar{\imath}$ should be approximated by $\alpha_{11}W_1$, where α_{11} is a constant. In this case, the probability that caste 1, in the absence of caste 2, would not solve the contingency is about $(1 - q_{11})^{\alpha_{11}W_1}$; and the probability that caste 2, in the absence of caste 1, would not solve the contingency is about $(1 - q_{12})^{\alpha_{12}W_2}$. Let us now assume that the castes operate independently of each other in meeting contingencies. That is, the presence of caste 2 does not enhance or diminish the capacity of a given worker belonging to caste 1 to deal with a single given contingency of type 1. This is probably not always the case, but on the basis of my own subjective impressions I believe it is close enough to the truth in most behavioral interactions to be accepted as an approximation. Where it applies, the cost due to contingency 1 in a given period of time is given as follows:

$$F_1 \doteq k_1x_1(1 - q_{11})^{\alpha_{11}W_1}(1 - q_{12})^{\alpha_{12}W_2} \qquad q_{11} > 0, \quad q_{12} > 0. \tag{2a}$$

And, symmetrically,

$$F_2 \doteq k_2x_2(1 - q_{21})^{\alpha_{21}W_1}(1 - q_{22})^{\alpha_{22}W_2} \qquad q_{21} > 0, \quad q_{22} > 0. \tag{2b}$$

By rearrangement of the two formulas, *contingency curves* are obtained:

Contingency Curve 1

$$W_1 \doteq \frac{\ln F_1 - \ln k_1x_1}{\alpha_{11} \ln (1 - q_{11})} - \frac{\alpha_{12} \ln (1 - q_{12})}{\alpha_{11} \ln (1 - q_{11})} W_2. \tag{3a}$$

Contingency Curve 2

$$W_1 \doteq \frac{\ln F_2 - \ln k_2x_2}{\alpha_{21} \ln (1 - q_{21})} - \frac{\alpha_{22} \ln (1 - q_{22})}{\alpha_{21} \ln (1 - q_{21})} W_2. \tag{3b}$$

Equations (3a) and (3b), already presented in abbreviated form as equations (1a) and (1b), provide the postulated linear relationships between W_1 and W_2 required to keep error in contingencies 1 and 2 down to a tolerable level. Equation (3a) gives the relationship for contingency 1, and (3b) for contingency 2.

These equations are based on one set of simple, explicit conditions. The object in deriving them, however, is not to make a guess about the precise relationships but, rather, to illustrate the general form they are expected to take. The only crucial result for the deductions to be made in Figures 2–7 is that the relationships can be put in a linear form. Equations (3a) and (3b) are of course in linear form, but they were derived from a particular hypothesis. It is easy, on the other hand, to show that the conditions can be greatly relaxed, or the graphical coordinates transformed, to provide a linear graphical representation of the kind to be used in Figures 2–7. For instance, suppose the number of contacts per caste per contingency were not linearly related to the weight W_i of the caste, that is, the number could not be expressed as $\alpha_{1i}W_i$, but was related in some other way, for example, as some logarithmic function of W_i. As long as the function is the same for the two castes, the contingency curves would still be linear. Next, even if the contingency curves are not linear algebraically, they can in a great many cases be made linear by transformations of the W_1 and W_2 axes. Finally, if the curves still cannot be made linear by any transformation, it should still be possible to conduct analyses on segments of the curves that are approximately linear.

In conclusion, we do not yet know the shape of real contingency curves, and to attempt a guess at their precise form would be inappropriate, except for illustrative purposes as undertaken here. On the other hand, it is a much more reasonable proposition that contingency curves can be presented as straight lines in a generalized graphical form, with or without transformation of the axes. When this is done, new insight can be gained into the possibilities of social organization in insects. But before the next step is taken, a brief discussion of certain special biological qualities of insect societies is needed.

Special Qualities of Insect Societies Put into the Model

The procedure of making the behavioral responses probabilistic may seem strange at first, but it has a sound biological rationale. To make this point clear, we might ask why the model was not constructed in a more straightforward manner by having members of different castes laboring at their assigned tasks through the day and night, producing to the utmost of their hexapodan capacities. Then the more conventional procedures of linear programming would have been easy to apply. But social insects simply do not behave that way. Contrary to folklore, most of the members of an insect colony are idle at any given moment, and the greater part of the life of each individual is spent in relative quietude. The members of colonies respond to contingencies, which can be identified as discrete, albeit very numerous, events, much as visualized in this model. Abundant documentation for the point is provided by studies cited in the recent reviews of Lindauer (1961) and Wilson (1966).

A second special feature that needs comment is the way cost is assessed.

In the model the colony is given a certain period of time to accomplish the task presented by the contingency; if it fails, it takes the penalty in reduced queen production. This would be the case for such contingencies as intrusion by a predator, which could result in the loss of one or more colony members, or self-fouling by a larva, which could result in the death of the larva. It is also easy to accomodate more complex, and perhaps more general, cost functions without seriously affecting the qualitative conclusions about to be made. For example, suppose the task remains undone until a member of the correct caste comes upon the contingency, that is, a contingency of type 1 requires an encounter by a member of caste 1. Then on the average the $[(\alpha_{11}W_1 + \alpha_{12}W_2)/\alpha_{11}W_1]$th worker encountering the contingency belongs to the correct caste (caste 1); and, where t is the average interval between encounters by workers of all castes, the mean time elapsing before the task is done is $t[(\alpha_{11}W_1 + \alpha_{12}W_2)/\alpha_{11}W_1]$. The cost would then be some function of this mean elapsed time. Actually, a reasonable estimate of cost as a function of caste ratios can only be accomplished empirically. It will be a necessary prelude to making this or any other programming model predictive for specific cases.

RESULTS

The optimal mix of castes is the one which gives the minimum summed *weights* of the different castes while keeping the combined cost of the contingencies at the maximum tolerable level. The manner in which the optimal mix is approached in evolution is envisaged as follows. Any new genotype that produces a mix falling closer to the optimum is also one that can increase its average net output of queens and males. In terms of energetics, the average number of queens and males produced per unit of energy expended by the colony is increased. Even though colonies bearing the new genotype will contain about the same adult biomass as other colonies, their average net output will be greater. Consequently, the new genotype will be favored in colony-level selection, and the species as a whole will evolve closer to the optimal mix.

The general form of the solution of the optimal mix problem for the case of two contingencies and two castes is given in Figure 2. It has been postulated that behavior can be classified into sets of responses in a one-to-one correspondence to a set of kinds of contingencies. Even if this conception only roughly fits the truth, it can be used to develop a theory of ergonomics. Not all kinds of contingencies occur frequently enough, or are important enough when they occur, to evoke the evolution of a distinct response. Those that do, I have suggested here, are handled by the colony in such a way that in the end they decrease fitness only by a certain threshold amount, with the same amount being permitted to each kind of contingency separately. From suitable modifications of the graphical model of Figure 2, a series of theorems can be drawn:

1. By inspection of Figures 3 and 4, it can be seen that, as long as the

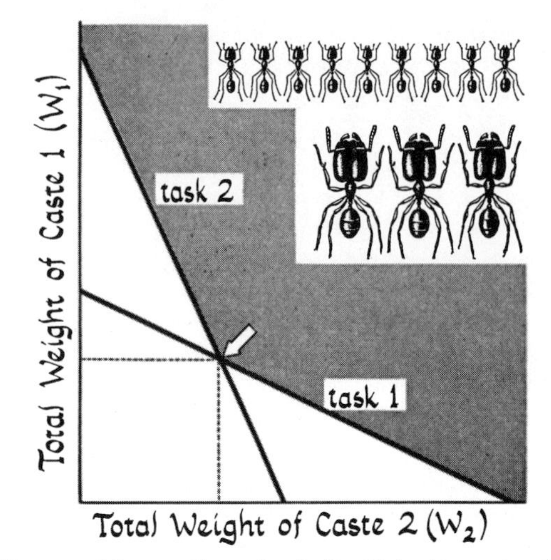

FIG. 2.—The case of two contingencies dealt with by two castes. The optimal mix for the colony, measured in terms of the respective total weights of all the individuals in each caste, is given by the intersection of the two curves (from equations 3a and 3b). Contingency curve 1, labeled "task 1," gives the combination of weights (W_1 and W_2) of the two castes required to hold losses in queen production to the threshold level due to contingencies of type 1; and contingency curve 2, labeled "task 2," gives the combination with reference to contingencies of type 2. The intersection of the two contingency curves determines the minimum value of $W_1 + W_2$ that can hold the losses due to both kinds of contingencies to the threshold level.

contingencies occur with relatively constant frequencies, it is of advantage for the species to evolve so that in each mature colony there is one caste specialized to respond to each kind of contingency. In other words, one caste should come into being which perfects the appropriate response, even at the expense of losing proficiency in its response to other kinds of contingencies.

2. Figure 5 demonstrates that if one caste increases in efficiency in the course of evolution, and the others do not, the proportionate total weight of the improving caste will decrease. In other words, the expected result of group selection is precisely the opposite of that of individual selection, which would be an increase in the more efficient form. The decrease in weight is proportional to the increase in specialization and (inversely) to the slope of the opposite contingency curve. More precisely, as seen in Figure 5, the increase in specialization of caste 1 with reference to task 1 is measured by the ratio of intersects a/a', while the decrease in weight is measured by b/b'. Then, by similar triangles, $a/\omega = b(\omega - \beta)$ and $a'/\omega = b'(\omega - b')$, and from these relations

$$\frac{a}{a'} = \frac{(\omega - \beta')}{(\omega - \beta)} \frac{b}{b'}.$$

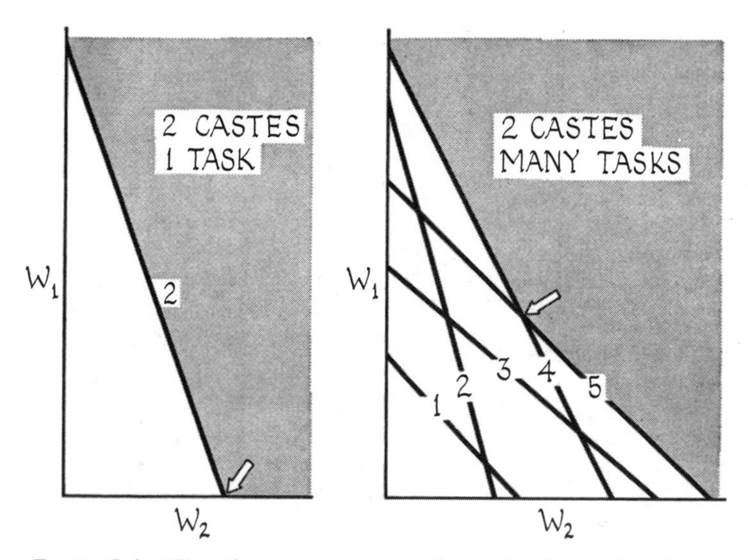

FIG. 3.—*Left:* When there are more castes than tasks, the number of castes will be reduced in evolution to equal the number of tasks. The surplus castes removed will be the least efficient ones (in this case, caste 1). *Right:* If there are more tasks than castes, the optimal mix of castes will be determined entirely by those tasks, equal or less in number to the number of castes, which deal with the contingencies of greatest importance to the colony in terms of average productivity of virgin queens and males (in this case, tasks 4 and 5).

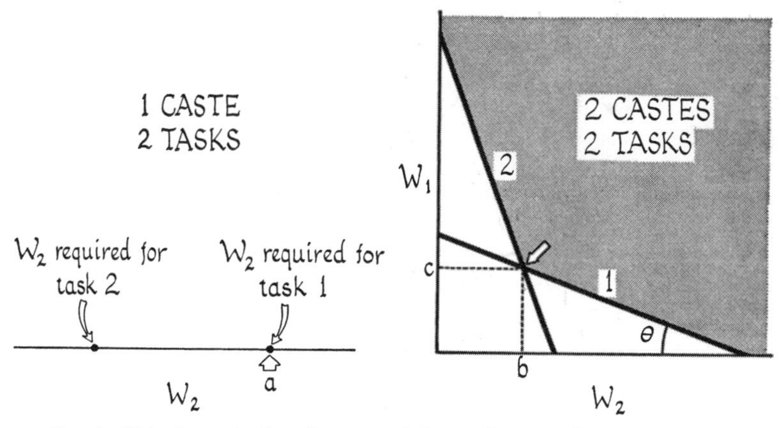

FIG. 4.—It is always to the advantage of the species to evolve new castes until there are as many castes as contingencies and each caste is specialized uniquely on a single contingency. This theorem can be substantiated readily from a comparison of the two graphs in this figure. With the addition of caste 1, the total weight of workers is changed from a to $b + c$. Since caste 1 specializes on task 1, θ is acute, therefore $a - b > c$ and $a > b + c$ for all a, b, and c.

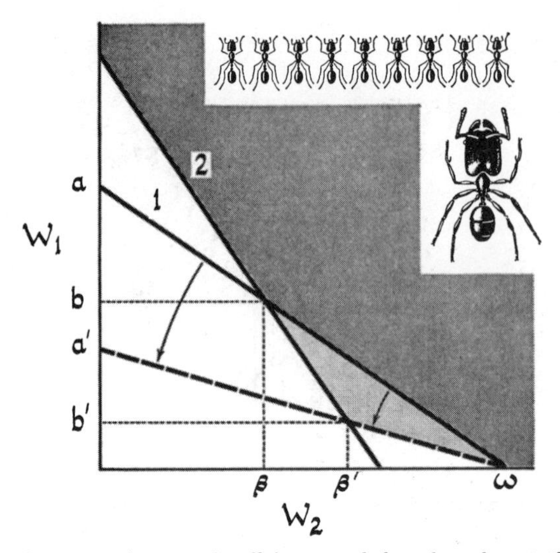

FIG. 5.—If one caste increases in efficiency, and the others do not, the proportionate total weight of the improving caste will decrease. Compare with Fig. 2.

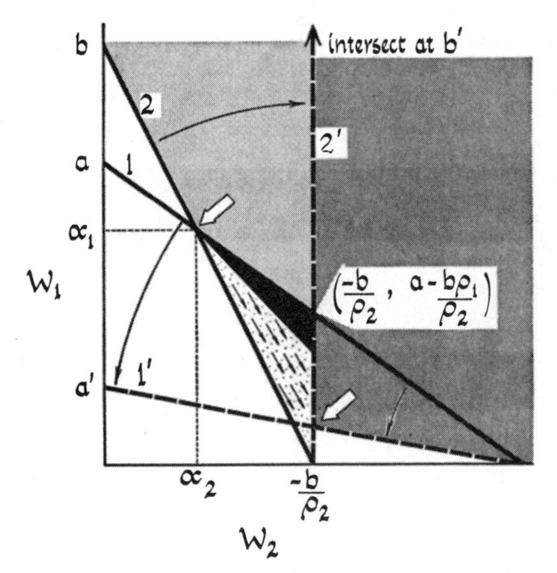

FIG. 6.—An increase in specialization of a given caste will result, under most conceivable circumstances, in increased efficiency and colony fitness. As specialization occurs through evolution, in this imaginary example in caste 1, the slope of the contingency 1 curve decreases, while the slope of the contingency 2 curve may or may not increase (but it does not decrease). (Further explanation in the text.)

The graphical presentation of Figure 6 demonstrates that there are conditions under which any increase in specialization, involving even a total loss of capacity in every other contingency, will increase colony fitness. Under certain other conditions, on the other hand, an increase in specialization at the favored task will, if accompanied by loss of capacity in other tasks beyond a certain degree, result in loss in colony fitness. But any increase in specialization not accompanied by a loss of capacity in other tasks will result in increased colony fitness. To be more specific, a change in specialization of a caste involves a shift in one or (more likely) both contingency curves. As shown in Figure 6, the intersection of the curves move to some point in the "sweep-out" triangle. A move into the stippled part of the triangle results in less total weight for both castes combined and hence greater ergonomic efficiency. A move into the blackened part of the triangle results in equal or greater total weight for both castes combined and equal or less ergonomic efficiency. Under certain conditions, explained in Appendix I, the evolving species loses colony efficiency if, while improving a caste at one task, it permits the caste to reduce ability at performing other tasks. But interestingly there are other conditions under which any specialization increases efficiency; that is, the black triangle does not exist.

The inferences from the models of Figures 5 and 6 can be summarized in the following simplified way: In a constant environment, caste determination should evolve so that each caste becomes increasingly specialized to its single assigned task.

3. If proliferation and divergence of castes are the expected consequences of selection at the colony level, why have they not reached greater heights throughout the social insects? In fact, these qualities vary greatly from group to group and even from species to species within the same taxonomic group. The only answer consistent with the theory is that, as in most evolving systems, the various levels reached by individual species are compromises between opposing selection pressures. The obvious pressure that must oppose proliferation and divergence is fluctuation of the environment. From Figure 7 we can see that a long-term change can eliminate a caste if the caste that supersedes it (by taking over its task through superior numbers) is not very specialized. In this example, contingency 2 has been increased in frequency (or importance), shifting the contingency curve to the right of the contingency 1 curve intercept of the W_2 axis. Consequently, the number of caste 2 workers required to take care of contingency 2 is also more than enough to take care of contingency 1. The presence of caste 1 now reduces colony fitness, and if the environmental change is of long duration, caste 1 will tend to be eliminated by colony-level selection. In this case the species tracks the environment to acquire a new optimal mix that just happens to eliminate the superseded caste. Thus if the critical features of the environment are changing at a rate slow enough to be tracked by the species but too fast to permit much

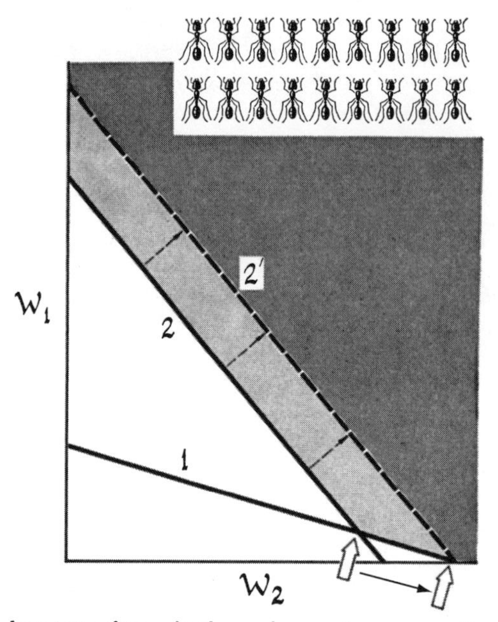

FIG. 7.—A long-term change in the environment can cause the evolutionary loss of a caste, even when the task to which the caste is specialized remains as common and important as ever.

specialization of individual castes, both the number and the degree of specialization of castes will be kept low.

At another level, the critical features of the environment may be changing too fast to be tracked genetically, yet too slow to provide each colony with a consistent average for the duration of its maturity. In this case, a mix of specialized castes would be inferior to a few generalized forms able to adapt to new circumstances.

In Figures 8 and 9 a relation is shown to exist between the degree of caste specialization and the magnitude of change in the optimal mix which is invoked by a given change in the environment. The castes represented in Figure 8 are relatively unspecialized. Task 2 is shown to become somewhat less common (or less important), resulting in a shift of the contingency curve toward the origin without a change in slope. As a consequence, the optimal mix changes from one comprised predominantly of caste 2 to one comprised predominantly of caste 1. In contrast, the castes represented in Figure 9 are highly specialized; and a shift in the contingency curve results in little change in caste ratios.

From the models presented in Figures 7–9, we can draw the conclusion that species with unspecialized castes will have on the average fewer castes and more variable caste ratios, and this effect will be enhanced in fluctuating environments. The more specialized the castes become, the more entrenched they become, in the sense that they are more likely to be represented in the optimal mix regardless of long-term fluctuations in the environment. Here we have another, peculiar theoretical result of group selection; for, in

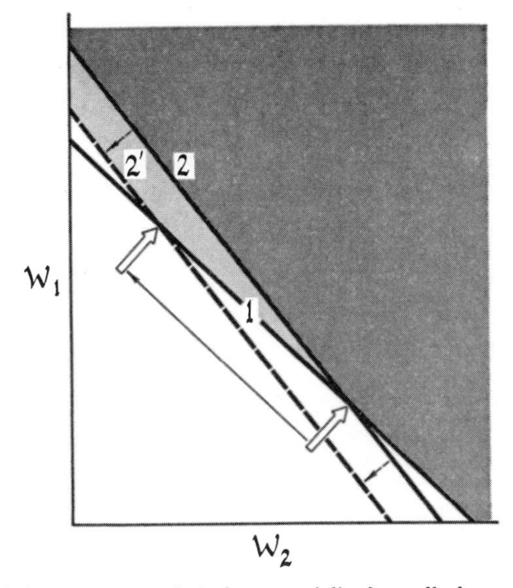

FIG. 8.—If the castes are relatively unspecialized, small changes in the environment will result in large changes in the optimal caste ratios.

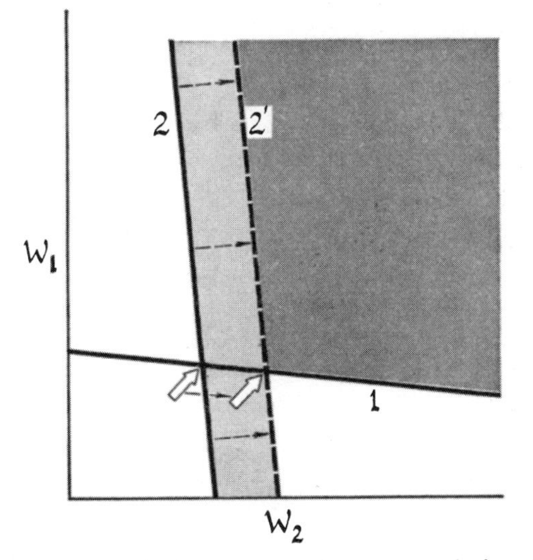

FIG. 9.—The more specialized the castes are in aggregate, the less change there is in the optimal mix.

classical population genetic theory, which entails individual selection, it is the generalized genotypes and species and not the specialized ones that are the more likely to survive in the face of long-term fluctuation in the environment.

EMPIRICAL EVIDENCE

No contingency curves of actual species have yet been drawn. At the moment, the required steps of defining contingencies and measuring their effects in natural populations seem technically formidable. Yet I can see no way of probing very deeply into the evolution of castes except by this means, or at least by comparable studies guided by an ergonomic theory equivalent to the one offered here.

Turning to empirical information already available, we find that temporal division of labor in insect societies has not been studied enough to permit a comparison of "castes," as broadly defined, among species and thus to identify evolutionary trends of relevance to the theory. In order to test the predictions of the ergonomic models in any manner, it is necessary for the moment to fall back on available data concerning the purely physical castes. Even this kind of evidence has an important limitation. Worker subcastes are known only from ants and termites. They are usually, although not always, produced by a partition of the workers into a soldier subcaste concerned primarily with defense and a minor worker (or pseudergate, etc.) subcaste devoted to the remaining functions. These special conditions make generalizations from physical castes to temporal castes an uncertain step.

Nevertheless, the evidence from the physical castes does seem compatible with the theory. We can recognize at least four phenomena that are better explained by the present formulation than by any other, earlier explanations.

First, it turns out that the soldier caste has been lost secondarily in some polymorphic ant lines, even where the defensive function surely still exists (see Wilson, 1953). The possibility of just such an occasional paradoxical development is predicted in Figure 7.

Second, physical castes are more frequent in tropical ant faunas than in temperate ant faunas. This correlation, which I do not believe has been investigated by previous authors, is documented in Appendix II. It is consistent with the notion that castes always tend to proliferate in evolution but are simultaneously being reduced in response to fluctuations in the environment, the degree of response being proportionate to the degree of fluctuation.

Third, the most specialized castes are found primarily in tropical genera and species. The bizarre soldiers of ant genera such as *Paracryptocerus*, *Pheidole (Elasmopheidole)*, *Acanthomyrmex*, *Zatapinoma*, *Camponotus (Colobopsis)*, and of termite genera such as *Nasutitermes*, *Mirotermes*, and *Capritermes* are all but limited to the tropics and subtropics. Polymorphism in temperate ant species, representing the less extreme members of *Pheidole, Solenopsis, Monomorium, Myrmecocystus,* and *Camponotus,* is predominantly of the simpler form produced by elementary allometry. This climatic correlation is consistent with the prediction from ergonomic theory that specialization in castes already in existence should increase indefinitely until countered by opposite selection pressures imposed by fluctuations in the environment.

Finally, it is my subjective impression—but principally from field observations and so far unsupported by quantitative data—that, among ant species, the more anatomically specialized the caste, the scarcer it is. If proven true, this relation would be consistent with the result given in Figure 5 and in fact would be difficult to explain by any other hypothesis.

SUMMARY

The analysis of caste ratios and their effect on colony efficiency can be approached by linear programming models. In the formulation offered here, account has been taken of certain general features of organization and behavior peculiar to insect societies; selection was assumed to be at the colony level; and the optimization goal was given as the maximum production of new queens and males by a mature colony whose size has an upper limit characteristic of the species. Even in their elementary form, the linear models here produced some interesting new conclusions, among which are the following:

1. Different kinds of contingencies, for which distinct behavioral responses are evolved (e.g., invasion by enemies, larval hunger, nest fouling), will be met by the colony as a whole in such a way that each kind of contingency causes about the same amount of reduction in average queen production.
2. Castes, including both physical variants and temporal behavioral stages, will tend to be proliferated in evolution until there is one and only one distinct caste specialized to meet each contingency.
3. The weight of workers belonging to a given caste at a given moment will increase in evolution according to the frequency and importance of the contingency to which it is specialized.
4. Under most, but not all, circumstances, it is of advantage to the colony progressively to increase the degree of specialization of each caste.
5. The more specialized the caste becomes, and in general the more efficient it becomes, the less will be its representation in the colony. This inverse relation between ability and numbers, which is a consequence of selection at the colony level, is the direct opposite of what would be predicted from selection at the level of the individual organism.
6. Proliferation of castes should be countered by those contingencies whose changes in frequency through time are poorly autocorrelated and paced so as to be difficult to track genetically. Certain changes in the environment can result in a caste being dropped, a result that at first seems paradoxical. This is due to the fact that, even though the caste is still the

best one to handle a contingency that continues to be important, its presence nonetheless results in lowered efficiency of the colony as a whole.

7. The more specialized a caste has become, the less likely it is to be dropped from the optimal mix due to fluctuation in the environment. This, too, is the opposite of what would be expected if selection were operating at the level of the organism rather than at the colony level.

8. Certain empirically determined qualities of physical castes in ants and termites are consistent with the result of this ergonomic theory and, for the moment at least, seem best explained by it.

ACKNOWLEDGMENT

I am very grateful to Mr. Daniel Simberloff for critically reading the manuscript and to Mrs. H. Lyman for preparing the illustrations. The author's research has been supported by grants from the National Science Foundation.

APPENDIX I

A Fuller Account of the Consequences of Caste Specialization

Following is a fuller explanation of the principles expressed under Figure 6.

Let ρ_1 be the slope of the contingency 1 curve before evolution and ρ_1' the slope afterward; and ρ_2 the slope of the contingency curve 2 before evolution and ρ_2' afterward. Specialization on the part of caste 1 means that it becomes more efficient at task 1, so that $\rho_1 \to \rho_1'$ must involve a decrease. On the other hand, the specialization to task 1 may or may not result in caste 1 becoming less efficient at task 2; if there is any change at all, $\rho_2 \to \rho_2'$ therefore involves an increase.

A mandatory decrease in ρ_1 and an optional increase in ρ_2 results in the intersection of contingency curves 1 and 2 moving down and to the right. A new value is reached which is one of a large convex set of values bounded by the triangle whose coordinates are (α_1, α_2), $[-b/\rho_2, a - (b\rho_1/\rho_2)]$, and some values near but not equal to $[-(b/\rho_2'), 0]$. The second and third apexes are set by the 2′ curve being rotated until it becomes parallel with the W_1 axis. This is what would happen in the extreme case of caste 1 losing all ability to perform function 2.

Now let us define the stippled part of the sweep-out triangle. Before evolution, the combined weight of the castes is $\alpha_1 + \alpha_2$, the sum of the coordinates of the upper left apex of the triangle. This apex is simply the intersection of the original contingency curves, so that α_1 and α_2 are related as follows:

$$\alpha_1 = a + \rho_1\alpha_2,$$
$$\alpha_1 = b + \rho_2\alpha_2.$$

Solving for α_1 and α_2 and then adding them,

$$\alpha_1 = \frac{\rho_2 a - \rho_1 b}{\rho_2 - \rho_1},$$

$$\alpha_2 = \frac{a - b}{\rho_2 - \rho_1},$$

$$\alpha_1 + \alpha_2 = \frac{a - b + \rho_2 a - \rho_1 b}{\rho_2 - \rho_1}.$$

If $\alpha_1 + \alpha_2$ is greater than the sum of the coordinates of the point within the sweep-out triangle reached by increased specialization, the colony has reduced the combined weight of caste 1 and 2 and thus increased efficiency. The coordinates and sum of coordinates in the triangle are given by

$$W_1 = a' + \rho_1' W_2,$$

$$W_1 = b' + \rho_2' W_2,$$

$$W_1 + W_2 = \frac{a' - b' + \rho' a' - \rho_1' b'}{\rho_2' - \rho_1'}.$$

And it follows that, when

$$\frac{a - b + \rho_2 a - \rho_1 b}{\rho_2 - \rho_1} > \frac{a' - b' + \rho_2' a' - \rho_1' b'}{\rho_2' - \rho_1'},$$

the increase in specialization results in increased ergonomic efficiency, while the reverse relation results in decreased efficiency. If a caste increases its capacity to perform one task without losing ability to perform the other task, there will always be an increase in ergonomic efficiency. This is intuitively apparent, and it can be confirmed by noting that under the given condition $b = b'$ and $\rho_2 = \rho_2'$; and since by definition $a > a'$ and

$$\rho_1 > \rho_1', \qquad \frac{a - b + \rho_2 a - \rho_1 b}{\rho_2 - \rho_1} > \frac{a' - b' + \rho_2' a' - \rho_1' b_1'}{\rho_2' - \rho_1'}$$

always.

Finally, when

$$\frac{a - b + \rho_1 b - \rho_2 a}{\rho_1 - \rho_2} > \frac{a - b\rho_1 + b}{\rho_2},$$

any increase in ability at task 1 increases ergonomic efficiency regardless of the loss of ability at task 2. This result is obtained as follows. The maximum sum of coordinates that can be obtained in any sweep-out triangle is at the outer apex of that particular triangle formed when the evolving caste loses all ability to perform the function to which it is not specialized. This extreme situation is illustrated in Figure 6. The value b', defined as the weight of caste 1 required to perform task 2 if left by itself, approaches infinity. The coordinates of the outer apex of the sweep-out triangle become

$$\left(-b/\rho_2,\ a - \frac{b\rho_1}{\rho_2}\right)$$

and their sum is

$$\frac{a - b\rho_1 + b}{\rho_2}.$$

If the sum of the coordinates of the original intersection of contingency curves is greater than this sum, then it is greater than the sum of coordinates of any other point in the triangle.

To summarize the conclusions of this argument, we can define widely ranging conditions under which an increase in caste specialization in an unchanging environment results in greater ergonomic efficiency, and much more limited conditions under which it does not. If we could measure the contingency curves in experiments, and they should prove linear, we could then predict the permissible evolutionary pathways in caste evolution.

APPENDIX II

Variation in Frequency of Species with Physical Worker Subcastes ("Polymorphic Species") in Different Ant Faunas around the World

The data in Table 1 were compiled from taxonomic studies published by various authors in the past 60 years. While the completeness of the collections on which the studies were based varied greatly, there is no reason to doubt that each represents a random sample of species as far as polymorphism is concerned.

An examination of these data reveals several trends, and lack of trends, of interest. In the New World, the frequency of polymorphic species drops off in both the north and south temperate zones. This is noteworthy because the genera and species are mostly different at the two ends. There is a sharp decline also in going from North Africa to the limits of the ant fauna in northern Europe. A slight decline is apparent in temperate South Africa. Also, a slight decline is seen from the southern to the northern Palaearctic region and from Australia to New Zealand. However, the remainder of the Indo-Australian area does not display the same climatic trend. Tropical Asia and Melanesia have a relatively low frequency of polymorphic species, although not so low as most of the marginal cold temperate regions around the world. There is a local *increase* going north into the Himalayas. There is no apparent trend either way across Australia.

In general, then, there is a decline in polymorphism with increasing coldness of climate everywhere in the world except in the Indo-Australian region. In the Himalayan region the reverse effect is seen.

Perhaps we should expect environments to be more fluctuating on islands and in deserts, and polymorphism to decline accordingly. However, the data show this not to be the case. The frequency of polymorphism on islands is consistent with that on the nearest continent at the same latitude. The faunas of deserts may even increase in polymorphism somewhat. But it appears that the overriding determinant on a global scale is latitude.

In conclusion, the latitude effect is probably clear enough to be regarded as consistent with the prediction from ergonomic theory. On the other hand, the lack of an island or desert effect is not consistent with it—yet does not oppose it.

TABLE 1

Locality	Climate	Total Number of Species	Number of Polymorphic Species	Percentage of Polymorphic Species	Authority
Neotropical:					
Amazonas (state), Brazil	Tropical	71	27	38.0	Wheeler (1923)
Galápagos	Tropical	19	6	31.6	Wheeler (1924*b*)
Honduras and Guatemala	Tropical	86	32	37.2	Mann (1922)
Haiti	Tropical	67	16	23.9	Wheeler and Mann (1914)
Puerto Rico	Tropical	60	10	16.7	Smith (1936)
Jamaica	Tropical	34	14	41.2	Wheeler (1917*b*)
Cuba	Tropical	74	17	23.0	Wheeler (1913)
Tucuman Prov., Argentina	Subtropical	130	45	34.6	Kusnezov (1953)
Hidalgo State, Mexico	Warm temperate	37	11	29.7	Wheeler (1914*a*)
East Patagonia and Tierra del Fuego	Cold temperate	19	2	10.5	Kusnezov (1959)
Nearctic:					
Florida Keys	Subtropical	41	12	29.3	Wilson (unpublished)
Welaka Reserve, central Florida	Warm temperate	76	13	17.1	Van Pelt (1956)
California	Subtropical to warm temperate	94	20	21.3	Mallis (1941)
North Carolina: coastal plain	Warm temperate	115	17	14.8	Carter (1962)
North Carolina: mountains	Warm to cold temperate	83	12	14.5	Carter (1962)
George Reserve, Michigan	Cold temperate	67	5	7.5	Talbot (in Wilson, 1959)
North Dakota	Cold temperate	83	8	9.6	G. C. Wheeler and J. Wheeler (1963)
Alaska	Cold temperate	12	1	8.3	Wheeler (1917*a*)
Western Palaearctic:					
Tassili des Ajjer, central Sahara	Tropical (desert)	45	13	29.0	Bernard (1953)
Canary Islands	Subtropical	55	11	20.0	Wheeler (1927*c*)
Madeira	Subtropical	15	2	13.3	Wheeler (1927*c*)
Mamora Forest, Morocco	Subtropical	41	12	29.3	Bernard (1953)
Balearics	Subtropical to warm temperate	26	7	26.9	Wheeler (1926)

TABLE 1.—*Continued*

Locality	Climate	Total Number of Species	Number of Polymorphic Species	Percentage of Polymorphic Species	Authority
France: Tourettes-sur-Loup, south coast	Warm temperate	37	8	21.6	Collingwood (1956)
France: Fontainebleau	Warm to cold temperate	30	1	3.3	Collingwood (1956)
South Lake Dist., England	Cold temperate	17	0	0	Collingwood and Satchell (1956)
Ireland	Cold temperate	16	0	0	Collingwood (1958)
Scottish Highlands	Cold temperate	18	0	0	Collingwood (1961*a*)
Sweden	Cold temperate	60	4	6.7	Forsslund (1957)
Baltic Amber (Eocene)	Tropical to temperate?	92	10	10.9	Wheeler (1914*b*)
Ethiopian:					
Congo	Tropical	318	91	28.6	Wheeler (1922)
Malagasy	Tropical	237	69	29.1	Wheeler (1922)
Imatong Mts., Sudan: lower slopes	Tropical	99	23	23.2	Weber (1943)
Imatong Mts., Sudan: 5,600 ft and higher	Warm to cold temperate	34	8	23.5	Weber (1943)
Rhodesia	Subtropical	33	5	15.2	Forel (1913*a*)
Natal, S. Africa	Warm temperate	72	15	20.8	Santschi (1914)
Oriental to Eastern Palaearctic:					
Borneo	Tropical	243	58	23.9	Wheeler (1919)
Krakatau Islands (1919–21)	Tropical	65	14	21.5	Wheeler (1924*a*)
Indochina	Tropical	65	14	21.5	Wheeler (1927*b*)
Formosa	Tropical to subtropical	68	17	25.0	Forel (1913*c*); Wheeler (1930)
China	Subtropical to cold temperate	191	39	20.4	Wheeler (1931)
Afghanistan	Mountainous, cold temperate	44	14	31.8	Collingwood (1961*b*)
Nanga Parbat, NW Himalayas	Mountainous, cold temperate	18	6	33.3	Eidmann (1942)
Japan (entire)	Warm to cold temperate	69	12	17.4	Wheeler (1928)

TABLE 1.—*Continued*

Locality	Climate	Total Number of Species	Number of Polymorphic Species	Percentage of Polymorphic Species	Authority
Japan: Sapporo	Cold temperate	28	6	21.4	Hayashida (1960)
Japan: Mt. Atusanupuri	Cold temperate	11	1	9.1	Hayashida (1959)
West China and Tibet	Mostly cold temperate	43	4	9.3	Eidmann (1941)
Turkestan	Cold temperate	39	7	17.9	Kusnezov (1926)
Hsingan Prov., North Manchuria	Cold temperate	7	0	0	Yasumatsu (1941)
Oceania and Australia:					
Busu River, New Guinea	Tropical	179	39	21.8	Wilson (1959)
Solomons	Tropical	123	19	15.4	Mann (1919)
Fiji	Tropical	88	25	28.4	Fullaway (1956)
Wallis-Futuna	Tropical	36	7	19.4	Wilson and Hunt (1967)
Polynesia	Tropical	35	5	14.3	Wilson and Taylor (1967)
New Caledonia	Subtropical	67	17	25.4	Emery (1914)
Lord Howe I.	Subtropical	14	1	7.1	Wheeler (1927*a*)
Norfolk I.	Subtropical	12	1	8.3	Wheeler (1927*a*)
Australia: Queensland	Tropical to subtropical	174	29	16.7	Forel (1915)
Australia: Everard and Musgrave Ranges, center of Australia	Warm temperate	30	10	33.3	Wheeler (1915)
Australia: Rottnest I.	Warm temperate	43	7	16.3	Wheeler (1934)
Australia: southwest	Warm temperate	72	15	20.8	Forel (1907)
New Zealand	Warm to cold temperate	31	4	12.9	Wilson and Taylor (1967)
Tasmania	Cold temperate	20	4	20.0	Forel (1913*b*)

NOTE.—In each region, localities are listed with the one farthest from the pole first, then proceed poleward, until the last entry in each category is closest to the pole.

LITERATURE CITED

Bernard, F. 1953. Les fourmis du Tassili des Ajjer (Sahara Central). Inst. Recherches Sahariennes Univ. D'Alger. 130 p.

Brian, M. V. 1965. Caste differentiation in social insects. Symp. Zool. Soc. London 14: 13–38.

———. 1966. Social insect populations. Academic Press, New York.

Carter, W. G. 1962. Ant distribution in North Carolina. J. Elisha Mitchell Sci. Soc. 78:150–204.

Collingwood, C. A. 1956. Ant hunting in France. Entomologist 89:105–108.

———. 1958. A survey of Irish Formicidae. Roy. Irish Acad., Proc., B(59):213–219.

———. 1961*a*. Ants in the Scottish Highlands. Scottish Natur. 70:12–21.

———. 1961*b*. Formicidae (Insecta) from Afghanistan. Vidensk. Medd. Dansk naturh. Foren. 123:51–79.

Collingwood, C. A., and J. E. Satchell. 1956. The ants of the South Lake District. J. Soc. Brit. Entomol. 5:159–164.

Eidmann, H. 1941. Zur Ökologie und Zoogeographie der Ameisenfauna von Westchina und Tibet. Z. Morphol. Ökologie Tiere 38:1–43.

———. 1942. Zur Kenntnis der Ameisenfauna des Nanga Parbat. Zool. Jahrb. (Abt Syst.) 75:239–266.

Emery, C. 1914. Les fourmis de la Nouvelle-Calédonie et des îles Loyalty. *In* F. Sarasin and J. Roux [ed.], Nova Caledonia, Zool. 1:393–439.

Forel, A. 1907. Formicidae. Die Fauna Südwest-Australiens. Ergeb. Hamburger südwest-australisch. Forschungreise 1905. W. Michaelsen and R. Hartmeyer [ed.]. Vol. 1, p. 263–310.

———. 1913*a*. Ameisen aus Rhodesia, Kapland usw. (Hym.) Deutsch. Entomol. Z. 1913: 203–225.

———. 1913*b*. Fourmis de Tasmanie et d'Australie récoltées par Mm. Lea, Frogatt, etc. Soc. Vaudoise Sci. Nat., Bull. 49:173–196.

———. 1913*c*. H. Sauter's Formosa—Ausbeute: Formicidae II. Arch. Naturgesch. (A)6: 183–202.

———. 1915. Results of Dr. E. Mjöbergs Swedish scientific expeditions to Australia 1910–1913. II. Ameisen. Arkiv. Zool. (Stockholm) 9:1–119.

Forsslund, K.-H. 1957. Catalogus insectorum Sueciae. XV. Hymenoptera: Fam. Formicidae. Opuscula Entomol. 22:70–78.

Fullaway, D. T. 1956. Checklist of the Hymenoptera of Fiji. Hawaiian Entomol. Soc., Proc. 16:269–280.

Hamilton, W. D. 1964. The genetical evolution of social behaviour. I, II. J. Theoretical Biol. 7:17–52.

Hayashida, K. 1959. Ecological distribution of ants in Mt. Atusanupuri, an active volcano in Akan National Park, Hokkaido. J. Fac. Sci. Hokkaido Univ. 14(6):252–260

———. 1960. Studies on the ecological distribution of ants in Sapporo and its vicinity. I, II. Insectes Sociaux 7:125–162.

Kusnezov, N. 1926. Die Entstehung der Wüstenameisenfauna Turkestans. Zool. Anzeiger 65:140–159.

———. 1953. Lista de las hormigas de Tucuman con descripcion de dos nuevos generos (Hymenoptera, Formicidae). Acta Zool. Lilloana 13:327–339.

———. 1959. La fauna de hormigas en el oeste de la Patagonia y Tierra del Fuego. Acta Zool. Lilloana 17:321–401.

Lindauer, M. 1961. Communication among social bees. Harvard Univ. Press, Cambridge, Mass.

Lüscher, M. 1961. Social control of polymorphism in termites. Symp. Roy. Entomol. Soc. London 1:57–67.

Mallis, A. 1941. A list of the ants of California with notes on their habits and distribution. Southern California Acad. Sci., Los Angeles.

Mann, W. M. 1919. The ants of the British Solomon Islands. Mus. Comp. Zool. Harvard, Bull. 63:273–391.
———. 1922. Ants from Honduras and Guatemala. U.S. Nat. Mus., Proc. 61:1–54.
Michener, C. D. 1961. Social polymorphism in Hymenoptera. Symp. Roy. Entomol. Soc. London 1:43–56.
Richards, O. W., and M. J. Richards. 1951. Observations on the social wasps of South America (Hymenoptera: Vespidae). Roy. Entomol. Soc. London, Trans. 102:1–170.
Santschi, F. 1914. Fourmis du Natal et du Zoulouland. Medd. Göteborgs Kungl. Vetenskaps. Vitterhetssamhälles Handl. 15:3–47.
Smith, M. R. 1936. The ants of Puerto Rico. J. Agr. Univ. Puerto Rico 20:819–875.
Van Pelt, A. F. 1956. The ecology of the ants of the Welaka Reserve, Florida (Hymenoptera: Formicidae). Amer. Midland Natur. 56:358–387.
Weaver, N. 1966. Physiology of caste determination. Annu. Rev. Entomol. 11:79–102.
Weber, N. A. 1943. The ants of the Imatong Mountains, Anglo-Egyptian Sudan. Mus. Comp. Zool. Harvard, Bull. 93:263–389.
Wheeler, G. C., and J. Wheeler. 1963. The ants of North Dakota. Univ. North Dakota Press, Grand Forks.
Wheeler, W. M. 1913. The ants of Cuba. Mus. Comp. Zool. Harvard, Bull. 54:477–505
———. 1914*a*. Ants collected by W. M. Mann in the State of Hidalgo, Mexico. J. New York Entomol. Soc. 22:37–61.
———. 1914*b*. The ants of the Baltic amber. Schrift. Phys.-ökon. Ges. Königsberg 15: 1–142.
———. 1915. Hymenoptera. [Collections by Capt. S. A. White in the Everard and Musgrave Ranges, Central Australia]. Roy. Soc. South Australia, Trans. 39:805–823.
———. 1917*a*. The ants of Alaska. Mus. Comp. Zool. Harvard, Bull. 61:15–22.
———. 1917*b*. Jamaican ants collected by Prof. C. T. Brues. Mus. Comp. Zool. Harvard, Bull. 61:457–471.
———. 1919. The ants of Borneo. Mus. Comp. Zool. Harvard, Bull. 63:43–147.
———. 1922. Ants of the American Museum Congo Expedition. A contribution to the myrmecology of Africa. Amer. Mus. Natur. Hist., Bull. 45:1–269.
———. 1923. Wissenschaftliche Ergebnisse der schwedischen entomologischen Reise des Herrn Dr. A. Roman in Amazonas 1914–1915. VII. Formicidae. Arkiv. Zool. 15:1–6.
———. 1924*a*. Ants of Krakatau and other islands in the Sunda Strait. Treubia 5:1–20.
———. 1924*b*. The Formicidae of the Harrison Williams Galapagos expedition. Zoologica 5:101–122.
———. 1926. Ants of the Balearic Islands. Folia Myrmecologica et Termitologica 1:1–6.
———. 1927*a*. The ants of Lord Howe Island and Norfolk Island. Amer. Acad. Arts Sci., Proc. 62:121–153.
———. 1927*b*. Ants collected by Professor F. Silvestri in Indochina. Lab. Gen. Agr. Portici, Boll. 20:83–106.
———. 1927*c*. The ants of the Canary Islands. Amer. Acad. Arts Sci., Proc. 63:93–120.
———. 1928. Ants collected by Professor F. Silvestri in Japan and Korea. Lab. Gen. Agr. Portici, Boll. 21:96–125.
———. 1930. Formosan ants collected by Dr. R. Takahashi. New England Zool. Club, Proc. 11:93–106.
———. 1931. A list of the known Chinese ants. Peking Natur. Hist. Bull. 5:53–81.
———. 1934. Contributions to the fauna of Rottnest Island, Western Australia. J. Roy. Soc. Western Australia 20:137–163.
Wheeler, W. M., and W. M. Mann. 1914. The ants of Haiti. Amer. Mus. Natur. Hist., Bull. 33:1–61.
Wilson, E. O. 1953. The origin and evolution of polymorphism in ants. Quart. Rev. Biol. 28:136–156.

———. 1959. Some ecological characteristics of ants in New Guinea rain forests. Ecology 40:437–447.

———. 1963. The social biology of ants. Annu. Rev. Entomol. 8:345–368.

———. 1966. Behaviour of social insects. Symp. Roy. Entymol. Soc. London 3:81–96

Wilson, E. O., and G. L. Hunt. 1967. The ant fauna of Futuna and the Wallis Islands, stepping stones to Polynesia. Pacific Insects (in press).

Wilson, E. O. and R. W. Taylor. 1967. The ants of Polynesia. Pacific Insect Monogr 14:1–109.

Yasumatsu, K. 1941. Ants collected by Mr. H. Takahasi in Hingan (Hsingan) North Province, North Manchuria (Hymenoptera, Formicidae). Natur. Hist. Soc. Formosa, Trans. 31:182–185.

1971

"The prospect for a unified sociobiology," *American Scientist* 59(4): 400–403

Sociobiology is commonly said to have begun as a structured discipline with the publication of my general work *Sociobiology: The New Synthesis* in 1975. In fact, it began in 1971 with *The Insect Societies,* an encyclopedic treatment of all forms of communal behavior known in insects to that time. The last chapter, reproduced here in modified form, proposes that similar, perhaps identical principles of communal organization are to be found in insect and vertebrate societies—up to and including non-human primates. Humans were not added until the 1975 synthesis.

Another common misconception holds that "sociobiology" is the belief (or doctrine, as critics once liked to call it) that human behavior is genetically determined, or at least genetically predisposed to follow certain forms or patterns.

Quite the contrary. Sociobiology, which was in formation before *The Insect Societies* and *Sociobiology: The New Synthesis,* is correctly defined as the systematic study of the biological basis of all forms of social behavior in all organisms. That meaning is the one employed by the many investigators who have taken the concept in new directions and to higher levels of sophistication.

The foundation disciplines of sociobiology are population biology, evolutionary theory, and behavioral biology. The consilient reasoning that led to the idea that I developed is as follows. A society is a population, and hence it has a demography, a group genetics, and a well-integrated information network. All these properties are shaped through evolution by natural selection and hence are analyzable in a manner parallel to that used for populations of solitary organisms.

The connection I inferred between the parameters of population biology and sociobiology are depicted in Figures 2 and 3.

❁ ARTICLE II

Edward O. Wilson

The Prospects for a Unified Sociobiology

This important branch of behavioral biology should be joined with population biology

When the same parameters and quantitative theory are used to analyze both termite colonies and troops of rhesus macaques, we will have a unified science of sociobiology. Can this really ever happen? As my own studies have advanced, I have been increasingly impressed with the functional similarities between insect and vertebrate societies and less so with the structural differences that seem, at first glance, to constitute such an immense gulf between them.

Consider for a moment termites and macaques. Both are formed into cooperative groups that occupy territories. The group members communicate hunger, alarm, hostility, caste status or rank, and reproductive status among themselves, by means of something on the order of ten to a hundred nonsyntactical signals. Individuals are intensely aware of the distinction between group-mates and nonmembers. Kinship plays an important role in group structure and has probably served as a chief generative force of sociality in the first place. In both kinds of society there is a well-marked division of labor, albeit with a much stronger reproductive component on the part of the insects. The

After completing his undergraduate work and an M.S. in biology at the University of Alabama, Edward O. Wilson came to Harvard University for the Ph.D. (1955) and has remained on the Harvard faculty, where he is now Professor of Zoology. Although his chief interest is the social insects, especially ants, he works on a variety of other subjects, including pheromones and theoretical and experimental biogeography. He is a member of the National Academy of Sciences and a Fellow of the American Academy of Arts and Sciences. This article is based on excerpts from The Insect Societies, *The Belknap Press of Harvard University Press, Cambridge, Mass., © 1971, President and Fellows of Harvard College. Address: The Biological Laboratories, 16 Divinity Ave., Harvard University, Cambridge, MA 02138.*

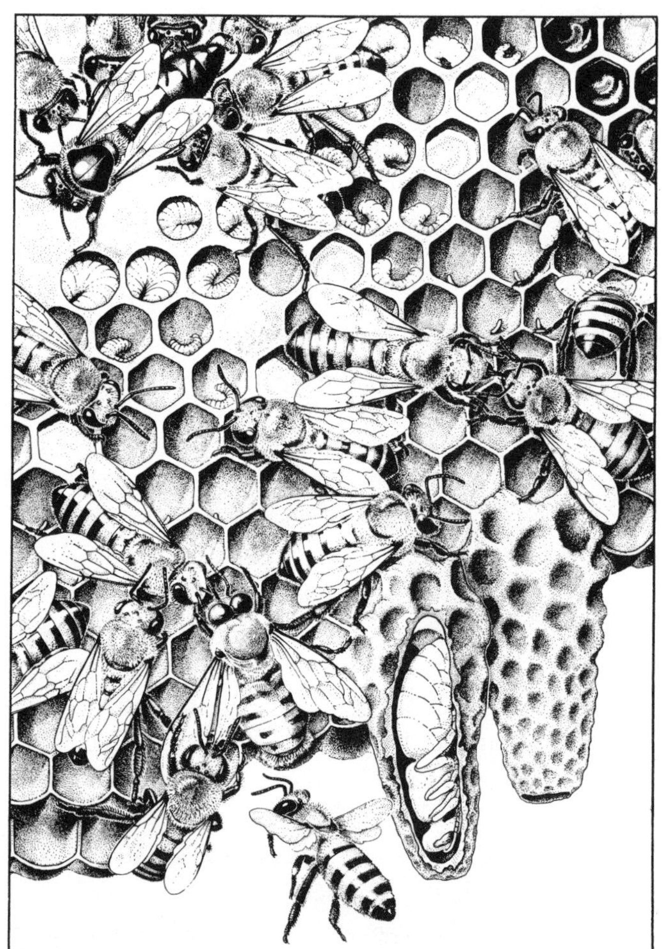

Figure 1. The impersonal quality of the bonds in higher social insects is exemplified by life in the honeybee hive. Although a high degree of order is sustained, including care of the mother queen (*upper left*), rearing of new queens (in royal cells, *lower right*), and frequent sharing of liquid food by regurgitation (*center right*), the worker bees have little or no knowledge of each other as individuals. Most live only one or two months in the midst of tens of thousands of nestmates. (Drawing by Sarah Landry.)

Reprinted from AMERICAN SCIENTIST, Vol. 59, No. 4, July-August 1971, pp. 400-403

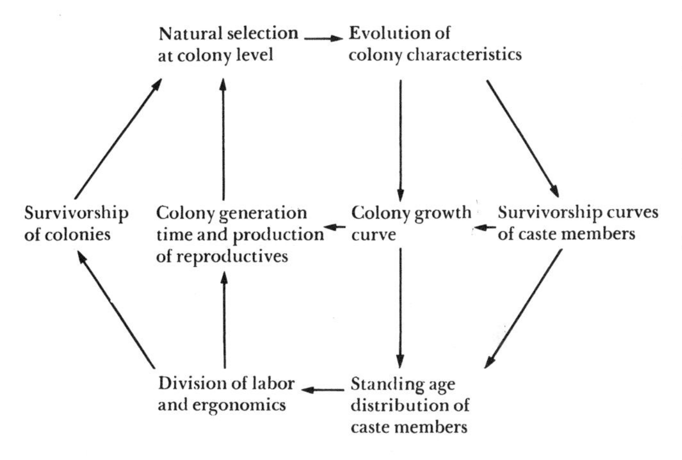

Figure 3. The causal relationships of principal phenomena in the biology of the higher social insects (ants, social bees, social wasps, termites). The exterior aspects of caste, communication, and other social phenomena represent adaptations that are fixed by natural selection at the level of the colony. Selection at the colony level is the differential survival and reproduction of closely related genotypes. In order to explain survival and reproduction fully, it is necessary to go beyond insect sociobiology into population ecology.

lined. The role of historical constraints has been examined recently by a number of authors in, for example, the volumes edited by Baker and Stebbins (1965), Waddington (1968, 1969), and Lewontin (1968). Levins (1968) has deliberately set out to develop deductive theories of the niche and population genetics in fluctuating environments. At the next level, MacArthur and Wilson (1967) have derived some of the population parameters and linked them to dispersal and distribution, while Hamilton (1964) has produced the first comprehensive theory of altruistic behavior based on first principles of population genetics. An increasing number of experimental biologists are devoting themselves to the study of behavioral genetics, which holds the key to the population consequences of behavioral modifiability (Parsons 1967). The remaining lacunae of evolutionary biology are nevertheless wide and treacherous, and relevant experimental investigation is in its earliest stages.

In the future development of sociobiology the insect societies must play a key role. They are so remote in phylogenetic origin from vertebrate societies as to resemble creations of some other world. They provide the comparative material every good theoretical scheme needs. Where the theory predicts convergence of a functional trait, its validity can be put to a convincingly stringent test by comparing insects and vertebrates. It would therefore be appropriate to augment this brief attempt at prophesy by identifying the basic differences that visibly exist, not between insects and vertebrates as organisms, but between the societies they form.

If we exclude man, with his unique language and revolutionary capacity for cultural transmission, the best organized societies of vertebrates can be distinguished by a single trait so overriding in its consequences that the other characteristics seem to flow from it. This is personal recognition among the members of the group. As a rule each adult animal knows and bears some particular relationship to every other member. Status is extremely important. Where dominance hierarchies exist, elaborate signaling is employed to implement them. Parent-offspring relationships are specific to individuals, tightly binding, and relatively long in duration. Within primate societies, personal groupings often ascend to the level of cliques, and, in the case of the macaques, to the "uncle" and "aunt" forms of parental care.

Group members spend large amounts of time and energy establishing and maintaining these multifarious personal bonds. Associated with the personal form of communication is a prolonged period of socialization in the young. By constant experience, much of it obtained in the form of play, the young animal learns its personal relationship to its parents and establishes an early status among its age-peers. Some division of labor exists: there are weakly defined "specialists" in defense, leadership, and foraging. But is not based upon morphologically defined castes, and it does not include reproductive neuters.

In contrast, the insect societies are for the most part impersonal. The small, relatively primitive colonies of bumblebees and *Polistes* wasps are organized in part by dominance hierarchies, and individuals appear to recognize one another to a limited extent. However, in other kinds of social insects personalized relationships play little or no role. The sheer size of the colonies and the short life of the members combine to make it inefficient, if not impossible, to establish individual bonds. The average adult honeybee, for example, lives for only about six weeks in the midst of a rapidly changing population of up to 80,000 workers. The army ant worker has only a few weeks or months to become acquainted with a million or more nestmates. In the more advanced societies even the relationship of the workers to the queen is impersonal. She is "recognized" by a small set of pheromones, which can be extracted and absorbed into dummy substitutes of inert material. While songbirds have individual calls that are identified by territorial neighbors and mammals use individual scent marks, the members of an insect colony employ signals that are for the most part uniform throughout the species. The one known exception is the colony odor, which is acquired at least in part from food and nesting material and is used to distinguish nestmates, all of them together, from members of other colonies. Socialization is minimal in insects, and play is apparently absent.

Yet the insect colony, as a unit, can equal or exceed the accomplishments of the nonhuman vertebrate society—as a unit. One need only think of the giant air-conditioned nests of the macrotermitines, the organized foraging expeditions of the army ants, and the swift recruitment of honeybees in response to pheromones and waggle dancing, to see the strength of this

details of organization have been evolved by an evolutionary optimization process of unknown precision, during which some measure of added fitness was given to individuals with cooperative tendencies—at least toward relatives. The fruits of cooperativeness depend upon the particular conditions of the environment and are available to only a minority of animal species during the course of their evolution.

From the specialist's point of view this comparison may at first seem facile—or worse. But it is out of such deliberate oversimplification that the beginnings of a general theory are made. The formulation of a theory of sociobiology offers one of the great manageable tasks of biology for the next twenty or so years. Let us try to guess part of its future outline and some of the directions it is most likely to lead animal behavior research.

It is my belief, shared in varying degrees by some other biologists (for example, Altmann and Altmann 1970; Brown and Orians 1970; Crook 1970; Hall 1965; Hamilton 1964; Williams 1966), that the evolution of social behavior cannot begin to be fully comprehended except through an understanding first of demography, which yields the vital information concerning population growth and age structure, and second of the genetic structure of the populations, which tells us what we need to know about effective population size in the genetic sense, the coefficients of relationship within the societies, and the amounts of gene flow between them. The principal goal of a general theory of sociobiology should be an ability to predict features of social organization from a knowledge of these population parameters combined with information on the behavioral constraints imposed by the genetic constitution of the species. It will be a chief task of evolutionary ecology in turn to derive the population parameters from a knowledge of the evolutionary history of the species and of the environment in which the most recent portion of that history has unfolded. This sequential relation between evolutionary studies, ecology, and sociobiology is represented in Figure 2, while a more detailed schema relating demography and the study of social insects is given in Figure 3.

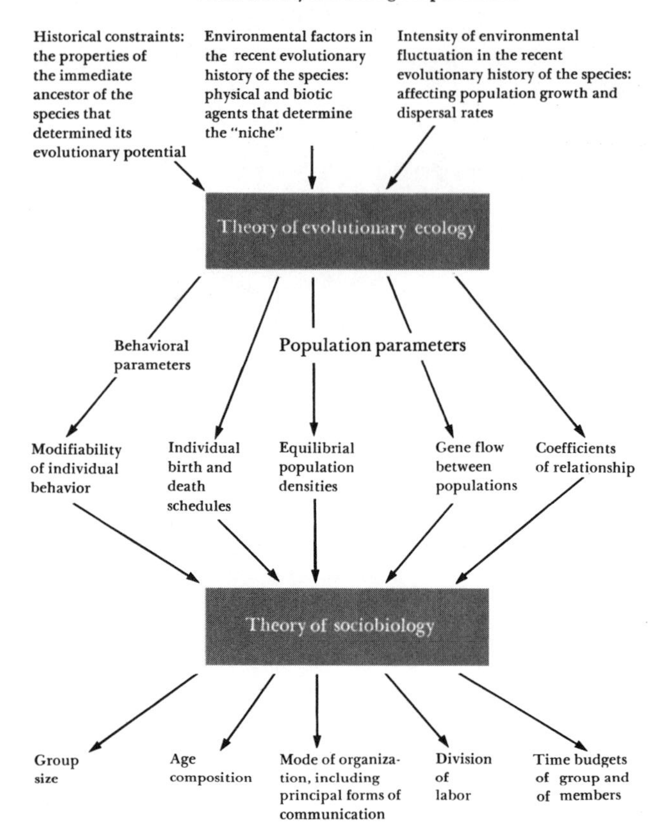

Notice that a special meaning of the word "theory" is implied in these outlines. Mature theory is conceived as being postulational-deductive, or neo-Cartesian, in nature, which means that it is built from models designed explicitly to test and extend our basic assumptions. This form of ratiocination has of course worked exceedingly well in the physical sciences and is just now beginning to have an influence throughout evolutionary biology. It proceeds roughly in three steps. First, empirical knowledge, both actual and desired, is organized into concepts which in their most desirable form are sets of measured variables. What we know, or think we know, is then expressed in precisely defined relations among the concepts. These are the postulates of our knowledge. In the second step of the postulational-deductive method the consequences of the postulates are deduced by means of models. New predictions about the real world are thereby contrived, and, in the third and final step, they are examined through experimentation and analysis in order to test and revise the original concepts and postulates. The predictions, insofar as they can be verified, provide postulates for the next, subordinate level of theory. This is the procedure that can best be followed to link evolutionary theory and ecology to population biology and thence to sociobiology.

Adding to the foundations provided by classical population genetics, evolutionary biology is now in the process of building fragments of such a theory by approximately the means just out-

generalization. The insects do it with programmed divisions of behavior among castes. Their great innovation in evolution, achieved over 100 million years ago, was the reproductive neuter, which fixed the colony as the unit of natural selection and removed the limits on the amount of caste differentiation that could occur among the colony members. Once this happened, it was possible to fashion entire organisms to perform difficult, highly specialized tasks and to combine them in mixtures capable of matching the feats of single vertebrates.

Had vertebrates gone the same route the results would have been much more spectacular. The individual bird or mammal, possessing a brain vastly larger than that of an insect, might have been shaped into a comparably better specialist. The vertebrate society then could have evolved into a considerably more complex, efficient unit—at the price, of course, of independent action on the part of its members. Vertebrates have instead remained chained to the cycle of individual reproduction. This forever enhances freedom on the part of the individual at the expense of efficiency on the part of the society. The dilemma of mankind is that technology and population growth have propelled us to the point where we could perhaps operate better as a society with termite-like altruism and regimentation, yet we cannot and must not forsake the primate individuality that brought us to the threshold of civilization in the first place.

The optimistic prospect for sociobiology can be summarized briefly as follows. In spite of the phylogenetic remoteness of vertebrates and insects and the basic distinction between their respective personal and impersonal systems of communication, these two groups of animals have evolved social behaviors that are similar in degree of complexity and convergent in many important details. This fact conveys a special promise that sociobiology can eventually be derived from the first principles of population and behavioral biology and developed as a single, mature science. The discipline can then be expected to increase our understanding of the unique qualities of social behavior in animals as opposed to those in man, and to bring us closer to the vision expressed in 1810 by Pierre Huber, one of the first sociobiologists:

This great attribute, which signifies unbounded wisdom, induces us to admire those laws by which providence rules the insect societies and reserves to herself their exclusive direction; and it shows us that in delivering man to his own guidance, she has subjected him to a great and heavy responsibility. If natural history only serves to prove this truth, it will have attained the most dignified end of which science may boast—that of endeavoring to ameliorate the human species by the examples it lays before us.

References

Altmann, S. A., and Jeanne Altmann. 1970. *Baboon Ecology: African Field Research.* University of Chicago Press. 220 pp.

Baker, H. G., and G. L. Stebbins, ed. 1965. *The Genetics of Colonizing Species.* Academic Press. xv + 588 pp.

Brown, J. L., and G. H. Orians. 1970. Spacing patterns in mobile animals. *Ann. Rev. Ecol. Syst.* 1:239–62.

Crook, J. H. 1970. The socio-ecology of primates. In J. H. Crook, ed., *Social Behaviour in Birds and Mammals.* Academic Press. xl + 492 pp. (pp. 103–66)

Hall, K. R. L. 1965. Social organization of the Old World monkeys and apes. *Symp. Zool. Soc. London* 14:265–89.

Hamilton, W. D. 1964. The genetical evolution of social behaviour. I, II. *J. Theoret. Biol.* 7:1–52.

Huber, P. 1810. *Recherches sur les moeurs des fourmis indigènes.* Paris: J. J. Paschoud. xvi + 328 pp.

Lewontin, R. C., ed. 1968. *Population Biology and Evolution.* Syracuse University Press. vii + 205 pp.

Levins, R. 1968. *Evolution in Changing Environments: Some Theoretical Explorations.* Princeton University Press. ix + 120 pp.

MacArthur, R. H., and E. O. Wilson. 1967. *The Theory of Island Biogeography.* Princeton University Press. xi + 203 pp.

Parsons, P. A. 1967. *The Genetic Analysis of Behaviour.* London: Methuen. ix + 174 pp.

Waddington, C. H., ed. 1968. *Towards a Theoretical Biology,* Vol. 1, *Prolegomena.* Aldine. ii + 234 pp.

Waddington, C. H., ed. 1969. *Towards a Theoretical Biology,* Vol. 2, *Sketches.* Aldine. ii + 351 pp.

Williams, G. C. 1966. *Adaptation and Natural Selection.* Princeton University Press. x + 307 pp.

Wilson, E. O. 1971. *The Insect Societies.* Belknap Press of Harvard University Press. x + 548 pp.

1975

"Slavery in ants," *Scientific American* 232(6): 32–36

Ants, which have occupied the land for over 100 million years (reaching to the roots of *Sphecomyrma* and other ancestral forms) and ruled as dominant organisms worldwide for the last half of that time, have evolved an astonishing number of extreme social adaptations. One is slavery, in which young individuals, usually pupae, are captured during organized raids on other colonies, then allowed by their captors to mature into the worker caste. The slave-maker workers on the other hand are specialized to be raiders more or less exclusively: they contribute little or nothing to nest construction, brood care, and other quotidian tasks in their own nests.

A few slave-making species raid nests of their own species, while retaining the ability to live independently if, for some reason, they have no slaves. Most, however, raid other species, and therefore the "slaves" can be more accurately thought of as livestock, albeit much more intelligent than their single-minded parasite captors.

In the course of my studies on chemical communication in ants, I learned how to extract and apply the trail and recruitment pheromones of slave-maker species. I was then able to draw the raiders out of their nest and lead them to other nests of my choosing. Alternatively, I could lead them to nowhere in particular, whereupon they milled around in confusion at the end of the artificial trail.

My discovery of the existence of "propaganda substances" employed by one of the ant species *(Formica subintegra),* revealed how raiders can scatter the adults of the invaded colony with a maximum of chemical noise and a minimum of combat. Finally, in this quite entertaining research program, I examined one of the slave members *(Leptothorax duloticus)* with the least evolved raiding behavior. When I took away the slaves of a colony to see if the raiders were still generalists enough to fill in for the absent domestic force, they proved able to assume this role, although with limited competence.

❋ ARTICLE 12

SLAVERY IN ANTS

Certain species of ants raid the nests of other species for ants to work in their own nest. Some raiding species have become so specialized that they are no longer capable of feeding themselves

by Edward O. Wilson

The institution of slavery is not unique to human societies. No fewer than 35 species of ants, constituting six independently evolved groups, depend at least to some extent on slave labor for their existence. The techniques by which they raid other ant colonies to strengthen their labor force rank among the most sophisticated behavior patterns found anywhere in the insect world. Most of the slave-making ant species are so specialized as raiders that they starve to death if they are deprived of their slaves. Together they display an evolutionary descent that begins with casual raiding by otherwise free-living colonies, passes through the development of full-blown warrior societies and ends with a degeneration so advanced that the workers can no longer even conduct raids.

Slavery in ants differs from slavery in human societies in one key respect: the ant slaves are always members of other completely free-living species that themselves do not take slaves. In this regard the ant slaves perhaps more closely resemble domestic animals—except that the slaves are not allowed to reproduce and they are equal or superior to their captors in social organization.

The famous Amazon ants of the genus *Polyergus* are excellent examples of advanced slave makers. The workers are strongly specialized for fighting. Their mandibles, which are shaped like miniature sabers, are ideally suited for puncturing the bodies of other ants but are poorly suited for any of the routine tasks that occupy ordinary ant workers. Indeed, when *Polyergus* ants are in their home nest their only activities are begging food from their slaves and cleaning themselves ("burnishing their ruddy armor," as the entomologist William Morton Wheeler once put it).

When *Polyergus* ants launch a raid, however, they are completely transformed. They swarm out of the nest in a solid phalanx and march swiftly and directly to a nest of the slave species. They destroy the resisting defenders by puncturing their bodies and then seize and carry off the cocoons containing the pupae of worker ants.

When the captured pupae hatch, the workers that emerge accept their captors as sisters; they make no distinction between their genetic siblings and the *Polyergus* ants. The workers launch into the round of tasks for which they have been genetically programmed, with the slave makers being the incidental beneficiaries. Since the slaves are members of the worker caste, they cannot reproduce. In order to maintain an adequate labor force, the slave-making ants must periodically conduct additional raids.

It is a remarkable fact that ants of slave-making species are found only in cold climates. Although the vast majority of ants live in the Tropics and the

RAID BY SLAVE-MAKING AMAZON ANTS of the species *Polyergus rufescens* (*light color*) against a colony of the slave species *Formica fusca* (*dark color*) is depicted. The *fusca* ants make their nest in dry soil under a stone. The raiding Amazon ants kill resisting

warm Temperate zones, not a single species of those regions has been implicated in any activity remotely approaching slavery. Among the ants of the colder regions this form of parasitism is surprisingly common. The colonies of many slave-making species abound in the forests of the northern U.S., and ant-slave raids can be observed in such unlikely places as the campus of Harvard University.

The slave raiders obey what is often called Emery's rule. In 1909 Carlo Emery, an Italian myrmecologist, noted that each species of parasitic ant is genetically relatively close to the species it victimizes. This relation can be profitably explored for the clues it provides to the origin of slave making in the evolution of ants. Charles Darwin, who was fascinated by ant slavery, suggested that the first step was simple predation: the ancestral species began by raiding other kinds of ants for food, carrying away their immature forms in order to be able to devour them in the home nest. If a few pupae could escape that fate long enough to emerge as workers, they might be accepted as nestmates and thus join the labor force. In cases where the captives subsequently proved to be more valuable as workers than as food, the raiding species would tend to evolve into a slave maker.

Although Darwin's hypothesis is attractive, I recently obtained evidence that territorial defense rather than food is the evolutionary prime mover. I brought together in the Harvard Museum of Comparative Zoology different species of *Leptothorax* ants that normally do not depend on slave labor. When colonies were placed closer together than they are found in nature, the larger colonies attacked the smaller ones and drove away or killed the queens and workers. The attackers carried captured pupae back to their own nests. The pupae were then allowed by their captors to develop into workers. In the cases where the newly emerged workers belonged to the same species, they were allowed to remain as active members of the colony. When they belonged to a different *Leptothorax* species, however, they were executed in a matter of hours. One can easily imagine the origin of slave making by the simple extension of this territorial behavior to include tolerance of the workers of related species. The more closely related the raiders and their captives are, the more likely they are to be compatible. The result would be in agreement with Emery's rule.

One species that appears to have just crossed the threshold to slave making is *Leptothorax duloticus*, a rare ant that so far has been found only in certain localities in Ohio, Michigan and Ontario. The anatomy of the worker caste is only slightly modified for slave-making behavior, suggesting that in evolutionary terms the species may have taken up its parasitic way of life rather recently.

In experiments with laboratory colonies I was able to measure the degree of behavioral degeneration that has taken place in *L. duloticus*. Like the Amazon ants, the *duloticus* workers are highly efficient at raiding and fighting. When colonies of other *Leptothorax* species were placed near a *duloticus* nest, the workers launched intense attacks until all the pupae of the other species had been captured.

In the home nest the *duloticus* workers were inactive, leaving almost all the ordinary work to their captives. When the slaves were temporarily taken away from them, the workers displayed a dramatic expansion in activity, rapidly taking over most of the tasks formerly carried out by the slaves. The *duloticus* workers thus retain a latent capacity for working, a capacity that is totally lack-

***fusca* workers by piercing them with their saberlike mandibles. Most of the Amazon ants are transporting cocoons containing the pupae of *fusca* workers back to their own nest. When the workers emerge from the cocoons, they serve as slaves. Two dead *fusca* workers that resisted lie on the ground. Two other workers have retreated to upper surface of the rock over the nest's entrance.**

33

ing in more advanced species of slave-making ants.

The *duloticus* workers that had lost their slaves did not, however, perform their tasks well. Their larvae were fed at infrequent intervals and were not groomed properly, nest materials were carried about aimlessly and were never placed in the correct positions, and an inordinate amount of time was spent collecting and sharing diluted honey. More important, the slaveless ants lacked one behavior pattern that is essential for the survival of the colony: foraging for dead insects and other solid food. They even ignored food placed in their path. When the colony began to display signs of starvation and deterioration, I returned to them some slaves of the species *Leptothorax curvispinosus*. The bustling slave workers soon put the nest back in good order, and the slave makers just as quickly lapsed into their usual indolent ways.

Not all slave-making ants depend on brute force to overpower their victims. Quite by accident Fred E. Regnier of Purdue University and I discovered that some species have a subtler strategy. While surveying chemical substances used by ants to communicate alarm and to defend their nest, we encountered two slave-making species whose substances differ drastically from those of all other ants examined so far. These ants, *Formica subintegra* and *Formica pergandei*, produce remarkably large quantities of decyl, dodecyl and tetradecyl acetates. Further investigation of *F. subintegra*

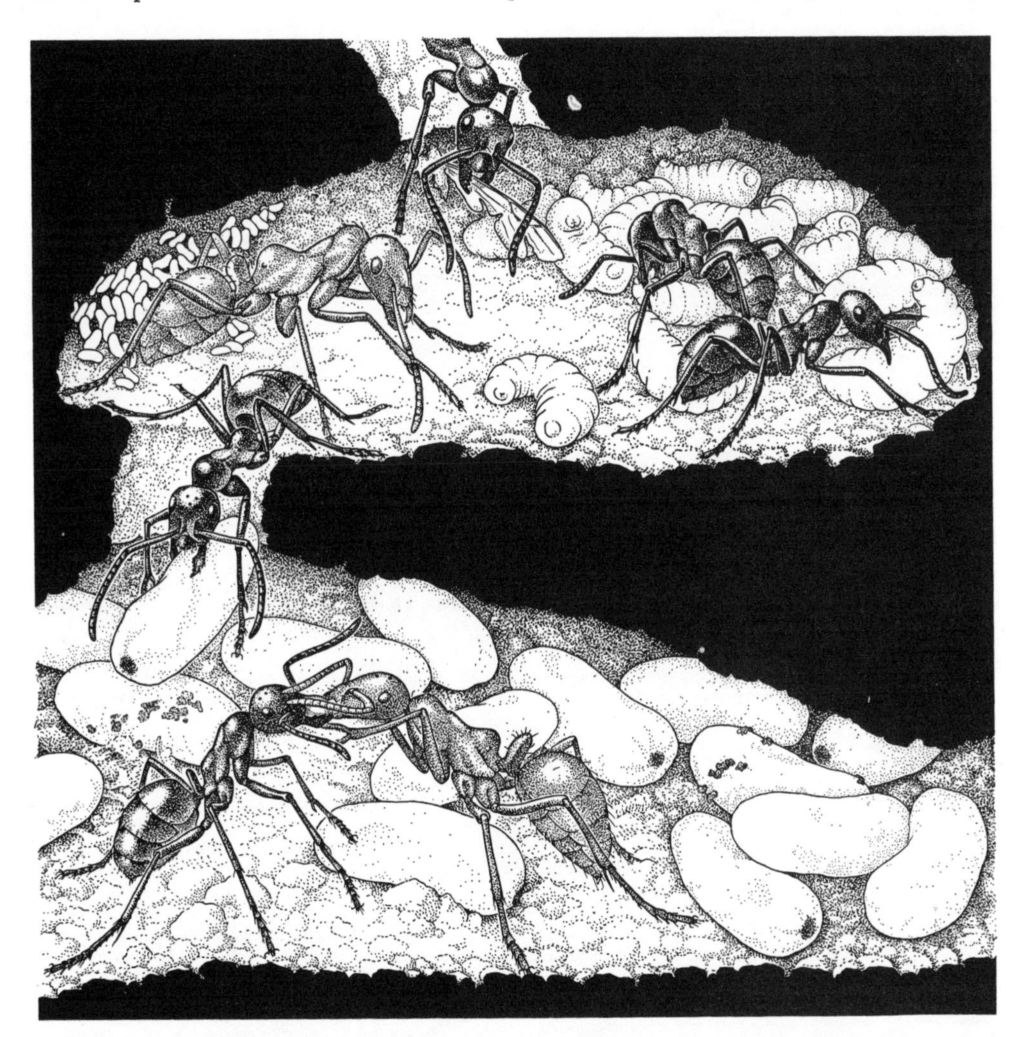

INTERIOR VIEW OF THE HOME NEST of a colony of Amazon ants shows *Formica fusca* slaves (*dark color*) performing all the housekeeping labor. At top center one of the slaves brings a fly wing into the nest for food. Other slave workers care for the small eggs, grublike larvae and cocoon-enclosed pupae of their captors. During the raiding season some of the pupae are likely to be those of *fusca* workers. The slave makers (*light color*) can do nothing more than groom themselves (*upper left*). In order to eat, the Amazon ants must beg slave workers to regurgitate liquid droplets for them (*lower left*). These ant species are found in Europe.

revealed that the substances are sprayed at resisting ants during slave-making raids. The acetates attract more invading slave makers, thereby serving to assemble these ants in places where fighting breaks out. Simultaneously the sprayed acetates throw the resisting ants into a panic. Indeed, the acetates are exceptionally powerful and persistent alarm substances. They imitate the compound undecane and other scents found in slave species of *Formica,* which release these substances in order to alert their nestmates to danger. The acetates broadcast by the slave makers are so much stronger, however, that they have a long-lasting disruptive effect. For this reason Regnier and I named them "propaganda substances."

We believe we have explained an odd fact first noted by Pierre Huber 165 years ago in his pioneering study of the European slave-making ant *Formica sanguinea.* He found that when a colony was attacked by these slave makers, the survivors of the attacked colony were reluctant to stay in the same neighborhood even when suitable alternative nest sites were scarce. Huber observed that the "ants never return to their besieged capital, even when the oppressors have retired to their own garrison; perhaps they realize that they could never remain there in safety, being continually liable to the attacks of their unwelcome visitors."

Regnier and I were further able to gain a strong clue to the initial organization of slave-making raids. We had made a guess, based on knowledge of the foraging techniques of other kinds of ants, that scout workers direct their nestmates to newly discovered slave colonies by means of odor trails laid from the target back to the home nest. In order to test this hypothesis we made extracts of the bodies of *F. subintegra* and of *Formica rubicunda,* a second species that conducts frequent, well-organized raids through much of the summer. Then at the time of day when raids are normally made we laid artificial odor trails, using a narrow paintbrush dipped in the extracts we had obtained from the ants. The trails were traced from the entrances of the nest to arbitrarily selected points one or two meters away.

The results were dramatic. Many of the slave-making workers rushed forth, ran the length of the trails and then milled around in confusion at the end. When we placed portions of colonies of the slave species *Formica subsericea* at the end of some of the trails, the slave makers proceeded to conduct the raid in a manner that was apparently the same

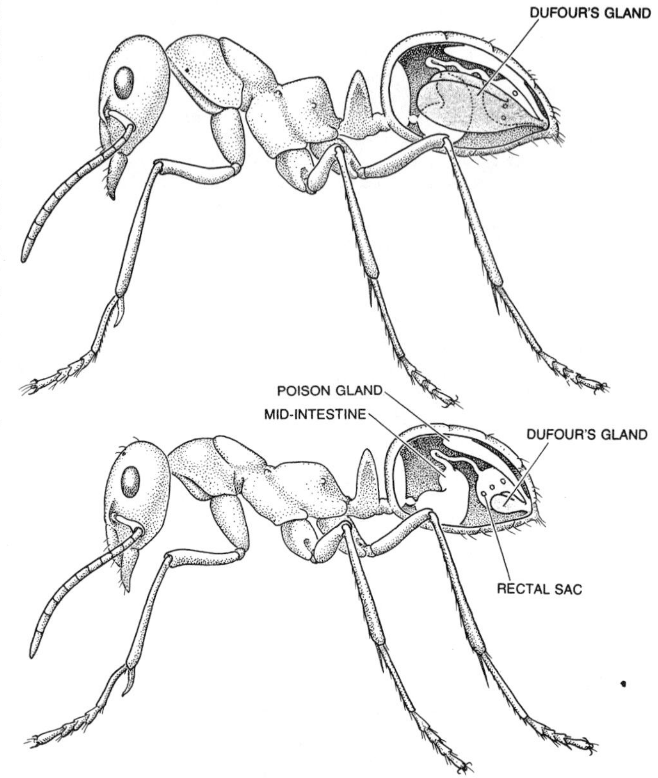

DUFOUR'S GLAND, which produces substances that serve as communication scents among ants, is much larger in the slave-making species *Formica subintegra* (*top*) than in the slave species *F. subsericea* (*bottom*). The *subsericea* ant releases its scent to alert its nestmates to the presence of danger. The *subintegra* sprays its secretions at resisting ants during slave raids. The secretions are so strong that they create panic in the colony being attacked.

in every respect as the raids initiated by trails laid by their own scouts. Studies of the slave-making species *Polyergus lucidus* and *Harpagoxenus americanus* by Mary Talbot and her colleagues at Lindenwood College provide independent evidence that raids are organized by the laying of odor trails to target nests; indeed, this form of communication may be widespread among slave-making ants.

The evolution of social parasitism in ants works like a ratchet, allowing a species to slip further down in parasitic dependence but not back up toward its original free-living existence. An example of nearly complete behavioral degeneration is found in one species of the genus *Strongylognathus,* which is found in Asia and Europe. Most species in this genus conduct aggressive slave-making raids. They have characteristic saber-shaped mandibles for killing other ants. The species *Strongylognathus testaceus,* however, has lost its warrior habits. Although these ants still have the distinctive mandibles of their genus, they do not conduct slave-making raids. Instead an *S. testaceus* queen moves into the nest of a slave-ant species and lives alongside the queen of the slave species. Each queen lays eggs that develop into workers, but the *S. testaceus* offspring do no work. They are fed by workers of the slave species. We do not know how the union of the two queens is formed in the first place, but it is likely that the parasitic queen simply induces the host colony to adopt her after her solitary dispersal flight from the nest of her birth.

Thus *S. testaceus* is no longer a real slave maker. It has become an advanced social parasite of a kind that commonly infests other ant groups. For example,

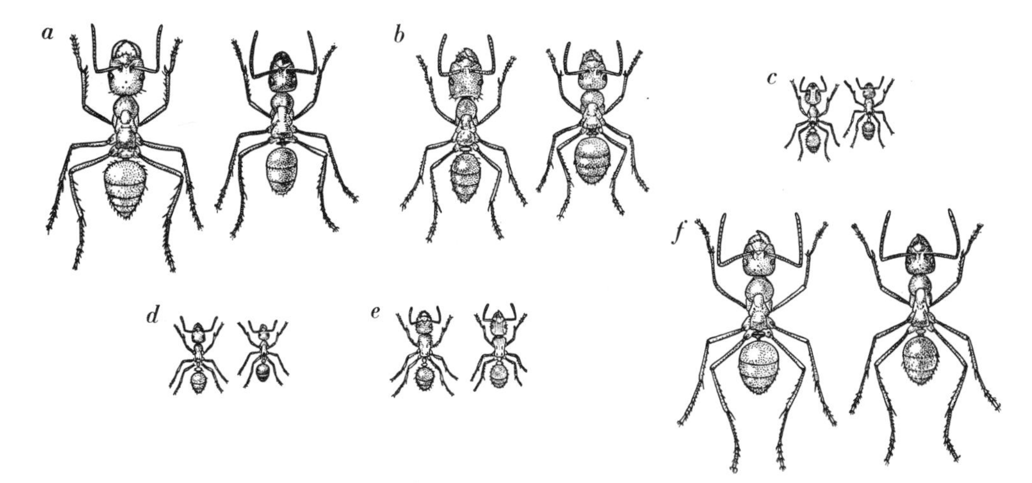

RESEMBLANCE of slave maker and slave was noted by an Italian myrmecologist, Carlo Emery, in 1909. In each pair of ants shown here the slave maker is on the left and the slave on the right. The species depicted are (*a*) *Polyergus rufescens* and *Formica fusca*, (*b*) *Rossomyrmex proformicarum* and *Proformica nasutum*, (*c*) *Harpagoxenus americanus* and *Leptothorax curvispinosus*, (*d*) *L. duloticus* and *L. curvispinosus*, (*e*) *Strongylognathus alpinus* and *Tetramorium caespitum* and (*f*) *F. subintegra* and *F. subsericea*.

many species of ant play host to parasites such as beetles, wasps and flies, feeding them and sheltering them [see "Communication between Ants and Their Guests," by Bert Hölldobler; SCIENTIFIC AMERICAN, March, 1971].

Does ant slavery hold any lesson for our own species? Probably not. Human slavery is an unstable social institution that runs strongly counter to the moral systems of the great majority of human societies. Ant slavery is a genetic adaptation found in particular species that cannot be judged to be more or less successful than their non-slave-making counterparts. The slave-making ants offer a clear and interesting case of behavioral evolution, but the analogies with human behavior are much too remote to allow us to find in them any moral or political lesson.

COLONY OF ANTS housed in a glass tube consists of the rare species *Leptothorax duloticus* and a slave species, *L. curvispinosus*. The *duloticus* ant, found in Ohio, Michigan and Ontario, has only recently become a slave maker. One of the *duloticus* workers can be seen in the center of the photograph; below it are three slave workers. The white objects are immature forms of both species. When the slave workers are removed, the *duloticus* workers attempt to carry out necessary housekeeping tasks but do so poorly.

1975

Sociobiology: the new synthesis (Belknap Press of Harvard University Press, Cambridge, MA), excerpts from "Man: from sociobiology to sociology," Chap. 27, pp. 547–575

These selections capture much of the essence of the final chapter of *Sociobiology: The New Synthesis,* which brought human behavior into the sociobiological framework. In this section of a book otherwise devoted to animals, I suggested that the social sciences would in time become part of an extended biology. That prophecy ignited a major controversy over human nature, which has since been resolved in favor of the arguments presented here. Human sociobiology has been largely taken over, as it should be, by psychologists, anthropologists, and others whose scholarly interests are centered in the social sciences. Since the 1990s the subject has been treated as a separate discipline called evolutionary psychology, but its links to general sociobiology remain strong.

Chapter 27 Man: From Sociobiology to Sociology

Let us now consider man in the free spirit of natural history, as though we were zoologists from another planet completing a catalog of social species on Earth. In this macroscopic view the humanities and social sciences shrink to specialized branches of biology; history, biography, and fiction are the research protocols of human ethology; and anthropology and sociology together constitute the sociobiology of a single primate species.

Homo sapiens is ecologically a very peculiar species. It occupies the widest geographical range and maintains the highest local densities of any of the primates. An astute ecologist from another planet would not be surprised to find that only one species of *Homo* exists. Modern man has preempted all the conceivable hominid niches. Two or more species of hominids did coexist in the past, when the *Australopithecus* man-apes and possibly an early *Homo* lived in Africa. But only one evolving line survived into late Pleistocene times to participate in the emergence of the most advanced human social traits.

Modern man is anatomically unique. His erect posture and wholly bipedal locomotion are not even approached in other primates that occasionally walk on their hind legs, including the gorilla and chimpanzee. The skeleton has been profoundly modified to accommodate the change: the spine is curved to distribute the weight of the trunk more evenly down its length; the chest is flattened to move the center of gravity back toward the spine; the pelvis is broadened to serve as an attachment for the powerful striding muscles of the upper legs and reshaped into a basin to hold the viscera; the tail is eliminated, its vertebrae (now called the coccyx) curved inward to form part of the floor of the pelvic basin; the occipital condyles have rotated far beneath the skull so that the weight of the head is balanced on them; the face is shortened to assist this shift in gravity; the thumb is enlarged to give power to the hand; the leg is lengthened; and the foot is drastically narrowed and lengthened to facilitate striding. Other changes have taken place. Hair has been lost from most of the body. It is still not known why modern man is a "naked ape." One plausible explanation is that nakedness served as a device to cool the body during the strenuous pursuit of prey in the heat of the African plains. It is associated with man's exceptional reliance on sweating to reduce body heat; the human body contains from two to five million sweat glands, far more than in any other primate species.

The reproductive physiology and behavior of *Homo sapiens* have also undergone extraordinary evolution. In particular, the estrous cycle of the female has changed in two ways that affect sexual and social behavior. Menstruation has been intensified. The females of some other primate species experience slight bleeding, but only in women is there a heavy sloughing of the wall of the "disappointed womb" with consequent heavy bleeding. The estrus, or period of female "heat," has been replaced by virtually continuous sexual activity. Copulation is initiated not by response to the conventional primate signals of estrus, such as changes in color of the skin around the female sexual organs and the release of pheromones, but by extended foreplay entailing mutual stimulation by the partners. The traits of physical attraction are, moreover, fixed in nature. They include the pubic hair of both sexes and the protuberant breasts and buttocks of women. The flattened sexual cycle and continuous female attractiveness cement the close marriage bonds that are basic to human social life.

At a distance a perceptive Martian zoologist would regard the globular head as a most significant clue to human biology. The cerebrum of *Homo* was expanded enormously during a relatively short span of evolutionary time (see Figure 27-1). Three million years ago *Australopithecus* had an adult cranial capacity of 400-500 cubic centimeters, comparable to that of the chimpanzee and gorilla. Two million years later its presumptive descendant *Homo erectus* had a capacity of about 1000 cubic centimeters. The next million years saw an increase to 1400-1700 cubic centimeters in Neanderthal man and 900-2000 cubic centimeters in modern *Homo sapiens*. The growth in intelligence that accompanied this enlargement was so great that it cannot yet be measured in any meaningful way. Human beings can be compared among themselves in terms of a few of the basic components of intelligence and creativity. But no scale has been invented that can objectively compare man with chimpanzees and other living primates.

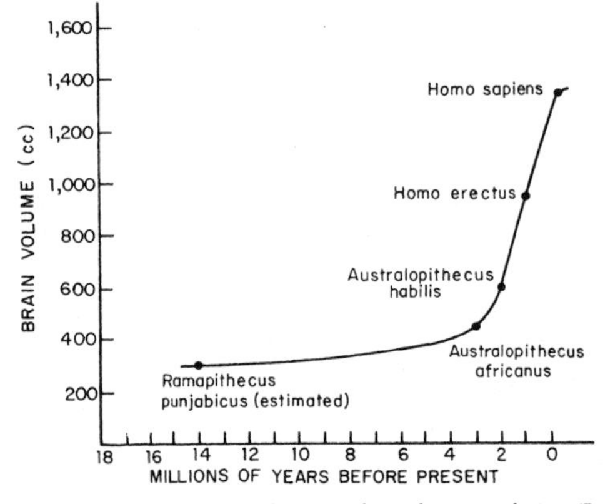

Figure 27-1 The increase in brain size during human evolution. (Redrawn from Pilbeam, 1972.)

We have leaped forward in mental evolution in a way that continues to defy self-analysis. The mental hypertrophy has distorted even the most basic primate social qualities into nearly unrecognizable forms. Individual species of Old World monkeys and apes have notably plastic social organizations; man has extended the trend into a protean ethnicity. Monkeys and apes utilize behavioral scaling to adjust aggressive and sexual interactions; in man the scales have become multidimensional, culturally adjustable, and almost endlessly subtle. Bonding and the practices of reciprocal altruism are rudimentary in other primates; man has expanded them into great networks where individuals consciously alter roles from hour to hour as if changing masks.

It is the task of comparative sociobiology to trace these and other human qualities as closely as possible back through time. Besides adding perspective and perhaps offering some sense of philosophical ease, the exercise will help to identify the behaviors and rules by which individual human beings increase their Darwinian fitness through the manipulation of society. In a phrase, we are searching for the human biogram (Count, 1958; Tiger and Fox, 1971). One of the key questions, never far from the thinking of anthropologists and biologists who pursue real theory, is to what extent the biogram represents an adaptation to modern cultural life and to what extent it is a phylogenetic vestige. Our civilizations were jerrybuilt around the biogram. How have they been influenced by it? Conversely, how much flexibility is there in the biogram, and in which parameters particularly? Experience with other animals indicates that when organs are hypertrophied, phylogeny is hard to reconstruct. This is the crux of the problem of the evolutionary analysis of human behavior. In the remainder of the chapter, human qualities will be discussed insofar as they appear to be general traits of the species. Then current knowledge of the evolution of the biogram will be reviewed, and finally some implications for the planning of future societies will be considered.

Plasticity of Social Organization

The first and most easily verifiable diagnostic trait is statistical in nature. The parameters of social organization, including group size, properties of hierarchies, and rates of gene exchange, vary far more among human populations than among those of any other primate species. The variation exceeds even that occurring between the remaining primate species. Some increase in plasticity is to be expected. It represents the extrapolation of a trend toward variability already apparent in the baboons, chimpanzees, and other cercopithecoids. What is truly surprising, however, is the extreme to which it has been carried.

Why are human societies this flexible? Part of the reason is that the members themselves vary so much in behavior and achievement. Even in the simplest societies individuals differ greatly. Within a small tribe of !Kung Bushmen can be found individuals who are acknowledged as the "best people"—the leaders and outstanding specialists among the hunters and healers. Even with an emphasis on sharing goods, some are exceptionally able entrepreneurs and unostentatiously acquire a certain amount of wealth. !Kung men, no less than men in advanced industrial societies, generally establish themselves by their mid-thirties or else accept a lesser status for life. There are some who never try to make it, live in run-down huts, and show little pride in themselves or their work (Pfeiffer, 1969). The ability to slip into such roles, shaping one's personality to fit, may itself be adaptive. Human societies are organized by high intelligence, and each member is faced by a mixture of social challenges that taxes all of his ingenuity. This baseline variation is amplified at the group level by other qualities exceptionally pronounced in human societies: the long, close period of socialization; the loose connectedness of the communication networks; the multiplicity of bonds; the capacity, especially within literate cultures, to communicate over long distances and periods of history; and from all these traits, the capacity to dissemble, to manipulate, and to exploit. Each parameter can be altered easily, and each has a marked effect on the final social structure. The result could be the observed variation among societies.

The hypothesis to consider, then, is that genes promoting flexibility in social behavior are strongly selected at the individual level. But note that variation in social organization is only a possible, not a necessary consequence of this process. In order to generate the amount of variation actually observed to occur, it is necessary for there to be multiple adaptive peaks. In other words, different forms of society within the same species must be nearly enough alike in survival ability for many to enjoy long tenure. The result would be a statistical ensemble of kinds of societies which, if not equilibrial, is at least not shifting rapidly toward one particular mode or another. The alternative, found in some social insects, is flexibility in individual behavior and caste development, which nevertheless results in an approach toward uniformity in the statistical distribution of the kinds of individuals when all individuals within a colony are taken together. In honeybees and in ants of the genera *Formica* and *Pogonomyrmex*, "personality" differences are strongly marked even within single castes. Some individuals, referred to by entomologists as the elites, are unusually active, perform more than their share of lifetime work, and incite others to work through facilitation. Other colony members are consistently sluggish. Although they are seemingly healthy and live long lives, their per-individual output is only a small fraction of that of the elites. Specialization also occurs. Certain individuals remain with the brood as nurses far longer than the average, while others concentrate on nest building or foraging. Yet somehow the total pattern of behavior in the colony converges on the species average. When one colony with its hundreds or thousands of members is compared with another of the same species, the statistical patterns of activity are about the same. We know that some of this consistency is due to negative feedback. As one requirement such as brood care or nest repair intensifies, workers shift their activities to compensate until the need is met, then change back again. Experiments have shown that disruption of the feedback loops, and thence deviation by the colony from the statistical norms, can be disastrous. It is therefore not surprising to find that the loops are both precise and powerful (Wilson, 1971a).

The controls governing human societies are not nearly so strong, and the effects of deviation are not so dangerous. The anthropological literature abounds with examples of societies that contain obvious inefficiencies and even pathological flaws—yet endure. The slave society of Jamaica, compellingly described by Orlando Patterson (1967), was unquestionably pathological by the moral canons of civilized life. "What marks it out is the astonishing neglect and distortion of almost every one of the basic prerequisites of normal human living. This was a society in which clergymen were the 'most finished debauchees' in the land; in which the institution of marriage was officially condemned among both masters and slaves; in which the family was unthinkable to the vast majority of the population and promiscuity the norm; in which education was seen as an absolute waste of time and teachers shunned like the plague; in which the legal system was quite deliberately a travesty of anything that could be called justice; and in which all forms of refinements, of art, of folkways, were either absent or in a state of total disintegration. Only a small proportion of whites, who monopolized almost all of the fertile land in the island, benefited from the system. And these, no sooner had they secured their fortunes, abandoned the land which the production of their own wealth had made unbearable to live in, for the comforts of the mother country." Yet this Hobbesian world lasted for nearly two centuries. The people multiplied while the economy flourished.

The Ik of Uganda are an equally instructive case (Turnbull, 1972). They are former hunters who have made a disastrous shift to cultivation. Always on the brink of starvation, they have seen their culture reduced to a vestige. Their only stated value is *ngag*, or food; their basic notion of goodness (*marangik*) is the individual possession of food in the stomach; and their definition of a good man is *yakw ana marang*, "a man who has a full belly." Villages are still built, but the nuclear family has ceased to function as an institution. Children are kept with reluctance and from about three years of age are made to find their own way of life. Marriage ordinarily occurs only when there is a specific need for cooperation. Because of the lack of energy, sexual activity is minimal and its pleasures are considered to be about on the same level as those of defecation. Death is treated with relief or amusement, since it means more *ngag* for survivors. Because the unfortunate Ik are at the lowest sustainable level, there is a temptation to conclude that they are doomed. Yet somehow their society has remained intact and more or less stable for at least 30 years, and it could endure indefinitely.

How can such variation in social structure persist? The explanation may be lack of competition from other species, resulting in what biologists call ecological release. During the past ten thousand years or longer, man as a whole has been so successful in dominating his environment that almost any kind of culture can succeed for a while, so long as it has a modest degree of internal consistency and does not shut off reproduction altogether. No species of ant or termite enjoys this freedom. The slightest inefficiency in constructing nests, in establishing odor trails, or in conducting nuptial flights could result in the quick extinction of the species by predation and competition from other social insects. To a scarcely lesser extent the same is true for social carnivores and primates. In short, animal species tend to be tightly packed in the ecosystem with little room for experimentation or play. Man has temporarily escaped the constraint of interspecific competition. Although cultures replace one another, the process is much less effective than interspecific competition in reducing variance.

It is part of the conventional wisdom that virtually all cultural variation is phenotypic rather than genetic in origin. This view has gained support from the ease with which certain aspects of culture can be altered in the space of a single generation, too quickly to be evolutionary in nature. The drastic alteration in Irish society in the first two years of the potato blight (1846-1848) is a case in point. Another is the shift in the Japanese authority structure during the American occupation following World War II. Such examples can be multiplied endlessly—they are the substance of history. It is also true that human populations are not very different from one another genetically. When Lewontin (1972b) analyzed existing data on nine blood-type systems, he found that 85 percent of the variance was composed of diversity within populations and only 15 percent was due to diversity between populations. There is no a priori reason for supposing that this sample of genes possesses a distribution much different from those of other, less accessible systems affecting behavior.

The extreme orthodox view of environmentalism goes further, holding that in effect there is no genetic variance in the transmission of culture. In other words, the capacity for culture is transmitted by a single human genotype. Dobzhansky (1963) stated this hypothesis as follows: "Culture is not inherited through genes, it is acquired by learning from other human beings . . . In a sense, human genes have surrendered their primacy in human evolution to an entirely new, nonbiological or superorganic agent, culture. However, it should not be forgotten that this agent is entirely dependent on the human genotype." Although the genes have given away most of their sovereignty, they maintain a certain amount of influence in at least the behavioral qualities that underlie variations between cultures. Moderately high heritability has been documented in introversion-extroversion measures, personal tempo, psychomotor and sports activities, neuroticism, dominance, depression, and the tendency toward certain forms of mental illness such as schizophrenia (Parsons, 1967; Lerner, 1968). Even a small portion of this variance invested in population differences might predispose societies toward cultural differences. At the very least, we should try to measure this amount. It is not valid to point to the absence of a behavioral trait in one or a few societies as conclusive evidence that the trait is environmentally induced and has no genetic disposition in man. The very opposite could be true.

In short, there is a need for a discipline of anthropological genetics. In the interval before we acquire it, it should be possible to characterize the human biogram by two indirect methods. First, models can be constructed from the most elementary rules of human behavior. Insofar as they can be tested, the rules will characterize the biogram in much the same way that ethograms drawn by zoologists identify the "typical" behavioral repertories of animal species. The rules can be legitimately compared with the ethograms of other primate species. Variation in the rules among human cultures, however slight, might provide clues to underlying genetic differences, particularly when it is correlated with variation in behavioral traits known to be heritable. Social scientists have in fact begun to take this first approach, although in a different context from the one suggested here. Abraham Maslow (1954, 1972) postulated that human beings respond to a hierarchy of needs, such that the lower levels must be satisfied before much attention is devoted to the higher ones. The most basic needs are hunger and sleep. When these are met, safety becomes the primary consideration, then the need to belong to a group and receive love, next self-esteem, and finally self-actualization and creativity. The ideal society in Maslow's dream is one which "fosters the fullest development of human potentials, of the fullest degree of humanness." When the biogram is freely expressed, its center of gravity should come to rest in the higher levels. A second social scientist, George C. Homans (1961), has adopted a Skinnerian approach in an attempt to reduce human behavior to the basic processes of associative learning. The rules he postulates are the following:

1. If in the past the occurrence of a particular stimulus-situation has been the occasion on which a man's activity has been rewarded, then the more similar the present stimulus-situation is to the past one, the more likely the man is at the present time to emit this activity or one similar to it.

2. The more often within a given period of time a man's activity rewards the behavior of another, the more often the other will perform the behavior.

3. The more valuable to a man a unit of the activity another gives him, the more often he behaves in the manner rewarded by the activity of the other.

4. The more often a man has in the recent past received a rewarding activity from another, the less valuable any further unit of that activity becomes to him.

Maslow the ethologist and visionary seems a world apart from Homans the behaviorist and reductionist. Yet their approaches are reconcilable. Homans' rules can be viewed as comprising some of the enabling devices by which the human biogram is expressed. His operational word is *reward*, which is in fact the set of all interactions defined by the emotive centers of the brain as desirable. According to evolutionary theory, desirability is measured in units of genetic fitness, and the emotive centers have been programmed accordingly. Maslow's hierarchy is simply the order of priority in the goals toward which the rules are directed.

The other indirect approach to anthropological genetics is through phylogenetic analysis. By comparing man with other primate species, it might be possible to identify basic primate traits that lie beneath the surface and help to determine the configuration of man's higher social behavior. This approach has been taken with great style and vigor in a series of popular books by Konrad Lorenz (*On Aggression*), Robert Ardrey (*The Social Contract*), Desmond Morris (*The Naked Ape*), and Lionel Tiger and Robin Fox (*The Imperial Animal*). Their efforts were salutary in calling attention to man's status as a biological species adapted to particular environments. The wide attention they received broke the stifling grip of the extreme behaviorists, whose view of the mind of man as a virtually equipotent response machine was neither correct nor heuristic. But their particular handling of the problem tended to be inefficient and misleading. They selected one plausible hypothesis or another based on a review of a small sample of animal species, then advocated the explanation to the limit. The weakness of this method was discussed earlier in a more general context (Chapter 2) and does not need repetition here.

The correct approach using comparative ethology is to base a rigorous phylogeny of closely related species on many biological traits. Then social behavior is treated as the dependent variable and its evolution deduced from it. When this cannot be done with confidence (and it cannot in man) the next best procedure is the one outlined in Chapter 7: establish the lowest taxonomic level at which each character shows significant intertaxon variation. Characters that shift from species to species or genus to genus are the most labile. We cannot safely extrapolate them from the cercopithecoid monkeys and apes to man. In the primates these labile qualities include group size, group cohesiveness, openness of the group to others, involvement of the male in parental care, attention structure, and the intensity and form of territorial defense. Characters are considered conservative if they remain constant at the level of the taxonomic family or throughout the order Primates, and they are the ones most likely to have persisted in relatively unaltered form into the evolution of *Homo*. These conservative traits include aggressive dominance systems, with males generally dominant over females; scaling in the intensity of responses, especially during aggressive interactions; intensive and prolonged maternal care, with a pronounced degree of socialization in the young; and matrilineal social organization. This classification of behavioral traits offers an appropriate basis for hypothesis formation. It allows a qualitative assessment of the probabilities that various behavioral traits have persisted into modern *Homo sapiens*. The possibility of course remains that some labile traits are homologous between man and, say, the chimpanzee. And conversely, some traits conservative throughout the rest of the primates might nevertheless have changed during the origin of man. Furthermore, the assessment is not meant to imply that conservative traits are more genetic—that is, have higher heritability—than labile ones. Lability can be based wholly on genetic differences between species or populations within species. Returning finally to the matter of cultural evolution, we can heuristically conjecture that the traits proven to be labile are also the ones most likely to differ from one human society to another on the basis of genetic differences. The evidence, reviewed in Table 27-1, is not inconsistent with this basic conception. Finally, it is worth special note that the comparative ethological approach does not in any way predict man's unique traits. It is a general rule of evolutionary studies that the direction of quantum jumps is not easily read by phylogenetic extrapolation.

Barter and Reciprocal Altruism

Sharing is rare among the nonhuman primates. It occurs in rudimentary form only in the chimpanzee and perhaps a few other Old World monkeys and apes. But in man it is one of the strongest social traits, reaching levels that match the intense trophallactic exchanges of termites and ants. As a result only man has an economy. His high intelligence and symbolizing ability make true barter possible. Intelligence also permits the exchanges to be stretched out in time, converting them into acts of reciprocal altruism (Trivers, 1971). The conventions of this mode of behavior are expressed in the familiar utterances of everyday life:

"Give me some now; I'll repay you later."

Table 27-1 General social traits in human beings, classified according to whether they are unique, belong to a class of behaviors that are variable at the level of the species or genus in the remainder of the primates (labile), or belong to a class of behaviors that are uniform through the remainder of the primates (conservative).

Evolutionarily labile primate traits	Evolutionarily conservative primate traits	Human traits
		SHARED WITH SOME OTHER PRIMATES
Group size		Highly variable
Group cohesiveness		Highly variable
Openness of group to others		Highly variable
Involvement of male in parental care		Strong
Attention structure		Centripetal on leading males
Intensity and form of territorial defense		Highly variable, but territoriality is general
		SHARED WITH ALL OR ALMOST ALL OTHER PRIMATES
	Aggressive dominance systems, with males dominant over females	Consistent with other primates, although variable
	Scaling of responses, especially in aggressive interactions	Consistent with other primates
	Prolonged maternal care; pronounced socialization of young	Consistent with other primates
	Matrilineal organization	Mostly consistent with other primates
		UNIQUE
		True language, elaborate culture
		Sexual activity continuous through menstrual cycle
		Formalized incest taboos and marriage exchange rules with recognition of kinship networks
		Cooperative division of labor between adult males and females

"Come to my aid this time, and I'll be your friend when you need me."

"I really didn't think of the rescue as heroism; it was only what I would expect others to do for me or my family in the same situation."

Money, as Talcott Parsons has been fond of pointing out, has no value in itself. It consists only of bits of metal and scraps of paper by which men pledge to surrender varying amounts of property and services upon demand; in other words it is a quantification of reciprocal altruism.

Perhaps the earliest form of barter in early human societies was the exchange of meat captured by the males for plant food gathered by the females. If living hunter-gatherer societies reflect the primitive state, this exchange formed an important element in a distinctive kind of sexual bond.

Fox (1972), following Lévi-Strauss (1949), has argued from ethnographic evidence that a key early step in human social evolution was the use of women in barter. As males acquired status through the control of females, they used them as objects of exchange to cement alliances and bolster kinship networks. Preliterate societies are characterized by complex rules of marriage that can often be interpreted directly as power brokerage. This is particularly the case where the elementary negative marriage rules, proscribing certain types of unions, are supplemented by positive rules that direct which exchanges must be made. Within individual Australian aboriginal societies two moieties exist between which marriages are permitted. The men of each moiety trade nieces, or more specifically their sisters' daughters. Power accumulates with age, because a man can control the descendants of nieces as remote as the daughter of his sister's daughter. Combined with polygyny, the system insures both political and genetic advantage to the old men of the tribe.

For all its intricacy, the formalization of marital exchanges between tribes has the same approximate genetic effect as the haphazard wandering of male monkeys from one troop to another or the exchange of young mature females between chimpanzee populations. Approximately 7.5 percent of marriages contracted among Australian aborigines prior to European influence were intertribal, and similar rates have been reported in Brazilian Indians and other preliterate societies (Morton, 1969). It will be recalled (Chapter 4) that gene flow of the order of 10 percent per generation is more than enough to counteract fairly intensive natural pressures that tend to differentiate populations. Thus intertribal marital exchanges are a major factor in creating the observed high degree of genetic similarity among populations. The ultimate adaptive basis of exogamy is not gene flow per se but rather the avoidance of inbreeding. Again, a 10 percent gene flow is adequate for the purpose.

The microstructure of human social organization is based on sophisticated mutual assessments that lead to the making of contracts. As Erving Goffman correctly perceived, a stranger is rapidly but politely explored to determine his socioeconomic status, intelligence and education, self-perception, social attitudes, competence, trustworthiness, and emotional stability. The information, much of it subconsciously given and absorbed, has an eminently practical value. The probe must be deep, for the individual tries to create the impression that will gain him the maximum advantage. At the very least he maneuvers to avoid revealing information that will imperil his status. The presentation of self can be expected to contain deceptive elements:

> Many crucial facts lie beyond the time and place of interaction or lie concealed within it. For example, the "true" or "real" attitudes, beliefs, and emotions of the individual can be ascertained only indirectly, through his avowals or through what appears to be involuntary expressive behavior. Similarly, if the individual offers the others a product or service, they will often find that during the interaction there will be no time or place immediately available for eating the pudding that the proof can be found in. They will be forced to accept some events as conventional or natural signs of something not directly available to the senses. (Goffman, 1959)

Deception and hypocrisy are neither absolute evils that virtuous men suppress to a minimum level nor residual animal traits waiting to be erased by further social evolution. They are very human devices for conducting the complex daily business of social life. The level in each particular society may represent a compromise that reflects the size and complexity of the society. If the level is too low, others will seize the advantage and win. If it is too high, ostracism is the result. Complete honesty on all sides is not the answer. The old primate frankness would destroy the delicate fabric of social life that has built up in human populations beyond the limits of the immediate clan. As Louis J. Halle correctly observed, good manners have become a substitute for love.

Bonding, Sex, and Division of Labor

The building block of nearly all human societies is the nuclear family (Reynolds, 1968; Leibowitz, 1968). The populace of an American industrial city, no less than a band of hunter-gatherers in the Australian desert, is organized around this unit. In both cases the family moves between regional communities, maintaining complex ties with primary kin by means of visits (or telephone calls and letters) and the exchange of gifts. During the day the women and children remain in the residential area while the men forage for game or its symbolic equivalent in the form of barter and money. The males cooperate in bands to hunt or deal with neighboring groups. If not actually blood relations, they tend at least to act as "bands of brothers." Sexual bonds are carefully contracted in observance with tribal customs and are intended to be permanent. Polygamy, either covert or explicitly sanctioned by custom, is practiced predominantly by the males. Sexual behavior is nearly continuous through the menstrual cycle and marked by extended foreplay. Morris (1967a), drawing on the data of Masters and Johnson (1966) and others, has enumerated the unique features of human sexuality that he considers to be associated with the loss of body hair: the rounded and protuberant breasts of the young woman, the flushing of areas of skin during coition, the vaso-dilation and increased erogenous sensitivity of the lips, soft portions of the nose, ear, nipples, areolae, and genitals, and the large size of the male penis, especially during erection. As Darwin himself noted in 1871, even the naked skin of the woman is used as a sexual releaser. All of these alterations serve to cement the permanent bonds, which are unrelated in time to the moment of ovulation. Estrus has been reduced to a vestige, to the consternation of those who attempt to practice birth control by the rhythm method. Sexual behavior has been largely dissociated from the act of fertilization. It is ironic that religionists who forbid sexual activity except for purposes of procreation should do so on the basis of "natural law." Theirs is a misguided effort in comparative ethology, based on the incorrect assumption that in reproduction man is essentially like other animals.

The extent and formalization of kinship prevailing in almost all human societies are also unique features of the biology of our species. Kinship systems provide at least three distinct advantages. First, they bind alliances between tribes and subtribal units and provide a conduit for the conflict-free emigration of young members. Second, they are an important part of the bartering system by which certain males achieve dominance and leadership. Finally, they serve as a homeostatic device for seeing groups through hard times. When food grows scarce, tribal units can call on their allies for altruistic assistance in a way unknown in other social primates. The Athapaskan Dogrib Indians, a hunter-gatherer people of the northwestern Canadian arctic, provide one example. The Athapaskans are organized loosely by the bilateral primary linkage principle (June Helm, 1968). Local bands wander through a common territory, making intermittent contacts and exchanging members by intermarriage. When famine strikes, the endangered bands can coalesce with those temporarily better off. A second example is the Yanomamö of South America, who rely on kin when their crops are destroyed by enemies (Chagnon, 1968).

As societies evolved from bands through tribes into chiefdoms and states, some of the modes of bonding were extended beyond kinship networks to include other kinds of alliances and economic agreements. Because the networks were then larger, the lines of communication longer, and the interactions more diverse, the total systems became vastly more complex. But the moralistic rules underlying these arrangements appear not to have been altered a great deal. The average individual still operates under a formalized code no more elaborate than that governing the members of hunter-gatherer societies.

Role Playing and Polyethism

The superman, like the super-ant or super-wolf, can never be an individual; it is the society, whose members diversify and cooperate to create a composite well beyond the capacity of any conceivable organism. Human societies have effloresced to levels of extreme complexity because their members have the intelligence and flexibility to play roles of virtually any degree of specification, and to switch them as the occasion demands. Modern man is an actor of many parts who may well be stretched to his limit by the constantly shifting demands of his environment. As Goffman (1961) observed, "Perhaps there are times when an individual does march up and down like a wooden soldier, tightly rolled up in a particular role. It is true that here and there we can pounce on a moment when an individual sits fully astride a single role, head erect, eyes front, but the next moment the picture is shattered into many pieces and the individual divides into different persons holding the ties of different spheres of life by his hands, by his teeth, and by his grimaces. When seen up close, the individual, bringing together in various ways all the connections he has in life, becomes a blur." Little wonder that the most acute inner problem of modern man is identity.

Roles in human societies are fundamentally different from the castes of social insects. The members of human societies sometimes cooperate closely in insectan fashion, but more frequently they compete for the limited resources allocated to their role-sector. The best and most entrepreneurial of the role-actors usually gain a disproportionate share of the rewards, while the least successful are displaced to other, less desirable positions. In addition, individuals attempt to move to higher socioeconomic positions by changing roles. Competition between classes also occurs, and in great moments of history it has proved to be a determinant of societal change.

A key question of human biology is whether there exists a genetic predisposition to enter certain classes and to play certain roles. Circumstances can be easily conceived in which such genetic differentiation might occur. The heritability of at least some parameters of intelligence and emotive traits is sufficient to respond to a moderate amount of disruptive selection. Dahlberg (1947) showed that if a single gene appears that is responsible for success and an upward shift in status, it can be rapidly concentrated in the uppermost socioeconomic classes. Suppose, for example, there are two classes, each beginning with only a 1 percent frequency of the homozygotes of the upward-mobile gene. Suppose further that 50 percent of the homozygotes in the lower class are transferred upward in each generation. Then in only ten generations, depending on the relative sizes of the groups, the upper class will be comprised of as many as 20 percent homozygotes or more and the lower class of as few as 0.5 percent or less. Using a similar argument, Herrnstein (1971b) proposed that as environmental opportunities become more nearly equal within societies, socioeconomic groups will be defined increasingly by genetically based differences in intelligence.

A strong initial bias toward such stratification is created when one human population conquers and subjugates another, a common enough event in human history. Genetic differences in mental traits, however slight, tend to be preserved by the raising of class barriers, racial and cultural discrimination, and physical ghettos. The geneticist C. D. Darlington (1969), among others, postulated this process to be a prime source of genetic diversity within human societies.

Yet despite the plausibility of the general argument, there is little evidence of any hereditary solidification of status. The castes of India have been in existence for 2000 years, more than enough time for evolutionary divergence, but they differ only slightly in blood type and other measurable anatomical and physiological traits. Powerful forces can be identified that work against the genetic fixation of caste differences. First, cultural evolution is too fluid. Over a period of decades or at most centuries ghettos are replaced, races and subject people are liberated, the conquerors are conquered. Even within relatively stable societies the pathways of upward mobility are numerous. The daughters of lower classes tend to marry upward. Success in commerce or political life can launch a family from virtually any socioeconomic group into the ruling class in a single generation. Furthermore, there are many Dahlberg genes, not just the one postulated for argument in the simplest model. The hereditary factors of human success are strongly polygenic and form a long list, only a few of which have been measured. IQ constitutes only one subset of the components of intelligence. Less tangible but equally important qualities are creativity, entrepreneurship, drive, and mental stamina. Let us assume that the genes contributing to these qualities are scattered over many chromosomes. Assume further that some of the traits are uncorrelated or even negatively correlated. Under these circumstances only the most intense forms of disruptive selection could result in the formation of stable ensembles of genes. A much more likely circumstance is the one that apparently prevails: the maintenance of a large amount of genetic diversity within societies and the loose correlation of some of the genetically determined traits with success. This scrambling process is accelerated by the continuous shift in the fortunes of individual families from one generation to the next.

Even so, the influence of genetic factors toward the assumption of certain *broad* roles cannot be discounted. Consider male homosexuality. The surveys of Kinsey and his coworkers showed that in the 1940's approximately 10 percent of the sexually mature males in the United States were mainly or exclusively homosexual for at least three years prior to being interviewed. Homosexuality is also exhibited by comparably high fractions of the male populations in many if not most other cultures. Kallmann's twin data indicate the probable existence of a genetic predisposition toward the condition. Accordingly, Hutchinson (1959) suggested that the homosexual genes may possess superior fitness in heterozygous conditions. His reasoning followed lines now standard in the thinking of population genetics. The homosexual state itself results in inferior genetic fitness, because of course homosexual men marry much less frequently and have far fewer children than their unambiguously heterosexual counterparts. The simplest way genes producing such a condition can be maintained in evolution is if they are superior in the heterozygous state, that is, if heterozygotes survive into maturity better, produce more offspring, or both. An interesting alternative hypothesis has been suggested to me by Herman T. Spieth (personal communication) and independently developed by Robert L. Trivers (1974). The homosexual members of primitive societies may have functioned as helpers, either while hunting in company with other men or in more domestic occupations at the dwelling sites. Freed from the special obligations of parental duties, they could have operated with special efficiency in assisting close relatives. Genes favoring homosexuality could then be sustained at a high equilibrium level by kin selection alone. It remains to be said that if such genes really exist they are almost certainly incomplete in penetrance and variable in expressivity, meaning that which bearers of the genes develop the behavioral trait and to what degree depend on the presence or absence of modifier genes and the influence of the environment.

Other basic types might exist, and perhaps the clues lie in full sight. In his study of British nursery children Blurton Jones (1969) distinguished two apparently basic behavioral types. "Verbalists," a small minority, often remained alone, seldom moved about, and almost never joined in rough-and-tumble play. They talked a great deal and spent much of their time looking at books. The other children were "doers." They joined groups, moved around a great deal, and spent much of their time painting and making objects instead of talking. Blurton Jones speculated that the dichotomy results from an early divergence in behavioral development persisting into maturity. Should it prove general it might contribute fundamentally to diversity within cultures. There is no way of knowing whether the divergence is ultimately genetic in origin or triggered entirely by experiential events at an early age.

Communication

All of man's unique social behavior pivots on his use of language, which is itself unique. In any language words are given arbitrary definitions within each culture and ordered according to a grammar that imparts new meaning above and beyond the definitions. The fully symbolic quality of the words and the sophistication of the grammar permit the creation of messages that are potentially infinite in number. Even communication about the system itself is made possible. This is the essential nature of human language. The basic attributes can be broken down, and other features of the transmission process itself can be added, to make a total of 16 design features (C. F. Hockett, reviewed by Thorpe, 1972a). Most of the features are found in at least rudimentary form in some other animal species. But the productivity and richness of human languages cannot be remotely approached even by chimpanzees taught to employ signs in simple sentences. The development of human speech represents a quantum jump in evolution comparable to the assembly of the eucaryotic cell.

Even without words human communication would be the richest known. The study of nonverbal communication has become a flourishing branch of the social sciences. Its codification is made difficult by the auxiliary role so many of the signals play to verbal communication. Categories of these signals are often defined inconsistently, and classifications are rarely congruent (see, for example, Renský, 1966; Crystal, 1969; Lyons, 1972). In Table 27-2 a composite arrangement is presented that I hope is both free of internal contradiction and consistent with current usage. The number of nonvocal signals, including all facial expressions, body postures and movement, and touch, probably number somewhat in excess of 100. Brannigan and Humphries (1972) have made a list of 136, which they believe is close to exhaustive. The number is consistent with the wholly independent estimate of Birdwhistle (1970), who believes that although the human face is capable of as many as 250,000 expressions, less than 100 sets of the expressions comprise distinct, meaningful symbols. Vocal paralanguage, insofar as it can be separated from the prosodic modifications of true speech, has not been cataloged so painstakingly. Grant (1969) recognized 6 distinct sounds, but several times this number would probably be distinguished by a zoologist accustomed to preparing ethograms of other primate species. In summary, all paralinguistic signals taken together almost certainly exceed 150 and may be close to 200. This repertory is larger than that of the majority of other mammals and birds by a factor of three or more, and it exceeds slightly the total repertories of both the rhesus monkey and chimpanzee.

Table 27-2 The modes of human communication.

I. Verbal Communication (Language): the utterance of words and sentences

II. Non-verbal Communication

- A. *Prosody:* tone, tempo, rhythm, loudness, pacing, and other qualities of voice that modify the meaning of verbal utterances
- B. *Paralanguage:* signals separate from words used to supplement or to modify language
    1. Vocal paralanguage: grunts, giggles, laughs, sobs, cries, and other nonverbal sounds
    2. Nonverbal paralanguage: body posture, motion, and touch (kinesic communication); possibly also chemical communication

Another useful distinction in the analysis of human paralanguage can be made between signals that are prelinguistic, defined as having been in service before the evolutionary origin of true language, and those that are postlinguistic. The postlinguistic signals are most likely to have originated as pure auxiliaries to speech. One approach to the problem is through the phylogenetic analysis of the relevant properties of primate communication. Hooff (1972), for example, has established the homologues of smiling and laughing in facial expressions of the cercopithecoid monkeys and apes, thus classifying these human behaviors among our most primitive and universal signals.

Human language, as Marler (1965) argued, probably stemmed from richly graded vocal signals not unlike those employed by the rhesus monkey and chimpanzee, as opposed to the more discrete sounds characterizing the repertories of some of the lower primates. Human infants can utter a wide variety of vocalizations resembling those of macaques, baboons, and chimpanzees. But very early in their development they convert to the peculiar sounds of human speech. Multiple plosives, fricatives, nasals, vowels, and other sounds are combined to create the 40 or so basic phonemes. The human mouth and upper respiratory tract have been strongly modified to permit this vocal competence (see Figure 27-2). The crucial changes are associated with man's upright posture, which may have provided the initial but still incomplete impetus toward the present modification. With the face directed fully forward, the mouth gave way to the upper pharyngeal space at a 90-degree angle. This configuration helped to push the rear of the tongue back until it formed part of the forward wall of the upper pharyngeal tract. Simultaneously the pharyngeal space and the epiglottis were both considerably lengthened.

These two principal changes, the shift in tongue position and lengthening of the pharyngeal tract, were responsible for the versatility in sound production. When air is forced upward through the vocal cords it generates a buzzing noise that can be varied in intensity and duration but not in the all-important qualities of tone that produce phoneme differentiation. The latter effect is achieved as the air passes up through the pharyngeal tract and mouth cavity and out through the mouth. These structures together form an air tube which, like any cylinder, serves as a resonator. When its position and shape are altered, the tube emphasizes different combinations of frequencies emanating from the vocal cords. The result, illustrated in Figure 27-2, is the sounds we distinguish as phonemes (see also Lenneberg, 1967, and Denes and Pinson, 1973).

However, the great advance in language acquisition did not come from the ability to form many sounds. After all, it is theoretically possible for a highly intelligent being to speak only a *single* word and still communicate rapidly. It need only be programmed like a

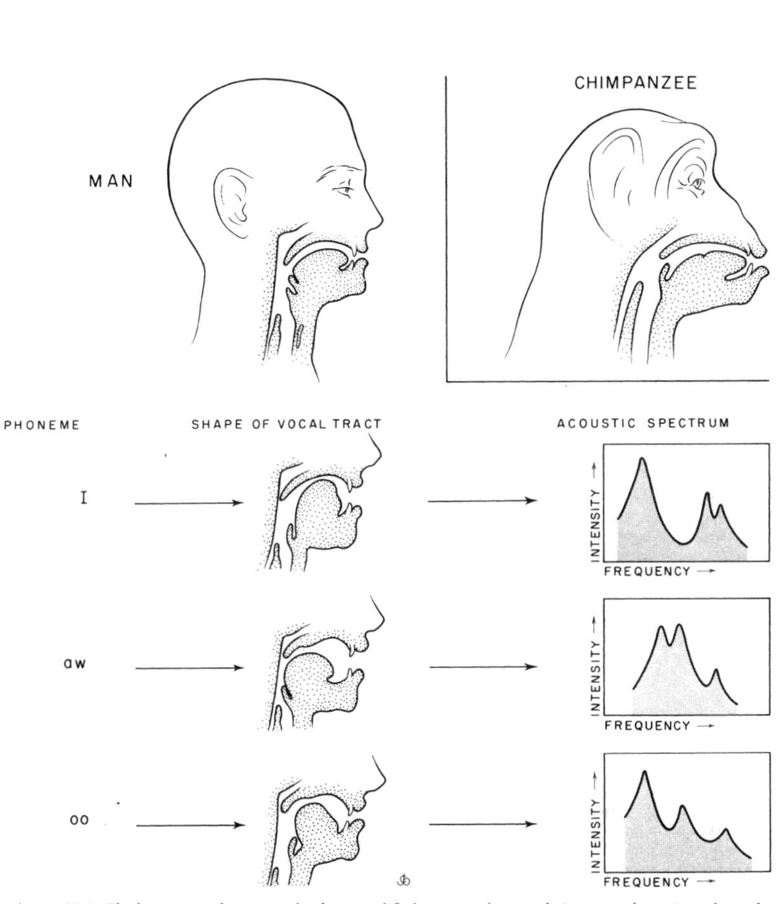

Figure 27-2 The human vocal apparatus has been modified in a way that greatly increases the variety of sounds that can be produced. The versatility was an essential accompaniment of the evolution of human speech. The upper diagrams show the ways in which man differs from the chimpanzee and other nonhuman primates: the angulation between the mouth and the upper respiratory tract is increased, the pharyngeal space is lengthened, and the back half of the tongue has come to form the front wall of the long tract above the vocal cords. The lower diagrams illustrate how movement of the tongue changes the shape of the air space to generate different sounds. (Modified from Howells, 1973, and Denes and Pinson, 1973.)

digital computer. Variation in loudness, duration, and pacing could be added to increase the transmission rate still more. It will be recalled that a single chemical substance, if modulated perfectly under ideal conditions, can generate up to 10,000 bits per second, far in excess of the capacity of human speech. Human languages gain their power instead from syntax, the dependence of meaning on the linear ordering of words. Each language possesses a grammar, the set of rules governing syntax. To truly understand the nature and origin of grammar would be to understand a great deal about the construction of the human mind. It is possible to distinguish three competing models that attempt to describe the known rules:

First Hypothesis: *Probabilistic left-to-right model.* The explanation favored by extreme behavioristic psychologists is that the occurrence of a word is Markovian, meaning that its probability is determined by the immediately preceding word or string of words. The developing child learns which words to link together in each appropriate circumstance.

Second Hypothesis: *Learned deep-structure model.* There exist a limited number of formal principles by which phrases of words are combined and juxtaposed to create various meanings. The child more or less unconsciously learns the deep structure of his own culture. Although the principles are finite in number, the sentences that can be generated from them are infinite in number. Animals cannot speak simply because they lack the necessary level of cognitive or intellectual ability, not because of the absence of any special "language faculty."

Third Hypothesis: *Innate deep-structure model.* The formal principles exist as suggested in hypothesis number two, but they are partially or wholly genetic. In other words, at least some of the principles emerge by maturation in an invariant manner. A corollary of this proposition is that much of the deep structure of grammar is widespread if not universal in mankind, notwithstanding the profound differences in surface structure and word meaning that exist between languages. A second corollary is that animals cannot speak because they lack this inborn language faculty, which is a qualitatively unique human property and not simply an outcome of man's quantitatively superior intelligence. The innate deep-structure model is the one that has come to be associated most prominently with the name of Noam Chomsky, and appears to be currently favored by most psycholinguists.

The probabilistic left-to-right model has already been eliminated, at least in its extreme version. The number of transitional probabilities a child would have to learn in order to compute in a language such as English is enormous, and there is simply not enough time in childhood to master them all (Miller, Galanter, and Pribram, 1960). Grammatical rules are actually learned very rapidly and in a predictable sequence, with the child passing through forms of construction that anticipate the adult form while differing significantly from it (Brown, 1973). This kind of ontogeny is typical of the maturation of innate components of animal behavior. Nevertheless, the similarity cannot be taken as conclusive evidence of a genetic program general to humanity.

The ultimate resolution of the problem, as Roger Brown and other developmental psycholinguists have stressed, cannot be achieved until deep grammar itself has been securely characterized. This is a relatively new area of investigation, scarcely dating beyond Chomsky's *Syntactic Structures* (1957). From the beginning it has been marked by a complicated, rapidly shifting argumentation. The basic ideas have been presented in recent reviews by Slobin (1971) and Chomsky (1972). Here it will suffice to define the main processes recognized by the new linguistic analysis. *Phrase structure grammar,* which is exemplified in Figure 27-3, consists of the rules by which sentences are built up in a hierarchical manner. Phrases can be thought of as modules that are substituted for other, equivalent modules or added *de novo* into sentences to change meanings. These elements cannot be split and the parts interchanged without creating serious difficulties. In the example "The boy hit the ball," "the ball" is intuitively such a unit. It can be easily taken out and replaced with some other phrase such as "the shuttlecock" or simply the word "it." The combination "hit the" is not such a unit. Despite the fact that the two words are juxtaposed, they cannot be easily replaced without creating difficulties for the construction of the entire remainder of the sentence. By observing the rules we all know subconsciously, the sentence can be expanded by the insertion of appropriately selected phrases: *After taking his position,* the *little* boy *swung twice and finally* hit the ball *and ran to first base.*

In short, phrase structure grammar decrees the ways in which phrases can be formed. It generates what has been called the deep structure of the word strings as opposed to the surface structure, or the mere order in which the individual words appear. But of course the sequences in which phrases and terminal words appear are crucial to the meaning of the sentence. "The boy hit the ball" is very different from "What did the boy hit?" even though the deep (phrase) structure is similar. The rules by which the deep structures are converted into surface structures by the assembling of phrases are called *transformational grammar.* A transformation is an operation that converts one phrase structure into another. Among the most basic operations are substitutions ("what" for "the ball"), displacement (placing "what" before the verb), and permutation (switching the positions of related words).

The psycholinguists have described, for English, both phrase structure and transformational grammar. The evidence does not appear

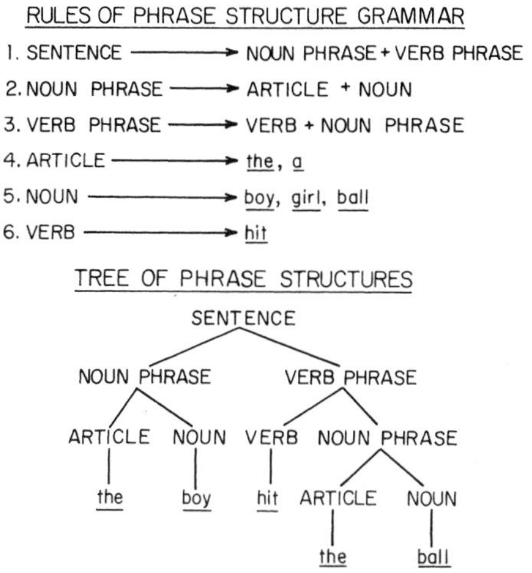

RULES OF PHRASE STRUCTURE GRAMMAR

1. SENTENCE → NOUN PHRASE + VERB PHRASE
2. NOUN PHRASE → ARTICLE + NOUN
3. VERB PHRASE → VERB + NOUN PHRASE
4. ARTICLE → the, a
5. NOUN → boy, girl, ball
6. VERB → hit

TREE OF PHRASE STRUCTURES

SENTENCE
NOUN PHRASE VERB PHRASE
ARTICLE NOUN VERB NOUN PHRASE
the boy hit ARTICLE NOUN
the ball

Figure 27-3 An example of the rules of phrase structure grammar in the English language. The simple sentence "The boy hit the ball" is seen to consist of a hierarchy of phrases. At each level one phrase can be substituted for another of equivalent composition, but the phrases cannot be split and their elements interchanged. (Based on Slobin, 1971.)

to be adequate, however, to choose between hypotheses two and three, in other words to decide whether the grammars are innately programmed or whether they are learned. The basic operations of transformation occur in all known human languages. However, this observation by itself does not establish that the precise rules of transformation are the same.

Is there a universal grammar? This question is difficult to answer because most attempts to generalize the rules of deep grammar have been based on the semantic content of one particular language. Students of the subject seldom confront the problem as if it were genuinely scientific, in a way that would reveal how concrete and soluble it might be. In fact, natural scientists are easily frustrated by the diffuse, oblique quality of much of the psycholinguistic literature, which often seems unconcerned with the usual canons of proposition and evidence. The reason is that many of the writers, including Chomsky, are structuralists in the tradition of Lévi-Strauss and Piaget. They approach the subject with the implicit world view that the processes of the human mind are indeed structured, and also discrete, enumerable, and evolutionarily unique with no great need to be referred to the formulations of other scientific disciplines. The analysis is nontheoretical in the sense that it fails to argue from postulates that can be tested and extended empirically. Some psychologists, including Roger Brown and his associates and Fodor and Garrett (1966), have adduced testable propositions and pursued them with mixed results, but the trail of speculation on deep grammar has not been easy to follow even for these skillful experimentalists.

Like poet naturalists, the structuralists celebrate idiosyncratic personal visions. They argue from hidden premises, relying largely on metaphor and exemplification, and with little regard for the method of multiple competing hypotheses. Clearly, this discipline, one of the most important in all of science, is ripe for the application of rigorous theory and properly meshed experimental investigation.

A key question that the new linguistics may never answer is when human language originated. Did speech appear with the first use of stone tools and the construction of shelters by the *Australopithecus* man-apes, over two million years ago? Or did it await the emergence of fully modern *Homo sapiens*, perhaps even the development of religious rites in the past 100,000 years? Lieberman (1968) believes that the date was relatively recent. He interprets the Makapan *Australopithecus* restored by Dart to fall close to the chimpanzee in the form of its palate and pharyngeal tract. If he is right, this early hominid might not have been able to articulate the sounds of human speech. The same conclusion has been drawn with respect to the anatomy and vocal capacity of the Neanderthal man (Lieberman et al., 1972), which if true places the origin of language in the latest stages of speciation in the genus *Homo*. Other theoretical aspects of the evolutionary origin of human speech have been discussed by Jane Hill (1972) and I. G. Mattingly (1972). Lenneberg (1971) has hypothesized that the capacity for mathematical reasoning originated as a slight modification of linguistic ability.

Culture, Ritual, and Religion

The rudiments of culture are possessed by higher primates other than man, including the Japanese monkey and chimpanzee (Chapter 7), but only in man has culture thoroughly infiltrated virtually every aspect of life. Ethnographic detail is genetically underprescribed, resulting in great amounts of diversity among societies. Underprescription does not mean that culture has been freed from the genes. What has evolved is the capacity for culture, indeed the overwhelming tendency to develop one culture or another. Robin Fox (1971) put the argument in the following form. If the proverbial experiments of the pharaoh Psammetichos and James IV of Scotland had worked, and children reared in isolation somehow survived in good health,

I do not doubt that they *could* speak and that, theoretically, given time, they or their offspring would invent and develop a language despite their never having been taught one. Furthermore, this language, although totally different from any known to us, would be analyzable by linguists on the same basis as other languages and translatable into all known languages. But I would push this further. If our new Adam and Eve could survive and breed—still in total isolation from any cultural influences—then eventually they would produce a society which would have laws about property, rules about incest and marriage, customs of taboo and avoidance, methods of settling disputes with a minimum of bloodshed, beliefs about the supernatural and practices relating to it, a system of social status and methods of indicating it, initiation ceremonies for young men, courtship practices including the adornment of females, systems of symbolic body adornment generally, certain activities and associations set aside for men from which women were excluded, gambling of some kind, a tool- and weapon-making industry, myths and legends, dancing, adultery, and various doses of homicide, suicide, homosexuality, schizophrenia, psychosis and neuroses, and various practitioners to take advantage of or cure these, depending on how they are viewed.

Culture, including the more resplendent manifestations of ritual and religion, can be interpreted as a hierarchical system of environmental tracking devices. In Chapter 7 the totality of biological responses, from millisecond-quick biochemical reactions to gene substitutions requiring generations, was described as such a system. At that time culture was placed within the scheme at the slow end of the time scale. Now this conception can be extended. To the extent that the specific details of culture are nongenetic, they can be decoupled from the biological system and arrayed beside it as an auxiliary system. The span of the purely cultural tracking system parallels much of the slower segment of the biological tracking system, ranging from days to generations. Among the fastest cultural responses in industrial civilizations are fashions in dress and speech. Somewhat slower are political ideology and social attitudes toward other nations, while the slowest of all include incest taboos and the belief or disbelief in particular high gods. It is useful to hypothesize that cultural details are for the most part adaptive in a Darwinian sense, even though some may operate indirectly through enhanced group survival (Washburn and Howell, 1960; Masters, 1970). A second proposition worth considering, to make the biological analogy complete, is that the rate of change in a particular set of cultural behaviors reflects the rate of change in the environmental features to which the behaviors are keyed.

Slowly changing forms of culture tend to be encapsulated in ritual. Some social scientists have drawn an analogy between human ceremonies and the displays of animal communication. This is not correct. Most animal displays are discrete signals conveying limited meaning. They are commensurate with the postures, facial expressions, and elementary sounds of human paralanguage. A few animal displays, such as the most complex forms of sexual advertisement and nest changing in birds, are so impressively elaborate that they have occasionally been termed ceremonies by zoologists. But even here the comparison is misleading. Most human rituals have more than just an immediate signal value. As Durkheim stressed, they not only label but reaffirm and rejuvenate the moral values of the community.

The sacred rituals are the most distinctively human. Their most elementary forms are concerned with magic, the active attempt to manipulate nature and the gods. Upper Paleolithic art from the caves of Western Europe shows a preoccupation with game animals. There are many scenes showing spears and arrows embedded in the bodies of the prey. Other drawings depict men dancing in animal disguises or standing with heads bowed in front of animals. Probably the function of the drawings was sympathetic magic, based on the quite logical notion that what is done with an image will come to pass with the real thing. This anticipatory action is comparable to the intention movements of animals, which in the course of evolution have often been ritualized into communicative signals. The waggle dance of the honeybee, it will be recalled, is a miniaturized rehearsal of the flight from the nest to the food. Primitive man might have understood the meaning of such complex animal behavior easily. Magic was, and still is in some societies, practiced by special people variously called shamans, sorcerers, or medicine men. They alone were believed to have the secret knowledge and power to deal effectively with the supernatural, and as such their influence sometimes exceeded that of the tribal headmen.

Formal religion *sensu stricto* has many elements of magic but is focused on deeper, more tribally oriented beliefs. Its rites celebrate the creation myths, propitiate the gods, and resanctify the tribal moral codes. Instead of a shaman controlling physical power, there is a priest who communes with the gods and curries their favor through obeisance, sacrifice, and the proffered evidences of tribal good behavior. In more complex societies, polity and religion have always blended naturally. Power belonged to kings by divine right, but high priests often ruled over kings by virtue of the higher rank of the gods.

It is a reasonable hypothesis that magic and totemism constituted direct adaptations to the environment and preceded formal religion in social evolution. Sacred traditions occur almost universally in human societies. So do myths that explain the origin of man or at the very least the relation of the tribe to the rest of the world. But belief in high gods is not universal. Among 81 hunter-gatherer societies surveyed by Whiting (1968), only 28, or 35 percent, included high gods in their sacred traditions. The concept of an active, moral God who created the world is even less widespread. Furthermore, this concept most commonly arises with a pastoral way of life. The greater

Table 27-3 The religious beliefs of 66 agrarian societies, partitioned according to the percentage of subsistence derived from herding. (From *Human Societies* by G. and Jean Lenski. Copyright © 1970 by McGraw-Hill Book Company. Used with permission.)

Percentage of subsistence from herding	Percentage of societies believing in an active, moral creator God	Number of societies
36–45	92	13
26–35	82	28
16–25	40	20
6–15	20	5

the dependence on herding, the more likely the belief in a shepherd god of the Judaeo-Christian model (see Table 27-3). In other kinds of societies the belief occurs in 10 percent or less of the cases. Also, the God of monotheistic religions is always male. This strong patriarchal tendency has several cultural sources (Lenski, 1970). Pastoral societies are highly mobile, tightly organized, and often militant, all features that tip the balance toward male authority. It is also significant that herding, the main economic base, is primarily the responsibility of men. Because the Hebrews were originally a herding people, the Bible describes God as a shepherd and the chosen people as his sheep. Islam, one of the strictest of all monotheistic faiths, grew to early power among the herding people of the Arabian peninsula. The intimate relation of the shepherd to his flock apparently provides a microcosm which stimulates deeper questioning about the relation of man to the powers that control him.

An increasingly sophisticated anthropology has not given reason to doubt Max Weber's conclusion that more elementary religions seek the supernatural for the purely mundane rewards of long life, abundant land and food, the avoidance of physical catastrophes, and the defeat of enemies. A form of group selection also operates in the competition between sects. Those that gain adherents survive; those that cannot, fail. Consequently, religions, like other human institutions, evolve so as to further the welfare of their practitioners. Because this demographic benefit applies to the group as a whole, it can be gained in part by altruism and exploitation, with certain segments profiting at the expense of others. Alternatively, it can arise as the sum of generally increased individual fitnesses. The resulting distinction in social terms is between the more oppressive and the more beneficent religions. All religions are probably oppressive to some degree, especially when they are promoted by chiefdoms and states. The tendency is intensified when societies compete, since religion can be effectively harnessed to the purposes of warfare and economic exploitation.

The enduring paradox of religion is that so much of its substance is demonstrably false, yet it remains a driving force in all societies. Men would rather believe than know, have the void as purpose, as Nietzsche said, than be void of purpose. At the turn of the century Durkheim rejected the notion that such force could really be extracted from "a tissue of illusions." And since that time social scientists have sought the psychological Rosetta stone that might clarify the deeper truths of religious reasoning. In a penetrating analysis of this subject, Rappaport (1971) proposed that virtually all forms of sacred rites serve the purposes of communication. In addition to institutionalizing the moral values of the community, the ceremonies can offer information on the strength and wealth of tribes and families. Among the Maring of New Guinea there are no chiefs or other leaders who command allegiance in war. A group gives a ritual dance, and individual men indicate their willingness to give military support by whether they attend the dance or not. The strength of the consortium can then be precisely determined by a head count. In more advanced societies military parades, embellished by the paraphernalia and rituals of the state religion, serve the same purpose. The famous potlatch ceremonies of the Northwest Coast Indians enable individuals to advertise their wealth by the amount of goods they give away. Rituals also regularize relationships in which there would otherwise be ambiguity and wasteful imprecision. The best examples of this mode of communication are the *rites de passage*. As a boy matures his transition from child to man is very gradual in a biological and psychological sense. There will be times when he behaves like a child when an adult response would have been more appropriate, and vice versa. The society has difficulty in classifying him one way or the other. The *rite de passage* eliminates this ambiguity by arbitrarily changing the classification from a continuous gradient into a dichotomy. It also serves to cement the ties of the young person to the adult group that accepts him.

To sanctify a procedure or a statement is to certify it as beyond question and imply punishment for anyone who dares to contradict it. So removed is the sacred from the profane in everyday life that simply to repeat it in the wrong circumstance is a transgression. This extreme form of certification, the heart of all religions, is granted to the practices and dogmas that serve the most vital interests of the group. The individual is prepared by the sacred rituals for supreme effort and self-sacrifice. Overwhelmed by shibboleths, special costumes, and the sacred dancing and music so accurately keyed to his emotive centers he has a "religious experience." He is ready to reassert allegiance to his tribe and family, perform charities, consecrate his life, leave for the hunt, join the battle, die for God and country. *Deus vult* was the rallying cry of the First Crusade. God wills it, but the summed Darwinian fitness of the tribe was the ultimate if unrecognized beneficiary.

It was Henri Bergson who first identified a second force leading to the formalization of morality and religion. The extreme plasticity of human social behavior is both a great strength and a real danger. If each family worked out rules of behavior on its own, the result would be an intolerable amount of tradition drift and growing chaos. To counteract selfish behavior and the "dissolving power" of high intelligence, each society must codify itself. Within broad limits virtually any set of conventions works better than none at all. Because arbitrary codes work, organizations tend to be inefficient and marred by unnecessary inequities. As Rappaport succinctly expressed it, "Sanctification transforms the arbitrary into the necessary, and regulatory mechanisms which are arbitrary are likely to be sanctified." The process engenders criticism, and in the more literate and self-conscious societies visionaries and revolutionaries set out to change the system. Reform meets repression, because to the extent that the rules have been sanctified and mythologized, the majority of the people regard them as beyond question, and disagreement is defined as blasphemy.

This leads us to the essentially biological question of the evolution of indoctrinability (Campbell, 1972). Human beings are absurdly easy to indoctrinate—they *seek* it. If we assume for argument that indoctrinability evolves, at what level does natural selection take place? One extreme possibility is that the group is the unit of selection. When conformity becomes too weak, groups become extinct. In this version selfish, individualistic members gain the upper hand and multiply at the expense of others. But their rising prevalence accelerates the vulnerability of the society and hastens its extinction. Societies containing higher frequencies of conformer genes replace those that disappear, thus raising the overall frequency of the genes in the metapopulation of societies. The spread of the genes will occur more rapidly if the metapopulation (for example, a tribal complex) is simultaneously enlarging its range. Formal models of the process, presented in Chapter 5, show that if the rate of societal extinction is high enough relative to the intensity of the counteracting individual selection, the altruistic genes can rise to moderately high levels. The genes might be of the kind that favors indoctrinability even at the expense of the individuals who submit. For example, the willingness to risk death in battle can favor group survival at the expense of the genes that permitted the fatal military discipline. The group-selection hypothesis is sufficient to account for the evolution of indoctrinability.

The competing, individual-level hypothesis is equally sufficient. It states that the ability of individuals to conform permits them to enjoy the benefits of membership with a minimum of energy expenditure and risk. Although their selfish rivals may gain a momentary advantage, it is lost in the long run through ostracism and repression. The conformists perform altruistic acts, perhaps even to the extent of risking their lives, not because of self-denying genes selected at the group level but because the group is occasionally able to take advantage of the indoctrinability which on other occasions is favorable to the individual.

The two hypotheses are not mutually exclusive. Group and individual selection can be reinforcing. If war requires spartan virtues and eliminates some of the warriors, victory can more than adequately compensate the survivors in land, power, and the opportunity to reproduce. The average individual will win the inclusive fitness game, making the gamble profitable, because the summed efforts of the participants give the average member a more than compensatory edge.

Ethics

Scientists and humanists should consider together the possibility that the time has come for ethics to be removed temporarily from the hands of the philosophers and biologicized. The subject at present consists of several oddly disjunct conceptualizations. The first is *ethical intuitionism,* the belief that the mind has a direct awareness of true right and wrong that it can formalize by logic and translate into rules of social action. The purest guiding precept of secular Western thought has been the theory of the social contract as formulated by Locke, Rousseau, and Kant. In our time the precept has been rewoven into a solid philosophical system by John Rawls (1971). His imperative is that justice should be not merely integral to a system of government but rather the object of the original contract. The principles called by Rawls "justice as fairness" are those which free and rational persons would choose if they were beginning an association from a position of equal advantage and wished to define the fundamental rules of the association. In judging the appropriateness of subsequent laws and behavior, it would be necessary to test their conformity to the unchallengeable starting position.

The Achilles heel of the intuitionist position is that it relies on the emotive judgment of the brain as though that organ must be treated as a black box. While few will disagree that justice as fairness is an ideal state for disembodied spirits, the conception is in no way explanatory or predictive with reference to human beings. Consequently, it does not consider the ultimate ecological or genetic consequences of the rigorous prosecution of its conclusions. Perhaps explanation and prediction will not be needed for the millennium. But this is unlikely—the human genotype and the ecosystem in which it evolved were fashioned out of extreme unfairness. In either case the full exploration of the neural machinery of ethical judgment is desirable and already in progress. One such effort, constituting the second mode of conceptualization, can be called *ethical behaviorism.* Its basic proposition, which has been expanded most fully by J. F. Scott (1971), holds that moral commitment is entirely learned, with operant conditioning being the dominant mechanism. In other words, children simply internalize the behavioral norms of the society. Opposing this theory is the *developmental-genetic conception* of ethical behavior. The best-documented version has been provided by Lawrence Kohlberg (1969). Kohlberg's viewpoint is structuralist and specifically Piagetian, and therefore not yet related to the remainder of biology. Piaget has used the expression "genetic epistemology" and Kohlberg "cognitive-developmental" to label the general concept. However, the results will eventually become incorporated into a broadened developmental biology and genetics. Kohlberg's method is to record and classify the verbal responses of children to moral problems. He has delineated six sequential stages of ethical reasoning through which an individual may progress as part of his mental maturation. The child moves from a primary dependence on external controls and sanctions to an increasingly sophisticated set of internalized standards (see Table 27-4). The analysis has not yet been directed to the question of plasticity in the basic rules. Intracultural variance has not been measured, and heritability therefore not assessed. The

Table 27-4 The classification of moral judgment into levels and stages of development. (Based on Kohlberg, 1969.)

Level	Basis of moral judgment	Stage of development
I	Moral value is defined by punishment and reward	1. Obedience to rules and authority to avoid punishment 2. Conformity to obtain rewards and to exchange favors
II	Moral value resides in filling the correct roles, in maintaining order and meeting the expectations of others	3. Good-boy orientation: conformity to avoid dislike and rejection by others 4. Duty orientation: conformity to avoid censure by authority, disruption of order, and resulting guilt
III	Moral value resides in conformity to shared standards, rights, and duties	5. Legalistic orientation: recognition of the value of contracts, some arbitrariness in rule formation to maintain the common good 6. Conscience or principle orientation: primary allegiance to principles of choice, which can overrule law in cases where the law is judged to do more harm than good

difference between ethical behaviorism and the current version of developmental-genetic analysis is that the former postulates a mechanism (operant conditioning) without evidence and the latter presents evidence without postulating a mechanism. No great conceptual difficulty underlies this disparity. The study of moral development is only a more complicated and less tractable version of the genetic variance problem (see Chapters 2 and 7). With the accretion of data the two approaches can be expected to merge to form a recognizable exercise in behavioral genetics.

Even if the problem were solved tomorrow, however, an important piece would still be missing. This is the *genetic evolution of ethics.* In the first chapter of this book I argued that ethical philosophers intuit the deontological canons of morality by consulting the emotive centers of their own hypothalamic-limbic system. This is also true of the developmentalists, even when they are being their most severely objective. Only by interpreting the activity of the emotive centers as a biological adaptation can the meaning of the canons be deciphered. Some of the activity is likely to be outdated, a relic of adjustment to the most primitive form of tribal organization. Some of it may prove to be *in statu nascendi,* constituting new and quickly changing adaptations to agrarian and urban life. The resulting confusion will be reinforced by other factors. To the extent that unilaterally altruistic genes have been established in the population by group selection, they will be opposed by allelomorphs favored by individual selection. The conflict of impulses under their various controls is likely to be widespread in the population, since current theory predicts that the genes will be at best maintained in a state of balanced polymorphism (Chapter 5). Moral ambivalency will be further intensified by the circumstance that a schedule of sex- and age-dependent ethics can impart higher genetic fitness than a single moral code which is applied uniformly to all sex-age groups. The argument for this statement is the special case of the Gadgil-Bossert distribution in which the contributions of social interactions to survivorship and fertility schedules are specified (see Chapter 4). Some of the differences in the Kohlberg stages could be explained in this manner. For example, it should be of selective advantage for young children to be self-centered and relatively disinclined to perform altruistic acts based on personal principle. Similarly, adolescents should be more tightly bound by age-peer bonds within their own sex and hence unusually sensitive to peer approval. The reason is that at this time greater advantage accrues to the formation of alliances and rise in status than later, when sexual and parental morality become the paramount determinants of fitness. Genetically programmed sexual and parent-offspring conflict of the kind predicted by the Trivers models (Chapters 15 and 16) are also likely to promote age differences in the kinds and degrees of moral commitment. Finally, the moral standards of individuals during early phases of colony growth should differ in many details from those of individuals at demographic equilibrium or during episodes of overpopulation. Metapopulations subject to high levels of r extinction will tend to diverge genetically from other kinds of populations in ethical behavior (Chapter 5).

If there is any truth to this theory of innate moral pluralism, the requirement for an evolutionary approach to ethics is self-evident. It should also be clear that no single set of moral standards can be applied to all human populations, let alone all sex-age classes within each population. To impose a uniform code is therefore to create complex, intractable moral dilemmas—these, of course, are the current condition of mankind.

Esthetics

Artistic impulses are by no means limited to man. In 1962, when Desmond Morris reviewed the subject in *The Biology of Art*, 32 individual nonhuman primates had produced drawings and paintings in captivity. Twenty-three were chimpanzees, 2 were gorillas, 3 were orang-utans, and 4 were capuchin monkeys. None received special training or anything more than access to the necessary equipment. In fact, attempts to guide the efforts of the animals by inducing imitation were always unsuccessful. The drive to use the painting and drawing equipment was powerful, requiring no reinforcement from the human observers. Both young and old animals became so engrossed with the activity that they preferred it to being fed and sometimes threw temper tantrums when stopped. Two of the chimpanzees studied extensively were highly productive. "Alpha" produced over 200 pictures, while the famous "Congo," who deserves to be called the Picasso of the great apes, was responsible for nearly 400. Although most of the efforts consisted of scribbling, the patterns were far from random. Lines and smudges were spread over a blank page outward from a centrally located figure. When a drawing was started on one side of a blank page the chimpanzee usually shifted to the opposite side to offset it. With time the calligraphy became bolder, starting with simple lines and progressing to more complicated multiple scribbles. Congo's patterns progressed along approximately the same developmental path as those of very young human children, yielding fan-shaped diagrams and even complete circles. Other chimpanzees drew crosses.

The artistic activity of chimpanzees may well be a special manifestation of their tool-using behavior. Members of the species display a total of about ten techniques, all of which require manual skill. Probably all are improved through practice, while at least a few are passed as traditions from one generation to the next. The chimpanzees have a considerable facility for inventing new techniques, such as the use of sticks to pull objects through cage bars and to pry open boxes. Thus the tendency to manipulate objects and to explore their uses appears to have an adaptive advantage for chimpanzees.

The same reasoning applies a fortiori to the origin of art in man. As Washburn (1970) pointed out, human beings have been hunter-gatherers for over 99 percent of their history, during which time each man made his own tools. The appraisal of form and skill in execution were necessary for survival, and they probably brought social approval as well. Both forms of success paid off in greater genetic fitness. If the chimpanzee Congo could reach the stage of elementary diagrams, it is not too hard to imagine primitive man progressing to representational figures. Once that stage was reached, the transition to the use of art in sympathetic magic and ritual must have followed quickly. Art might then have played a reciprocally reinforcing role in the development of culture and mental capacity. In the end, writing emerged as the idiographic representation of language.

Music of a kind is also produced by some animals. Human beings consider the elaborate courtship and territorial songs of birds to be beautiful, and probably ultimately for the same reasons they are of use to the birds. With clarity and precision they identify the species, the physiological condition, and the mental set of the singer. Richness of information and precise transmission of mood are no less the standards of excellence in human music. Singing and dancing serve to draw groups together, direct the emotions of the people, and prepare them for joint action. The carnival displays of chimpanzees described in earlier chapters are remarkably like human celebrations in this respect. The apes run, leap, pound the trunks of trees in drumming motions, and call loudly back and forth. These actions serve at least in part to assemble groups at common feeding grounds. They may resemble the ceremonies of earliest man. Nevertheless, fundamental differences appeared in subsequent human evolution. Human music has been liberated from iconic representation in the same way that true language has departed from the elementary ritualization characterizing the communication of animals. Music has the capacity for unlimited and arbitrary symbolization, and it employs rules of phrasing and order that serve the same function as syntax.

Territoriality and Tribalism

Anthropologists often discount territorial behavior as a general human attribute. This happens when the narrowest concept of the phenomenon is borrowed from zoology—the "stickleback model," in which residents meet along fixed boundaries to threaten and drive one another back. But earlier, in Chapter 12, I showed why it is necessary to define territory more broadly, as any area occupied more or less exclusively by an animal or group of animals through overt defense or advertisement. The techniques of repulsion can be as explicit as a precipitous all-out attack or as subtle as the deposit of a chemical secretion at a scent post. Of equal importance, animals respond to their neighbors in a highly variable manner. Each species is characterized by its own particular behavioral scale. In extreme cases the scale may run from open hostility, say, during the breeding season or when the population density is high, to oblique forms of advertisement or no territorial behavior at all. One seeks to characterize the behavioral scale of the species and to identify the parameters that move individual animals up and down it.

If these qualifications are accepted, it is reasonable to conclude that territoriality is a general trait of hunter-gatherer societies. In a perceptive review of the evidence, Edwin Wilmsen (1973) found that these relatively primitive societies do not differ basically in their strategy of land tenure from many mammalian species. Systematic overt aggression has been reported in a minority of hunter-gatherer pepples, for example the Chippewa, Sioux, and Washo of North America and the Murngin and Tiwi of Australia. Spacing and demographic balance were implemented by raiding parties, murder, and threats of witchcraft. The Washo of Nevada actively defended nuclear portions of their home ranges, within which they maintained their winter residences. Subtler and less direct forms of interaction can have the same result. The !Kung Bushmen of the Nyae Nyae area refer to themselves as "perfect" or "clean" and other !Kung people as "strange" murderers who use deadly poisons.

Human territorial behavior is sometimes particularized in ways that are obviously functional. As recently as 1930 Bushmen of the Dobe area in southwestern Africa recognized the principle of exclusive family land-holdings during the wet season. The rights extended only to the gathering of vegetable foods; other bands were allowed to hunt animals through the area (R. B. Lee in Wilmsen, 1973). Other hunter-gatherer peoples appear to have followed the same dual principle: more or less exclusive use by tribes or families of the richest sources of vegetable foods, opposed to broadly overlapping hunting ranges. Thus the original suggestion of Bartholomew and Birdsell (1953) that *Australopithecus* and the primitive *Homo* were territorial remains a viable hypothesis. Moreover, in obedience to the rule of ecological efficiency, the home ranges and territories were probably large and population density correspondingly low. This rule, it will be recalled, states that when a diet consists of animal food, roughly ten times as much area is needed to gain the same amount of energy yield as when the diet consists of plant food. Modern hunter-gatherer bands containing about 25 individuals commonly occupy between 1000 and 3000 square kilometers. This area is comparable to the home range of a wolf pack but as much as a hundred times greater than that of a troop of gorillas, which are exclusively vegetarian.

Hans Kummer (1971), reasoning from an assumption of territoriality, provided an important additional insight about human behavior. Spacing between groups is elementary in nature and can be achieved by a relatively small number of simple aggressive techniques. Spacing and dominance within groups is vastly more complex, being tied to all the remainder of the social repertory. Part of man's problem is that his intergroup responses are still crude and primitive, and inadequate for the extended extraterritorial relationships that civilization has thrust upon him. The unhappy result is what Garrett Hardin (1972) has defined as tribalism in the modern sense:

> Any group of people that perceives itself as a distinct group, and which is so perceived by the outside world, may be called a tribe. The group might be a race, as ordinarily defined, but it need not be; it can just as well be a religious sect, a political group, or an occupational group. The essential characteristic of a tribe is that it should follow a double standard of morality—one kind of behavior for in-group relations, another for out-group.
>
> It is one of the unfortunate and inescapable characteristics of tribalism that it eventually evokes counter-tribalism (or, to use a different figure of speech, it "polarizes" society).

Fearful of the hostile groups around them, the "tribe" refuses to concede to the common good. It is less likely to voluntarily curb its own population growth. Like the Sinhalese and Tamils of Ceylon, competitors may even race to outbreed each other. Resources are sequestered. Justice and liberty decline. Increases in real and imagined threats congeal the sense of group identity and mobilize the tribal members. Xenophobia becomes a political virtue. The treatment of nonconformists within the group grows harsher. History is replete with the escalation of this process to the point that the society breaks down or goes to war. No nation has been completely immune.

Early Social Evolution

Modern man can be said to have been launched by a two-stage acceleration in mental evolution. The first occurred during the transition from a larger arboreal primate to the first man-apes (*Australopithecus*). If the primitive hominid *Ramapithecus* is in the direct line of ancestry, as current opinion holds, the change may have required as much as ten million years. *Australopithecus* was present five million years ago, and by three million years B.P. it had speciated into several forms, including possibly the first primitive *Homo* (Tobias, 1973). As shown in Figure 27-1, the evolution of these intermediate hominids was marked by an accelerating increase in brain capacity. Simultaneously, erect posture and a striding, bipedal locomotion were perfected, and the hands were molded to acquire the precision grip. These early men undoubtedly used tools to a much greater extent than do modern chimpanzees. Crude stone implements were made by chipping, and rocks were pulled together to form what appear to be the foundations of shelters.

The second, much more rapid phase of acceleration began about 100,000 years ago. It consisted primarily of cultural evolution and must have been mostly phenotypic in nature, building upon the genetic potential in the brain that had accumulated over the previous millions of years. The brain had reached a threshold, and a wholly new, enormously more rapid form of mental evolution took over.

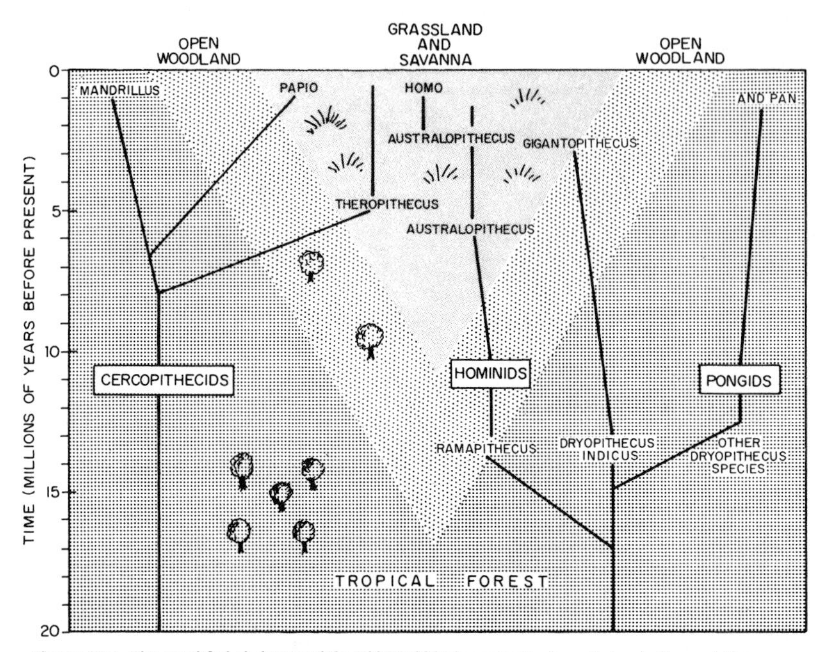

Figure 27-4 This simplified phylogeny of the Old World higher primates shows that only three existing groups have shifted from the forest to the savanna. They are the baboons (*Papio*), the gelada monkey (*Theropithecus gelada*), and man. (Based on Napier and Napier, 1967, and Simons and Ettel, 1970.)

This second phase was in no sense planned, and its potential is only now being revealed.

The study of man's origins can be referred to two questions that correspond to the dual stages of mental evolution:

———What features of the environment caused the hominids to adapt differently from other primates and started them along their unique evolutionary path?

———Once started, why did the hominids go so far?

The search for the prime movers of early human evolution has extended over more than 25 years. Participants in the search have included Dart (1949, 1956), Bartholomew and Birdsell (1953), Etkin (1954), Washburn and Avis (1958), Washburn et al. (1961), Rabb et al. (1967), Reynolds (1968), Schaller and Lowther (1969), C. J. Jolly (1970), and Kortlandt (1972). These writers have concentrated on two indisputably important facts concerning the biology of *Australopithecus* and early *Homo*. First, the evidence is strong that *Australopithecus africanus*, the species most likely to have been the direct ancestor of *Homo*, lived on the open savanna. The wear pattern of sand grains taken from the Sterkfontein fossils suggests a dry climate, while the pigs, antelopes, and other mammals found in association with the hominids are of the kind usually specialized for existence in grasslands. The australopithecine way of life came as the result of a major habitat shift. The ancestral *Ramapithecus* or an even more antecedent form lived in forests and was adapted for progression through trees by arm swinging. Only a very few other large-bodied primates have been able to join man in leaving the forest to spend most of their lives on the ground in open habitats (Figure 27-4). This is not to say that bands of *Australopithecus africanus* spent all of their lives running about in the open. Some of them might have carried their game into caves and even lived there in permanent residence, although the evidence pointing to this often quoted trait is still far from conclusive (Kurtén, 1972). Other bands could have retreated at night to the protection of groves of trees, in the manner of modern baboons. The important point is that much or all of the foraging was conducted on the savanna.

The second peculiar feature of the ecology of early men was the degree of their dependence on animal food, evidently far greater than in any of the living monkeys and apes. The *Australopithecus* were catholic in their choice of small animals. Their sites contain the remains of tortoises, lizards, snakes, mice, rabbits, porcupines, and other small, vulnerable prey that must have abounded on the savanna. The man-apes also hunted baboons with clubs. From analysis of 58 baboon skulls, Dart estimated that all had been brought down by blows to the head, 50 from the front and the remainder from behind. The *Australopithecus* also appear to have butchered larger animals, including the giant sivatheres, or horned giraffes, and dinotheres, elephantlike forms with tusks that curved downward from the lower jaws. In early Acheulean times, when *Homo erectus* began employing stone axes, some of the species of large African mammals became extinct. It is reasonable to suppose that this impoverishment was due to excessive predation by the increasingly competent bands of men (Martin, 1966).

What can we deduce from these facts about the life of early man? Before an answer is attempted, it should be noted that very little can be inferred directly from comparisons with other living primates. Geladas and baboons, the only open-country forms, are primarily vegetarian. They represent a sample of at most six species, which differ too much from one another in social organization to provide a baseline for comparison. The chimpanzees, the most intelligent and socially sophisticated of the nonhuman primates, are forest-dwelling and mostly vegetarian. Only during their occasional ventures into predation do they display behavior that can be directly correlated with ecology in a way that has meaning for human evolution. Other notable features of chimpanzee social organization, including the rapidly shifting composition of subgroups, the exchange of females between groups, and the intricate and lengthy process of socialization (see Chapter 26), may or may not have been shared by primitive man. We cannot argue either way on the basis of ecological correlation. It is often stated in the popular literature that the life of chimpanzees reveals a great deal about the origin of man. This is not necessarily true. The manlike traits of chimpanzees could be due to evolutionary convergence, in which case their use in evolutionary reconstructions would be misleading.

The best procedure to follow, and one which I believe is relied on implicitly by most students of the subject, is to extrapolate backward from living hunter-gatherer societies. In Table 27-5 this technique is made explicit. Utilizing the synthesis edited by Lee and DeVore (1968; see especially J. W. M. Whiting, pp. 336-339), I have listed the most general traits of hunter-gatherer peoples. Then I have evaluated the lability of each behavioral category by noting the amount of variation in the category that occurs among the nonhuman primate species. The less labile the category, the more likely that the trait displayed by the living hunter-gatherers was also displayed by early man.

What we can conclude with some degree of confidence is that primitive men lived in small territorial groups, within which males were dominant over females. The intensity of aggressive behavior and the nature of its scaling remain unknown. Maternal care was prolonged, and the relationships were at least to some extent matrilineal. Speculation on remaining aspects of social life is not supported either way by the lability data and is therefore more tenuous. It is likely that the early hominids foraged in groups. To judge from the behavior of baboons and geladas, such behavior would have conferred some protection from large predators. By the time *Australopithecus* and early *Homo* had begun to feed on large mammals, group hunting almost certainly had become advantageous and even necessary, as in the African wild dog. But there is no compelling reason to conclude that men did the hunting while women stayed at home. This occurs today in hunter-gatherer societies, but comparisons with other primates offer no clue as to *when* the trait appeared. It is certainly not essential to conclude a priori that males must be a specialized hunter class. In chimpanzees males do the hunting, which may be suggestive. But in lions, it will be recalled, the females are the providers, often working in groups and with cubs in tow, while the males usually hold back. In the African wild dog both sexes participate. This is not to suggest that male group hunting was not an early trait of hominids, only that there is no strong independent evidence to support the hypothesis.

This brings us to the prevailing theory of the origin of human sociality. It consists of a series of interlocking models that have been fashioned from bits of fossil evidence, extrapolations back from extant hunter-gatherer societies, and comparisons with other living primate species. The core of the theory can be appropriately termed the *autocatalysis model.* It holds that when the earliest hominids became bipedal as part of their terrestrial adaptation, their hands were freed, the manufacture and handling of artifacts was made easier, and intelligence grew as part of the improvement of the tool-using habit. With mental capacity and the tendency to use artifacts increas-

Table 27-5 Social traits of living hunter-gatherer groups and the likelihood that they were also possessed by early man.

Traits that occur generally in living hunter-gatherer societies	Variability of trait category among nonhuman primates	Reliability of concluding early man had the same trait through homology
Local group size:		
Mostly 100 or less	Highly variable but within range of 3–100	Very probably 100 or less but otherwise not reliable
Family as the nuclear unit	Highly variable	Not reliable
Sexual division of labor:		
Women gather, men hunt	Limited to man among living primates	Not reliable
Males dominant over females	Widespread although not universal	Reliable
Long-term sexual bonding (marriage) nearly universal; polygyny general	Highly variable	Not reliable
Exogamy universal, abetted by marriage rules	Limited to man among living primates	Not reliable
Subgroup composition changes often (fission-fusion principle)	Highly variable	Not reliable
Territoriality general, especially marked in rich gathering areas	Occurs widely, but variable in pattern	Probably occurred; pattern unknown
Game playing, especially games that entail physical skill but not strategy	Occurs generally, at least in elementary form	Very reliable
Prolonged maternal care; pronounced socialization of young; extended relationships between mother and children, especially mothers and daughters	Occurs generally in higher cercopithecoids	Very reliable

ing through mutual reinforcement, the entire materials-based culture expanded. Cooperation during hunting was perfected, providing a new impetus for the evolution of intelligence, which in turn permitted still more sophistication in tool using, and so on through cycles of causation. At some point, probably during the late *Australopithecus* period or the transition from *Australopithecus* to *Homo*, this autocatalysis carried the evolving populations to a certain threshold of competence, at which the hominids were able to exploit the antelopes, elephants, and other large herbivorous mammals teeming around them on the African plains. Quite possibly the process began when the hominids learned to drive big cats, hyenas, and other carnivores from their kills (see Figure 27-5). In time they became the primary hunters themselves and were forced to protect their prey from other predators and scavengers. The autocatalysis model usually includes the proposition that the shift to big game accelerated the process of mental evolution. The shift could even have been the impetus that led to the origin of early *Homo* from their australopithecine ancestors approximately two million years ago. Another proposition is that males became specialized for hunting. Child care was facilitated by close social bonding between the males, who left the domiciles to hunt, and the females, who kept the children and conducted most of the foraging for vegetable food. Many of the peculiar details of human sexual behavior and domestic life flow easily from this basic division of labor. But these details are not essential to the autocatalysis model. They are added because they are displayed by modern hunter-gatherer societies.

Although internally consistent, the autocatalysis model contains a curious omission—the triggering device. Once the process started, it is easy to see how it could be self-sustaining. But what started it? Why did the earliest hominids become bipedal instead of running on all fours like baboons and geladas? Clifford Jolly (1970) has proposed that the prime impetus was a specialization on grass seeds. Because the early pre-men, perhaps as far back as *Ramapithecus*, were the largest primates depending on grain, a premium was set on the ability to manipulate objects of very small size relative to the hands. Man, in short, became bipedal in order to pick seeds. This hypothesis is by no means unsupported fantasy. Jolly points to a number of convergent features in skull and dental structure between man and the gelada, which feeds on seeds, insects, and other small objects. Moreover, the gelada is peculiar among the Old World monkeys and apes in sharing the following epigamic anatomical traits with man: growth of hair around the face and neck of the male and conspicuous fleshy adornments on the chest of the female. According to Jolly's model, the freeing of the hands of the early hominids was a preadaptation that permitted the increase in tool use and the autocatalytic concomitants of mental evolution and predatory behavior.

Later Social Evolution

Autocatalytic reactions in living systems never expand to infinity. Biological parameters normally change in a rate-dependent manner to slow growth and eventually bring it to a halt. But almost miraculously, this has not yet happened in human evolution. The increase in brain size and the refinement of stone artifacts indicate a gradual improvement in mental capacity throughout the Pleistocene. With the appearance of the Mousterian tool culture of *Homo sapiens neanderthalensis* some 75,000 years ago, the trend gathered momentum, giving way in Europe to the Upper Paleolithic culture of *Homo s. sapiens* about 40,000 years B.P. Starting about 10,000 years ago agriculture was invented and spread, populations increased enormously in density, and the primitive hunter-gatherer bands gave way locally to the relentless growth of tribes, chiefdoms, and states. Finally, after A.D. 1400 European-based civilization shifted gears again, and knowledge and technology grew not just exponentially but superexponentially (see Figures 27-6, 27-7).

There is no reason to believe that during this final sprint there has been a cessation in the evolution of either mental capacity or the predilection toward special social behaviors. The theory of population genetics and experiments on other organisms show that substantial changes can occur in the span of less than 100 generations, which for man reaches back only to the time of the Roman Empire. Two thousand generations, roughly the period since typical *Homo sapiens* invaded Europe, is enough time to create new species and to mold them in major ways. Although we do not know how much mental evolution has actually occurred, it would be false to assume that modern civilizations have been built entirely on capital accumulated during the long haul of the Pleistocene.

Since genetic and cultural tracking systems operate on parallel tracks, we can bypass their distinction for the moment and return to the question of the prime movers in later human social evolution in its broadest sense. Seed eating is a plausible explanation to account for the movement of hominids onto the savanna, and the shift to big-game hunting might account for their advance to the *Homo erectus* grade. But was the adaptation to group predation enough to carry evolution all the way to the *Homo sapiens* grade and farther, to agriculture and civilization? Anthropologists and biologists do not consider the impetus to have been sufficient. They have advocated the following series of additional factors, which can act singly or in combination.

Sexual Selection

Fox (1972), following a suggestion by Chance (1962), has argued that sexual selection was the auxiliary motor that drove human evolution all the way to the *Homo* grade. His reasoning proceeds as follows. Polygyny is a general trait in hunter-gatherer bands and may also have been the rule in the early hominid societies. If so, a premium would have been placed on sexual selection involving both epigamic display toward the females and intrasexual competition among the males. The selection would be enhanced by the constant mating provocation that arises from the female's nearly continuous sexual receptivity. Because of the existence of a high level of cooperation within the band, a legacy of the original *Australopithecus* adaptation, sexual selection would tend to be linked with hunting prowess, leadership, skill at tool making, and other visible attributes that contribute to the success of the family and the male band. Aggressiveness was constrained and the old forms of overt primate dominance replaced by complex social skills. Young males found it profitable to fit into the group, controlling their sexuality and aggression and awaiting their turn at leadership. As a result the dominant male in hominid societies was most likely to possess a mosaic of qualities that reflect the necessities of compromise: "controlled, cunning, cooperative, attractive to the ladies, good with the children, relaxed, tough, eloquent, skillful, knowledgeable and proficient in self-defense and hunting." Since positive feedback occurs between these more sophisticated social traits and breeding success, social evolution can proceed indefinitely without additional selective pressures from the environment.

Multiplier Effects in Cultural Innovation and in Network Expansion

Whatever its prime mover, evolution in cultural capacity was implemented by a growing power and readiness to learn. The network of contacts among individuals and bands must also have grown. We can postulate a critical mass of cultural capacity and network size in which it became advantageous for bands actively to enlarge both. In other words, the feedback became positive. This mechanism, like sexual selection, requires no additional input beyond the limits of

Figure 27-5 At the threshold of autocatalytic social evolution two million years ago, a band of early men (*Homo habilis*) forages for food on the African savanna. In this speculative reconstruction the group is in the act of driving rival predators from a newly fallen dinothere. The great elephantlike creature had succumbed from exhaustion or disease, its end perhaps hastened by attacks from the animals closing in on it. The men have just entered the scene. Some drive away the predators by variously shouting, waving their arms, brandishing sticks, and throwing rocks, while a few stragglers, entering from the left, prepare to join the fray. To the right a female sabertooth cat (*Homotherium*) and her two grown cubs have been at least temporarily intimidated and are backing away. Their threat faces reveal the extraordinary gape of their jaws. In the left foreground, a pack of spotted hyenas (*Crocuta*) has also retreated but is ready to rush back the moment an opening is provided.

The men are quite small, less than 1.5 meters in height, and individually no match for the large carnivores. According to prevailing theory, a high degree of cooperation was therefore required to exploit such prey; and it evolved in conjunction with higher intelligence and the superior ability to use tools. In the background can be seen the environment of the Olduvai region of Tanzania as it may have looked at this time. The area was covered by rolling parkland and rimmed to the east by volcanic highlands. The herbivore populations were dense and varied, as they are today. In the left background are seen three-toed horses (*Hipparion*), while to the right are herds of wildebeest and giant horned giraffelike creatures called sivatheres. (Drawing by Sarah Landry; prepared in consultation with F. Clark Howell. The reconstruction of *Homotherium* was based in part on an Aurignacian sculpture; see Rousseau, 1971.)

social behavior itself. But unlike sexual selection, it probably reached the autocatalytic threshold level very late in human prehistory.

Increased Population Density and Agriculture

The conventional view of the development of civilization used to be that innovations in farming led to population growth, the securing of leisure time, the rise of a leisure class, and the contrivance of civilized, less immediately functional pursuits. The hypothesis has been considerably weakened by the discovery that !Kung and other hunter-gatherer peoples work less and enjoy more leisure time than most farmers. Primitive agricultural people generally do not produce surpluses unless compelled to do so by political or religious authorities (Carneiro, 1970). Ester Boserup (1965) has gone so far as to suggest the reverse causation: population growth induces societies to deepen their involvement and expertise in agriculture. However, this explanation does not account for the population growth in the first place. Hunter-gatherer societies remained in approximate demographic equilibrium for hundreds of thousands of years. Something else tipped a few of them into becoming the first farmers. Quite possibly the crucial events were nothing more than the attainment of a certain level of intelligence and lucky encounters with wild-growing food plants. Once launched, agricultural economies permitted higher population densities which in turn encouraged wider networks of social contact, technological advance, and further dependence on farming. A few innovations, such as irrigation and the wheel, intensified the process to the point of no return.

Warfare

Throughout recorded history the conduct of war has been common among tribes and nearly universal among chiefdoms and states. When Sorokin analyzed the histories of 11 European countries over periods of 275 to 1,025 years, he found that on the average they were engaged in some kind of military action 47 percent of the time, or about one year out of every two. The range was from 28 percent of the years in the case of Germany to 67 percent in the case of Spain. The early chiefdoms and states of Europe and the Middle East turned over with great rapidity, and much of the conquest was genocidal in nature. The spread of genes has always been of paramount importance. For

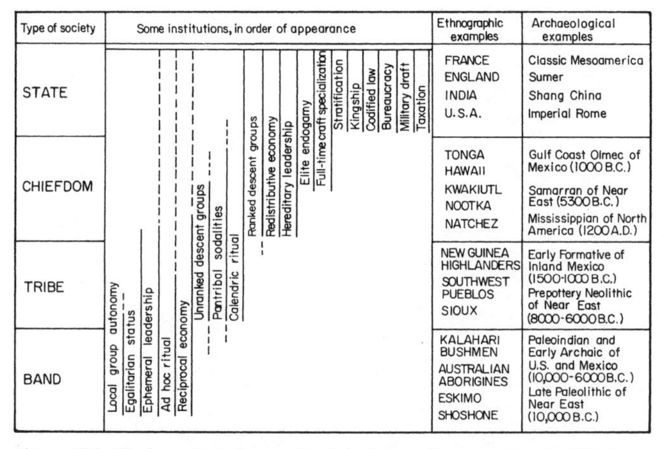

Figure 27-6 The four principal types of societies in ascending order of sociopolitical complexity, with living and extinct examples of each. A few of the sociopolitical institutions are shown, in the approximate order in which they are interpreted to have arisen. (From Flannery, 1972. Reproduced, with permission, from "The Cultural Evolution of Civilizations," *Annual Review of Ecology and Systematics,* Vol. 3, p. 401. Copyright © 1972 by Annual Reviews, Inc. All rights reserved.)

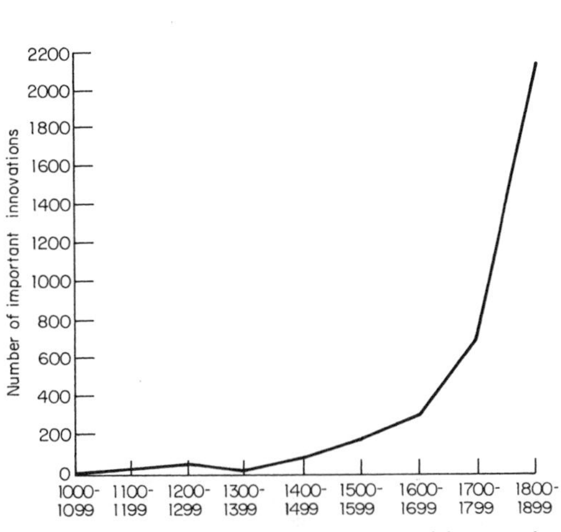

Figure 27-7 The number of important inventions and discoveries, by century, from A.D. 1000 to A.D. 1900. (From Lenski, 1970; after Ogburn and Nimkoff, 1958. Compiled from L. Darmstaedter and R. DuBois Reymond, *4000 Jahre-Pionier-Arbeit in den Exacten Wissenschaften*, Berlin, J. A. Stargart, 1904.)

example, after the conquest of the Midianites Moses gave instructions identical in result to the aggression and genetic usurpation by male langur monkeys:

> Now kill every male dependent, and kill every woman who has had intercourse with a man, but spare for yourselves every woman among them who has not had intercourse. (Numbers 31)

And centuries later, von Clausewitz conveyed to his pupil the Prussian crown prince a sense of the true, biological joy of warfare:

> Be audacious and cunning in your plans, firm and persevering in their execution, determined to find a glorious end, and fate will crown your youthful brow with a shining glory, which is the ornament of princes, and engrave your image in the hearts of your last descendants.

The possibility that endemic warfare and genetic usurpation could be an effective force in group selection was clearly recognized by Charles Darwin. In *The Descent of Man* he proposed a remarkable model that foreshadowed many of the elements of modern group-selection theory:

> Now, if some one man in a tribe, more sagacious than the others, invented a new snare or weapon, or other means of attack or defence, the plainest self-interest, without the assistance of much reasoning power, would prompt the other members to imitate him; and all would thus profit. The habitual practice of each new art must likewise in some slight degree strengthen the intellect. If the invention were an important one, the tribe would increase in number, spread, and supplant other tribes. In a tribe thus rendered more numerous there would always be a rather greater chance of the birth of other superior and inventive members. If such men left children to inherit their mental superiority, the chance of the birth of still more ingenious members would be somewhat better, and in a very small tribe decidedly better. Even if they left no children, the tribe would still include their blood-relations, and it has been ascertained by agriculturists that by preserving and breeding from the family of an animal, which when slaughtered was found to be valuable, the desired character has been obtained.

Darwin saw that not only can group selection reinforce individual selection, but it can oppose it—and sometimes prevail, especially if the size of the breeding unit is small and average kinship correspondingly close. Essentially the same theme was later developed in increasing depth by Keith (1949), Bigelow (1969), and Alexander (1971). These authors envision some of the "noblest" traits of mankind, including team play, altruism, patriotism, bravery on the field of battle, and so forth, as the genetic product of warfare.

By adding the additional postulate of a threshold effect, it is possible to explain why the process has operated exclusively in human evolution (Wilson, 1972a). If any social predatory mammal attains a certain level of intelligence, as the early hominids, being large primates, were especially predisposed to do, one band would have the capacity to consciously ponder the significance of adjacent social groups and to deal with them in an intelligent, organized fashion. A band might then dispose of a neighboring band, appropriate its territory, and increase its own genetic representation in the metapopulation, retaining the tribal memory of this successful episode, repeating it, increasing the geographic range of its occurrence, and quickly spreading its influence still further in the metapopulation. Such primitive cultural capacity would be permitted by the possession of certain genes. Reciprocally, the cultural capacity might propel the spread of the genes through the genetic constitution of the metapopulation. Once begun, such a mutual reinforcement could be irreversible. The only combinations of genes able to confer superior fitness in contention with genocidal aggressors would be those that produce either a more effective technique of aggression or else the capacity to preempt genocide by some form of pacific maneuvering. Either probably entails mental and cultural advance. In addition to being autocatalytic, such evolution has the interesting property of requiring a selection episode only very occasionally in order to proceed as swiftly as individual-level selection. By current theory, genocide or genosorption strongly favoring the aggressor need take place only once every few generations to direct evolution. This alone could push truly altruistic genes to a high frequency within the bands (see Chapter 5). The turnover of tribes and chiefdoms estimated from atlases of early European and Mideastern history (for example, the atlas by McEvedy, 1967) suggests a sufficient magnitude of differential group fitness to have achieved this effect. Furthermore, it is to be expected that some isolated cultures will escape the process for generations at a time, in effect reverting temporarily to what ethnographers classify as a pacific state.

Multifactorial Systems

Each of the foregoing mechanisms could conceivably stand alone as a sufficient prime mover of social evolution. But it is much more likely that they contributed jointly, in different strengths and with complex interaction effects. Hence the most realistic model may be fully cybernetic, with cause and effect reciprocating through subcycles that possess high degrees of connectivity with one another. One such scheme, proposed by Adams (1966) for the rise of states and urban societies, is presented in Figure 27-8. Needless to say, the equations needed to translate this and similar models have not been written, and the magnitudes of the coefficients cannot even be guessed at the present time.

In both the unifactorial and multifactorial models of social evolution, an increasing internalization of the controls is postulated. This shift is considered to be the basis of the two-stage acceleration cited earlier. At the beginning of hominid evolution, the prime movers were external environmental pressures no different from those that have guided the social evolution of other animal species. For the moment, it seems reasonable to suppose that the hominids underwent two adaptive shifts in succession: first, to open-country living and seed eating, and second, after being preadapted by the anatomical and mental changes associated with seed eating, to the capture of large mammals. Big-game hunting induced further growth in mentality and social organization that brought the hominids across the threshold into the autocatalytic, more nearly internalized phase of evolution. This second stage is the one in which the most distinctive human qualities emerged. In stressing this distinction, however, I do not wish to imply that social evolution became independent of the environment. The iron laws of demography still clamped down on the spreading hominid populations, and the most spectacular cultural advances were impelled by the invention of new ways to control the environment. What happened was that mental and social change came to depend more on internal reorganization and less on direct responses to features in the surrounding environment. Social evolution, in short, had acquired its own motor.

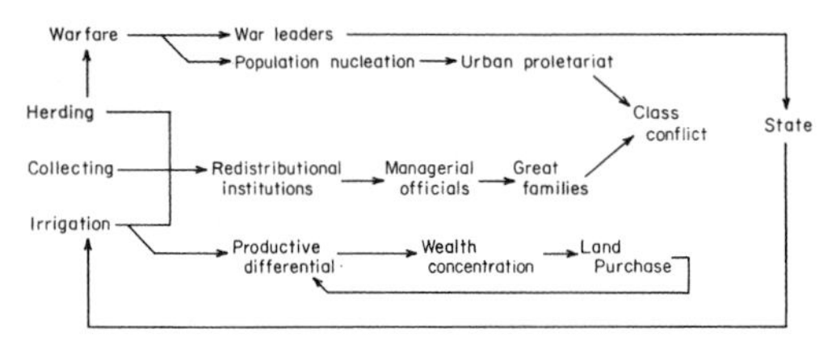

Figure 27-8 A multifactorial model of the origin of the state and urban society. (From Flannery, 1972; based on Adams, 1966. Reproduced, with permission, from "The Cultural Evolution of Civilizations," *Annual Review of Ecology and Systematics,* Vol. 3, p. 408. Copyright © 1972 by Annual Reviews, Inc. All rights reserved.)

The Future

When mankind has achieved an ecological steady state, probably by the end of the twenty-first century, the internalization of social evolution will be nearly complete. About this time biology should be at its peak, with the social sciences maturing rapidly. Some historians of science will take issue with this projection, arguing that the accelerating pace of discoveries in these fields implies a more rapid development. But historical precedents have misled us before: the subjects we are talking about are more difficult than physics or chemistry by at least two orders of magnitude.

Consider the prospects for sociology. This science is now in the natural history stage of its development. There have been attempts at system building but, just as in psychology, they were premature and came to little. Much of what passes for theory in sociology today is really labeling of phenomena and concepts, in the expected manner of natural history. Process is difficult to analyze because the fundamental units are elusive, perhaps nonexistent. Syntheses commonly consist of the tedious cross-referencing of differing sets of definitions and metaphors erected by the more imaginative thinkers (see for example Inkeles, 1964, and Friedrichs, 1970). That, too, is typical of the natural history phase.

With an increase in the richness of descriptions and experiments, sociology is drawing closer each day to cultural anthropology, social psychology, and economics, and will soon merge with them. These disciplines are fundamental to sociology *sensu lato* and are most likely to yield its first phenomenological laws. In fact, some viable qualitative laws probably already exist. They include tested statements about the following relationships: the effects of hostility and stress upon ethnocentrism and xenophobia (LeVine and Campbell, 1972); the positive correlation between and within cultures of war and combative sports, resulting in the elimination of the hydraulic model of aggressive drive (Sipes, 1973); precise but still specialized models of promotion and opportunity within professional guilds (White, 1970); and, far from least, the most general models of economics.

The transition from purely phenomenological to fundamental theory in sociology must await a full, neuronal explanation of the human brain. Only when the machinery can be torn down on paper at the level of the cell and put together again will the properties of emotion and ethical judgment come clear. Simulations can then be employed to estimate the full range of behavioral responses and the precision of their homeostatic controls. Stress will be evaluated in terms of the neurophysiological perturbations and their relaxation times. Cognition will be translated into circuitry. Learning and creativeness will be defined as the alteration of specific portions of the cognitive machinery regulated by input from the emotive centers. Having cannibalized psychology, the new neurobiology will yield an enduring set of first principles for sociology.

The role of evolutionary sociobiology in this enterprise will be twofold. It will attempt to reconstruct the history of the machinery and to identify the adaptive significance of each of its functions. Some of the functions are almost certainly obsolete, being directed toward such Pleistocene exigencies as hunting and gathering and intertribal warfare. Others may prove currently adaptive at the level of the individual and family but maladaptive at the level of the group—or the reverse. If the decision is taken to mold cultures to fit the requirements of the ecological steady state, some behaviors can be altered experientially without emotional damage or loss in creativity. Others cannot. Uncertainty in this matter means that Skinner's dream of a culture predesigned for happiness will surely have to wait for the new neurobiology. A genetically accurate and hence completely fair code of ethics must also wait.

The second contribution of evolutionary sociobiology will be to monitor the genetic basis of social behavior. Optimum socioeconomic systems can never be perfect, because of Arrow's impossibility theorem and probably also because ethical standards are innately pluralistic. Moreover, the genetic foundation on which any such normative system is built can be expected to shift continuously. Mankind has never stopped evolving, but in a sense his populations are drifting. The effects over a period of a few generations could change the identity of the socioeconomic optima. In particular, the rate of gene flow around the world has risen to dramatic levels and is accelerating, and the mean coefficients of relationship within local communities are correspondingly diminishing. The result could be an eventual lessening of altruistic behavior through the maladaption and loss of group-selected genes (Haldane, 1932; Eshel, 1972). It was shown earlier that behavioral traits tend to be selected out by the principle of metabolic conservation when they are suppressed or when their original function becomes neutral in adaptive value. Such traits can largely disappear from populations in as few as ten generations, only two or three centuries in the case of human beings. With our present inadequate understanding of the human brain, we do not know how many of the most valued qualities are linked genetically to more obsolete, destructive ones. Cooperativeness toward groupmates might be coupled with aggressivity toward strangers, creativeness with a desire to own and dominate, athletic zeal with a tendency to violent response, and so on. In extreme cases such pairings could stem from pleiotropism, the control of more than one phenotypic character by the same set of genes. If the planned society—the creation of which seems inevitable in the coming century—were to deliberately steer its members past those stresses and conflicts that once gave the destructive phenotypes their Darwinian edge, the other phenotypes might dwindle with them. In this, the ultimate genetic sense, social control would rob man of his humanity.

It seems that our autocatalytic social evolution has locked us onto a particular course which the early hominids still within us may not welcome. To maintain the species indefinitely we are compelled to drive toward total knowledge, right down to the levels of the neuron and gene. When we have progressed enough to explain ourselves in these mechanistic terms, and the social sciences come to full flower, the result might be hard to accept. It seems appropriate therefore to close this book as it began, with the foreboding insight of Albert Camus:

> A world that can be explained even with bad reasons is a familiar world. But, on the other hand, in a universe divested of illusions and lights, man feels an alien, a stranger. His exile is without remedy since he is deprived of the memory of a lost home or the hope of a promised land.

This, unfortunately, is true. But we still have another hundred years.

1975, 2000

"Sociobiology at century's end," the foreword to the twenty-fifth anniversary edition of *Sociobiology: The New Synthesis* (Belknap Press of Harvard University Press, Cambridge, MA), pp. v–viii

This essay was written in 2000 to introduce the twenty-fifth anniversary and otherwise unchanged reissue of *Sociobiology: The New Synthesis.* It tells some of the story of the response to this work, together with my own perception of the significance of the discipline. Other, fuller accounts of this history can be found in my memoir *Naturalist* (Island Press, Washington, DC, 2000) and Ullica Segerstråle's admirably exhaustive *Defenders of the Truth: The Battle for Science in the Sociobiology Debate and Beyond* (Oxford University Press, 2000).

❖ ARTICLE 14

Sociobiology at Century's End

Sociobiology was brought together as a coherent discipline in *Sociobiology: The New Synthesis* (1975), the book now reprinted, but it was originally conceived in my earlier work *The Insect Societies* (1971) as a union between entomology and population biology. This first step was entirely logical, and in retrospect, inevitable. In the 1950s and 1960s studies of the social insects had multiplied and attained a new but still unorganized level. My colleagues and I had worked out many of the principles of chemical communication, the evolution and physiological determinants of caste, and the dozen or so independent phylogenetic pathways along which the ants, termites, bees, and wasps had probably attained advanced sociality. The idea of kin selection, introduced by William D. Hamilton in 1963, was newly available as a key organizing concept. A rich database awaited integration. Also, more than 12,000 species of social insects were known and available for comparative studies to test the adaptiveness of colonial life, a great advantage over the relatively species-poor vertebrates, of which only a few hundred are known to exhibit advanced social organization. And finally, because the social insects obey rigid instincts, there was little of the interplay of heredity and environment that confounds the study of vertebrates.

During roughly the same period, up to 1971, researchers achieved comparable advances in population biology. They devised richer models of the genetics and growth dynamics of populations, and linked demography more exactly to competition and symbiosis. In the 1967 synthesis *The Theory of Island Biogeography*, Robert H. MacArthur and I (if you will permit the continued autobiographical slant of this account) meshed principles of population biology with patterns of species biodiversity and distribution.

It was a natural step then to write *The Insect Societies* at the close of the 1960s as an attempt to reorganize the highly eclectic knowledge of the social insects on a base of population biology. Each insect colony is an assemblage of related organisms that grows, competes, and eventually dies in patterns that are consequences of the birth and death schedules of its members.

And what of the vertebrate societies? In the last chapter of *The Insect Societies*, entitled "The Prospect for a Unified Sociobiology," I made an optimistic projection to combine the two great phylads:

> In spite of the phylogenetic remoteness of vertebrates and insects and the basic distinction between their respective personal and impersonal systems of communication, these two groups of animals have evolved social behaviors that are similar in degree and complexity and convergent in many important details. This fact conveys a special promise that sociobiology can eventually be derived from the first principles of population and behavioral biology and developed into a

single, mature science. The discipline can then be expected to increase our understanding of the unique qualities of social behavior in animals as opposed to those of man.

The sequel in this reasoning is contained in the book before you. Presented in this new release by Harvard University Press, it remains unchanged from the original. It provides verbatim the first effort to systematize the consilient links between termites and chimpanzees, the goal suggested in *The Insect Societies*, but it goes further, and extends the effort to human beings.

The response to *Sociobiology: The New Synthesis* in 1975 and the years immediately following was dramatically mixed. I think it fair to say that the zoology in the book, making up all but the first and last of its 27 chapters, was favorably received. The influence of this portion grew steadily, so much so that in a 1989 poll the officers and fellows of the international Animal Behavior Society rated *Sociobiology* the most important book on animal behavior of all time, edging out even Darwin's 1872 classic, *The Expression of the Emotions in Man and Animals.* By integrating the discoveries of many investigators into a single framework of cause-and-effect theory, it helped to change the study of animal behavior into a discipline connected broadly to mainstream evolutionary biology.

The brief segment of *Sociobiology* that addresses human behavior, comprising 30 out of the 575 total pages, was less well received. It ignited the most tumultuous academic controversy of the 1970s, one that spilled out of biology into the social sciences and humanities. The story has been told many times and many ways, including the account in my memoir, *Naturalist,* where I tried hard to maintain a decent sense of balance; and it will bear only a brief commentary here.

Although the large amount of commotion may suggest otherwise, adverse critics made up only a small minority of those who published reviews of *Sociobiology.* But they were very vocal and effective at the time. They were scandalized by what they saw as two grievous flaws. The first is inappropriate reductionism, in this case the proposal that human social behavior is ultimately reducible to biology. The second perceived flaw is genetic determinism, the belief that human nature is rooted in our genes.

It made little difference to those who chose to read the book this way that reductionism is the primary cutting tool of science, or that *Sociobiology* stresses not only reductionism but also synthesis and holism. It also mattered not at all that sociobiological explanations were never strictly reductionist, but interactionist. No serious scholar would think that human behavior is controlled the way animal instinct is, without the intervention of culture. In the interactionist view held by virtually all who study the subject, genomics biases mental development but cannot abolish culture. To suggest that I held such views, and it was suggested frequently, was to erect a straw man—to fabricate false testimony for rhetorical purposes.

Who were the critics, and why were they so offended? Their rank included the last of the Marxist intellectuals, most prominently represented by Stephen Jay Gould and Richard C. Lewontin. They disliked the idea, to put it mildly, that human nature could have any genetic basis at all. They championed the opposing view that the developing human brain is a tabula rasa. The only human nature, they said, is an indefinitely flexible mind. Theirs was the standard political position taken by Marxists from the late 1920s forward: the ideal political economy is socialism, and the tabula rasa mind of people can be fitted to it. A mind arising from a genetic human nature might not prove conformable. Since socialism is the supreme good to be sought, a tabula rasa it must be. As Lewontin, Steven Rose, and Leon J. Kamin frankly expressed the matter in *Not in Our Genes* (1984): "We share a commitment to the prospect of the creation of a more socially just—a socialist—society. And we recognize that a critical science is an integral part of the struggle to create that society, just as we also believe that the social function of much of today's science is to hinder the creation of that society by acting to preserve the interests of the dominant class, gender, and race."

That was in 1984—an apposite Orwellian date. The argument for a political test of scientific knowledge lost its strength with the collapse of world socialism and the end of the Cold War. To my knowledge it has not been heard since.

In the 1970s, when the human sociobiology controversy still waxed hot, however, the Old Marxists were joined and greatly strengthened by members of the New Left in a second objection, this time centered on social justice. If genes prescribe human nature, they said, then it follows that ineradicable differences in personality and ability also might exist. Such a possibility cannot be tolerated. At least, its discussion cannot be tolerated, said the critics, because it tilts thinking onto a slippery slope down which humankind easily descends to racism, sexism, class oppression, colonialism, and—perhaps worst of all—capitalism! As the century closes, this dispute has been settled. Genetically based variation in individual personality and intelligence has been conclusively demonstrated, although statistical racial differences, if any, remain unproven. At the same time, all of the projected evils except capitalism have begun to diminish worldwide. None of the change can be ascribed to human behavioral genetics or sociobiology. Capitalism may yet fall—who can predict history?—but, given the overwhelming evidence at hand, the hereditary framework of human nature seems permanently secure.

Among many social scientists and humanities scholars a deeper and less ideological source of skepticism was expressed, and remains. It is based on the belief that culture is the sole artisan of the human mind. This perception is also a tabula rasa hypothesis that denies biology, or at least simply ignores biology. It too is being replaced by acceptance of the interaction of biology and culture as the determinant of mental development.

Overall, there is a tendency as the century closes to accept that *Homo sapiens* is an ascendant primate, and that biology matters.

The path is not smooth, however. The slowness with which human sociobiology (nowadays also called evolutionary psychology) has spread is due not merely to ideology and inertia, but also and more fundamentally to the traditional divide between the great branches of learning. Since the early nineteenth century it has been generally assumed that the natural sciences, the social sciences, and the humanities are epistemologically disjunct from one another, requiring different vocabularies, modes of analysis, and rules of validation. The perceived dividing line is essentially the same as that between the scientific and literary cultures defined by C. P. Snow in 1959. It still fragments the intellectual landscape.

The solution to the problem now evident is the recognition that the line between the great branches of learning is not a line at all, but instead a broad, mostly unexplored domain awaiting cooperative exploration from both sides. Four borderland disciplines are expanding into this domain from the natural sciences side:

Cognitive neuroscience, also known as the brain sciences, maps brain activity with increasingly fine resolution in space and time. Neural pathways, some correlated with complex and sophisticated patterns of thought, can now be traced. Mental disorders are routinely diagnosed by this means, and the effects of drugs and hormone surges can be assessed almost directly. Neuroscientists are able to construct replicas of mental activity that, while still grossly incomplete, go far beyond the philosophical speculations of the past. They can then coordinate these with experiments and models from cognitive psychology, thus drawing down on independent reservoirs from yet another discipline bridging the natural and social sciences. As a result, one of the major gaps of the intellectual terrain, that between body and mind, may soon be closed.

In human genetics, with base pair sequences and genetic maps far advanced and near completion, a direct approach to the heredity of human behavior has opened up. A total genomics, which includes the molecular steps of epigenesis and the norms of reaction in gene-environment interaction, is still far off. But the technical means to attain it are being developed. A large portion of research in molecular and cellular biology is devoted to that very end. The implications for consilience are profound: each advance in neuropsychological genomics narrows the mind-body gap still further.

Where cognitive neuroscience aims to explain *how* the brains of animals and humans work, and genetics how heredity works, evolutionary biology aims to explain *why* brains work, or more precisely, in light of natural selection theory, what adaptations if any led to the assembly of their respective parts and processes. During the past 25 years an impressive body of ethnographic data has been marshaled to test adaptation hypotheses, especially those emanating from kin-selection and ecological optimization models. Much of the research, conducted by both biologists and social scientists, has been reported in the journals *Behavioral Ecology and Sociobiology, Evolution and Human Behavior* (formerly *Ethology and Sociobiology*), *Human Nature,* the *Journal of Social and Biological Structures,* and others, as well as in excellent summary collections such as *The Adapted Mind: Evolutionary Psychology and the Generation of Culture* (Jerome H. Barkow, Leda Cosmides, and John Tooby, eds., 1992) and *Human Nature: A Critical Reader* (Laura Betzig, ed., 1997).

As a result we now possess a much clearer understanding of ethnicity, kin classification, bridewealth, marriage customs, incest taboos, and other staples of the human sciences. New models of conflict and cooperation, extending from Robert L. Trivers' original parent-offspring conflict theory of the 1970s and from ingenious applications of game theory, have been applied fruitfully to developmental psychology and an astonishing diversity of other fields—embryology, for example, pediatrics, and the study of genomic imprinting. Comparisons with the social behavior of the nonhuman primates, now a major concern of biological anthropology, have proven valuable in the analysis of human behavioral phenomena that are cryptic or complex.

Sociobiology is a flourishing discipline in zoology, but its ultimately greatest importance will surely be the furtherance of consilience among the great branches of learning. Why is this conjunction important? Because it offers the prospect of characterizing human nature with greater objectivity and precision, an exactitude that is the key to self-understanding. The intuitive grasp of human nature has been the substance of the creative arts. It is the ultimate underpinning of the social sciences and a beckoning mystery to the natural sciences. To grasp human nature objectively, to explore it to the depths scientifically, and to comprehend its ramifications by cause-and-effect explanations leading from biology into culture, would be to approach if not attain the grail of scholarship, and to fulfill the dreams of the Enlightenment.

The objective meaning of human nature is attainable in the borderland disciplines. We have come to understand that human nature is not the genes that prescribe it. Nor is it the cultural universals, such as the incest taboos and rites of passage, which are its products. Rather, human nature is the epigenetic rules, the inherited regularities of mental development. These rules are the genetic biases in the way our senses perceive the world, the symbolic coding by which our brains represent the world, the options we open to ourselves, and the responses we find easiest and most rewarding to make. In ways that are being clarified at the physiological and even in a few cases the genetic level, the epigenetic rules alter the way we see and intrinsically classify color. They cause us to evaluate the aesthetics of artistic design according to elementary abstract shapes and the degree of complexity. They lead us differentially to acquire fears and phobias concerning dangers in the ancient environment of humanity (such as snakes and heights), to communicate with certain facial expressions and forms of body language, to bond with infants, to bond conjugally, and so on across a wide spread of categories in behavior and thought. Most of these rules are evidently very ancient, dating back millions of years in mammalian ancestry. Others, like the ontogenetic steps of linguistic development in children, are uniquely human and probably only hundreds of thousands of years old.

The epigenetic rules have been the subject of many studies during the past quarter century in biology and the social sciences, reviewed for example in my extended essays *On Human Nature* (1978) and *Consilience: The Unity of Knowledge* (1998), as well as in *The Adapted Mind*, edited by Jerome L. Barkow et al. (1992). This body of work makes it evident that in the creation of human nature, genetic evolution and cultural evolution have together produced a closely interwoven product. We are only beginning to obtain a glimmer of how the process works. We know that cultural evolution is biased substantially by biology, and that biological evolution of the brain, especially the neocortex, has occurred in a social context. But the principles and the details are the great challenge in the emerging borderland disciplines just described. The exact process of gene-culture coevolution is the central problem of the social sciences and much of the humanities, and it is one of the great remaining problems of the natural sciences. Solving it is the obvious means by which the great branches of learning can be foundationally united.

Finally, during the past quarter century another discipline to which I have devoted a good part of my life, conservation biology, has been tied more closely to human sociobiology. Human nature—the epigenetic rules—did not originate in cities and croplands, which are too recent in human history to have driven significant amounts of genetic evolution. They arose in natural environments, especially the savannas and transitional woodlands of Africa, where *Homo sapiens* and its antecedents evolved over hundreds of thousands of years. What we call the natural environment or wilderness today was home then—the environment that cradled humanity. Before agriculture the lives of people depended on their intimate familiarity with wild biodiversity, both the surrounding ecosystems and the plants and animals composing them.

The link was, on a scale of evolutionary time, abruptly weakened by the invention and spread of agriculture and then nearly erased by the implosion of a large part of the agricultural population into the cities during the industrial and postindustrial revolutions. As global culture advanced into the new, technoscientific age, human nature stayed back in the Paleolithic era.

Hence the ambivalent stance taken by modern *Homo sapiens* to the natural environment. Natural environments are cherished at the same time they are subdued and converted. The ideal planet for the human psyche seems to be one that offers an endless expanse of fertile, unoccupied wilderness to be churned up for the production of more people. But Earth is finite, and its still exponentially growing human population is rapidly running out of productive land for conversion. Clearly humanity must find a way simultaneously to stabilize its population and to attain a universal decent standard of living while preserving as much of Earth's natural environment and biodiversity as possible.

Conservation, I have long believed, is ultimately an ethical issue. Moral precepts in turn must be based on a sound, objective knowledge of human nature. In 1984 I combined my two intellectual passions, sociobiology and the study of biodiversity, in the book *Biophilia* (Harvard University Press). Its central argument was that the epigenetic rules of mental development are likely to include deep adaptive responses to the natural environment. This theme was largely speculation. There was no organized discipline of ecological psychology that addressed such a hypothesis. Still, plenty of evidence pointed to its validity. In *Biophilia* I reviewed information then newly provided by Gordon Orians that points to innately preferred habitation (on a prominence overlooking a savanna and body of water), the remarkable influence of snakes and serpent images on culture, and other mental predispositions likely to have been adaptive during the evolution of the human brain.

Since 1984 the evidence favoring biophilia has grown stronger, but the subject is still in its infancy and few principles have been definitively established (see *The Biophilia Hypothesis*, Stephen R. Kellert and Edward O. Wilson, eds., Island Press, 1993). I am persuaded that as the need to stabilize and protect the environment grows more urgent in the coming decades, the linking of the two natures—human nature and wild Nature—will become a central intellectual concern.

December 1999
Cambridge, Massachusetts

1975

"Human decency is animal," *New York Times Magazine* October 12, pp. 38–50

This article on human sociobiology written for a popular audience has continued in circulation, mostly in classrooms, for more than 30 years. Its popularity is undoubtedly due to the argument that nature is not all red in tooth and claw, but leavened—at least in families and societies—by altruism and kindness. Sociobiology, which is built on the theory of genetic evolution by natural selection, has demonstrated how moral precepts and other binding processes of society can arise from the superiority of genes favoring them. If a group is more tightly supportive of individuals and as a result more prosperous than groups less supportive, then genes favoring such behavior will spread through the population as a whole, even if they prescribe some amount of self-sacrifice on the part of group members.

In a forest in Uganda, a troop of gorillas conduct their lives with the relaxation, amiability and cooperation that are typical of their society. Aggression in the group exists, but less so than in human society. At the left, beyond the dominant silverback male, two female gorillas watch 2-year-old twins at play. In the right

HUMAN DECENCY IS ANIMAL

Hawks and baboons are not usually heroic, but altruistic dolphins and charitable chimps point to a kindly strain in our genes

By Edward O. Wilson

During the American wars of this century, a large percentage of Congressional Medals of Honor were awarded to men who threw themselves on top of grenades to shield comrades, aided the rescue of others from battle sites at the price of certain death to themselves, or made other, often carefully considered but extraordinary, decisions that led to the same fatal end. Such altruistic suicide is the ultimate act of courage and emphatically deserves the country's highest honor. It is also

Edward O. Wilson, professor of zoology at Harvard University, is the author of "Sociobiology: The New Synthesis."

foreground a female cradles a nursing infant, while another grooms a 3-year-old. At the far right, another female carries a 2-year-old on her back as she feeds. There is a great deal of variation of facial features; this is believed to be used by the gorillas themselves in recognizing individual group members.

only the extreme act that lies beyond the innumerable smaller performances of kindness and giving that bind societies together. One is tempted to leave the matter there, to accept altruism as simply the better side of human nature. Perhaps, to put the best possible construction on the matter, conscious altruism is a transcendental quality that distinguishes human beings from animals. Scientists are nevertheless not accustomed to declaring any phenomenon off limits, and recently there has been a renewed interest in analyzing such forms of social behavior in greater depth and as objectively as possible.

Much of the new effort falls within a discipline called sociobiology, which is defined as the systematic study of the biological basis of social behavior in every kind of organism, including man, and is being pieced together with contributions from biology, psychology and anthropology. There is of course nothing new about analyzing social behavior, and even the word "sociobiology" has been around for some years. What is new is the way facts and ideas are being extracted from their traditional matrix of psychology and ethology (the natural history of animal behavior) and reassembled in compliance with the principles of genetics and ecology.

In sociobiology, there is a heavy emphasis on the comparison of societies of different kinds of animals and of man, not so much to draw analogies (these have often been dangerously misleading, as when aggression is compared directly in wolves and in human beings) but to devise and to test theories about the underlying hereditary basis of social behavior. With genetic evolution always in mind, sociobiologists search for the ways in which the myriad forms of social organization adapt particular species to the special opportunities and dangers encountered in their environment.

A case in point is altruism. I doubt if any higher animal, such as a hawk or a baboon, has ever deserved a Congressional Medal of Honor by the ennobling criteria used in our society. Yet minor altruism does occur frequently, in forms instantly understandable in human terms, and is bestowed not just on offspring but on other members of the species as well. Certain small birds, robins, thrushes and titmice, for example, warn others of the approach of a hawk. They crouch low and emit a distinctive thin, reedy whistle. Although the warning call has acoustic properties that make it difficult to locate in space, to whistle at all seems at the very least unselfish; the caller would be wiser not to betray its presence but rather to remain silent and let someone else fall victim.

When a dolphin is harpooned or otherwise seriously injured, the typical response of the remainder of the school is to desert the area immediately. But, sometimes, they crowd around the stricken animal and lift it to the surface, where it is able to continue breathing air. Packs of African wild dogs, the most social of all carnivorous mammals, are organized in part by a remarkable division of labor. During the denning season, some of the adults, usually led by a dominant male, are forced to leave the pups behind in order to hunt for antelopes and other prey. At least one adult, normally the mother of the litter, stays behind as a guard. When the hunters return, they regurgitate pieces of meat to all that stayed home. Even sick and crippled adults are benefited, and as a result they are able to survive longer than would be the case in less generous societies.

Other than man, chimpanzees may be the most altruistic of all mammals. Ordinarily, chimps are vegetarians, and during their relaxed foraging excursions they feed singly in the uncoordinated manner of other monkeys and apes. But, occasionally, the males hunt monkeys and young baboons for food. During these episodes, the entire mood of the troop shifts toward what can only be characterized as a manlike state. The males stalk and chase their victims in concert; they also gang up to repulse any of the victims' adult relatives which oppose them. When the hunters have dismembered the prey and are feasting, other chimps approach to beg for morsels. They touch the meat and the faces of the males, whimpering and *hoo*ing gently, and hold out their hands—palms up—in supplication. The meat eaters sometimes pull away in refusal or walk off. But, often, they permit the other animal to chew directly on the meat or to pull off small pieces with its hands. On several occasions, champanzees have actually been observed to tear off pieces and drop them into the outstretched hands of others—an act of generosity unknown in other monkeys and apes.

Adoption is also practiced by chimpanzees. Jane Goodall has observed three cases at the Gombe Stream National Park in Tanzania. All involved orphaned infants taken over by adult brothers and sisters. It is of considerable interest, for more theoretical reasons to be discussed shortly, that the altruistic behavior was displayed by the closest possible relatives rather than by experienced females with children of their own, females who might have supplied the orphans with milk and more adequate social protection.

In spite of a fair abundance of such examples among vertebrate creatures, it is only in the lower animals and in the social insects particularly, that we encounter altruistic suicide comparable to man's. A large percentage of the members of colonies of ants, bees and wasps are ready to defend their nests with insane charges against intruders. This is the reason that people move with circumspection around honeybee hives and yellowjacket burrows, but can afford to relax near the nests of solitary species such as sweat bees and mud daubers.

The social stingless bees of the tropics swarm over the heads of human beings who venture too close, locking their jaws so tightly onto tufts of hair that their bodies pull loose from their heads when they are combed out. Some of the species pour a burning glandular secretion onto the skin during these sacrificial attacks. In Brazil, they are called *cagafogos* ("fire defecators") The great entomologist William Morton Wheeler described an encounter with the "terrible bees," during which they removed patches of skin from his face, as the worst experience of his life.

Honeybee workers have stings lined with reversed barbs like those on fishhooks. When a bee attacks an intruder at the hive, the sting catches in the skin; as the bee moves away, the sting remains embedded, pulling out the entire venom gland and much of the viscera with it. The bee soon dies, but its attack has been more effective than if it withdrew the sting intact. The reason is that the venom gland continues to leak poison into the wound, while a bananalike odor emanating from the base of the sting incites other members of the hive into launching Kamikaze attacks of their own at the same spot. From the point of view of the colony as a whole, the suicide of an individual accomplishes more than it loses. The total worker force consists of 20,000 to 80,000 members, all sisters born from eggs laid by the mother queen. Each bee has a natural life span of only about 50 days, at the end of which it dies of old age. So to give a life is only a little thing, with no genes being spilled in the process.

My favorite example among the social insects is provided by an African termite with the orotund, technical name *Globitermes sulfureus*. Members of this species' soldier caste are quite literally walking bombs. Huge paired glands extend from their heads back through most of their bodies. When they attack ants and other enemies, they eject a yellow glandular secretion through their mouths; it congeals in the air and often fatally entangles both the soldiers and their antagonists. The spray appears to be powered by contractions of the muscles in the abdominal wall. Sometimes, the contractions become so violent that the abdomen and gland explode, spraying the defensive fluid in all directions.

Sharing a capacity for extreme sacrifice does not mean that the human mind and the "mind" of an insect (if such exists) work alike. But it does mean that the impulse need not be ruled divine or otherwise transcendental, and we are justified in seeking a more conventional biological explanation. One immediately encounters a basic problem connected with such an explanation: Fallen heroes don't have any more children. If self-sacrifice results in fewer descendants, the genes, or basic units of heredity, that allow heroes to be created can be expected to disappear gradually from the population. This is the result of the narrow mode of Darwinian natural selection: Because people who are governed by selfish genes prevail over those with altruistic genes, there should be a tendency over many generations for selfish genes to increase in number and for the human population as a whole to become less and less capable of responding in an altruistic manner.

How can altruism persist? In the case of the social insects, there is no doubt at all. Natural selection has been broadened to include a process called kin selection. The self-sacrificing termite soldier protects the rest of the colony, including the queen and king which are the soldier's parents. As a result, the soldier's more fertile brothers and sisters flourish, and it is *they* which multiply the altruistic genes that are shared with the soldier by close kinship. One's own genes are multiplied by the greater production of nephews and nieces. It is natural, then, to ask whether the capacity for altruism has also evolved in human beings through kin selection. In other words, do the emotions we feel, which on occasion in exceptional individuals climax in total self-sacrifice, stem ultimately from hereditary units that were implanted by the favoring of relatives during a period of hundreds or thousands of generations? This explanation gains some strength from the circumstance that during most of mankind's history the social unit was the immediate family and a tight network of other close relatives. Such exceptional cohesion, combined with a detailed awareness of kinship made possible by high intelligence, might explain why kin selection has been more forceful in human beings than in monkeys and other mammals.

To anticipate a common objection raised by many social scientists and others, let me grant at once that the intensity and form of altruistic acts are to a large extent culturally determined. Human social evolution is obviously more cultural than genetic. The point is that the underlying emotion, powerfully manifested in virtually all human societies, is what is considered to evolve through genes. This sociobiological hypothesis does not therefore account for differences among societies, but it could explain why human beings differ from other mammals and why, in one narrow aspect, they more closely resemble social insects.

In cases where sociobiological explanations can be tested and proved true, they will, at the very least, provide perspective and a new sense of philosophical ease about human nature. I believe that they will also have an ultimately moderating influence on social tensions. Consider the case of homosexuality. Homophiles are typically rejected in our society because of a narrow and unfair biological premise made about them: Their sexual preference does not produce children; therefore, they cannot be natural. To the extent that this view can be rationalized, it is just Darwinism in the old narrow sense: Homosexuality does not directly replicate genes. But homosexuals *can* replicate genes by kin selection, provided

Thus came cooperation: Two million years ago, a band of men, foraging for food, fight off animals which also want the newly fallen, elephantlike dinothere. Since these men were less than five feet tall, they had to pull together to keep their prey.

they are sufficiently altruistic toward kin.

It is not inconceivable that in the early, hunter-gatherer period of human evolution, and perhaps even later, homosexuals regularly served as a partly sterile caste, enhancing the lives and reproductive success of their relatives by a more dedicated form of support than would have been possible if they produced children of their own. If such combinations of interrelated heterosexuals and homosexuals regularly left more descendants than similar groups of pure heterosexuals, the capacity for homosexual development would remain prominent in the population as a whole. And it has remained prominent in the great majority of human societies, to the consternation of anthropologists, biologists and others.

Supporting evidence for this new kin-selection hypothesis does not exist. In fact, it has not even been examined critically. But the fact that it is internally consistent and can be squared with the results of kin selection in other kinds of organisms should give us pause before labeling homosexuality an illness. I might add that if the hypothesis is correct, we can expect homosexuality to decline over many generations. The reason is that the extreme dispersal of family groups in modern industrial societies leaves fewer opportunities for preferred treatment of relatives. The labor of homosexuals is spread more evenly over the population at large, and the narrower form of Darwinian natural selection turns against the duplication of genes favoring this kind of altruism.

A peacemaking role of modern sociobiology also seems likely in the interpretation of aggression, the behavior at the opposite pole from altruism. To cite aggression as a form of social behavior is, in a way, contradictory; considered by itself, it is more accurately identified as antisocial behavior. But, when viewed in a social context, it seems to be one of the most important and widespread organizing techniques. Animals use it to stake out their own territories and to establish their rank in the pecking orders. And because members of one group often cooperate for the purpose of directing aggression at competitor groups, altruism and hostility have come to be opposite sides of the same coin.

Konrad Lorenz, in his celebrated book "On Aggression," argued that human beings share a general instinct for aggressive behavior with animals, and that this instinct must somehow be relieved, if only through competitive sport. Erich Fromm, in "The Anatomy of Human Destructiveness," took the still dimmer view that man's behavior is subject to a unique death instinct that often leads to pathological aggression beyond that encountered in animals. Both of these interpretations are essentially wrong. A close look at aggressive behavior in a variety of animal societies, many of which have been carefully studied only since the time Lorenz drew his conclusions, shows that aggression occurs in a myriad of forms and is subject to rapid evolution.

We commonly find one species of bird or mammal to be highly territorial, employing elaborate, aggressive displays and attacks, while a second, otherwise similar, species shows little or no territorial behavior. In short, the case for a pervasive aggressive instinct does not exist.

The reason for the lack of a general drive seems quite clear. Most kinds of aggressive behavior are perceived by biologists as particular responses to crowding in the environment. Animals use aggression to gain control over necessities—usually food or shelter—which are in short supply or likely to become short at some time during the life cycle. Many species seldom, if ever, run short of these necessities; rather, their numbers are controlled by predators, parasites or emigration. Such animals are characteristically pacific in their behavior toward one another.

Mankind, let me add at once, happens to be one of the aggressive species. But we are far from being the most aggressive. Recent studies of hyenas, lions and langur monkeys, to take three familiar species, have disclosed that under natural conditions these animals engage in lethal fighting, infanticide and even cannibalism at a rate far above that found in human beings. When a count is made of the number of murders committed per thousand individuals per year, human beings are well down the list of aggressive creatures, and I am fairly confident that this would still be the case even if our episodic wars were to be averaged in. Hyena packs even engage in deadly pitched battles that are virtually indistinguishable from primitive human warfare. Here is some action in the Ngorongoro Crater as described by Hans Kruuk of Oxford University:

"The two groups mixed with an uproar of calls, but within seconds the sides parted again and the Mungi hyenas ran away, briefly pursued by the Scratching Rock hyenas, who then returned to the carcass. About a dozen of the Scratching Rock hyenas, though, grabbed one of the Mungi males and bit him wherever they could—especially in the belly, the feet and the ears. The victim was completely covered by his attackers, who

Altruism: Four dolphins assist another which has been struck by an electroharpoon. "They crowd around the stricken animal and lift it to the surface, where it is able to continue breathing."

proceeded to maul him for about 10 minutes while their clan fellows were eating the wildebeest. The Mungi male was literally pulled apart, and when I later studied the injuries more closely, it appeared that his ears were bitten off and so were his feet and testicles, he was paralyzed by a spinal injury, had large gashes in the hind legs and belly, and subcutaneous hemorrhages all over. . . . The next morning, I found a hyena eating from the carcass and saw evidence that more had been there; about one-third of the internal organs and muscles had been eaten. Cannibals!"

Alongside ants, which conduct assassinations, skirmishes and pitched battles as routine business, men are all but tranquil pacifists. Ant wars, incidentally, are especially easy to observe during the spring and summer in most towns and cities in the Eastern United States. Look for masses of small blackish brown ants struggling together on sidewalks or lawns. The combatants are members of rival colonies of the common pavement ant, *Tetramorium caespitum*. Thousands of individuals may be involved, and the battlefield typically occupies several square feet of the grassroots jungle.

Although some aggressive behavior in one form or another is characteristic of virtually all human societies (even the gentle !Kung Bushmen until recently had a murder rate comparable to that of Detroit and Houston), I know of no evidence that it constitutes a drive searching for an outlet. Certainly, the conduct of animals cannot be used as an argument for the widespread existence of such a drive.

In general, animals display a spectrum of possible actions, ranging from no response at all, through threats and feints, to an all-out attack; and they select the action that best fits the circumstances of each particular threat. A rhesus monkey, for example, signals a peaceful intention toward another troop member by averting its gaze or approaching with conciliatory lip-smacking. A low intensity of hostility is conveyed by an alert, level stare. The hard look you receive from a rhesus when you enter a laboratory or the primate building of a zoo is not simple curiosity—it is a threat.

From that point onward, the monkey conveys increasing levels of confidence and readiness to fight by adding new components one by one, or in combination: The mouth opens in an apparent expression of astonishment, the head bobs up and down, explosive ho's! are uttered and the hands slap the ground. By the time the rhesus is performing all of these displays, and perhaps taking little forward lunges as well, it is prepared to fight. The ritualized performance, which up to this point served to demonstrate precisely the mood of the animal, may then give way to a shrieking, rough-and-tumble assault in which hands, feet and teeth are used as weapons. Higher levels of aggression are not exclusively directed at other monkeys.

Once, in the field, I had a large male monkey reach the hand-slapping stage three feet in front of me when I accidentally frightened an infant monkey which may or may not have been a part of the male's family. At that distance, the male looked like a small gorilla. My guide, Professor Stuart Altmann of the University of Chicago, wisely advised me to avert my gaze and to look as much as possible like a subordinate monkey.

Despite the fact that many kinds of animals are capable of a rich, graduated repertory of aggressive actions, and despite the fact that aggression is important in the organization of their societies, it is possible for individuals to go through a normal life, rearing offspring, with nothing more than occasional bouts of play-fighting and exchanges of lesser hostile displays. The key is the environment: Frequent intense display and escalated fighting are adaptive responses to certain kinds of social stress which a particular animal may or may not be fortunate enough to avoid during its lifetime. By the same token, we should not be surprised to find a few human cultures, such as the Hopi or the newly discovered Tasaday of Mindanao, in which aggressive interactions are minimal. In a word, the evidence from comparative studies of animal behavior cannot be used to justify extreme forms of aggression, bloody drama or violent competitive sports practiced by man.

This brings us to the topic which, in my experience, causes the most difficulty in discussions of human sociobiology: the relative importance of genetic vs. environmental factors in the shaping of behavioral traits. I am aware that the very notion of genes controlling behavior in human beings is scandalous to some scholars. They are quick to project the following political scenario: Genetic determinism will lead to support for a status quo and continued social injustice. Seldom is the equally plausible scenario considered: Environmentalism will lead to support for authoritarian mind control and worse injustice. Both sequences are highly unlikely, unless politicians or ideologically committed scientists are allowed to dictate the uses of science. Then anything goes.

That aside, concern over the implications of sociobiology usually proves to be due to a simple misunderstanding about the nature of heredity. Let me try to set the matter straight as briefly but fairly as possible. *What the genes prescribe is not necessarily a particular behavior but the capacity to develop certain behaviors and, more than that, the tendency to develop them in various specified environments.* Suppose that we could enumerate all conceivable behavior belonging to one category—say, all the possible kinds of aggressive responses — and for convenience label them by letters. In this imaginary example, there might be exactly 23 such responses, which we designate A through W. Human beings do not and cannot manifest all the behaviors; perhaps all societies in the world taken together employ A through P. Furthermore, they do not develop each of these with equal facility; there is a strong tendency under most possible conditions of child rearing for behaviors A through G to appear, and consequently H through P are encountered in very few cultures. It is this *pattern* of possibilities and probabilities that is inherited.

To make such a statement wholly meaningful, we must go on to compare human beings with other species. We note that hamadryas baboons can perhaps develop only F through J, with a strong bias toward F and G, while one kind of termite can show only A and another kind of termite only B. Which behavior a particular human being displays depends on the experience received within his own culture, but the total array of human possibilities, as opposed to baboon or termite possibilities, is inherited. It is the evolution of this pattern which sociobiology attempts to analyze.

e can be more specific about human patterns. It is possible to make a reasonable inference about the most primitive and general human social traits by combining two procedures. First, note is made of the most widespread qualities of hunter-gatherer societies. Although the behavior of the people is complex and intelligent, the way of life to which their cultures are adapted is primitive. The human species evolved with such an elementary economy for hundreds of thousands of years; thus, its innate pattern of social responses can be expected to have been principally shaped by this way of life. The second procedure is to compare the most widespread hunter-gatherer qualities with similar behavior displayed by the species of langurs, colobus, macaques, baboons, chimpanzees, gibbons and other Old World monkeys and apes that, together, comprise man's closest living relatives.

Where the same pattern of traits occurs in man—and in most or all of the primates—we conclude that it has been subject to relatively little evolution. Its possession by hunter-gatherers indicates (but does not prove) that the pattern was also possessed by man's immediate ancestors; the pattern also belongs to the class of behaviors least prone to change even in economically more advanced societies. On the other hand, when the behavior varies a great deal among the primate species, it is less likely to be resistant to change.

The list of basic human patterns that emerges from this screening technique is intriguing: (1) The number of intimate group members is variable but normally 100 or less; (2) some amount of aggressive and territorial behavior is basic, but its intensity is graduated and its particular forms cannot be predicted from one culture to another with precision; (3) adult males are more aggressive and are dominant over females; (4) the societies are to a large extent organized around prolonged maternal care and extended relationships between mothers and children, and (5) play, including at least mild forms of contest and mock-aggression, is keenly pursued and probably essential to normal development.

We must then add the qualities that are so distinctively ineluctably human that they can be safely classified as genetically based: the overwhelming drive of individuals to develop some form of a true, semantic language, the rigid avoidance of incest by taboo and the weaker but still strong tendency for sexually bonded women and men to divide their labor into specialized tasks.

In hunter-gatherer societies, men hunt and women stay at home. This strong bias persists in most agricultural and industrial societies and, on that ground alone, appears to have a genetic origin. No solid evidence exists as to when the division of labor appeared in man's ancestors or how resistant to change it might be during the continuing revolution for women's rights. My own guess is that the genetic bias is intense enough to cause a substantial division of labor even in the most free and most egalitarian of future societies.

Responsibility: In Tanzania, a wild dog, home from the hunt, prepares to regurgitate fresh pieces of meat to pups. All the adults participate fully in the upbringing of the young.

As shown by research recently summarized in the book "The Psychology of Sex Differences," by Eleanor Emmons Maccoby and Carol Nagy Jacklin, boys consistently show more mathematical and less verbal ability than girls on the average, and they are more aggressive from the first hours of social play at age 2 to manhood. Thus, even with identical education and equal access to all professions, men are likely to continue to play a disproportionate role in political life, business and science. But that is only a guess and, even if correct, could not be used to argue for anything less than sex-blind admission and free personal choice.

Certainly, there are no a priori grounds for concluding that the males of a predatory species must be a specialized hunting class. In chimpanzees, males are the hunters; which may be suggestive in view of the fact that these apes are by a wide margin our closest living relatives. But, in lions, the females are the providers, typically working in groups with their cubs in tow. The stronger and largely parasitic males hold back from the chase, but rush in to claim first share of the meat when the kill has been made. Still another pattern is followed by wolves and African wild dogs: Adults of both sexes, which are very aggressive, cooperate in the hunt.

The moment has arrived to stress that there is a dangerous trap in sociobiology, one which can be avoided only by constant vigilance. The trap is the naturalistic fallacy of ethics, which uncritically concludes that what is, should be. The "what is" in human nature is to a large extent the heritage of a Pleistocene hunter-gatherer existence. When any genetic bias is demonstrated, it cannot be used to justify a continuing practice in present and future societies. Since most of us live in a radically new environment of our own making, the pursuit of such a practice would be bad biology; and like all bad biology, it would invite disaster. For example, the tendency under certain conditions to conduct warfare against competing groups might well be in our genes, having been advantageous to our Neolithic ancestors, but it could lead to global suicide now. To rear as many healthy children as possible was long the road to security; yet with the population of the world brimming over, it is now the way to environmental disaster.

Our primitive old genes will therefore have to carry the load of much more cultural change in the future. To an extent not yet known, we trust—we insist—that human nature can adapt to more encompassing forms of altruism and social justice. Genetic biases can be trespassed, passions averted or redirected, and ethics altered; and the human genius for making contracts can continue to be applied to achieve healthier and freer societies. Yet the mind is not infinitely malleable. Human sociobiology should be pursued and its findings weighed as the best means we have of tracing the evolutionary history of the mind. In the difficult journey ahead, during which our ultimate guide must be our deepest and, at present, least understood feelings, surely we cannot afford an ignorance of history. ■

1976

"Behavioral discretization and the number of castes in an ant species," *Behavioral Ecology and Sociobiology* 1(2): 141–154

The research reported here attempted to objectively measure the number of worker subcastes in an ant species. My focus was on discretization—the dividing of labor roles that have boundaries sharp enough for the roles to be counted in discrete numbers. This effort was a sequel to the formulation of my earlier theory of adaptive demography, which attributed the shaping of demographic traits of colony populations to natural selection at the colony level. In this follow-up paper I took into account the existence in ants of temporal castes, orthogonal to physical castes, rising not from anatomical differences but from the ageing of individual colony members. In other words, each worker ant can and usually does pass through two or more castes during its lifetime. Temporal castes are almost universal in ants and other eusocial insects. They invariably follow this timeline statistically: upon emergence as adults, they tend first to care for the queen and brood; then, over a period of days or weeks, they move to new locations to participate in other internal tasks such as nest construction and repair; and finally, they travel away from the nest to forage for the colony. It has been well known since the 1920s that this sequence is followed only loosely, such that during times of severe labor shortage nurses tend to begin foraging early and foragers to resume nursing.

Two subsequent investigators, C. Tofts and N. R. Franks (*Trends in Ecology and Evolution* 7: 346–349, 1992) have suggested that there is no innate physiological program to guide the centrifugal spread by aging workers away from the queen and brood; there is only the working of chance movements. In this "foraging-for-work" model, workers wander until they find a task available, then attend to it until it is finished, or nestmates take over. By stochastic dispersion, in the manner of randomly diffusing molecules, older workers eventually reach the outdoors and commence to forage. This model, however, cannot be reconciled with the behavior of most ant species, especially those such as members of the genus *Pheidole,* which possess large, complex societies. Workers have innate programs that guide the timeline of labor division even though they do not strictly determine it. The evidence includes dominance-driven division of labor among workers of some ponerine ants and the failure of young workers in many species to move to foraging even when nest exits are only a few centimeters from where they begin the adult stage. In some species also, newly eclosed adults travel within nests to larval piles immediately after they emerge from pupae sequestered in separate nest chambers. Finally, the genetic programming of a temporal sequence in honeybees has been demonstrated to be a finely orchestrated interplay of hormones, exocrine gland development, and labor roles of workers of unstressed colonies that unfolds during the 45–60 days of their adult longevity.

Behav. Ecol. Sociobiol. 1, 141–154 (1976)

Behavioral Ecology and Sociobiology

Behavioral Discretization and the Number of Castes in an Ant Species

Edward O. Wilson

Museum of Comparative Zoology Laboratories, Harvard University, Cambridge, Massachusetts, USA

Summary. 1. Two extreme possibilities in the evolution of temporal castes can be envisaged. First, workers can undergo changes in responsiveness to various kinds of stimuli in a strongly discordant manner as they grow older, so that each task is addressed by a distinctly different frequency distribution of workers belonging to different age groups. Because these age-frequency distributions change almost gradually from one task to another in covering many such tasks, the resulting temporal caste system is referred to as continuous. At the opposite extreme, the aging worker can undergo changes in responsiveness to different stimuli in a highly concordant manner, so that all of the tasks are attended by one or relatively few frequency distributions of workers belonging to different age groups. The resulting temporal caste system is referred to as discrete, and the evolutionary process leading to it is called behavioral discretization (Fig. 1).

2. The temporal system of the minor worker caste of *Pheidole dentata* proves to be much closer to the discrete state, although it is not extreme in form (Figs. 3, 4). On the basis of ethograms constructed of stressed and unstressed colonies in which the approximate ages of the minor workers were known, it is possible to recognize five discrete female castes: the queen, a single temporal subcaste of the major worker, and three temporal subcastes of the minor worker. These are the elements which can now be employed in ergonomic analyses of the species' caste system.

Introduction

A basic but hitherto unsolved problem of insect sociobiology is the exact number of castes found in each colony of a given species. The exact analysis of ergonomics depends on the specification of this parameter. Until the present time an imaginary number has been set for the purposes of model building and the analysis pursued as a largely theoretical exercise (Wilson, 1968; Oster, 1976).

The problem can be partly solved where physical castes exist and are discrete enough simply to be counted. Even when variation is continuous it is often possible to equate castes with the distinct modes of multimodal size-frequency distributions. But this solution takes us only half the way. Most eusocial insects, and virtually all those characterized by advanced traits in communication, queen-worker differentiation, and other important social qualities, also display temporal polyethism: the worker changes roles, usually progressing from nurse to forager, as it grows older (reviews by Wilson, 1971; Michener, 1974; Schmidt 1974). The temporal shift is ordinarily complex. Earlier I showed that for the purposes of optimization studies it is useful to define age groups as equivalent to physical castes (Wilson, 1968). The castes can then be conveniently defined as discrete age-size groups. The present report examines the intriguing question of whether in the course of evolution eusocial species themselves have discretized age-size groups or whether they have opted for more complex continuous systems. Data are then presented to show that in at least one ant species (*P. dentata*) discretization has occurred, and fits what appear to be peculiar spatial arrangements in the colony.

Results

1. The Alternatives Open to Evolving Ant Species

Figure 1 presents the two extreme alternatives open to an ant species in the process of evolving temporal castes. The ageing period depicted is that which occurs from the moment of the adult worker's eclosion from the pupal skin to the moment of its death by senescence. The worker's entire life span is arbitrarily divided into six periods (the number was chosen to conform with the six periods used in the experimental work to be reported later). The worker is envisaged as undergoing physiological change with age such that its responsiveness to each of various contingencies is altered out of phase with reference to the alteration of the other contingencies. Suppose that T_1 is the responsiveness to a misplaced egg: the curve indicates that when the worker is very young (age I) it is likely both to be in the vicinity of the egg and to react by picking the egg up and putting it on an egg pile. Its location and/or behavioral responsiveness change as it ages in such a way that its probability of response to the contingency drops off rapidly after ages I or II.

Let us now consider the possibilities. In the upper half of Figure 1, labeled Model 1, the response curves to four contingencies (T_1 through T_4) are all out of phase. The curve of response to T_1 (misplaced egg) is different from the curve of response to T_2 (say, a hungry larva), and so on. As a result, the ensemble of age groups, represented on the right-hand side by the frequency distribution of workers in different age groups that attend to task T_1, is different from that attending to T_2, and so on. As the number of contingencies is increased, and their response curves are all made discordant, there will be one age-group ensemble for each task. *Let us now define an age-group ensemble as a caste, a temporal caste to be exact, and state that in Model 1 there is a caste for each task.* However, the distinction of age-group ensembles will soon be blurred as more tasks are added. The overlap in the age-group frequency

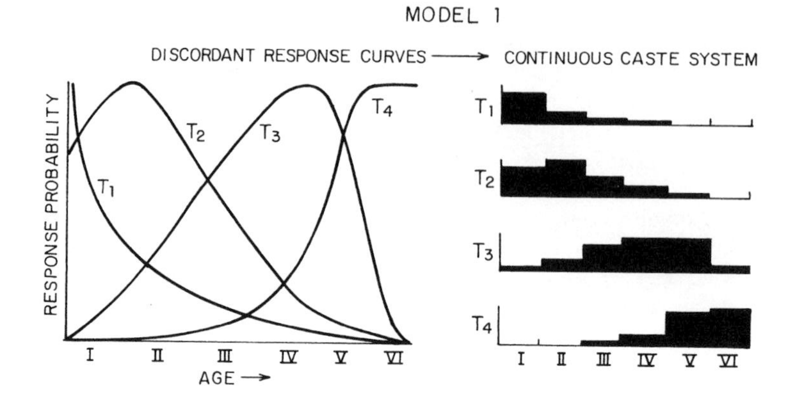

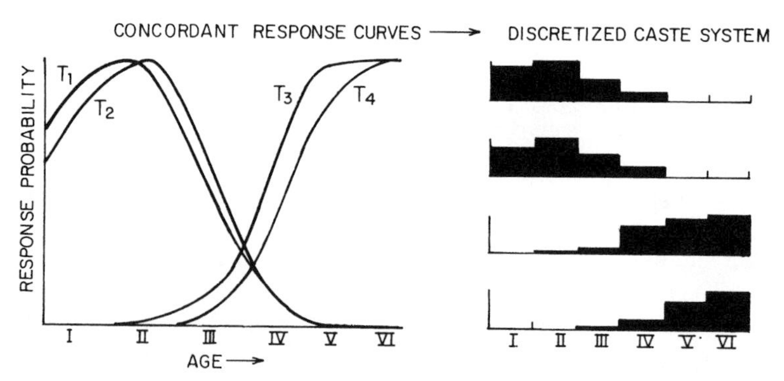

Fig. 1. The two extreme alternatives open to an ant species in the evolution of temporal castes. The age of adult workers is arbitrarily divided into six periods. In the first model (*upper*) the responsiveness of the worker to each of four contingencies (T_1 through T_4) changes markedly out of phase with reference to its responsiveness to the other contingencies. As a consequence each of the four contingencies are met by a distinct ensemble of age groups (temporal castes) which are represented on the right by the frequency distributions of workers in different age groups attending to the contingency. If the number of contingencies were increased substantially, the overlap of the age-group ensembles would increase to a corresponding degree and the resulting system would approach a continuous transition. In the second model (*lower*) the response curves are clustered into groups that are approximately in phase, resulting in more than one contingency being attended to by the same age-group ensemble (caste). If the number of contingencies were increased substantially, the number of age-group ensembles would remain small

curves is so extensive that after ten or so contingencies are added, the system becomes effectively continuous. For this reason I suggest that such an arrangement be called a *continuous caste system.*

The approach to continuity in Model 1 is marked by complexity and subtlety. The evolving ant species can easily adjust the programming of individual worker responsiveness to attain discordance, which in turn yields one caste specialized for each task. Thus only a relatively elementary alteration in physiology is needed to produce a complex caste system.

Next consider the alternative option, depicted in the lower half of Figure 1. Here various of the response curves are concordant, or at least approximately so. As a result the same statistical ensemble of workers attends to more than one task. As the number of tasks increases, the number of castes does not keep pace; conceivably it could remain low, say corresponding to as few as 2 or 3 distinct ensembles. Thus the species has chosen to operate with a *discrete caste system*—comprised of a relatively few, easily recognized age-group ensembles. The evolutionary process leading to such a system can be called "behavioral discretization." It is attained through physiological alterations as potentially simple as those that yield continuous cyste systems.

2. An Analysis of Temporal Castes

Which evolutionary route have ant species actually followed? We know that the behavior and physiology of workers change profoundly with age, and that this shift is a widespread if not universal pattern in the 12,000 or more living ant species. In general, workers begin their adult lives nursing immature forms and gradually progress to the status of full-time foragers. Many aspects of the change have been documented in detail in the genera *Formica* (Otto, 1958; Dobzrańska, 1959) and *Myrmica* (Ehrhardt, 1931; Weir, 1958; Cammaerts-Tricot, 1974, 1975), but the data are still not complete enough or of such a character as to distinguish the presence or absence of discretization in temporal castes.

In order to make this distinction I selected an ant species, *P. dentata,* which has proved especially favorable for laboratory studies of caste (see Wilson, 1975, 1976a). The analysis consisted of three sequential stages, which are (1) construction of ethograms, (2) age determination, and (3) measurement of temporal division of labor.

a) Behavioral Repertories. The study began with the cataloging of behaviors of the workers caste of *P. dentata.* Over a period of a week, a total of 14 h was devoted to the accumulation of 1,406 records of separate acts by a single colony. Although the acts were recorded at random within a given region of the nest or nest vicinity under surveillance, the region itself was not chosen randomly; instead, the brood area was consistently favored. The result, then, is not a precise frequency diagram but rather a list of the great majority of activities in which these ants engage.

The repertories are presented in Table 1. Recognition is made of the fact that the workers of *P. dentata* are divided into two very different subcastes unconnected by intermediate forms: the minor workers, which have the ordinary body proportions of myrmicine worker ants and which conduct most of the foraging, nest construction, nursing of immature stages, and other quotidian tasks; and the major workers, which have massive heads filled mostly with the adductor muscles of the mandibles and which are anatomically and behavio-

Table 1. Relative frequencies of behavioral acts by the two physical castes of the ant *P. dentata* in an undisturbed colony. *N*, total number of behavioral acts recorded in each column

Behavioral act	Frequency of behavioral acts	
	minor workers (*N*=1,222)	major workers (*N*=204)
Self-grooming	0.18003	0.56373
Allogroom adult		
Minor worker	0.04992	0
Major worker	0.00573	0
Alate or mother queen	0.01146	0
Brood care		
Carry or roll egg	0.01391	0
Lick egg	0.00245	0
Carry or roll larva	0.12357	0
Lick larva	0.09984	0.02941
Assist larval ecdysis	0.00409	0
Feed larva solid food	0.00573	0
Carry or roll pupa	0.03601	0
Lick pupa	0.01882	0
Assist eclosion of adult	0.00818	0
Regurgitate		
With larva	0.02128	0
With minor worker	0.03764	0.22059
With major worker	0.00573	0
With alate or mother queen	0.00327	0
Forage	0.12111	0.02941
Feed outside nest	0.04337	0.01471
Carry food particles inside nest	0.05237	0
Feed inside nest	0.05810	0.01471
Lick meconium	0.00573	0
Carry dead nestmate	0.01882	0.04902
Carry or drag live nestmate	0.00246	0
Eat dead nestmate	0.06383	0.07843
Handle nest material	0.00655	0
Totals	1.0	1.0

rally specialized for colony defense. The major workers also store more liquid food in their crops on a per-gram basis with respect to their body weight and can therefore also be regarded as a storage subcaste. The repertories reveal strong behavioral differences between the two subcastes, as well as an overall behavioral impoverishment of the major workers. Both are qualities found generally in such completely dimorphic ant species. The minor workers were observed in the performance of 26 kinds of behavioral acts. This number is arbitrary to an extent, in the sense that some of the categories could be combined or

Table 2. Relative frequencies of responses by the two physical castes of the ant *P. dentata* to stress. The responding individuals of both castes were recorded up to the moment that 200 minor workers had been counted

Stress imposed on colony	Number of minor workers responding	Number of major workers responding
Assault: a white card was placed on top of the nest, which was then tapped and shaken; the ants swarming over the card were then counted	200	1
Assault: ten fire ants (*Solenopsis geminata*) were placed 20 cm from the nest, and the *Pheidole* emerging from the nest to fight were counted	200	47
Burial: soil and leaf litter were dumped over the nest entrances, and the ants engaged in digging the colony out were counted	200	0
Exposure: a large quantity of larvae and other immature stages were dumped from the nest onto the floor of the foraging arena; the ants that retrieved them were counted	200	8

subdivided to make a lower or higher number. Through the employment of the Fagen-Goldman method based on the fitting of the frequency data to a lognormal Poisson distribution (Fagen and Goldman, 1976), the true number of behavioral acts in the minor worker repertory was estimated to be 27, with a 95% confidence interval of [26, 28]. These numbers are comparable to those based on repertories of the worker caste of monomorphic *Leptothorax* species and of the minor worker subcaste of *Zacryptocerus varians* (see Wilson, 1976b). The major workers of *P. dentata* were observed performing 8 kinds of behavioral acts; the true number was estimated to be 9, with a confidence interval of [8, 10]. Unlike the minor workers, the major workers do not appear to undergo significant age-related changes within this limited repertory.

In order to assess division of labor further between the two subcastes, a colony was subjected to four kinds of major stress encountered in nature. The results, given in Table 2, again demonstrate strong differences between the two forms. Major workers apparently never participate in excavation, even when the colony has been buried. They respond strongly to invasion by fire ants, which are among the chief enemies of *P. dentata* in nature (but much less strongly to other kinds of ants; see Wilson, 1976a). Surprisingly, major workers hardly respond at all to a mechanical disturbance of the surface of the nest, the kind of stimulus that would be associated with the approach of a vertebrate. However, many of the minor workers swarm out excitedly and attack any alien object they encounter. When the brood is suddenly exposed, which would occur if the nest were broken apart, a few major workers join in the retrieval effort. However, they are proportionately less represented than the minor workers, and they quit sooner, even when brood pieces still lie about exposed.

b) Age Determination. The second step in the study was to devise a way of estimating the age of individual minor workers. Advantage was taken of the fact that the integument of individuals is clear yellow when the ants first eclose, and gradually darkens to a deep blackish-brown as the ant grows older. With practice I was able to separate workers into six color classes, which were then labeled I (lightest and youngest) through VI (darkest and apparently oldest).

In order to estimate the true ages of workers in these six stages, 21 minor worker pupae were segregated in artificial nests with groups of stage-VI workers. The latter workers were chosen to be the companions because their already dark color made them easy to distinguish from the young workers under observation. Following their eclosion the young workers were then classified each day according to color. During the period of the observation the temperature varied from 21° C to 35° C and averaged about 28° C. A description of each color phase and its average duration (to the nearest day) are given in the following list. With the exception of one worker that remained in stage V for 10 days (instead of the average of 6), the 21 workers changed through all color phases within three days of each other.

Stage I. Body uniformly clear yellow. Newly eclosed; duration of stage: 0 to 2 days after eclosion.

Stage II. Body mostly clear yellow but gaster ("abdomen") a slightly contrasting light yellowish-brown. Duration of stage: 2 to 7 days after eclosion.

Stage III. Thorax clear yellow, head and gaster a slightly contrasting light yellowish-brown. Duration of stage: 7 to 9 days after eclosion.

Stage IV. Thorax clear yellow to light brownish-yellow, head and gaster a contrasting medium to dark brown; occasionally the head is also light yellowish-brown. Duration of stage: 9 to 10 days after eclosion.

Stage V. Head and gaster dark brown, thorax a slightly contrasting medium brown. Duration of stage: 10 to 16 days after eclosion.

Stage VI. Body nearly uniformly dark brown, with at least extensive patches of dark brown covering most of the pronotum and mesonotum of the thorax. Duration of stage: 16 days after eclosion to death.

As soon as the behavioral studies had been completed, the entire adult population of the main observation colony was censused. The physical castes present were 6 queens (5 alate, virgin individuals born in the nest, plus a single dealate, mother queen), 1,093 minor workers, 82 major workers. The proportions of age groups among the minor workers are given in Figure 2. It will be

N = 1093	40	166	229	116	207	335
STAGE	I	II	III	IV	V	VI
MEAN AGE (DAYS)	1	4	8	10	13	>18
DURATION OF STAGES (DAYS)	2	5	2	1	6	>10

Fig. 2. The mean age, duration, and representation in the main observation colony of the six color stages of the *P. dentata* minor worker

noted that these proportions do not correspond closely to the relative durations of the age periods, as would be expected in a continuously growing colony with a steady rate of oviposition by the queen. In fact, brood development is not uniform; eggs appear to be laid in surges, and occasional peak periods of eclosion have been noted. The present study, as suggested by the data in Figure 2, was conducted when the number of young adult workers (Stages II-III) was relatively high.

c) Measurement of Temporal Division of Labor. In the third and final stage of analysis, information gained from the ethograms and aging studies were employed in a further study of the division of labor by the various age groups of minor workers. Additional ethograms were now compiled, taking into account the ages of the individuals. The results, based on 2,331 behavioral acts recorded during a five-day period, are presented in Figure 3. In two categories, "defend nest" and "excavate nest", stresses identical to those described in Table 2 were applied. Otherwise, the colony was observed in an undisturbed condition. The colony used in the study was in a mature, vigorous state, having recently produced a crop of winged virgin queens. Two conditions of a fleeting nature happened to exist during the study: the brood contained few larvae of intermediate size and, as just noted in the discussion of the adult census data of Figure 2, there was a relatively large number of younger workers. However, some "unusual" circumstance or other will always prevail in particular colonies, since oviposition and brood rearing are not uniform through time. The conditions that were encountered did not appear of such a nature as to influence the results in any important way.

Frequency data were taken for all 26 categories of behavior (reduced to 25 by lumping "regurgitation with the queen" and "allogroom the queen" into one category called "attend mother queen") plus two other categories, "guard nest entrance" and "guard food site" which are based on the location of workers rather than behavior as such. Still one other, "retrieve prey" was not in the original ethogram because of its rarity, but was added for this stage of the analysis. Thus a total of 28 categories was considered. *The data indicate discretization of the minor workers into three temporal castes; the relatively small size of this set conforms to Model 2 of Figure 1.* Fifteen of the 28 categories are represented in Figure 3. The remaining 13 categories can be characterized as follows:

1. Five (self-grooming, allogrooming minor and major workers, regurgitating with minor and major workers) are engaged in by all the minor workers, without apparent age bias. They do not contribute in any evident way to temporal division of labor, although if the direction of flow during regurgitation could have been ascertained, it is possible that an age-dependent pattern would have been detected.

2. Three categories (feed larva solid food, carry dead nestmate, eat dead nestmate) conform to pattern B, while three (feed outside nest, feed inside nest, carry food particles inside nest) conform to pattern C.

3. Two categories (assist larval ecdysis, lick meconium) were observed on too few instances to permit an assessment.

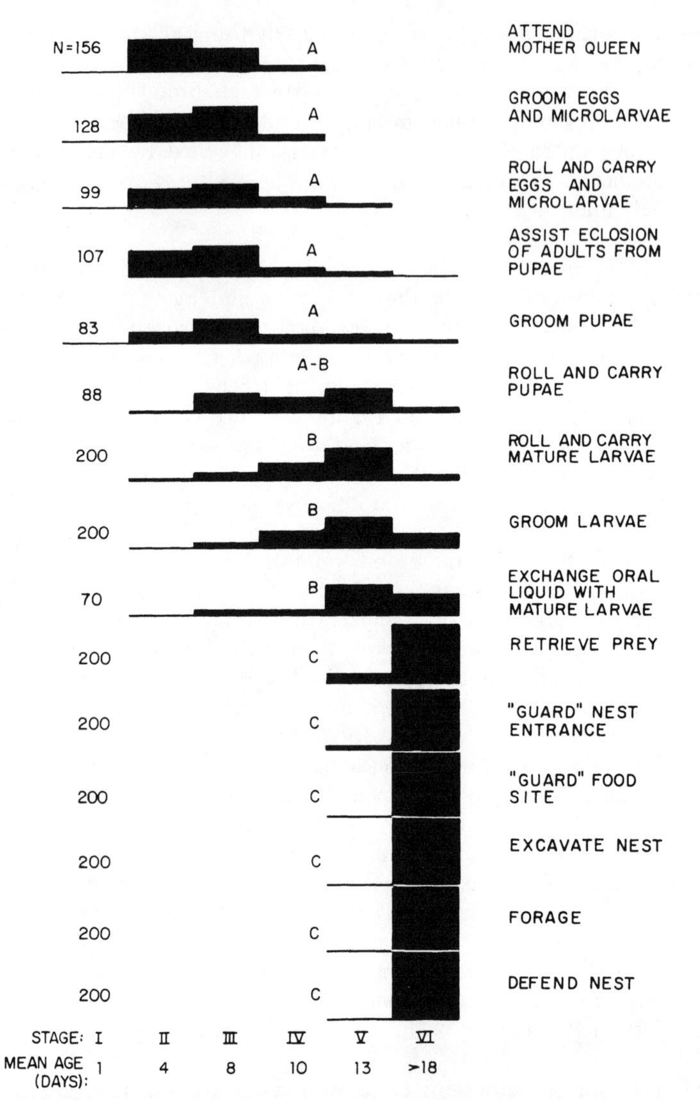

Fig. 3. The proportions of workers of the six age groups attending to all of the principal tasks are given in a series of histograms. The number of observed performances of each task, totaled through all of the age groups, are given on the left. The age groups (I-VI) and the average age of workers in each are given at the bottom. The histograms are classified into three groups (*A*, *B*, *C*), which are then identified as the temporal castes

In summary, 20 of the 28 behavioral categories are divided among three discrete age-group ensembles. Of the remaining 8 categories, 5 are conducted by workers with no evident age bias and can be eliminated from the analysis; 2 others are based on insufficient data and cannot be assessed. This leaves

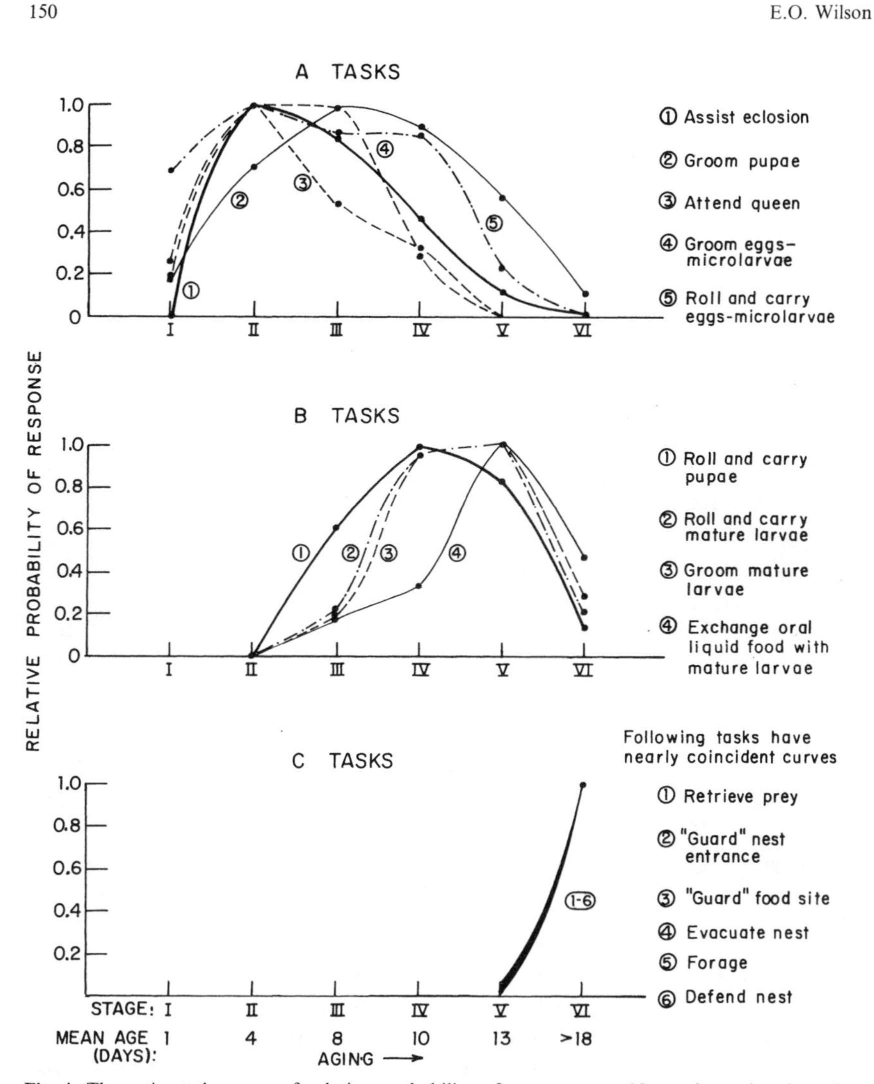

Fig. 4. The estimated curves of relative probability of response, to 15 contingencies, by minor workers of different ages

only one category which is known to be attended to by a fully intermediate ensemble, in this case (roll and carry pupae) intermediate between ensembles *A* and *B*. One might argue for recognizing a fourth caste for the last category, instead of calling it intermediate. However, this seems excessive in view of the fact that there is only one category; also individual workers performing the tasks were seen frequently to switch to the *A* or *B* tasks.

As suggested in the original model (Model 2 in Fig. 1), such discretization can be achieved by concordant response curves. Although these curves could not be drawn from data taken in the present study, their general form can be deduced from the age-group frequency data of Figure 3. The "relative probability of response" is defined as the probability that a worker in a given age group will attend to a particular contingency relative to the probability that a worker in some other designated age group will attend to a similar contingency. The age group designated for comparison is arbitrarily selected to be the most responsive one; as a result the "relative probability of response" can range from zero to one. The measure is derived in two steps as follows:

$$\textit{Probability of response}\text{ by a member of a given age group} = \frac{\text{Proportion of workers attending to the task that belong to the age group}}{\text{Proportion of workers in the total colony population that belong to the age group}}$$

$$\textit{Relative probability of response}\text{ by a member of the age group} = \frac{\text{Probability of response by a member of the age group}}{\text{Probability of response by a member of the age group with the highest such probability}}$$

In Figure 4 the curves of the relative probabilities of response are given with reference to the 15 behavioral categories for which frequency data were given previously (in Fig. 3).

Discussion

Why have the minor workers of *P. dentata* been discretized into just three temporal castes? The question can now be more precisely phrased: why have certain tasks been so closely coupled with each other in the response patterns of aging workers? For example, the age group of workers that assist the eclosion of others from pupal skins is nearly or exactly the same age group that attends the mother queen but it is strongly different from the age group that attends mature larvae. The age group that searches for food is nearly or exactly the same that excavates the nest but wholly distinct from the workers that nurse the first instar larvae. Furthermore, direct observations showed that the same individual workers shift from one task to another within the purview of their age group. If specialization within a temporal caste occurs, for example if one *A*-caste worker devotes itself mostly to the queen and another to the eggs, such specialization is at best very weak.

I would like to suggest that the observed discretization of temporal castes is an adaptation that increases *spatial efficiency*. It is obviously more efficient for a particular ant grooming a larva to regurgitate to it as well, or for the worker standing "guard" at the nest entrance to join in excavation when the entrance is buried. The other juxtapositions make equal sense when the spatial arrangement of the colony as a whole is considered. The queen, eggs, first instar larvae (microlarvae), and pupae are typically clustered together and apart from the older larvae, although the positions are constantly being shifted and pupae in particular are often segregated for varying periods of time well away from other immature stages. Thus the *A*-ensemble of workers can efficiently care for all of these groups, moving from egg to pupa to queen with a minimum of travel. The mean free path of a patrolling worker, to put the matter another way, is minimized by such versatility; it utilizes the least amount of energy in travelling from one contingency to another. It makes equal sense for *A* workers to assist the eclosion of adults from the pupae, since the latter developmental stage is already under their care.

It can be shown that in a purely abstract system, species of social insects should proliferate specialists until there is one caste for each distinguishable task (Wilson, 1968). In other words the twenty or so age ensembles should be evenly spread out to minimize overlap. The present study has revealed, on the other hand, that the advantages of spatial efficiency as a countervailing force has helped restrict the actual number of castes to less than 20% of the maximum conceivable.

A second constraint may exist: the small size of the worker's brain and the shortness of its life (especially true in *P. dentata*) could limit its ability to discriminate among stimuli. Thus if there were a single common attractant pheromone on the surface of pupae, eggs, and newly eclosed adults, the *A*-ensemble of workers could groom all three as virtually a single response to one stimulus. In the end we may discover that a great deal of complex social activity has been generated by the evolution of a relatively small number of pheromones made adequately efficient by the process of the discretization of temporal castes.

Trivers (personal communication) has suggested still another possible constraint on caste proliferation. As shown by Trivers and Hare (1976) it is to the advantage of the worker caste to control the sex ratio of the newly developed reproductive castes. For example, in species such as *P. dentata,* where workers are physiologically unable to produce offspring of their own, the inclusive fitness of the workers will be maximized when the ratio of investment is 3:1 in favor of new queens as opposed to new males. Workers having the most frequent contacts with brood are given the most opportunities to control the ratio of investment. Thus, temporal castes might evolve in such a way that foragers still come directly back to the brood and participate in their care; or, in a still more precise manner, workers nursing larvae might find it advantageous to monitor the eggs and pupae, and vice versa. The result would be less discretization of temporal castes. However, it turns out that in *P. dentata* older workers surrender much of their control over the brood. The foragers (mostly *C* individuals, 16 days and older) participate to a greatly diminished degree in larval care and contribute almost nothing to the care of eggs and larvae. Younger workers specializing on larvae (the *B* individuals) pay less attention to pupae and far less attention to eggs. The youngest workers seldom leave the eggs and queen to care for the larvae. In short, the evolutionary process of discretization does not appear to have been significantly diminished in intensity, nor the details of division of labor altered, in ways that facilitate control of the ratio of investment. The same statement is even more true with respect to the major worker ("soldier") subcaste, which has virtually nothing to do with the immature stages from eclosion to death.

The number of castes in *P. dentata* appears to be five: the queen, a single temporal subcaste of the major worker, and three temporal subcastes of the minor worker. For purposes of ergonomic analysis each of these can be treated as a full caste. Purists may wish to add the males as a sixth caste, but for reasons given elsewhere (Wilson, 1971) this extension creates more problems than it solves. It might also be argued that each and every age-group ensemble attending a task which is distinguishable from other ensembles at any level—say, statistically different at the 95% confidence level—deserves recognition as a caste. In this way, for example, workers grooming pupae can be distinguished (just barely) from workers grooming eggs and microlarvae. The counter-argument, on which the present analysis has been brought to a conclusion, is that major differences among ensembles are what matter in the economy of the colony, and that clusters of such ensembles are the features of interest, even if some variation can be demonstrated among the ensembles within each cluster. In this study visual assessment of clustering was deemed sufficient. More complex, ambiguous sets of data may demand formal clustering analysis of the kind used in numerical taxonomy.

It will be of paramount interest in the future to learn whether other species of social insects with temporal division of labor have undergone discretization. If they have, ergonomic analysis should prove easier than originally anticipated. Are the number of discretized castes small, as in *P. dentata*? And does the pattern of division of labor among them reflect the particular ecological adaptations of each species? Both of these questions seem tractable at the present time.

Acknowledgments. I thank Professor Robert M. Fagen for conducting the repertory estimations. The research reported here has been supported by grant number GB-40247 from the National Science Foundation.

References

Cammaerts-Tricot, M.-C.: Production and perception of attractive pheromones by differently aged workers of *Myrmica rubra* (Hymenoptera: Formicidae). Insectes Soc. **21**, 235–247 (1974)

Cammaerts-Tricot, M-C.: Ontogenesis of the defence reactions in the workers of *Myrmica rubra* L. (Hymenoptera: Formicidae). Anim. Behav. **23**, 124–130 (1975)

Dobrzańska, J.: Studies on the division of labour in ants genus *Formica*. Acta Biol. Exper. **19**, 57–81 (1959)

Ehrhardt, S.: Über Arbeitsteilung bei *Myrmica*- und *Messor*-Arten. Z. Morphol. Ökol. Tiere **20**, 755–812 (1931)

Fagen, R.S., Goldman, R.: Behavioural catalogue analysis methods. Anim. Behav., in press (1976)

Michener, C.D.: The social behavior of the bees: a comparative study. Cambridge, Mass.: Belknap Press of Harvard Univ. Press 1974

Oster, G.: Modelling social insect populations, I: ergonomics of foraging and population growth in bumblebees. Am. Naturalist, in press (1976)

Otto, D.: Über die Arbeitsteilung im Staate von *Formica rufa rufopratensis minor* Gössw. und ihre verhaltensphysiologischen Grundlagen, ein Beitrag zur Biologie der Roten Waldameisen. Wiss. Abh. Dt. Akad. Landw.-Wiss. Berl. **30**, 1–169 (1958)

Schmidt, G. (ed.): Sozialparasitismus bei Insekten. Stuttgart: Wissenschaftliche Verlagsges. mbH 1974

Trivers, R.L., Hare, H.: Haplodiploidy and the evolution of the social insects. Science **191**, 249–263 (1976)

Weir, J.S.: Polyethism in workers of the ant *Myrmica.* Insectes Soc. **5**, 97–128; 315–339 (1958)
Wilson, E.O.: The ergonomics of caste in the social insects. Am. Naturalist **102**, 41–66 (1968)
Wilson, E.O.: The insect societies. Cambridge, Mass.: Belknap Press of Harvard Univ. Press 1971
Wilson, E.O.: Enemy specification in the alarm-recruitment system of an ant. Science **190**, 798–800 (1975)
Wilson, E.O.: The organization of colony defense in the ant *Pheidole dentata* Mayr. Behav. Ecol. Sociobiol. **1**, 63–81 (1976a)
Wilson, E.O.: A social ethogram of the Neotropical arboreal ant *Zacryptocerus varians.* Anim. Behav., in press (1976b)

Received December 17, 1975

Edward O. Wilson
Department of Biology
Harvard University
MCZ Laboratories
Oxford Street
Cambridge, Mass. 02138, USA

1976

"The organization of colony defense in the ant *Pheidole dentata* Mayr (Hymenoptera: Formicidae)," *Behavioral Ecology and Sociobiology* 1(1): 63–81

Ants are among the most aggressive animals on Earth. The colonies of many species raid those of their own and other species as a matter of course and destroy them if possible. I can say, with a touch of hyperbole, that if ants had nuclear weapons the world would end in a week. A key part of the communication repertory of virtually all ant species is alarm signaling, with which workers alert nestmates to a breach in the nest wall or the approach of enemies. In a large fraction of species, they simultaneously lay trails to recruit nestmates to the site of the disturbance.

While studying the biology of *Pheidole dentata,* a common ant of the southern United States, I discovered what may be the most elaborate defense strategy currently known in the social insects. Colonies employ a graduated series of primary and fallback tactics in response to increasing threat from enemies. The series starts with simple recruitment of soldiers (large-headed major workers) by minor worker scouts. If the combined force fails to eliminate the threat, the soldiers retreat to form a tightening circle around the nest entrance. If that fails, the entire colony scatters outward in an "every ant for herself" evacuation of the nest. This final maneuver is triggered by chaotic alarm-defense signaling throughout the nest interior, constituting what appears to be context-specific positive feedback at the social level.

Of equal interest in the *Pheidole* defense strategy is the phenomenon of enemy specification, characterized here for the first time in ants and possibly in social insects in general. Fire ant colonies, which nest all around the much smaller colonies of *Pheidole dentata,* are among this species' deadliest enemies. Minor workers of *Pheidole dentata* foraging outside the nests are far more sensitive to the pressure of fire ants than to other ant species. In fact, they are hypersensitive: a single fire ant worker encountered close to a *Pheidole* nest is enough to send minor workers into a frenzy of alarm-recruitment, drawing soldiers out in a prolonged search for intruders.

Behav. Ecol. Sociobiol. 1, 63–81 (1976)

Behavioral Ecology and Sociobiology

The Organization of Colony Defense in the Ant *Pheidole dentata* Mayr (Hymenoptera: Formicidae)*

Edward O. Wilson

Museum of Comparative Zoology Laboratories,
Harvard University, Cambridge, Massachusetts, USA

Summary. 1. Colonies of *Pheidole dentata* employ a complex strategy of colony defense against invading fire ants. Their responses can be conveniently divided into the following three phases: (1) at low stimulation, the minor workers recruit nestmates over considerable distances, after which the recruited major workers ("soldiers") take over the main role of destroying the intruders; (2) when the fire ants invade in larger numbers, fewer trails are laid, and the Pheidole fight closer to the nest along a shorter perimeter; (3) when the invasion becomes still more intense, the Pheidole abscond with their brood and scatter outward in all directions (Figs. 1, 4).

2. Recruitment is achieved by a trail pheromone emitted from the poison gland of the sting. Majors can distinguish trail-laying minors that have just contacted fire ants, apparently by transfer of the body odor, and they respond by following the trails with more looping, aggressive runs than is the case in recruitment to sugar water. Majors are superior in fighting to the minors and remain on the battleground longer.

3. The first phase of defense, involving alarm-recruitment, is evoked most strongly by fire ants and other members of the genus *Solenopsis*; the presence of a single fire ant worker is often sufficient to produce a massive, prolonged response (Figs. 2, 5, 6). In tests with *Solenopsis geminata*, it was found that the Pheidole react both to the odor of the body surface and to the venom, provided either of these chemical cues are combined with movement. Fire ants, especially *S. geminata*, are among the major natural enemies of the Pheidole, and it is of advantage for the Pheidole colonies to strike hard and decisively when the first fire ant scouts are detected. Other ants of a wide array of species tested were mostly neutral or required a large number of workers to induce the response. The alarm-recruitment response is not used when foragers are disturbed by human hands or inanimate objects. When such intrusion results in a direct mechanical disturbance of the nest, simulating the attack of a vertebrate, both minor and major workers swarm out and attack without intervening recruitment.

* This research has been supported by National Science Foundation Grant No. GB 40247.

Introduction

The conventional view of defense in ants has been one of relatively complex individual behavior but simple colony organization. The single ant was seen to be sensitive to a wide range of stimuli, including the alarm pheromones of its nestmates, and to respond with generalized forms of aggression or retreat. In cases where the worker caste is divided into minor and major subcastes, the latter was regarded as being the more prone to combat under most or all circumstances. Defense at the colony level was not interpreted to be as well organized as, say, recruitment or caste determination (see review in Wilson, 1971).

All of these generalizations are probably incorrect, at least for some ant species. In the course of studies on the division of labor in the myrmicine *P. dentata* Mayr, I discovered that colony defense is at least as complicated and precisely organized as the most advanced, better known forms of social behavior. This article will show that colonies of *P. dentata* employ a flexible strategy in dealing with invading ants, one that is qualitatively different from defense against vertebrates. The strategy consists of three phases initiated in sequence by an increasing magnitude of the challenge: destruction of scouts and small enemy forces well away from the nest entrances, followed by reduction of the defense perimeter so that fighting occurs closer to the nest until the enemy is eliminated by attrition, and, finally, when the colony is attacked by still larger numbers, the abandonment of the nest premises in a sudden exodus. Most of the defense is orchestrated by the use of the trail pheromone by the minor worker caste. As demonstrated in an earlier, preliminary report (Wilson, 1975), this alarm-recruitment system is evoked only by a narrow range of ant species, the foremost of which are the fire ants and other members of the genus *Solenopsis*. It will be argued that such selectivity, which in effect constitutes enemy specification, is an adaptation to the particular environment in which *P. dentata* lives.

Materials and Culturing Methods

Pheidole is one of the several most abundant, diverse, and geographically widespread ant genera of the world. *P. dentata* is a medium-sized form which is abundant in woodland over most of the southern United States (Creighton, 1950). Like other members of the genus, it is characterized by a sharp division of the worker caste into small-headed minor workers, which forage for food and conduct the other quotidian tasks of the colony, and large-headed major workers, or "soldiers" as they are often called, which function primarily in defense. The majors also have proportionately more distensible crops and hence store larger quantities of liquid food. But the large head, with its massive adductor muscles and clipper-like mandibles, is highly specialized for the fighting role of this caste.

Entire colonies, collected in the field in the Tallahassee region of northern Florida, were placed in plexiglas containers 28 cm × 45 cm and 16 cm deep, the sides of which were coated by Fluon GP-1 (ICI America, Inc., Stamford, Conn.) to prevent escape. The floor of each container served as the foraging space and experimental arena for a single colony. The ants were permitted to move into test tubes 148 mm long with 23 mm inner diameter, kept moist by tight cotton plugs that trap water at the bottom of the tubes. This simple arrangement required a minimum of subsequent care, yet permitted close inspection of the behavior of the entire colony both inside and outside the nests. The colonies flourished when maintained on a mixture of Bhatkar diet (Bhatkar and Whitcomb, 1970) and freshly killed insects, eventually rearing large numbers of winged queens and males.

In the behavioral tests the nest tubes were placed at one end of the container, and the remainder of the floor of the container was kept clear for observations of foraging and colony defense. Alien ants were introduced at a point on the floor approximately 25 cm from the Pheidole tube nests. A positive response was defined as the recruitment of major workers by minor workers to this area. The number recruited was defined as the maximum number seen beyond a line 20 cm from the end of the container (hence, a 20 × 28 cm sector around the invaders) for 30 min following the introduction minus the maximum number of majors seen in the same area during the 15 min immediately preceding the introduction.

Results

1. The Three Phases of Colony Defense

Phase 1 (Distant Alarm-Recruitment). When workers of the native fire ant *S. geminata* are placed within 25 cm of the nest tubes of a laboratory *P. dentata* colony, they soon encounter foraging Pheidole minor workers. Some of the foragers grapple with the intruders, while others flee and travel in irregular loops through the surrounding area. Within seconds, some of the minor workers run swiftly back and forth to the nest, dragging the tips of their abdomens over the ground. The trail pheromone thus deposited attracts both minor and major workers from the nest in the direction of the invaders. The majors have never been seen to lay trails. Their glandular anatomy is very different from that of the minors, and artificial trail bioassays (see Wilson, 1959) employed with them suggest that no part of their body contains a trail pheromone. Thus the communication is unilateral.

Upon arriving at the battle scene the major workers become highly excited, snapping at the fire ants with their powerful mandibles and soon chopping them to pieces (see Figs. 1, 5). The recruited minor workers also join the fighting, but they are less persistent and remain in the area for much shorter periods of time. As a result the majors increase in proportion, and for all but the more transient invasions they eventually outnumber the minors, despite the fact that they constitute only 8–20 percent of the worker population in the great majority of nests. The majors remain in the battle area for an hour or more after the last Solenopsis has been dispatched, restlessly patrolling back and forth (see Fig. 2). Often a single Solenopsis worker is enough to evoke the full response, which brings ten or more Pheidole majors into the field. This is one of the first examples documented of an alarm-recruitment system in ants, a phenomenon previously well known in termites (see review in Wilson, 1971). A second, independently discovered case has recently been reported in *Myrmica rubra* by Cammaerts-Tricot (1974, 1975).

Fig. 1. The first phase of colony defense by the ant *P. dentata*. After contacting fire ant workers near the nest, minor workers of Pheidole run back and forth to the nest, dragging the tips of their abdomens over the ground and laying odor trails (*upper left*). The trail pheromone attracts both minor and major workers to the battle ground. The majors are especially effective in destroying the invaders, which they chop to pieces with their powerful, clipper-shaped mandibles. Some of the Pheidole are themselves crippled or killed by the venom of the fire ants. (Original drawing by Sarah Landry)

Phase 2 (Close Defense). If the initial invading force strongly outnumbers the foraging minors, the colony goes directly into the second phase of defense. If, on the other hand, the invaders increase gradually in number, the colony passes from the first phase into the second. In the second phase few, if any, minor workers lay odor trails to the outer foraging area. Some grapple with the fire ants, but most simply flee from the area back toward the nest. As a result, the fire ants soon advance close to the Pheidole nest. Here they encoun-

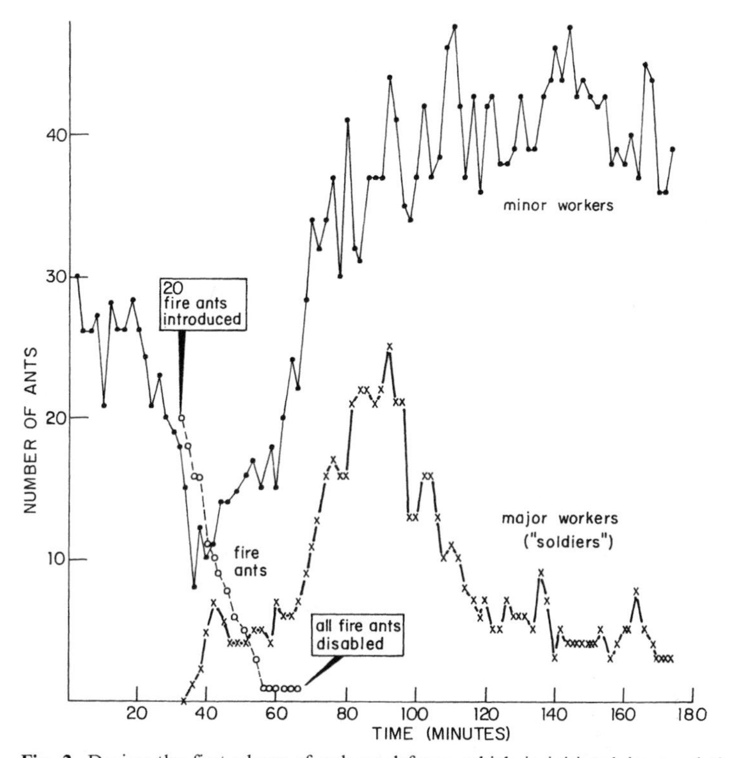

Fig. 2. During the first phase of colony defense, which is initiated by a relatively small force of invading fire ants, the number of minor workers in the 28 × 20 cm area around the invaders decreases, while that of the major workers increases. The majors restlessly patrol the area long after the last fire ant has been destroyed. In the typical experiment reported here, 20 workers of the fire ant *S. geminata* were introduced in a group 25 cm from the Pheidole tube nests

ter and are challenged by majors resting in and around the nest entrances. Some of the Pheidole minors also lay trails at close quarters, bringing still more majors to the conflict. The mass of fighting majors, assisted by a few minors, form a short, tight defense formation around the nest entrances (see Fig. 3).

Phase 3 (Absconding). If fire ants continue to invade the nest area in spite of the efforts of the Pheidole majors, excitement spreads through the workers remaining in the nest. An increasing number of minor workers race back and forth, some picking up and carrying packets of eggs, larvae, and pupae. Others leave the nest and stand or run around in the immediate vicinity of the nest entrances. Then, during a period of only a few minutes, the minors start leaving the nest with pieces of brood in their mandibles. Some remain near the nest entrances, but more and more run rapidly away in various directions, so that the colony as a whole scatters outward. The queen joins the exodus, breaking through the fighting masses of fire ants and Pheidole majors to flee on her own (see Fig. 4). Under laboratory conditions at least, the colony slowly returns

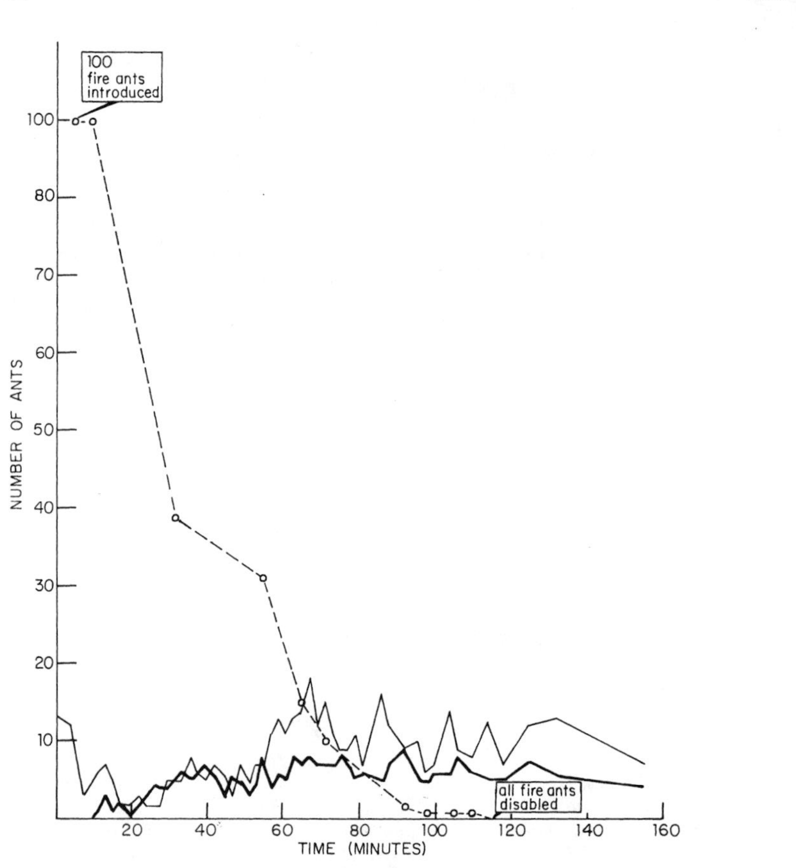

Fig. 3. The second phase of colony defense is initiated when a relatively large number of fire ants are introduced near the nest. In this case 100 *S. geminata* were placed 25 cm from the Pheidole nest tubes (cf. Fig. 1). The number of minor workers (*thin line*) decreases as in the first phase, but now few if any lay odor trails. As a result a smaller number of major workers (*thick line*) are attracted away from the nest. Most of the fighting occurs close to the nest entrances rather than in the 20×28 cm outer area from which the above data were obtained

to the nest after the fire ants are destroyed or leave. But when a large fire ant colony is permitted to attack a Pheidole confined to its nest area, it wipes out the defenders and removes and eats their brood.

2. *Analysis of the First Phase of Defense*

The Recruitment Process. Some of the minor workers that have just contacted fire ants run homeward, dragging the tips of their abdomens over the ground. Nestmates that encounter these individuals or cross their paths run excitedly along the paths, usually in a direction away from the nest. It is a reasonable

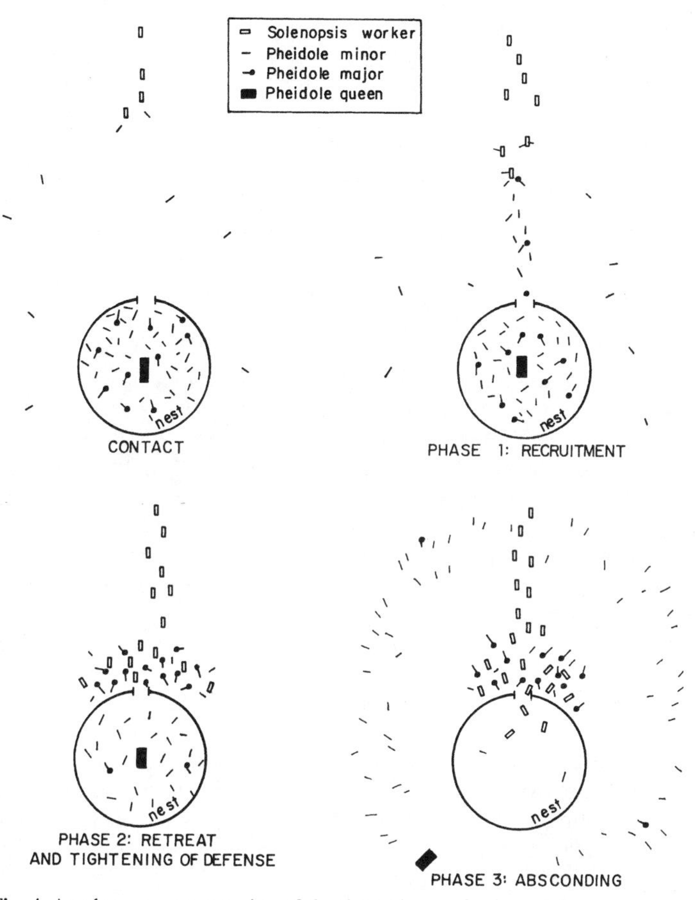

Fig. 4. An abstract representation of the three phases of colony defense

presumption, therefore, that odor trails are being laid from the vicinity of the invaders. In time substantial numbers of minor and major workers can be seen travelling back and forth along the exact paths taken by some of the recruiter ants. No other reasonable explanation of the alarm-recruitment process seems possible except that it is based in large part on the deposition of a trail pheromone.

The source of the trail pheromone was sought by means of the artificial trail bioassay. First, it was established that trails made from ethanol extracts of the whole bodies of minor workers were followed by large numbers of both minor and major workers. Similar whole extracts of major workers failed to cause this response. Next, several of the most likely source organs were dissected one at a time from the abdomens of single workers, washed in insect Ringer's solution, crushed on the tip of a birch applicator stick, and smeared in artificial trails running from one of the nest entrances outward for a distance of 35 cm into the foraging area. The number of workers following the trail for at least half its distance was then recorded. The results, presented in Table 1, appear conclusive. The trail pheromone is concentrated in the poison gland of the minor worker. The paired poison glands proper and the vesicle into which they empty were assayed together. As expected, no trace of the pheromone was detected in the poison gland of the major worker. This structure is proportionately larger than in the minor worker and differently shaped, and it almost certainly serves a different function.

Although the experimenter can easily induce trail-following with the use of the pheromone alone, other elements are present in alarm recruitment that make it qualitatively different from ordinary recruitment. When minor workers of *P. dentata* discover a drop of sugar water, for example, they first feed—often to repletion—and then return to the nest laying a trail. The speed at which they run is no greater than that seen during foraging. When they encounter nestmates they often turn toward them briefly, and a very brief mutual antennation may ensue that appears no different from that exchanged between workers on other occasions. No other forms of tactile communication have been observed. Both minor and major workers follow the trail closely, moving calmly at a rate not greatly exceeding that seen during foraging. The ratio of major to minor workers responding is close to or somewhat below that occurring in the colony population at large (see Table 2); the latter ratio ordinarily falls between 10 and 20 major workers per 100 minor workers.

Alarm recruitment contains differences at every step. After encountering fire ants, Pheidole minor workers lay odor trails while running swiftly. When they encounter other minor workers, their contact consists at most of a fleeting antennation. When major workers are contacted, however, the trail-layers rush at them, strongly vibrating their bodies for a fraction of a second. The major workers respond by rushing outward in short, irregular loops, following the odor trail only loosely and for brief intervals. Their mandibles are held more

Table 1. Response of Pheidole colonies to artificial trails made of various crushed abdominal organs of minor and major workers. A positive response was recorded when a worker ran along the trail for a distance of at least 18 cm. The ranges (and means) of numbers of minors and majors responding are given

Source of artificial trail		Number of replicates	Number following trail	
Caste	Organ		Minor workers	Major workers
Minor worker	Hindgut	6	0	0
	Dufour's gland	7	0	0
	Poison gland + vesicle	7	24–136 (71)	1–22 (8)
Major worker	Hindgut	6	0	0
	Dufour's gland (absent in this caste)	–	–	–
	Poison gland + vesicle	6	0	0

widely open than usual, and the majors rush at any objects they encounter, including their own nestmates. Fire ants are invariably attacked. Minor workers also move in looping movements but are not noticeably more aggressive. As shown in Table 2, the proportion of major workers responding to alarm recruitment is higher than those responding to sucrose recruitment and in the nest population at large. In 27 of the 32 comparisons made, the difference was significant at the 99 percent level or higher. To be sure that the difference was truly qualitative and not simply due to an indifference to sucrose due to overfeeding, I recorded the ratios of majors and minors in colonies that were first freshly fed and then starved for a period of eleven days. The results, given in Table 2, show that the difference in response persists in hungry colonies.

How do the major workers know that they are being recruited to fire ants and not to sucrose water? The vibration of the recruiters could provide a tactile clue. However, it is equally possible that the vibration merely enhances the transmission of an odor cue. The latter hypothesis is supported by the fact that minor workers do not receive the vibrating movement and yet seem to recognize that the object of the recruitment is different. The critical cue may be nothing more than odor accidentally transferred from the fire ants onto the defending Pheidole minor workers during the initial contact. To test this possibility, ten minor workers were seized with flexible forceps as soon

Table 2. The number of minor workers and major workers responding to odor trails laid by minor workers from sucrose solution and ten fire ants, respectively, on successively longer periods after the last feeding. Each count was terminated when 100 minor workers had been recorded

Days since last feeding	Caste responding	Colony No. 1		Colony No. 2		Colony No. 3	
		Sucrose solution	Fire ants	Sucrose solution	Fire ants	Sucrose solution	Fire ants
1	Minor	100	–	100	100	100	100
	Major	6	–	1	23	1	16
2	Minor	100	100	100	100	100	100
	Major	4	30	16	44	6	13
3	Minor	100	100	100	100	100	100
	Major	6	39	25	54	21	23
4	Minor	100	100	100	100	100	100
	Major	13	46	13	40	9	51
7	Minor	100	100	100	100	100	100
	Major	9	18	14	37	20	16
8	Minor	100	100	100	100	100	100
	Major	5	72	21	38	10	17
9	Minor	100	100	100	100	100	100
	Major	5	45	13	39	8	32
11	Minor	100	100	100	100	100	100
	Major	8	37	12	27	9	45

as they contacted fire ants, then dropped into the midst of groups of minor and major workers resting outside the nest entrance. In every case, the first nestmates to encounter the introduced workers rushed toward them in what appeared to be an initially hostile move; in two instances the workers were even grasped by the appendages and dragged about briefly before being released. As controls, ten other minor workers were seized and transported by forceps in a fashion identical in every detail except that they had not contacted a fire ant worker. These individuals were accepted by nestmates with neither a display of hostility nor even an outward show of interest. It is reasonable to conclude at least tentatively that the combined stimuli of pheromone trails and the odor of fire ants are crucial to triggering the distinctive response of the Pheidole alarm-recruitment system. It remains to be learned whether the vibration of the minor worker's body, or other, still unsuspected signals, play auxiliary roles.

Duration on the Battlefield. As the fighting proceeds, the proportion of major workers on the battlefield gradually increases (see Figs. 2, 3). In extreme cases only major workers remain in the end. When the colony absconds, most of the majors stay behind to fight. Furthermore, recruited majors remain in the battle area for much longer periods of time than do recruited minors. In one typical episode 30 majors, chosen at random as they crossed into a 100 cm^2 square around ten fire ants, remained in the square from 4 to 470 secs, averaging 70.5 secs per stay; 14, or 47 percent, remained for more than 50 secs. In contrast, 30 minors chosen at random stayed 2 to 64 secs, averaging 17.2 secs per stay, with only 2 individuals, or 7 percent, remaining over 50 secs.

Fighting Ability. The Pheidole majors are far superior to the minor workers in their ability to disable and to kill fire ant workers. They seize the fire ants by an appendage, the neck, or some portion of the petiole or postpetiole, then cut through the body part with their powerful mandibles. When faced by a phalanx of scurrying, snapping majors, few fire ants remain mobile for more than a minute or two. The Pheidole minor workers attempt the same tactic, but they are less frequently successful. Members of both Pheidole castes are crippled and killed by the bites and venom of the fire ants, but the majors suffer a much lower casualty rate. These differences in performance are exempli-

Table 3. Results of one-on-one combat between each of the two castes of *P. dentata* against minor workers of the fire ant *S. geminata*

	Both separate unhurt	Both crippled or killed	Pheidole crippled or killed, Solenopsis unhurt	Solenopsis crippled or killed, Pheidole unhurt
Pheidole minor worker vs. Solenopsis minor worker	20	5	36	11
Pheidole major worker vs. Solenopsis minor worker	12	7	4	60

fied by the results of 155 randomly chosen one-on-one combats shown in Table 3. The probability that these differences could be due to chance alone is less than one in a million.

3. The Second and Third Phases of Defense

When minor workers encounter large numbers of fire ants, a few grapple with the intruders but most scatter outward in all directions. Only a minority of those fleeing return directly to the nest, and proportionately few odor trails are laid even by these individuals. Consequently the fire ants are able to move quickly to the vicinity of the Pheidole nest. There they encounter stiffening resistance as more minor workers are encountered, a larger number of trails are laid, and a growing force of major workers is assembled. Thus the second phase of defense is only an extension of the first conducted in another location. It may seem paradoxical that when faced with more enemies the colony should recruit majors less efficiently. But this weaker response applies only to the territory farthest from the nest. The result of importance is that the majors are committed to battle close to the nest, where in fact they are now most needed.

If the second phase of defense begins to break down, the colony absconds, which constitutes the final, desperate, yet very effective maneuver. The following events lead to absconding. As the fire ants crowd closer to the nest entrances, an increasing number of Pheidole minor workers lay odor trails all the way back to the brood areas. As a result the excitement of the nest workers increases, and many begin to pick up pieces of brood and carry them back and forth. Meanwhile, the defense perimeter continues to shrink as Pheidole majors are immobilized in combat and more Pheidole minor workers are either immobilized or flee from the nest area after contacting fire ants. Activity within the brood area builds up, sometimes at an apparently exponential rate, and minor workers then begin to run out of the nest, many laden with pieces of brood. At first there is a strong tendency for the refugees to loop back and reenter the nest, but if they continue to encounter fire ants they break away entirely and flee outward. No particular direction is taken by individuals during the exodus, so that the colony ends up scattered over a wide area. The queen also departs under her own power. Later, after settling down, she attracts minor workers who cluster around her. No direct physical transport of one adult by another has been observed. When the fire ants are removed, the Pheidole adults slowly return to their nest and reoccupy it.

In summary, there appear to be two processes contributing to the relative suddenness and *en masse* quality of absconding:

1) As fire ants press in, more and more Pheidole of both castes are killed, while other minor workers simply desert the area. This leads to a steadily decreasing ratio of defenders to invaders and a shrinking of the defense perimeter. A point is reached in which the nest workers are contacting fire ants at a sufficiently high rate to cause them to seize brood pieces and leave. This final, critical level is approached steeply.

Table 4. Effectiveness of thirty ant species in evoking the alarm-recruitment response of *P. dentata*. A positive response was recorded when major workers of *P. dentata* were recruited by the minor workers. Also shown is the change in the number of majors in the battle area. Species indicated by an asterisk (*) occur as native inhabitants in at least part of the range of *P. dentata*

Species	Number of invading workers	Number of repli-cations	Number of positive responses	Change in number of majors in battle area: range (mean)
Subfamily Ponerinae				
**Leptogenys manni*	1 or 2	5	0	−2 to 0 (−1)
Subfamily Dorylinae				
**Neivamyrmex opacithorax*	10	16	11	−1 to 16 (7)
**N. opacithorax*	1	5	0	−1 to 1 (0)
Subfamily Pseudomyrmecinae				
**Pseudomyrmex elongatus*	10	5	0	−1 to 1 (0)
Subfamily Myrmicinae				
**Myrmica* sp.	5	3	0	−1 to 1 (0)
**Stenamma* sp.	10	3	0	−1 to 1 (0)
**Pogonomyrmex badius*	3	5	0	−1 to 2 (1)
**P. badius*	10	1	1	41
P. occidentalis	2	5	0	−1 to 3 (1)
**Aphaenogaster rudis*	10	5	1	−2 to 6 (1)
Tetramorium caespitum	10	5	0	0 to 2 (1)
T. caespitum	100	1	1	15
Atta sexdens, minor workers	10	5	0	0 to 1 (0)
Zacryptocerus varians, minors + majors	10	5	0	−2 to 2 (0)
**Pheidole dentata*	1	18	0	−2 to 5 (0)
**P. dentata*	10	26	4	−3 to 11 (2)
**P. dentata*	100	6	3	−3 to 63 (17)
**P. dentata*	200	4	4	10 to 57 (32)
P. desertorum	10	1	0	2
**P. floridana*	10	3	0	0 to 1 (1)
Xenomyrmex floridanus	10	5	0	−1 to 3 (1)
**Crematogaster atkinsoni*	10	1	0	0
**C. minutissima*	10	5	0	0
**C. minutissima*	100	1	0	3
**Monomorium minimum*	10	10	1	−5 to 4 (0)
**M. minimum*	50	2	2	6 to 10 (8)
M. pharaonis	10	4	3	0 to 9 (5)
Solenopsis fugax	1	5	2	−1 to 8 (2)
S. fugax	10	5	5	7 to 12 (11)
**S. geminata*	1	24	19	−1 to 37 (9)
**S. geminata*	10	33	33	3 to 72 (27)
**S. geminata*	100	6	6	4 to 31 (20)
**S. geminata*	200	2	2	3 to 10 (6)
**S. molesta*	10	5	4	1 to 32 (11)
**S. picta*	10	5	2	−1 to 10 (3)
**S. xyloni*	10	5	5	3 to 68 (41)
Subfamily Dolichoderinae				
**Tapinoma sessile*	10	3	0	0 to 1 (0)

Table 4 (continued)

Species	Number of invading workers	Number of repli-cations	Number of positive responses	Change in number of majors in battle area: range (mean)
Subfamily Formicinae				
Brachymyrmex depilis	10	1	0	1
Lasius alienus	10	3	0	2 to 4 (3)
Acanthomyops interjectus	5	3	3	5 to 7 (6)
A. interjectus	1	10	3	0 to 10 (3)
Camponotus fraxinicola	10	3	0	0 to 2 (1)
C. planatus	5	3	0	−2 to 1 (0)

2) Simultaneously, there is an exponential increase in the rate at which minor workers run back and forth between the vicinity of the nest entrance and the brood chambers. Since many lay odor trails, there appears to be a buildup in the concentration of the pheromone.

It has been possible to induce part of the absconding process by exposing the ants to their own volatile substances. In two replications I first immobilized 50 minor workers of *P. dentata* by freezing, then crushed them simultaneously between two glass slides approximately 5 cm above the nest entrances. The action caused moderate excitement, with both minor and major workers emerging from the nest. The ants were apparently attracted by odor, since many lifted and waved their antennae back and forth in the direction of the slides. They spread over the nest tubes and in the surrounding area; some dispersed outward still farther, trying to climb the walls of the arena. These actions closely resembled the early stages of absconding, although neither brood carrying nor departure of the queen were included. The movement of glass slides in an identical manner but without the crushing of ants did not induce attraction or dispersal.

4. Enemy Specification

Perhaps the most surprising discovery made during the entire analysis was the specificity of the alarm-recruitment response by *P. dentata*. In addition to the native fire ant *S. geminata*, five other species of *Solenopsis* tested in the laboratory were found to evoke alarm recruitment when introduced into the foraging arenas in small numbers. These are the red imported fire ant *Solenopsis invicta*, the native fire ant *Solenopsis xyloni*, and the "thief ants" *Solenopsis fugax, Solenopsis molesta*, and *Solenopsis picta*. The last three forms are members of the subgenus *Diplorhoptrum,* recognized as taxonomically distinct from the fire ants of the genus *Solenopsis*, and they are among the smallest of all ants. Their effectiveness in producing the response is thus even more impressive. None of the 24 other ant species tested thus far, representing 20 non-solenopsidine genera and six of the eight living subfamilies, proved effective at comparable numbers (see Table 4, and Fig. 5).

The closest approach to an exception is *Pogonomyrmex badius*, which is ineffective when the number of invaders is three, but evokes the response when the number is increased to ten. But compared to the Solenopsis and most of the other ant species tested, *P. badius* is gigantic; a single minor worker

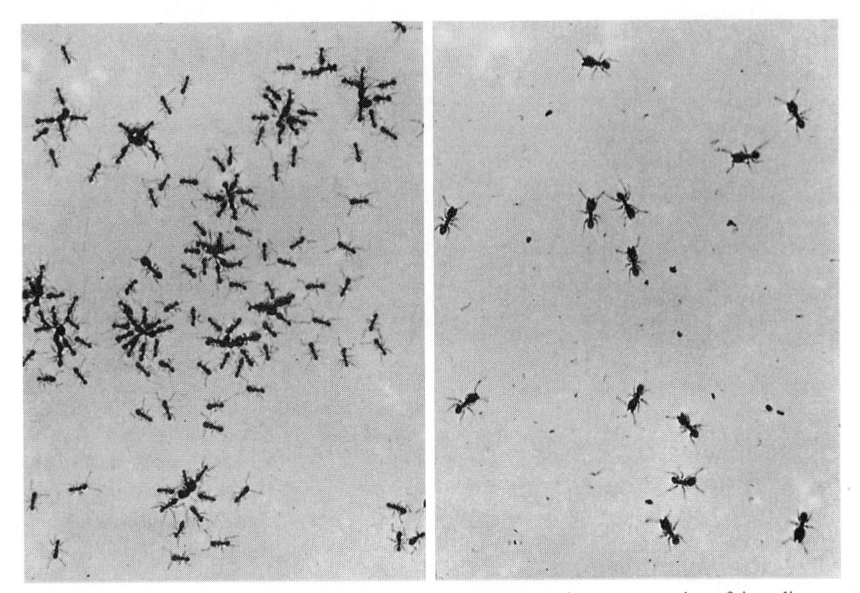

Fig. 5. Different modes of defense by the ant *P. dentata* against two species of invading ants. *Left:* workers of *T. caespitum* are pinioned and disabled mostly by Pheidole minor workers, which have not recruited major workers to their aid. The few majors present were stragglers in the area. *Right:* in response to the same number of workers of the fire ant *S. geminata*, the Pheidole minor workers have already destroyed the invaders and are patrolling the battle area. The scene has been largely abandoned by the minor workers. (From Wilson, 1975)

weighs about 6.88 mg, compared with 0.33 mg in the case of *P. dentata*, 0.38 mg in *S. geminata*, and only 0.06 mg in *S. molesta* (based on the wet weights of ten workers of each species). Relatively mild responses were also obtained by ten workers of the army ant *Neivamyrmex opacithorax*, the myrmicine *Monomorium pharaonis*, and formicine *Acanthomyops interjectus*, but they were well below those produced by fire ants. There may be a level at which any ant species can cause the alarm-recruitment response. One hundred minor workers from alien colonies of *P. dentata* were sufficient to cause it, as were 100 *Tetramorium caespitum*. However, 100 workers of *Crematogaster minutissima* still failed to evoke the response. In general, the numbers of non-solenopsidine ants comparable in size to *P. dentata* and *S. geminata* required to evoke the response is more than ten times that for *S. geminata*.

A more detailed comparison of the effects of two species is presented in Fig. 6. It can be seen that a single fire ant worker causes recruitment, and ten workers substantial recruitment, but 100 of these invaders reduces the magnitude of the response. The decline under more intense pressure has been explained previously; the colony is entering Phase 2 of its defense as more of the minor workers pull back, and fighting is increasingly restricted to the immediate nest vicinity. The data illustrate another phenomenon for which the author has no explanation: a marked variation among *P. dentata* colonies in the pattern of response to fire ants. Considerable differences exist in the intensity of response

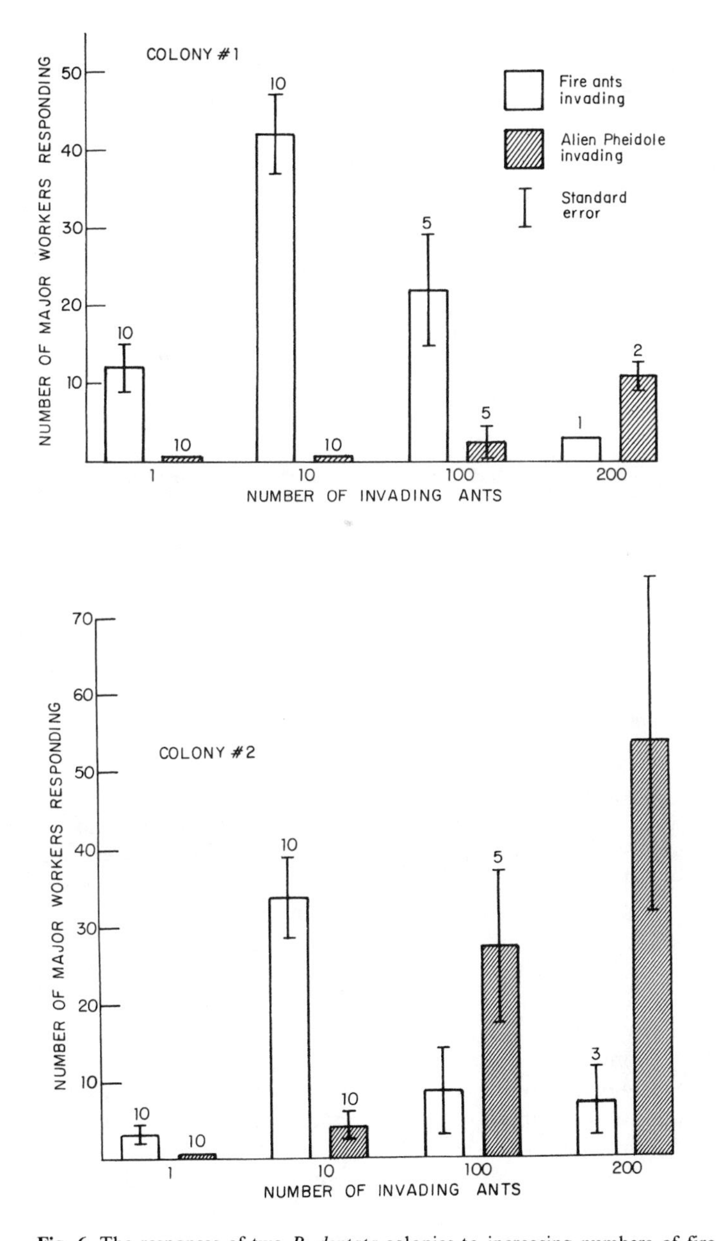

Fig. 6. The responses of two *P. dentata* colonies to increasing numbers of fire ants (*S. geminata*) and minor workers of *P. dentata* from alien colonies. The number of replications are given at the tops of the bars. A second trial in colony no. 1 utilizing 200 fire ant workers, not indicated here, resulted in most of the Pheidole colony absconding. These two colonies were selected because they represent different patterns of response to increasing numbers of invaders. (Fig. based on colony no. 1 from Wilson, 1975)

to particular numbers of invaders. Some variation of this kind has also been displayed by single colonies tested repeatedly over a period of weeks, but it does not equal the variation among many colonies noted in the laboratory on a single day.

In order to determine whether the response specificity is a widespread phenomenon, 15 colonies collected at four localities in Leon and Wakulla Counties, northern Florida, were challenged at successive intervals with ten minor workers of *S. geminata* and ten minor workers from alien colonies of *P. dentata*. All of the colonies responded positively to the Solenopsis, while only two were activated by the alien Pheidole. Moreover, the latter two colonies recruited many more major workers to the Solenopsis than to the Pheidole.

An effort was made to localize the cues by which *P. dentata* recognize Solenopsis.

The procedure followed was closely similar to that employed in the testing of various species of ants. The tests were conducted on *P. dentata* minor workers in the same zone of the foraging arena. Corpses of *S. geminata* tested without movement were simply placed on the floor of the arena shortly after being killed by freezing and rewarmed to room temperature. Movement was achieved in other tests by holding the object to be tested in the tips of fine forceps (which forceps were previously washed in ethanol and water and dried) and rapidly vibrating the object back and forth through a 1-cm arc by hand movement. During the vibration the object was directed at one minor worker after another, striking it lightly and allowing it to attack and depart at will. Bare forceps were also used to seize the legs of the minor workers and to pull them back and forth during the 1-cm movement. In order to test the effectiveness of *S. geminata* venom, the forceps tips were smeared with venom "milked" from three Solenopsis minor workers. Milking was accomplished by breaking off the abdomens of the Solenopsis workers and smearing the droplets of venom directly onto the forceps as the droplets exuded from the tips of the extended stings.

The results of these experiments, which are presented in Table 5, can be summarized as follows. It was found that freshly killed *S. geminata* were ineffective, even in large numbers, and whether intact or crushed. Also, steel forceps and wooden dummies do not cause the response, even when violently agitated among the Pheidole foragers—even to the extent of crippling and killing some of them. But single, freshly killed *S. geminata* workers are effective when held in forceps and agitated. The abdomen alone is more effective than the head or mesosoma alone. Wooden dummies treated with *S. geminata* venom and agitated cause the response, but so do live *S. geminata* workers with abdomens (and hence venom) removed. Thus either contact with venom or the odor of the body surface is sufficient, providing the chemical stimulus is associated with movement. The differences on which these conclusions are drawn are significant to at least the 95% level, and in most instances to the 99% level.

The specificity of response to the body odor of *Solenopsis* is further indicated by the following result. As shown in Table 4, *P. dentata* responds with relatively modest vigor to the presence of ten *M. pharaonis*. The venom of this species has been shown to have an especially repellent effect against other ants (Hölldobler, 1973). When the abdomens of ten workers were removed, and hence the venom, the workers failed to evoke alarm-recruitment in two replications. Hence, unlike the stimuli provided by *S. geminata*, the body odor of *M. pharaonis* was not sufficient by itself to trigger the response.

Table 5. The response of *P. dentata* colonies to various objects laid motionless on the arena floor or held in forceps and agitated. The object of the tests was the localization of the cues by which *P. dentata* recognize fire ants (*S. geminata*) and initiate alarm recruitment

Objects tested	Procedure	Number of replications	Number of positive responses	Number of Pheidole major workers responding
Controls:				
Wooden square 3 mm wide	Agitated 4 min	10	0	0 to 1 (0)
Forceps	Tips agitated and used to seize Pheidole during 4-min periods	9	1	0 to 2 (1)
Tetramorium caespitum (1 worker)	Fresh corpse held in forceps and agitated 4 min	6	0	−1 to 1 (0)
Experimentals:				
Solenopsis geminata (10 entire workers)	Fresh corpses placed on arena floor	10	0	−1 to 1 (0)
S. geminata (10 workers with abdomens removed)	Live workers with abdomens removed placed on arena floor	10	10	7 to 28 (15)
S. geminata (1 entire worker)	Fresh corpse held in forceps and agitated 4 min	10	7	0 to 10 (6)
S. geminata (1 head)	Head of corpse held in forceps and agitated 4 min	10	4	−1 to 5 (1)
S. geminata (1 mesosoma)	Mesosoma of corpse held in forceps and agitated 4 min	10	2	−2 to 2 (0)
S. geminata (1 abdomen, incl. pedicel)	Abdomen (incl. pedicel) held in forceps and agitated 4 min	10	7	3 to 18 (8)
Forceps with *S. geminata* venom	Tips smeared with venom and agitated 4 min	10	10	5 to 23 (14)

Discussion: The Ecology of Alarm Recruitment

The author was at first baffled by the discovery of enemy specification. However, further consideration of the ecology of *P. dentata* removed the mystery, and the following explanation now seems reasonably well established. The distribution of the native fire ant *S. geminata*, to which *P. dentata* responds so strongly, broadly overlaps that of *P. dentata* through the southern United States. The two species occur together in many of the same habitats, particularly open mixed pine-hardwood forests. To some extent they even utilize the same nest sites, principally logs and stumps in an advanced stage of decay. *S. geminata* forms the largest, most aggressive colonies of any potential competing native ant species. It also employs a swift, precise trail system which is initiated by scouts when they discover food or new nest sites (Wilson, 1962).

When *S. geminata* colonies are given close access to *P. dentata* colonies in the laboratory, they launch swift, overwhelming attacks. Each fire ant colony contains tens of thousands of workers—in nature, this number often exceeds 100,000—which enjoy a decisive numerical advantage over the several thousand workers that comprise a typical *P. dentata* colony.

On two occasions I allowed wars between Solenopsis and Pheidole colonies to proceed to a conclusion in the laboratory. The nest areas of the two species were connected by a glass tube 80 cm long supported on Erlenmeyer flasks which were in turn set alongside the nest tubes. In both cases *S. geminata* scouts quickly climbed the flasks, ran along the bridges, and descended to the Pheidole nests, all within 15 min and before Pheidole scouts could begin to climb the flasks on their own side. The Solenopsis laid odor trails back and forth, and the invading forces engaged the Pheidole minors and majors as the Pheidole colony proceeded through the first two phases of colony defense. In one case the Pheidole were quickly overwhelmed, and they absconded only 28 min after the bridge was laid in place. Unable to escape over the arena walls, the fleeing Pheidole were then methodically destroyed by the fire ants. Within 65 min all of the Pheidole adults, including the queen, were dead, and the fire ants were carrying off the captured brood. The fire ants subsequently ate the Pheidole brood, but they merely collected the Pheidole adult corpses in piles and abandoned them. They also occupied the deserted nest tubes. Hence the Solenopsis gained both food and nest space as the rewards of their assault. In the second experiment the Pheidole held their own for a while, destroying virtually all of the fire ants that descended to their nest area. However, within 2 h the tide of battle turned, and at 2 h 14 min the Pheidole absconded. The invaders were beginning to destroy the survivors and to carry away the brood when the experiment was halted in order to save the colony.

Clearly, it is essential for Pheidole colonies to halt invasions by fire ants in the earliest stages. It is of advantage to strike hard and fast whenever a fire ant scout is discovered near the nest. The danger is sufficient to commit major workers to destroy even a single intruder and to search the surrounding area for the presence of additional scouts. This observation is apparently sufficient to explain the seeming overreaction to single fire ants which has been observed in the first phase of colony defense (see Fig. 6).

The slight departures from genus specificity in the alarm-recruitment response also appear to have rational explanations. At an early stage in the study it was predicted that *P. dentata* would react to army ants of the genus *Neivamyrmex*, which are predators of other ants that occur in the same habitats as the Pheidole. This was subsequently shown to be the case, although the response did not prove to be as strong as that displayed toward fire ants (see Table 4). It was also anticipated that species of *Monomorium* might evoke the response, because they are phylogenetically the closest to Solenopsis of all the North American ants. This also proved to be true to a limited degree. The moderately positive response to *A. interjectus* was more surprising. *Acanthomyops* is a northern genus with a distribution that barely overlaps the northern fringes of the *P. dentata* range. It is possible that the relatively large size of these formicine ants and the large quantities of citronellal and other defensive secretions they expel are the key stimuli.

Finally, it should be noted that alarm-recruitment appears to be limited to other ants. When foraging minor workers are disturbed by human hands or inanimate objects they usually attempt to flee and only rarely attack. They have never been observed to recruit nestmates. When the nest is shaken or penetrated by human hands or an inanimate object, in a manner simulating an attack by a vertebrate predator, a large percentage of both the minor and major workers rush out, swarm over the offender, and attempt to bite it. The total effect is slightly painful to a human being and might serve as a deterrent to smaller vertebrates. Other minor workers seize pieces of brood and retreat deeper into the nest. If the disturbance continues and is serious enough, the colony leaves. There is no evidence of alarm recruitment during the entire process. Thus basically different defense strategies are employed against vertebrates as opposed to other ants.

It will be interesting to learn to what extent alarm-recruitment systems occur in other ant species in addition to pure recruitment systems, as is the case in *P. dentata*, and whether they are specifically directed at principal enemies. The significance of the present study is the demonstration that the entire defense strategy of an ant, including its alarm-recruitment system, is a sequence of complex events triggered by varying forms and intensities of stress. Also, the threshold of response varies according to the identity of the invading enemy in a way that appears appropriate to the ecology of the species.

Acknowledgements. Appreciation is expressed to Professor Walter R. Tschinkel and other members of the Department of Biological Science, Florida State University, as well as to personnel of the Tall Timbers Research Station, for their hospitality and assistance during the author's field research.

References

Bhatkar, A., Whitcomb, W.H.: Artificial diet for rearing various species of ants. Florida Entomol. **53**, 229–232 (1970)

Cammaerts-Tricot, M.-C.: Piste et phéromone attractive chez la fourmi *Myrmica rubra*. J. comp. Physiol. **88**, 373–382 (1974)

Cammaerts-Tricot, M.-C.: Ontogenesis of the defence reactions in the workers of *Myrmica rubra* L. (Hymenoptera: Formicidae). Animal Behaviour **23**, 124–130 (1975)

Creighton, W.S.: The ants of North America. Bull. Mus. Comp. Zool., Harv. **104**, 1–585 (1950)

Hölldobler, B.: Chemische Strategie beim Nahrungserwerb der Diebsameise (*Solenopsis fugax* Latr.) und der Pharaoameise (*Monomorium pharaonis* L.). Oecologia (Berl.) **11**, 371–380 (1973)

Wilson, E.O.: Source and possible nature of the odor trail of fire ants. Science **129**, 643–644 (1959)

Wilson, E.O.: Chemical communication among workers of the fire ant *Solenopsis saevissima* (Fr. Smith), parts 1–3. Animal Behaviour **10**, 134–164 (1962)

Wilson, E.O.: The insect societies. Cambridge, Mass.: Harvard University Press 1971

Wilson, E.O.: Enemy specification in the alarm-recruitment system of an ant. Science **190**, 798–800 (1975)

Received September 30, 1975

Edward O. Wilson
Harvard University
Department of Biology
MCZ Laboratories
Oxford Street
Cambridge, Mass. 02138, USA

1977

"The number of queens: an important trait in ant evolution," *Naturwissenschaften* 64: 8–15 (with Bert Hölldobler, first author)

Whether an ant colony, or any animal colony, has a single queen responsible for reproduction, as opposed to multiple queens, may seem of relatively minor importance. However, as Bert Hölldobler and I show in this first systematic examination of the subject, it has deep implications for our understanding of the evolution of social structure, ecology, and biodiversity in the ants. The difference between a single or multiple queens determines how fast a colony can grow, the size it can reach at maturity, and how much territory it can occupy. Such might seem to be a universal goal, but it is not. There are many niches open to ant species that are narrow, entailing specialized nest sites and food, and these restrictions require multiple small colonies serviced by more than a single queen. Other niches, including those of the army ants, require relatively huge colonies regulated by the presence of a single, highly productive queen. Much of the social evolution of ants entails such permutations of adaptation to the environment by the regulation of colony size and multiplicity of queens.

The Number of Queens: An Important Trait in Ant Evolution*

Bert Hölldobler and Edward O. Wilson
Department of Biology, Museum of Comparative Zoology Laboratories, Harvard University, Cambridge, Massachusetts 02138, U.S.A.

The pervasive social and ecological differences between ant colonies that have a single queen and those that have multiple queens are defined. The evolutionary tendencies which lead to polygyny and the adaptive significance of multiple queens are examined. The discussion of the ecological consequences of polygyny and monogyny leads to a deeper understanding of territoriality, spacing and species packing in ants.

Introduction

Colonies of higher social insects, in particular termites, ants, bees, or wasps, are by definition divided into a fertile "queen" caste and a supportive worker caste which has reduced fertility or is wholly infertile. The number of queens per colony varies a great deal between species and sometimes even within species. Entomologists make an elementary distinction between colonies that have a single queen (monogynous) and those that have multiple queens (polygynous). In this article we will briefly review previous literature and advance some new evidence to show that the distinction is actually fundamental, with consequences that ramify into various aspects of the life cycle, population structure, and mode of competition of ant species.

It is useful to begin with a note on terminology. Monogyny refers simply to the possession by a colony of a single queen, polygyny to the possession of multiple queens [1]. The founding of a colony by a single queen is referred to as haplometrosis; when multiple queens found a colony the condition is called pleometrosis [2–4]. The term metrosis can be used to refer generally to this biological variable.[1] Monogyny can be primary, meaning that the single queen is also the foundress; or it can be secondary, meaning that multiple queens start a colony pleometrotically, but only one survives. In a symmetric fashion, polygyny can be primary (multiple queens persist from a pleometrotic association) or secondary (the colony is started by a single queen and others are added later by adoption or fusion with other colonies). Finally, the mode of colony founding varies enormously. It can be accomplished by swarming, a process also called budding, hesmosis, or sociotomy, in which one or more reproductive forms depart with a force of supporting workers; or independent, in which case the reproductive form makes the attempt alone (see Fig. 1). If independent, the colony founding is also in many instances claustral, meaning that the queen seals herself off in a chamber and rears the first brood in isolation. This is the dominant mode of independent colony formation in ants, although the queens of such primitive forms as *Myrmecia* and *Amblyopone,* as well as those of a very few more advanced species (in *Manica* and *Acromyrmex*), still forage outside their cells for food.

Evolutionary Tendencies in Metrosis and Gyny

We will argue that elementary properties in the organization of insect societies generally bias species toward monogyny in the course of evolution by natural selection, and that the tendency is reversed only when special ecological constraints are imposed on the species. Two such properties exist. First, queens of all kinds of social insects should prefer to retain personal reproductive rights and surrender none to their sisters or daughters, because they are more closely related to their daughters and sons than to their nieces, nephews, and grandoffspring. Second, workers should prefer to have only one queen as the colony progenitrix. In species with colonies of small to mo-

* Dedicated to Professor Dr. Karl von Frisch on the occasion of his 90th birthday

[1] Wilson [3], following W.M. Wheeler, used the terms gyny and metrosis interchangeably as suffixes, but we here recognize the usefulness of the distinction made by earlier authors

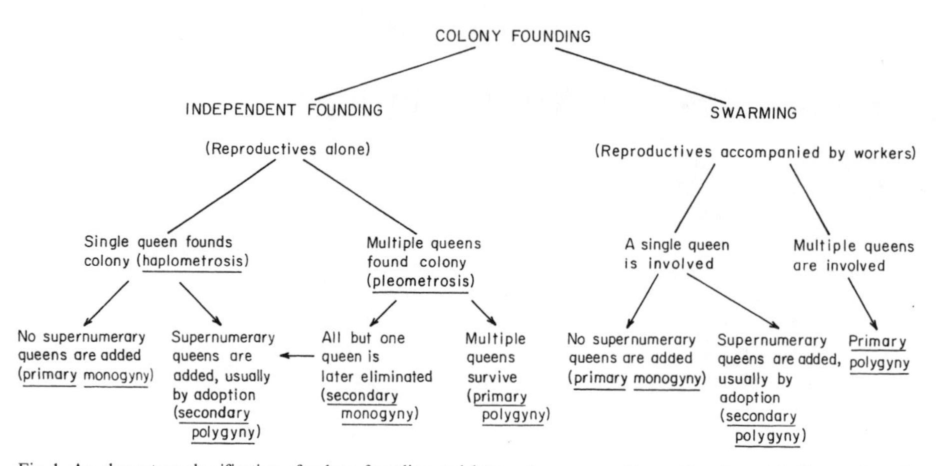

Fig. 1. An elementary classification of colony founding and later colony composition, with reference to the number of queens

derate size, the rate of colony growth, and hence the amount of colony genetic fitness, is limited primarily by the number of workers and not by the number of queens, one queen usually being able to supply as many eggs as a worker force can rear [5, 6]. Since extra queens would then be an unnecessary energetic burden on the colony, an especially significant factor during the colony's early growth, it should be of advantage to the workers to eliminate them. In short, one can reasonably expect the dominant queen and the workers to conspire to eliminate supernumerary queens during the early phase of colony growth and thus to attain a state of monogyny.

Under a wide range of conceivable conditions, independent colony foundation should be a second trait favored by natural selection. If entire colonies can be started by one or a few queens, mother colonies producing such females can deploy far more of them over greater distances than otherwise comparable colonies that reproduce by swarming. Each swarm drains off a part of the original worker force, and its dispersal range is limited by the difficulties inherent in mass orientation and mobility. Swarming is likely to be of advantage only if the survival rate of queens is overwhelmingly greater when they are accompanied by workers than when they proceed alone.

In addition, haplometrosis can be expected to be the preferred mode of independent colony foundation. Unless circumstances give a large advantage to founding in groups, each queen should attempt to start a colony well away from all possible rivals. The tendency just cited toward the restitution of monogyny by both the dominant queen and the first worker force means that a queen choosing to become a member of a group of n founding queens has a $1/n$ chance of surviving to be the nest queen.

Finally, colony foundation should be claustral whenever possible. The highest mortality of social-insect workers occurs during foraging trips [3], and it is probable that the same is true of founding queens forced to leave their nests in order to search for food.

In sum, a logical examination of the more abstract general properties of insect societies leads us to expect that natural selection will lead species to monogyny, haplometrosis, and claustral nest founding. Yet deviations from this expected pattern are many and exceedingly diverse. Their close examination in a more theoretical context, can, we believe, shed light on the evolution of other aspects of social behavior.

Differences between Ants and Wasps

Let us begin with a comparison of two of the major groups of social Hymenoptera. Most ant species are monogynous and obligatorily haplometrotic. Within most major phyletic lines, polygyny and swarming appear to be evolutionarily derived conditions. Among the wasps, in contrast, only the temperate-zone species of *Vespa* and *Vespula* are known to be obligatorily haplometrotic. *Belonogaster, Mischocyttarus,* and *Polistes* are sometimes haplo- and sometimes pleometrotic, while most or all of the Polybiini, containing 20 of the 26 known social wasp genera, are polygynous and reproduce by swarming [7, 8].

We suggest the following simple explanation for the difference. When an ant colony moves to a new nest site, for example following a disturbance by a predator, it must walk to the new site. The workers are wingless, and the queen must travel on the ground with them. Thus queens can afford to engage in claustral colony foundation. They shed their wings following the nuptial flight, and in all but a few primitive species they histolyze their alary muscles to nurture the first worker brood within a completely closed cell. Since they never need to take flight again, the queens are able to take advantage of the surplus energy available in the alary muscles. Both wide dispersal and independent, claustral colony founding are within their reach. Since these are the optimum techniques under almost all conceivable conditions, most ant species have evolved to acquire them.

When wasp colonies are disrupted, they fly to a new nest site. Lengthy ground travel is not only unnecessary but would be disadvantageous for insects so fully adapted to life in the air. Because the queens must fly with them, they must "stay in shape" by not losing their flight muscles. The nest queens of *Vespa* lose the power of flight as they become older, presumably because of the weight of their ovaries, but this is the exception rather than the rule in social wasps. Thus wasp queens are less well equipped to be solitary foundresses. Since independently founding individuals find it disadvantageous to convert the alary muscles into energy for the brood, they must engage in the risky process of foraging for food. It appears to follow that wasp species should be more likely to rely on pleometrosis or even swarming, which is in fact the case.

Primary versus Secondary Polygyny in Ants

In a sizeable minority of ant genera and species, pleometrosis is at least an optional mode of colony foundation. In other words, some queens attempt to start colonies singly after the nuptial flight while others of the same species simultaneously form groups for the same purpose. When this phenomenon was examined in the laboratory in *Lasius flavus* [9], *Solenopsis invicta* [10], and *Myrmecocystus mimicus* [11], it was found that groups have a higher survival rate and bring their first broods to maturity more quickly than do single queens. Because groups of multiple queens of all three species are found in nature, it is reasonable to conclude that pleometrosis conveys a selective advantage.

After completing their nuptial flights, queens of *Lasius neoniger* and *Solenopsis molesta* alight on the ground in what appears to be a random distribution. Afterward they shed their wings and search for suitable nest sites during short excursions on foot [12]. Let us suppose that this pattern is followed generally in pleometrotic ant species, and that queens associate with some probability greater than zero whenever they contact one another on the ground. Then the

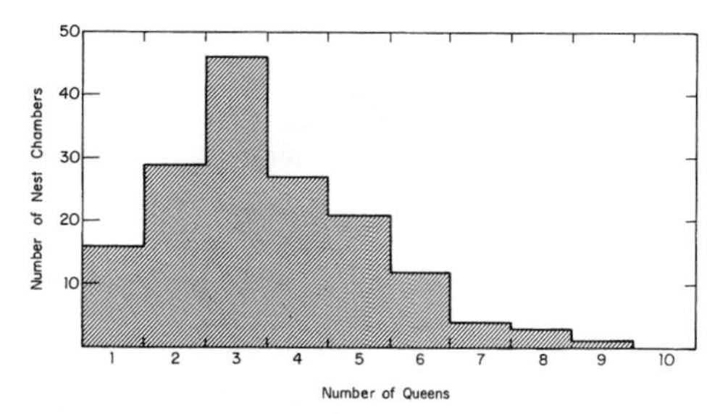

Fig. 2. The frequency of group size in colony-founding queens of *Myrmecocystus mimicus*. The array is consistent with a Poisson distribution with the zero term truncated, as predicted by an assumption of random dispersal followed by free association of queens that meet on the ground

frequencies of the numbers of queens in various groups should fit a truncated Poisson distribution, with the "quadrat" width being roughly half the distance over which queens attract each other. This appears to be the case in data recently gathered for *Myrmecocystus mimicus* in the field in Arizona (see Fig. 2; the probability of drawing a distribution by chance alone as close as the one shown is greater than 0.80).

In a wide variety of ant genera, including *Aphaenogaster, Camponotus, Formica, Leptothorax, Myrmica, Prenolepis, Solenopsis,* and *Tapinoma,* species exist in which monogyny of established colonies is the rule but polygynous colonies are also frequent. The following data supplied by Talbot [13] for *Aphaenogaster rudis* are typical for this class of species: 52 colonies had a single queen, 10 colonies had 2 queens, 3 colonies had 3 queens, 3 colonies had 4 queens, and one colony had 15 queens. If the last datum is discounted, the distribution is close to a truncated Poisson and is not inconsistent with an origin through the random association of pleometrotic queens. Other, similar data have been published by several authors [14–17]. Unfortunately, it was not determined in most of these studies whether all of the queens had been inseminated. In some species of the genus *Leptothorax* the supernumerary queens often remain uninseminated [18].

Yet pleometrosis and primary polygyny cannot be proved by such data alone. The species involved also display marked colony territoriality. In the cases of *Lasius flavus, Myrmecocystus mimicus,* and *Solenopsis invicta,* also territorial species, pleometrotic laboratory groups revert to monogyny when the first brood matures. The *Lasius* queens fight with one another and then break apart into single-queen units. Those of *Myrmecocystus* form dominance hierarchies, with the supernumerary individuals eventually being

Naturwissenschaften 64, 8–15 (1977)

driven out by the workers. When multiple *Solenopsis invicta* queens are introduced to queenless workers, the latter usually execute all but one. In *Camponotus herculeanus* and *C. ligniperda,* large colonies often contain several queens, but these individuals are intolerant of one another and maintain territories within the diffuse nests [19].

Remarkably, we know of no case of pleometrosis leading smoothly to polygyny. In one replicate out of 23, a pair of *Myrmecocystus mimicus* queens persisted in harmony during two years of colony growth—then one was ejected. In a single replicate out of twenty, the workers of a laboratory colony of *Solenopsis invicta* kept two queens that had been introduced to them in a queenless state. The queens were still coexisting after one year [10]. It is possible that the oligogynous and polygynous conditions just cited in colonies of *Aphaenogaster, Camponotus, Formica,* and other territorial ant species is derived from such pleometrosis. Studies of pleometrotic groups have not been undertaken, either in the field or laboratory. But if the transition proves to occur, we suspect that some of the supernumerary queens will be ejected in later stages of colony growth, and that the frequency distribution of queen numbers will be modified in form. In short, there appears to be a "bottleneck" at the time of the first brood maturation which in many ant species results in a reduction in the number of queens, most commonly to a monogynous state.

Secondary polygyny, in contrast to primary polygyny, has been well documented. It occurs commonly in socially parasitic species, where mating occurs in or near the host nest and the host workers accept inseminated queens [3, 4]. It is also the rule in ant species in which no colony boundaries exist and local populations are comprised of networks of intercommunicating aggregations of workers, brood, and fertile queens.[2] Examples from the latter, important category are *Formica polyctena,* members of the *Formica exsecta* group, the Argentine ant *Iridomyrmex humilis,* Pharaoh's ant *Monomorium pharaonis,* and *Myrmica ruginodis* ("*microgyna*" form). Such species have been referred to as unicolonial, or more precisely as forming unicolonial populations, as opposed to the more frequent multicolonial ant species in which intercolony recognition and aggression are the rule [3]. Although the build-up of queens usually results from the re-adoption of newly inseminated individuals following their nuptial flights from the same nest, the initial state of polygyny in the arboreal ant *Pseudomyrmex venefica* comes from the fusion of small haplometric colonies [20].

Elmes [5] has argued that the queens of secondarily polygynous species live essentially as parasites on the colonies. They contribute fitness to the colonies, of course, in that they are the progenitrices of new aggregates necessary for overall population growth. But with reference to the growth rate of local aggregates, they represent an extra energetic load without being essential for a full oviposition rate. Elmes therefore views the population of queens as an entity that grows semi-independently from the population of workers. This viewpoint appears to be valid, as we noted earlier, because in general only a single queen is required to furnish a full supply of eggs for the workers to nurture. Both populations will be limited in a logistic manner by their own numbers and the numbers of the other caste, in which the parameters are caste-specific. Using modified Leslie-Gower equations for host-parasite populations, Elmes concluded that the frequency of aggregates ("colonies") with various numbers of queens should fit a lognormal distribution, which proves to be the case in *M. rubra.* A lognormal distribution, or more precisely a lognormal Poisson distribution, is also the expected result if the newly fecundated queens return randomly to the ground but then are recruited by the colonies with an efficiency that is not constant but rather a random variable. The latter result would accrue, for example, if the recruitment power of a colony were simply a linear function of its size.

The Adaptive Significance of Secondary Polygyny

Attention therefore focusses on secondary polygyny, a minority phenomenon that is evolutionarily derived, polyphyletic, and has certain apparent intrinsic disadvantages. When a phenomenon displays this kind of pattern, the biologist is justified in searching for unusual ecological circumstances that have promoted its deviant form of evolution.

It appears to us that fully polygynous species, i.e., species that are unicolonial or at least comprised mostly of colonies with multiple queens, fall into two sets characterized by very different adaptation syndromes. (1) The first set is specialized on exceptionally short-lived nest sites. Such species are opportunistic in the sense employed by ecologists—they occupy local sites that are too small or unstable to support entire large colonies with life cycles and behavioral patterns dependent on monogyny. (2) The second adaptation is specialization on habitats—entire habitats, as opposed merely to nest sites—that are long-lived, patchily distributed, and large enough to sup-

[2] The "colonies" of such species typically occupy multiple nest sites; they are, to use current terminology, polydomous. However, the correlation is very weak. Many monogynous ant species are also polydomous, while a few polygynous ones are monodomous

port large populations. The two forms of specialization are not mutually exclusive; some ant species, for example *Iridomyrmex humilis* and *Pheidole megacephala,* possess both.

Consider first the class of opportunistic nesters. Colonies of *Tapinoma melanocephalum, T. sessile, Paratrechina bourbonica,* and *P. longicornis* often occupy tufts of dead grass, plant stems, temporary cavities beneath detritus in urban environments, and other local sites that sometimes remain habitable for only a few days or weeks. Some species of *Cardiocondyla* excavate shallow tunnels and chambers in patches of soil of such places as the edges of palm trunks, sidewalks, and street gutters. Colonies of *Monomorium pharaonis* and *Tapinoma melanocephalum* invade houses to occupy cracks in walls, the lining of instrument cases, spaces in piles of discarded clothing, and similarly unlikely microhabitats. Colonies of these species are characterized by extreme vagility—a readiness to move when only slightly disturbed and the ability swiftly to discover new sites and to organize emigrations. Their colonies are also typically broken into subunits that occupy different nest sites and exchange individuals back and forth along odor trails. We suggest that it is the latter quality that gives polygyny a premium in opportunistic nesting. Because of the inevitable frequent fragmentation of the colonies, subunits probably lose contact with one another for long periods of time and occasionally forever. By having enough reproductive females to service most or all of the subunits means that the colony as a whole can exploit the rapidly fluctuating environment in which it lives. Other kinds of ants are not fragmented in this manner, and consequently a single queen suffices as the colony progenitrix.

The second major class of fully polygynous ant species contains only a few species, but they include the great majority of examples that have been both well studied and do not qualify as opportunistic nesters. Thus, almost all fully polygynous species the natural history of which is known to us are accounted for by the simple classification proposed here. The habitats favored by species in the second category are first of all patchily distributed; they have distinctive qualities and are more or less isolated from each other. They are also extensive enough to support substantial populations of ants. Hence propagules of species that are specially adapted for such places encounter a potential bonanza when they succeed in colonizing one. The habitats are also relatively long-lived, giving a premium to the type of slow but thorough occupation made possible by polygyny and budding.

For example, the Allegheny mound-builder *Formica exsectoides* typically occupies persistent grassy or heath-like clearings. Such habitats are relatively scarce and patchily distributed, and many are fully occupied by dense unicolonial populations of *F. exsectoides.* The *microgyna* form of *Myrmica ruginodis* shows a similar preference for scattered, very stable open habitats in England, some of which are known to have persisted at the same localities for as long as 200 years. The *macrogyna* form, which is haplometrotic and monogynous, favors less stable but more widespread habitats [21]. *Pseudomyrmex venefica* is specialized to occupy species of swollen-thorn *Acacia* that grow for long periods of time in areas of slow floristic succession. Because the acacias are able to expand into extensive thorn forests, the *P. venefica* colonies have the opportunity to build large unicolonial populations over a period of many years. Single populations may contain 20 million or more workers, rivalling *Dorylus* ants for the possession of the largest ant "colonies" in the world [20]. "Tramp" species, those ants distributed widely by human commerce and living in close association with man, are typically polygynous. In a sense they can be said to have been preadapted for patchy but persistent and species-poor habitats created within man-made environments. Some of the best known species are comprised of unicolonial populations and spread largely or entirely by budding off of groups of workers accompanied on foot by inseminated queens. Examples include *Monomorium pharaonis, Pheidole megacephala, Iridomyrmex humilis,* and *Wasmannia auropunctata.*

The isolated-habitat specialists, as distinguished from the nest-site opportunists, are species that have difficulty locating their preferred habitats, but once having found a suitable place, are opposed by relatively less competition from colonies of their own or other ant species. Thus it is to the advantage of the founding colony to spread out as a continuous unicolonial population, occupying all of the habitable nest sites and foraging areas. In contrast, most other ant species have no trouble finding a suitable habitat, but such places are typically already saturated with other ant colonies. The best strategy of these ants is to send out large numbers of flying queens capable of founding new colonies independently. A tiny fraction of the propagules will locate some of the rare and constricted nest sites and foraging areas left unfilled; the colonies they produce will not find it profitable to try to spread outward through the surrounding occupied territories.

The full life cycle of unicolonial populations remains to be worked out. Among the interesting variables to be studied as functions of the age and condition of the populations are the queen:worker ratios and the behavior of queens during nuptial flights. Scherba [22] reported that most of the queens of *Formica opaciventris,* a unicolonial species closely related to

 Naturwissenschaften 64, 8–15 (1977)

F. exsectoides, mate within a few meters of their nests of origin and return at once, while a small minority, originating entirely from mounds near the edge of the population, fly out of sight in a direction away from the population. We predict that in the early stages of population growth, the queen: worker ratio of unicolonial species will start high (immediately following haplo- or pleometrosis, or adoption by an alien host species), drop to a steady state as the population comes to saturate the habitat, then rise again as the habitat quality declines and a higher premium is thereby placed on dispersal away from the habitat. It also seems likely that the fraction of queens engaging in dispersal flights will increase when the habitat is saturated, and increase still more as the habitat declines.

An interesting phenomenon for which no explanation yet exists is the frequent coexistence of pairs of closely related species, of which one is monogynous and the other polygynous. Among the examples we have noted in the literature and from personal observation include the *"microgyna"* and *"macrogyna"* forms of *Myrmica ruginodis* already described, two sibling species of *Pseudomyrmex flavidula* in southern Florida, two apparent species placed under *Crematogaster minutissima, Dorymyrmex insana* and *D. flavopectus* [23], *Formica incerta* and *F. nitidiventris* [14], and two apparently distinct species of *Formica neorufibarbis* in the White Mountains of New Hampshire.

Finally, we have used a combination of logic and a synthesis of very incomplete data to arrive at the hypothesis of the duality of selective forces behind full polygyny. It is hoped that the set of generalizations associated with this explanation will be recognized as preliminary and used primarily to make a more precise definition of the many unanswered questions in what remains a largely unexplored area of ant biology.

The Ecological Consequences of Polygyny

Having reviewed the possible ecological prime movers that lead to monogyny or polygyny, let us now consider some of the consequences of these adaptations.

Monogyny is closely associated with colony distinctness. Each colony occupies a separate nest site, and its workers avoid or attack the members of other colonies encountered during foraging. The basis of discrimination is the colony odor. It is one of the enduring mysteries of biology that ant workers can almost instantly distinguish nestmates from alien workers belonging to the same species. We still do not know whether *every* colony has its own odor, meaning that abundant species would contain many millions of such olfactory signatures, or whether there are a lesser number of odor groups, such that a small fraction of the colonies might be joined in experiments without resulting hostility. Accidental chemical differences in the diet and nest material might contribute to the colony odor, as demonstrated in bumblebees and honeybees [3, 24]. It is just as likely that genetic differences play a role, as indicated by recent experiments on the primitively eusocial bee *Lasioglossum zephyrum* [25, 26]. If that is the case, the simplest conceivable mechanism is for the queen to provide the essential ingredients, because genetic variety among the workers would easily increase the diversity of odors to a confusing and possibly unmanageable level. Elaborate genetic controls are not needed to generate the amount of odor diversity observed in nature. If, in the simplest conceivable case, two alleles influenced odor at each locus, only ten such loci would generate $3^{10} = 59{,}049$ diploid combinations; three alleles at ten loci could yield $9^{10} = 3.49 \times 10^9$ such combinations. Monogyny would make such a system easily operable; polygyny would tend to break it down. It therefore may be significant that the total erasure of colony boundaries that occurs in the unicolonial populations of such species as *Monomorium pharaonis, Iridomyrmex humilis* and *Formica exsectoides* is associated with an extreme development of polygyny.

The loss of colony territories in unicolonial species has some important ecological implications. Unicolonial species are notable for their high local abundance and the degree to which they appear to dominate the environment. Introduced populations of *Pheidole megacephala, Wasmannia auropunctata,* and *Iridomyrmex humilis* extirpate many other kinds of ants, although it is not known whether species occurring within the native ranges of the unicolonial ants have evolved competitive resistance to the point of being able to coexist [27–29]. Habitats containing populations of *Formica exsectoides* have notably sparse ant faunas, and the related *F. exsecta* and *F. opaciventris* are known to exclude some other territorial ants, including other species of *Formica,* by aggression [30, 31]. It is not known to what extent the general occurrence of unicolonial species in species-poor habitats is a cause and to what extent it is an effect. Put as a question: have the unicolonial ants simply adapted to habitats that are species-poor in the first place, or does the formation of supercolonies provide them with a decisive competitive edge in habitats that would otherwise be species-rich? This problem seems eminently tractable to field analysis.

It is possible that gyny and the nature of territoriality also affect the patterns of species diversity and quite possibly the mechanism of speciation itself. Local faunas consisting of multicolonial ant populations, which are comprised in turn chiefly of monogynous and oligogynous colonies, seem to be more susceptible to increases in within-habitat species diversity,[3] for three reasons. First, as previously noted, they appear to suppress other ant species less severely than do unicolonial populations, permitting the buildup of larger numbers of coexisting species. Second, because the colonies are much smaller in size, multicolonial species are able to specialize more on nest sites and food. For example, a colony of one species might do well with a single hollow stem, while a colony of another species is able to occupy the space beneath a nearby stone. Other species can be differentiated according to food, some preying exclusively on small arthropods, others mostly tending scale insects, and so forth. The relatively huge populations of unicolonial species cannot afford a narrowing of their niche. In order to survive they must remain broad generalists, which brings them into competition with a broad range of specialists as well as other generalists.

The third reason why multicolonial species can be expected to exhibit higher within-habitat diversity is more subtle. Many studies have shown that colonies of ants are hostile to a degree inversely proportional to the degree of similarity to their competitors. That is, they are most aggressive to other colonies of the same species, somewhat less to other species in the genus, and least of all to forms that not only belong to other genera but differ strongly in size and behavior [34, 35]. This being the case, within-habitat diversity can be expected to be enhanced by the divergence of colony odors between closely related species. Suppose that two newly formed, cognate species have arisen by genetic divergence during geographical isolation, and have recently come into contact along the boundaries of their respective geographical ranges. Suppose further that the species are sufficiently divergent in ecological requirements so that neither would exclude the other by means of non-aggressive preemption of resources. Yet the two forms are likely to remain in separate geographical ranges so long as their colony odors are too similar, causing interspecific territorial exclusion to be as strong as intraspecific exclusion. The two species can penetrate one another's range only if one or both undergoes a divergence in the species-specific components of the colony odor. This would be a form of character displacement comparable to the divergence of identifying songs by which territorial bird species penetrate one another's ranges [36]. The result is an increase in the within-habitat species diversity.

The penetration of territories belonging to alien monogynous species is made easier by the very fact that territorial colonies repel one another so effectively—to the extent that colonies are overdispersed in their statistical pattern of distribution. In many kinds of ants, such as members of the genus *Pogonomyrmex*, this pattern is geometrically highly regular and is maintained by high levels of intercolonial aggression [35]. If colonies of other species are different enough in colony odor to escape such aggressive response, and if they are also distinct enough in their foraging habits not to be replaced through competitive exclusion, they might easily slip into nest sites located between those of the resident species.

We therefore hypothesize that the high degree of odor specificity associated with monogyny and intercolony territoriality lends itself to interspecific recognition, the reduction of interspecific territorial aggression by means of that recognition, and an increase in numbers of species that can coexist in the same habitat. Polygynous ant species, and particularly those that are unicolonial, have surrendered some of this discriminatory power. As a result they remain aggressive toward a broader range of species, and it is not surprising to find very few unicolonial species occupying the same habitat, as well as an overall decrease in species diversity. Although it is theoretically possible for unicolonial ant species to build up between-habitat diversity by means of specialization on habitats as opposed to niches within habitats, very little appears to have occurred in nature—quite possibly due to the preemption of most kinds of habitats by monogynous, multicolonial ant species.

The preparation of this article, and some of the original research reported in it, has been supported by National Science Foundation Grants Nos. BMS75-06447 and BNS73-00889.

1. Reuter, O.M.: Lebensgewohnheiten und Instinkte der Insekten bis zum Erwachen der sozialen Instikte. Berlin: R. Friedländer u. Sohn 1913
2. Wasmann, E.: Biol. Centralbl. *30*, 453 (1910). See also Wheeler, W.M.: Colony-founding among ants. Cambridge, Mass.: Harvard Univ. Press 1933
3. Wilson, E.O.: The insect societies. Cambridge, Mass.: Belknap Press of Harvard Univ. Press 1971
4. Buschinger, A., in: Sozialpolymorphismus bei Insekten (Schmidt, G., ed.). Stuttgart: Wissenschaftliche Verlagsges. 1974
5. Elmes, G.W.: J. Animal Ecol. *35*, 761 (1972)
6. Wilson, E.O.: Ann. Entomol. Soc. Amer. *67*, 781 (1974)
7. Evans, H.E., West Eberhard, M.J.: The wasps. Ann Arbor, Michigan: Univ. of Michigan Press 1970
8. Spradbery, J.P.: Wasps. London: Sidgwick & Jackson 1973
9. Waloff, N.: Insectes Sociaux *4*, 391 (1957)
10. Wilson, E.O.: Symp. Roy. Entomol. Soc. London *3*, 81 (1966)
11. Hölldobler, B., Bry, W.: unpublished observations
12. Wilson, E.O., Hunt, G.L.: Ecology *47*, 485 (1966)
13. Talbot, M.: Ann. Entomol. Soc. Amer. *44*, 302 (1951)

[3] For general accounts of species diversity patterns, see [32, 33]

14. Talbot, M.: Ecology *24*, 31 (1943); *29*, 316 (1948); *38*, 449 (1957)
15. Talbot, M.: Ann. Entomol. Soc. Amer. *38*, 365 (1945)
16. Headley, A.E.: ibid. *36*, 743 (1943); *42*, 265 (1949)
17. Markin, G.P., et al.: ibid. *66*, 803 (1973)
18. Buschinger, A.: Insectes Sociaux *15*, 217 (1968)
19. Hölldobler, B.: Z. Angew. Entomol. *49*, 337 (1962)
20. Janzen, D.H.: J. Animal Ecol. *42*, 727 (1973)
21. Brian, M.V., Brian, A.D.: Evolution *10*, 280 (1955)
22. Scherba, G.: J. N.Y. Entomol. Soc. *69*, 71 (1961)
23. Nickerson, J.C., et al.: Ann. Entomol. Soc. Amer. *68*, 1083 (1975)
24. Michener, C.D.: The social behavior of the bees. Cambridge, Mass.: Belknap Press of Harvard Univ. Press 1974
25. Bell, W.J.: J. Comp. Physiol. *93*, 195 (1974)
26. Barrows, E.M., Bell, W.J., Michener, C.D.: Proc. Nat. Acad. Sci. U.S.A. *72*, 2824 (1975)
27. Smith, M.R.: Circular U.S. Dept. Agriculture *387*, 1 (1936)
28. Way, M.J.: Bull. Entomol. Res. *44*, 669 (1953)
29. Levins, R., Pressick, M.L., Heatwole, H.: Amer. Scientist *61*, 463 (1973)
30. Scherba, G.: J. N.Y. Entomol. Soc. *72*, 231 (1964)
31. Pisarski, B.: Ekologia Polska *20*, 111 (1972)
32. Whittaker, R.H.: Taxon *21*, 213 (1972)
33. MacArthur, R.H.: Geographical ecology. New York: Harper & Row 1972
34. Carroll, C.R., Janzen, D.H.: Annu. Rev. Ecol. Syst. *4*, 231 (1973)
35. Hölldobler, B.: Behav. Ecol. Sociobiol. *1*, 3 (1976)
36. Murray, B.G.: Ecology *52*, 414 (1971)

Received May 14, 1976

1980

"The ethical implications of human sociobiology," *The Hastings Center Report* 10(6) (December): 27–29

The controversy over my inclusion of human beings in *Sociobiology: The New Synthesis* had at least one salutary effect on me. I was made to understand the necessity of learning more about cognitive and developmental psychology and to think more deeply than I had prior to 1975 about the relation of genetic and cultural evolution. I also saw the opportunity to further explore the implications of evolution in the origin of religion and ethics. The result was the extended essay, *On Human Nature,* published in 1978. This work was very well received. It won the 1979 Pulitzer Prize in General Nonfiction and suffered surprisingly little controversy. I think it fair to say that it was the first systematic application of sociobiology to psychology and thus a key founding document of the discipline later called evolutionary psychology.

The present essay, which is based on a part of *On Human Nature,* is typical of my expositions on this and related subjects in that period. It describes the support that biological reasoning can give to moral philosophy based on rational secularism.

Will Biology Transform the Humanities?

Two basic questions have always pervaded the work of the Hastings Center. The first, more general, concerns the proper relationship between the sciences and the humanities. Much of our work is at just that intersection, and vexing questions of disparate methodologies, perspectives, and concepts constantly arise. The second, more specific, bears on the pertinence of scientific theories and data for the development both of moral theories and of practical moral judgments. That issue surfaced in some of our earlier work on genetic counseling and prenatal diagnosis, in more recent work on health policy, and is central to a present project on the relationship between social science knowledge and the formation of public policy.

On October 12, 1978, The Hastings Center organized a conference, held at and supported by The Rockefeller Foundation, that brought both of those problems to the fore. The broad topic was the way in which, if at all, the humanities should try to incorporate the new findings of the biological sciences in their perennial attempt to understand human nature and human values. A number of scientists, not always immodestly, have recently suggested that these findings must inevitably alter the way in which the humanities have traditionally viewed human nature. The question before the conference was: is that so? As it turned out, a more concrete focus of the conference was that of sociobiology, a relatively new discipline greatly stimulated by the work of Professor E.O. Wilson of Harvard. For it is in that new discipline that some of the fresh claims of knowledge offering a challenge to the humanities have been most pronounced. The presentations collected here underscore that focus of the conference.

The late John Knowles, M.D., then President of the Rockefeller Foundation, was an active participant in the conference, as were Joel Colton, director of the Humanities Division at the Foundation; other Foundation staff members; and a number of other invited guests. It was a lively occasion, captured only in part by the essays presented here. **—Daniel Callahan**

The Ethical Implications of Human Sociobiology
by EDWARD O. WILSON

Sociobiology is not, as the popular press often describes it, a hypothesis about the genetic determination of human behavior. A well-established scientific discipline, it is the systematic study of the biological basis of all forms of social behavior.[1] Sociobiology is distinguished from ethology by its emphasis on the study of societies as populations; it is to a large degree an extension of traditional population biology and evolutionary theory. Ethology, which provides some of the factual information used in sociobiology, has remained primarily concerned with the behavioral responses of individual organisms, the evolution of those responses as part of species-specific patterns, and the ways that these patterns of behavior enable organisms to cope with and adapt to particular environments.

By contrast, sociobiology is concerned more with the biology of whole groups, especially those aspects of demography, genetic kinship, hierarchy, and communication that constitute the species-specific qualities of societies. What is new about sociobiology is the way that it has extracted information previously scattered through ethology, comparative psychology, and natural history, and reassembled it on a foundation of ecology and genetics studied at the population level. By this procedure students of the subject have been able to show how societies adapt to particular environments, in ways that increase the overall genetic fitness of the group members.

General sociobiology has been concerned primarily with animal societies. Aside from an ordinary number of intramural disputes over specific models and pieces of evidence, the discipline has had a relatively normal development as the newest branch of evolutionary biology. What is in contention—the source of the media's version of the "sociobiology controversy"—is the application of biological analysis to human social behavior. Much of the critics' initial opposition has been misdirected. Sociobiological theory does not rest on any assumptions that human behavior must be genetically controlled. It leaves open the possibility that a highly intelligent species can abandon genetic programs and come to depend more or less exclusively on cultural evolution (although it has always seemed unlikely to me

EDWARD O. WILSON *is Frank B. Baird, Jr., Professor of Science at Harvard University, and the author of* Sociobiology: The New Synthesis *and* On Human Nature.

that even the most intelligent species could make a total shift and survive for very long).

It is therefore not the content of general sociobiology that has been disputed but rather the evidence that substantial genetic constraints exist and the proposition that the theory of sociobiology is the best means of organizing and explaining the evidence. The data concerning genetic constraints have been summarized elsewhere.[2] For the moment let me simply note the five categories into which the evidence can be conveniently divided: (1) the specificity of human social behavior, numerous details of which distinguish the human species from the thousands of other social creatures on earth; (2) the close resemblance of some behavior to that of other Old World primates, as one might expect from knowledge of our ancestry based on the fossil record and present-day comparisons of anatomy, physiology, and biochemistry; (3) the conformity of at least some of the basic patterns of human sociality to evolutionary theory; (4) the preservation of elementary, but still surprisingly complex, mammalian patterns of behavior in genetically caused retardation; (5) the now abundant studies, some of them conclusive enough to persuade even the most determined skeptics, of a genetic component in the variation among the members of a population. The component can be found in a wide array of behavioral traits that affect social behavior, including word fluency, memory, the timing of language and Piagetian stages of intellectual development, psychomotor skills, extroversion and introversion, homosexuality, the timing of heterosexual activity, and some psychoses and neuroses.

With only a slight effort, students of human biology have adduced what I consider overwhelming evidence of substantial genetic constraints on cognition and social behavior. These take the form of learning rules (the tendency to learn one response as opposed to another when faced with identical reinforcement schedules), programmed fine-tuning of mood and mental set by hormones and neurotransmitters, and limitations on the ranges in the variation of thresholds in sensory reception. Each of the constraints is mediated by molecular and cellular processes of the kind known to be modifiable by mutations.

The final behavioral effects, and the mutations from which they stem, are sorted out by the process of natural selection, leading to genetic evolution by altering the statistical properties of the populations as a whole. The chain of causation runs from genes to physiological process to species-specific behavioral responses to a limited array of possible social organizations. And there is a simultaneous feedback: genes increase or decline in frequency according to their success in existing environments, including those altered by the social organization.

On existing empirical grounds it can therefore be defensibly argued that our deepest emotions, our most general sense of what is right or wrong, and the main trajectory of our intellectual development are species-specific, and guided by human genes; that they are not just the product of culture and personal history. Because the principles of sociobiology are derived from ecology and genetics, are testable, and generate rich and complex effects, this quite orthodox extension of biology promises eventually to explain many, if not all, the most distinctive forms of human social behavior. It may even accomplish this with more detail and precision than other, prevailing explanatory schemes including psychoanalysis, behaviorism, and Marxism. The essential flaw in these alternative systems is the lack of grounding of their theories in biology.

In considering the role of biology in social theory, one should bear in mind that the evolution of mankind occurred over a period of three to five million years. What we are today is the product of a very long episode of organic evolution in preliterate, economically simple societies. It should thus come as no great surprise to learn that our social behavior is still moderately or even strongly constrained by our biological past. It is also reasonable to suppose, as a working hypothesis, that cultural evolution is largely an elaboration of underlying biological imperatives.

In addition to misunderstanding the theory, the criticisms of human sociobiology are inappropriate in another sense. We think nothing of formulating the principles of human genetics and human biochemistry from studies of Norway rats, fruit flies, and colon bacteria. But when sociobiologists recommend the same time-honored point of departure for human social behavior, many scholars balk. They ask: is there nothing more to human existence? Can the mind ever, in any manner, be mapped onto 250,000 DNA triplets?

I wish to suggest that the biological research program will eventually make a detailed, if not a point-by-point, correspondence possible. But success in the enterprise will not prove restrictive, oppressive, or dehumanizing. Obviously history matters. And so does environment. Phenotypes—outward appearances—are the outcome of an immensely complex interaction of genes and environment along developmental pathways whose outlines are as yet only dimly perceived, and the whole process is so complex that individual free will probably will remain forever invulnerable. But at another species-wide level, deeper biological knowledge can tell us a great deal about ourselves that can be grasped in no other way: where we come from, why and not just how we act the way we do—in other words, the ultimate meaning of the human condition.

In the twentieth century the social sciences and humanities have been dominated by the tradition of philosophical dualism and the strategy of disciplinary autonomy. Insofar as they have sought to adduce general, scientific principles—the nomothetic as opposed to idiographic approach—they have deliberately avoided building models upon premises taken from the supporting antidisciplines.[3] Sociology has remained aloof from psychology, economics from ecology, anthropology from zoology, and all the social sciences from evolutionary biology. The social sciences have failed to produce a convincing, predictive general theory for the simple reason that they lack an axiomatically formulated and testable prime mover based on a study of the origin of the human condition. I have suggested that the prime mover is the biologically designed central nervous system. As one illustration of the applicability of biology to social theory, let me cite a short list of generalizations about aggression. These have been derived from animal studies but are relevant to a broad array of topics in the social sciences.[4]

1. *Aggression is a biologically eclectic phenomenon.* It consists of a set of often very different behaviors that are under separate neural and endocrine controls and can therefore evolve independently of each other.

2. *Aggression is not based on universally or even widely occurring instincts.* It evolves as one or more idiosyncratic traits in response to competition among members of the same species for a limited resource. The competition is most frequently for food or space, in which case aggressive behavior serves as one of the devices that confers equity or superiority in the struggle. But competitive behavior can also evolve as nonaggressive or avoidance responses. And where the size of populations is limited by emigration, predators, parasites, or deterioration of the physical environment, competition may play no significant role. Thus we find species of animals in which aggressive behavior among members is unknown. However, *Homo sapiens*, like most or all other higher Old World primates, is one of the innately aggressive species.

3. *When mates become a limiting resource, intrasexual aggression often evolves.* Such competition develops most frequently among males, but in special circumstances females contend for access to males. In mammals, competition is generally among males; and there is a close correlation between the average size difference between males and females and the degree of polygyny, that is, the number of females sequestered by each successful male. In humans the average size difference is relatively small, and the degree of polygyny is correspondingly modest.

4. *In theory at least, significant genetic changes can occur under moderate selection pressure within as few as ten generations.* Single alleles (forms of genes on a chromosome site) can be mostly substituted in this short period of time, and relatively rapid shifts can probably also be effected in the frequencies of complex sets of polygenes. Thus the human species is not permanently fixed within a species-specific pattern of aggressive responses.

5. *The innate patterns of aggressive behavior in mammals and other higher vertebrates are not automatic, "push-button" responses but are species-specific programs.* Each species has a distinctive array of responses that its members are capable of developing, and a much larger array exists beyond their developmental capabilities. Some species, as noted, are incapable of any aggressive repertory. Given a knowledge of the species, the pattern of responses for individuals can be predicted with greater or lesser accuracy depending upon the knowledge of their environmental history. Human beings fit this rule; they are strongly predisposed to respond with unreasoning hatre·' to external threats and to escalate their hostility sufficiently to overwhelm the threat by a respectably wide margin of safety. In at least one specialized form of aggressive behavior—territoriality—hunter-gatherers and other economically primitive societies conform to ecological theory based on animal studies.

If the origins of human sociality and morality are truly biological, they raise three dilemmas for ethics.[5] The first is that no species, including *Homo sapiens*, has any purpose beyond the neurophysiological censors and motivators created by its genetic history. These biological structures constitute innate emotional guides. Rational decision-making, however sublime in complexity, is subject to these innate guides. Thus decisions for the future must be based on past genetic history. The second dilemma is even more basic. We cannot obey all the species-specific censors and motivators blindly (some, such as proneness to violent aggression under a wide array of stresses, have become destructive even to personal genetic fitness); yet in choosing which to reinforce culturally and which to circumvent, we must consult these very same guides.

The third dilemma is the most critical of all. Because genetic components of behavior can be altered by selection in as few as ten generations, our species can theoretically alter its basic nature within centuries. The capacity to alter the foundation of human nature, and thence the source of our species-specific morality, raises significant philosophical issues. The ability to change our innate emotional guides calls into question the validity of the distinction commonly drawn between the deontological and consequentialist views of morality. It also challenges the traditional belief that we cannot deduce values from facts or moral prescriptions from scientific information.

REFERENCES

[1]See Edward O. Wilson, *Sociobiology: The New Synthesis* (Cambridge, Mass.: Belknap Press of Harvard University Press, 1975). I wrote this book in order to help define a then inchoate discipline. For a brief history of the term "sociobiology," see my article, "What is Sociobiology?" *Society* 15 (September/October, 1978), 10-14.

[2]Edward O. Wilson, *On Human Nature* (Cambridge, Mass.: Harvard University Press, 1978); I. Eibl-Eibesfeldt, "Human Ethology—Concepts and Implications for the Sciences of Man," *The Behavioral and Brain Sciences* 2 (December 1979), 1-57.

[3]For a definition of antidisciplines, see Edward O. Wilson, "Biology and the Social Sciences," *Daedalus* 106 (1977), 127-40.

[4]The documentation can be found in *Sociobiology*, pp. 88-90, 242-97. For human applications, see W. H. Durham, "Resource Competition and Human Aggression. Part I: A Review of Primitive War," *Quarterly Review of Biology* 51 (1976), 385-415; and Rada Dyson-Hudson and E. A. Smith, "Human Territoriality: An Ecological Assessment," in N. Chagnon and W. Irons, eds., *Evolutionary Biology and Human Social Behavior* (North Scituate, Mass.: Duxbury Press, 1979).

[5]For a further consideration of the three dilemmas, see *On Human Nature*, pp. 1-13, 195-209. Other, excellent discussions of the significance of evolutionary biology for ethical philosophy are to be found in G. S. Stent, ed., *Morality as a Biological Phenomenon*, (Berlin: Abakon Verlagsgesellschaft, 1978).

1980

"Caste and division of labor in leaf-cutter ants (Hymenoptera: Formicidae: *Atta*), II. The ergonomic optimization of leaf cutting," *Behavioral Ecology and Sociobiology* 7(2): 157–165

During the 1970s, I conducted an extensive study on caste systems and division of labor in leafcutter ants of the genus *Atta.* Members of this genus, which abound in the tropical and subtropical zones of the New World, exhibit the most complex ecological adaptations based on social organization in all of the ants. They are also ideal laboratory animals. Colonies flourish in plastic observation nests on nothing more than fresh leaves and flowers and are mostly unaffected by close observation. They chew the vegetation to a pulp, which is then enriched with enzymes excreted from their own bodies. They next mold the resulting paste into a comblike substrate on which they raise a symbiotic fungus. In addition to plant sap obtained from freshly cut vegetation, the fungal mycelia are their sole source of nutriment.

The study reported here used laboratory colonies in an attempt to answer the following question: How good are castes specialized for different tasks at what they do? Are they the best caste for the job? I chose foraging for the test. The workers sallying from the nest each day in both field and laboratory to harvest vegetation are medium-sized, with a modal head width of 2.2–2.8 mm, the exact number depending on the hardness of the vegetation. With the "pseudomutant" technique invented for this purpose, I was able to measure precisely the performance and energy consumption of the 2.2- to 2.8-mm size class against other size classes and concluded that the workers devoting themselves to foraging are indeed adaptively optimal for the needs of the colony. Other conceivable adaptations would produce different optima.

❉ ARTICLE 20

Behav. Ecol. Sociobiol. 7, 157–165 (1980)

Behavioral Ecology and Sociobiology

Caste and Division of Labor in Leaf-Cutter Ants (Hymenoptera: Formicidae: *Atta*)

II. The Ergonomic Optimization of Leaf Cutting

Edward O. Wilson

Museum of Comparative Zoology Laboratories, Harvard University, 26 Oxford Street, Cambridge, Massachusetts 02138, USA

Received December 10, 1979 / Accepted February 20, 1980

Summary. Leaf cutting was selected for an evaluation of ergonomic efficiency in the fungus-growing ant *Atta sexdens* because it is performed largely by medias (head width 1.8–2.8 mm), which attend to relatively few other functions and hence are less likely to be evolutionarily compromised by the demands of competing tasks (Fig. 1).

Three alternative a priori criteria of evolutionary optimization were envisioned that are consistent with natural selection theory: the reduction of predation by means of defense and evasion during foraging, the minimization of foraging time through skill and running velocity during foraging, and energetic efficiency, which must be evaluated with reference to both the energetic construction costs of new workers and the energetic cost of maintenance of the existing worker force.

In order to measure the performance of various size groups within the *A. sexdens* worker caste in isolation, I devised the 'pseudomutant' technique: in each experiment, groups of foraging workers were thinned out until only individuals of one size class were left outside the nest. Measurements were then made of the rate of attraction, initiative in cutting, and performance of each size group at head-width intervals of 0.4 mm (Figs. 2, 3, and 7). Other needed measurements were made in body weight, oxygen consumption, and running velocity (Figs. 5, 6, and 8).

The size-frequency distribution of leaf cutters in the *A. sexdens* conforms closely to the optimum predicted by the energetic efficiency criterion for harder forms of vegetation, such as rhododendron leaves. The distribution is optimum with reference to both construction and maintenance costs. The difference between the predicted and actual modal size groups specializing on leaf cutting is 10% or less of the total size range of the *sexdens* worker caste.

A model was next constructed in which attraction and initiative were allowed to 'evolve' genetically to uniform maximum levels. The theoretical maximum efficiency levels obtained by this means were found to reside in the head-width 2.6–2.8 mm size class, or 8% from the actual maximally efficiency class (head width 2.2–2.4 mm). In the activity of leaf cutting, *A. sexdens* can therefore be said to be not only at an adaptive optimum but also, within at most a relatively narrow margin of error, to have been optimized in the course of evolution.

Introduction

How well organized is an ant colony? This question is simple in tone but potentially profound in meaning. It can be translated into the following more operational form: is the colony as efficient in its basic operations as natural selection can make it, without some basic change in the ground plan of anatomy and behavior? The answer, deduced species by species, is of importance not only for sociobiology, but for general evolutionary theory (Oster and Wilson 1978). At stake may be our very understanding of the process of natural selection.

I have selected the *Atta* leaf-cutter ants for analysis of efficiency and optimization because they possess caste systems and patterns of division of labor that rank among the most complex found in ants (Wilson 1980). They provide an unusual number of opportunities to measure and evaluate performance. As understanding of the species increases, we can hope to understand better the forces that have set an upper limit on the social evolution of ants.

I have, for the moment, further narrowed the choice of tasks to leaf cutting, for two reasons. First, the work is performed by the ants outside the nest and is therefore experimentally more easily managed. Second, the great majority of leaf cutters fall within

0340-5443/80/0007/0157/$01.80

an intermediate size range of workers (head width 1.8–2.8 mm) that engage in few other tasks (Wilson 1980); the evolution of the workers within this size segment with reference to leaf cutting is consequently less likely to have been compromised by necessities imposed by other modes of specialization.

The ideal way in which to test the natural selection hypothesis and to estimate the degree of optimization is to first write a list of all conceivable optimization criteria, deduced a priori from a knowledge of the natural history of the species. The next step is to conduct experiments to determine which of the criteria has been most closely approached, and to what degree. Finally, with the results in hand, the theoretician can alter behavioral and anatomical parameters in simulations in order to judge whether the species is capable of still further optimization by genetic evolution. If the approach actually taken by the species cannot be significantly improved by the simulations, we are justified in concluding that the species has not only been shaped in this particular part of its repertory by natural selection, but that it is actually on top of an adaptive peak. In other words, it is both at an optimum and evolutionarily optimized; it is doing as well as its genotype permits, and the genotype cannot evolve over short distances to produce a better optimum. This is the sequence taken in the analysis of leaf cutting in *Atta*.

The Optimization Criteria

The criteria envisioned a priori were the following:

1) Evasion or Defense. If predation were an overwhelmingly important mortality factor and hence a principal cost item in the energy budget of the growing colony (see Oster and Wilson 1978 p 169), we could expect to see evidence that the foraging caste was molded so as to place an effective defensive technique ahead of other criteria. Thus the leaf cutters might be the smallest size class capable of cutting leaves, regardless of their degree of skill or energetic efficiency. The size of these workers would enable them to run in inconspicuous, even partially hidden columns – a method employed by many other ant species, such as *Solenopsis* fire ants and *Labidus* army ants. Or the leaf cutters might consist of very large, soldierlike forms capable of fending off most enemies. Alternatively, a mixed technique might be employed: small foragers guarded by a large soldier caste; this is in fact the method employed by some other ant species, including members of the genus *Pheidole*.

2) Skill and Running Velocity. Pure skill is the rate of cutting measured by the length of the cut or the weight of material cut, regardless of the degree of attraction, initiative, or energetic efficiency. Running velocity determines how quickly the foragers can retrieve the material. These criteria are expected to be dictated by natural selection if there is a premium on speed, due for example to predator pressure or shortness of time in which the vegetation can be harvested.

3) Energetic Efficiency. This is the criterion to be expected if the foragers are acting as a relatively untroubled economic system, attempting to maximize their net energetic yield in an environment not excessively dominated by predators and with a relatively plentiful food resource. Skill and running speed continue to matter a great deal, but attraction to the vegetation and initiative in cutting would be expected to be adjusted in evolution so as to enable workers to cut the maximum amount of vegetation per unit of energy expended. There exist in turn two categories of energy expenditure. The first is *construction cost*: the number of calories required to rear a worker of a given size class from egg to adult; this quantity is reasonably assumed to be a linear function of the dry weight of the adult. Construction cost can be expected to rise to the status of a major factor in the energy budget if mortality is high. Consider the extreme imaginable case in which every leaf-cutter worker dies following its first foraging trip; then a very large part of the cost of leaf cutting must come from the heavy construction costs of daily forager replacement. The second category of energy expenditure is *maintenance cost*: the rate of metabolic energy consumption by a worker ant. In the extreme imaginable case, in which foraging workers live for the life of the queen, maintenance cost would be paramount and construction cost minor. In the intermediate case, where active members of the leaf-cutting class survive for periods of weeks or months, maintenance will outweigh construction, but to a degree that has not been determined. The prudent course in evaluating energetic yield is to measure both categories of cost, that is, both construction and maintenance, and treat them as a mix.

In order to evaluate adherence to the three competing criteria of optimization, both in the real species and in potentially evolvable species, it is necessary to measure the following properties for each size class in turn: tendency to leave the nest on foraging expeditions, attraction to fresh vegetation, initiative, pure skill in leaf cutting, performance in leaf cutting, energetic efficiency by the two cost criteria (construction and maintenance), speed and cost of running back and forth between the nest and vegetation, and differential mortality. In the present study, all of these properties have been measured across the size range of *Atta sexdens* workers except for the last, differential mortality. I have not been able to estimate the death rate of workers under natural conditions. Nevertheless, I believe that it is a relatively minor factor, for the reason that foraging workers in the laboratory are long-lived and do not appear, at least under these special conditions, to vary significantly in mortality rate as a function of size. On the other hand, the role of other factors, particularly cutting skill, does vary so greatly with size as to outweigh greatly the effects of all but the most striking differential mortality.

Materials and Methods

The study was conducted with a colony of *Atta sexdens* (Linné) from Timehri, Guyana, about eight years of age and containing the original mother queen and approximately 7,000 workers. The colony was maintained in a closed series of clear plastic chambers 14 cm × 19 cm and 9 cm high. Foraging workers were allowed to leave the chambers through a 26-mm-wide plastic tube that opened onto the floor of a foraging platform. The platform, measuring 32 cm × 40 cm, was the bottom of an inverted enameled metal tray, which was set in turn in a larger, upright tray partly filled with water to provide a moat. During the routine maintenance of the colony, a wide variety of vegetable materials was given to the foraging workers, including the leaves and flowers of several plant species as well as wheat germ and other dry cereals. As the colony expanded, the ants filled one chamber after another with characteristic spongelike masses of degraded and processed vegetable material, through which the whitish symbiotic fungus (*Rozites*) grew abundantly. In the experiments of leaf cutting, petals of American roses (ForeverYours variety) were used to represent soft vegetation, while the thick, leathery leaves of overwintering ornamental rhododendron were used to represent hard vegetation. Most vegetation utilized by *A. sexdens* in the field falls between these two extremes. In order to standardize the conditions, the ants were offered ten 22-mm-wide circles of the petals or leaves freshly cut with a large cork borer. They were then observed during the ensuing period of 15 min.

The following measurements were taken. *Attraction* is the fraction of workers on the foraging table that are on or at least touching the vegetation sample; the figure is the average of three counts taken 2, 7, and 12 min, respectively, after the vegetation sample is placed on the table. *Initiative* is the number of cuts started per ant per minute by the ants attracted to the vegetation sample. More precisely, it was defined as the number of pieces of vegetation cut free by the ants divided by 15 times the average number of workers observed on the vegetable sample at 2, 7, and 12 min. *Skill* is the length of the cut made by one ant in mm/s, where the measurements apply only to the interval of time in which the ant was actually cutting. *Performance* is the rate of cutting in either mm of cut per ant per minute or mg of dry weight of vegetation cut per ant per minute, averaged over the 15-min interval following introduction of the vegetable sample and for all the workers in the foraging force, regardless of whether they were attracted to the sample or not. *Energetic efficiency by the first criterion* is the dry weight in mg of vegetation cut per mg of ant per minute during the first 15 min following introduction of the vegetable sample and for all workers in the foraging force, regardless of whether they were attracted to the sample or not. *Energetic efficiency by the second criterion* is the dry weight in mg of vegetation cut by the foraging force in the first 15 min, per μl of oxygen consumed by resting or slowly moving workers at 30° C/min. Note that the oxygen consumption was not taken while the ants were cutting the leaves, an activity that consumes only a tiny fraction of their time in nature, but while they were at rest or slowly moving about in a confined space.

Respiration was measured with Scholander manometers of a design modified by C.M. Williams and Edward Seling. Ants were placed singly or in groups of up to ten individuals of the same size in tubes 18 mm in diameter. A screen floor separated them from a layer of saturated KOH solution at the bottom of the tube, so that they had a cylindrical space 18 mm in diameter and 35 mm long in which to move. As they consumed oxygen, the reduction in volume was marked by an advancing column of dyed water drawn forward in a connecting tube by the reduced pressure. At the end of 24 h, the volume was restored by injecting oxygen with a syringe – until the column of dyed water reached the original level – and the volume of consumed oxygen was read off as the amount of advance by the plunger in the cylinder. To ensure that the apparatus was airtight, it was submerged in a beaker of water after the apparatus had been sealed. All tests were run at 30° ± 2° C.

The heart of the experimental procedure is the creation of what I have loosely called *pseudomutants*. These can be defined as colonies or fractions of colonies in which the age-sex ratios have been altered in chosen ways by differentially screening out members of various age-sex classes. Thus the foraging *A. sexdens* workers were all older individuals and mostly medias, in the head-width range 1.8–2.8 mm, accompanied by a few smaller and larger individuals. As this foraging force emerged onto the platform each morning, I removed all of the workers except those of the size class (say, head width 1.0–1.2 mm or head width 2.2–2.4 mm) whose performance I wished to study on that occasion. Thus it was possible to create a 'pseudomutant', a colony which for the moment fielded a worker force consisting exclusively of minors or medias of a narrowly restricted size range, as though its caste ratios had been altered by a genetic mutation. The performance of each size class could be evaluated separately, without the complicated interaction effects caused by the presence and activity of other size classes. Moreover, the same colony – and even part of the same set of foraging workers – was used throughout, minimizing variance due to overall colony size and health.

In the report to follow, the frequency distribution of the size classes is referred to as the *caste distribution function* (CDF), to align the experimental results with the models of Oster and Wilson (1978). The unmodified foraging force is then said to possess a natural CDF, while the 'pseudomutant' force altered by restriction is called a synthetic CDF. The size measurement used is the *head width* (HW) of ant systematics, defined as the greatest width of the head that can be obtained by measuring at right angles to the longitudinal axis of the head while the head is viewed full face.

Results

As illustrated by the data in Fig. 1, the population of *A. sexdens* workers that left the laboratory nest to forage has nearly the same size-frequency mode as the populations that cut various kinds of vegetation (HW 2.2–2.4 mm), although the range is slightly greater (HW 1.2–4.0 mm as opposed to 1.4–3.8 mm). However, both explorer and cutter samples were very

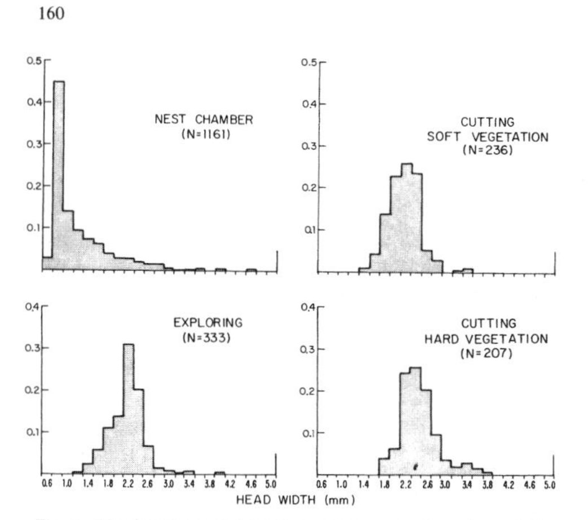

Fig. 1. The frequency distributions shown are of workers of the *A. sexdens* laboratory colony collected at random from a nest chamber during a period of nonforaging (hence representative of the colony as a whole), other workers sampled while exploring the foraging platform outside the nest, and still others sampled while cutting vegetation. Soft vegetation was represented by rose petals, hard vegetation by rhododendron leaves (see additional size-frequency data in Wilson 1980)

different from the overall colony population, which had a mode of 0.8 mm and proportionately few workers in the HW 2.2–2.4 mm range. The population cutting soft vegetation had a somewhat smaller overall head width than that cutting hard vegetation, but the modal difference is less than 10%; additional data on variation as a function of vegetation type are given in Wilson (1980). In general, workers in the exploring and foraging population are substantially larger in average size than those drawn randomly from the total nest population. They consist mostly of medias in the 1.8–2.8 mm range. The trait is probably a general one in this and closely related species. Authors who have observed *A. sexdens* and other *Atta* species in the wild refer consistently to medias as the leaf-cutter caste (references in Wilson 1980). In the film 'Millions of years ahead of man' (Taunus Film, GMBH, Wiesbaden, West Germany), workers of *A. sexdens* are shown foraging and cutting leaves in the field in Brazil; from the proportions of the head and body, I have subjectively estimated them to fall in the size range HW 1.8–2.8 mm. Samples of *A. sexdens* from Panama (Ancon Hill, Balboa, Gamboa) collected while cutting a wide variety of types of vegetation in the field also conform to the size distribution of *A. sexdens* leaf cutters in the laboratory. All of four samples had modes at HW 2.2 mm. Comparable but more variable results were obtained with ten colonies of *Atta cephalotes* sampled in the field at three localities in Ecuador. Cherrett has presented data on the size of foraging *A. cephalotes* workers in Guyana rain forest. With the aid of allometry curves, I have transformed his data, which are based on the length of the hind femur, into frequency curves based on the standard head-width measurement. The mode of workers collecting flower parts is HW 2.1 mm and that of workers collecting leaves is HW 2.4 mm, a result closely consistent with curves obtained in my own studies with both *A. cephalotes* and *A. sexdens*.

Using the 'pseudomutant' method, I tested all size groups in the HW 1.0–3.6 mm range at 0.4-mm intervals to determine the degree of their attraction to vegetation, initiative in cutting, cutting skill, and overall performance. As noted previously, two types of vegetation were chosen to represent different degrees of hardness: rose petals as a form of very soft vegetation, and the thick, tough evergreen leaves of ornamental rhododendron as a form of relatively hard vegetation. The results are presented in Figs. 2–4. A further estimate displayed in Figs. 2 and 3 is energetic efficiency measured by the criterion of construction costs; this measure is based on the dry weight of leaves cut in mg per mg of dry weight of the ants, taken as a function of worker size (HW). The conversion of the linear size measurement (HW) to dry weight of the entire ant, required for the estimate of energetic efficiency by the construction criterion, is given in Fig. 5. In Fig. 4, the energetic efficiency by the maintenance criterion is given. The measure used is the amount of dry-weight vegetation cut per µl oxygen consumed per mg dry weight of worker at rest in 30° C ambient air.

The oxygen consumption rate of workers used to evaluate maintenance cost is presented in Fig. 6. In this illustration, the curves are fitted to the power equation with reference to the standard linear measure (that is, head width), in order to facilitate estimates of energetic efficiency with reference to maintenance costs across the worker size range. However, to make the data more accessible to physiologists, the data can also be expressed in the more conventional form relating oxygen consumption to total body dry weight, as follows:

$$y = 3.52\ w^{0.79}$$

where y is O_2 consumption in µl per individual ant per hour at 30° C, and w is the total dry weight of the individual ant. These parameters are consistent with those obtained by Jensen and Nielsen (1975) for nine species of European ants. All of their combined data fit the following equation:

$$y = 1.95\ w^{1.01}$$

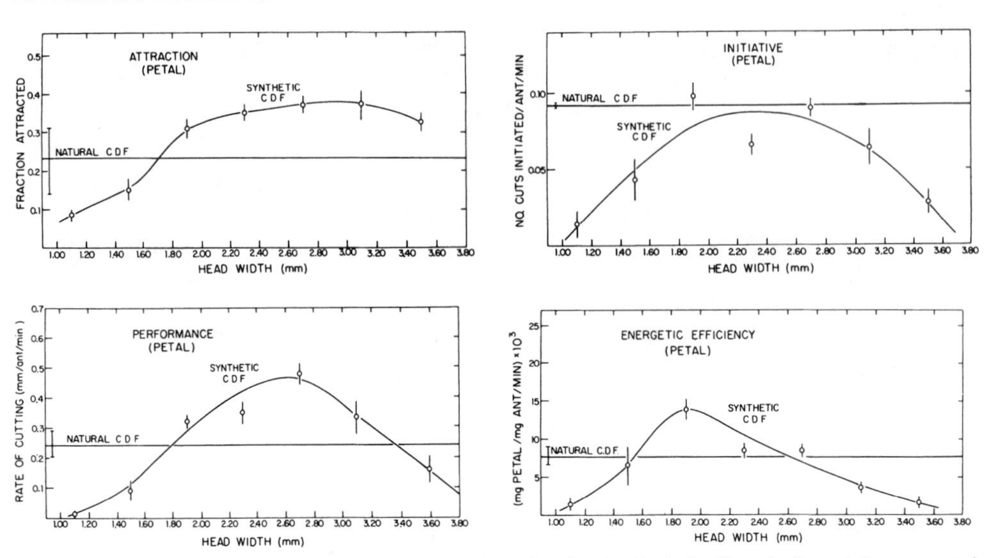

Fig. 2. The capabilities of isolated ensembles of *A. sexdens* workers of varying sizes in the handling of soft vegetation, represented by rose petals. Energetic efficiency shown here is based on the criterion of construction costs, estimated from the total dry body weight. The *natural CDF* (caste distribution function) is an unmodified group of foragers; the *synthetic CDFs* are the groups reduced to consist only of members of single size classes

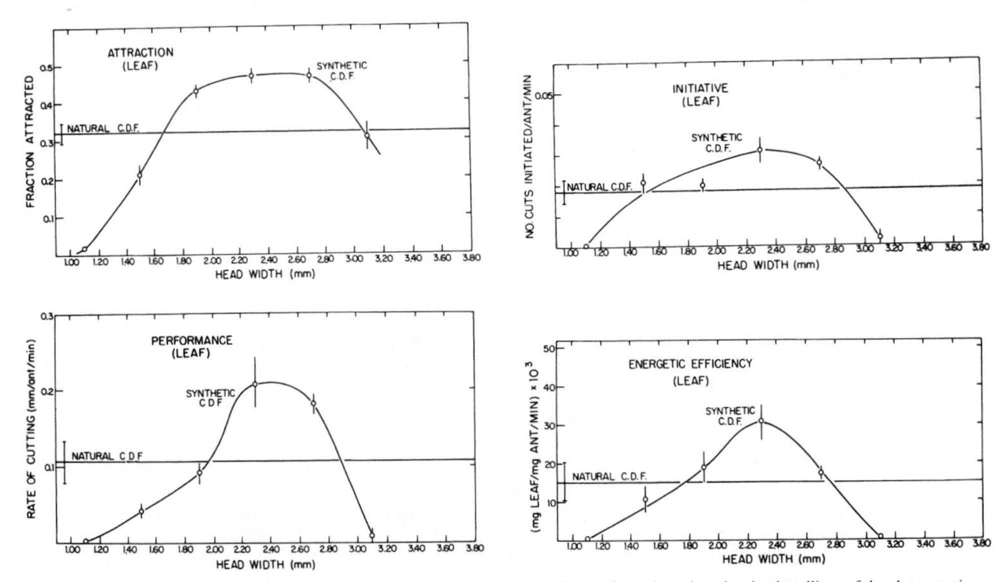

Fig. 3. The capabilities of isolated ensembles of *A. sexdens* foraging workers of varying sizes in the handling of hard vegetation, represented by overwintered rhododendron leaves. Energetic efficiency shown here is based on the criterion of construction costs, estimated from the total dry body weight

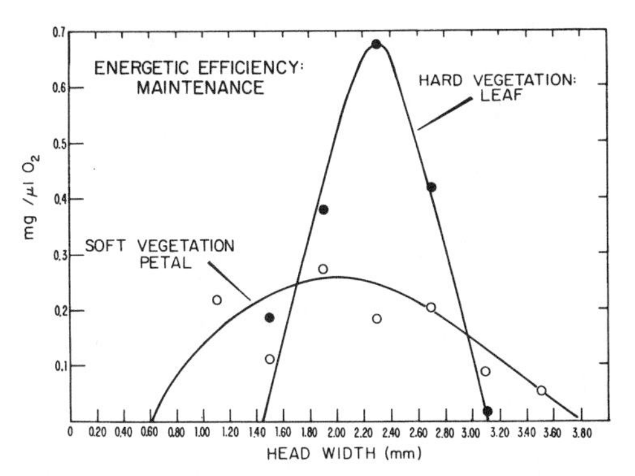

Fig. 4. Energetic efficiency estimated on the basis of maintenance. The measure used is the dry weight of vegetation cut per mg dry weight of worker per unit time, divided by the oxygen consumed by mg dry weight of resting workers in µl per unit time, at 30° C (oxygen consumption data are displayed in Fig. 6)

The *A. sexdens* values fall well within the total range of parameter values displayed among these European species, and the measurements obtained are also consistent with the response of the European species to the higher temperature employed (30° C). It is of interest that the power coefficient is significantly less than 1.0 in *A. sexdens*, so that maintenance cost falls off with an increase in size.

The skill of workers in cutting the two extreme forms of vegetation as a function of head width is illustrated in Fig. 7. These data reveal several notable features. First, the curves for petal cutting and leaf cutting are parallel, allowing the inference that the slope ($a=0.1$) is relatively invariant for all forms of vegetation. Second, the curves are linear or nearly so; as the head width increases by a given proportion, the cutting skill increases by an approximately equal proportion. The fact that skill so closely reflects linear dimensions, rather than area or volume of the head, may be of still wider significance. Skill might well depend primarily on the length of the mandibles, which is related to head width as follows:

$$y=0.11+0.34x$$

where y is length of the mandibular blade and x is head width (based on measurements of 23 workers in the range HW 0.6–3.8 mm).

Finally, and this is a feature of unusual importance in the ecology of *A. sexdens*, all size classes are at least theoretically capable of cutting the softest vegetation, although the smallest (HW 0.6–1.0 mm) have not been observed to do so, and even though their skill – estimated by an extrapolation of the curve – would be very low. On the other hand, rhododendron leaves can be cut only by ants with HW over 1.6 mm. By comparing the two skill curves, it is apparent that most fresh vegetation of moderate to very

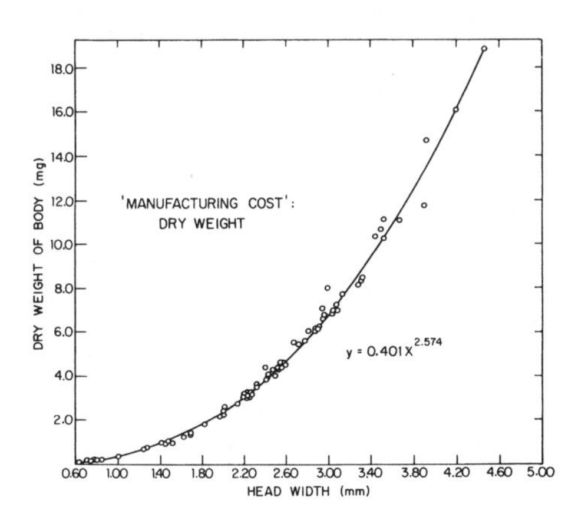

Fig. 5. The relation of head width to body dry weight in workers of *A. sexdens*, used in the analysis of energetic efficiency

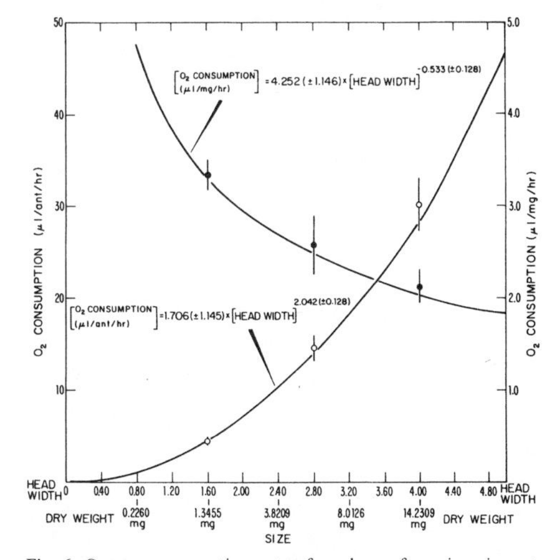

Fig. 6. Oxygen consumption rate of workers of varying sizes at 30° C, given as a function of both total body dry weight and per mg of dry weight

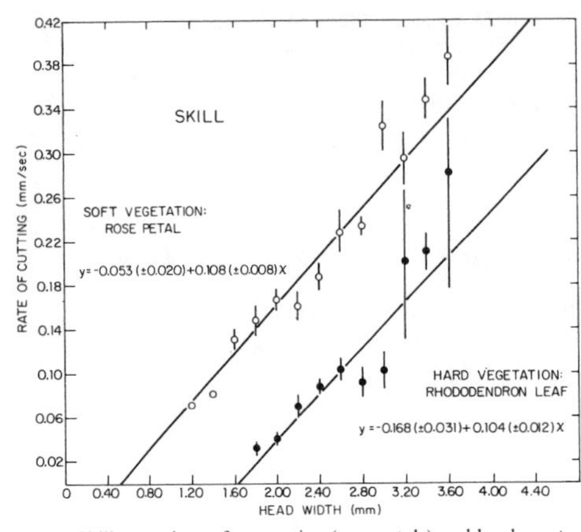

Fig. 7. Skill at cutting soft vegetation (rose petals) and hard vegetation (rhododendron leaves) as a function of head width

hard consistency could not be handled by workers with head widths less than a head width of about 1.4 mm. The inference is consistent with the fact that relatively few workers in the numerically dominant size class of head width 0.6–1.4 mm leave the nest on foraging expeditions.

Finally, the running speed of workers in the various linear size classes is evaluated in Fig. 8. It is an interesting fact that velocity changes very little over a wide range of size variation, and an even more remarkable fact that the *average* velocity of burdened and unburdened running changes not at all as a function of linear size. The result can be interpreted as an adaptation to maintain an even traffic flow in the foraging columns, so that workers of large size do not overtake and trip over their smaller nestmates. Without doubt this is a result of the uniform velocity; the smooth flow of the *A. sexdens* foraging column in both the laboratory and the field is an impressive sight. However, the data are also consistent with a more general, canonical rule (see, for example, Alexander 1971): because the work performed per unit time is a constant times body mass ($a \times M$), and the work done during running is equal to the kinetic energy expended, which is equal to another constant times the square of velocity, multiplied by the mass ($b \times M \times V^2$), it follows that

$$aM = bMV^2$$

$$V = \left(\frac{a}{b}\right)^{\frac{1}{2}}$$

In other words, velocity should be a constant independent of body size.

Discussion

The conclusions suggested by the data in this ergonomic study can be stated very simply. In its performance of leaf cutting, *Atta sexdens* does not depend upon an 'evasive' caste as part of a defensive strategy, the possibility suggested by the first a priori criterion. To be specific, it does not commit the smallest size class (HW 1.6 mm, see Fig. 9) capable of cutting most forms of vegetation. Similarly, colonies of the species do not depend upon their soldiers for leaf cutting, the possibility raised as an alternative defensive strategy. In fact these individuals, with HW exceeding about 3.4 mm, rarely cut leaves at all (see Fig. 1). Furthermore, *A. sexdens* has not utilized the workers with the greatest pure skill. These too would be the largest individuals (Fig. 7), which do little or no cutting.

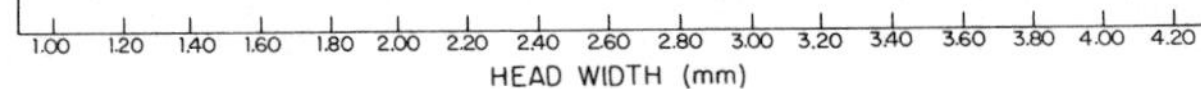

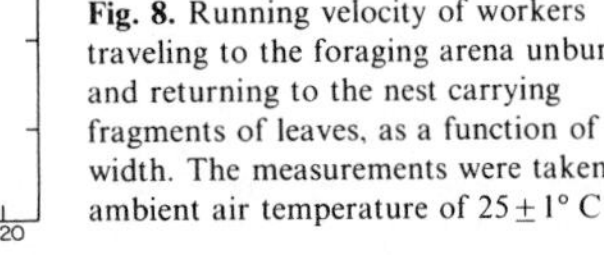

Fig. 8. Running velocity of workers traveling to the foraging arena unburdened and returning to the nest carrying fragments of leaves, as a function of head width. The measurements were taken in an ambient air temperature of $25 \pm 1°$ C

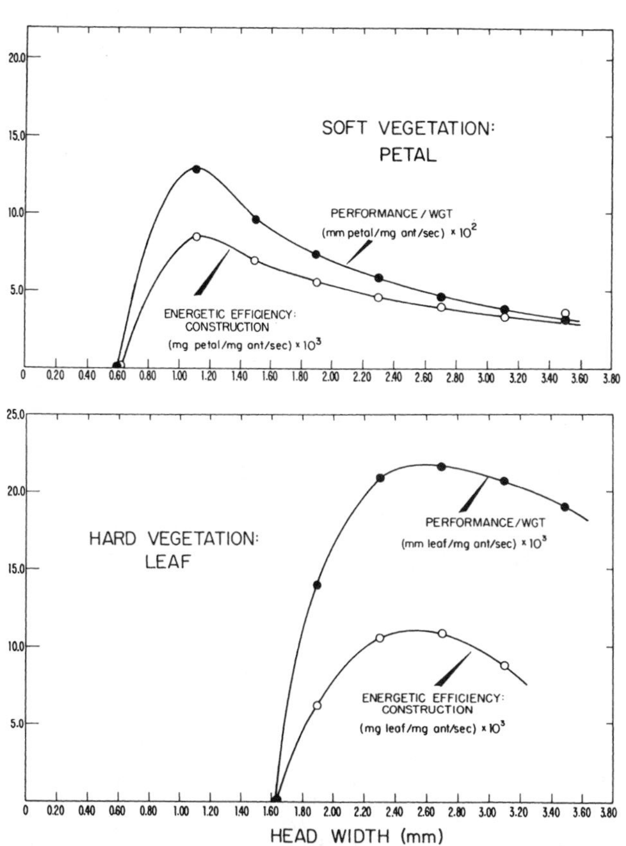

Fig. 9. The theoretical maximum performances and energetic efficiencies (construction criterion) of various worker size classes. These are the values obtained if every one of the workers in all size classes of the foraging force were attracted to the vegetation and all initiated cutting, but continued to display the size-dependent skill characteristic of the species. The numbers are consequently the skill in mg dry weight of vegetation cut per ant divided by the body dry weight of one ant, given for each size class in turn. (Compare with Fig. 10)

What *A. sexdens* has done is to commit the size classes that are energetically the most efficient, by both the criterion of the cost of construction of new workers (Figs. 2 and 3) and the criterion of the cost of maintenance of workers (Fig. 4). The two energetic criteria, cost and maintenance, turn out to be virtually indistinguishable; by optimizing with respect to one, the species at least approximately optimizes with respect to the other.

Moreover, *A. sexdens* has adapted more closely to hard vegetation, represented in the experiments by rhododendron leaves, than to soft vegetation, represented by rose petals. There has been a remarkably close match with the requirements posed by this kind of material, which can be seen by comparing the size-frequency distributions of foragers in the laboratory (Fig. 1) with the energetic efficiencies estimated by the pseudomutant technique (Figs. 2 and 3). The maximum energetic efficiency of rhododendron-leaf cutting is obtained by workers within HW 2.2–2.6 mm, which is also the mode of the size-frequency distribution of workers cutting rhododendron and most other forms of vegetation. Since the full size range of the worker caste is HW 0.6–5.4 mm, it is reasonable to say that the species is accurate to within $(2.6-2.2)/(5.4-0.6)=0.083$, or about 10%, of the energetic optimum. This can be taken as the upper limit; the ants may have done better than 10%.

Why has the species fitted its innate polyethism curve more closely to hard vegetation than to soft vegetation? The answer, I believe, can be deduced from the skill curves presented in Fig. 7. The smallest workers capable of cutting rhododendron leaves are about HW 1.6 mm. Since most of the vegetation collected by free-living colonies of *A. sexdens* consists of at least moderately tough leaves (John Wenzel, personal communication), the colony will field a partly idle work force unless it utilizes workers over HW 1.6 mm.

In summary, the evidence clearly shows that *A.*

sexdens are operating at or close to the optimum defined by the criterion of energy harvesting. But are they *optimized*? This is a question seldom addressed in the literature of behavioral ecology. Translated into operational terms, it can be restated as follows: is there any direction in which *A. sexdens* can evolve further – short of a major anatomical or behavioral reorganization – so as to commit another size group that is even more energetically efficient than HW 2.2–2.6 mm? If not, then the species can be said to be truly optimized in an evolutionary sense. It sits atop an adaptive peak (see further discussion in Wilson 1975 p 24).

In order to answer the question of true evolutionary optimization at least in part, I treated the *A. sexdens* foraging system as a problem in design, considering which features might be changed readily in evolution and which cannot. On a priori grounds, the features most readily altered are behavioral, especially attraction to vegetation and initiative in cutting. These belong to broad categories of behavior that have changed most readily in genetic experiments on laboratory populations of other insect species (see, for example, Dobzhansky et al. 1972; Ehrman and Parsons 1976). The feature less readily altered is skill, which is a linear function of size, based on a head and mandibular form conventional for ants generally, and hence likely to be an evolutionarily stable trait. The relation between size and skill probably could not be significantly altered without large changes in the allometry of head shape, and even then all that might be affected is the slope of the allometric curve – with little final alteration redounding in the final relation between size and skill.

Consequently, I conducted a thought experiment in which attraction and initiative were allowed to evolve to uniform maximum levels for all size classes. The skill curves (Fig. 7) were not altered. Then new, theoretical maximum curves for performance and energetic efficiency, drawn across all size classes, were calculated on the basis of the empirical data for skill, dry body weight, and oxygen consumption. The results, given in Figs. 9 and 10, show that energetic efficiency in cutting petals is relatively insensitive to variation in worker size. But efficiency with reference to hard leaves remains very sensitive. It also does not shift very much: only from HW 2.2–2.4 mm to HW 2.6–2.8 mm, or about 8% of the total worker size range. Thus *A. sexdens* can be said to be within approximately 10% of the theoretical maximum level of energetic efficiency attainable by short-range evolution. It lies on or very close to the local adaptive peak. While it might be possible to design an imaginary leaf-cutting ant with higher levels of efficiency than *A. sexdens*, and in that sense identify a distant, higher adaptive peak, there appears to be no higher peak

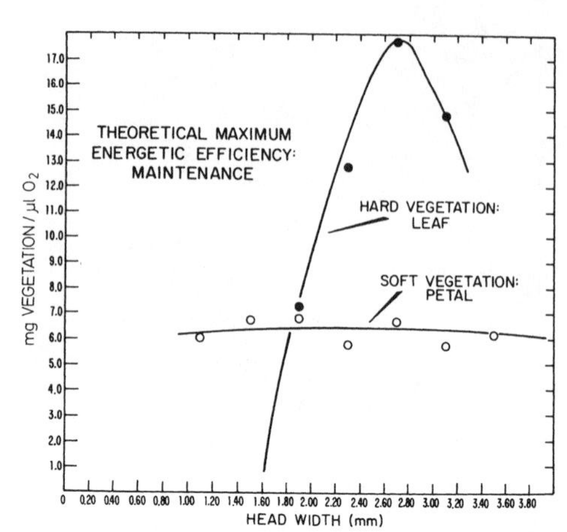

Fig. 10. The theoretical maximum energetic efficiencies (maintenance criterion), derived according to the same criteria employed to obtain the theoretical maxima with respect to the construction criterion (see Fig. 9). The values given are the skill in mg dry weight vegetation cut per worker divided by the oxygen consumption of one worker for each size class in turn

in the close vicinity of the *A. sexdens* genotype, and it is therefore likely that the species has stabilized with reference to its leaf-cutting behavior and performance.

Acknowledgements. I wish to thank Laurie Burnham and John Wenzel for collecting samples of *Atta cephalotes* and *A. sexdens* in the field, Donna J. Howell and Thomas A. McMahon for advice on the general principles of animal locomotion, and Edward Seling and Carroll M. Williams for advice on insect respiration and the loan of the manometers they devised to measure it. The author's research has been supported by National Science Foundation grant DEB77-27515.

References

Alexander RMcN (1971) Size and shape. Arnold, London

Cherrett JM (1972) Some factors involved in the selection of vegetable substrate by *Atta cephalotes* (L.) (Hymenoptera: Formicidae) in tropical rain forest. J Anim Ecol 41:647–660

Dobzhansky T, Levene H, Spassky B (1972) Effects of selection and migration on geotactic and phototactic behaviour of *Drosophila*. III. Proc R Soc (London) Biol 180:21–41

Ehrman L, Parsons PA (1976) The genetics of behavior. Sinauer Associates, Sunderland MA

Jensen TF, Nielsen MG (1975) The influence of body size and temperature on worker ant respiration. Nat Jutlandica 18:21–25

Oster GF, Wilson EO (1978) Caste and ecology in the social insects. Princeton University Press Princeton

Wilson EO (1975) Sociobiology: the new synthesis. Belknap Press of Harvard University Press, Cambridge

Wilson EO (1980) Caste and division of labor in leaf-cutter ants (Hymenoptera: Formicidae: *Atta*). I. The overall pattern in *A. sexdens*. Behav Ecol Sociobiol 7:143–156

1982

"Précis of *Genes, Mind, and Culture*," *The Behavioral and Brain Sciences* 5(1): 1–37

In 1979–80 I spent much of my time working with Charles J. Lumsden, a theoretical biophysicist, in an attempt to construct a theory on the relation between genetic evolution and cultural evolution. The result was *Genes, Mind, and Culture* (1981). In this work we addressed (and coined the term) gene-culture coevolution by exploring the chain of causal links from genes to neural and mental development, to mind, and thence to culture. The natural selection of epigenetic rules was also illustrated and modeled as the key to the link between genetic and cultural evolution.

In 1982 we followed this work with a popularized version, *Promethean Fire* (Harvard University Press, Cambridge). Much later, in *Consilience: The Unity of Knowledge* (Knopf, New York, 1998), I developed additional subjects based on the same theme.

A brief summary of *Genes, Mind, and Culture* is presented here, with critiques from specialists in several relevant disciplines. Although the book was generally well received, and a few (surprisingly few) researchers have worked on different aspects of what must be granted to be an important subject, very little progress has been made in the past twenty years.

THE BEHAVIORAL AND BRAIN SCIENCES (1982) 5, 1–37
Printed in the United States of America

Précis of *Genes, Mind, and Culture*

Charles J. Lumsden
Department of Medicine, University of Toronto, Toronto, Ontario, Canada M5S 1A8

Edward O. Wilson
Museum of Comparative Zoology, Harvard University, Cambridge, Mass. 02138

Abstract: Despite its importance, the linkage between genetic and cultural evolution has until now been little explored. An understanding of this linkage is needed to extend evolutionary theory so that it can deal for the first time with the phenomena of mind and human social history. We characterize the process of gene–culture coevolution, in which culture is shaped by biological imperatives while biological traits are simultaneously altered by genetic evolution in response to cultural history. A case is made from both theory and evidence that genetic and cultural evolution are inseverable, and that the human mind has tended to evolve so as to bias individuals toward certain patterns of cognition and choice rather than others. With the aid of mathematical models we trace the coevolutionary circuit: The genes prescribe structure in developmental pathways that lay down endocrine and neural systems, imposing regularities in the development of cognition and behavior; these regularities (loosely labeled "epigenetic rules") translate upward into holistic patterns of culture, which can be predicted in the form of probability density distributions (ethnographic curves); natural selection acts within human history to favor certain epigenetic rules over others; and the selection alters the frequencies of the underlying genes. The effects of genetic and cultural changes reverberate throughout the circuit and are consequently tested with the passage of each life cycle. In addition to modeling gene–culture coevolution, we apply methods from island biogeography and information theory to examine the cultural capacity of the genes, the factors determining the magnitude of cultural diversity, and the possible reasons for the uniqueness of the human achievement.

Keywords: behavior; cognition; culture; ethnography; evolution; genes; genetics; mind; sociobiology

Mind and the linkage between genes and culture

We undertook the studies that led to *Genes, Mind, and Culture* (Lumsden & Wilson 1981b; henceforth *Genes*) because we regard the linkage between biological and cultural evolution to be one of the great unsolved but tractable problems of contemporary science. Many philosophers and scientists still consider the gap between the biological and social sciences to be a permanent discontinuity, grounded in epistemology and reinforced by a fundamental difference in goals on the part of specialists (e.g., Bock 1980; Medawar 1981). We view it instead as occupied by a largely unknown evolutionary process – a complicated, fascinating interaction in which culture is generated and shaped by biological imperatives while biological properties are simultaneously altered by genetic evolution in response to cultural history. We have suggested that this process be called *gene–culture coevolution* (Lumsden & Wilson 1980).

Any effort to cross this no-man's-land between the natural and social sciences is certain to encounter forbidding technical problems and to come under an extraordinary amount of critical cross-fire. But even a small amount of progress will more than compensate for the difficulties. The reason is that the key to the social sciences is the appropriate characterization of human nature. Each of the disciplines of the social sciences is ultimately shaped by its perception of the core properties of human mind and behavior. If these properties can somehow be specified in a form that makes it possible to discover laws of mind and culture, even in crude form, the contents of the constituent disciplines of the social sciences might be reformulated to some degree and explained by a common theory. Moreover, in order to accommodate the peculiarities of human mind and cultural history, evolutionary theory itself will have to be reshaped in novel and interesting ways.

The basic conception

We define a *society* as a group of individuals belonging to the same species and organized in a cooperative manner; the diagnostic criterion is reciprocal communication of a cooperative sort, extending beyond mere sexual activity (see also E. O. Wilson 1975). A key property of *culture* used in our approach is its origin as the resultant of all the artifacts, behavior, institutions, and mental concepts transmitted through learning by the members of a society, and the holistic patterns they form. It has been conventional in anthropology to use static terms such as *culture trait* to refer to this array of perceivable features of culture; we suggest the term *culturgen* in order to capture the dynamic, generative role these entities actually play. Considerable care is devoted to the formal characterization of the culturgen in the first chapter of the book, in order to relate it to the more familiar definition of artifact (a kind of culturgen) employed by some archaeologists and to permit culturgens to be further described in terms of knowledge

 0140-525X/82/010001-37/$06.00/0

structures in long-term memory. In some instances, such as the adoption versus rejection of a particular ritual or sexual practice, the relevant culturgens are denumerable and few in number. On the other hand, variation among relevant culturgens can be continuous (as, for example, in the array of all conceivable color classifications). The partitioning of such a continuum into categories is sometimes a practical step. We have taken care to show that the categories can be made as coarse or as fine as required to reach the desired level of precision, and that continuous variation can be treated per se. We stress that culturgens are *perceivable features* of the integrated cultural system; their existence does not mean that culture is composed of discrete, isolated units.

During socialization the culturgens are processed according to what we have loosely labeled the *epigenetic rules*. These rules are the genetically underwritten peripheral sensory filters, interneuron coding processes, and more centrally located cognitive procedures of perception, learning, and decision making. They affect the probability of using one culturgen as opposed to another.

It follows that the transfer of a set of competing culturgens among individuals in any evolving system belongs to one or the other of three classes. The first is pure genetic transmission, in which learning – and perhaps even teaching – takes place but only one culturgen is permitted by the epigenetic rules (hence, an entirely genetic culture is a theoretical possibility). The second is pure cultural transmission, in which no innate mechanisms pattern the acquisition and use of any culturgen in competition with others (the tabula rasa state). The third is gene–culture transmission, in which innate epigenetic rules, underwritten by genes and modulated by culture, discriminate multiple culturgens and are more likely to use some rather than others in guiding the assembly of the human mind.

Contrary to an impression held by many biologists and social scientists, most and possibly all forms of human culture are processed through gene–cultural transmission rather than pure cultural transmission. There have been detailed studies of the development of choice among very distinct culturgens. In some of these the evidence for an innate bias favoring the selection of certain stimuli over others is very strong. This evidence is in accord with our purely theoretical considerations concerning the evolutionary process. The tabula rasa state tends to be unstable, and simple models can estimate the mean time to departure from a tabula rasa state resulting from the incorporation of nonuniform epigenetic rules. The results suggest that such a shift should occur most rapidly when the number of dimensions used to distinguish culturgens is small. It will also be enhanced when the ability to distinguish culturgens within any one dimension is small, and when the number of discoverable culturgens that are advantageous or disadvantageous in genetic natural selection is large.

This brings us to the process of gene–culture coevolution envisaged in *Genes*. The genes prescribe culture-modulated epigenetic rules, which use the information latent in culturgens to guide the assembly of the human mind in certain patterns as opposed to others. The collective choices in cognition and behavior made by the members of the society create the culture and the social fabric. Genetic variation exists in the epigenetic rules, contributing to at least part of the variance of cognitive and behavioral traits within the population. Evolutionary mechanisms such as mutation and migration become relevant to the coevolutionary process. In particular, natural selection can ensue: Individuals whose choices favor their survival and reproduction – and that of their kin – within the contemporary culture transmit more genes to future generations, and as a consequence the population as a whole tends to shift toward the epigenetic rules and the forms of cognition and behavior favored by the rules. The coevolutionary circuit is thus completed: The epigenetic rules (or the genes, or the culture, depending on the starting point chosen) make themselves by creating effects that reverberate around the circuit, finally to be tested by natural selection with each passage through the life cycle. To take a simple example, the avoidance of sibling incest appears to be based on an epigenetic rule in which sexual interest is neutralized by close domestic proximity during the first six years of life. Avoidance of incest confers a demonstrable advantage in the viability of offspring; children born of incestuous marriages have a much higher level of homozygosity and probability of suffering from one or more debilitating genetic defects. Thus the culturgens consisting in incest avoidance and the associated taboos are facilitated by the epigenetic rule, and they in turn favor the genes that generate the mental predisposition against incest. The two inheritance systems, genetic and cultural, can be said to have coevolved.

When pure cultural transmission is taken into account as one of the three possible patterns of transmission, it can be seen that genetic and cultural evolution are inseverably coupled. Even in the unlikely event that a species could evolve to the tabula rasa state, such uniform perception and information processing would require genetic fine tuning of the cellular apparatus at each stage through the full sequence of cognition. In other words, pure cultural transmission is simply a special case of epigenesis, which in this sense is no less under genetic control than the pure genetic form of cultural transmission.

Four conditions are necessary for the analysis of gene–culture coevolution: that nonuniform epigenetic rules exist and be investigable in such a way as to test gene–culture coevolutionary models; that some variance in the expression of the epigenetic rules be heritable; that cultural practice affect genetic fitness; and that links exist between genetically controlled processes at the molecular and cellular level and the epigenetic rules. All of these conditions have been documented in contemporary human populations.

The epigenetic rules

Each epigenetic rule affecting cognition and behavior constitutes one or more elements of a complex sequence of events occurring throughout the nervous system. In Chapters 2 and 3 of *Genes* we review all of the information known to us in the literature of cognitive and developmental psychology that pertains not just to stages of mental development but to *choices* made at various stages. Because psychologists do not often lay out a range of options and measure the relative frequency of choices, the data needed to advance gene–culture theory are relatively few. We nevertheless found 12 studies that appeared to contain sufficient data to be either definitive or strongly suggestive. The categories of cognition or behavior addressed by the studies are dietary preference in infants, proportionate representation of vision and the other principal senses in vocabularies, color classification, phoneme formation in the development of language, preference for certain geometric designs, preference for normally composed facial features, the choice of facial expressions to denote basic emotions, the forms of mother–infant bonding and communication, the mode of carrying infants and other intermediate-sized objects, the form of the fear-of-stranger response, phobias, and incest avoidance (see review in *Genes*, pp. 35–98).

The following examples are representative of the choice phenomena. Newborn infants choose most kinds of sugars over water in the following order of preference: sucrose, fructose, lactose, and then glucose (Maller & Desor 1974). They also discriminate among tastes that are acid, salty, and bitter, and react by twisting their faces into the distinctive expressions employed by adults in responding to strong and unpleasant tastes (Chiva 1979). Young infants prefer to gaze at certain geometric patterns, including those with curved lines, elements that touch, and arrays that are nonlinear instead of latticed (Fantz, Fagan & Miranda 1975). The tendency for even adults to learn one thing as opposed to another is perhaps most dramatically illustrated by the phobias,which typically emerge full-blown after only a single unpleasant experience and are exceptionally difficult to eradicate. It is a remarkable fact that phobias are easily evoked by many of the greatest dangers of mankind's ancient environment, including closed spaces, heights, thunderstorms, running water, snakes, and spiders. Of equal significance, phobias are rarely evoked by the greatest dangers of modern technological society, such as guns, knives, automobiles, and electric sockets (Seligman 1972).

Other epigenetic rules, more difficult to specify by means of overt choice experiments, exist in the preferred ways the mind evaluates culturgens and makes decisions. The disparate components used in judging other people, for example, are summed additively and without prominent interaction effects. Thus when the estimations of level-headedness, sophistication, boldness, and good-naturedness are combined pairwise (at least by persons in industrialized Western societies), they contribute to the overall impression in a nearly pure additive fashion, even though different weights are assigned to each (N. H. Anderson 1979).

We found it useful for the purposes of gene–culture coevolutionary theory to divide the elements of the developmental rules into two classes: primary epigenetic rules, which range from sensory filtering to perception; and secondary epigenetic rules, which include the procedures of feature evaluation and decision making through which individuals are predisposed to use certain culturgens in preference to others. Many, perhaps most, categories of cognition and overt behavior are channeled by combinations of these two classes of rules. The distinction between primary and secondary rules is crude and to some extent arbitrary. We have employed it as a convenient and refinable taxonomy through which the basic relevant processes can be evolutionarily classified for the first time.

That the epigenetic rules have a genetic basis is strongly indicated by the circumstance that so many are relatively inflexible and appear during early childhood. In addition, pedigree analysis and standard comparisons of fraternal and identical twins, in some instances strengthened by longitudinal studies of development, have yielded evidence of genetic variance in virtually every category of cognition and behavior investigated by these means, including some that either constitute epigenetic rules or share components with them (e.g., McClearn & DeFries 1973; R. S. Wilson 1978). Single alleles have been identified that selectively affect certain cognitive abilities (Ashton, Polovina, & Vandenberg 1979) and the ability to discriminate certain odorants (Amoore 1977). It has also become apparent that mutations at a single locus can result in profound but highly specific changes in the architecture and operation of brain tissues such as mammalian neocortex, which modify locomotion, perception, and choice (Caviness & Rakic 1978).

The translation from mind to culture

A key step in the modeling of gene–culture coevolution is the translation from the activity of the individual mind to the social and cultural patterns that characterize the society as a whole. A society is a system in flux, in which the proportions of its members using different patterns of culturgens can change. Moreover, the probabilistic nature of human choice and decision, together with uncertainties about environment and social structure, limits the precision of theoretical prediction. But one can aim at a theory that both generates probability distributions, such as those constructed empirically in cross-cultural ethnographic surveys, and predicts the likelihood of observing given cultural patterns of action and organization in a specific society. (For example, in 0.4 of the societies 0.5 of the members express culturgen c_1 and 0.5 express c_2, where c_1 and c_2 are learnable belief systems; in 0.05 of the societies 0.9 of the members express c_1 and 0.1 express c_2; and so on through all possible patterns of c_1, c_2 abundance.) We call this distribution an *ethnographic curve*.

In most cases the ethnographic curve cannot be expected to depend on individuals' fixed bias values. Information about the culture modulates even the moment-by-moment activity of the epigenetic rules. Thus the proportion of other members using a given culturgen (for example, choosing outbreeding over incest, one color classification over another, and so forth) often affects the probabilities that an individual will adopt the culturgen. We have made such relationships explicit in the theory, and have gone on to build first models of them by means of the assimilation function $v_{ij}(\xi)$, where the probability per unit time for a switch from use of culturgen i to use of culturgen j is a function of macrocultural pattern variable ξ, defined so that $\xi = -1$ when no one possesses culturgen j and $+1$ when all possess it. The assimilation functions express the operation of secondary epigenetic rules of valuation and decision. Our fragmentary information indicates that they vary greatly among behavioral categories, and their effects amplify differences among the social patterns that are produced. We have suggested explicit functions to which data from studies in social psychology can be fitted.

When the members of the society are combined by the theory through ξ and related elements of the social fabric, mathematical techniques can be used to calculate ethnographic curves for the population. In *Genes*, we have concentrated on simple models involving homogeneous, egalitarian social groups. Indeed, the general properties of human socialization are favorable to a first exploration of the translation process by means of such models. In many societies culture is systematically passed along not just by the nuclear family but by a much broader array of relatives and parent surrogates (Williams 1972a). This condition ensures a relatively uniform exposure of younger people to most culturgens, especially in the smaller societies that prevailed throughout most of human gene–culture coevolution. Exchanges of information channeled through guilds and other preference groups represent higher-order inhomogeneities that the formal structure of the theory also includes.

Two findings of general significance have emerged from our elementary translation models. The first is that even small differences in the epigenetic rules, as reflected in the assimilation function, are magnified in the resulting ethnographic patterns. Even differences as low as 0.02 in the intrinsic bias toward competing culturgens can translate upward to produce ethnographic curves that differ qualitatively in their structure. An amplification law has been derived that provides an exact and simple relation between the magnitude of the intrinsic bias, the values of ξ selected for analysis, and the amplification observed in the ethnographic curves (see Equation 4-42). Where bias has been measured in a variety of developmental studies, the values are mostly an order of magnitude greater than those required to produce easily detectable effects in ethnographic data (Table 4-1).

The second finding is a corollary of the first: Even when the underlying epigenetic rules and assimilation functions are rigidly constrained by genetic prescription, they can generate wide cultural diversity (see, for example, Figure 4-23). Fine tuning in these innate parameters can create large shifts in the dependent social patterns. Additional variation arises from the probabilistic nature of ethnographic distributions resulting from continuing flux in the decisions of individuals. As hundreds or thousands of individual decisions are made throughout the society in a given interval of time, with some running in opposite directions, the culture as a whole tends to drift. Even when a steady state is attainable, different cultures observed at the same time can be expected to vary substantially. A single culture can also move back and forth through time. In either situation the result is an ethnographic curve that may be centered on the idealized steady state but be spread far on either side. This finding gains added importance from the light it sheds on the relation of cultural diversity to biological determinism. Until recently it has been routinely argued by many writers that the existence of diversity proves that human social behavior is free of genetic influence. We have shown that, on the contrary, a large amount of diversity is expected in most cases of gene–culture transmission, and under appropriate conditions the amount can be predicted from knowledge of the biologically based epigenetic rules in mental development. The linkage we have proposed is not that of genetic determinism, that is, between genetic diversity and cultural diversity, although where genetic diversity exists in the epigenetic rules it can be expected to increase cultural diversity. Cultural diversity arises rather from the remarkable range of choices and decisions open to each individual. The mind grows as it picks its way through the space of choices, but it does so equipped with knowledge and preferences it formulates under the influence of the epigenetic rules. The epigenetic rules, although modulated by culture, are under genetic control.

Although suitable data concerning the epigenetic rules exist in a few developmental studies and can be used to construct rough ethnographic curves, we know of no case in which information on both individual development and ethnography is sufficient to examine the full linkage between the two levels. Perhaps the closest approach is encountered in the category of brother–sister incest. It is possible to make an approximation of the assimilation function, which on the basis of developmental studies is considered to be relatively insensitive to social context. We have derived the relatively narrow range of ethnographic curves in which brother–sister incest should lie. The available ethnographic measures, which are largely anecdotal, appear to fall within the predicted range. Another case chosen for analysis is village fissioning in the Yanomamö Indians, where the assimilation function seems to approach a step form based on the tension between kinship ties and intravillage strife. A third example is fashion in women's dress, which fluctuates with a periodicity of about 100 years. The rebound apparently arises from the conflict between competition for status through innovation and exclusiveness and the tendency to adhere to natural body form. These three case histories illustrate the great variety to be found in the assimilation functions and their deduced ethnographic effects, as well as the explanatory scope to be expected of useful gene–culture theories. Although their details are still very incomplete, they demonstrate that developmental psychology and human sociobiology can be unified with the remainder of the social sciences both conceptually and formally. They also illustrate the kinds of data required to adapt such theory to programs of empirical research.

The gene–culture adaptive landscape

After analyzing the step between the epigenetic rules and culture, we turn (in Chapter 5) to the examination of the full coevolutionary circuit. We felt that it would be desirable to describe gene–culture coevolution in the language of theoretical population genetics and hence to keep human sociobiology aligned closely with evolutionary biology while bringing it into contact with the social sciences. In so doing, we were able to emphasize more clearly the similarities and differences between pure genetic evolution and gene–culture coevolution.

We demonstrate that an adaptive topography in a simple type of gene–culture coevolution exists and can be described with general dynamical equations. The topography resembles that of the conventional genetic landscape, except that in addition to the genetic fitness vector of individuals it contains an epigenetic rule vector. Its characterization permits an evaluation of the effect on individual fitness of the joint possession of particular genotypes and particular culturgens.

With the existence theorem established, we next consider gene–culture coevolution as it occurs not just under static conditions but in environments that change in space and time. The true adaptive landscape is more accurately described as a viscous seascape, in which peaks subside into depressions and populations must move toward more complex solutions that integrate gene–culture fitness through time. When the environments are coarse-grained, that is, when changes occur on the order of once per generation or less, the average gene–culture fitness can be crudely modeled by a geometric mean. The maximum potential fitness curve is a hyperbola that touches the real fitness space at a point prescribing relatively unselective epigenetic rules. When the environments are fine-grained, shifting frequently within single generations, the average gene–culture fitness resembles an arithmetic mean and the maximum potential fitness curve is a straight line that intersects one or the other of the ends of the real fitness space. The result is the genetic prescription of relatively specific epigenetic rules. These models lead to a general rule: The more uniform the environment encountered by individuals in a particular life stage through many generations, the more specific will be the epigenetic rules evolved for that life stage. This prediction is supported by the fact that the richest arrays of selective epigenetic rules have been detected during early childhood and in other circumstances in which encounters are uniform and predictable. However, the correlation is based on a possibly biased sample, and the rule cannot be considered to be established.

Haldane-Jayakar analysis has been applied to the model in order to incorporate a simple form of history. In one case examined, two usage bias curves are prescribed by *AA* and *aa* respectively in a diallelic system with complete dominance, where the *aa* curve is selectively superior generation after generation but is more vulnerable to occasional catastrophes. If the arithmetic gene–culture fitness mean of *aa* is greater than one and the geometric mean less than one - a not unlikely combination - the two alleles will be maintained in a polymorphic mixture. Thus the idiosyncratic histories of populations affect gene–culture coevolution in ways that cannot be assessed by classical analysis alone. A new interpretation can be placed on the propensity to acquire and retain certain destructive culturgens in spite of a newly acquired rational understanding of their peril.

In this class of models, if the mean value of the inherited bias (or absence of bias) is at the adaptive maximum, phenotypic variation in the predisposition has no influence on fitness. Only the mean values, and no other statistical properties in the bias curves, enter into the gene–culture coevolutionary landscape. General conclusions based on the simpler case of invariant bias curves are therefore preserved. In addition, conclusions based on the single-locus model can be extended to many systems of polygenic inheritance generating continuous variation. Because of the adaptive neutrality of phenotypic variance, the variance can drift in such a way as to become broader or narrower, creating greater or lesser amounts of behavioral heterogeneity within populations. The variance will tend to broaden maximally in equiprobable bias curves and minimally in curves with high selectivity. If some degree of selectivity other than the contemporaneous mean value becomes more adaptive, an increase in phenotypic variation will lead to an earlier attainment of this new, more adaptive bias by the population. There will be a tendency to incorporate the bias (and hence shift to a new epigenetic rule) through genetic assimilation, followed by a return to selective neutrality of the variance.

The full coevolutionary circuit

Following the establishment of the most general properties of the coevolutionary process, we proceed to create a model that specifies the actual changes in gene and culturgen frequencies. Our account (in Chapter 6) begins with the principles of epigenesis in the nervous system. Among the most important of these is the generalization that mutations inaugurate changes not in single neurons but in the field gradients and properties of cell form; these gradients and properties in turn affect whole populations of neurons. Only relatively few such developmental rules are needed to reach precision in neuronal form, location, and connectivity. Although these rules are under the control of polygenes, the introduction of single new alleles can induce major modifications in specific features of brain structure and behavior.

The exact rules of pattern formation in the human brain are still unknown, but some of their general features can be inferred from the results of studies of epigenesis in other organisms. In some cases the complex form and patterning of cells appear to depend upon the diffusion of inducing substances. The patterns are canalized by the interactions of genes. Genetic assimilation is most rapid when this canalization is less restrictive, or else when extreme environmental events occur frequently enough to generate many phenotypes markedly different from the population norm. It follows that the polygenic control of traits is a theater of unusual opportunity for gene–culture coevolution. By cultural experimentation and the continual exploration of new environments, species are more likely to test the potential of the genes that prescribe epigenetic rules of cognitive development and to produce novel results. When the behaviors eventually emitted also confer selective advantage, the genes tend to shift into different frequency distributions that prescribe epigenetic rules biased toward the new responses.

Neuronal and cognitive development are tightly linked processes that continue unbroken from the early embryo to full maturity. Learning can be regarded as cognitive epigenesis within and beyond the womb. Recent studies on long-term memory have cast considerable light on the assembly of the cognitive structures that form the schemata by which the mind recalls information, evaluates new contingencies, solves problems, and directs motor activity. We use this new conception to relate culturgens to mental activity and by this means to evaluate their genetic fitness as the outcome of learning and reasoning.

We then construct a model that includes the main steps around the entire coevolutionary circuit. A life cycle is stipulated in which the offspring are socialized by both their age peers and the parental generation. The young learn and evaluate all the culturgens of the society by means of exploration and observation. At some future date they use this stored information to exploit the environment. In particular, individuals learn and later choose between two culturgens under the influence of epigenetic rules that are prescribed by two alleles. Variation is permitted in the degree of bias in the epigenetic rules, the amount of sensitivity to peer and parental usage, and the function by which resources garnered during exploitation of the environment are converted into genetic fitness.

The time scales of genetic change in this type of coevolution, along with a diversity of interesting effects (some of them previously unsuspected), have been revealed by investigation of the model. The tabula rasa state, in which no innate bias exists in culturgen choice, is again shown to be unstable, easily replaced by any one of a virtually unlimited range of biasing epigenetic rules. Sensitivity to usage pattern increases the rate of evolution in epigenetic biasing and hence the genetic assimilation of culturgen use. This catalytic effect might have contributed to the rapid evolutionary increase in human brain size associated with the onset of gene–culture coevolution. Culture slows the rate of genetic evolution, but coevolution still occurs quickly enough for the genes to track many forms of cultural evolution. We have summarized our inferences by a 1,000-year rule: The alleles of epigenetic rules favoring more successful culturgens can largely replace competing alleles within as few as 50 generations, or on the order of 1,000 years in human history. It is thus possible that substantial genetic evolution occurred during historical times and continues even today.

Finally, gene–culture coevolution of this type entails frequency-dependent selection. Under certain conditions, and in particular when genetic fitness declines after a certain amount of a resource has been harvested, selection leads to such intermediate evolutionary states as stable genetic polymorphisms or chaotic fluctuations in gene frequency. This finding cautions against the application of overly facile optimization reasoning to human evolution. It also suggests the existence of a novel mechanism that may have enhanced genetic diversity, division of labor, and individuality among human beings.

The biogeography of the mind

In Chapter 7 we develop a new set of concepts concerning the evolution of mind within a gene–culture framework. The idea of steady-state cultural diversity is introduced. We refer in this case to the evolution of local cultures that become static in the number of culturgens used, rather than to any notion of ultimate limits in global culture.

In order to evaluate steady states as well as deduce rates of change at various nonequilibrium diversities, we view the minds in a society as an "archipelago." Using procedures that have proved successful in island biogeography (that is, the analysis of the diversity of animal and plant species on islands), we envisage new culturgens being adopted and old ones being lost or placed in passive store at rates that depend on the current culturgen diversity level. A regular rise in diversity with population size across societies – the area effect – can be predicted and documented. The decrease in diversity as a consequence of isolation – the distance effect – can also be projected, but data are inadequate to measure its magnitude.

The analogies to island biogeography also permit an evaluation of the turnover rate of culturgens, as well as estimates of the expected duration time of individual culturgens in various categories. It is possible to recognize not only similarities between culturgenic and biotic assemblages but also important differences, especially in the case of the organizational traits of societies. A fuller consideration of these differences is likely to enhance rather than diminish the value of biogeographic analysis in the study of cultural diversity.

Culturgen packing is another key property affecting the richness of cultures. The immigration and extinction rates of culturgens, and hence of cultural diversity within individual societies, are determined by the number of culturgens that can be packed into the same cognitive system. This property is in turn the result of three capacities in cognition: the ability to discriminate among culturgens belonging to particular categories, categorization and recall in long-term memory, and valuation of the stimuli associated with each culturgen. Examination of these processes must be the first step toward the ultimate goal of characterizing the packing process. We have paid particular attention to discrimination, since it has been the best characterized in experimental psychology. Models are developed in which it is possible to estimate the optimum degrees of ambiguity in signal reception and the numbers of signals clustered into single culturgens.

We relate valuation to culturgen packing by considering the number of schema ensembles that can be linked together in stable configurations through value learning. The larger the linked ensembles, or "polymers," the greater the potential cultural diversity.

The study of diversity leads back to the question of the uniqueness of human-level culture. We ask: Out of the hundreds of millions of animal species that have existed during the past 600 million years, why has man alone attained this overwhelmingly successful ability? We have considered simple hypotheses concerned only with ecological specialization and rejected them on the basis of evidence from comparative zoology. We believe it necessary to seek internal constraints on the origin of advanced culture as well. From a priori considerations it appears that the cost of cognitive processing in terms of genetic fitness is a monotone rising function of cognitive and behavioral flexibility. If this conjecture is correct, it follows that the benefit curves are nonmonotone. They are likely to vary greatly in form according to social structure and environment but to possess general properties that make the evolutionary pathways to advanced flexibility few and relatively inaccessible.

Finally, we have raised the theoretical question of whether the level of civilization created from human culture might be duplicated entirely by genetic programming. In other words, we have explored whether genes and culture are partners of convenience or partners of necessity. Using a basic argument from information theory, we show that if the genome is regarded as a blueprint from which the structures of neural circuits and cognitive schemata are specified directly, it is not possible to program a human-sized brain and language structure solely with the amount of DNA found in human beings and other organisms. *For human beings, genes and cultures are partners of necessity.* It is nevertheless possible to prescribe some knowledge structures, as well as rules in the growth and differentiation of neurons, that reduce the amount of genetic specification to some intermediate level. This amount cannot be exactly estimated at present, because of the lack of an adequate developmental theory of brain structure.

The role of gene–culture coevolution in social theory

The theory of gene–culture coevolution is a development in sociobiology and evolutionary theory that is meant to create an internally consistent network of explanation between biology and the social sciences. It is designed to include all cultural systems, from the protoculture of macaques and chimpanzees to the advanced culture of human beings, as well as forms of culture hitherto conceived only in the imagination. Hence we have spoken of the goal of a comparative social theory, within which the study of human behavior is embedded. To some this concept may seem premature, but in fact it is only a general theoretical program of a kind routinely conceived in the natural sciences. What makes it different is the fact that it is addressed in part to mind and culture, which have traditionally remained outside the ambit of the natural sciences.

The success of the gene–culture theory we have developed, or of any other formulation that may spring up to replace it, will depend on its capacity to perform three services. First, it must derive rigorous propositions that are the unexplained axioms of other theories in the social sciences. Second, we require that it achieve a level of predictiveness and testability greater than that provided by other modes of explanation, or at least that it subsume the exact phenomenological models of disciplines such as economics and anthropology so as to make the underlying assumptions identical. Finally, it must suggest new questions and problems, as well as identify previously unknown properties and laws to be woven into a network of verifiable explanation from genes through the mind to culture.

Although the technical and conceptual difficulties facing such an enterprise are formidable, at stake is nothing less than the understanding of the true nature of man and of the mechanism that created him. These matters are extraordinary in their significance and require the attention of scholars from both the natural and the social sciences. *Genes, Mind, and Culture* is the most complete statement in this new field to date. We emphasize, however, the still rudimentary nature of gene–culture theory and its empirical base, and look forward to the progress that can result from the greater involvement of investigators who believe, as we do, that the gap between the natural and social sciences can now be closed.

Open Peer Commentary

Commentaries submitted by the qualified professional readership of this journal will be considered for publication in a later issue as Continuing Commentary on this article.

From genes to mind to culture: Biting the bullet at last

David P. Barash
Departments of Psychology and Zoology, University of Washington, Seattle, Wash. 98195

Thanks to our friends the physicists (among whom Charles Lumsden is – or used to be – numbered), we now have a pretty good and ever-growing understanding of molecules, atoms, and the various subatomic particles. Yet imagine, if you will, how odd it would be if no one worked on the interface between the basic building blocks of matter and the actual structure found in the world around us; that is, what if we had no physical chemistry, crystallography, or biochemistry? Such a void is almost inconceivable today. But it has long existed in the study of behavior: on the one hand we have evolutionary biology, notably population, developmental, and molecular genetics – the building blocks of the life sciences – and on the other we have anthropology, sociology, and social, developmental, and cognitive psychology – the actual structure of our own social behavior. Like ships passing in the night, evolutionary biology and the social sciences have rarely even taken serious notice of each other, although admittedly, many introductory psychology texts give an obligatory toot of the Darwinian horn somewhere in the first chapter, occasionally even mentioning genes as well, before passing on to discuss human behavior as though it were determined only by environmental factors.

But as prophesied by E. O. Wilson in 1975, the new discipline of sociobiology has been drawing the conceptual and empirical links between biology and social behavior, *Homo sapiens* included. In this regard, an important endeavor in human sociobiology has been the evaluation of whether, in fact, the behavior of human beings tends to accord with predictions based on sociobiology's central principle of fitness maximization. But very few students have had the multidisciplinary training or perhaps the intellectual courage to bite the bullet and face one of the central questions posed by such work: Just what is the nature of the connection leading from genes to mind to culture? In *Genes, Mind, and Culture,* Lumsden & Wilson (L & W) have done just this, and they are to be congratulated.

Their effort is as audacious as it is overdue. They have integrated an impressive array of data and theory from behavioral and population genetics, neurobiology, developmental and cognitive psychology, sociology, anthropology, and even social history, attempting to point toward the underlying dynamics of the human behavioral system. They show how genetic effects can (and have) readily been taken for purely cultural transmission, and emphasize that diversity and flexibility are not incompatible with genetic influence. They also demonstrate mathematically how biases in "epigenetic rules" may be amplified during the translation from gene to culture, so that dramatic cultural differences are not necessarily incompatible with the genetic as well as the psychic unity of humankind.

It is instructive to compare the L & W effort with the only other well described (and equally, if not more mathematical) theory of cultural transmission, recently concretized by the geneticists L. Cavalli-Sforza and M. Feldman (1981). Cavalli-Sforza and Feldman develop an approach that is likely to be significantly less controversial, and probably less interesting as well. (The difference reminds me of a comment by the sociobiologically oriented anthropologist Napoleon Chagnon at a 1978 AAAS symposium on sociobiology: "I would rather be wrong about something important than right about something trivial.") By considering only pure cultural transmission - whatever that is - and explicitly ignoring individual differences and gene-environment interactions, Cavalli-Sforza and Feldman achieve something that may be cleaner and more elegant than the L & W approach, but it is probably too far removed from the guts of biology to stimulate the social sciences to much debate or any real synthesis.

It seems inconceivable that human behavior is rigidly programmed by our genotype, and yet equally unlikely that the two are totally uncoupled. L & W trace one of several possible connection schemes in the no-man's-land in between. As with any audacious effort, theirs can easily be criticized, especially since very few figures break the horizon in the demilitarized zone between biology and the social sciences. Despite the opacity (at least to me!) of much of their mathematics, even in this there is room for disagreement, since a fundamental assumption throughout is that Markov processes are appropriate to the assumption of different culturgens, whereas in fact the reality is likely to be more complex. Similarly, their genetic models are avowedly too simple, based on diallelic, single locus systems with at most only three possible genotypes. But as with any modelling effort, we start with the simple, see how far it takes us, and then either complicate or discard it as it gets tested against reality. The data available thus far are certainly suggestive and lead to the hope that more will shortly be forthcoming, so that tests and possible falsification can be carried out. In the meanwhile, as Darwin said when he first read Malthus, at last we have something to work with!

Stalking the wild culturgen

Arthur L. Caplan
Hastings Center, Hastings-on-Hudson, N.Y. 10706

If *Genes, Mind, and Culture* (henceforth *Genes*) is to be properly assessed, then it surely must be seen as an effort to respond to a number of recent criticisms of the sociobiological research program. This program argues that social behavior can only be explained by theories in biology and the social sciences that recognize the importance of kin selection, reciprocal altruism, and the competitive nature of parent-child reproductive efforts (E. O. Wilson 1975). Among the most trenchant criticisms mounted against this suggestion has been the claim that the link between genes and social behavior is far too tenuous, reticulate, and complex to make the sociobiological approach worthwhile in the study of the behavior of *homo sapiens* and other "advanced" species (Burian 1978; Gould 1977a; Sociobiology Study Group 1977).

Lumsden & Wilson (L & W) attempt to meet this criticism in the present volume. They admit that human social behavior is best understood as the end product of a host of interactions among genes, genotypes, phenotypes, and culture. They admit that earlier sociobiological allusions to such metaphors as culture being kept "on a leash" by genes did not suffice as an accurate depiction or explanation of the genesis of social behavior.

If the gene-behavior connection was the Achilles' heel of earlier sociobiological accounts of human behavior, then *Genes* is intended as an enormous prosthetic. The authors endeavor to describe and analyze systematically the nature of gene-culture interaction, using such apparatuses as epigenesis, bias curves, life cycles, and coevolutionary circuits.

The choice of the term "prosthetic" to describe L & W's enterprise is based upon my view that the effort is, in many ways, somewhat artificial and unnatural. This claim is not motivated by the appearance in their exposition of complex mathematics and ornate technical models which are peppered throughout the book and seem to have galled a number of reviewers (Leach 1981; Lewontin 1981; Medawar 1981). If the nature of gene-culture interaction is as complex as the critics of sociobiology have insisted that it is, then it will undoubtedly require complex descriptions, models, laws, and lemmas for its analysis.

My problems with the analyses of behavioral epigenesis proferred in *Genes* center around two key conceptual issues - the definition of culturgen, and the degree to which the cultural products of *homo sapiens* can be said to meet the conditions described by L & W as "necessary for the analysis of gene-culture coevolution" (p. 26).

Culturgens are introduced into the analysis of gene-culture coevolution to serve as the operational units of culture. They are intended to be the cultural equivalents of the physical traits studied by various biologists interested in organic phenotypes. L & W define culturgens as

> a relatively homogeneous set of artifacts, behaviors, or mentifacts (mental constructs having little or no direct correspondence with reality) that either share without exception one or more attribute states selected for their functional importance or at least share a consistently recurrent range of such attribute states within a given polythetic set. (p. 27)

Definitions dependent upon the use of polythetic sets of attributes to individuate and identify classes have come in for a good deal of hard sledding in recent discussions of the species concept in evolutionary biology (Ghiselin, 1974; 1981; Hull 1976; 1978; Rosenberg 1980). While I think there are sound reasons for rejecting the objections that have been raised against the class-based definition of species (Caplan 1980; 1981a; 1981b), there seem to be a number of problems confronting L & W's attempt to use this type of definition in the analysis of culturgens.

L & W's definition presupposes that it is possible to identify and individuate the "attributes" that are to be used to group artifacts, behaviors, and what they term "mentifacts" into various kinds of culturgens. But without some prior assumptions and hypotheses about what constitutes a "significant" attribute of culture, it will be impossible to simply "look and see" what traits ought to be the object of study in cultural evolution. The issue of how one defines key attributes, distinctive attributes, or defining attributes in forming polythetic sets that can serve as the objects of biocultural explanations is one that must not be glossed over lightly. This is particularly so in view of the low esteem in which L & W hold the descriptions and generalizations currently in use in the social sciences. We are never told why anyone should accept the existing categories and classifications of various forms of cultural behavior as valid. In the absence of some sort of theoretical framework for identifying and individuating certain attributes as important or distinctive, there is no reason to accept such culturgenic clusters as incest avoidance, village fissioning, and depth of décolletage in women's dresses as *basic* cultural units. This is not to say that such units could not be defined and defended. But L & W fail to do so adequately in their text.

Indeed, the failure to define clearly the theoretical framework used to individuate culture and thought into culturgens leads to other serious problems in the overall assessment of the utility of the various models and mechanisms proposed in *Genes* for explaining cultural evolution.

L & W look to the functional similarity of attributes in defining culturgens. But surely functional similarity is merely one criterion or test by which the validity or utility of a particular classification of entities can be assessed.

In studying phenotypic traits, biologists use such diverse criteria as similarity of form, similarity of function, heritability, polarity of phylogenetic sequence, malleability in the face of environmental change (both natural and experimental), and consistency with other existing and accepted trait classifications (Bock & von Wahlert 1965). Each of these criteria provides independent support for the legitimacy of treating a particular group of attributes as a unit. However, no such independent criteria are adduced in L & W's analysis of the culturgen. Without independent criteria for corroborating the classification of certain attributes into culturgens, L & W are open to the charge that the "biases" they see in culturgen distributions in various populations are as much a function of peculiarities in their definition as they are the function of diverse epigenetic rules at work among individuals and populations.

The problem of individuation becomes particularly acute in attempts to dissect the units, if such there be, of mentation. Many philosophers have noted that the amenability to analysis in physical terms of such mental states as purposes, hopes, beliefs, intentions, motives, and reasons is problematic at best (Davidson 1963; Kim 1978). It is not at all evident that the spatiotemporal criteria of individuation useful in analyzing the external world will suffice for individuating our internal mental worlds. L & W merely presume that univocal individuation criteria will do, but sufficient uncertainty exists as to whether mental phenomena are all classifiable as events, objects, exemplifications, or some mix of these that their presumptive suggestion seems unwarranted. It is one thing to argue for the reification of the mind, and quite another to provide adequate means for divvying up its contents.

Perhaps the greatest difficulty confronting the concept of a culturgen is that it is not clear that culturgens must be particulate units. For Darwinian evolution to occur, the units of heredity must remain constant and relatively immune to environmental alteration if selection is to pick and choose among genotypes. As Darwin's critics noted nearly a century ago, if interbreeding resulted in the blending of hereditary materials, the efficacy of the mechanisms involved in natural selection would be greatly diminished.

We know through years of experimental work and microscopic study that there is little direct action between genetic materials and the environment and that there are no interactions that can be shown to have a directional or Lamarckian effect on genes (E. O. Wilson 1975). But the situation is very different for units of culture or culturgens. The accuracy and reliability of transmission among these units are unknown. Moreover, there are many reasons for suspecting that culturally based habits, customs, behaviors, and personal ideas are malleable entities open to both directional and nondirectional environmental influences. Even if it is possible to parse human thought and behavior into classes of attributes or units, these units cannot be presumed a priori to resemble the buffered, particulate units so dear to generations of population biologists. While it may be true that cultural variability exists, that it can affect fitness, and that it in part results from genetic and physiological sources, these facts are not sufficient to prove that cultural evolution is amenable to Darwinian analysis in terms of natural selection. If the epigenetic rules posited by L & W are open programs (Mayr 1974b) rather than rigid, closed systems, then Darwinism may not be the best theoretical approach to take in analyzing behavior. If culturgens are more akin to putty than chromosomes, then Lamarckianism or other non-Darwinian evolutionary schemas may be necessary for the analysis of culture.

L & W provide an abundance of provocative claims and hypotheses in *Genes, Mind, and Culture*. However, the value of their proposals is contingent upon the clarity and soundness of their definition of the units of culture. I think there are some reasons for questioning the adequacy of the culturgen concept and, thus, some reasons for skepticism about the degree to which epigenetic constraints limit cultural evolution.

The epigenetic connection between genes and culture: Environment to the rescue

William R. Charlesworth
Institute of Child Development, University of Minnesota, Minneapolis, Minn. 55455

This is a very interesting book which is bound to elicit a wide range of reactions. Those who have concerned themselves with the relationship between genes and culture are finally offered the missing connective – development, or as Lumsden & Wilson (L & W) label it, "epigenesis." Unfortunately, the term "epigenesis" has a number of different meanings. In its general sense it refers to changes over ontogenetic time involving the gradual diversification and differentiation of an initially undifferentiated entity (a zygote, for example). Specific use of the term, however, often goes beyond description of developmental changes to include the determinants of the changes as well (Gould 1977b; Hess 1970; Moltz 1963). Herein lies the crucial question: Is behavioral epigenesis a process predictably controlled by maturational factors predetermined by genetic factors, or is it a probabilistic process contingent on not totally predictable outcomes of genotype–environment interactions (Gottlieb 1970)?

L & W define epigenesis as processes of interaction between genes and environment that ultimately result in traits of the organism. At first glance this definition fits the probabilistic definition. But L & W go further and postulate epigenetic rules as devices that channel or constrict such interactions, and these rules are "ultimately genetic in basis." Thus their view of epigenesis is fundamentally one of genetic determination. This view brings down on them charges of "unjustified hereditarian advocacy" (Cloninger & Yokoyama 1981, p. 571) and also weakens their credibility among developmentalists, the majority of whom feel that environmental factors play at least as crucial a role in behavioral development as genetic factors.

This weakening of credibility is unfortunate because, in my estimation, L & W are actually vastly improving conditions for falsifying claims about the connections between genes and culture. Such a move should be especially welcome by developmentalists because they are in the best position to identify the causal mechanisms involved in such construction (L & W's second requirement for understanding gene–culture coevolution, p. 16) and therefore of sharpening tests of any particular hypothesis.

The immediate question, then, is whether L & W can make their case more credible to developmentalists without distorting their own position. I think they can. In my estimation L & W would be convincing to many developmentalists if they put more emphasis upon environmental or ecological factors in their formulation. For example, it is generally recognized among developmentalists that behavior and cognition are constrained not only by maturational processes associated with physical growth but also by environmental constraints, such as socialization and educational practices, as well as by constraints contingent upon the developing child's behavior. An example of the latter is learning to walk, an almost universally regular event which takes the toddler away from the primary caretaking setting, thereby putting a constraint on parental influences on the child. Of course, enculturation practices and walking are geared to biological maturation, but this does not mean that variations in them have no effect on the child's future behavior and consequently on his level of adaptation. I am sure the authors recognize this, but they give the general impression (through omissions, mostly) that environmental or child-induced regularities are not very important.

This impression is especially strong when L & W define basic terms. For example, their definition of epigenetic rules seems to imply that only two classes of terms are important – genotype and phenotype. This goes against the well-known working principle of the reaction range, namely, that variations in phenotype are contingent upon variations in genotype–environment interactions, not upon variations in genotype only or in environment only. That L & W are aware of this is clear from their discussion of idiosyncratic early experiences leading to adult behavioral deviations, many of which negatively influence fitness (p. 69). It is curious that they do not develop this point, because critical or sensitive periods during early ontogenesis (when enculturation is presumably most effective) can be viewed as evolutionary adaptations favored by regularly occurring environmental pressures.

L & W could have strengthened their case further by putting more emphasis upon the functional value of epigenetic rules in aiding organisms to adapt to environmental changes. Their discussion of early cessation of lactase production in oriental populations (p. 69) is a good example. Instead of insisting solely upon a rule of parsimony when discussing the evolution of epigenetic rules governing lactase regulation, L & W could have entertained the thesis that lactase production in adults may have evolved in populations where sheep and goats were domesticated for milk products, thereby opening up a new food niche that favored genotypes for adult lactase production (Gottesman & Heston 1972; McCracken, 1971). It is important to search for environmental pressures whenever a phenotypic difference ascribed to a particular epigenetic rule is identified. Otherwise, one is telling only half the story of behavioral (and therefore cultural) evolution. I make the same argument for such a search in an attempt to develop an ethological approach to the study of intelligence (Charlesworth 1979). While intelligence can be viewed as an evolved genetic disposition (varying across individuals), it can also be viewed as an adaptive feature of behavior shaped by the various kinds and numbers of environmentally induced problems the individual faces over ontogeny. Johnston (1981) takes an analogous position on learning, arguing for studying the ecology of learning and the function it serves the individual in adapting to the environment. In other words, emphasis upon behavior relative to its ecological function expands rather than weakens the evolutionary viewpoint.

Despite their lack of emphasis upon environmental factors, L & W in my estimation have made a significant step forward because they have dared to suggest connections in an area where almost none presently exist. Whether these connections exist mostly in their imaginations is not as crucial at the present state of our knowledge as believing that they exist, and that they are theoretically important and therefore worth studying. In my opinion, L & W have imagined some very interesting things. The next step is obvious – a longitudinal, interdisciplinary research program inspired by the book's enthusiasm and guided by its leading concepts, modified and enriched by constructive commentary. The results will be interesting.

Epigenesis and culture

Robert Fagen
Department of Animal Biology, University of Pennsylvania, Philadelphia, Pa. 19104

Genes, Mind, and Culture is a work of serious, committed scholarship in a vastly important and uncharted area – the interaction of cultural and biological processes to produce mind and society. Lumsden & Wilson's (L & W's) synthesis of several disparate disciplines, their uncompromising standards, and their dedicated pursuit of difficult truths distinguish their book from all other works on the topic of biological roots of culture that have so far appeared. L & W have written a scholarly book in the classic style. Courageously, they choose to view human culture and its evolution from sociobiological perspectives; these originate in a biological analysis of development in terms of evolved interactions between genes and environments.

L & W's central argument is that epigenetic mechanisms in behavioral development link biological and cultural evolution. They cite abundant and compelling evidence for this view in contemporary research in the behavioral and social sciences, as well as in artificial intelligence. I, for one, find their narrative of explorations into these fields fascinating. It stimulated me to consult the original sources and to read more deeply in these areas, especially in child psychology and artificial-intelligence models of children's cognitive development. L & W do indeed give a representative assessment of the current state-of-the-art in these disciplines. Where any consensus at all may be said to exist, L & W state it fairly, whether or not they happen to find it convincing. In more controversial areas, they consistently cite elegant and sophisticated empirical research generally (though often indirectly) supporting their hypothesis that epigenetic mechanisms in human cognitive development are far more pervasive and complex than previously believed.

Is this book useful? By all means. It presents a very complete theoretical model of behavioral development in intelligent animals, a topic of considerable current interest in evolutionary behavioral biology; it offers fascinating new ways of looking at human cultural diversity and social-political history; and it may even inspire others to found arsenals of practical techniques for education, psychiatry, advertising, and marketing research.

L & W have undeniably written a fascinating and controversial book, but are they right? Have they made the kinds of assumptions that one really ought to make about human nature in the process of modelling genes, culture, and the human mind? Because they have given careful attention to contemporary research on human behavior, their assumptions are difficult to fault as a whole. Whether revision of specific elements of the system will generate predictions different from those of L & W's models remains to be seen. However, I have a strong hunch that their major conclusions are correct. Better to abandon advocacy and systematically vary the central assumptions of these models as current knowledge (and uncertainty) dictate.

L & W have raised the level and mode of discourse on genes and cultures. Their analyses establish formal models of hypotheses in this field as a prerequisite for constructive criticism and dialogue. From now on, anything less will simply miss the point. Could the more intuitive sorts of critics prefer verbal potshots simply because the advantage is theirs only so long as the battle is joined in the domain of the irrational and the emotional?

Unlike many books in human science whose finale is a grand apotheosis of humanity in a safe society and a better world, L & W's work ends by affirming interdisciplinary studies of culture rather than by affirming the human condition. This is obviously a very different sort of book, destined not to uplift its readers' souls by flattering their egos with moist and tepid platitudes, but rather to advance and expand a given area of scholarly inquiry by confronting it with new and difficult ideas. L & W's efforts to do so are a total success.

On mechanisms of cultural evolution, and the evolution of language and the common law

Michael T. Ghiselin
Department of Biology, University of Utah, Salt Lake City, Utah 84112

In *Genes, Mind, and Culture* Lumsden & Wilson (L & W) argue for a strong linkage between biological and cultural evolution. Traditional evolutionary biologists should not find such a thesis uncongenial. That behavior plays an important role in evolution was well appreciated by Darwin (1871), and has been cogently argued by Mayr (1960; 1974a) among others. If, for example, an animal learned to eat a new kind of food, and if the pursuit of it led that animal into a different environment, it stands to reason that a new set of selection pressures might come into play. By all analogy one would expect cultural behavior to be subject to like principles. The question is just how much of the data of cultural diversity and evolution need to be explained in terms of such processes. Probably few would deny that the general faculties, say, for the use of language, are thus subject to natural selection. Powerful selective forces ought to control the gene ratios that affect our ability to understand and generate sentences. On the other hand, students of cultural evolution are not particularly interested in this aspect of the problem. It seems far less probable that genetic models will tell us much of real interest about why English differs from French.

Theories of linguistic evolution should nonetheless have much in common with theories of biological evolution. This is not because genes control the ontogeny of organisms. Rather, it is because cultural units have a formal structure not unlike that of biological ones, and are subject to similar laws of nature. Natural selection is only a particular subset of the more general process that Campbell (1974) calls "blind variation and selective retention."

Although they do consider "holistic properties," L & W use an approach that tends to treat integrated units larger than organisms as if they were classes (see Ghiselin 1981). Such an approach is apt to give the impression that the only "real" units are those at lower hierarchical levels. But different processes and different laws and principles become important when we deal, say, with such individuals as *Homo sapiens* and French. Speciation theory includes such matters as the founder principle and allopatry. I have been much impressed by the similarities between the study of geographical races and clines on the one hand, and dialect geography on the other. Indeed such parallels between linguistic and biological evolution are discussed at considerable length in the older literature on comparative linguistics. L & W do excellent work in this tradition when they apply island biogeography theory to cultural evolution. However, all of this seems far removed from the chromosomes, and has little to do with linkage between genes and culture.

Various aspects of linguistic evolution are not obviously coupled to the genes. Linguists have long recognized that there is an analogue of genetic mutation in the imperfect replication that occurs when words are transmitted from one generation to the next. The same fundamental principle is involved, but there are different material substrata and the frequency of such mutation has different causes and different laws. Again, consider the possibility that a phonemic change makes two words into homonyms, and that this leads to misunderstandings. There are two likely outcomes. The first is "competitive exclusion" – one word ceases to be used (Darwin 1859). The other is "character displacement" – the competing words get modified (Brown & Wilson 1956). Many changes in language occur in response to social circumstances the particulars of which are quite accidental – such as "donkey" coming to replace "ass." Such phenomena suggest that a lot of features of culture are in one sense adaptively neutral. The parts of culture are coadapted as a result of their interaction, but it is indifferent to the organism which particular coadapted system evolves.

For some aspects of linguistic evolution there is very little reason to invoke much of what is called "hard wiring." Indeed, the forces operative are largely economic ones that pervade behavior quite generally. L & W (pp. 320–21) suggest that there is an economic balance between ambiguity and cost in the production of cultural products. This principle is well known to linguists as Zipf's Law (Zipf 1968). We are sloppy or precise in our pronunciation according to an equilibrium set by saving effort on the one hand and being understood on the other. It hardly seems necessary for much detailed genetical programming to exist for this sort of behavior to occur. On the face of it, all that is required is a propensity to communicate and a propensity to save work.

The changes that occur in language would thus result from a kind of artificial selection, with what Dunn (1970) has called a "culture pool." It would follow that the properties of the selective agent will influence the course of such evolution. The question nonetheless remains unanswered of just how often the sort of reciprocal interaction hypothesized by L & W occurs. With respect to the coevolution of *Homo sapiens* and domesticated species, it is clear that dogs have been much altered by artificial selection. But have people been modified by dogs?

The selective processes that change cultures may be rather different from what one might expect from a straightforward mechanism of cognition plus choice such as that hypothesized by L & W. In this connection some recent studies on the evolution of the common law are of considerable interest (Cooter & Kornhauser 1980; Goodman 1978; Priest 1977). According to conventional wisdom, adaptive changes in law occur through a rational choice on the part of judges. The new model proposes that there is no change in the law without litigation (much as a gene, to be selected, must be phenotypically expressed). People litigate only when they are dissatisfied with the existing condition of the law. Judges then vary the law – but in a manner that is fortuitous or "random" in the sense that a mutation is. If people are satisfied with the change, they stop litigating. If, on the other hand, the change displeases them, they litigate all the more. This model has some intriguing features. For one thing it proposes a very simple and straightforward mechanism of choice: Either one is satisfied with the law or one is not. Not much reasoning is invoked, just trial and error. Yet the result could be the sort of effect that one might hope for from a wise and prescient legislator. If cultural evolution results largely from such a trial and error process, one might wonder how much selective advantage really accrues to an organism opting for my particular "culturgen." In other words, suppose that we do not evaluate a "culturgen" as such, but merely find out whether life in general seems better or worse without it. We would then be making a rather different judgment from the sort hypothesized by L & W. Drop the ending off a word, and see if you are misunderstood.

Even in that most ratiocinative of human activities, science, it is remarkable how little thinking is necessary. Run a hypothesis up the flagpole and see if anybody shoots. Intellectuals tend to see more cognition in the world than the data really warrant, and not just when they are writing articles for the *Journal of Anthropomorphic and Teleological Zoology.*

Genes for general intellect rather than particular culture

Howard E. Gruber
Institute for Cognitive Studies, Rutgers University, Newark, N.J. 07102

It has not been possible for Lumsden & Wilson (L & W) to present their case, or for me to evaluate it, without reference to the human plight. As I see it, they join a wave of intellectual pessimism that is not entirely unjustified in this thermonuclear age.

Can we as a species change our ways of thinking and acting more in the next 25 years than we did in the last 10,000? This is what we require to avoid a thermonuclear ***finis*** to the present discussion (see Thomas 1981a; 1981b). In pronouncing their "1,000-year rule" L & W also stress the centrality of rate of change in their argument. In particular, which of three time scales, we may ask, is relevant to the emergence and establishment of genetic changes that can significantly affect culture?

Omega-time. The 10^5–10^6 years in which the body and brain of humanity have evolved from earlier hominids. On this scale, genetic change is irrelevant to our present needs. The intellectual and emotional equipment *homo sapiens* used in constructing its immense variety of cultures is what we must use to rescue ourselves now. While this is not entirely a pleasant thought, the very fact that this animal has produced so many diverse cultures, *some* of them peaceable, gives some ground for hope.

Beta-time. The 10^2–10^3 years that L & W claim are necessary and sufficient to produce significant genetic changes to account for known cultural variations – and that must by the same token be relied on to effect any changes to come. If L & W are right, then the necessary changes are out of reach, and we are lost. The note of hope suggested by existing cultural diversity disappears in this time perspective, since cultural change is linked to genetic change. The dominant civilizations, those that will destroy the species, are locked into their culturgens, locked onto their targets – ourselves.

Alpha-time. Suppose that effective genetic – and consequently "culturgenetic" – change were possible on an even shorter time scale, within a generation. This seems biologically impossible, but with the onrush of genetic engineering, the idea ought at least to be considered. Kenneth B. Clark (1971) had something equivalent in mind when in his 1971 APA presidential address he recommended that psychoactive, antiaggression drugs be administered to world leaders. I suppose that he would have settled for gene substitutions having the same effect.

In 1971 Clark's proposal seemed outlandish and pessimistic. A statement criticizing it began, "the normal human brain is beautiful." That phrase still gives me hope, and I must say that I am sad that some of my colleagues have now invented a time scale for culturgenetic change even more pessimistic than Clark's. In alpha-time we can at least dream: ***if only our leaders would submit to the administration of these miracle drugs. . . .***

In beta-time, L & W's time frame, no such dream is possible.

Of course, the implication of a gloomy future does not decide scientific issues which must be addressed in their own right. How does an innovation come about? How is an innovation disseminated? How is an innovation assimilated by a population (or by a culture)? L & W make use of the mathematics of population genetics to show how dissemination and assimilation of innovations might follow functions corresponding to those found in the spread and evolution of genetic changes. Even supposing that they have made this case convincingly, it does not go far to support their underlying point that cultural change is genetically regulated. Many profoundly different processes can have the same function form. L & W restrict themselves almost entirely to biocultural decisions among given alternatives. Almost nothing is said about the process of making an innovation in the first place. Yet invention is one of the prime facts of all human existence.

In contrast, in another recent work the anthropologist Peter Wilson (1980) has argued that *homo sapiens* represents the extreme thus far attained in a general trend in hominid evolution – a trend toward openness and ever increasing diversity. But what is this openness? At the cognitive level it consists in curiosity and inventiveness – both technical and social. Rather than prescribing a range of particular outcomes, the genes for culture may be only those for a wide-ranging and powerful intellect.

How could this view be squared with the occurrence of such cultural universals as incest avoidance? The general mammalian preference, or need, for varied stimulation may be enough to account for an initial bias against couplings between littermates, nestmates, housemates, fellow members of the same kibbutz. The sexual stimulus value of a potential partner may be akin to all other perceptual phenomena, subject to the laws of perceptual adaptation. In situations of choice this may be all the bias that is necessary to account for what is actually known about spontaneous gravitation toward outmates rather than inmates of the same social unit.

The life of Charles Darwin is a case in point. He married his first cousin, Emma Wedgwood. All his life he was very close to her branch of the Wedgwood family. Her father was almost a father surrogate for Charles. But Charles did not slide directly into his union with Emma. First he went away on that famous five-year circumnavigation. Only after this long "vacation from cousinhood" did he court and marry Emma (Gruber 1981).

This "perceptual adaptation" hypothesis is just as biological as the notion that a particular avoidance pattern is carried directly in the genes. But it draws on the most general cognitive abilities and propensities of mammals and especially humans. The aim of this argument is not to deny the existence of incest taboos, but to provide a baseline of sexual choice behavior biased against incestuous pairings, without invoking anti-incest genes (see Feffer 1981 and Fischer & Watson 1981 for similar cognitive-developmental approaches to the Oedipus complex, avoiding the notion of pro-Oedipal genes). Widely different mechanisms account for outbreeding. Plants avoid self-fertilization by nonjuxtaposition of male and female sex organs on the same plant, by use of insects as go-betweens, and so on. To accomplish similar evolutionary ends, animals must have evolved entirely different mechanisms – certainly not insect go-betweens. Even nonliterate peoples accumulate considerable lore about heredity.

Let us do the following thought-experiment. Imagine an evolutionary situation in which genes favoring general intellect are competing with genes favoring some cultural trait that is adaptive in a particular environment. Whenever the environment changed (as it always does) the genes for intellect would gain.

It seems plausible to propose that a powerful intellect may be our species' major tool for the production of social inventions, rather than genes or even culturgens dictating particular behaviors.

The "culturgen": Science or science fiction?

C. R. Hallpike
Department of Anthropology, McMaster University, Hamilton, Ontario L8S 4L9, Canada

As a social anthropologist unsympathetic to tabula rasa theories of the mind I can appreciate the significance and use for cultural theory of what Lumsden & Wilson (L & W) have to say about the epigenetic rules of human perception and cognition. But to explain their origins or deduce their actual impact on culture is quite another matter, and in these respects their book is, in my view, totally unconvincing. One of its major defects is the concept of "culturgen," and I shall devote the rest of this review to examining its deficiencies.

The theory of gene–culture coevolution rests on the attempt to "decompos[e] . . . social behavior into objective functional units" (p. 362), to find "the operational unit of culture" (p. 27); this unit is presented as the "culturgen." A number of disciplines employ concepts of roughly this type: the atom and molecule of chemistry, the phoneme and morpheme of linguistics, the stimulus and response of behaviorist psychology, and so on. But in all these cases the relevant units are so defined that it is possible to decide what is or is not a unit, how units are to be distinguished from one another, and therefore, the number and kind of units present in any empirically describable situation. There is, of course, absolutely no point in positing basic units unless clear rules are given for identifying them. To what extent does the concept of culturgen meet these requirements of definitional adequacy? The problem begins with the definition:

> A culturgen is a relatively homogeneous set of artifacts, behaviors, or mentifacts (mental constructs having little or no direct correspondence with reality) that either share without exception one or more attribute states selected for their functional importance or at least share a consistently recurrent range of such attribute states within a given polythetic set. (p. 27)

This definition is so vague and all-encompassing that a culturgen could be any discriminable aspect of human thought or behavior whatsoever: It is as though the "thing" were to be proposed as the basic unit of physics. Not surprisingly, therefore, one finds in the book that the examples of culturgens comprise an extraordinary ragbag of oddments: carpenters' tools, dress fashions, marriage arrangements, food items, ten-second slowdowns by individual drivers causing traffic jams, speech, a color classification, 6,000 attributes of camels among Arabs, and so on. While the definition includes the requirement that a culturgen have "functional importance," we cannot say that something has functional importance until we can first define what that something is. So, for example, how many, and what, culturgens are there in: "the Roman Catholic concept of God," "marriage," "gift," "cardinal integer," "the Liberal Party of Canada"? I can see no sign that L & W would be able to answer such questions, or have even reflected on the problems that answering them would involve; yet, if a biologist cannot precisely define a phenotypic trait, for example, any argument about its selective value must be worthless. One can only conclude that the concept of culturgen fails to meet the minimal criteria of definitional adequacy required by its proposed status as an "objective functional unit."

Second, one may also ask why, in the case of culture, we should even expect to find any such basic unit as the culturgen or why, for that matter, there should be only *one* such unit. Despite L & W's claim in their accompanying *BBS* précis that culturgens' "existence does not mean that culture is composed of discrete, isolated units," the culturgen does in fact boil down to just this. For, we are told in the book that culturgens "can be ultimately delimited as *portions of knowledge structures*. As such they contain fragments of information about culture, environment, and social action, and they constitute the raw material used by the epigenetic rules in the construction of the mind" (p. 349, my italics). Even if one ignores this statement, it is clear that throughout the book culturgens are treated as discrete entities with frequencies, distributions, loss rates, and the like, and subject, like genes, to natural selection. Indeed, the ideal type of the culturgen is the artifact of the archaeologist, and the authors express the hope that this model will be generally applicable to the rest of culture. But, despite their appeal to archaeological artifacts (a highly specialized and unrepresentative case), L & W's real reason for introducing the concept of culturgen is not derived from any well-founded social theory at all, but is simply that, if one is trying to explain culture on the basis of a neo-Darwinian theory of natural selection, it is highly inconvenient *not* to have a concept like culturgen, quantifications of which can be treated as continuously variable over time.

The hypothesis that societies and cultures are composed of such units is vulnerable to at least two basic objections. The first is that sociocultural phenomena are not just aggregates or heaps of elements but structures, with rules, relationships, and transformations of these, at differing hierarchical levels. Structures include elements, but they do not have portions, as in "portions of knowledge structures," in the same sense that a wall has bricks. L & W also pay far too little attention to the autonomy of social and cultural structures, which are not solely the product of mind, and are also accommodated to reality. More generally, the concept of culturgen rests on a social theory of individualist reductionism, as, for example, in "the key to the social sciences is the appropriate characterization of human nature" (accompanying précis). Human nature is undoubtedly a vital ingredient in the explanation of sociocultural forms, but to claim that it is *the* ingredient assumes that culture and society are *nothing more* than the aggregate of the decisions and acts of individual human beings. Here again, one can see that this assumption, too, is almost necessitated by L & W's model of gene–culture coevolution, and provides an essential basis for the culturgen; but if they wish to be taken seriously by social scientists they will have to demonstrate by evidence and argument that naive individualism is an adequate theory of culture. Mere assumption is not good enough.

In conclusion, the culturgen is a concept that is introduced because the authors' theory requires it, not because of adequate research into the nature of culture and society. Its definition is vacuous, and since propositions including it are unverifiable, neither it, nor theories employing it, can claim scientific status.

A too simple view of population genetics

Daniel L. Hartl
Department of Genetics, Washington University, School of Medicine, St. Louis, Mo. 63110

I agreed sight unseen to review the population-genetic content of *Genes, Mind, and Culture: The Coevolutionary Process* (henceforth *Genes*), hoping that it would provide new insights into the relationships between genes and behavior, genes and culture, and culture and evolution. I was disappointed. What population genetics is found in the book is embodied in Chapters 5 and 6 (mainly Chapter 5), but it is merely a reinterpretation of classical formulations in terms of culturgens. The principal assumption is that genes can influence virtually any aspect of phenotype, including behavioral patterns; to the extent that differences in fitness can arise because of differences in behavioral patterns, changes in allele frequency can be mediated by natural selection acting on such differences. Moreover, the dynamics of allele-frequency change will be the same as if natural selection were acting with an identical intensity on any other phenotypic attribute related to fitness. But such conclusions are self-evident; they are embodied in the assumptions and require only correct logic to emerge.

This sort of idealism is widespread not only in contemporary theoretical population genetics but also in theoretical branches of certain other fields. In his recent autobiography, John Kenneth Galbraith (1981, p. 514) remarks as follows:

> Much economic discussion – theoretical model-building it is called – proceeds within the framework of a larger assumption that is never examined. "We assume a competitive market." The validity of the result then depends not on congruity with what exists, with reality, but on whether it derives in a valid way from the assumption.

Now, if you replace the word "economic" with "population genetics" and the phrase "We assume a competitive market" with the phrase "We assume that genetic variation exists for all traits and that all traits are acted upon directly by natural selection at all times," you will have a reasonably accurate description of much contemporary theory in population genetics.

The population genetics of Lumsden & Wilson is not so much wrong as unquestioned, entirely within the framework of the above paraphrase of Galbraith. However, if one interprets the purpose of theory as being to tell us, not what is true, but only what *could* be true, then this sort of theory serves a valid and useful scientific purpose. (On the other hand, Galbraith goes on to point out a danger – somebody might actually begin to believe this stuff; worse, it might be converted into social or economic policy.) In this restricted sense of the purpose of theoretical population genetics, *Genes* does not go far enough; it gives us no alternatives. Normal variability in behavior, if influenced significantly by genes at all, is not influenced by single genes of major effect; such traits are quantitative traits, and thus the theory of quantitative genetics is appropriate, not classical single-locus population genetics. Moreover, in any realistic consideration of genes, mind, and culture, one must take into account the intensity of selection on traits because the intensity of selection can be so small relative to a population's effective size that allele-frequency changes are dominated by random genetic drift. In any serious discussion of what *could* be true, random genetic drift cannot be ignored.

Another phenomenon that cannot be ignored relates to correlated responses, which are inevitable concomitants of selection involving quantitative traits. It simply must be recognized that certain traits, in particular certain behavioral traits, may not result from natural selection acting directly on the trait itself, but rather may result accidentally from genetic correlations with traits that are subject to the direct action of natural selection. For example, play behavior in juvenile mammals need not necessarily be a direct result of natural selection; it could as well be an indirect result arising from a correlated response to selection for particular behavioral traits in adults. [See also Smith: "Does Play Matter" *BBS* 5(1) 1982.]

In short, the problem with the population genetics in *Genes* is that it is oversimplified and takes itself too seriously. Oversimplification is not so much wrong as misleading, but the danger is that somebody might actually believe it.

Concepts of development in the mathematics of cultural change

Timothy D. Johnston
North Carolina Division of Mental Health, Research Branch, Dorothea Dix Hospital, Raleigh, N.C. 27611

The point of departure for Lumsden & Wilson's (L & W's) *Genes, Mind, and Culture* is the recognition that human social institutions are the product of both cultural and biological evolutionary processes, so that the key to their understanding lies in the interaction between these two modes of evolution. That is a sensible proposition that has motivated several generations of research by physical anthropologists and provides the raison d'être for nonhuman animal studies attempting to elucidate human evolution.

Many students of human evolution, including Darwin (1871), Muller (1959), Dobzhansky (1962), Hockett and Ascher (1964), and Parker and Gibson (1979), have struggled with the problem of reconciling the biological and cultural sides of human nature. The most noticeable difference between L & W's theory of gene–culture coevolution and the work of earlier writers lies in their extensive use of mathematical formalism. The way they have chosen to present their highly mathematical theory makes me wonder whom they have in mind as the intended audience for this book. Its subject matter will be of greatest interest to anthropologists and evolutionary biologists, but I doubt that many of these will have the mathematical sophistication necessary to evaluate the theory properly. Although L & W claim (p. xii) that their argument can be followed by means of sections that summarize the mathematics, for the most part these sections present only the conclusions and not the argument, so the skeptic is more or less obliged to tackle the mathematics. But the reader without professional training in mathematics (such as myself) will find this a daunting task. Little attempt is made to help such a reader gain an understanding of what the equations signify; and unless one has a fairly well developed mathematical intuition, much of the argument remains somewhat opaque.

The book is profusely illustrated with graphs and diagrams, but for the most part I found these to be of little help; many of them are so cryptically labeled as to be even less clear than the equations they portray. In fact, some of them appear to be incorrectly labeled: the caption to Figure 7-7 (p. 330), for example, asserts that several selective costs increase with developmental flexibility (which seems to be correct; see Johnston, in press), but the figure itself clearly shows these costs as a decreasing function of flexibility. Encountering this figure late in the book made me wonder whether my difficulties interpreting earlier figures might have arisen from similar errors.

One persistent problem that arises in trying to interpret L & W's mathematics stems from the fact that many variables are only defined implicitly, by inclusion in one or more equations, so that many important elements of the theory never receive explicit elucidation and discussion. For example, Equation 4-36 (p. 127) introduces, without definition, two parameters, a_2 and a_3. Two pages later, we learn that these are drawn from a set of "adjustable parameters," which characterize changes that occur in the rate of transition between cultural options (or "culturgens") as a result of learning, cultural shifts, and natural selection. In the absence of any further interpretation or discussion, it does not take a particularly unsympathetic reader to wonder whether these parameters might not be sufficiently adjustable to allow subsequent derivations to accord conveniently with the authors' theoretical expectations. What do the parameters measure in the real cultural world? When do they get adjusted and what do the adjustments correspond to? Since minor changes in these parameters are said to have important evolutionary consequences (p. 129), they demand a lot more explicit attention than they receive.

The authors could have made things a little easier for their readers by defining their variables and functions in a glossary, but the glossary they do provide (which includes only five of the innumerable variables in the book) is mostly concerned with the definition of terms like "ecology," "learning," and "mammal," with which most readers will surely be familiar. (Incidentally, a glossary that includes those terms, but omits terms such as "stoichiometry," "virilocal," and "Dirac delta function" is, to say the least, curiously selective. Again I wonder, whom is this book aimed at?)

In terms of the style and presentation of its argument, I regret that I can find almost nothing to recommend this book. Too much of the mathematics seems designed to impress rather than to instruct an audience whose mathematical sophistication is likely to be on a nonprofessional level. Certainly the authors make little effort to persuade their readers that the effort of mastering all these equations will be repaid by deeper insights into the evolutionary processes that they model. The book would have been more valuable if they had limited their mathematical exposition and devoted more attention to the conceptual foundation on which their argument rests.

However, conceptual explication does not appear to be one of L & W's more pressing concerns, and as a result, the reader is asked to make some rather dramatic leaps of faith so that the mathematical exposition can proceed. The conceptual fuzziness of their theory is particularly evident in their treatment of individual development, or epigenesis. Although L & W are, I think, right in making development the focus of their theory, they can hardly be said to have done justice to the problems this focus raises. Although one would never suspect it from their discussion, the role of development in evolution has received a considerable amount of attention, much of it quite recently. To cite only a few sources, Alberch, Gould, Oster and Wake (1979); Baldwin (1902); Bates (1979); Bonner (1965); deBeer (1958); Fishbein (1976); Gould (1977b); Ho and Saunders (1979); Parker and Gibson (1979); Waddington (1957); and Zuckerkandl (1976) have all made serious attempts, from different points of view, to understand the relation between development and evolution, yet none of these authors receives more than a passing mention. (In fact, only Waddington and Fishbein are cited, the latter in a footnote.)

L & W's developmentally based model of evolution is thus highly idiosyncratic. It rests largely on their concept of an epigenetic rule, an entity that is involved in determining the choices that individuals make among alternative culturgens such as marriage customs, agricultural practices, and dress styles. Epigenetic rules "are the genetically determined procedures that direct the assembly of the mind" (p. 7), a formulation that implies a strong genetic determination espoused by only a very few developmental theorists. L & W's developmental theory has much in common with that of classical ethologists such as Lorenz (1950; 1956; 1965) and Eibl-Eibesfeldt (1979), and if they are aware of the numerous strong arguments against this way of thinking about development (e.g. Gottlieb 1970; 1976; Hailman 1967; 1979; Hinde 1970; Kuo 1967; Lehrman 1953; 1970; Schneirla 1956; 1966) L & W do not reveal it. In suggesting that the authors have based their theory of development on a strong genetic determinism, I do not mean that they envisage development as a simple mapping of alternative genetic arrangements onto alternative cultural patterns. Rather, they see the genes as having a strong and pervasive influence on those patterns through the action of epigenetic rules, and those rules are genetically *determined*, not just genetically influenced. Although their glossary defines epigenesis as an "interaction between genes and the environment" (itself a questionable formulation; Ho & Saunders 1979, p. 579), they more often use the term to imply a genetic prescription of structure and behavior. For example, in discussing the development of the brain they write: "The traditional view that canalization produces a vertebrate brain made up largely of vast but haphazardly connected nerve nets, with fine structure beyond the capacity of the genes, is no longer acceptable. . . . Epigenesis achieves a previously unsuspected precision of neuronal form, location, and connectivity" (p. 241).

Epigenetic rules determine a set of "innate biases" in culturgen choice (pp. 133 ff.) which constrain the possibility for cultural change and are themselves changed only as a result of natural selection. These biases, and their updating during development by reference to the "adjustable parameters" mentioned earlier, play a central role in the model of gene-culture translation presented in Chapter 4. The concept of innateness has worried developmental theorists for a long time; there are cogent arguments for abandoning the concept altogether, or at least sharply restricting its use (e.g. Hailman 1979; Lehrman 1970), but even those who continue to employ the term generally recognize that its use raises some difficult problems (e.g. Jacobs 1981). At the very least, assessing innateness requires controlled-rearing experiments of a kind that cannot be performed with human subjects, so the attribution of "innate preferences" to humans must rely on highly indirect evidence. None of these difficulties appears to worry L & W, for whom any regularity in development seems adequate grounds for a judgment of innateness. Although they shy away from defining "innate" in their glossary, the entry for "epigenetic rule" is revealing; it begins: "*Any* regularity during epigenesis" (p. 370; emphasis added). Thus they can confidently assert, on the basis of cross-cultural evidence alone, that there is an innate tendency for women to carry babies on their left side, even going so far as to give a quantitative estimate for the degree of innateness (Table 4-1, p. 146). But regularities in development are the result of both genetic and environmental constancies, and of complex interactions between the two. Any attempt to model the evolution of such regularities, as in L & W's theory of gene–culture coevolution, requires far more attention to the nature of those interactions than they have provided. A simple division into "innate biases" and "updating functions" (Equation 4-38, p. 133) in no way reflects the complexity of developmental processes as revealed by current theory and experiment in the field (see Cairns 1979, for a recent review of much of the pertinent literature in human social development). Concluding their discussion of innate biases in gene–culture translation, L & W write: "These examples are limited principally to initial enculturation, but it is reasonable to suppose that the innate biases carry over into later transitions between the culturgens" (pp. 145–47). That "reasonable supposition" elides several decades of careful developmental study. L & W are sorely in need of a theory of development whose conceptual sophistication matches the mathematical sophistication of their evolutionary theory. Mathematical analysis is an excellent tool for obtaining a close focus on complex conceptual problems, but a close focus on fuzz only gives more fuzz.

How, then, ought we to proceed with the kind of undertaking that L & W have in mind? Frankly, I doubt that our current understanding of development, and of the relationship between developmental and evolutionary processes, can support mathematical models of this degree of complexity. L & W apparently believe that the developmental relations between genes and cultural phenomena such as marriage customs and agricultural practices are understood well enough in principle and only require elucidation in detail to be made amenable to the sort of modeling they espouse. Even the most cursory survey of the literature on behavioral development would have revealed that there are no grounds for that belief. Before models of this kind can play a useful role in the study of evolution, they must rest on far more solid empirical and conceptual foundations than are presently available. L & W are right to call for more detailed studies of human development from an evolutionary standpoint, but such analysis cannot be based on the inadequate and outmoded concepts of development to which they subscribe. [See also Johnston: "Contrasting Approaches to a Theory of Learning" *BBS* 4(1) 1981.]

From genes to culture: The missing links

Joseph K. Kovach
Research Department, The Menninger Foundation, Topeka, Kans. 66601

Had I not read the latest discussion of evolution and sociobiology in *BBS* (Plotkin & Odling-Smee 1981 and the accompanying twenty-six commentaries), and Medawar's (1981) delightfully nasty review of Lumsden & Wilson's (L & W's) book, I would not have summoned the interest and courage to comment here in so few words. Plotkin et al. put me at ease: There is no *unified theory of life in all of its manifestations*, after all, and I suspect there never will be. Medawar emboldened me to hum my way through most of the formulas, derivations, and symbolic bravado L & W offer in their book in lieu of data. Thus unburdened of concern for the merits and demerits of sociobiology, and well prepared to skip L & W's unpalatable obfuscations, I discovered a brave and provocative book. It has captured my interest by its daring shortcuts, leaps and lapses of logic, and confusions of purpose.

L & W argue that the time has come to unite social and biological sciences under the canopy of population genetics and the theory of evolution. They claim to have found the fine texture of this union in the coevolution of genes and cultures, as it is supposedly revealed in variably distributed culture units and choices that are guided in individual development by epigenetic rules (genetically fixed sensory filters and preadapted processes of perception, motivation, learning, and cognition). They believe they have solved in the single gesture of this formulation the problem of differences between evolution and development, and the attendant mysteries of human nature. Because the nature of human nature is at the heart of this book, let me start commenting in earnest with this matter.

Human nature has been regarded throughout history as a product of two fundamentally different forces. One, traditionally, if disparagingly, referred to as "the flesh," is man's animal nature. Another, traditionally anchored in the Judeo-Christian view of similarity between the images of Man and God, is the human potential for transcendental good and progress through grace and experience. No two ideas have greater currency in our culture than these. They have dominated discussions of theology, philosophy, art, and literature, and they live on in modern sciences as well.

The idea of man's perfectibility and progress lives on in positivistic approaches, especially radical and logical behaviorism and social determinism. Its proponents have tended to attribute all significant variations of behavior to environmental causes.

The idea of man's biological limitations lives on in the psychologies that arose from Darwin's theory of evolution. It has penetrated the field through two major avenues: through the theories and researches of comparative psychology and ethology, and through Freud's psychoanalytic theory of instinct. Practitioners of these fields have tended to approach human behavior from perspectives of instinct, genetic determination, and evolution.

But the problem of inborn limitations and transcendental potentialities of human nature has been also closely linked throughout history to another puzzling problem. Human beings reflecting on the conflicting forces that govern their behavior have always been puzzled by this question: How can bodily actions that are observable and localizable in time and space be controlled by thoughts, feelings, and wishes that are neither observable, nor localizable, nor categorizable under any other common notion of a physical object or process? This question too has been with us since antiquity, at least since the time of the early Greek philosopher Anaxagoras; yet it is also far from being settled to everyone's satisfaction (see Fodor 1981; Popper & Eccles 1977; Sperry 1980).

Entries in the long list of proposed solutions to the mind-body problem date back to Plato's dualism and Aristotle's monism. These solutions and the many others that followed them down the centuries have all been rooted in fundamental assumptions about cause-effect relationships in nature. Here too science is no exception. Its fundamental assumption has been that there exist uniform continuities between causes and effects in all phenomena of nature, including the phenomena of life. This implies that a valid statement of fact arising from a particular set of observations must not contradict another valid statement of fact about the same or any other set of observations. The various branches of science are united by this assumption.

Because the Platonic and Cartesian ideas that mind and body are of fundamentally different substances do not agree with science's notion of continuity in causation, and perhaps more important, because dualism has not led to a productive science, modern science has either rejected the mind as an agent of causation or accepted in one form or another the Aristotelian view that mind and body are but a single substance. I think L & W are on secure scientific ground when, by implication, they claim that the unity of mind and body is a matter of their coevolution and is thereby inseparable from the nature–nurture problem. Had they remained with this problem of coevolution of mind and body, and had they left the topics of culture and historical determinism alone, or to another book, I could now exclaim, "How right they are!" But L & W misconceive the ways of science when they claim, again by implication, that the mind–body unity is a ground for the union of methods of social and biological sciences. It seems to me altogether out of bounds, however, to extend this idea to historical analyses and ideological issues. What L & W say about epigenetic rules and development of culturgens (aside, of course, from their statistical gerrymandering), it seems to me, applies to the coevolution of genes and mind, not genes and culture. Let me elaborate these points.

The study of evolution deals with overt variations of organisms within and across generations and species. Its task is to trace phenotypic variations to variations of genotype, ecological niche, ontogenetic experience, and the forces and mechanisms that ensure the transmission of related information from generation to generation, including natural selection, genetic inheritance, epigenetic rules of individual development, learning, cognition, and cultural inheritance. A complete list of topics under the purview of evolutionary investigations would pretty much cover all the current facts, theories, and questions of the life sciences, for which evolution is indeed the single unifying principle. The nature–nurture and mind–body problems would most certainly be on such a list. Yet, these problems long antedate the theory of evolution. They are rooted in a long tradition of concern with the mainspring of human conduct. For that reason they touch on moral, ethical, artistic, philosophical, ideological, and political considerations that are beyond the ken of science. This explains, but does not excuse, why the boundaries of science, philosophy, and ideology are so blurred when it comes to the discussion of mind–body and nature–nurture problems (see Skinner 1971 and E. O. Wilson 1978, for examples). This also explains, but still does not excuse, why the discussions of these problems take on the air of urgency, finality, and all-inclusiveness exemplified by L & W's book.

That the self-image of man has been as much a guiding force in the development of science as it has been in the development of ethical, philosophical, and ideological principles is self-evident. That the knowledge science generates does influence man's image of himself is likewise self-evident. In this sense, then, science, philosophy, and ideology stem from common roots and are subject to reciprocal feedback and shared mechanisms of social control. Yet the facts, theories, and procedures of science, including the facts of evolution and the logic of scientific inquiry, can and must remain free of ideological considerations. Choosing and justifying particular social actions by reference to science unavoidably bring ethics and ideology into play.

By placing evolution at the center of the search for the origins of the mind, and the mind at the center of the development of culture, we can and should remain focused on purely scientific issues. Within the context of L & W's book these issues, it seems to me, are: (1) What is the origin of variations among species-typical capacities for culture? (2) What is the origin of variations among human cultures? And (3) what is the origin of individual variations within a human culture? It seems to me that L & W confuse these questions. Ostensibly their idea of coevolution of genes and cultures is addressed to all three of them, yet I am able to relate it only to number 1: to the sources of the human capacity for culture. Let me explore this point a little further.

That man's evolutionary history and individual constitution impose limits on his ontogenetic development seems to me self-evident. That culture, any culture, develops within such limits is also self-evident. The problem then is: Are there constitutional differences among human beings that would predispose certain groups of individuals to acquire certain cultures? If we take L & W's derivations of the coevolution of genes and cultures seriously, including and especially the 1000-year limit on significant evolutionary change, and the "leash principle," the answer is yes. In terms of L & W's theorizing, cross-culturing (and fostering), let's say, a Jivaro and a !Kung Bushman baby from birth should result in serious developmental difficulties. It should require resolution of their divergent epigenetic rules and incongruous cultural environments. We have not the slightest evidence of such difficulties, but evidence to the contrary abounds – evidence of quick acculturation, not only in babies but in adults, in initially alien environments. All human cultures seem quite capable of assimilating all the normal constitutional variations of humanity! It is not genes and cultures but genes and mind (or, better still, genes and the human propensity for culture) that have evolved jointly.

As we examine the changing complexities of behavior along the ascending scale of evolution we cannot escape noting that the capacities for ontogenetic acquisition of new behaviors (and for the neural coding and storing of new information) have progressively superseded the primacy of genetic adaptation (of the role of genetically fixed action patterns and stimulus-response associations). Yet the capacity to learn is also a product of evolution. To understand the evolution of the mind we first need to eliminate the many unknowns in the evolutionary progression of vertebrate learning. While the mind is the link between genes and culture, it is but a bag of its own missing links.

The history of science is scattered with concepts that promised new ways to find the missing links of the mind. Newton's explanation of the physical world quickly led to La Mettrie's reduction of the mind to simple mechanistic principles. The birth of chemistry and understanding of the changing attributes of compound matter led to the "mental chemistry" of John Stuart Mill. Darwin's theory of evolution legitimized the monistic view of unity between mind and body. The set of new ideas that now promises a new approach to the mind comes from a new understanding of information – from such concepts as negentropy, information content of organized matter, informational codes, information flow, and the hardware-software distinction in coding and processing information.

The most notable accomplishments that have emerged to date from the new understanding of information are the genetic code, the birth of new technologies in computation, communication, and now genetic engineering, and the development of the "cognitive sciences" (computational theory, cybernetics, linguistics, and the study of human and artificial intelligence). This "information revolution" rests on the recognition of simple yet pervasive cognitive capacities of matching and discriminating isomorphic structures that characterize the replicating and self-organizing processes of living matter. We find these capacities in the molecular processes of DNA and RNA synthesis, in protein synthesis, in immunological reactions, in the cellular interactions and organizations of complex structures during embryonic and postnatal morphogenesis, and in the epigenetic developments and episodic operations of the central nervous system. Studying the processes of matching and discrimination may thus be the most promising new tool for studying the missing links of the mind; I think L & W have correctly perceived this. I for one have welcomed this tool in the task of unraveling the hardware–software dimensions and gene–environment interactions implicated in the brain's handling of information. But L & W's promise of linking population genetics, developmental neurobiology, and cultural anthropology, and doing it with the help of statistical estimates of the matching and discrimination of heteromorphic "culturgens," seems to me but a pie in the sky.

Top-down guidance from a bottom-up theory

Geoffrey R. Loftus
Department of Psychology, University of Washington, Seattle, Wash. 98195

It is not uncommon in psychological circles to hear the claim that "psychology is a young science." This statement is typically made defensively, often in response to the criticism that psychology, and the social sciences in general, lack internal unity and common scientific goals.

Such a criticism is a telling one. The field of psychology is broken into a variety of subareas – developmental, cognitive, social, personality, and physiological, among others – and communication among workers in these subareas is minimal. Experiments are typically performed with the notion that the results of a given research endeavor will provide another brick in the mythical edifice whose construction is somehow supposed to constitute a common goal. The architectural plans for this edifice, however, are rarely glimpsed or even considered important. It is therefore a noteworthy event when a work appears in which an overall scheme for the social sciences is proposed, as appears to be the case in *Genes, Mind, and Culture* (henceforth *Genes*). This particular book may not, and probably will not, represent the ultimate unifying structure. What is important, though, is that it might, and it at least represents a serious attempt to do so.

Given the potential of *Genes* for providing some guidance to research in the social sciences it seems that two issues should be given primary consideration. First, presentation of the ideas should be designed so as to be maximally comprehensible to all workers whose intellectual sympathy the authors hope to capture. Second, the data forming the empirical foundations for the theoretical claims must be evaluated very critically. In the remainder of this review, I remark on both of these issues.

The top-down-bottom-up distinction. The ideas in *Genes* are difficult to assimilate, particularly if one has only a passing acquaintance with the major issues and tenets of sociobiology. To their credit, Lumsden & Wilson (L & W) anticipate this problem to some degree, and they have provided some clever pedagogical devices to smooth the waters. I applaud, for example, their efforts to at least partially separate their core ideas from the mathematical framework within which these ideas are formalized. I found, nonetheless, that after reading a third of the book I was in the uncomfortable position of trying to absorb a substantial amount of theory and data without really having formulated a clear idea of what it was all supposed to mean or how it was all supposed to fit together. What seemed to be most sorely lacking was a clear initial delineation of a reasonable alternative to L & W's own position. Indeed, it is not until page 176 that such an alternative is casually suggested, under the heading "Can Culture Have a Life of Its Own?" My reaction after reading this section was: "Aha! The critical issue we're dealing with here is a distinction between those systems that operate top-down on the one hand versus those that operate bottom-up on the other."

It is at this point that the reader (at least this one) comes to realize that a perfectly reasonable a priori view of a culture is that it operates top-down – that culture is "a virtually independent entity that grows, proliferates, and bends the members of society to its own imperatives" (p. 176). In this same passage, we simultaneously acquire an encapsulated version of the opposite, bottom-up, view – that espoused by L & W, who

> have perceived culture as the product of a myriad of personal cognitive acts that are channeled by the innate epigenetic rules. The "invisible hand" in this marketplace of culturgens has been made visible by characterizing the epigenetic rules at the level of the person and translating them upward to the social level through the procedures of statistical mechanics.

When one comes to understand this passage, the main point of the book becomes considerably clearer than it had been previously, since we now realize that the arguments being made and the data presented can potentially serve to disconfirm a particular, interesting, alternative view rather than merely provide support for the authors' position. If the top-down–bottom-up dichotomy had been explicitly spelled out in the beginning of the book, however, the reader would have had an easier time from the outset. L & W provide earlier hints of such a spelling out (as with a brief description on page 20 of the excellent *Micromotives and Macrobehavior* by their colleague Thomas Schelling), but these hints are not nearly sufficient.

Cautionary notes on psychological data. L & W review a massive number of psychological experiments whose results they then cite as support for their position. Unfortunately, however, another shortcoming of the young science of psychology is that the quality of its research methodology along with the robustness and replicability of its data is far from exemplary. Yet L & W describe experiment after experiment in a completely uncritical way as if the validity of each were entirely unquestioned and as if the pervasive squabbles and debates within psychology did not exist. In short, the empirical foundations of the L & W theory are not as solid as they might appear to be from a casual reading of the book. A review such as the present one is obviously not the place for an extensive critique of all the data described in *Genes*, but such a critique should certainly constitute a primary focus in a future, major evaluation of the theory that L & W have offered us.

The power of reduction and the limits of compressibility

Hubert Markl
Fakultät für Biologie, Universität Konstanz, D-775 Konstanz, W. Germany

The cause of reductionism is a noble one: It has proven to be the only means of acquiring reliable knowledge about the causal structure behind the bewildering richness of phenomena. *Reducere* need not mean to diminish or depreciate; in its original sense it means to lead back to the roots, to derive phenomena from their sources. Its rationale is that nothing occurs without cause and that to explain what exists means to define the conditions for its existence. In order to succeed, it is not enough simply to describe the conditions necessary for the existence of a class of phenomena, that, for example, known as "culture." One must also specify conditions sufficient to explain all the extant variants within the class. The final proof of reductionist success is therefore to arrive at a theory from which phenomena follow by necessity, in other words, a theory with predictive power. Although physics has had to give up on this deterministic claim at the microlevel, it has convincingly fulfilled it for the macroworld. Why then should not the biological and social sciences be able to produce comparable explanations for the phenomena of the living world?

Genes, Mind, and Culture is a breathtakingly bold attempt to provide such an explanation for man's biological-cultural existence. The result of this tour de force attempt to untie the Gordian knot by mathematically modelling its causal structure may not be the solution to this problem, but to date it seems to be the most powerful tool of thought for approaching a solution. Although Lumsden & Wilson (L & W) may not have the answer, they show us how to ask promising questions.

Why must their theory of gene–culture coevolution – like any biological theory – nevertheless ultimately fail to develop the explanatory power of reductionist physical theories? Because the biological and social sciences deal with ontogenetically and historically individualized entities: individual organisms, demes, cultures, species, or ecosystems. Individualized entities cannot be replaced in a theoretical model by arbitrarily interchangeable, identical entities without a damaging loss of essential characteristics, namely, those that cannot be neglected if we want to explain the entities' performance. Another way of stating this is that individualized entities are unique, and their unique properties are not a *quantité négligeable*.

For this reason, theories about living systems allow at best probabilistic predictions about classes of phenomena – never deterministic predictions about the behavior of individual organisms, species, or cultures (a situation analogous to the physical microworld!). To put it differently: Individualized entities are unique; the properties and behavior of unique entities cannot be reliably predicted; it is at best possible to "retrodict" the causes of their unique existence.

Why should this be so? L & W make it abundantly clear with their demonstration that genes alone cannot completely specify the properties of the brain, let alone the behavior controlled by it. Genes provide the ground rules for the organism's epigenetic self-construction in interaction and feedback with the environment. Since every individual of a sexually reproducing species has a unique genome, and since no two individuals will ever have exactly the same developmental environment, some of that variation being stochastic "noise," it follows that all living systems are so underdetermined as to never be even approximately "compressible" into general models or laws that try to explain their behavior.

Therefore gene–culture coevolutionary theory (like any biological theory) is a strong predictor of generalities but weak at specifics; it is strong at explaining possible causes of human behavioral diversity, but weak at deriving existing outcomes of this process, for example, the specific place of a given society on a particular ethnographic curve. It is limited by the indeterminacy (we may also say freedom) of epigenetically self-organizing systems. This freedom from complete predetermination enables living systems to evolve. It also has its costs, one of them being that living systems and their evolution resist fully reductionist explanation.

This by no means belittles the importance and power of the theory of gene–culture coevolution. In addition to being the most comprehensive attempt to model the complex interactions between the biological and the cultural evolution of man, which takes into account what we know about man's biology, culture, and mind, it is the only theory formulated in a way that allows it to be empirically tested and further developed according to the hypothetico-deductive scientific method. Even where L & W will have erred – as without doubt they will have – they will have done it in a way that is bound to lead to an advance in our knowledge about human nature.

Moving now from a general evaluation to specifics, I find three major problems with this approach (in addition to inevitable minor quibbles, such as, for instance, the fact that L & W don't even bother to discuss why nearly all cultures found it necessary to impose the severest possible sanctions against a behavior – incest – the avoidance of which, according to L & W, is one of the prime examples of a biological imperative that is safeguarded by a well-nigh unfailing epigenetic rule).

1. Despite its intimidating mathematical armor, the axiomatic core of gene–culture coevolution theory is rather soft; the definition of some basic concepts, on which the models operate, is exceedingly (though admittedly) fuzzy. Accordingly, some of the conclusions are so vague and so open to very different interpretations as to nearly immunize the theory against any possible objections, while other results, although reached by awe-inspiring theoretical effort, are anything but unexpected. Thus, the result that incest will rarely be found as regular behavior in human cultures comes as little surprise if we feed into the model that people don't like to practice it. More important: the model does not help us to understand why some cultures *do* make it a regular practice.

Or, to take another example, L & W consider the extremes of pure genetic or pure cultural transmission of information in a social species. Anything in between is gene–culture transmission subject to analysis by gene–culture coevolutionary theory. An effort is made to show that the tabula rasa state of pure cultural transmission (i.e. one developing free from guidance by epigenetic rules) is unstable. However, this is hardly news, since the way L & W apply the term "epigenetic rule" – including, for instance, defining our trichromatic color vision by an epigenetic rule – makes it analytically true for purely logical reasons that there can be no tabula rasa man or any tabula rasa organism, except perhaps angels, because to have body means to have a body with defining properties. Such a broad notion of tabula rasa could be rejected with little mathematics by simply stating that man can't fly because he lacks wings! Therefore it must be true that man is a product of gene–culture coevolution in this sense, since nobody in his right mind can maintain that he is fully genetically determined either.

All this becomes interesting only when we try to learn just how restrictive the ensemble of epigenetic rules is for the development of human culture. I wonder whether proving that man cannot be tabula rasa is not "breaking down open doors" after Lorenz, Piaget, and Chomsky (or actually even Immanuel Kant and Charles Darwin). That epigenetic rules range from very strict to very lax must be true (considering how broadly they are defined), but this begs the question (as L & W of course admit). One reader may conclude that gene–culture theory demonstrates "biological imperatives" (L & W's words) at work in human behavior, while another can claim that – except in some special cases like incest avoidance or mother–infant bonding – culture is held on such a long leash by human nature as to let it virtually develop according to its own rules. This breadth of interpretability may make it a grand theory, but grand theories are always in danger of becoming weak theories, with little explanatory power.

I found notions like "culturgen" or "epigenetic rule" conceptually useful, but applied with such protean versatility as to make them lose much of the discriminatory power a good term should have. Just as in Midas's hands everything turned to gold, it seems whatever a human being does, makes, thinks, values, obeys, or even discriminates can turn into a culturgen, if it can be perceived, memorized, and, preferably, communicated. In the beginning I felt that epigenetic rules could be what ethologists call "genetic constraints on" or "dispositions for" learning, but it seems that almost any process causally involving genes can be an epigenetic rule, from "make rhodopsin" to "thou shalt not kill."

I may be dull-witted or else these notions need restrictive sharpening if we don't want to lose them in the same all-explaining semantic morass where much of psychoanalytic, structuralist, or Marxist terminology has ended – which would be an eternal shame.

2. My next problem is that I miss in L & W's gene–culture theory a sufficiently explicit treatment of the role of cultural tradition in the coevolutionary process ("tradition" does not even turn up in the index). Tradition is of course central to any culture because it is the ensemble of "culturgens" that are kept together and inherited in a package from generation to generation by imitation and formal teaching, thereby giving a culture its specific character of adaption and adaptiveness. Although learned, it is quite different from learning from individual experience and from insight. Tradition is a culture's second store of selected wisdom (the first being the genome), and just as information from the first store can shape and constrain human behavior by epigenetic rules, so tradition shapes and constrains the behavior of the developing child by what we might call "tradigenetic" rules.

Clearly, tradigenetic rules cannot enlarge the range of possible behavior specified by the epigenetic rules; they can only maintain or restrict this range. This being so, tradigenetic rules can even replace postnatally active epigenetic rules as long as teaching and imitation occur sufficiently quickly and efficiently. Thus, they become a kind of buffer between the selective impact of the environment and the genome, and should therefore have important consequences for gene–culture coevolution. They could shield the genetic information store from direct selective pressures and consequently relax epigenetic constraints on behavior. In fact, provided that tradigenetic shaping is efficient enough, it should follow that the fittest genotypes are those that least restrain tradigenetic rule flexibility, since tradition can be more easily and swiftly changed according to the perception of traditions and to innovative insights.

Thus, a system with tradition cannot degenerate toward an unstable tabula rasa state by relaxing epigenetic control since tradition can instruct no less reliably and adaptively than epigenetic rules. Tradition makes the costs of freedom bearable and its benefits accessible. It also follows that epigenetic rules should be most prominently in control of behavior immediately after birth (because tradigenetic shaping needs time) and with respect to rarely occurring situations, for which imitation cannot prepare and teaching may lack the necessary repetitive reinforcing experience. This seems to be borne out by what we know of human behavior. The most important epigenetic rule could then be: Learn and trust the experience of your culture as you find it embodied in its tradition! This should be most important in coping with rapid changes of the cultural environment as they have occurred over the last 10,000 years. Since L & W show that 1,000 years may suffice to change epigenetic rules profoundly, the unleashing of our behavior from the genes under the guarding guidance of tradition could easily have taken place in this period, if our hunter and gatherer ancestors had happened to be under tighter genetic control. To model this, however, a group selective approach would seem inevitable, an extension already foreseen by L & W.

Teaching and imitation can also allow an epigenetically poorly prepared animal to jump over to the adaptive peak of successful innovators – provided that its behavior is not restrained from this by excessively strict epigenetic rules but rather prepared for it by flexible ones *combined* with high motivation to learn from successful example. Both these arguments seem to work in the direction of relaxing epigenetic and strengthening tradigenetic control of human behavior. If this is encompassed in the L & W model of gene–culture coevolution, I find its consequences insufficiently spelled out in the book.

3. Finally, I want to address one aspect touched upon briefly in the last chapter of *Genes, Mind, and Culture*. I find it difficult to see why deeper insight into the innate dictates of epigenetic rules could help us to find and agree on universal goals of our behavior, that is, how it could give us moral guidance.

Either we can continue to trust the magnificent coadapting feedback of gene–culture coevolution to enable us to face approaching environmental challenges in the future, or, if we can't trust this to do the job in time, we must guide ourselves to agreed-upon goals by the best strategies we can think of, independent of what our epigenetic rules may incline us to do. What is incorporated in the epigenetic rules is of necessity past wisdom. If this wisdom is not good enough for the future, how can we benefit from knowing this for mastering our unpredictable future? Of course, we gain knowledge about ourselves, which has a value in itself and satisfies our curiosity. It seems to me that gene–culture theory does not show any hitherto undiscovered way around the gap between "is" and "ought." Of course, human nature as determined by epigenetic rules may not allow us to change our behavior sufficiently in time to survive; but we will have to try anyway, to find out! One could even read another message from gene–culture theory: Let us change our ways as we find it necessary and possible, and let our biological nature follow suit.

Toward a natural science of human culture

Roger D. Masters
Department of Government, Dartmouth College, Hanover, N.H. 03755

Genes, Mind, and Evoution represents a fundamental contribution to the growing body of theory and empirical research linking the biological and social sciences. Despite an austere style and difficult material – certain to repel many readers – Lumsden & Wilson (L & W) have taken a major step toward analyzing human social systems from the perspective of evolutionary biology. Their terminology is so hard to grasp, however, that I have found it necessary to reconceptualize their main points in ordinary language.

The L & W theory of "gene–culture coevolution" attempts to link individual development, social processes, and evolutionary trends into a single framework. I see the essence of their approach in terms of five fundamental concepts: *individual development* (or "epigenesis"), *within-group variation*, *between-group variation*, *hierarchy* (or different levels of analysis), and *feedback*. Their theory is more accessible if each of these concepts is considered separately.

1. Individual development (epigenesis). In contrast to the persistent habit of distinguishing between genetic and environmental "causes," L & W stress *interactions* between the organism's genome and its setting throughout the life cycle. Since the innate "reaction ranges" of different traits are known to vary, sometimes permitting wide phenotypic plasticity and sometimes narrowly determining the range of viability, the concept of "epigenetic rules" makes it possible to analyze individual development (ontogenesis) without prejudging the role of genetic and cultural factors. Summarizing a mass of data on both "primary" epigenetic rules (establishing the amount of variation in perception and fundamental neurological processes) and "secondary" ones (relating to more complex cognitive tasks), L & W move from the programmatic assertions of either genetic or environmental determinism toward a more probabilistic, empirically grounded theory of human cognition.

2. Within-group variation. Individual thought and behavior can thus be treated as a phenotypic expression of innate potentialities, sometimes narrowly constrained and sometimes extraordinarily open. But this means that variation within any human group can be seen as the product of differential responses to the environment, reinforced to varying degrees by shifts in the "epigenetic" potentialities ("bias curves") of individuals. Often, however, a primary factor will be the responses of others in the group, in other words, the social context. In its simplest form, this leads to the L & W model of the probabilities that one of two mutually exclusive "culturgens" (whether practices or "ideas") will be adopted within a single group. While their approach is by no means limited to such a simplifying case, this model is itself a big step forward – if only because it permits the junction between similar models of social behavior (such as the Prisoner's Dilemma in rational choice theory) and evolutionary theories of greater complexity (Axelrod & Hamilton 1981; Hirshleifer 1980; Margolis 1981; Masters 1982, in press; Maynard Smith 1978).

3. Between-group variation. L & W's greatest contribution, however, is the way they approach between-group variation (differences from one entire society or culture to another) as distinguished from within-group differences (variation from one individual to another). Put in elementary terms, the L & W theory of "gene–culture coevolution" focuses on the percentage of human populations in which any given percentage of individuals shares a particular practice or trait. It therefore concerns the frequency distribution of frequency distributions – not the raw data. Most social science consists of formal models, Weberian ideal types, statistical or factual description, ad hoc theory, and prescriptive advice; as a result, it is usual to find either analysis of a particular trait within one or more societies (as if within-group variation were all that mattered), or typological classification of societies (as if between-group differences could best be approached by specifying kinds or categories of social systems).

L & W elaborate a more scientific approach to the differences from one society to another by focusing on the percentage of populations in which a given percentage of individuals exhibit a trait – that is, on what they call "ethnographic curves." Their terminology (e.g., emphasis on "epigenetic translation") may hide the simplicity and importance of this shift in conceptualization. Since the social sciences have – with very few exceptions (e.g., Whiting 1969) – failed to adopt such a probabilistic approach to between-group variation of social behavior, it is little wonder that existing theories seem fundamentally unscientific and "arbitrary" (Rosenberg 1980; Tiersky 1980).

4. Hierarchy. The relationship between individual developmental processes, within-group variation, and between-group variation can and must be "hierarchical" in nature. As with many other problems in both logic and empirical science, such a hierarchical approach often disentangles what otherwise seem intractable controversies (P. W. Anderson 1972; Pattee 1973). In particular, from this perspective it is neither necessary nor even possible to seek narrowly reductionist or deterministic "causes" for complex social phenomena. Even where the "epigenetic rules" produce a high probability of conformity (as with the avoidance of sibling incest), L & W show the frequency with which behavior contradicting the norm is to be expected. How could it be otherwise, since even strong selection in favor of genetic canalization can only produce a high probability of individual phenotypic expression – and since some residue of within-group variation is to be expected on most complex traits?

5. Feedback. In their analysis of the feedback between levels (which they call the "coevolutionary circuit"), L & W develop propositions such as the "1,000-year rule," specifying conditions in which a strong developmental "bias" can be genetically established in human populations over periods as short as 50 generations or 1,000 years. Much attention will be directed to such conclusions. Curiously enough, critics to date have been less disposed to stress the equally interesting – and perhaps more important – parameters that favor diversity in the "epigenetic rules" or predispositions. But more important than any one of these conclusions is the general principle they entail: A hierarchical model linking individual developmental processes (epigenesis), social variation, and evolutionary change must include feedback between different levels of analysis.

There are doubtless shortcomings in the specific mathematical formalizations presented by L & W and in their treatment of particular empirical phenomena. More than one reviewer has already stressed the sheer difficulty of the work (e.g. Lewin 1981; Medawar 1981). It is far from clear that feedback is as effective in producing genetic change among humans – at least for behaviorally relevant epigenetic rules – as is sometimes implied by L & W's discussion of the "coevolutionary circuit" (Masters 1975). But I would argue that such caveats, while necessary, are insufficient as an overall assessment of one of the most challenging and significant monographs in contemporary social science.

Mind and the linkage between genes and culture

J. Maynard Smith
School of Biological Sciences, University of Sussex, Falmer, Brighton BN1 9QG, England

I will confine myself to commenting on Lumsden & Wilson's (L & W's) mathematical models, and in particular on the model of the "coevolutionary circuit," which is the most general one, and hence the one on which L & W's book must be judged. The model is based on four basic assumptions:

i. individuals have biases, making them more likely to adopt one behavioural trait than another;

ii. individuals differ genetically in the strength and direction of these biases;

iii. the trait adopted by an individual is influenced by the behaviour of other members of the society;

iv. the fitness (expected number of offspring) of an individual is influenced by the behavioural trait adopted.

I don't think many behavioural scientists will doubt assumptions (i) or (iii). Assumption (iv) is at least sometimes true; consider, for example, the traits of taking heroin, practising birth control, or hang gliding. There will be argument about assumption (ii). It will no doubt be asserted that there is no single case in which it is known that a difference in a human gene causes a difference in behaviour. This is at the same time true (if one excludes cases of physical or mental defect) and irrelevant. All that is needed if models of this kind are to be relevant is that there should be some additive genetic variance for the behavioural trait in question. This, too, is hard to establish, basically because one cannot treat man like a laboratory animal. It would, however, be very odd if it were not true; experience suggests that there is additive genetic variance for almost every characteristic that has been studied (including behavioural ones), and I see no reason why our species should be an exception. However, I can see good reasons why the heritability of human behavioural traits may prove to be small.

Perhaps the main virtue of the mathematical models in L & W's book is that they make it easier to identify the assumptions that are being made. The authors claim that the model predicts "some remarkable phenomena." I do not think it makes any predictions that are not more or less self-evident. I will explain this by considering in turn the main conclusions as listed in the book. To simplify exposition, I shall refer to the favourable and unfavourable traits as "not smoking" and "smoking" respectively.

i. "Pure tabula rasa is an unlikely state" (p. 290). That is, a genotype with equal a priori probabilities of smoking or not smoking is not stable against invasion. This seems self-evident. If there exists an allele making its carriers less likely to smoke, and if smoking reduces fitness, then that allele will spread.

ii. "Sensitivity to usage patterns increases the rate of genetic assimilation" (p. 290). That is, an allele making smoking less likely will spread more rapidly if there is a tendency for individuals to copy their peers. This *is* a counterintuitive result. Thus consider first a pair of alleles, *A* and *a*, such that *AA* and *Aa* never smoke and *aa* always do; that is, no cultural affects and extreme genetic determinism. *A* will then replace *a*, and it will do so as rapidly as possible; any cultural effects can only slow things down. Why, then, do L & W conclude that culture speeds things up? I have spent several weeks finding out. It turns out that the effect depends on very special features of the model. In particular, it is assumed that there is a "cost" associated with copying one's peers. If this cost is omitted, culture slows down genetic change. To get an accelerating effect of culture, costs have to be incorporated in a particular way, and cultural effects have to be weak; even then, the accelerating effect is small. I have no doubt that, in practice, the common sense conclusion – that cultural effects slow down genetic change – is the correct one. This point will be dealt with in more detail in a review in *Evolution* (in preparation).

iii. "Culture slows the rate of genetic evolution" (p. 294). This seems to be a direct contradiction of (ii), and to be correct. However, it turns out to be based on the following conclusions from the simulations: a "tabula rasa," 50-50 genotype is replaced by a "nonsmoking" genotype more slowly than a "smoking" genotype would be. Obviously so, but what has this to do with culture?

iv. "Changes in gene frequency during the coevolutionary process can nevertheless be rapid" (p. 295). This is the basis of the "1,000-year rule" which L & W regard as their major finding. In what sense does it follow from the model? Clearly, if genes influence behaviour, if cultural effects are weak (as they are in the simulation), and if fitness differences are large (in the simulation, individuals adopting the favourable trait accumulate "resources" five times as fast as those adopting the unfavourable trait), then the genetic constitution of the population will change substantially in 1,000 years. And if not, not. The conclusion does not depend on the cultural components of the model: It follows from the assumptions of strong selection and high heritability.

v. "Gene-culture coevolution can promote genetic diversity" (p. 297). This arises because, in the model, fitnesses are "frequency-dependent" – that is, individuals with rarer traits are fitter. This can promote genetic diversity. This is correct, but the idea is not particularly new.

To sum up, the models are useful in clarifying the assumptions being made in sociobiological theories. In my view, they would have served this purpose better if they had been simpler; the models have sacrificed simplicity without a corresponding gain in realism. I do not think any startling conclusions follow from the models. The "1,000-year rule" reduces to the assertion that if there is high heritability of behaviour, and strong selection, then the genetic constitution of populations will change rapidly. It does not depend at all on the cultural components of the model.

Genes, mind, and emotion

Robert Plutchik

Department of Psychiatry, Albert Einstein College of Medicine of Yeshiva University, Bronx, N.Y. 10461

Genes, Mind, and Culture is a highly original attempt to provide a theory of the relations between genetic codes and social behavior. Whether or not it succeeds will probably not be known for one or two generations. This is partly because of the formidable mathematical models and partly because there is as yet little empirical data that bear directly on the tenets or derivatives of the theory. My response to the book focuses on two general issues: possible ambiguities or obscurities in the text, and relevance of the theory to the subject of emotions, my own special area of interest.

One important issue concerns the specification of the rules for analysis and synthesis of culturgens. Consider two examples. Lumsden & Wilson (L & W) suggest that the epigenetic rules of valuation and decision making lead to economic conservatism and cultural neophobia. They also suggest that bigotry and group aggression may stem from the interaction of the fear-of-strangers response, the proneness to associate with groups in the early stages of social play, and the intellectual tendency to dichotomize continua. What is not clear is what guidelines one can use to make judgments about what cultural patterns may be derived (synthesized) from a given set of epigenetic rules (assuming we know them). Conversely, given a cultural pattern, how does one go about determining (analyzing) the epigenetic rules that may have interacted to produce this pattern? If, as the authors say, "the structure of the mind . . . cannot be inferred by unaided intuition" (p. 353), how can it be inferred?

One is reminded of psychoanalytic theory, which often makes plausible interpretations of historical events (fixations, traumas, Oedipal conflicts, etc.) given a particular observed behavior such as pedophilia or fear of horses. However, as Freud and others have noted, it is much more difficult to predict the outcome of these same historical events. Knowing that a child has had a difficult birth, has suffered early deprivations, has had a major Oedipal conflict, and so on, allows only the weakest of predictions concerning the future lifestyle and psychopathology of this individual. Assuming that detailed knowledge will eventually be available about epigenetic rules, how precisely will that enable us to predict the existence of particular social institutions?

Another question concerns the meaning of the term "epigenetic rules." L & W describe the organization of color and sound by means of a circumplex. One may add that emotions and personality traits can also be conceptualized in terms of a circumplex (Plutchik 1980). Is the circumplex an epigenetic rule or is it a derivative of certain underlying epigenetic rules? If the latter is true, what form might these rules take? Similarly, is the "natural" tendency to avoid strangers an epigenetic rule or a derivative of epigenetic rules? Are the Moro reflex and the startle response epigenetic rules or culturgens?

The topic of emotion is an important one for psychology as well as for gene–culture coevolution theory. Unfortunately, relatively little formal attention is devoted to this topic, and the term emotion is not defined in the glossary. However, the definition of a culturgen sounds remarkably like the definition of emotion: "a homogeneous set of... behaviors... that... share... attribute states [of] functional importance" (p. 27). Later L & W say that "each epigenetic rule affecting behavior comprises one or more elements of a complex sequence of events that occur at sites distributed throughout the nervous system" (p. 36). This also sounds like a partial definition of the term "emotion." This is further supported by the statement that secondary epigenetic rules also pertain to emotional responses, as illustrated by an infant's fear of strangers or by its avoidance of bitter flavors. The question thus arises: Is an emotion a culturgen or an epigenetic rule?

Another use of the term "emotion" is found in L & W's discussion of memory. In one context they say that the permanent quality of long-term memory is "achieved through repetition and strong reinforcement, especially that which engenders emotion" (p. 62). This appears inconsistent with the well-known fact that strong emotion seems to disrupt memory and learning. In a different context they suggest that the mind grows like a polymer by forming semantic networks in long-term memory. Some of the nodes of the network are emotionally indifferent and form core stimulus associations of the culturgens; others are emotionally "hot" and provide the basis for the valuations people put on events. As an illustration, mother–infant bonding reflects the existence of a positive emotional node, while snakes and heights reflect negative emotional nodes.

This second use of the term "emotion" appears inconsistent with the findings of Osgood, May & Miron (1975) that every concept has an evaluative element. Does Osgood et al.'s finding represent a kind of diffusion process from certain nodes to all others or is there something innately emotional about all words?

One may also ask why lower animals show many of the same basic patterns of emotion that are seen in humans and why emotions can be described in the same functional terms across all phylogenetic levels (Plutchik 1980). We should remember that organisms without a limbic system also show emotional patterns. Is it possible that emotions reflect the existence of certain general organism–environment relations that are so widespread throughout the various phyla that a small group of epigenetic rules evolved early in the history of life, and that these rules have been so adaptive functionally, that they have been maintained in a conservative fashion – like DNA and amino acids?

It seems to me that emotions must take a central role in gene–culture coevolution theory in view of the point L & W make that their theory focuses on individual fitness as well as group fitness. Emotions are individual attempts to achieve homeostatic balances in interpersonal relations, and to some degree they influence and regulate the course of these relations. As I explain in my own writings (Plutchik 1980), emotions are intimately related to such group phenomena as dominance, territoriality, and identity, and emotions largely determine the choices one makes. The evidence suggests that cognitions are in the service of emotions.

These remarks deal with only a few of the ideas included or implicit in the rich structure of this fundamentally important book. To be most convincing, however, the theory should provide us with some surprises (another emotion), just as relativity theory predicted the displacement of light rays in the vicinity of large gravitational fields. This has not yet been clearly done.

Are there culturgens?

Alexander Rosenberg
Philosophy and Social Science, Syracuse University, Syracuse, N.Y. 13210

Seeing how the question of the existence of "culturgens" is to be answered is crucial to setting the standards against which a theory like Lumsden & Wilson's (L & W's) is to be measured. It shows that the standards of success they have set themselves are both unattainable and irrelevant to the real prospects of gene–culture (GC) theory, propects that seem to me to be far brighter than the prospects of alternative accounts of the distribution and transmission of culture.

Consider the parallel question, Are there phenotypes? This is not the question, Are there hereditary traits? – one presumably answered in the affirmative independently of any commitment to Mendel's theory and its successors. The former question asks whether these traits, or *others*, perhaps undescribed in Mendel's time, behave in accordance with *something* like Mendelian principles of segregation and assortment, and are caused by something like what his successors called genes. The answer to the question was not given by the natural history of Mendel's own time: The existence and identity of phenotypes cannot be read off the observational descriptions of flora and fauna, either longitudinally or cross-sectionally. They can be identified only by presuming the truth of the theory in which they figure. What is more, traits at one time identified as phenotypes did not remain so as genetics developed. Indeed the polypeptide products today so identified are as far from and as foreign to Mendel's candidates for phenotype as possible, and these molecular products certainly do not behave in accordance with *his* laws, nor are their genetic determinants much like Mendel thought they would be. Finally, the entrenchment of the notion of phenotype rested as much on the simultaneous development of another theory, equally foreign to conventional natural history (though it consumed the data but not the assumptions and theories of 19th-century natural history and geology): the theory of natural selection. The question of whether there are phenotypes is the question of whether Mendel's theory was on the right track. And the affirmative answer is a product of experimental successes inspired by Mendel's theory, and by its subsequent coupling with another independently elaborated theory, equally foreign to the received science of its day.

Mutatis mutandis, Are there culturgens? is *not* the question of whether some cultural traits, hitherto identified and canonically described by contemporary social science, can be regimented into the sorts of relations reflected in assimilation functions, amplification rules, usage bias curves, or, arguably, ethnographic curves. If it turned out that the curves are all flat and the rules too weak to effect or explain transmission and distribution of traits identified by previous work in social science, this would be no more reason to surrender GC theory than the fact that the most ordinarily noticeable traits of flora and fauna do not assort as phenotypes is reason to reject Mendelian genetics. If, however, the curves are flat and the rules too weak for bona fide culturgens, then the theory would be in serious difficulty. But how do we identify bona fide culturgens? Only by employing GC theory to motivate experiments and field studies, by consuming the anectodal data of conventional social science (but not its theories and explanations), and perhaps by making use of developments in other, nonsocial sciences, such as neurophysiology. We certainly cannot answer the question by determining whether traits identified and individuated in accordance with the dictates of conventional social science have distinctive assimilation functions, strong amplification rules, epigenetically determined bias usage curves, or ethnographic curves that change at the rate L & W require. At most these disciplines will provide anecdotal information, whose best case usage will be no better than L & W's use of anthropological reports of incest avoidance, and their provisional identification of incest choice as reflecting a culturgen.

L & W are aware of the fluid nature of any identification of culturgens, as revealed in their own shifts in its characterization throughout *Genes, Mind, and Culture* (henceforth *Genes*). Initially, and in the glossary, culturgens are described as the basic units of culture. This description is purely formal, and will only prove useful if something like their theory turns out to be well confirmed and to have considerable explanatory power. More specifically, they describe culturgens as sets of artifacts, behaviors, or "mentifacts," that is, as sets of *particular* objects, thoughts, or acts. But it shortly becomes clear that culturgen is a far more theoretical notion, whose instances are not observable occurrents, but the dispositions, discriminative abilities, and preferences that give rise to sets of occurrent artifacts, mentifacts, and behaviors. Subsequently these dispositions are given a neurological reading. Culturgens come finally to be described as structures of linked nodes in long-term memory, constructed in the neural architecture by the interaction of epigenetic rules and environmental contingencies. Given their ultimate neurological characterization, coupled with the relatively underdeveloped state of contemporary neurological theory, it is clear that culturgens can at present only be identified indirectly, by appeal to their functions in shaping behavior. But the behavior to which such identification must appeal cannot be intentionally described. For we know already that intentional descriptions cannot be systematically linked to physical ones; that intentional descriptions are not reducible, or, as L & W would have it, "compressible," into physical and, in particular, neurophysiological ones (see Searle 1980 for an example of arguments to this effect). Indeed this irreducibility of intentional descriptions of human behavior (in terms like belief, desire, action, and their cognates) explains why the inevitably intentional disciplines, the conventional social sciences, have been unable to generate theories that reflect the increasing predictive and explanatory findings characteristic of natural science. Intentional states are nomologically anomalous. Because they are nomologically anomalous, and because culturgens are nomologically homogeneous (as L & W's system of lawful regularities among culturgens of course requires), it follows that conventional social sciences cannot uncover or properly describe culturgens, just because they are irretrievably intentional.

These considerations about the nature and methodological status of culturgens make unacceptably stringent, indeed superfluous because logically unattainable, the criteria of success for GC theory that L & W lay down for themselves. These criteria, as expounded in *Genes* (p. 343) and in the authors' précis that accompanies these commentaries, are threefold: First, the fundamental, underived laws of conventional social sciences should be deducible as theorems in an axiomatic presentation of GC theory. Second, the explanatory power and predictive content of GC theory should at least match that of the current social sciences; at least in L & W's words, "it [should] subsume the exact phenomenological models of disciplines such as economics and anthropology so as to make the underlying assumptions identical." Third, GC theory must suggest new problems, and identify new parameters and laws in a network of verifiable explanations from genes through minds to culture. The first of these requirements cannot be satisfied, the second is either trivially satisfiable or impossible to satisfy, and the third, which is the only criterion of success GC theory need realize, cannot be realized compatibly with the other two.

Since there is nothing worthy of the name "underived law" in conventional social science, any theory that implied the false or vacuous general statements of the nomothetic social sciences could only be saddled with their defects. As for the ideographic social sciences (the disciplines of "thick" descriptions; *Genes* p. 351), they eschew the search for laws and so cannot be linked with GC theory in the way L & W envision. In any case, since both sorts of disciplines are intentional, neither could be linked to GC theory in the way the first criterion requires. Thus, for example, L & W envision the explanation of the optimizing models of economics and economic anthropology by GC theory via its ability to account for individual motivation (i.e. intention) by bridging psychology and biology (p. 352). But this is logically impossible, on a neurological or any physicalistic reading of culturgens. Accordingly it cannot be required of GC theory that it entail as theorems the axioms of the social sciences as presently constituted. If the explanatory and predictive power of conventional social science is considered negligible, then any theory, including GC theory, can satisfy L & W's second stricture on explanatory and predictive power. On the other hand, if their theory satisfies the first requirement and is therefore saddled with the burden of the social sciences, then GC theory will be incapable of transcending the limits on scientific significance to which conventional social science long ago doomed itself. The latter interpretation of the purport of the second requirement seems most textually plausible, for L & W demand that GC theory attain increased power, "at least" by "subsum[ing] the exact phenomenological models of [conventional social sciences], so as to make the underlying assumptions identical." GC theory cannot do this if it hopes to transcend their explanatory and predictive power, as L & W demand.

GC theory stands a good chance of meeting the third of L & W's demands on it, but only on the condition that they forgo the first and second requirements for success. GC theory may well suggest new problems. Indeed, it must, if, like Mendelian genetics, it is on the right track. It may well generate new hypotheses about the parameters and the laws governing culturgens and their consequences. But it cannot do this unless it remains rigorously biological (i.e. nonintentional) in its conceptual machinery. And remaining rigorously biological will prevent GC theory from answering most of the questions about actual human behavior and its meaning that are conventional social sciences' stock in trade. GC theory is an account of the forces governing the intergenerational transmission and intragenerational distribution of culturgens. At most such a theory can tell us about the dispositional determinants of culture, but at a level of description too far removed from conventional descriptions of culture to answer the social scientists' sorts of question. Thus, the new problems, the new laws, the new parameters will not be ones current social scientists recognize, for GC theory is a sharp departure from their modes of thought, one that holds out far greater promise, because it is backed by theories that we know independently to govern human phenomena. GC theory will not be able to generate new results if it must drag along the dead weight of contemporary social scientists' expectations and theoretical commitments. For then it will be forced up the same blind alley in which they have found themselves, thus failing to meet this third requirement. And the surest means of ensuring this sorry conclusion is to demand that GC theory encompass these disciplines by satisfying the first and second of L & W's requirements.

Frankly, I suspect that L & W realize that these two requirements are really not to be taken seriously, but that their presence in the last chapter of *Genes* represents its authors' attempt to conciliate – if not co-opt – potentially hostile social scientists; I suspect that they believe that there is little that current theory in social science can contribute to the assessment of their research program, but that they are eager to avoid unnecessary and fruitless controversy of the sort that has too long bedeviled sociobiology. Their praise for visions so inimical to their own naturalistic approach as hermeneutics, structuralism, or psychoanalysis can only be explained in these terms.

Epigenesis: The newer synthesis?

Glendon Schubert
Department of Political Science, University of Hawaii at Manoa, Honolulu Hawaii 96822

It is disheartening to find so much that is wrong in the statement of a thesis that is so fundamentally right. Lumsden & Wilson (L & W) certainly merit praise for having argued so forcefully the proposition that the relationship between human genes and human culture is both dynamic and recursive; and more particularly, for their emphasis upon the brain as the bridge between genes and culture, for our species. My concern with this aspect of L & W's model and presentation is not that they have gone too far, but rather that they are nowhere near radical enough. Their characteristically linear mode of thinking is exemplified by their suggested "three-step improvement" upon simplistic gene–culture interactionalism: "from genes to epigenesis, from epigenesis to individual behavior, and from individual behavior to culture" (p. 343). A far better and also much more realistic approach would posit transactional relationships among physiological systems (including the cellular level at which gene-cytoplasmic effects are manifested in animo acid production), neurological systems (including the brain), and the components of human culture (as both internalized by the brain and externalized by the brain in its effects upon human physiology).

Throughout the book, however, the authors' reach exceeds their grasp, as in the first sentence (p. ix) where they preempt as their own "the first attempt to trace development all the way from genes through the mind to culture"; again, when they claim (p. 230) that their Chapter 5 presents the "first concrete[1] model of the coupling between genetic and cultural evolution"; and that (p. 256) "gene-culture theory... goes far beyond conventional notions." But such huffing claims constitute a less than generous admission of these authors' indebtedness to the prior[2] insights, experimental research, and mathematical formulations of the late Conrad Hal Waddington who, not 50 generations but a full generation ago, formulated the key concepts of epigenesis, developmental canalization, and genetic assimilation, upon which L & W's thesis depends.

The major arguments in the early chapters of the book exemplify culture as archeological artifacts, whereas the later chapters, in which more sophisticated versions of human culture are sometimes mentioned, deal with culture in contemporary terms on a merely verbal level – and the latter is an activity that social scientists and humanists probably can do just as well as physicists and entomologists. Frequent mention, for example, is made of the relatively specific epigenetic "rule" against incest, which is said to be illustrated by the culturally established brother and sister relationships experienced by children brought up in kibbutzim, who manifest a striking aversion for heterosexual intercourse or marriage among each other. This is said to be the consequence of "an automatic sexual inhibition between persons who lived intimately together... to the age of six" (p. 86). There certainly is cultural (i.e., literary) evidence to the contrary, involving dizygotic heterosexual twins (Mann 1938); but given Wilson's long-standing interest and expertise in haploid breeding systems, which were also the subject of Hamilton's (1964a; 1964b) primogenitive articles on inclusive fitness, the preoccupation of this book with brother-sister incest (e.g., pp. 147–58) seems odd. The evidence from human culture, contemporary as well as historical, suggests a much higher probability of parent-offspring incest (an especially, of course, father–daughter), where the coefficient of relatedness is just as high as between even full siblings: This implies just as high a risk of negative genetic consequences. One would therefore expect that it should provide an example, on both scores, of greater interest and utility to these authors than the alternative that they proffer.

There are some patent asymmetries, notwithstanding L & W's emphasis upon rules, in their application of them to genetic, as distinguished from cultural, data. Binary classification, for example, is quite rational – one is tempted to say "eufunctional" – for the authors themselves to use when they "wish to emphasize the power and flexibility of two- 'culturgen' models, whose utility parallels the two-allele case in theoretical population genetics" (p. 274; and cf. the cell-color dichotomies discussed at p. 48); but when social scientists do something equivalent, then "there is a nonrational proneness to use two-part classifications in treating socially important arrays, such as in-group versus out-group... and so forth" (p. 95).

Clarence Day (1920) provided some intriguing speculations about what humans would be like behaviorally if they had evolved as super-felids instead of as super-primates (and cf. Marais 1969); and in their scenario for what genetic determinism really would imply for humans, L & W (p. 331) describe a society *with* which (at least) a specialist in the behavior of social insects ought to feel right at home (and cf. Marais 1937). Such a robot civilization is portrayed as archetypical in contrast to the tabula rasa notion of human consciousness, which L & W attribute to most social scientists; and this leaves their own "gene–culture" interaction theory to occupy all of the middle ground between the "pure genetic" and the "pure cultural" approaches (p. 99). In practice, however, virtually all of their mathematical modeling (as distinguished from graphic displays of more qualitatively based ideas) clusters around the genetic pole of their continuum, where they examine binary choice between cultural units, each of which is assumed to be monotonically controlled by a single and corresponding monogene. And the further they get from perception, the more frequently (and vociferously) they justify their inability to say anything explicit (about genetic effects upon learned behavior) as resulting from the underdevelopment of population or behavior genetics in relation to other fields such as developmental neurobiology or developmental psychology. But the latter complaints also cut the ground from under their sometimes guarded (e.g., pp. 300–301) but usually hyperbolic (pp. 295–96) claims in behalf of their so-called 1,000-year rule: that human culture can have positive feedback upon its own genetic and epigenetic bases within as few as 50 human generations. The models from which this deduction is made are built in terms of the (for humans) tremendously oversimplified assumptions of dichotomous choice, and large and randomly mating populations "in order to exploit the deterministic equations of population genetics" (p. 266). Nothing in the models takes into account the empirical parameters mentioned elsewhere (pp. 197–200) of 100,000 human genes, mostly acting epistatically with both polygenic and pleiotropic effects; and of which only 1,200 have been identified and barely 210 have been mapped. Given those parameters and their absence, in any case, from the gene–culture coevolutionary model, it is a difficult feat even in imagination to contemplate how the "1,000-year rule" can be tested with data on humans, during any future in which any of the latter are likely to be reading the L & W book.

The book includes its fair share of vacuous remarks, beginning quite early on with the invocation (p. 6) of "sheer drive" as the explanation why chimpanzees, who lack it, cannot conceptualize language like humans. The authors subsequently speak (p. 330) of the genus *Homo* having overcome "the resistance to advanced cognitive evolution by the cosmic good fortune of being in the right place at the right time." But this kind of poetic remark could be applied equally to dinosaurs, army ants, sharks, or indeed to any other (temporarily) well adapted species; certainly the remark explains nothing, or at least nothing new. Soon thereafter readers are reassured (p. 345) that "the inseverable linkage between genes and culture does not also chain mankind to an animal level": of course, the authors mean to imply "of mind"; but their statement remains a fatuous one for a biologist to make. Similarly, at the very beginning of the book (p. 2) and throughout it passim, the authors speak of "the epigenetic rules feeding on" this and that; in this initial instance (and also at pp. 272 and 349) it is on "information derived from culture and physical environment." One suspects that in so doing, the authors have not taken the time to think through the implications of what some sensitive readers might consider to be a disgusting alimentary metaphor of information processing, the logical output of which is, of course, a great deal of excrement in one form or another.

An unhappy feature of this book is found in the authors' penchant for gobbledegook: pretentious and dysfunctional neologisms. We can briefly note "heterarchy" (p. 108); and "social contagion" (p. 113) with all of its pathological undertones, notwithstanding the ready availability of such accepted ethological concepts as "mood convection" which offers the additional advantage of already having been introduced into the research literature of political and social science (e.g., Masters 1976, pp. 225–26). L & W's designation of perception as "primary epigenetic rules," and of learning as "secondary epigenetic rules" (p. 36), is a more glaring example of taking feckless (if not reckless) liberties with the English language. Perception is a well-established and active field in physiological psychology; and so are the parallel fields of learning (in psychology), socialization (in sociology and political science), and acculturation (in social and cultural anthropology). Calling perception "primary epigenetic rules," and learning "secondary epigenetic rules," certainly makes no contribution to present empirical knowledge. These proposals are likewise of most dubious value as theoretical contributions, particularly in the use of the concept "rules" in relation to the effect of rules upon human behavior; and this is a matter about which political scientists, like lawyers, can be presumed to have some specialized professional knowledge (including rules applied to the behavior of research scientists; Schubert 1967a; 1967b; 1975). L & W's emphasis passim upon epigenetic *rules* (in lieu of the Waddingtonian concept of epigenetic *processes*) would, if taken seriously, be likely to encourage a giant step backward toward premature closure, especially if the actual network relationships in the brain turn out to be more open, more variable, and more rapidly changing than the circumstances defined (or associated) with "rule-ordered" modes of canalizing behavior (Davidson & Davidson 1980; Geist 1978; Pribram 1979; 1980; Roederer 1978).

L & W's misuse of "reification" constitutes another example of semantic obfuscation: For reasons that they nowhere divulge, they decided to call conceptualization reification. Social scientists, including psychologists, use "reification" pretty much as it is defined in *Webster's New International Dictionary* (2d ed. unabridged, p. 2100, col. 2), to mean hypostatization, or the regarding of an abstraction or mental construction *as though it were* a material thing, discrete and objective. These authors choose – in a patently arbitrary manner – to describe as reification "the operations of the human mind [that] incorporate (1) the production of concepts and (2) the continuously shifting reclassification of the world" (p. 5), a usage in which they persist throughout the book, with unfortunate consequences for their communication with literate readers. They refer to a metaphor in which "the archipelago-society can be treated as one insular ensemble;... as though it were a single space open for occupation by competing culturgens" (p. 306): now *that* is reification in the usual sense (although L & W do not, of course, call it that); indeed, the authors instead reveal that "language is the means whereby the culturgens are labeled" (p. 253). Another good example of reification in the conventional sense is found in their explanation of the especially powerful role that mentifacts play in culturgen packing: Mentifacts "are the nearly pure creations of the mind, the reveries, fictions, and myths that have little connection with reality but take on a vigorous life of their own and can be transmitted from one generation to the next" (p. 316). There, as Humpty Dumpty said, is glory for you.

L & W's understanding of Marxism (pp. 354–56) appears to be informed exclusively in terms of Western culture, but there are lessons in this regard to be learned from the East as well. Like the late (and until recently great) Mao Tse-tung, L & W are enthralled with the idea of making a Great Leap Forward, in the authors' case in cultural understanding of the biological basis for human social behavior; and this leads them to deplore the "balkanization" of the social sciences, which labor under a "crippling state of affairs" (p. 345). Like Mao, they would have been better advised to have taken a different page from the Little Red Book, seeking instead to "Let a Thousand Flowers Bloom."

NOTES

1. Describing the human brain and nervous system as "concrete" is a less than felicitous way to suggest its most important and (in the context of the quotation) transcendent function in human development and evolution.

2. Waddington himself would not necessarily have taken an Olympian view of Wilson's cultural assimilation; (see Waddington 1975, p. viii and Chapter 10).

Collaboration between biology and the social sciences: A milestone

Joseph Shepher
Department of Sociology and Anthropology, Haifa University, Haifa 31999, Israel

I do not know whether the international scientific establishment will rightly appreciate this work, but I am convinced that a student of the development of human knowledge hundreds of years hence will see in it a milestone on the way to the fertile collaboration of biology and the social sciences that paved the way to a true understanding of human nature. Moreover he will probably point out that the publication of Lumsden & Wilson's (L & W's) book redeemed to social sciences from the most sterile period of their history.

Those who have tried to apply sociobiological theory to the basic questions of one of the social sciences have frequently asked: How is an obvious biological predisposition translated into an empirically verifiable biased cultural frequency distribution? How does it happen that in an overwhelmingly large majority of cultures certain matings become defined as incestuous and are avoided or forbidden? The ultimate cause, prevention of inbreeding, was obvious, but it was not clear how people who did not understand anything about genetics acted as if they did. Although it was clear that the asymmetry in human parental investment should result in polygynous tendencies in the human male, no preference for polyandry was manifest in the female. It was difficult to explain exactly how the biased frequency distribution of human cultures, resulting in more than 70 percent polygyny and only 1 percent polandry came into being (Daly & Wilson 1978; Murdock 1967).

Then, too, biological predispositions have evolved during millions of years through Darwinian individual selection, but concomitantly cultures have evolved through Lamarckian group selection. The almost generally accepted reification of the group in the social sciences obscured the process of crystallization of individually and biologically evolved predispositions into cultural traits accepted and adhered to by groups and transmitted through learning, teaching, and even indoctrination. Moreover, the Durkheimian verdict, "every social fact has to be explained by another social fact," created a formidable taboo that not only prevented the understanding of many a human universal, but raised an inpenetrable barrier between the social and the life sciences. This formidable barrier has been removed by L & W's *Genes, Mind, and Culture*.

Nevertheless the authors could have improved the book by collaborating with an experienced social scientist and a cognitive psychologist. Whereas it is true that "no word burdened with a history, as Nietzsche said, can be defined perfectly," most social scientists will have a difficult time with the definition of a culturgen proposed on page 27. Since without an effective operationalization of this central concept, the three "services" that gene–culture theory is meant to perform are hardly feasible, much work has to be done before we can reach the degree of testability and predictiveness that the authors call for in their summary. Thus L & W's example on page 28 might be compared with length of skirts, but to compare these two examples with patterns of political organization or religious belief systems would be somewhat tenuous.

The selective incorporation into the long-term memory of epigenetic rules, according to their dependence on environmental circumstances (p. 70), explains the extent to which the resulting flexibility and conscious deliberation are used in dealing with the exigencies of everyday life. Thus we are flexible in applying different forms of economic behavior and our deliberations are conscious, but we are very rigid in our use of deep grammar, incest avoidance, and the consumption of certain high caloric foods, and unaware of the relation of those patterns to our genetic fitness. How is our brain constructed that it allows for such selectivity?

The use of the concept "biased ethnographic curves" will open a new era in comparative cross-cultural anthropology. Instead of the Durkheimian cat that wants to catch its own tail, the epigenetic rules will form a solid axis for this important research activity. No more quest for sterile associations between two or more sets of cultural traits (Textor 1967, pp. 189–298), but a testing and refining of the definition of epigenetic rules – or revealing heretofore hidden ones. Preliminary work in that spirit (Chagnon & Irons 1979) will be refined and deepened, and a promising collaboration between anthropologists, sociologists, and psychologists will result.

A bully pulpit

L. B. Slobodkin
Department of Ecology and Evolution, SUNY-Stony Brook, N.Y. 11794

The tone of this book is reminiscent of G. B. Shaw's prefaces to his plays, in which he briskly cuts through all complexities and proclaims both his own clever solutions and their originality. An Edwardian criticism of Shaw applies to Lumsden & Wilson (L & W) as well:

> He is one of those gifted observers who can always see through a brick wall. But the very fact that a man can see through a brick wall means that he can not see the brick wall. . . . Flesh and blood are quite invisible to Mr. Shaw. He thinks because he can not see them they do not exist, and that he is to be accepted as a realist. I need hardly point out to my readers that he is mistaken. (Behrman 1960, p. 167)

Of course, these were Edwardian readers. Since then we have had more practice in believing in the vision of seers of this type. Only recall the "orgones" of Reich and the "engrams" of L. Ron Hubbard. These are now joined by the "culturgens" of L & W.

I used quotation marks for two reasons. First, the above-listed authors made up these words. Second, a string of letters can move as a mature word in a scientific context – free of the training wheels of quotation marks – only if it is clear that there is a referent for that word in the real world.

L & W define "culturgen" as the "operational unit of culture," as if culture were composed of operational units. Maynard Smith (quoted in Lewin 1981) doubts the usefulness and interest of a theory based on a particulate view of culture. L & W discount this objection by stating (in the accompanying précis) that: "culturgens are perceivable features of the integrated cultural system; their existence does not mean that culture is composed of discrete, isolated units." That is, L & W are apparently asserting that culture is not definable as an assemblage of countable units. Why would Maynard Smith and Medawar, both distinguished biologists, make such a simple error? Perhaps because L & W find (pp. 314–15) that "cultural diversity depends on the number of culturgens that can be incorporated into the mind." They also provide a sample of culturgen counts: "Indian tribes of California produced 3,000 to 6,000 artifact types . . . the armed forces of the United States landed 500,000 artifact types at Casablanca in World War II. . . . Arabs [recognize] 6,000 attributes of camels." (p. 317) How many attributes of each of the 500,000 artifact types did the GIs recognize at Casablanca?

I conclude that "culturgens" are at least sometimes countable, in their manifestation as artifacts, or as attributes, or as mental constructs. It is still not clear to me how to make such a count for myself. I am not helped by the assertion (in the précis): "The categories [used as culturgens?] can be made as coarse or as fine as required to reach the desired level of precision." Desired to what end?

The "theory of culturgens" rests on "simple" mathematical models of culturgen innovation, in which "culturgens" and the "number of qualities of culturgens" are enumerated. Is a "quality of a culturgen" then not a "culturgen"?

The term "simple" also has a very special meaning of being suitable for manipulation by the mathematical tools that come to hand. It does not mean that a simple culture is being discussed. Certainly a society in which "the distinctness of the culturgens from one another is a random variable and . . . [they] are created at a constant rate" (p. 14) is not a simple one from an anthropological or political viewpoint.

Since the mathematical theory is built on modifications of this kind of assumption, the primary value it might have had would have been to generate insights that were themselves of great merit. If this had occurred all else would have been forgiven.

Among the claimed insights are that L & W have made two findings of general significance. The first, stated as a theorem and its corollary, seems to be that if two human populations differ genetically they may (theorem) or may not (corollary) differ culturally (see the précis). L & W have also discovered that there exists a coevolutionary circuit between genetics and culture, and that they are being original in that they "deal for the first time with the phenomena of mind and human social history." L & W do not acknowledge Plotkin and Odling Smee (1981) in this context, and they do not use the results of Cavalli-Sforza and Feldman (1981).

Perhaps the most mysterious thing about the book is the amount of attention it has been given. Without deliberately searching, two strongly negative reviews have come to my attention, one by a Nobel laureate developmental biologist (Medawar 1981) and one by one of the most brilliant living population geneticists (Lewontin 1981). A third review refers to the "sophistic presentation" and "serious flaws" (Cloninger & Yokoyama 1981). There was also a news article in *Science* (Lewin 1981) that began "There is nothing modest about *Genes, Mind, and Culture.*" These should be read for critical comments that space limitations forbid discussing here.

Why this concern with a fuzzy work which is basically a sermon with loose mathematical trappings? The *Science* news report states: "[It] is certain to attract a good deal of attention because the Harvard University Press is promoting it unusually vigorously" (Lewin 1981).

Apparently the pressure of the Harvard Press creates a bully pulpit.

Resistance to biological self-understanding

Pierre L. van den Berghe
Department of Sociology, University of Washington, Seattle, Wash. 98195

The ultimate black box in understanding human behavior and culture is, of course, that prodigious product of the last couple of million years of evolution known as the human brain. That organ has achieved a clear consciousness of its own existence, but only the faintest glimmerings of *what* makes it tick (and that, only in the last century or so). We can arrive at reasonably satisfactory accounts, descriptions, analyses, and even explanations and predictions about the *outcomes* of mental processes: behaviors, artifacts, historical events, cultures. Yet, we are still frustratingly ignorant of just *how* these products get processed through our little 1.5-liter black box.

The computers crudely mimic some of the simplest functions of the mind, and give us the illusion that a mind that creates "artificial intelligence" also understands the real thing. But, of course, a computer is to the mind as the phonograph record is to human language: an ingenious but limited piece of mimicry. Fortunately, we can reproduce a human brain with great ease and enjoyment through copulation, so we may well ask why we futilely pursue far more cumbersome and less satisfactory methods. Perhaps the answer is in the frustrating elusiveness of the Socratic precept: Know thyself.

Such also was, I think, the motivation for the Lumsden & Wilson (L & W) book. Its great merit is that it tries to bore peepholes into our cranial black box, and squarely asks the right questions about what happens between genes and behavior (including culture). Thus, it fills a great gap in much of the animal behavior literature, whether of behaviorist, ethological, or sociobiological orientation, and it points to ways of incorporating the vast research tradition of brain physiology into the "new synthesis" of sociobiology.

At the same time, *Genes, Mind, and Culture* (henceforth *Genes*) makes us painfully aware of how far we are from a real understanding of the complex linkages and feedback loops between these three levels of human social reality. With all their zeal and talent, L & W can only dredge up scraps of highly suggestive, but far from conclusive, evidence for the epigenetic rules of culturgens: incest avoidance, color terminology, dietary preferences of infants, and a few others. (Freedman's [1979] work on neonates also gives some interesting data on possible intergroup variability in epigenetic rules.) Nor do L & W's attempts to model these relationships seem to take us much beyond their verbal formulations at this point. (Admittedly, my foggy understanding of their mathematics may be at fault here, rather than any intrinsic limitation thereof.)

Despite its limitations, *Genes* is a pioneering effort of great importance, and by far the most suggestive of the half-dozen or so recent formulations of gene–culture coevolution. The limitations are attributable far more to the present state of knowledge than to the authors. Notwithstanding the paucity of hard evidence, I find their statement of the problem convincing because what evidence exists points consistently in the same direction and leaves little room for alternative explanations. Of the three main theoretical alternatives to account for human behavior and culture posited by L & W (pure genetic transmission, pure cultural transmission, and gene–culture coevolution), it seems to me that we can safely rule out the first two.

Our cognitive and emotional processes are primed to learn and do certain things, and to avoid other things. As culture, in the last analysis, is the modal expression of the cumulative behavior of individuals, it cannot be divorced from either its genetic substratum or the environment in which behavioral phenotypes are expressed. Add to our extraordinary capacity to adapt through learning the further ability to transmit learning vicariously through language, and the extraordinary achievements of human culture are fully accounted for in materialist, evolutionary terms. Finally, as culturgens have fitness consequences for individuals, and for the rise and fall of the societies people form, culturgens therefore feed back on genetic evolution. The loop is closed: Genes and culture coevolve.

In a nutshell, this is what L & W say (though they say much of interest beyond this capsule summary). That such a cautious, indeed, almost trite formulation should continue to be found controversial by so many social scientists and even biologists defies rational explanation. Perhaps our mind is programmed *not* to understand itself after all, for if we knew ourselves we would find ourselves insufferable. We are back to another Socratic question: whether we would rather be satisfied pigs or dissatisfied humans. Insofar as it is the mark of the humanist to opt for the latter alternative, we should all be grateful to L & W for asking inportant questions, and for making so many people mad at them.

Information, feedback, and transparency

Robert Van Gulick
Department of Philosophy, Rutgers University, New Brunswick, N.J. 08903

The coevolutionary approach advocated by Professors Lumsden and Wilson (L & W) offers a commendably interactive view of the relation between genetics and culture. Their model is genuinely reciprocal, and the arguments they offer on its behalf make a surprisingly plausible case for the occurrence of rapid culturally induced changes in the human gene pool. However, their inclination to describe and conceptualize the relevant interactions within a nonintentional framework of the sort appropriate for modeling dynamic processes in the physical and nonmental sciences unduly restricts the range of feedback and informational relations that they discuss in their account of coevolution.

Consider, for example, how they characterize the tabula rasa state and distinguish it from other cases of cultural transmission. It is defined as one of "pure cultural transmission in which no innate mechanisms pattern the acquisition of any culturgen in competition with others." At the opposite extreme of their division are cases of pure genetic transmission in which all members of a society are genetically constrained to acquire the same culturgen. Given such extremes, it is not surprising that L & W believe most actual cases involve an intermediate sort of gene–culture transmission. We should not restrict our attention to mere variablility in the output of the transmission process, however, for we are concerned here with *learning* processes. Thus what most requires explanation is the existence of processes that are highly variable but adaptively attuned or fitted to the particular learning environment. It is not variability per se that is striking, but variability mirroring fitness that is improbable from the viewpoint of physical science and thus in need of special explanation.

Commentary/Lumsden & Wilson: Genes, mind, and culture

The term "tabula rasa" suggests an absence or decrease of genetically transmitted information. But if we recognize that the variable outputs of a plastic learning process are generally adapted to individual conditions, we must acknowledge that providing the organism with an "open program" requires building more genetically transmitted information into it rather than less (Lorenz 1965, pp. 44ff, 79–82; 1973, pp. 64–66). This information will be carried in the learning mechanisms that complete the program on the basis of the organism's individual interaction with its social and natural environment. The epigenetic rules must be understood as carrying a great deal of information about the environment if they are to have any chance of directing an open process of development to an adaptive result.

Insofar as the term "tabula rasa" carries misleading implications about the nature of open programming, one must also question the theoretical value of the ordering that it anchors at one extreme. Since the progression from genetically fixed to open programs does not involve a decrease of genetically transmitted information, there is no good reason to describe only intermediate cases as ones of gene–culture transmission. Relevant genetically transmitted information will have to be present to at least as great a degree in the extreme cases, which show the greatest differential sensitivity to the particular context of enculturation.

I should add that to my knowledge it was Konrad Lorenz (Lorenz 1965, pp. 79–82; 1973, p. 81) who first clearly recognized that learning does not involve a nature–nurture trade-off. He saw that in sophisticated organisms the learning mechanisms are of necessity the greatest carriers of genetically transmitted information. The accumulation of information is like that of wealth: The more you want a creature to be able to acquire on its own, the more you have to give it at the start. Given the many insightful and creative observations made by Lorenz on the very relations among culture, learning, and evolution to which L & W address themselves, it is puzzling that his work receives no mention in their text or in their extensive list of references. Lorenz's flair for speculation should not lead to his neglect by those investigating the interactions among genetics, evolution, and culture.

Also relevant to coevolutionary theory are the distinctions Lorenz draws among the different sorts of feedback mechanisms involved in natural selection and individual learning. Natural selection operating on the genetic level is restricted to the buildup of information by trial and success (Lorenz 1973, pp. 83–84). No information is gained from failures, and unsuccessful trials leave no record to prevent repetition. Moreover, genetic innovation is fortuitous and not under feedback control from either successes or failures. The process of accumulating information in the genome is thus restricted. It has nonetheless succeeded in developing organisms with a capacity for individual learning, and in so doing it has made available to the information-acquisition process forms of feedback beyond those to which it remains restricted by its own functional structure. Nervous systems, unlike genes, can build up information by trial and error; mistakes, once met, can be avoided. Moreover, the range of new variations to be tried and tested can be greatly constrained and selectively projected on the basis of feedback from past interactions with the environment. Foresight joins chance as a source of novelty.

Of course, as L & W remind us, the ontogenesis of the nervous system, including its self-modification by learning mechanisms, must occur under the direction of epigenetic rules that have been shaped by natural selection. An animal capable of learning through operant conditioning can adapt its behavior to novel stimuli, but only insofar as the genetically based feedback structure of its learning mechanism carries information about the environment, enabling it to discriminate reliably those of its behaviors that produce beneficial or harmful consequences. The information carried in such a structure should be regarded as a genetically determined mentifact. Though the information is opaquely embedded in a conditioning mechanism, it nonetheless constitutes a structure or filter, which conceptualizes and organizes the organism's informational dealings with its environment. However, as L & W note, the more fitness depends on fine variations in the environmental features, the more transparent to the organism must be the information it possesses about the relevant aspects of the environment (the transparency principle). This forces a move toward nervous system structures that function as representations of the environment and thus make explicit the information opaquely embedded in less flexible learning- or behavior-regulating mechanisms. Thus, even in precultural organisms, there is a natural development toward reified mentifacts insofar as nervous system structures arise that function symbolically and as representations.

Individual learning can thus overcome many of the functional and temporal factors that constrain the acquisition of information via genetic selection. Nonetheless, individual learning restricts knowledge accumulation to the life span and experience of a single organism. Communication and culture are required to free further the process of knowledge accumulation. Communication allows one organism to transfer individually acquired information to another. It makes possible a sort of Lamarckian evolution allowing for the inheritance of individually acquired information. Simple transfers of this kind can take place even in species without a capacity for genuine language. Experienced ravens can condition their fellows to regard as dangerous some genetically neutral stimulus by exhibiting a genetically fixed danger response in the presence of the stimulus (Lorenz 1973, pp. 157–59). An inexperienced raven can come to recognize a local predator without ever having observed the animal doing anything dangerous or threatening. The case is clearly one in which learned information is transferred, but the transfer is restricted to a generalization of the fixed conditioning mechanism and can operate only in the actual presence of the stimulus.

Genuine language allows for the reification of mentifacts and frees the transfer of acquired information from dependence on the adventitious presence of proximity of the relevant stimulus. The linguistically mediated transfer of mentifacts thus leads to an enormous acceleration in the accumulation of knowledge. This acceleration results not only from simple accretion but from epistemic snowballing. Linguistic reification makes possible not only the retention of individually acquired information, but also underlies the whole social enterprise of constructing, testing, and modifying theories. Theories as evolving representations of the environment can be subjected to critical examination and test within the framework of standards and data made available by the scientific enterprise itself. As we increase our knowledge of what the world is like, we also improve our methods for finding out more about it. The capacity for linguistic representation of abstract thought allows knowledge to become reflexive. Our methods and techniques for acquiring information about the world can themselves be made transparent to observation, analysis, and modification. The rapid development of modern science demonstrates how drastically the growth of knowledge can accelerate when such thresholds are crossed.

Science may not be quite so rational and objective a matter as logical empiricist philosophers of science supposed in their accounts of confirmation and theory choice. But the acceptance or rejection of a scientific hypothesis is still a far more rational and self-critical process than the sort of contagion model L & W refer to approvingly to describe the spread of such culturgens as the use of hybrid corn. Adopting a naturalistic attitude toward scientific method may displace the view of science as operating according to some canons of a priori reason, but it also implies that the self-corrective character of the theoretical enterprise should and does apply to our methods of scientific investigation and evaluation as well. Reason having been dethroned, we can be scientific about reasoning.

None of the above is epistemological news. The point in retelling a familiar story has been to put the coevolutionary circuit in perspective, as one among many forms of informational feedback. None of what I've recounted is incompatible with the existence of the sorts of coevolutionary processes L & W describe. Human populations may today be undergoing genetic selection in interaction with cultural changes. And it seems likely that our particular human cognitive processes are arbitrary and idiosyncratic in ways that reflect our evolutionary history and that merit careful investigation. Nonetheless, given the diversity of rapid and sophisticated self-correcting mechanisms that characterize the modern enterprise of acquiring knowledge, there seems little role to be played by the slow information-collecting mechanism of genetic selection that originally set the whole self-propelling process in motion.

Thus, in the cognitive sphere, claims of cultural autonomy come to more than a mere humanistic resistance to seeing human activity brought within the scope of natural science. Natural selection has freed the mind by giving it methods to acquire knowledge more powerful than its own. So provident a parent surely would not begrudge us our autonomy, and we would be ungrateful to deny the openness of our possibilities.

Genes, mind, and culture; A turning point

Thomas Rhys Williams
Department of Anthropology, George Mason University, Fairfax, Va. 22030

The prevailing anthropological view of the relationships between an evolving human biology and a developing human cultural behavior system recognizes the validity of the concept of a "social fact" as defined by Durkheim, emphasizes the differences between human heredity and learning, stresses the crucial importance of using archaeological and historical data in forming general statements concerning the origins and causes of human behavior, and avoids the fundamental error of 19th-century scholars, who arranged data of human biological and cultural evolution in schemes without analyzing individual cases that would have proved particular schemes incorrect.

The general response by anthropologists to Wilson's earlier works (E. O. Wilson 1975; 1978) was to severely question his failures to pay close attention to this general anthropological viewpoint in the study of what evolutionary biologists now term "coevolution" (see Harris 1978; Montagu 1980; Sahlins 1976). The new work by Lumsden & Wilson (L & W) will continue to concern many anthropologists, since they will be able to find (*i*) failures to provide relevant supporting social facts, (*ii*) failures to use readily available archaeological and historical data, and (*iii*) the presentation of an evolutionary scheme relating human biology, culture, and "mind" that appears to ignore individual cases that would have challenged L & W's views.

It has not helped the work of anthropologists concerned with understanding evolutionary relationships between human biology and culture to have had to deal in recent years with the view of Dawkins (1976) that humans are nothing more than "containers" for genes seeking expression, or the statements of Cavalli-Sforza and Feldman (1973; 1981) concerning the way culture is supposedly transmitted, in which they generally ignore the existence of an extensive literature of more than 50 years of ethnographic study of the cultural transmission process, as well as nearly all of the anthropological efforts to create theoretical and conceptual understanding of that process. In short, anthropologists feel put upon by the growing numbers of evolutionary biologists, population geneticists, assorted types of ethologists, zoologists, and astrophysicists who have turned in recent years to devising schemes and offering generalizations concerning the evolution of human biology, culture, mind and human nature without knowing an *Australopithecus robustus* from *Homo habilis* or Iroquois cousin terminology from a bifurcate collateral kinship system. Put another way, anthropologists tend to be made uneasy by the way persons trained in other disciplines now seem to be wandering selectively through the data of anthropology, while ignoring generally accepted methods, theory, and concepts for the use of such materials.

However, as an anthropologist with an abiding interest in the study of the development of a human nature (Williams 1959) and an ongoing concern with understanding the evolutionary origins, development, and current form and functioning of the cultural transmission process and with the ways human biology and culture have come to be dynamically related (Williams 1972a; 1972b; 1975; 1981) I find the L & W work to be an epochal one. I learned much from reading it.

The main point of the work, that genetic and cultural evolution are "inseverable" processes, will be accepted easily by anthropologists concerned with the study of human biological and cultural evolution since this has been a key theoretical outlook in anthropology for more than two decades. However, I believe that the general conclusion reached by L & W from their demonstration of the inseverability of human genetic and cultural evolution, which states that the "human mind" has tended to evolve so as to bias individuals toward certain genetically fixed patterns of cognition and choice over others, will be rejected by most anthropologists as not demonstrated; Sahlins (1976) and Montagu (1980) can be read to discover why many anthropologists are likely to come up with the response. And I would venture to guess that L & W's use of mathematical models to trace what they term the "coevolutionary circuit" will greatly jar the sensibilities of anthropologists, who still generally follow Wiener's (1950) view that scientists do not have the right to give the impression of a precise mathematical analysis of complex and difficult to understand events and situations unless they use a language they fully understand and can apply correctly. Short of this, Wiener noted, a purely descriptive account of the gross, or whole, appearance of phenomena is really more accurate. I expect that my anthropological colleagues will have some difficulty accepting the various mathematical expressions used by L & W to describe the "coevolutionary circuit" because they will keep recalling certain social facts, data concerning human learning and cultural transmission, and archaeological and historical data, that call into question the "epigenetic rules" (e.g., genes prescribe structure in developmental pathways that lay down endrocrine and neural systems which in turn impose regularities on human cognition and behavior) formulated mathematically by L & W.

Despite this, I find L & W's work an epochal one because, to my knowledge, it marks the first serious effort by evolutionary biologists to really make an attempt to use critically the language – that is, the concepts, theory, and methods – of contemporary anthropology and to try to incorporate social facts and data of culture history and archaeology into a scheme for understanding the evolution of human biology and culture. They are to be commended for their effort. I will return to their work again and again in my research.

This does not mean that I will not be disturbed by having to refer to a work concerned with the evolution of human mind that has only one reference (a passing rejection of a hypothesis) to the work of Freud or those who have since followed in his giant footsteps. Or that I will not be put off each time I use the work when I note L & W's confusion – a fundamental one – between human genetic processes and learning. And this does not mean that I will not be greatly concerned by the casual way that a gratuitous sexism creeps into otherwise informative work; it is now time that evolutionary biologists who insist on talking about humans recognize that the descriptive language for biogenetic processes among crickets, gall wasps, or geese simply will not do when describing such processes in humans, if that language leaves the impression of an unalterable biological or social inferiority on the part of human females.

Given this, I would say that not to have read L & W's work is to have missed a turning point in what can become a growing entente cordiale between what, in effect, have become warring camps. This work takes us all a very long way from the point where, only five years ago, it was possible for a very distinguished senior anthropologist to describe Wilson's work (1975) as reading "like a mixture of science fiction and the 19th century" (MacPherson 1976). I do not know whether this is the result that L & W had in view when they set out to write their work, but this is what I think they have accomplished; and I, for one, am glad of it. Now, as evolutionary biologists learn the language and prevailing view in anthropology concerning the evolutionary interrelations between human biology and culture, just as anthropologists have long since learned the language and viewpoints of evolutionary biology, a new synthesis will indeed emerge.

The place of mind, and the limits of amplification

Joachim F. Wohlwill
College of Human Development, Pennsylvania State University, University Park, Penn. 16802

By a strange set of coincidences, the invitation to act as a commentator on this volume arrived on the same day as the issue of *Science* containing Skinner's (1981) article "Selection by consequences," as well as the presentation, on Cronkite's Universe program, of the problems created by the synthesis, through genetic engineering, of a growth-accelerating hormone. However far apart, each of these provides significant sidelights on central issues raised by *Genes, Mind, and Culture*.

Although it may appear questionable to compare a four-page, necessarily superficial treatment such as Skinner's with the highly formalized, systematic theory presented by Lumsden & Wilson (L & W), there are some interesting similarities, as well as revealing differences between the two. Like the authors of the volume under discussion, Skinner is concerned with interrelations among phenomena at the level of genetic functioning, individual behavior, and cultural patterns and institutions. Skinner contents himself with suggesting a single mechanism, selection, operating in homologous fashion in the three realms. L & W have the much more ambitious aim of tracing the functional relations and interdependencies among these three classes of phenomena. If their aim is thus incomparably more far-reaching, it is hardly surprising that they should fall short of its realization in diverse ways. From the perspective of this commentator (a developmental psychologist), one major shortcoming of the book resides in the treatment accorded to the intermediary term "mind," and more particularly to the role of cognitive development, nominally emphasized by the authors as mediating the coevolutionary process that is at the heart of their theory.

According to L & W, "the epigenetic rules emphasized in our development of gene–culture theory are the cognitive mechanisms that ultimately produce the transition probabilities within the state matrix" (p. 186). They proceed to elaborate the notion of "node-link structures," representing a set of interconnected core concepts that evolve during the course of the individual's experience in a given culture and suggest that these structures "can be defined as the form in which culturgens reside in long term memory" (p. 246). But it is not at all clear how these cognitive structures serve to mediate the gene–culture relationships that form the basis of the "coevolutionary circuit" postulated in the theory. Indeed, the authors stress the subtlety, complexity, and affectivity of the cognitive representations of the basic units of their theory, the "culturgens" (defined earlier [p. 27] as a "relatively homogeneous set of artifacts, behaviors, or mentifacts" – surely a terminological looseness that invites confusion at the outset). The "mentifacts" just alluded to represent a further source of cognitively based indeterminacy, representing reveries, fictions, and myths transmitted from one generation to the next. Neither they, nor the node-link structures of concepts more generally, appear to mediate behavior, at least that embodied in the gene–culture system. Instead, one is left with the impression that these cognitive phenomena represent reflections of the culture in the individual, and thus consequences, rather than determinants of the coevolutionary circuit.

It might seem uncharitable to castigate L & W unduly in this regard, given the difficulties that major cognitive-psychological theorists, from Tolman on down, have had in linking cognition to behavior. But the matter is of some importance here, because of the explicit behavioral focus in the formulation of the epigenetic rules that are considered to underlie the gene–culture relation. Indeed, the illustrations given for these rules, which are thought to provide the basis for the genetically determined biases that modulate the process of gene–culture coevolution, include specific preferences for foods, visual patterns (e.g., human faces), and the like, as well as such behavior patterns as mother–infant bonding, incest avoidance, and particular fears and phobias. All these appear quite divorced from the cognitive structures invoked in later chapters of the book in the discussion of mind and "mentifacts."

Particularly revealing in this connection is L & W's reference to preferences for complexity, since these could have served to link the culturgen bias to cognitive structure. But the problem is considered mainly with regard to the choices of infants, with little interest in the developmental changes over later years related to the child's cognitive development, or in the information-processing implications of complexity preference, even when studied in adults. Indeed, in spite of repeated reference to cognitive development, to "the ontogeny of mind," to Piaget and other developmental psychologists, there is little explicit treatment of ontogenetic change in this book, as opposed to cultural evolution and differentiation.

On a more positive note, the amplification principle that L & W introduce to account for the progressive magnification of some incipient genetic bias or differential through cultural transmission and elaboration deserves special mention. It might be noted that the principle applies far more broadly than the authors appear to recognize – the feeding upon itself of some differential in behavior appears as a general phenomenon in much of differential psychology, notably in the realm of individual differences in intelligence (Wohlwill 1980). Applied to the action of genetic factors, it points to a pervasive interaction between such factors and environmental forces that would seem to render the linear conception underlying heritability coefficients altogether meaningless.

Concerned as they are with cultural evolution, L & W appear less interested in the emergence of intragroup differences in behavior; rather, they concentrate on the choice between alternative culturgens on the part of members of a group, which is a process resulting in increasingly stable behavior patterns within that group. That situation appears to fit best the case of incest avoidance between brother and sister. In the case of the second example, the fissioning of a tribal village, the choice between the responses appears to involve the opposite of amplification: The more one of the behaviors, that is, staying with the tribe, has been chosen, the more probable it becomes that subsequent choices will shift to the alternative mode, that is, leaving the tribe, because of the negative consequences of mushrooming group size. Finally, the case of historical changes in fashion involves an oscillatory process, veering back and forth between relative extremes (e.g., of skirt length), presumably reflecting a cybernetic process that hardly seems to fit a simple, positive-feedback-based amplification principle.

Rather, we appear to be dealing with an interweaving of amplification and damping processes that act to stabilize both the behavioral norm characterizing a group and the individual differences around that norm. An interesting aspect of this process is that the culture may serve both of these functions at the same time, acting in some ways to reinforce differences between individuals or groups (e.g., between the sexes in athletic achievement), and in others to counteract them (e.g., through deliberately instituted compensatory experiences; see Wohlwill 1980). Which brings us back, finally, to the Cronkite broadcast, which focused on the dilemma created by the availability of synthesized forms of the human growth hormone for individuals who by their genetic constitution were destined to be relatively short (through not suffering from any pathological condition of dwarfism). To the extent that the culture will tolerate, or perhaps even encourage, the use of this drug (a matter of debate at present among physicians), it will clearly act in a markedly differences-damping fashion on a basic genetic trait - stature. Whether or not such a societally chanelled biological change will affect genetic fitness, the case appears to illustrate a direct action of culture on a basic physical characteristic of man that would hardly fit the gene–culture translation model sketched out by L & W. Admittedly, this represents a rather special case, but when translated into the behavioral domain, it represents a far more commonplace process, one that sociobiology needs sooner or later to come to terms with.

Authors' Response

Genes and culture, protest and communication

Charles J. Lumsden[a] and Edward O. Wilson[b]
[a] *Department of Medicine, University of Toronto, Toronto, Ontario, Canada M5S 1A8 and* [b] *Museum of Comparative Zoology, Harvard University, Cambridge, Mass. 02138*

Certain key points *not* made by the 23 reviewers are at least as important as the points they do stress. The reviewers do not deny that biological and cultural evolution are somehow coupled, and that the nature of the coupling is one of the great remaining problems of science. Because its solution could bridge the biological and social sciences and shed light on the origin of mind, a special concentration on this problem seems well justified.

The reviewers do not identify any systems of explanation that are competitive to the theory of gene–culture coevolution under discussion. **Schubert** implies that it has been done before, but does not say where or by whom. Similarly, **Barash** suggests that competing schemes are conceivable, but does not provide examples. **Markl** goes so far as to cite the work of Chomsky, Piaget, and their collaborators. However, as we noted in *Genes, Mind, and Culture* (henceforth *Genes*), these investigators have defined certain epigenetic rules, not constructed an explicit theory of the coupling between genetic and cultural evolution. **Slobodkin** mentions the valuable work of Cavalli-Sforza and Feldman (1981) on cultural transmission (which, contrary to his assertion, is reviewed in our book); but these authors did not attempt to treat the mind or incorporate cognitive science, nor did they formulate a theory that treats the evolution of either the epigenetic rules of mental development or macrocultural structure. Such a conception of gene–culture coevolution is unique to *Genes;* the coevolutionary circuit is spelled out in explicit detail in this book for the first time. We stress this point not to raise questions of priority, but to remove the possibility of the kind of objection commonly encountered in social theory, that in some way this mode of reasoning "has been tried before," has already been subjected to all sorts of scholarly rebuttal, and is therefore dismissible by reference to the history of ideas.

While accepting this point, **Caplan** nevertheless undervalues the gene–culture coevolution program as an attempt "to prove that cultural evolution is amenable to Darwinian analysis in terms of natural selection . . . [which] may not be the best theoretical approach to take in analyzing behavior." We go much further than that simple prescription, arguing that cultural form and change are the result of gene–culture coevolution, a hitherto unrecognized mechanism. It is likely that natural selection figures prominently in the aspects of this process affecting *biological* change, and we have constructed some of the test models accordingly (Chapters 5 and 6). But the basic scheme is much more than an extension of Darwinism and a prosthetic device added to conventional sociobiology. It incorporates the epigenetic rules of mental development and cultural innovation, as well as the basic principles of socialization and group behavior. Of equal importance, the gene–culture translation models (Chapter 4) are not dependent on natural selection processes, while the full coevolutionary equations (Chapter 6) can be extended as needed to accommodate non-Darwinian changes in gene frequency.

Are culturgens real? Most of the reviewers either approve of our concept of the culturgen as the basic unit of culture used by the epigenetic rules or at least let it pass without comment. **Rosenberg** correctly recognizes that the precise definition of the culturgen is not crucial to the success of a theory of gene–culture coevolution. But others (**Caplan, Hallpike, Shepher, Slobodkin**) see the unit as a source of trouble. To them, arbitrariness in the definition of the culturgen jeopardizes the theory.

These critics forget that a great deal of science is built on units of very complex or even continuous variation. For example, most of morphology, developmental biology, population biology, and ecosystems research is conducted in this manner. We have suggested that culturgenic variation can be described either as a continuum or as a set of discrete elements (see *Genes*, Figure 3-2, p. 55) and that models within gene–culture coevolutionary theory can be based on continuous mathematical functions approximating culturgenic variation. Moreover, it is of great theoretical and practical importance that some culturgens *are* discrete and readily conceptualized in the procedure followed in our elementary test models. They include the many binary choices made by people, such as the case of brother–

sister incest versus outbreeding used in our first analysis. Other culturgenic categories, such as hand position in tool use and kin classifications, are broader but still composed of finite arrays of denumerable entities. We have shown how to cluster such elements into groups of a size chosen to facilitate study at any desired level of precision (see Figure 3-12, p. 98).

Some of the critics (**Caplan, Hallpike**) appear to confuse the possibility of arbitrariness in the definition of culturgen with arbitrariness or flexibility in the selection by human investigators of methods for sorting identified culturgens into sets or families. If our approach is correct, the existence of epigenetic rules provides a natural, nonarbitrary criterion for a culturgen, namely, the feature or set of features in the culture that is recognized by epigenetic rules and so used in the assembly of the mind. Thus, as we pointed out in Chapter 6 of *Genes*, the resulting knowledge structures of the mind can in turn be viewed as the abstract embodiments of culturgens, the relationship being provided by the epigenetic rules. Moreover, we believe that classification methods of the type reviewed in Appendix 1-1 will be useful in categorizing culturgens and searching for natural affinities among them. Despite their assertions that this cannot be the case, the critics offer no proof or argument to support their position. Certainly the weight of the evidence, some of which we touch on in Appendix 1-1, is against them. The existence and usefulness of powerful methods of culturgen classification constitute empirical problems and cannot be prescribed by aprioristic reasoning of the kind used by Caplan and other critics.

In summary, the concept of the culturgen appears to us to present no difficulty for gene–culture coevolutionary theory. Examples already exist of small numbers of competing discrete culturgens that can serve as the basis for the first test models of the theory. The goal of delineating more complex types of culturgens must also be vigorously pursued.

The models: Too simple – or too complex? As is pointed out by **Caplan** models of the gene–culture coevolutionary process must be relatively detailed if, as earlier critics of human sociobiology have argued, the essential processes of mind and culture are distinct from and possibly more complicated than ordinary biological evolution. **Maynard Smith** proposes that there exists an upper limit to useful complexity in models (an unremarkable idea in itself) and suggests that we have already crossed this threshold. He offers no supporting argument, but apparently expects the inference to be gathered from his expressed lack of astonishment at the several highly specific technical results that he selects as the sole topics of his critique. Such verbal blandishments after the fact seem unconvincing. Repeatedly he asks what it all has to do with culture. Maynard Smith is implying that the "novelty value" of the technical results is relative to the set of all possible sources of genetic change. If this were true, one might indeed find it natural to inquire what news there is in the result that significant microevolutionary change can occur in a few dozen generations. But this perspective is surely the unproductive one, because what one wants is the perspective Maynard Smith avoids by means of his meticulously blinkered view: These specific technical results have exploratory value relative to the *subuniverse*, heretofore completely unstudied, of gene–culture coevolutionary models of the type based on our hypotheses. For this new class of models it is of significant interest to learn that genetic (and cultural) change can occur at given rates and in given directions, and to trace within the gene–culture linkage the effects that make this possible. Moreover, the fact that rates and other properties prove to be similar or dissimilar to those discovered in other, very different, classes of purely genetic models adds to the intriguing search for deep links among the areas of population biology.

In an early interview (Lewin, 1981) **Maynard Smith** covered the opposite pole to complexity by stating that our models were too simple to handle the process of gene–culture coevolution. We can only conclude that Maynard Smith's real views on this matter fall somewhere between judgments of simplicity and complexity, but given the apparent internal contradictions in his position, his lack of supporting argument or suggestion of superior methods, and his assiduous avoidance of the main conceptual and empirical issues faced in the book, his comments are unproductive.

Commenting on the models, **Hartl** reports that we have achieved "merely" a reformulation of traditional genetic theory with culturgens added. This is comparable to saying that population genetics is merely natural selection theory with genes added. In fact, the models as presented in Chapters 4 and 6 of *Genes*, despite their crudity, achieve the first unification of mathematical structures basic to population biology with mathematical structures fundamental to social theory. We have shown that in order to incorporate culturgens, it is necessary to transform individual decisions into cultural patterns, estimate the effects of cultural patterns on individual decisions and selection pressure, and incorporate substantial information on socialization and cognition – none of which has to our knowledge been previously addressed by population geneticists.

Hartl thinks that our test models are somehow too simple because "normal variability in behavior, if influenced significantly by genes at all, is not influenced by single genes of major effect; such traits are quantitative traits, and thus the theory of quantitative genetics is appropriate, not classical single-locus population genetics." The weight of recent experimental evidence, which we review in *Genes* (and which he does not refute), is against Hartl. Single genes of major effect in behavior are common enough to justify both their biological interest and their significant value in pilot models, while the number of polygenes in quantitative systems is often 10 or fewer (see, for example, various reports in Thompson & Thoday 1979). Hartl is also concerned that we did not incorporate genetic drift and pleiotropism into the models. This is true – exploratory first models do not accomplish all things at once. Hartl does not make clear the fact that the mathematical structure of gene–culture theory can be expanded to accommodate these and other phenomena in more detailed models. The accuracy of predictions in the models will be the test of how judiciously the elements were chosen.

On metaphysics. Doubt flourishes among a few of the reviewers over whether a true theory of human sociobiology can ever be made to work. **Markl** suggests that gene–culture coevolutionary theory, or any biological theory, will ultimately fail to develop the explanatory power of "reductionist physical theories." He does not supply an argument about why this pessimistic assessment is true or even likely. The center of Markl's position is the remarkable assumption that gene–culture theory *must* be weaker because people and cultures are unique. But physics deals with planets and apples, galaxies and protons, all of which have uniqueness and idiosyncracies due to history or boundary conditions. Physics draws strength from the realization that many key phenomena depend upon properties that are similar among the entities considered. What Markl and other writers with a similar philosophical viewpoint (e.g., Bock 1980; Sahlins 1976; and **Kovach** in this issue) have failed to establish is why human sociobiology must lack deep explanatory power because the uniqueness in organic systems arises by ontogeny and biological evolution. On the contrary, we believe that the theory of gene–culture coevolution, even in its primitive state, demonstrates the opposite. If valid, this formulation, and ultimately biological theory in general, will gain explanatory power to the extent that they can identify general mechanisms at work – such as gene–culture coevolution – and isolate the properties that such systems share.

Ghiselin, Kovach, and many others writing prior to this *BBS* symposium have urged that variation in culture itself (Ghiselin stresses language) is intrinsically more interesting than the supposed general properties of gene–culture coevolution. This is the equivalent of saying that planetology is intrinsically superior to physics. The U.S. and Soviet space probes have disclosed that each planet and moon is unique, sometimes to a spectacular degree. Scientists might have been satisfied with a detailed description of the planets in the style of natural history and taxonomy. But physics was needed to show how the diversity originated in the first place, while techniques based on physical theory were necessary to design the flights on which the greater part of our descriptive knowledge is based. In a similar manner, both the common features of cultures and their diversity await an explanation of the underlying mechanisms of coupled genetic and cultural evolution.

The theory: Too hereditarian? Human sociobiology is hagridden by the persistent belief that it is committed to the hereditarian end of some kind of gene-to-environment scale along which all theories can be positioned. Thus **Charlesworth** and **Johnston** worry about what they perceive to be our unjustified hereditarian advocacy and genetic determinism. But we have presented an interactionist formulation that makes the simple gene–environment scale obsolete. Gene–culture coevolutionary theory does not a priori exclude the evolution of any degree of developmental bias in particular epigenetic rules of mental development, from the perfect tabula rasa state to total restriction to a single phenotype. It seeks to predict and explain why certain degrees of biological constraint on the structure of real minds have occurred. The important features are in the epigenetic rules themselves, which feed on culture in order to assemble the mind. The importance of environment prevades every part of the book, in its synthesis of data from cognitive psychology, its pilot models, and its more qualitative discussions.

Williams suggests that we have not "demonstrated" a genetic basis for the epigenetic rules. He is apparently arguing that data do not dictate a unique choice of theory, and that steps of creative hypothesis forming are involved. In this regard we consider the theory of gene–culture coevolution to be particularly strong and deep. The data available to us (and reviewed in Chapters 2, 3, and 6) seem to point overwhelmingly to the existence of biologically based epigenetic rules. Furthermore, constraint in mental development appears to be a general if not universal property in all categories of cognition and behavior thus far examined. Some degree of constraint is also indicated readily and naturally from the test models that consider the alternative possibility of a tabula rasa. As many reviewers have noted, gene–culture coevolutionary theory gains added power because it is testable. Anthropologists and others who do not like the idea of specific epigenetic rules, especially those channeling mental development, can attempt to falsify it. It will also be incumbent on them to create an alternative theory of gene–culture coevolution with equivalent depth and explanatory power.

Several reviewers, including **Johnston** and **Plutchik,** have expressed reservations about the definition of the epigenetic rules and the amount of mathematical theory based on existing knowledge of mental development. Without doubt cognitive developmental psychology is in its infancy, and the models of gene–culture coevolution should be responsive to each important new discovery. On the other hand, our conception of cognitive epigenesis is in accord with the recent primary literature on the subject. It is almost entirely in advance of the review and theoretical articles cited by Johnston, most of which date from 1970 and earlier and most of whose authors had no inkling of what was to come in the choice experiments and other cognitive developmental studies of the 1970s. **Loftus** raises the possibility that the data we incorporated into *Genes* was selected uncritically. This is not the case. It is a fact that because of their subject matter and state of development the sophistication of experiments across the subfields of psychology is highly variable. However, in each area on which we reported, a serious effort was made to locate and report only the most significant and reliable data available.

Epigenetic rules, emotion, and intention. Several questions about the concept of the epigenetic rules have been raised by **Plutchik.** He asks whether such phenomena as perceptual circumplexes and fear of strangers are epigenetic rules or the result of such rules. The answer is that they are the result of such rules. The rules themselves are somewhat more abstract. To envisage them correctly, one must, to continue the example, consider the space of all possible perceptual circumplexes or the space of all possible patterns of fear of strangers. The epigenetic rules are then represented as mathematical functions (typically probability densities or related objects) on these spaces. In this way we can begin to

Response/Lumsden & Wilson: Genes, mind, and culture

incorporate into gene-culture theory the regularities and constraints involved in the development of each specific cognitive domain. In turn, the mathematical objects model the action of what in the developing organism are actual mechanisms or submechanisms of neural and mental development.

We share **Plutchik**'s conviction that the subject of emotion is a critical but little explored area of gene-culture coevolutionary theory. We anticipate a productive convergence of current research on emotion (e.g., Plutchik 1980) and gene-culture coevolutionary theory, although the entire subject of emotion is still in a nascent and rapidly changing state. In particular, we know of no evidence to support Plutchik's statement that emotion always disrupts the operation of memory. On the contrary, a strong emotional experience can facilitate recall under some circumstances. It is likely that various differing measures of disruption and reinforcement caused by emotion are in fact components of epigenetic rules.

We also disagree with **Rosenberg** concerning the irreducibility of intentional descriptions of human behavior (such as belief and desire) to nonintentional form such as that employed by neurobiology. Contrary evidence is supplied in the form of the rapidly increasing number of theories and models in the fields of artificial intelligence and cognition theory. Not only do many investigators in artificial intelligence believe that such a connection is conceivable, they are actively trying to achieve it in the form of computer simulations of mental activity. This connection is considered one of the central problems of artificial intelligence research (McCorduck 1979).

Style and other miscellany. Many reviewers have found *Genes* a difficult book to read. Although some would prefer to think that we are just poor writers, the real reason is that the human sciences to which it is addressed are still a tower of Babel. Most social scientists have refused to acknowledge the relevance of biology and therefore find its idioms and rhythms alien, even when the subject under discussion is human social behavior. Most biologists in turn have little interest in the problems of the social sciences. If efforts to understand social organization do not touch on subjects already important to them and also match the level of sophistication reached within their own narrow specialities, they hold the enterprise in contempt. They resemble William Hopkins, one of the chief critics of evolution, of whom Darwin said that he was "so much opposed because his course of study has never led him to reflect much on such subjects as geographical distribution, classification, homologies, &c., so that he does not feel it a relief to have some kind of explanation" (Darwin 1860).

We believe that a correctly formulated human science must inevitably contain a juxtaposition of previously specialized knowledge that will at first seem improper and discordant to scholars in the newly contiguous disciplines. *Genes* was written in this eclectic manner and has received a great deal of criticism for it. An anthropologist who finishes a paragraph on the familiar topics of culture and socialization and in the next paragraph plunges into a mathematical model of population genetics may be understandably disoriented, no less than the biologist who is directed abruptly out of genetics into cognitive psychology. The emotional discomfort of cognitive dissonance is everywhere felt; we ourselves have suffered it, and we expect it to occur in others who wish to press on into the subject of gene-culture coevolution. At the same time we insist, as we did in *Genes*, that the juxtapositions are natural and even inevitable.

We note that geneticists did not complain much about our explanations of genetics, cognitive psychologists about the accounts of their subject, and so forth. In fact, within each discipline we were careful to present material in a straightforward, clarified manner. The difficulty in *Genes* comes not from its prose but from its presentation of information in unfamiliar mixes and compounds. This difficulty will fade in the mind of the reader as the constituent disciplines become more familiar.

Genes is also grounded in the basic mode of the natural sciences, in which mathematics is deemed indispensable for structuring hypotheses, communicating arguments, formulating models, and constructing predictions. We can think of no theory in any field of science that achieves depth, clarity, explanatory power, and predictive strength without involving essential mathematical structures. If one exists, then none of the critics who lamented the book's mathematical content points it out as a counterexample to our approach. We doubt that they will be able to find one. It is foolish to pretend that mathematics is a secondary flourish to be used only when a scientific field has reached some ultimate state of verbal purity. Moreover, we can detect no signs of progress toward deep theory by those who offer instead only dogmatic assertions implying that the sciences of man are impervious to mathematical reasoning. And it is unfortunate that the appearance of mathematical sections blinded some of the reviewers, who avoided the central arguments and confined themselves to reservations over terminology and the philosophy of science.

One unusual response deserves special mention. **Slobodkin** employs a rarely seen anachronism: an appeal to authority. Rather than deal with the substance of gene-culture coevolutionary theory, he refers the reader to strongly negative reviews by "a Nobel laureate developmental biologist" (Peter Medawar, actually an immunologist) and "one of the most brilliant living population geneticists" (Richard Lewontin, of Harvard University). He notes that a third, largely favorable review in *Science* cites a "sophistic presentation" and "serious flaws." Slobodkin then offers the opinion that the only reason the book has received so much attention is that it is being vigorously promoted by Harvard University Press. This all sounded familiar to us. We recalled that Darwin's *Origin of Species* was bitterly attacked by the foremost comparative anatomist of the day (Richard Owen) and America's most brilliant zoologist (Louis Agassiz, of Harvard University), not to mention one of the founders of geology (Adam Sedgwick). Various distinguished reviewers condemned parts of Darwin's book as variously insincere, obscurantist, and seriously flawed (see Hull 1973). Samuel Haughton declared (on February 9, 1859) that the only reason the Darwin-Wallace essays received any notice at all was that they were promoted by Charles Lyell and the Geological Society. If

Slobodkin's logic is correct, it seems to follow that our own work is not without hope. (We have, incidentally, answered the largely ad hominem criticisms by Medawar and Lewontin elsewhere; see Lumsden & Wilson 1981a; 1981c).

Larger questions. Like sociobiology in general, *Genes, Mind, and Culture* has evoked strong and astonishingly diverse responses in part because it acts as a Rorschach test of the philosophical positions, professional interests, and even the ethical beliefs of the scholars who scrutinize it. The theory of gene–culture coevolution – in whatever form it will eventually take – does represent the growing edge of biology as it spreads toward the social sciences and, no less, the growing edge of the social sciences as they expand in the direction of biology. As the explorers and reigning shamans view each other clearly across that previously vast chasm, a great deal of communication and accommodation will be required. At stake is nothing less than the perception of the limits of the methods of natural science, the intellectual sovereignty of many disciplines, and the ethics-determining premises of human nature.

We have seen nothing in the reviews that challenges the basic conception of gene–culture coevolution. No alternative mechanism for the evolution of mind has been suggested, nor have the key steps of the coevolutionary circuit we propose been seriously challenged. Some reviewers doubt the feasibility of the explicit research program suggested in the book. Their concerns center on the definition of the culturgen, but this concept, as we believe we have demonstrated, is easily handled with reference to the model-building role to which it has been assigned. Nevertheless, the enterprise faces many kinds of obstacles. We foresee serious challenges for experimentalists in the elucidation of epigenetic rules in a form suitable for the realistic gene–culture translation models (Chapter 4). We are sensitive to the complexity in the interaction of epigenetic processes across categories of cognition and behavior, as in the relation of emotion and various forms of learning. We view as key mysteries the consequences of various possibilities in the still largely unknown relations of genes and brain structure and, still more, of brain structure and cognitive development. Our pilot models can be applied directly to hunter-gatherer and primitively agricultural societies, in which most gene–culture coevolution did in fact occur. We believe that they can reveal a great deal about the origin and evolution of the mind. But the usefulness of this first generation of models fades as one approaches the more advanced forms of societies, in which institutions are vastly more complex and economics and culture can be powerfully manipulated by a small minority.

All of these looming problems are technical, in the sense that they affect phenomenological theory at the level of the test models together with the techniques of field and experimental research. The fundamental theory, which was constructed with reference to our understanding of biological evolution and the key facts of cognition and human social behavior, awaits testing by means of these and other models, rather than aprioristic denials by skeptics who think that no such formulation can ever be true. Perhaps the fundamental theory will be set aside – but not solely by skepticism based on intuition. Only a competing explanation of mental and cultural evolution that more accurately accounts for biology and society will work. Meanwhile, as the antecedent sociobiology controversy showed, the topic is too important simply to ignore. We don't make this claim because we have worked on the subject; rather we have tried to produce a theory because we believe the claim.

References

Alberch, P.; Gould, S. J.; Oster, G. F. & Wake, D. B. (1979) Size and shape in ontogeny and phylogeny. *Paleobiology* 5:296–317. [TDJ]

Amoore, J. E. (1977) Specific anosmia and the concept of primary odors. *Chemical Senses and Flavor* 2:267–81. [ta CJL]

Anderson, N. H. (1979) Algebraic rules in psychological measurement. *American Scientist* 67:555–63. [ta CJL]

Anderson, P. W. (1972) More is different. *Science* 177:393–96. [RDM]

Ashton, G. C., Polovina, J. J. & Vandenberg, S. G. (1979) Segregation analysis of family data for 15 tests of cognitive ability. *Behavior Genetics* 9:329–47. [ta CJL]

Axelrod, R. & Hamilton, W. D. (1981) The evolution of cooperation. *Science* 211:1390–96. [RDM]

Baldwin, J. M. (1902) *Development and evolution*. London: Macmillan. [TDJ]

Bates, E. (1979) *The emergence of symbols*. New York: Academic Press. [TDJ]

Behrman, S. N. (1960) *Portrait of Max, an intimate memoir of Sir Max Beerbohm*. New York: Random House. [LBS]

Bock, K. (1980) *Human nature and history: A response to sociobiology*. New York: Columbia University Press. [tar CJL]

Bock, W. & von Wahlert, G. (1965) Adaptation and the form-function complex. *Evolution* 19:269–99. [ALC]

Bonner, J. T. (1965) *Size and cycle*. Princeton; Princeton University Press. [TDJ]

Brown, W. L., Jr. & Wilson, E. O. (1956) Character displacement. *Systematic Zoology* 5:49–64. [MTG]

Burian, R. M. (1978) A methodological critique of sociobiology. In: *The sociobiology debate*, ed. A. Caplan, pp. 376–95. New York: Harper & Row. [ALC]

Cairns, R. B. (1979) *Social development: The origins and plasticity of interchanges*. San Francisco: Freeman. [TDJ]

Campbell, D. T. (1974) Evolutionary epistemology. In: *The philosophy of Karl Popper*, vol. 1, ed. P. A. Schlipp, pp. 413–63. La Salle, Ill.: Open Court Press. [MTG]

Caplan, A. (1980) Have species become déclassé? In: *PSA-1980*, ed. P. Asquith & R. Giere, pp. 71–82. East Lansing, Mich.: Philosophy of Science Association. [ALC]

(1981a) Back to class: A note on the ontology of species. *Philosophy of Science* 48:130–40. [ALC]

(1981b) Pick your poison: Historicism, essentialism, and emergentism in the definition of species. *Behavioral and Brain Sciences* 4:285–86. [ALC]

Cavalli-Sforza, L. & Feldman, M. (1973) Models for cultural inheritance, I: Group mean and within-group variation. *Theoretical Population Biology* 4:42–55. [TRW]

(1981) *Cultural transmission; A quantitative approach*. Princeton: Princeton University Press. [DPB, r CJL, LBS, TRW]

Caviness, V. S., Jr. & Rakic, P. (1978) Mechanisms of cortical development: A view from mutations in mice. *Annual Review of Neuroscience* 1:297–326. [ta CJL]

Chagnon N. A. & Irons, W., eds. (1979) *Evolutionary biology and human social behavior*. North Scituate, Mass.: Duxbury Press. [JS]

Charlesworth, W. R. (1979) Ethology: Understanding the other half of intelligence. In: *Human ethology: Claims and limits of a new discipline*, ed. M. von Cranach; K. Foppa; W. Lepenies & D. Ploog, pp. 491–529. Cambridge: Cambridge University Press. [WRC]

Chiva, M. (1979) Comment la personne se construit en mangeant. *Communications* (Ecole des hautes etudes en sciences sociales – Centre d'Etudes Transdisciplinaires, Paris) 31:107–18. [ta CJL]

Clark, K. B. (1971) The pathos of power: A psychological perspective. *American Psychologist* 26:1047–57. [HEG]

Cloninger, C. R. & Yokoyama, S. (1981) The channeling of social behavior (*Review of genes, mind, and culture*, Lumsden, C. J. and Wilson, E. O.). *Science* 213:749–51 [WRC, r CJL, LBS]

Cooter, R. & Kornhauser, L. (1980) Can litigation improve the law without the help of judges? *Journal of Legal Studies* 9:139–63. [MTG]

Daly, M. & Wilson, M. (1978) *Sex, evolution and behavior*. North Scituate, Mass.: Duxbury Press. [JS]

References/Lumsden & Wilson: Genes, mind, and culture

Darwin, C. (1859) *On the origin of species by means of natural selection, or the preservation of favoured races in the struggle for life.* London: John Murray. [MTG]

(1860) Letter to Asa Gray, 22 July 1860. In: *The life and letters of Charles Darwin*, ed. F. Darwin. London: John Murray, 1887. [rCJL]

(1871) *The descent of man, and selection in relation to sex.* London: John Murray. [MTG, TDJ]

Davidson, D. (1963) Actions, reasons and causes. *Journal of Philosophy* 60: 685-700. [ALC]

Davidson, J. & Davidson, R., eds. (1980) *The psychobiology of human consciousness.* New York: Plenum Press. [GS]

Dawkins, R. (1976) *The selfish gene.* New York: Oxford University Press. [TRW]

Day, C. (1920) This simian world. Reprinted in *The best of Clarence Day.* New York: Knopf, 1956. [GS]

deBeer, G. (1958) *Embryos and ancestors.* 3rd. ed. London: Oxford University Press. [TDJ]

Dobzhansky, T. (1962) *Mankind evolving.* New Haven: Yale University Press. [TDJ]

Dunn, F. L. (1970) Cultural evolution in the late Pleistocene and Holocene of Southeast Asia. *American Anthropologist* 72:1041-54. [MTG]

Eibl-Eibesfeldt, I. (1979) Human ethology - concepts and implications for the sciences of man. *Behavioral and Brain Sciences* 2:1-57. [TDJ]

Fantz, R. L.; Fagan, J. F. III & Miranda, S. B. (1975) Early visual selectivity: As a function of pattern variables, previous exposure, age from birth and conception, and expected cognitive deficit. In: *Infant perception: From sensation to cognition*, vol. 1, ed. L. B. Cohen & P. Salapatek, pp. 249-345. New York: Academic Press. [ta CJL]

Feffer, M. (1981) *The structure of Freudian thought.* New York: International Universities Press. [HEG]

Fischer, K. W. & Watson, M. W. (1981) Explaining the Oedipus conflict. In: *New directions for child development: Cognitive development*, no. 12, ed. K. W. Fischer, pp. 79-92. San Francisco: Jossey-Bass. [HEG]

Fishbein, H. D. (1976) *Evolution, development, and children's learning.* Pacific Palisades, Calif.: Goodyear. [TDJ]

Fodor, J. A. (1981) The mind-body problem. *Scientific American* 244(1): 114-23. [JKK]

Freedman, D. G. (1979) *Human sociobiology.* New York: Free Press. [PLB]

Galbraith, J. K. (1981) *A life in our times.* Boston: Houghton Mifflin. [DLH]

Geist, V. (1978) *Life strategies, human evolution, environmental design.* New York: Springer. [GS]

Ghiselin, M. T. (1974) A radical solution to the species problem. *Systematic Zoology* 23:536-54. [ALC]

(1981) Catgories, life, and thinking. *Behavioral and Brain Sciences* 4: 269-83. [ALC, MTG]

Goodman, J. C. (1978) An economic theory of the evolution of common law. *Journal of Legal Studies* 7:393-406. [MTG]

Gottesman, I. I. & Heston, L. (1972) Human behavioral adaptations: Speculations on their genesis. In: *Genetics, environment, and behavior: Implications for educational policy*, ed. L. Ehrman; G. S. Omenn & E. Caspari, pp. 105-22. New York: Academic Press. [WRC]

Gottlieb, G. (1970) Conceptions of prenatal behavior. In: *Development and evolution of behavior*: Essays in memory of T. C. Schneirla, ed. L. R. Aronson, E. Tobach, D. S. Lehrman & J. S. Rosenblatt, pp. 111-37. San Francisco: Freeman. [WRC, TDJ]

(1976) The roles of experience in the development of behavior and the nervous system. In: *Neural and behavioral specificity*, ed. G. Gottlieb, pp. 25-54. New York: Academic Press. [TDJ]

Gould, S. J. (1977a) *Ever since Darwin.* New York: W. W. Norton. [ALC]

(1977b) *Ontogeny and phylogeny.* Cambridge, Mass.: Harvard University Press. [WRC, TDJ]

Gruber, H. E. (1981) *Darwin on man: A psychological study of scientific creativity.* 2d ed. Chicago: University of Chicago Press. [HEG]

Hailman, J. P. (1967) The ontogeny of an instinct. *Behaviour Supplement* 15:1-159. [TDJ]

(1979) The ethology behind human ethology. *Behavioral and Brain Sciences* 2:35-36. [TDJ]

Hamilton, W. (1964a) The genetical evolution of social behaviour. I. *Journal of Theoretical Biology* 7:1-16. [GS]

(1964b) The genetical evolution of social behavior. II. *Journal of Theoretical Biology* 7:17-52. [GS]

Harris, M. (with E. O. Wilson) (1978) Encounter: The envelope and the wig. *Sciences* 18:9-15; 27-28. [TRW]

Hess, E. (1970) Ethology and developmental psychology. In: *Carmichael's manual of child psychology*, ed. P. H. Mussen, pp. 1-38. New York: John Wiley. [WRC]

Hinde, R. A. (1970) *Animal behaviour.* 2d. ed. New York: McGraw-Hill. [TDJ]

Hirshleifer, J. (1980) Evolutionary models in economics and law: Cooperation versus conflict strategies. UCLA Department of Economics working paper 170. [RDM]

Ho, M. W. & Saunders, P. T. (1979) Beyond neo-Darwinism - an epigenetic approach to evolution. *Journal of Theoretical Biology* 78:573-91. [TDJ]

Hockett, C. F. & Ascher, R. (1964) The human revolution. *Current Anthropology* 5:135-68. [TDJ]

Hull, D. L. (1973) *Darwin and his critics.* Cambridge, Mass.: Harvard University Press. [rCJL]

(1976) Are species really individuals? *Systematic Zoology* 25:174-91. [ALC]

(1978)A matter of individuality. *Philosophy of Science* 45:335-60. [ALC]

Jacobs, J. (1981) How heritable is innate behaviour? *Zeitschrift für Tierpsychologie* 55:1-18. [TDJ]

Johnston, T. D. (1981) Contrasting approaches to a theory of learning. *Behavioral and Brain Sciences* 4:125-73. [WRC]

(in press) Selective costs and benefits in the evolution of learning. In: *Advances in the study of behavior*, vol. 12, ed. J. S. Rosenblatt; C. Beer; R. A. Hinde & M.-C. Busnel. New York: Academic Press. [TDJ]

Kim, J. (1978) Supervenience and nomological incommensurables. *American Philosophical Quarterly* 15:149-56. [ALC]

Kuo, Z.-Y. (1967) *The dynamics of behavior development.* New York: Random House. [TDJ]

Leach, E. (1981) Biology and social science: Wedding or rape? *Nature* 291: 267-68. [ALC]

Lehrman, D. S. (1953) A critique of Konrad Lorenz's theory of instinctive behavior. *Quarterly Review of Biology* 28:337-63. [TDJ]

(1970) Semantic and conceptual issues in the nature-nurture problem. In: *Development and evolution of behavior*, ed. L. R. Aronson; E. Tobach; D. S. Lehrman & J. S. Rosenblatt, pp. 17-52. San Francisco: Freeman. [TDJ]

Lewin, R. (1981) Cultural diversity tied to genetic differences. *Science* 212: 908-10. [r CJL, RDM, LBS]

Lewontin, R. C. (1981) Sleight of hand. *Sciences*: 23-26. [ALC, LBS]

Lorenz, K. Z. (1950) The comparative method in studying innate behavior patterns. *Symposia of the Society for Experimental Biology* 4:221-68. [TDJ]

(1956) The objectivistic theory of instinct. In: *L'Instinct dans le comportement des animaux et de l'homme*, ed. P.-P. Grasse, pp. 51-76. Paris: Masson. [TDJ]

(1965) *Evolution and the modification of behavior.* Chicago: University of Chicago Press. [TDJ, RVG]

(1973) *Behind the mirror.* New York: Harcourt, Brace, Jovanovich. [RVG]

Lumsden, C. J. & Wilson, E. O. (1980) Translation of epigenetic rules of individual behavior into ethnographic patterns. *Proceedings of the National Academy of Sciences, U.S.A.* 77:4382-86. [ta CJL]

(1981a) Genes & culture. *New York Review of Books*, September 24. [r CJL]

(1981b) *Genes, mind, and culture: The coevolutionary process.* Cambridge, Mass.: Harvard University Press.

(1981c) Genes, mind, and ideology. *The Sciences* 21:6-8. [r CJL]

McClearn, G. E. & DeFries, J. C. (1973) *Introduction to behavioral genetics.* San Francisco: Freeman. [ta CJL]

McCorduck, P. (1979) *Machines who think.* San Francisco: Freeman. [r CJL]

McCracken, R. (1971) Lactase deficiency: An example of dietary evolution. *Current Anthropology* 12:479-517. [WRC]

MacPherson, M. (1976) Sociobiology: Scientists at odds. *Washington Post.* November 21, pp. 81-84. [TRW]

Maller, O. & Desor, J. A. (1974) Effect of taste on ingestion by human newborns. In: *Fourth symposium on oral sensation and perception*, ed. J. Bosma, pp. 279-311. Washington, D.C.: Government Printing Office. [ta CJL]

Mann, T. (1938) The blood of the Walsungs. In *Stories of three decades*, pp. 297-319. New York: Knopf. [GS]

Marais, E. (1937) *The soul of the white ant.* New York: Dodd Mead. [GS]

(1969) *The soul of the ape.* New York: Atheneum. [GS]

Margolis, H. (1981) A new model of rational choice. *Ethics* 91:265-79. [RDM]

Masters, R. D. (1975) Politics as a biological phenomenon. *Social Science Information* 14:7-63. [RDM]

(1976) The impact of ethology on political science. In: *Biology and politics*, ed. A. Somit, pp. 197-233. The Hague: Mouton. [GS]

(1982, in press) Is sociobiology reactionary? The political implications of inclusive fitness theory. *Quarterly Review of Biology* 57. [RDM]

Maynard Smith, J. (1978) The evolution of behavior. *Scientific American* 239:176-92. [RDM]

Mayr, E. (1960) The emergence of evolutionary novelties. In: *The evolution of life*, ed. S. Tax, pp. 349-80. Chicago: University of Chicago Press. [MTG]

(1974a) Behavior programs and evolutionary strategies. *American Scientist* 62:650-59. [MTG]

(1974b) Teleological and teleonomic: A new analysis. In: *Methodological and historical essays in the natural and social sciences*, ed. R. S. Cohen & M. W. Wartofsky, pp. 91-117. Dordrecht: D. Reidel. [ALC]

Medawar, P. B. (1981) Stretch genes. *New York Review of Books*, July 16, pp. 45–48. [ALC, JKK, ta CJL, RDM, LBS]
Moltz, H. (1963) Imprinting: An epigenetic approach. *Psychological Review* 70:123–38. [WRC]
Montagu, M. F. A., ed. (1980) *Sociobiology examined*. New York: Oxford University Press. [TRW]
Muller, H. J. (1959) The guidance of human evolution. *Perspectives in Biology and Medicine* 3:1–43. [TDJ]
Murdock, G. P. (1967) *Ethnographic atlas*. Pittsburgh: Pittsburgh University Press. [JS]
Osgood, C. E., May, W. H., & Miron, M. S. (1975) *Cross-cultural universals of affective meaning*. Urbana: University of Illinois Press. [RP]
Parker, S. T. & Gibson, K. R. (1979) A developmental model for the evolution of language and intelligence in early hominids. *Behavioral and Brain Sciences* 2:367–408. [TDJ]
Pattee, H. H., ed. (1973) *Hierarchy theory*. New York: Braziller. [RDM]
Plotkin, H. C. & Odling-Smee, F. J. (1981) A multiple-level model of evolution and its implications for sociobiology. *Behavioral and Brain Sciences* 4: 225–68. [JKK, LBS]
Plutchik, R. (1980) *Emotion: A psychoevolutionary synthesis*. New York: Harper & Row. [r CJL, RP]
Popper, K. & Eccles, J. C. (1977) *The self and its brain: An argument for interactionism*. Berlin: Springer International. [JKK]
Pribram, K. (1979) Behaviourism, phenomenology, and holism in psychology. *Journal of Social and Biological Structures* 2:65–72. [GS]
(1980) The role of analogy in transcending limits in the brain sciences. *Daedalus* 109:19–38. [GS]
Priest, G. L. (1977) The common law process and the selection of efficient rules. *Journal of Legal Studies* 6:65–82. [MTG]
Roederer, J. (1978) On the relationship between human brain functions and the foundations of physics, science, and technology. *Foundations of Physics* 8:423–38. [GS]
Rosenberg, A. (1980) *Sociobiology and the preemption of social science*. Baltimore: Johns Hopkins University Press. [ALC, RDM]
Sahlins, M. (1976) *The use and abuse of biology*. Ann Arbor: University of Michigan Press. [r CJL, TRW]
Schneirla, T. C. (1956) The interrelationships of the "innate" and the "acquired" in instinctive behavior. In: *L'Instinct dans le comportement des animaux et de l'homme*, ed. P.-P. Grasse, pp. 387–452. Paris: Masson. [TDJ]
(1966) Behavioral development and comparative psychology. *Quarterly Review of Biology* 41:283–303. [TDJ]
Schubert, G. (1967a) Academic ideology and the study of adjudication. *American Political Science Review* 61:106–29. [GS]
(1967b) Ideologies and attitudes, academic and judicial. *Journal of Politics* 29:3–40. [GS]
(1975) *Human jurisprudence*. Honolulu: University Press of Hawaii. [GS]
(1981) The sociobiology of political behavior. In: *Sociobiology and human politics*, ed. E. White, pp. 193–238. Lexington, Mass.: D. C. Heath. [GS]
Searle, J. (1980) Minds, brains, and programs. *Behavioral and Brain Sciences* 3:417–58. [AR]
Seligman, M. E. P. (1972) Phobias and preparedness. In: *Biological boundaries of learning*, ed. M. E. P. Seligman & J. L. Hager, pp. 451–60. New York: Appleton-Century-Crofts. [ta CJL]
Skinner, B. F. (1971) *Beyond freedom and dignity*. New York: Knopf. [JKK]
(1981) Selection by consequences. *Science* 213:501–4. [JFW]
Sociobiology Study Group (1977) Sociobiology – A new biological determinism. In: *Biology as a social weapon*, ed. Ann Arbor Science for the People Editorial Collective, pp. 133–50. Minneapolis: Burgess. [ALC]
Sperry, R. W. (1980) Mind-brain interaction: Mentalism, yes; dualism no. *Neuroscience* 5:195–206. [JKK]
Textor, R. B. (1967) *A cross cultural summary*. New Haven: HRAF Press. [JS]
Thomas, L. (1981a) Unacceptable damage (book review of *Hiroshima and Nagasaki: The physical, medical, and social effects of the atomic bombings* by the Committee for the Compilation of Materials on Damage Caused by the Atomic Bombs in Hiroshima and Nagasaki). *New York Review of Books*, vol. 28, no. 14, pp. 3–8. [HEG]
(1981b) *Unforgettable fire: Pictures drawn by atomic bomb survivors*, ed. Japan Broadcasting Corporation. New York: Pantheon. [HEG]
Thompson, J. N., Jr. & Thoday, J. M., eds. (1979) *Quantitative genetic variation*. New York: Academic Press. [r CJL]
Tiersky, R. (1980) Confusion and uncertainty in comparative politics. Johns Hopkins University Bologna Center, Occasional paper 30. [RDM]
Waddington, C. H. (1957) *The strategy of the genes*. New York: Macmillan. [TDJ,GS]
(1960) *The ethical animal*. London: George Allen & Unwin. [GS]
(1962) *The nature of life*. New York: Atheneum. [GS]
(1975) *The evolution of a evolutionist*. Ithaca, N.Y.: Cornell University Press. [GS]
Whiting, J. W. M. (1969) Effects of climate on certain cultural practices. In: *Environment and cultural behavior*, ed. A. P. Vayda, pp. 416–55. Garden City, N.Y.: Natural History Press. [RDM]
Wiener, N. (1950) Some maxims for biologists and psychologists. *Dialectica* 4:22–27. [TRW]
Williams, T. R. (1959) The evolution of a human nature. *Philosophy of Science* 26:1–13. [TRW]
(1972a) *Introduction to socialization: Human culture transmitted*. St. Louis: C. V. Mosby. [ta CJL, TRW]
(1972b) The socialization process: A theoretical perspective. In: *Primate socialization*, ed. F. Poirier, pp. 207–60. New York: Random House. [TRW]
(1975) On the origin of the socialization process. In: *Socialization and communication in primary groups*, ed. T. R. Williams, pp. 233–49. The Hague: Mouton. [TRW]
(1981) *Socialization*. Englewood Cliffs, N.J.: Prentice-Hall (in press). [TRW]
Wilson, E. O. (1975) *Sociobiology: The new synthesis*. Cambridge, Mass.: Harvard University Press. [DPB, ALC, ta CJL, TRW]
(1978) *On human nature*. Cambridge, Mass.: Harvard University Press. [JKK,TRW]
Wilson, P. J. (1980) *Man, the promising primate: The conditions of human evolution*. New Haven: Yale University Press. [HEG]
Wilson, R. S. (1978) Synchronies in mental development: An epigenetic perspective. *Science* 202:939–48. [ta CJL]
Wohlwill, J. F. (1980) Cognitive development in childhood. In: *Constancy and change in human development*, ed. O. G. Brim & J. Kagan, pp. 359–444. Cambridge, Mass.: Harvard University Press. [JFW]
Zipf, G. K. (1968) *The psychobiology of language*. Cambridge, Mass.: MIT Press. [MTG]
Zuckerkandl, E. (1976) Programs of gene action and progressive evolution. In: *Molecular anthropology*, ed. M. Goodman, R. E. Tashian & J. H. Tashian, pp. 387–447. New York: Plenum Press. [TDJ]

1984

"The relation between caste ratios and division of labor in the ant genus *Pheidole* (Hymenoptera: Formicidae)," *Behavioral Ecology and Sociobiology* 16(1): 89–98

If the theory of adaptive demography I presented in 1968 is correct, and the members of different worker subcastes (such as minors and majors) are regulated by the innate schedules of birth and death of colony members in a manner that optimizes colony fitness, it follows that the more specialized each subcaste becomes, the fewer should be the individuals maintained by colonies to address the task for which the subcaste is specialized. *Pheidole* is an ant genus well suited to tests of this prediction. Its many species vary greatly in their characteristic ratio of majors to minors. In this study, combining behavioral and caste-ratio data in social insects with ergonomic theory, the prediction appears to have been confirmed.

Behav Ecol Sociobiol (1984) 16:89–98

Behavioral Ecology and Sociobiology

The relation between caste ratios and division of labor in the ant genus *Pheidole* (Hymenoptera: Formicidae)

Edward O. Wilson
Museum of Comparative Zoology Laboratories, Harvard University, Cambridge, Massachusetts 02138, USA

Received March 2, 1984 / Accepted March 20, 1984

Summary. Ten species of *Pheidole*, representing as many species groups from various localities in North and South America, Asia, and Africa, were analyzed to probe for possible relationships between caste ratios and division of labor.

Minor workers are behaviorally almost uniform among the species, but major workers vary in repertory from 4 to 19 behavioral acts (Table 1, Fig. 2). The major repertory size increases significantly across the species with the percentage of majors in the worker force (Fig. 3). This trend is consistent with the basic prediction of ergonomic optimization models under an assumption of colony-level selection. There is also a trend toward reduction of behavioral repertory with increase of size in the major relative to the minor, a second relation expected from theory, but the data are not sufficient to reach statistical significance.

When the minor: major ratio was lowered to below 1:1 (from the usual 3:1 to 20:1, according to species), in three widely different species (*guilelmimuelleri, megacephala, pubiventris*), the repertory size increased by 1.4–4.5X and the rate of activity by 15–30X (Table 1, Figs. 4–6). The change occurred within 1 h of the ratio change and was reversed in comparably short time when the original ratio was restored.

This abrupt and important shift in behavior permitted the major workers to serve as an emergency stand-by caste, available to be summoned to a nearly full repertory when the minor worker caste was depleted. The majors also restored 75% or more of the missing minor workers' activity rate under laboratory conditions. Their transformation allowed continued oviposition by the queen and the rearing of larvae to the adult stage.

In line with these findings, a distinction is made between programmed "elasticity" in the repertory of individual workers and castes and the "resiliency" of the colony as a whole, which depends upon the pattern of caste-specific elasticity.

Introduction

The myrmicine ant genus *Pheidole* possesses several outstanding advantages for experimental and phylogenetic studies of social behavior and ecology. It is first of all one of the most prevalent ant genera in the world. It rivals the formicine genus *Camponotus* in number of species and extent of geographic range, being virtually worldwide outside the very cold temperate and polar regions. It is also the dominant ant genus overall in numbers of colonies and individual workers in most localities within its vast range (Wilson 1976a).

Second, with the exception of several parasitic species, *Pheidole* is consistently dimorphic (see Fig. 1). The well-defined major caste differs from species to species in size, degree of morphological specialization, behavior, and numerical representation within the colony. (Members of the *tepicana* group, comprising about 1% of all known New World species, have two or more major castes.) The minor workers take care of most of the quotidian tasks of the colony, while the major workers are specialized either for seed milling, abdominal food storage, defense, or some combination of these functions (Buckingham 1911; Bruch 1916; Wheeler 1928; Wilson 1976b, c; Droual 1983; Calabi et al. 1983). Finally, most species of *Pheidole* have proved relatively easy to culture in the laboratory.

These traits together make *Pheidole* an especially felicitous group in which to analyze caste roles and division of labor. Correlative studies can

Fig. 1. Portion of a colony of the dimorphic ant *Pheidole pubiventris* from Brazil. Minor and major workers are seen with brood in various stages of development

be conducted on large samples of species, even into the hundreds, and caste ratios are readily manipulated to create "pseudomutant" colonies for optimization analysis (see Wilson 1980).

In this article I report the results of two related studies on division of labor in *Pheidole*. Estimates are given of the repertory size of ten species, representing an equal number of species groups and originating from various localities in North and South America, Asia, and Africa. The minor and major castes are compared across the sample in order to test a basic prediction of earlier ergonomic theory: that the more anatomically specialized a caste, the smaller should be its behavioral repertory as well as its numerical representation within the colony (Wilson 1968; Oster and Wilson 1978).

Next, experiments are described in which major workers were stressed by progressive reduction in the minor worker force. The research was designed to probe the extent to which major workers alter their behavior in compensation to care for themselves and the other colony survivors. A parallel study was conducted with the minor workers. My initial purpose was to learn whether the ants expand their repertory when the opposing caste is reduced and contract it when the other caste is increased.

When addressing the various possibilities it is useful to make a distinction between these responses of ant workers at two levels of organization. The first can be called *elasticity*: the degree of change by expansion or contraction of the behavior of individuals or castes as a reaction to major alterations in the environment. Elasticity is a more appropriate term than, say, plasticity or flexibility, because as will be seen from the present study, the repertory stretches back and forth in a predictable manner in response to outside "forces" in the form of stimuli from other castes. Elasticity in turn determines the *resiliency* of the colony as a whole, that is, the degree to which the colony responds to environmental alterations through changes in the behavioral repertories of the constituent castes and individuals (see for example, Wilson 1983). In order to make quantitative descriptions of homeostatic responses by colonies, it is necessary to identify the feedback loops that contribute to their resiliency. A key determining feature of these loops is the magnitude of the plasticity of individual repertories.

Materials and methods

In the following list the names of the species employed in the study are preceded by the abbreviations used to identify them in the illustrations; they are followed by the numbers of behavioral acts observed in each of the two castes (minor; major) in "normal", i.e., unmodified colonies housed in laboratory nests.

dent. *P. dentata*: St. Marks, Wakulla Co., Florida. Mixed warm temperate woodland. (1,222; 208)

dist. *P. distorta* or a closely related species: Fazenda Esteio, 87 km N of Manaus, Brazil. Primary lowland rain forest. (1,140; 83)

emb. *P. embolopyx*: Fazenda Esteio, Brazil. Primary lowland rain forest (1,166; 590)

guil. *P. guilelmimuelleri*: Fazenda Esteio, Brazil. Primary lowland rain forest. (1,542; 58)

hort. *P. hortensis*: Sri Lanka. Lowland rain forest. (Data from Calabi et al. 1983). (3,685; 255)

meg. *P. megacephala*: Bush Key, Dry Tortugas, Florida; a pantropical tramp species, ultimately African in origin. Red mangrove woodland. (1,418; 483)

mend. *P. mendicula*: Fazenda Esteio, Brazil. Primary lowland rain forest. (1,077; 175)

min. *P. minutula*: Fazenda Esteio, Brazil. Primary lowland rain forest, a symbiotic plant ant that nests in swollen leaf bases of melostome shrubs belonging to the genus *Maieta*. (1,535; 656)

pub. *P. pubiventris*: Fazenda Esteio, Brazil. Primary lowland rain forest. (1,286; 233)

sp. A *P. sp. indet.*, possibly undescribed; a medium-sized, shining dark brown species with elongate propodeal spines. Fazenda Esteio, Brazil. Primary lowland rain forest. (1,255; 267)

Colonies of *Pheidole dentata* and *P. megacephala* were housed in moistened test tubes placed on the floor of fluon-lined plastic tubs, the procedure described in an earlier account (Wilson 1976b). The Brazilian species were kept in a special form of artificial nest designed to simulate the natural forest site while permitting the entire colony to be observed constantly while both in the brood chambers and during foraging. Plaster of Paris was poured to a depth of 2 cm in the bottom of a rectangular clear plastic container 22 × 30 cm and 10 cm deep (smaller containers were used for the smallest *Pheidole* species, *P. mendicula* and *P. minutula*). As the plaster of Paris set, I carved 10–12 chambers into its surface that were roughly similar in size and proportions to the natural nest chambers of the colony to be cultured. In the case of medium-sized species living in pieces of rotting wood the chambers were typically ovoid or circular in shape and 1–4 cm across; hence I cut chambers in the artificial nest about 2 × 3 cm and 1 cm deep. The artificial nest chambers were connected by galleries 5 mm wide and deep and covered tightly by a rectangle of glass plate 10 × 18 cm across and 3 mm thick. Two to four exit galleries led from beneath the glass to the remainder of the plaster of Paris surface, which served as the foraging arena. Fragments of decaying wood and leaves from the vicinity of the original nest were scattered over the surface to add to the "naturalness" of the microenvironment. Colonies were fed twice weekly with cubes of Bhatkar diet (see Bhatkar and Whitcomb 1970) and fragments of freshly killed insects, usually mealworms (*Tenebrio molitor*) and cockroaches (*Nauphoeta cinerea*) offered in small dishes. The plaster of Paris was moistened by watering once or twice weekly, but the ants were also given an emergency retreat in the form of a test tube 14.8 cm long with 2.3 cm inner diameter; the tube interior was kept constantly moist by tight cotton plugs that trapped water at the bottom.

Behavioral repertories were accumulated by the ethological classification and methods employed earlier using *Pheidole dentata* and other ant species (see for example Wilson 1976b, c). Complete repertory sizes were estimated by the Fagen-Goldman method, in which the frequency data are fitted to a lognormal Poisson distribution and asymptotic repertory sizes and their standard errors estimated (Fagen and Goldman 1977). The total number of behavioral acts recorded in this study was 18,334, of which 3,940 came from the published research of Calabi et al. (1983) on *P. hortensis*.

Observations were made with a 10–70X swing-arm dissecting microscope, at 26° ± 2° C. Combinations of workers with different caste ratios were set up in the following manner: the nest queen and a small complement of minor and major workers were transferred to a second artificial nest made to resemble as much as possible the one from which they originated. Also added were a set of eggs, small larvae, large larvae, and pupae in the same ratios and proportions (brood piece/worker) that existed in the original colony. The sub-colonies were then given a period of 24 h before experiments began. Thereafter, the worker and brood proportions were modified by the addition or subtraction of individuals, and those left behind were given one hour to reorganize. Observations were made during the ensuing 10 h through periods of 10–30 min for a total observation period per day of 1 h. In a few cases when activity was exceptionally high and continuous, the observation periods were shortened to either 15 or 30 min. Two to four experiments with each caste combination were performed, with different combinations taken in random sequence to minimize habituation and other special time effects.

Voucher specimens of all the species have been placed in the ant collection of the Museum of Comparative Zoology, Harvard University.

Results

Species differences. The full worker repertories of three representative species of *Pheidole* are presented in Table 1 (the repertories of the other species are available upon request from the author). The first point of interest, reinforced by inspection of the repertory sizes of all ten species given in Fig. 2, is that very little variation occurs in the minor worker caste across species. The minor workers perform virtually all of the quotidian tasks of the major workers; they even engage in defense, seed milling, and food storage by means of abdominal repletion, although less dramatically and effectively than the major workers specializing in these tasks. The similarity of the minor worker repertories across species is probably even greater than these observational data reveal, because the few behavioral acts seen in one species and not another are all relatively uncommon acts that almost certainly would be added if either observation time were prolonged or the nest conditions were changed slightly. Examples include grooming of males and alate queen (which were absent in some of the laboratory nests), licking the nest wall, and anal trophallaxis. In effect, what these additions would do would be to move the lower-scoring species such as *dentata* and *hortensis* closer to the others. But even as it is, the range in repertory size across species is 40% (*hortensis* to *minutula* and *pubiventris*), and only a minority of the pairs of the species show repertory size differences significant at the 95% confidence level.

Differences do exist among the species in the way minor workers build nests, respond to alarm and defense pheromones, and forage (see for example comparisons in Wilson 1976b and Droual 1983). But overall, the basic repertories are remarkably similar across a wide range of species groups. Also, as revealed in Fig. 2, no significant relation exists between repertory size and total body size, even though head width varies through a factor of 2.1X (from *mendicula* to *distorta*), and the brain volume can safely be assumed to range over the square to cube of that amount – approximately, 5X or more.

In contrast, variation in the major worker repertories was dramatic and at least as strong as in any other polymorphic genus thus far studied. A typical low-repertory species is *P. guilelmimuelleri*, for which the full data are given in Table 1 under the column "major in normal colony." The majors possess one of the smallest repertories known in ants, being limited to self-grooming, eating, and defense. They attack intruders within the nest entrance gallery and brood chambers, and can be effective fighters under these circumstances, but do not venture outside the nest (at least, under laboratory conditions) except during colony emigrations. In sharp contrast is the behavior of the major workers of *P. pubiventris*, a species that occurs sympatrically with *P. guilelmimuelleri* in the Amazonian forests. Their repertory (see column labeled "major in normal colony") includes a large part of the minor responses, although the majors seldom engage in brood care and were never seen

Table 1. Repertories of minor and major castes of three *Pheidole* species, in normal (unmodified) laboratory colonies and when in subcultures with various experimentally modified caste ratios. The number given for each behavior is the fraction devoted to the behavior among the total number of acts observed in a given caste

Behavior	Species and caste combination										
	P. guilelmimuelleri				*P. megacephala*			*P. pubiventris*			
	Minor in normal colony	Minor alone	Major in normal colony	Major alone	Minor in normal colony	Major in normal colony	Major alone	Minor in normal colony	Minor (1 minor: 3 major)	Major in normal colony	Major alone
Self-grooming	0.278	0.330	0.569	0.441	0.314	0.680	0.540	0.159	0.290	0.442	0.405
Allogroom adult:											
Minor worker	0.110	0.108	–	–	0.166	0.018	–	0.051	0.020	–	–
Major worker	0.020	–	–	0.034	0.028	–	0.091	0.017	0.024	0.004	0.018
Dealate queen	0.003	0.003	–	0.027	0.015	–	0.012	0.014	0.003	–	0.002
Alate queen	–	–	–	–	–	–	–	0.010	0.017	–	–
Male	0.026	0.020	–	–	–	–	–	–	–	–	–
Brood care:											
Carry or roll egg	0.026	0.018	–	0.039	0.025	–	0.029	0.012	0.006	–	0.043
Lick egg	0.012	0.019	–	0.019	0.014	–	0.017	0.004	0.004	–	0.034
Carry or roll larva	0.103	0.090	–	0.049	0.101	–	0.023	0.098	0.024	0.026	0.056
Lick larva	0.132	0.135	–	0.126	0.071	–	0.062	0.182	0.172	0.082	0.109
Feed larva solids	0.008	0.007	–	0.002	0.004	–	0.006	0.005	0.011	–	0.001
Assist ecdysis to pupa	0.002	0.016	–	–	–	–	–	0.005	0.002	–	0.006
Carry or roll pupa	0.021	0.006	–	0.041	0.023	–	0.043	0.030	0.009	–	0.057
Lick pupa	0.028	0.015	–	0.109	0.032	–	0.081	0.061	0.058	–	0.134
Assist eclosion of adult	0.009	0.006	–	–	0.004	–	–	0.022	0.013	–	–
Handle meconium	0.006	0.005	–	0.003	0.002	–	–	0.019	0.002	–	–
Regurgitate with											
Larva	0.007	0.004	–	–	0.013	–	0.008	0.019	0.061	0.039	0.017
Minor worker	0.008	0.015	–	–	0.057	0.236	–	0.028	0.019	0.129	–
Major worker	–	–	–	–	0.014	–	0.017	0.010	0.009	0.009	0.010
Dealate queen	0.005	0.002	–	–	0.005	–	–	0.001	0.001	–	–
Alate queen	–	–	–	–	–	–	–	0.002	0.008	–	–
Male	–	0.001	–	–	–	–	–	–	–	–	–
Forage	0.045	0.036	–	0.002	0.020	0.031	0.014	0.037	0.057	0.021	–
Feed outside nest	0.006	0.002	–	–	0.030	0.009	–	0.025	0.010	–	–
Lay odor trail	0.003	0.003	–	–	0.004	–	–	0.002	0.004	–	–
Carry food particles	0.042	0.028	0.017	0.004	0.006	–	0.017	0.030	0.029	0.056	0.010
Feed inside nest	0.053	0.069	–	0.015	0.008	–	0.010	0.080	0.080	0.107	0.018
Carry adult nestmate	0.005	0.002	–	0.007	0.001	–	0.002	0.003	–	–	–
Aggression (drag or attack)	–	–	–	–	–	–	–	0.004	–	–	–
Carry dead adult	–	0.004	–	0.001	0.006	–	–	0.005	0.005	0.017	–
Carry dead larva or pupa	0.001	–	–	–	0.002	–	–	0.002	0.004	–	–
Feed on adult nestmate	–	0.004	–	–	0.007	–	–	0.017	0.014	0.013	–
Feed on larva or pupa	0.004	–	–	–	0.008	–	0.014	0.017	0.001	–	–
Handle nest material	0.030	0.046	–	0.054	0.009	–	0.012	0.011	0.038	–	0.053
Lick wall of nest	–	–	–	–	0.004	–	–	0.011	0.003	–	–
Chew on seed	0.005	0.004	0.069	–	–	0.004	–	0.005	0.002	0.043	0.010
Antennal tipping	0.002	0.001	–	–	0.006	0.022	–	0.004	0.001	–	–
Dispose of infrabuccal pellet	0.001	0.001	–	–	–	–	–	–	–	–	–
Guard nest entrance	–	–	0.345	0.025	–	–	–	–	–	–	0.015
Anal "trophallaxis"	–	–	–	–	–	–	–	–	0.003	0.013	0.001
Observed repertory size	30	30	4	18	30	8	18	35	34	14	19
Estimated true repertory size (±S.E.)	30±1	31±1	4±1	18±1	30±1	8±1	18±1	36±2	35±2	14±2	19±1
Number of separate acts observed	1,542	1,339	58	1,196	1,418	225	483	1,286	1,037	233	819

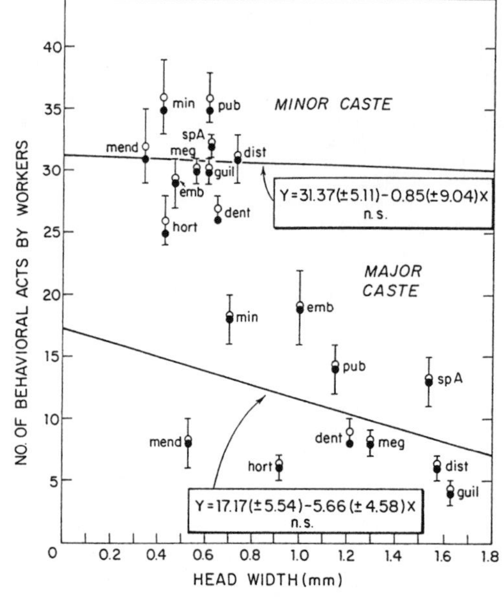

Fig. 2. The relation between absolute body size (measured as maximum head width, frontal view and exclusive of compound eyes) and repertory size. *Filled circles*: observed number of behavioral acts. *Open circles and brackets*: mean with standard error of behavioral acts estimated by fitting the data to a log-normal Poisson distribution. *Each pair of circles and brackets* refers to a different species of *Pheidole*. The full scientific names and colony origins are given at the beginning of the section on "Materials and methods." The head widths given are the averages of 3 to 5 measurements for each caste

to handle nest material. *P. megacephala*, the third species presented in Table 1, is intermediate in repertory size.

The size of the major worker repertory declines slightly with body size among the ten species studied (Fig. 2), although the trend is not statistically significant and no safe judgment can be made about it with existing data. Curiously, the major worker repertory size *increases* with the ratio of minor worker head width to major worker head width, as follows:

$$Y = 0.88\ (\pm 11.62) + 19.48\ (\pm 23.28)X$$

But this trend is also less than statistically significant, as can be seen by inspection ($P > 0.4$).

A very significant trend is found in the relation between the percentage of the worker force consisting of major workers and the major worker repertory size, as displayed in Fig. 3. In simple words, the less the majors do, the fewer there are.

The last two correlations, should they persist with the addition of more *Pheidole* species, are in

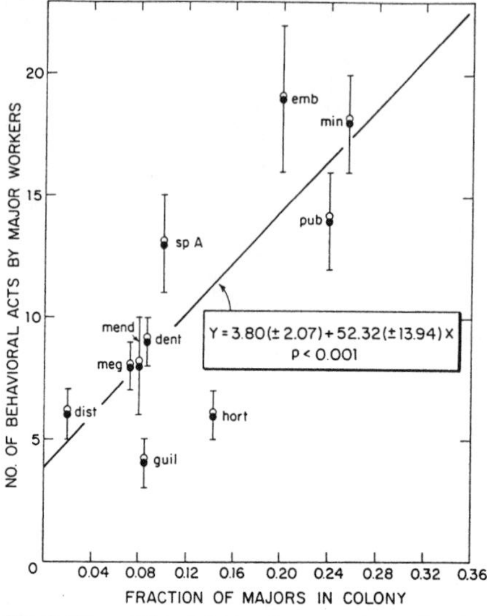

Fig. 3. The relation between the percentage of major workers in colonies and the repertory size of the major caste. Conventions as in Fig. 2

accord with a basic prediction from ergonomic theory: as a caste becomes more specialized in anatomy – as reflected in the larger size relative to minor workers – and as it becomes more specialized in behavior, these two changes together should cause a reduction in repertory size, accompanied by greater efficiency in the remaining behavioral acts and hence less representation in the worker force (Wilson 1968; Oster and Wilson 1978).

Behavioral elasticity. The repertory of the minor worker caste proved to be insensitive to the minor: major ratios. In other words they were inelastic. When all major workers were removed from the *P. guilelmimuelleri* colony, the repertory of the minor workers did not expand in compensation (see Table 1, "minor alone" column). When the ratio of minors to majors was changed from 3:1 to 1:3 in a *P. pubiventris* subculture (with the nest queen and proportionately reduced brood also present), the minor worker repertory was not compressed in compensation; see the "minor (1 minor:3 major)" column in Table 1.

Also, the *rates* of activity of minor workers did not change with the caste ratios. A subculture of *P. guilelmimuelleri* was set up with 12 majors, 12 minors, the nest queen, and proportionately reduced brood. In three replicates, the rate of self-grooming in acts/individual/h was 8.92, 9.85, and

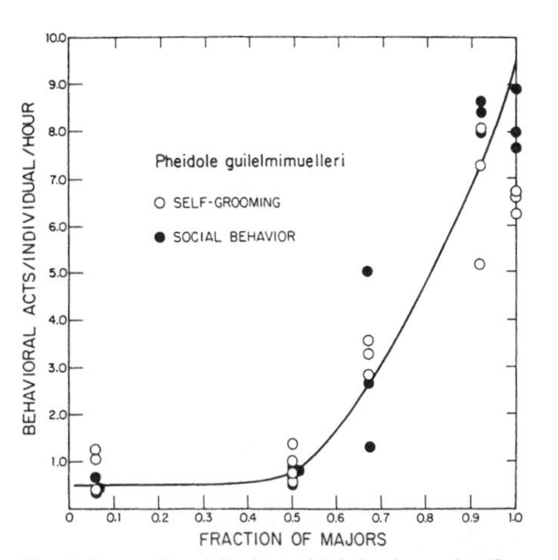

Fig. 4. Rates of activity in social behavior and self-grooming as functions of the proportion of majors in artificially constituted subcultures, in the South American species *Pheidole guilelmimuelleri.* Social behavior is defined as all of the repertory except for self-grooming, feeding, and individual waste disposal

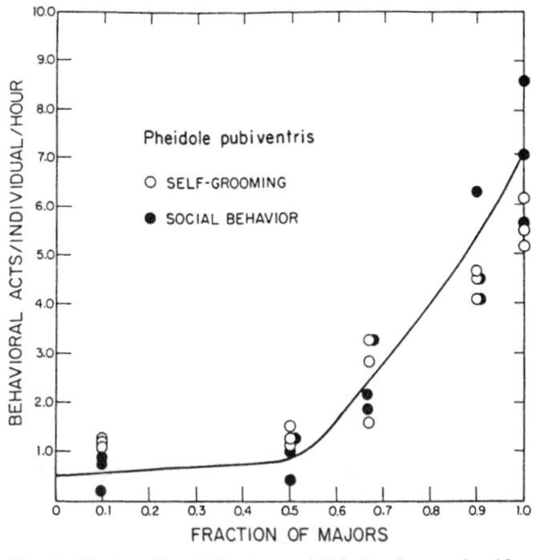

Fig. 5. Rates of activity in social behavior and self-grooming as functions of the proportion of majors in artificially constituted subcultures, in the South American species *Pheidole pubiventris.* Definitions of terms as in Fig. 4

11.67 respectively ($\bar{x}=10.15$). The rate of social behavior, that is, all acts except self-grooming, feeding, and personal waste disposal, was 10.79, 12.00, and 15.38 respectively ($\bar{x}=12.72$). A second subculture was established with 12 minor workers alone. Self-grooming rates were 11.00, 13.33, and 14.00 respectively ($\bar{x}=12.78$); and social behavior rates were 11.00, 13.33, and 13.33 respectively ($\bar{x}=12.55$). The rates of self-grooming and social behavior did not differ significantly between the two subcultures. Similar negative results were obtained with subcultures of *P. pubiventris.*

The major workers, on the other hand, proved strongly sensitive to changes in caste ratio and in a predictable, patterned manner. When major workers of *guilelmimuelleri, megacephala,* and *pubiventris* were placed in subcultures with nest queens and proportionately reduced amounts of brood but no minors, they expanded their repertories by 1.4X to 4.5X according to species (Table 1), and increased the rate of social activity by approximately 15X to 30X, again according to species (Figs. 4–6). The pattern of activity increase was remarkably similar across the three species, as shown by a comparison of the curves in Figs. 4–6. In each the rate rose abruptly when the minor:major ratio passed 1:1, as though a threshold exists at this level. Also, the transformation occurred within an hour or less. In other words, the majors did not require a long-term hormonal mediation or learning experience to become more like minor workers.

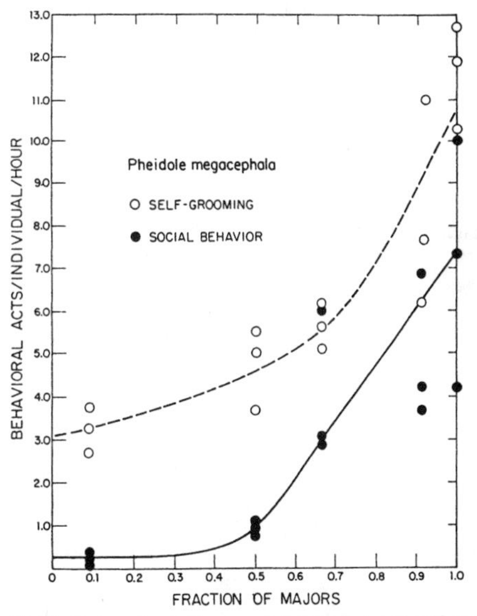

Fig. 6. Rates of activity in social behavior and self-grooming as functions of the proportion of majors in artificially constituted subcultures, in the African species *Pheidole megacephala.* Definitions of terms as in Fig. 4

In order to learn whether the rate increase was truly due to a change in the minor:major ratio and not to a change in the absolute numbers of workers present, in other words some form of "group effect," I repeated the experiment with subcultures of *P. pubiventris* containing 24 major workers alone instead of 12 major workers alone. The rates of self-grooming in three replicates were 3.00, 3.50, and 4.25 acts/individual/h respectively ($\bar{x}=3.58$), and the rates of social behavior were 4.42, 5.58, and 7.08 acts/individual/h respectively ($\bar{x}=5.69$). The self-grooming rates were slightly lower than in the groups of 12 (compare with Fig. 5), but the rates of social behavior were not significantly different. Also, the repertories expanded to a comparable extent in the group of 24 as opposed to the group of 12.

Ergonomic resiliency. The data for *P. guilelmimuelleri* and *P. pubiventris* were consulted, and additional data recorded, to answer the following question: to what extent does the sharp increase in social repertory size and rate of social activity in the majors left in solitude compensate for the reduced contribution due to the loss of the minor workers? Put another way, suppose that a *Pheidole* colony were deprived of most or all of its minor workers, a circumstance not at all unlikely when the colony is small, say consisting of 20–100 minor and major workers. How fully will the major workers make up for the loss? We seek a more quantitative measure of ergonomic resiliency, in other words the capacity of the colony as a whole to respond to a traumatic change in its caste structure.

The surprising answer is that the major workers make up most but not quite all of the repertory of the minor workers (see also behavioral repertories in Table 1); and in the case of *P. guilelmimuelleri* and *P. pubiventris*, for which appropriate measurements were taken, the major workers attain 75–85% of the full rate of social activity of the colony prior to the removal of the minor workers.

The essential data are the following. When 12 majors of *P. guilelmimuelleri* were placed with 12 minors, the mother queen, and brood in stages of development and quantity proportionate to those in the full colony, they averaged only 0.67 acts of social behavior/individual/h. When the minors were removed (and brood diminished proportionately, by one-half), the social activity of the majors increased to 8.20 acts/individual/h, a factor of 12X. When in the presence of 12 majors, the queen, and the same brood, the 12 minor workers performed at the rate of 12.72 acts/individual/h; the combined major and minor activity in the small colony is thus 0.67 + 12.72 or 13.39 acts/individual/h. Hence the 12X increase in the rate of activity of the majors in the absence of the minors brought the group to 8.20/13.39 = 61% of the full activity.

But note that the major workers deprived of the minors had only half the amount of brood to attend. When this reduced amount is taken into account by subtracting it from the activity of the combined 12 minors and 12 majors in the unreduced colony, the 12X rise in major activity is seen to be substantially greater than 61%. The adjusted rate of combined activity (before the minors were removed but with half the amount of brood care) is now 0.67 + 10.30 = 10.97 acts/individual/h, and the ergonomic compensation due to the 12X rise in major activity is 8.20/10.97 = 75% of full activity.

A comparable result was obtained with *P. pubiventris*. The 12 majors increased their rate from 0.89 acts/individual/h (when with 12 minors plus queen and proportionate brood) to a rate of 7.25 acts/individual/h when the minors and half the brood were removed, a factor of 8X. The rate of activity of minors in the undisturbed group (12 minors plus 12 majors) was 12.53 acts/individual/h and the combined rate for the two castes was therefore 13.42 acts/individual/h. The 8X increase in major activity when the minors were removed thus made up 7.25/13.42 = 54% of the original activity. But again, the majors deprived of the minors had only half the amount of brood to attend. The adjusted rate of combined activity (before the minors were removed, but with only half the amount of brood care) is 0.89 + 7.70 = 8.59 acts/individual/h, and the 8X rise in major activity is 7.25/8.59 = 84% of full activity.

Equivalent measurements of minor worker activity were not taken for *P. megacephala*, but it was my impression from recording the data for the behavioral repertories that the rate of minor activity is comparable to that of *P. pubiventris* and hence the compensation of major activity (due to an 8X increase in activity; see Fig. 6) is close to the 75–85% range demonstrated for *P. guilelmimuelleri* and *P. pubiventris*.

In addition to compensating for most of the *rate* of social activity, the *pubiventris* and *megacephala* majors also expanded their repertories to perform over 80% of the tasks performed by the minors. The majors of *guilelmimuelleri* did not expand their repertory as much. For example, no case was recorded of their regurgitating to larvae, whereas this was a common behavior in *pubiventris* and *megacephala*. Even so, it can be said that in all three of the *Pheidole* species, the majors serve very well as an emergency stand-by caste.

But – can all-major groups actually save the colony by sustaining the queen and rearing brood on their own? In order to answer this question,

96

I first set up five experimental groups of 50 *P. megacephala* with one nest queen each. As controls I composed (under identical conditions) five groups of 45 minor workers and five major workers each with a nest queen, simulating small natural colonies with normal caste ratios, as well as five nest queens alone, with no worker force at all. None of the groups, either experimental or control, was given brood. All of the queens in the "pure major colonies" and "natural" controls produced substantial masses of eggs during the following 3 weeks; none of the workerless queens produced eggs. The probability that the difference in presence or absence of eggs between the pure major and workerless units was due to sampling alone is $2^{-4}=0.02$. The probability that the average *number* of eggs produced is due to chance was less than 10^{-4}. It was concluded that the presence of the major workers results in increased oviposition by the queen. Both the control and experimental subcultures began to decline at this point, possibly due to the small group size (*megacephala* colonies are normally very populous), and the experiment was discontinued.

The following additional experiment was conducted with *P. pubiventris* in order to track the remainder of the brood cycle: three nests were set up with ten major workers and four worker larvae and one queen larva in each. As controls three more nests were set up with ten minor workers and a similar group of larvae. No queens were included in any of these subcultures. During the following month all of the groups produced pupae, and all but one produced at least one adult. It was concluded that majors alone are able to rear larvae to maturity and thus are capable of saving colonies largely or entirely depleted of the minor caste. This result was also obtained by Wheeler and Nijhout (1984), who reared worker larvae of *Pheidole bicarinata* to maturity with all-major groups.

The correlates of self-grooming. Since there is a close correlation between the rate of self-grooming and the rate of performance of other acts, including mostly social behavior, the question arises as to whether majors deprived of all other members of their colony would decrease social behavior and simultaneously decrease self-grooming. If this were to occur, we could reasonably postulate a direct causal relation: fewer social acts somehow induce less frequent self-grooming.

The following experiment revealed that this is not the case. Eleven major workers of *P. guilelmimuelleri* were isolated in subculture as previously described but this time without the queen, minor workers, and brood. Their behavior was recorded during three one-hour periods. Behavior other than self-grooming did decline significantly ($P<0.01$); the rate was 18–40 acts/individual/h ($\bar{x}=31.7$), approximately one-third that of the same major workers when kept with the queen and proportionately reduced brood (but no minor workers) with a range of 92–107 ($\bar{x}=98.3$). The decline was almost entirely due to the canceling of all forms of queen and brood care. At the same time there was little or no reduction in behaviors unrelated to these two categories. Thus allogrooming was 8–15 ($\bar{x}=15.6$), compared to 1–17 ($\bar{x}=6.3$) when the queen and brood were present. Nest construction was 5–14 ($\bar{x}=9.0$), compared to 8–13 ($\bar{x}=10.7$) when the majors were accompanied by the queen and brood. In both cases $P>0.1$. Self-grooming also remained at approximately the same high level: 83–103 acts/individual/h ($\bar{x}=90.3$), compared with 75–80 ($\bar{x}=78.3$) when the queen and brood were present; $P>0.1$. It was concluded that there is no direct relationship between the rates of self-grooming and social behavior. The inference is consistent with the more general conclusion that majors become active in all behavioral categories when minors are few or absent and relatively inactive when many minors are present.

Discussion

The remarkable variation in the repertory size of major workers among the ten species of *Pheidole* can be arrayed along a primary ethocline, with one end being the display of most of the 25 or more behaviors performed by minor workers (*embolopyx*) and the other end being only four relatively specialized behaviors (*guilelmimuelleri*). The phylogenetic directionality of this ethocline, whether from many to few or the reverse (or even from the middle outward in two directions), cannot be ascertained with present information, but a reduction in repertory seems the more likely trend within a recently evolved genus such as *Pheidole*.

Looked at more closely (Table 2), the ethocline is seen to consist of three trends toward the small-repertory end. The majors are limited in varying degrees to one or the other of three specialties: (1) seed milling (*distorta, guilelmimuelleri*); (2) guarding food found by the minor workers and assisting in its dismemberment away from the nest (*dentata*, sp. A); and (3) guarding the nest entrances while also serving as repletes in food storage (*hortensis, mendicula*). All three kinds of specialists tend to drop allogrooming and brood care, and no species is yet known in which the major workers specialize in brood care, queen atten-

Table 2. The performance of principal categories of social behavior by the major caste of *Pheidole* species, in undisturbed colonies with normal caste ratios

Species	Estimated repertory size	Social behaviors (principal categories)							
		Forage	Brood care	Chew seeds	Allo-groom adults	Carry foood particles	Guard nest entrance	Guard food	Handle nest material
embolopyx	19	X	X	X	X	X	X	X	
minutula	18	X	X		X	X	X		
pubiventris	14	X	X	X	X	X			
sp. A	13	X	X	X				X	
dentata	8	X	X					X	
megacephala	8	X		X	X				
mendicula	8		X			X	X		X
distorta	6			X	X				
hortensis	6	X					X	X	
guilelmimuelleri	4			X		X			

dance, or nest excavation. Many more species will need to be studied to learn whether these trends hold generally through the genus *Pheidole*, and whether other, still unsuspected trends exist.

Why should the percentage of major workers drop as specialization increases? As noted in earlier models of ergonomic optimization and adaptive demography (Wilson 1968; Oster and Wilson 1978), this relation is predictable providing the following two conditions are true: natural selection occurs at the level of the colony as a whole and not at the level of the individual colony members, and specialization actually results in greater efficiency and higher colony fitness. Because workers of *Pheidole* – unlike those of most other kinds of ants – entirely lack ovaries, the possibility for individual selection is precluded and selection affecting caste specialization must occur at the colony level. The second condition is also partly met. My studies of seed milling in colonies of *P. distorta, P. guilelmimuelleri*, and *P. pubiventris* have shown that the majors open mustard and lettuce seeds more quickly than the minors, which indeed have difficulty milling the seeds at all. Also, the majors of *P. mendicula* appear clearly to have more distended abdomens during repletion than minors, although this impression has not yet been tested by appropriate measurements.

In general, the hypothesis of colony-level selection is favored by the evidence but cannot be entirely accepted until energetic yield and colony fitness are measured, both in colonies with major workers and without them.

The elasticity of the major caste was a surprising discovery. These large-headed, often sluggish insects are galvanized into a nearly full minor-worker repertory when their representation in the colony rises above 50%. The transformation takes place within an hour or less, showing that the full behavior program is latent in the brain, a program waiting to be summoned as an emergency program.

In fact, the data show that the majors do serve as an effective emergency stand-by caste, substituting for the missing minor workers in most of their quotidian roles and at 75% or more of the normal activity of this smaller caste. The majors are able to take care of the queen and to rear new minor workers through from at least as far back as the larval stage. In other words, the majors can bring the entire colony through crises in which the minor worker caste is severely depleted, either during periods of starvation and queen infertility (majors are longer lived than minors and hence more likely to persist until the fertility is resumed) or because of heavy differential mortality during foraging.

Then – why don't the major workers perform the full repertory and at the higher rates of which they are capable *all* the time and not just during emergencies? The answer may well have to do with energetic efficiency. Each major has several times the dry weight of each minor, with correspondingly higher energetic costs during construction and maintenance, yet my observations show that they are no more mechanically efficient than minor workers in performing tasks and may well be inferior. It would appear that it is to the advantage of the colony on the whole for majors to stay with narrow repertories at which they are maximally efficient and to assume additional tasks only during emergencies.

The swift transition of the major workers from low to high activity when minor workers are removed and back again to low activity when the minor workers are returned suggests the existence of special stimuli readily perceived by the ants. The nature of these stimuli remains to be investigated.

Acknowledgements. Appreciation is expressed to Thomas Lovejoy and Woodruff Benson for assistance in the field in Brazil, and to Stephen Bartz and Bert Hölldobler for valuable suggestions on the manuscript. The research was supported by National Science Foundation grant BSR-81-19350.

References

Bhatkar A, Whitcomb WH (1970) Artificial diet for rearing various species of ants. Fla Entomol 53:229–232

Bruch C (1916) Contribución al estudio de las hormigas de la Provincia de San Luis. Rev Mus La Plata, Argentina 23:291–354

Buckingham E (1911) Division of labor among ants. Proc Am Acad Art Sci 56:425–507

Calabi P, Traniello JFA, Werner MH (1983) Age polyethism: its occurrence in the ant *Pheidole hortensis*, and some general considerations. Psyche 90:395–412

Droual R (1983) The organization of nest evacuation in *Pheidole desertorum* Wheeler and *P. hyatti* Emery (Hymenoptera: Formicidae). Behav Ecol Sociobiol 12:203–208

Fagen RM, Goldman RN (1977) Behavioural catalogue analysis methods. Anim Behav 25:261–274

Oster GF, Wilson EO (1978) Caste and ecology in the social insects. Princeton University Press, Princeton, NJ

Wheeler D, Nijhout F (1984) Soldier determination in *Pheidole bicarinata:* inhibition by adult soldiers. J Insect Physiol 30:127–135

Wheeler WM (1928) Mermis parasitism and intercastes among ants. J Exp Zool 50:165–237

Wilson EO (1968) The ergonomics of caste in the social insects. Am Nat 102:41–66

Wilson EO (1976a) Which are the most prevalent ant genera? Stud Entomol 19:187–200

Wilson EO (1976b) The organization of colony defense in the ant *Pheidole dentata* Mayr (Hymenoptera: Formicidae). Behav Ecol Sociobiol 1:63–81

Wilson EO (1976c) Behavioral discretization and the number of castes in an ant species. Behav Ecol Sociobiol 1:141–154

Wilson EO (1980) Caste and division of labor in leaf-cutter ants (Hymenoptera: Formicidae: *Atta*). II. The ergonomic optimization of leaf cutting. Behav Ecol Sociobiol 7:157–165

Wilson EO (1983) Caste and division of labor in leaf-cutter ants (Hymenoptera: Formicidae: *Atta*). III. Ergonomic resiliency in foraging by *A. cephalotes*. Behav Ecol Sociobiol 14:47–54

1985

“The sociogenesis of insect colonies,” *Science* 228: 1489–1495

In this brief synthesis I brought together several of the guiding themes of insect sociobiology under the rubric of “sociogenesis,” defined as the development of colonies through their life span and the mechanisms of growth and communication that regulate the development. Also included is a reprise of the theory of adaptive demography, introduced in 1968 (see “The ergonomics of caste in the social insects,” this volume). Adaptive demography addresses the determination of caste proportion and division of labor in different species according to the particular environmental pressures they have encountered through evolutionary time.

28 June 1985, Volume 228, Number 4707 SCIENCE

The Sociogenesis of Insect Colonies

Edward O. Wilson

Summary. Studies on the social insects (ants, bees, wasps, and termites) have focused increasingly on sociogenesis, the process by which colony members undergo changes in caste, behavior, and physical location incident to colonial development. Caste is determined in individuals largely by environmental cues that trigger a sequence of progressive physiological restrictions. Individual determination, which is socially mediated, yields an age-size frequency distribution of the worker population that enhances survival and reproduction of the colony as a whole, typically at the expense of individuals. This "adaptive demography" varies in a predictable manner according to the species and size of the colony. The demography is richly augmented by behavioral pacemaking on the part of certain castes and programmed changes in the physical position of colony members according to age and size. Much of what has been observed in these three colony-level traits (adaptive demography, pacemaking, and positional effects) can be interpreted as the product of ritualization of dominance and other forms of selfish behavior that is still found in the more primitive insect societies. Some of the processes can also be usefully compared with morphogenesis at the levels of cells and tissues.

Together with flight and metamorphosis, colonial life was one of the landmark events in the evolution of the insects and evidently served as a source of their ecological success. Preliminary studies indicate that approximately one-third of the entire animal biomass of the Amazonian terra firme rain forest may be composed of ants and termites, with each hectare of soil containing in excess of 8 million ants and 1 million termites (*1, 2*). On the Ivory Coast savanna the density of ants is 20 million per hectare, with one species, *Camponotus acvapimensis*, alone accounting for 2 million (*3*). Such African habitats are often visited by driver ants (*Dorylus* spp.), single colonies of which occasionally contain more than 20 million workers (*4*). And the driver ant case is far from the ultimate. A "supercolony" of the ant *Formica yessensis* on the Ishikari Coast of Hokkaido was reported to be composed of 306 million workers and 1,080,000 queens living in 45,000 interconnected nests across a territory of 2.7 square kilometers (*5*).

The environmental impact of these insects is correspondingly great. In most terrestrial habitats ants are among the leading predators of insects and other small invertebrates (*3, 6, 7*), and leafcutter ants (*Atta* spp.) are species for species the principal herbivores and most destructive insect pests of Central and South America (*8*). *Pogonomyrmex* and other harvester ants compete effectively with mammals for seeds in deserts of the southwestern United States (*9*). Other ants move approximately the same amount of soil as earthworms in the woodlands of New England, and they surpass them in tropical forests. Both are exceeded in turn by termites, which also break down a large part of the vegetable litter and diffuse the products through the humus (*10, 11*).

Edward O. Wilson is Frank B. Baird, Jr., Professor of Science and Curator in Entomology, Museum of Comparative Zoology, Harvard University, Cambridge, Massachusetts 02138. This article is based on the lecture of the 1984 Tyler Prize for Environmental Achievement, delivered at the University of Southern California on 24 May 1984.

The Reasons for Success

In general, the most abundant social insects are the evolutionarily more advanced groups of ants and termites, in other words, those with the highest percentage of derived traits in anatomy and physiology as well as the more populous and complexly organized societies (*6, 12, 13*). What is the real origin of this competitive advantage in the environment as a whole? At the risk of oversimplification, it can be said that entomologists have come to recognize three qualities as being most important. First, coordinated groups conduct parallel as opposed to serial operations and hence make fewer mistakes, especially when labor is divided among specialists. If different cadres of workers in an ant colony simultaneously forage for food, feed the queen, and remove her eggs to a safe place, they are more likely as a whole to complete the operation than if they perform the steps in repeated sequences in the manner of solitary insects (*13*). Second, groups can concentrate more energy and force at critical points than can single competitors, using sheer numbers to construct nests in otherwise daunting terrain, as well as to defend the young, and to retrieve food more effectively. Finally, there is caste: in ways that vary among species, the food supply is stabilized by the use of larvae and special adult forms to store reserves in the form of fat bodies and nutrient liquids held in the crop, while defense, nest construction, foraging, and other tasks are mostly accomplished by specialists (*14*).

The aim of much of contemporary research on social insects is to identify more fully the mechanisms by which colony members differentiate into castes and divide labor—and to understand why certain combinations of mechanisms have produced more successful products than others. The larger hope is that more general and exact principles of biological organization will be revealed by the meshing of comparable information from developmental biology and sociobiology. The definitive process at the level of the organism is morphogenesis, the set of procedures by which individual cells or cell populations undergo changes in shape or position incident to organismic development (*15*). The definitive process at the level of the colony is sociogenesis, the procedures by which individuals undergo changes in caste, behavior, and physical location incident to colonial development. The question of interest for general biology is the nature of the similarities between morphogenesis and sociogenesis.

The study of social insects is by necessity both a reductionistic and holistic enterprise. The behavior of the colony can be understood only if the programs and positional effects of the individual members are teased apart, ultimately at the physiological level. But this information makes full sense only when the patterns of colonial behavior of each species are examined as potential idiosyncratic adaptations to the natural environment in which the species lives. At both levels social insects offer great advantages over ordinary organisms for the study of biological organization. Although no higher organism can be readily dissected into its constituent parts for study and then reassembled, this is not the case for the insect colony. The colony can be fragmented into any conceivable combination of sets of its members, manipulated experimentally, and reconstituted at the end of the day, unharmed and ready for replicate treatment at a later time. The technique is used for analysis of optimization in social organization as follows: the colony is modified by changing caste ratios, as though it were a mutant. The performance of this "pseudomutant" is then compared with that of the natural colony and other modified versions. The same colony can be turned repetitively into pseudomutants in random sequences on different days, eliminating the variance that would otherwise be due to between-colony differences (*16*). At a still higher level of explanation, that of the ecosystem, the large numbers of species of various kinds of social insects (more than 1000 each in the ant genera *Camponotus* and *Pheidole* alone) give a panoramic view of the evolution of colonial patterns and make correlative analysis of adaptation more feasible.

Principles of Sociogenesis

In all species of social insects thus far studied, caste differences among colony members have proved to be principally or exclusively phenotypic rather than genetic. The environmental factors in each instance belong to one or more of the following six categories: larval nutrition (which is especially important in ants); inhibition caused by pheromones or other stimuli from particular castes (the key factor in many kinds of termites); egg size and hence quantity of nutrients available to the embryo; winter chilling; temperature during development; and age of the queen (*6, 17, 18*). Phenotypic caste determination is similar to restriction during cell differentiation. That is, the growing individual reaches one or more decision points at which it loses some of its potential, and this diminution continues progressively until it reaches the final decision point, where it is determined to the caste it will occupy as an adult. For example, in the ant genus *Pheidole* the restriction to either the queen line or worker line occurs in the egg; then larvae in the worker line become committed to development as either minor or major workers in the fourth and final instar. The cues affecting these two decisions, which include nutrition, winter chilling of queens, and inhibitory pheromones, are mediated to the developing tissue by juvenile hormone (*19, 20*).

The differentiation of the colony members into physical castes is supplemented in the great majority of social species by a regular progression on the part of most workers through different work roles during aging. In this way the individual belongs not only to one physical caste but to a sequence of temporal castes as it passes through its life-span. By far the most common sequence is for the worker to join in the care of the queen or immature stages shortly after it emerges into the adult stage, then to participate in nest building, and, finally, to forage outside the nest for food. Temporal castes are a derived trait in evolution, having become most clearly demarcated in species with the largest societies. They are typically weak or absent in anatomically primitive species with small colony populations (*6, 21*).

Although individual workers are flexible with respect to caste at the start of their personal development in the egg stage, the colony as a whole is rigidly limited to a single array of castes. Each species also has a particular size-frequency distribution of adult workers (*13, 22, 23*). Workers in the ant genus *Pheidole*, for example, are divided into two subcastes, the minors and the majors, by size and body proportions. Among ten species selected for their taxonomic diversity, the majors were found to range from 3 percent in *Pheidole distorta* to 25 percent in *Pheidole minutula* (*23*). A lesser amount of variation exists among colonies belonging to the same species, and recent work suggests indirectly that some of the variation is genetic. Seven colonies of *Pheidole dentata* raised under uniform laboratory conditions through three brood cycles maintained relatively constant major worker percentages, and these levels varied significantly among the colonies, from approximately 5 to 15 percent (*24*).

The size-frequency distribution can also persist through relatively long periods of geological time. A fragment of a colony of the extinct weaver ant *Oecophylla leakeyi* preserved intact from the African Miocene (the only fossil insect society collected to date) proved to have the same distinctive pattern as the two living species of the genus, *Oecophylla longinoda* and *Oecophylla smaragdina*. In particular, the frequency curve was sharply bimodal, with the major workers somewhat more numerous than the minors and with a small number of medias connecting the two moieties. The allometry, or disproportionate variation in body parts, is also similar between the extinct and living species (*25*).

These several lines of evidence have led to the hypothesis of adaptive demography (*13, 26*), which can be summarized as follows. The vast majority of insect, vertebrate, and other animal populations evolve primarily through selection at the level of the individual organism. As a consequence, survivorship curves and natality schedules are directly adaptive, whereas the age-frequency distribution of the population as a whole emerges as an epiphenomenon. In the advanced social insects, in contrast, selection occurs primarily at the level of the colony, with workers mostly or entirely eliminated from reproduction and colonies competing against one another as compact units. Colonies whose members possess the most effective age-frequency distribution are more likely to survive and to reproduce, regardless of the fate of individual colony members. It is generally believed that the workers will increase the replication of genes identical to their own by promoting the physical well-being of the colony, even if they sacrifice themselves to achieve this end. Hence the age-frequency distribution of the colony members is directly subject to natural selection. Survivorship and natality schedules are indirectly subject to natural selection, in the sense of being shaped according to the effect they have on the age-frequency distribution of the colony as a whole.

The adaptive demography hypothesis has begun to be tested by both correlative analysis and experimentation. For example, linear programming models predict that as a caste specializes, its members should decrease in proportion within the colony membership (*26*). This relation does hold among the species of *Pheidole* so far studied: the repertory size of the major caste is correlated significantly across species with the percentage of the majors in the worker force. Put another way, as the majors perform fewer tasks and devote more time proportionately to roles for which they are anatomically specialized, they become scarcer in the colony population (*23*).

And yet the major workers of *Pheidole* retain a remarkable flexibility. When the minor-major ratio was experimentally reduced to below 1:1 in three widely different species of the genus, the majors increased the number of kinds of acts they performed by as much as 4.5 times and their rate of activity 15 to 30 times. The change occurred within 1 hour of the ratio change and was reversed in comparably short time when the original ratio was restored. Thus the major workers were found to respond in a manner reminiscent of the genome of a somatic cell. Under normal circumstances most of their brain programs are silent: the active repertory is limited in a fashion appropriate to the tasks for which the majors are anatomically specialized. But when an emergency arises a much larger program is quickly summoned, the majors supply about 75 percent of the activity of the missing minors, and as a result the colony continues to feed and grow (*23*).

A second line of evidence of adaptive demography has been provided by studies of the leafcutter ant *Atta cephalotes*. New colonies of *Atta*, like those of most kinds of ants, are founded by single queens after the nuptial flights. These individuals dig a shaft into the ground, then eject a wad of symbiotic fungus from their mouths onto the ground and fertilize the hyphae with droplets of feces. During the next 6 weeks they rear the first brood of workers with reserves from their own bodies while bringing the small garden to flourishing condition. The queens have only enough ovarian yolk and other storage materials to rear one small group to maturity. In order for the colony to survive thereafter, the workers must range in size from a head width of 0.8 mm, which is small enough to culture the fungus, through 1.6 mm, which is just large enough to cut fresh leaves for the fungal substrate. It turns out that the first brood of workers possess a nearly uniform frequency distribution from 0.8 through 1.6 mm, which comes close to maximizing the number of individuals and at the same time achieves the minimum size range required to grow the fungus on which the colony depends (*27*).

As the leafcutter population expands afterward, the size-frequency distribution of the workers changes in dramatic fashion. The range is increased at both ends and the curve becomes strongly skewed toward the media and major worker classes (Fig. 1). An interesting question then arises: suppose that by some misadventure most of the population of a leafcutter colony were destroyed, reducing it to near the colony-founding state. Would the size-frequency distribution of new workers produced by the colony be characteristic of the beginning stage, or would it remain at the older stage? In other words, which is the more important in the ontogeny of the caste system, the size of the colony or its age? If age were more important, causing much of the available energy to be invested in workers larger than the minimum required to harvest leaves, the colony would be imperiled because of a

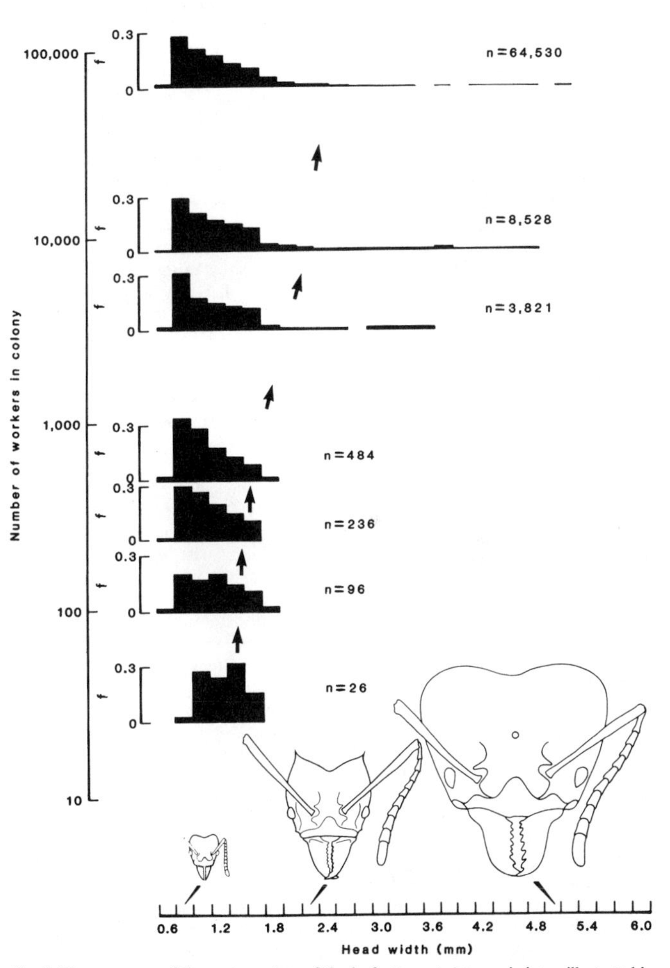

Fig. 1. The ontogeny of the caste system of the leafcutter ant *Atta cephalotes*, illustrated by seven representative colonies collected in the field or reared in the laboratory. The worker caste is differentiated into subcastes by continuous size variation associated with disproportionate growth in various body parts. The number of workers in each colony (*n*) is based on complete censuses; *f* is the frequency of individuals according to size class. The heads of three sizes of workers are shown in order to illustrate the disproportionate growth. Modified from Wilson (*27*).

shortage of the small gardener classes. The creation of just one new major worker, possessing a body weight 300 times that of a gardener worker, would bankrupt the already impoverished colony. In order to provide an answer, I selected four colonies 3 to 4 years old and with about 10,000 workers and reduced the population of each to 236, giving them an artificially imposed juvenile size-frequency distribution. The worker pupae produced at the end of the first brood cycle possessed a size-frequency distribution like that of small, young colonies rather than larger, older ones. Thus colony size is more important than age, and "rejuvenated" colonies are prevented from extinguishing themselves through an incorrect investment of their resources (*27*).

Such programmed resiliency implies the existence of control mechanisms operating at the level of the colony during population growth. An increasing fraction of the research on social insects is now being directed at the discovery of such mechanisms. This work has begun to reveal a fascinating pattern of feedback loops, pacemakers, and positional effects.

An example of negative feedback is provided by the events leading to the fission of honeybee colonies. The queen secretes a "queen substance," *trans*-9-keto-2-decenoic acid, which under most circumstances inhibits the construction of royal cells by the workers and hence the rearing of new queens (*28*). However, in large, freely growing colonies this pheromone must be supplemented by a second substance, the footprint pheromone, which is secreted in relatively large amounts from glands in the fifth tarsal segment of the queen. When bee colonies become overcrowded, the queen is unable to walk along the bottom edges of the comb, where the royal cells are ordinarily built. As a result the inhibition fails in that zone, the cells are built, and the colony reproduces. With the population density now reduced to below threshold density, the queen is able to resume her inhibitory control (*29*).

Most such controls are negative and hence contribute to physiological stability and smooth growth cycles within the colony. What appear to be properties of positive feedback and explosive chain reactions nevertheless do occur during nest evacuation in a few species. When attacking fire ant workers press closely on nests of the ant *Pheidole dentata*, the defending minor workers start laying odor trails back into the brood area. This causes excited movement through the nest and further bouts of recruitment. At the height of this expanding activity the workers and queen suddenly scatter from the nest and seek individual cover. When the fire ants are then experimentally removed, the *Pheidole* adults return to the nest and reoccupy it (*30*).

The coordination of activity is still imperfectly understood. Although the typical insect society is not quite the "feminine monarchie" envisioned by early entomologists (*31*), it is also much more than a republic of specialists. According to the species, certain immature stages and castes function as pacemakers and coordinators of colony activity. Ant larvae are specially effective in initiating foraging and nest construction by the adult workers. In army ants (*Eciton*), the hatching of larvae triggers the monthly nomadic cycle during which the entire colony marches to a new location daily (*32*). But in the great majority of other species thus far studied it is the queen that provides the maximum regulation. In more primitive societies, such as those of bumblebees (*Bombus*) and paper wasps (*Polistes*), she physically dominates her daughters and other females occupying the nests, prevents them from laying eggs, and by these actions forces most into foraging and other nonreproductive tasks. Such influence can transcend simple displacement. For instance, the presence of the queen of *Polistes fuscatus*, probably a typical species at this evolutionary level, increases and synchronizes overall worker activity (*33*). In carpenter ants (*Camponotus*), the mother queen is the principal source of the nest odor (*34*). When she is removed, the workers, now in a more chaotic state, fall back on odor cues emanating from their own bodies (*35*).

Workers of social insects move to different positions with reference to the queen and brood according to their ages. This pattern is usually centrifugal: soon after the worker emerges from the pupa into the adult stage, it attends the queen and immature stages, then drifts toward the outer chambers to assist in nest construction, and finally devotes itself primarily to foraging outside the nest. The progression is accompanied by physiological change. The details vary greatly among species, and even among members of the same colony, but in general the ovaries reach maximum development early in adult life, along with fat bodies and exocrine glands devoted to nutritive exchange (*6, 17, 36–38*). Afterward these tissues regress more than enough to counterbalance the growth of exocrine glands associated with nest construction and foraging, so that the worker declines overall in weight. Mortality due to accidental causes increases sharply among workers when they commence foraging. But this attrition has far less effect on the size-and-age structure of the worker population than if individuals commenced foraging early in life, because the natural life-span is curtailed in any case past the onset of foraging by physiological senescence. In the best documented case, the honeybee worker born in early summer typically begins foraging at 2 to 3 weeks of adult life and dies from senescence by 10 weeks into this period (*39*).

The workers of advanced insect societies are not unlike cells that emigrate to new positions, transform into new types, and aggregate to form tissues and organs. With relatively small adjustments in response thresholds according to size and age, intricate new patterns are created at the level of the colony. In the fungus-growing termite *Macrotermes subhyalinus*, for example, 90 percent of the foragers are large major workers past 30 days of age. Younger major and minor workers accept the grass collected by these foragers, consume it, and pass the partly digested material out into the fungus comb. Workers of various castes older than 30 days eat the fungus comb and produce the final feces (*40*). In the leafcutting ant *Atta sexdens* most of the fresh vegetation is gathered by workers of intermediate size (which, incidentally, achieve the highest net energetic yield of all the size groups). The material is then converted into new fungus substrate within the nest by an assembly-line operation that penetrates ever more deeply into the combs: successively smaller workers cut the leaves into tiny fragments, chew them into pulp, stick the processed lumps onto the growing combs, and transfer strands of fungi onto this newly prepared substrate. Finally, the smallest workers of all care for the proliferating fungus, virtually strand by strand (*16, 41*).

Such patterns are in fact much more intricate than a description of sequences alone indicates. In the ant *Pheidole dentata* and the honeybee *Apis mellifera* the tasks are broken into sets that are linked not by the similarity of the behaviors performed but by the proximity of the objects to which they are directed, thus reducing the travel time and energy expenditure of the individual workers (Figs. 2 and 3). The similarities between the two patterns can only be due to convergent evolution, since ants and bees arose during Mesozoic times from widely different stocks of aculeate wasps (*42*).

The Imperfection of Insect Societies

Although insects as a whole originated at least 350 million years ago, higher social insects did not appear until the Jurassic Period, roughly 200 million years ago, and they began an extensive evolutionary radiation only in the late Cretaceous and early Tertiary Periods, about 100 million years later (*42*). Even then, advanced social organization originated in as few as 13 stocks, 12 within the aculeate Hymenoptera (ants, bees, and wasps) and one in the cockroach-like orthopteroids that produced the termites (*6*).

Two possible explanations for this evolutionary conservatism have emerged from more detailed studies of individual colony members. The first is that the small size of the insect brain and the heavy reliance of social forms on chemical signaling place inherent limits on the amount of information flow through the colonies. This circumstance leads to frequent near-chaotic states and the dependence on colony decision-making by *force majeure*, a statistical preponderance of certain actions over others that lead to a dynamic equilibrium rather than clean binary choices (*6, 13, 43, 44*). Thus when released from threshold concentrations of the queen inhibitory pheromones, some honeybee workers build royal cells while a smaller number of workers set out to dismantle them. The final result is an equilibrial number of cells sufficient for the rearing of new queens (*44*).

On the other hand, a few mechanisms are coming to light that sharpen the precision of mass response and bring it closer to binary action. Markl and Hölldobler (*45*) reported the existence of "modulatory communication" in ants, a form of signaling in one channel that alters the threshold of response in another. For example, when harvester ants of the genus *Novomessor* encounter large food objects they make sounds by scraping together specialized surfaces on the thin postpetiole and adjacent abdominal segment. This stridulation does not cause an overt behavioral change in nestmates but raises the probability that they will release short-range recruitment chemicals. The overall result is a speeding and tightening of the coordination process.

The second force inhibiting social evolution, at least in the case of hymenopterans, is the substantial conflict among individuals for reproductive privileges. Dominance rank orders, once thought to be confined to simply organized societies of halictine bees, bumblebees, and polis-

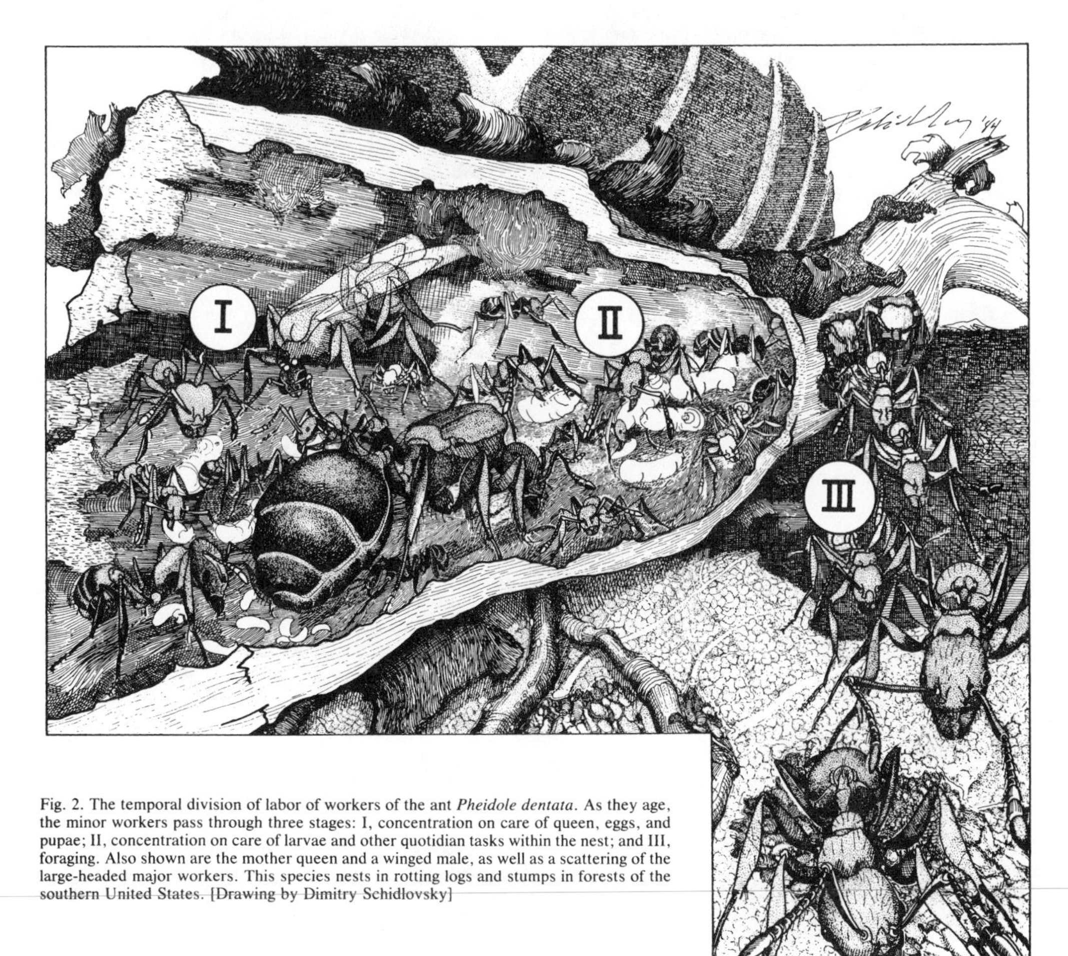

Fig. 2. The temporal division of labor of workers of the ant *Pheidole dentata*. As they age, the minor workers pass through three stages: I, concentration on care of queen, eggs, and pupae; II, concentration on care of larvae and other quotidian tasks within the nest; and III, foraging. Also shown are the mother queen and a winged male, as well as a scattering of the large-headed major workers. This species nests in rotting logs and stumps in forests of the southern United States. [Drawing by Dimitry Schidlovsky]

tine wasps, as well as associations of queens of a few kinds of ants [*Nothomyrmecia, Myrmecocystus*, and *Eurhopalothrix* (*46*)] have also been discovered in the workers of some species of ants as well (*47*). West-Eberhard has argued that competition among workers is more pervasive among advanced societies than has been recognized and that selection at the level of the individual has consequently played a key role in the division of labor (*36, 48*). She explains the centrifugal pattern of temporal castes (Figs. 2 and 3) as the product of such selection. The individual worker, by staying close to the brood chambers while still young and while her personal reproductive value is highest, maximizes her potential to contribute personal offspring. But as death approaches and fertility declines because of senescence, the optimum strategy for contributing genes to the next generation is to enhance colony welfare through more dangerous occupations such as defense and foraging, thus producing more brothers and sisters as opposed to personal offspring. By this criterion, Porter and Jorgensen (*37*) were correct to call foraging harvester ants the "disposable" caste. Hölldobler (*49*) has recently described what may be the ultimate case: aging workers of the Australian tree ant (*Oecophylla smaragdina*) occupy special "barracks nests" around the periphery of the main nest area. They stand idle most of the time and are among the first defenders to enter combat during territorial battles with other tree ant colonies.

Individual selection appears likely to have inhibited the refinement of social behavior, especially in the earliest stages of the evolution. Indeed, there is evidence that species of the bee genus *Exoneurella*, trading production of siblings for the production of offspring, have returned from primitive sociality back to a more nearly solitary state (*50*). Yet there does appear to be a point of no return in the rise of sociality. When colonies become very complex, organized by an intricate caste system and highly coordinated group movements, the advantages of queenlike behavior on the part of workers is diminished and may even disappear. In a few advanced ant genera, such as *Pheidole* and *Solenopsis*, the workers no longer even possess ovaries (*51*).

The pattern emerging from comparative studies suggests that as reproductive competition has declined during the elaboration of sociogenesis, dominance interactions have been ritualized to serve as part of the communicative signals dividing labor. In the more complex societies of bees and wasps, overt aggression is replaced by queen pheromones, but the inhibition of the ovaries of the subordinates and their induction into worker roles remain essentially the same (*6, 14*). Also, traces of aggressive and subordinate interactions persist in ritual form. The workers of stingless bees either hurriedly withdraw from the area when the

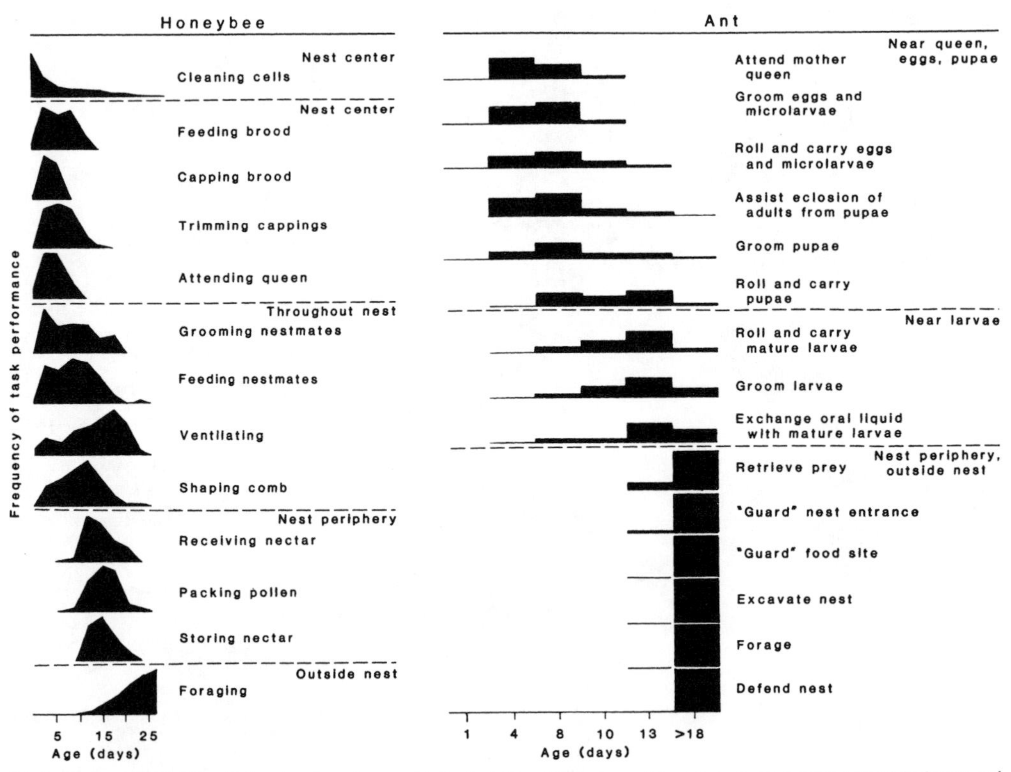

Fig. 3. The temporal division of labor, based on changes of behavior in the adult workers with aging, is shown in the ant *Pheidole dentata* and honeybee *Apis mellifera*: the insects shift from one linked set of tasks to another as they move their activities outward from the nest center (see Fig. 2). The similarities between the two species are convergent and believed to be adaptive. The sum of the frequencies in each histogram is 1.0. Adapted from Wilson (*54*) and Seeley (*55*).

queen approaches, thus clearing a path for her, or else they mock-attack, then bow to her head, and finally swing to her side to become part of the retinue (*52*). Ritualized dominance interactions may also be important between sterile workers. Major workers of the ant *Pheidole pubiventris* turn away from minor workers when they encounter them around the brood, thus yielding most of the care of the immature forms to these smaller nestmates. This aversion neatly divides colony labor into several principal categories (*23*).

Although seldom acknowledged in the literature, regulatory mechanisms are often found lacking even when they are intuitively anticipated by the investigator. For example, the major workers of *Pheidole dentata* are specialized for response against fire ants and other members of the genus *Solenopsis*, but when colonies are stressed continually with these enemies the major-minor ratio remains the same.

In other words, there is no increase in the defense expenditure in the face of a major threat (*24*). Leafcutter workers with head widths from 1.8 through 2.2 mm are responsible for most of the foraging, but when members of this important caste are removed experimentally, the colonies fail to compensate for the loss by increasing representation of the size class in later broods. The result is a reduction in energetic efficiency through two brood cycles (*53*).

On the whole, insect societies display impressive degrees of complexity and integrity on the basis of what appear to be relatively few sociogenetic processes. The mechanisms that do exist, together with their strengths, precision, and phylogenetic distribution, constitute a subject in an early and exciting period of investigation. Of comparable importance are the expected mechanisms that do not exist, so that investigators are likely to pay closer attention to them than has been the case in the past. As the full pattern becomes clearer, it may be possible to compare sociogenesis with morphogenesis in a way that leads to a more satisfying general account of biological organization.

References and Notes

1. L. Beck, *Amazoniana* **3**, 69 (1971).
2. F. J. Fittbau and H. Klinge, *Biotropica* **5**, 2 (1973).
3. J. Levieux, in *The Biology of Social Insects*, M. D. Breed, C. D. Michener, H. E. Evans, Eds. (Westview, Boulder, 1982), pp. 48–51.
4. A. Raignier and J. Van Boven, *Ann. Mus. R. Congo Belg. Tervuren* **2**, 1 (1955).
5. S. Higashi and K. Yamauchi, *Jpn. J. Ecol.* **29**, 257 (1979).
6. E. O. Wilson, *The Insect Societies* (Harvard Univ. Press, Cambridge, Mass., 1971).
7. R. L. Jeanne, *Ecology* **60**, 1211 (1979).
8. N. A. Weber, *Gardening Ants: The Attines* (American Philosophical Society, Philadelphia, 1972); J. M. Cherrett, in *The Biology of Social Insects*, M. D. Breed, C. D. Michener, H. E. Evans, Eds. (Westview, Boulder, 1982), pp. 114–118.
9. D. Davidson, J. H. Brown, R. S. Inouye, *BioScience* **30**, 233 (1980).
10. W. H. Lyford, *Harv. For. Pap.* **7**, 1 (1963).
11. T. Abe, in *The Biology of Social Insects*, M. D. Breed, C. D. Michener, H. E. Evans, Eds. (Westview, Boulder, 1982), pp. 71–75.
12. E. O. Wilson, *Stud. Entomol.* **19**, 187 (1976).
13. G. F. Oster and E. O. Wilson, *Caste and Ecology in the Social Insects* (Princeton Univ. Press, Princeton, N.J., 1978); J. M. Herbers, *J. Theor. Biol.* **89**, 175 (1981).
14. The large literature on the advantages of social life, much of it based on experimental studies, is reviewed by Wilson (*6*) and investigators in (*13*), as well as by C. D. Michener, *The Social Behavior of the Bees: A Comparative Study* (Harvard Univ. Press, Cambridge, Mass., 1974); H. R. Hermann, Ed., *Social Insects* (Academic Press, New York, 1979–1982), vols. 1–4; T. Seeley and B. Heinrich, in *Insect Thermoregulation*, B. Heinrich, Ed. (Wiley, New York, 1981), pp. 159–234.
15. See, for example, N. K. Wessells, *Tissue Interaction and Development* (Benjamin, Menlo Park, Calif., 1977).
16. E. O. Wilson, *Behav. Ecol. Sociobiol.* **7**, 157 (1980).
17. M. V. Brian, in *Social Insects*, H. R. Hermann, Ed. (Academic Press, New York, 1979), vol. 1, pp. 121–222.
18. J. de Wilde and J. Beetsma, *Adv. Insect Physiol.* **16**, 167 (1982).
19. D. E. Wheeler and H. F. Nijhout, *J. Insect Physiol.* **30**, 127 (1984).
20. L. Passera and J.-P. Suzzoni, *Insectes Soc.* **26**, 343 (1979).
21. For example, the very primitive ant *Amblyopone pallipes* appears to lack temporal castes completely [J. F. A. Traniello, *Science* **202**, 770 (1978)].
22. M. I. Haverty, *Sociobiology* **2**, 199 (1977); in *The Biology of Social Insects*, M. D. Breed, C. D. Michener, H. E. Evans, Eds. (Westview, Boulder, 1982), p. 251.
23. E. O. Wilson, *Behav. Ecol. Sociobiol.* **16**, 89 (1984).
24. A. B. Johnston and E. O. Wilson, *Ann. Entomol. Soc. Am.* **78**, 8 (1985).
25. E. O. Wilson and R. W. Taylor, *Psyche* **71**, 93 (1964).
26. E. O. Wilson, *Am. Nat.* **102**, 41 (1968); J. M. Herbers, *Evolution* **34**, 575 (1980).
27. E. O. Wilson, *Behav. Ecol. Sociobiol.* **14**, 55 (1983).
28. C. G. Butler and R. K. Callow, *Proc. R. Entomol. Soc. London* **B43**, 62 (1968).
29. Y. Lenski and Y. Slabezki, *J. Insect Physiol.* **27**, 313 (1981).
30. E. O. Wilson, *Behav. Ecol. Sociobiol.* **1**, 63 (1976).
31. C. Butler, *The Feminine Monarchie* (Barnes, Oxford, 1609).
32. T. R. Schneirla, *Army Ants: A Study in Social Organization*, H. R. Topoff, Ed. (Freeman, San Francisco, 1971).
33. H. K. Reeve and G. J. Gamboa, *Behav. Ecol. Sociobiol.* **13**, 63 (1983).
34. N. F. Carlin and B. Hölldobler, *Science* **222**, 1027 (1983).
35. ———, personal communication.
36. M. J. West-Eberhard, in *Natural Selection and Social Behavior*, R. D. Alexander and D. W. Tinkle, Eds. (Chiron, New York, 1981), pp. 3–17.
37. S. D. Porter and C. D. Jorgensen, *Behav. Ecol. Sociobiol.* **9**, 247 (1981).
38. T. D. Seeley, *ibid.* **11**, 287 (1982).
39. M. Rockstein, *Ann. Entomol. Soc. Am.* **43**, 152 (1950); S. F. Sakagami and H. Fukuda, *Res. Popul. Ecol.* **10**, 127 (1968).
40. S. Badertscher, C. Gerber, and R. H. Leuthold, *Behav. Ecol. Sociobiol.* **12**, 115 (1983).
41. E. O. Wilson, *ibid.* **7**, 143 (1980).
42. F. M. Carpenter and H. R. Hermann, in *Social Insects*, H. R. Hermann, Ed. (Academic Press, New York, 1979), vol. 1, pp. 81–89.
43. P. Hogeweg and B. Hesper, *Behav. Ecol. Sociobiol.* **12**, 271 (1983).
44. D. H. Baird and T. D. Seeley, *ibid.* **13**, 221 (1983).
45. H. Markl and B. Hölldobler, *ibid.* **4**, 183 (1978).
46. S. H. Bartz and B. Hölldobler, *ibid.* **10**, 137 (1982); B. Hölldobler and R. W. Taylor, *Insectes Soc.* **30**, 384 (1983); E. O. Wilson, *ibid.* **30**, 408 (1985).
47. B. J. Cole, *Science* **212**, 83 (1981); N. Franks and E. Scovell, *Nature (London)* **304**, 724 (1983).
48. M. J. West-Eberhard, *Proc. Am. Philos. Soc.* **123**, 222 (1979).
49. B. Hölldobler, *Biotropica* **15**, 241 (1983).
50. C. D. Michener, *Kansas Univ. Sci. Bull.* **46**, 317 (1965).
51. E. O. Wilson, *J. Kansas Entomol. Soc.* **51**, 615 (1978).
52. S. F. Sakagami, in *Social Insects*, H. R. Hermann, Ed. (Academic Press, New York, 1982), vol. 3, pp. 361–423.
53. E. O. Wilson, *Behav. Ecol. Sociobiol.* **15**, 47 (1983).
54. ———, *ibid.* **1**, 141 (1976).
55. T. D. Seeley, *ibid.* **11**, 287 (1982).

56. I am grateful to D. M. Gordon, B. Hölldobler, T. Seeley, and D. Wheeler for critical readings of the manuscript. Supported by a series of grants from the National Science Foundation, the latest of which is BSR 81-19350.

1985

"Between-caste aversion as a basis for division of labor in the ant *Pheidole pubiventris* (Hymenoptera: Formicidae)," *Behavioral Ecology and Sociobiology* 17(1): 35–37

Colonies of the ant genus *Pheidole* possess two sterile worker subcastes: small-headed minor workers, which attend to brood care and all other ordinary tasks of the colony, and large-headed major workers, which are specialized for defense. When the nest is attacked by enemies, the minor workers assist the majors in defense. I posed the following question: Do major workers have a similar capacity to expand their repertories in emergencies?

To find the answer, I removed varying numbers of minor workers, which are the numerically prominent caste in undisturbed *Pheidole* colonies. When the ratio of minors to majors dropped to 1:1 or lower, some of the majors quickly moved to join the larvae and began to care for them. The second question then followed: How do majors "know" that there is a shortage of minor worker nurses? It turns out that they do not have to make an estimate or carry any memorized information in their heads. Instead, they operate by a simple off-and-on rule: when majors approach a cluster of larvae with minors on it, they turn away. And they do so abruptly, in a clear and active avoidance of the combination of larvae and minors. When there are no minors with the larvae, however, the majors proceed onto the pile and commence nursing behavior.

The context-controlled aversive response is an example of a very simple decision rule of individual behavior that is translated into a major adaptive response at the level of the colony.

Behav Ecol Sociobiol (1985) 17:35–37

Behavioral Ecology and Sociobiology

Between-caste aversion as a basis for division of labor in the ant *Pheidole pubiventris* (Hymenoptera: Formicidae)

Edward O. Wilson

Museum of Comparative Zoology, Harvard University, Cambridge, Massachusetts 02138, USA

Received June 19, 1984 / Accepted October 22, 1984

Summary. When deprived of minor workers under experimental conditions, major workers of the ant *Pheidole pubiventris* dramatically increase their repertory and rate of activity, and the change is due in good part to the greater attention they pay the brood. When minor workers are reinstated in appropriate numbers, the majors reduce their attention to the immature stages to the ordinary, low levels. Their response consists of the active avoidance of minors while in the vicinity of the immature stages. However, majors do not turn from other majors near the brood as much as they do from the minors, and they do not avoid minors at all while in other parts of the nest. In addition, minors do not avoid either minors or majors anywhere in the nest. The result is a striking division of labor with reference to brood care.

Introduction

Major workers of *Pheidole*, which comprise from about 2 to 30% of the adult worker force according to species, display a very limited behavioral repertory and a low rate of activity in comparison with the minor workers. But when the proportion of majors is increased to 50% or more, both the repertory size and rate of activity increase dramatically, almost to the same levels as those of the minors. When the original proportion is restored, the repertory and rate drop to their original levels. Both transitions, that is, from low to high and high to low, occur within an hour or less. Thus the majors serve as an emergency stand-by caste (Wilson 1984).

This striking pattern provides an opportunity to examine the basis of division of labor between the two castes. When the percentage of majors is set between 60 and 80%, both minor and major workers are relatively active, and they are both numerous enough for frequent interactions to occur. During earlier studies of three species (*guilelmimuelleri, megacephala, pubiventris*) I noticed that one of the changes occurring in the majors as minors are restored is the early abandonment of the brood (eggs, larvae, pupae). Moreover, the desertion appeared to be based at least in part on an active avoidance of minors by majors when members of the two castes meet on or close to the brood pile. Accordingly, I extended the study of *P. pubiventris* in a way that allowed me to measure the magnitude of the avoidance in both castes and its dependence on caste ratios and location within the nest.

Methods

A colony of *P. pubiventris* collected near Manaus, Brazil, was cultured until it reached mature size, during which virgin queens were being steadily produced. The techniques of culturing and observation were the same as reported earlier (Wilson 1984). Subcultures consisted of a dealate queen, 40 workers varying in major/minor ratio, and a quantity of brood proportionate to that in the source colony (10 larvae of various sizes, 5 worker and queen pupae, 5 eggs).

Most of the subcultures were set up with either 25 or 75% major workers, in order to allow a comparison of behavior under these two conditions. At the former proportion (25%) the majors remained largely displaced from the brood piles and their behavioral rates were low. At 75% the majors were active on the brood pile and frequently contacted minor workers there.

As the adults approached each other inside the nest chambers, a record was made of the location of the encounter (on the brood or <1 mm from the nearest brood piece versus >1 mm from the brood), the position of the two workers at the moment of antennal contact by one or both, and the response of the ant making the antennal contact. Responses were classified according to one or the other of two categories: "turn away," in which the ant making the contact took at least one turn of 90 degrees or more before travelling more than four body lengths; and "stay" or "proceed," the latter response

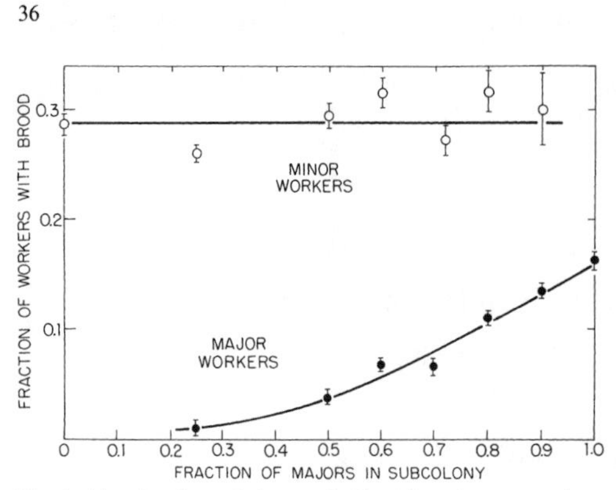

Fig. 1. The fraction of the total *Pheidole pubiventris* minor worker force and the fraction of the total major worker force on or within 1 mm of the brood are given as functions of the percentage of the workers composed of majors. The difference in the slopes of the two curves is due to the avoidance of minors by majors

being a trajectory of four body lengths with turns of less than 90%. The differences in proportions of responses were evaluated by the χ^2-test with two classes.

Table 1. Numbers of major or minor workers, respectively, giving one or the other of two responses (stay or proceed vs turn away) to minor or major workers, either near or apart from the brood. Significant inter-row or inter-column differences are indicated as *($P<0.05$) and **($P<0.01$)

Fraction of majors in colony	Near brood		Apart from brood
	Stay or proceed vs turn away		Stay or proceed vs turn away
A. Majors encountering minors			
0.25	(7 vs 35)	–**–	(52 vs 3)
	**		
0.75	(54 vs 67)	–**–	(63 vs 8)
B. Minors encountering majors			
0.25	(25 vs 0)		(39 vs 3)
0.75	(100 vs 8)		(64 vs 7)
C. Majors encountering majors			
0.25	(insufficient data)		(28 vs 4)
0.75	(49 vs 14)	–*–	(141 vs 14)
D. Minors encountering minors			
0.25	(90 vs 3)		(16 vs 0)
0.75	(100 vs 11)		(49 vs 1)

Results

As shown in Fig. 1, the percentage of minor workers (out of the total minor worker force) located on the brood remained constant as the minor/major ratio was varied. The minor worker curve is $y = -0.055\ (\pm 1.182) + 2.122(\pm 4.225)x$, with the slope not deviating significantly from zero. In contrast, the percentage of major workers falls off rapidly as minor workers are added: the curve for the range of abscissal values 0.5–1.0 fitted to a straight line is $y = 0.381\,(\pm 0.039) + 3.852(\pm 0.371)x$; the probability that the slope is zero is less than 0.001.

The data in Table 1 show that the difference in location between the two castes is due to the aversion of majors to minors on or close to the brood pile. Moreover, the response is specific to that location. The principal results leading to this conclusion are as follows:

(1) Majors turn from minors more at the brood pile than away from it.

(2) Majors also turn more from other majors when on or near the brood, but this aversion is less than that toward minors. (This relative aversion disappears when the adult population consists entirely of majors – as revealed in separate tests not displayed in Table 1, when 147 majors stayed or proceeded, while only 16 turned away.) Also, majors do not avoid other majors when away from the brood.

(3) Minors do not display the above pattern of aversion toward either caste at either location.

Discussion

It is generally assumed that much of the division of labor in social insects is based on caste differences in attraction to stimuli and in thresholds of response to the stimuli, although supporting evidence is surprisingly scarce in the literature (see for example Wilson 1978). The present study shows that division of labor can also be based on caste-specific aversions. The question then arises: is the mechanism of the kind documented in *Pheidole pubiventris* exceptional in occurrence, perhaps linked to the unusual behavioral flexibility of the major caste? The ability of the majors to serve as emergency stand-by auxiliaries depends on some kind of rapid recognition of castes as well as on their location in the nest, and hence the tendency to adjust the entire pattern of response after receipt of the correct stimuli. A simple aversion is one mechanism to achieve this result, and it will be interesting to learn whether it occurs in other categories of division of labor, in other species of social insects, under similar circumstances.

The mechanism of recognition was not established experimentally during this study, but it is almost certainly olfactory rather than tactile in nature. The majors showed the clearest responses after making direct antennal contact with the minors or at least coming within less than a millimeter of them. On the other hand they did not play the antennae over the bodies of the nestmates sufficiently well to obtain clues concerning distinctive shapes of major as opposed to minor workers.

Acknowledgements. The research was supported by National Science Foundation grant BSR 81-19350. I am grateful to Bert Hölldobler and Diana Wheeler for critical readings of the manuscript.

References

Wilson EO (1978) Division of labor in fire ants based on physical castes (Hymenoptera: Formicidae: *Solenopsis*). J Kans Entomol Soc 51:615–636

Wilson EO (1984) The relation between caste ratios and division of labor in the ant genus *Pheidole* (Hymenoptera: Formicidae). Behav Ecol Sociobiol 16:89–98

1987

"The earliest known ants: an analysis of the Cretaceous species and an inference concerning their social organization," *Paleobiology* 13(1): 44–53

When the Russian entomologist Gennedy Dlussky and I, and others later, were studying sphecomyrmines, the first ants discovered from the Mesozoic Era, we faced a problem: Were these primitive forms really ants, or were they, as Dlussky believed, a separate line of nonsocial wingless wasps? In the research reported here, I compared the anatomy of sphecomyrmine workers known at the time with contemporary ants and wingless solitary wasps to conclude that the wingless Mesozoic specimens were indeed true worker ants. I also associated what are likely the winged queens and males of the same or similar species. This interpretation came to be favored by later investigators.

Paleobiology, 13(1), 1987, pp. 44–53

The earliest known ants: an analysis of the Cretaceous species and an inference concerning their social organization

E. O. Wilson

Abstract.—The known Cretaceous formicoids are better interpreted from morphological evidence as forming a single subfamily, the Sphecomyrminae, and even a single genus, *Sphecomyrma,* rather than multiple families and genera. The females appear to have been differentiated as queen and worker castes belonging to the same colonial species instead of winged and wingless solitary females belonging to different species. The former conclusion is supported by the fact that the abdomens of workers of modern ant species and extinct Miocene ant species are smaller relative to the rest of the body than is the case for modern wingless solitary wasps. The wingless Cretaceous formicoids conform to the proportions of ant workers rather than to those of wasps (Figs. 1–2) and hence are reasonably interpreted to have lived in colonies.

The Cretaceous formicoids are nevertheless anatomically primitive with reference to modern ants and share some key traits with nonsocial aculeate wasps. They were distributed widely over Laurasia and appear to have been much less abundant than modern ants.

E. O. Wilson. Museum of Comparative Zoology, Harvard University, Cambridge, Massachusetts 02138

Accepted: October 14, 1986

Introduction

This article attempts to resolve a problem in systematics that bears significantly on the origin and early evolution of the ants and hence the antiquity of advanced social behavior in insects generally. Studies of four mid-Cretaceous amber specimens have established the presence of the ant subfamily Sphecomyrminae across a wide portion of present-day North America: *Sphecomyrma freyi* from New Jersey (Wilson et al. 1967a,b) and *S. canadensis* from Alberta (Wilson 1985a). Dlussky (1975, 1983) has described an important additional collection of ant-like forms from the Upper Cretaceous of the Taymyr Peninsula (extreme north-central Siberia), southern Kazakh S.S.R., and the Magadan region of extreme eastern Siberia. He erected 10 new genera to accommodate this material. In his more recent article, he also created a new family, the Armaniidae, to accommodate some of the genera while elevating the Sphecomyrminae to family rank (hence, Sphecomyrmidae) to receive others. This classification is summarized in Table 1.

Dlussky's taxonomic interpretation, which is based on careful and accurate descriptions of the new material, has sweeping consequences for our conception of the origin of the ants. First, it presents a picture of an extensive radiation of ants or ant-like forms by the early part of the Upper Cretaceous—in other words, by no later than 80 ma B.P. Second, Dlussky suggested that the Sphecomyrmidae are not true ants or even precursors of the Formicidae but a closely related side branch, while the Armaniidae are the true ancestors of the ants. He doubted that either the Sphecomyrmidae or the Armaniidae were eusocial; in other words he questioned whether they possessed the most advanced mode of social organization in which distinct queen and worker castes form overlapping adult generations and care for the developing young (see Wilson 1971).

Because of the light that the early fossils can shed on the origin of the ants and their distinctive social systems, which are matters hitherto largely unexplored due to the exclusively eusocial status of modern ant species, I decided to reexamine closely the morphological and biogeographic evidence provided by the Cretaceous material. I have arrived at a wholly different conclusion from that of Dlussky. In essence, the differences among the fossils cannot support the separation of the two new families, the Sphecomyrmidae and Armaniidae, from the Formicidae. It is difficult to justify even the recognition of any genus other than *Sphecomyrma* on the

0094-8373/87/1301-0003/$1.00

TABLE 1. Interpretation by Dlussky (1983) of the Mesozoic (Cretaceous) or antlike forms, classified to genus, with localities and approximate dates. Spelling of formation and estimation of dates follow van Eysinga (1978).

Taxon	Sex	Locality	Formation and age before present (million years B.P.)
Superfamily Formicoidea			
Family Sphecomyrmidae Dlussky 1983			
Sphecomyrma Wilson and Brown 1967	Female (wingless)	New Jersey, U.S.A.; Alberta, Canada	New Jersey: Santonian (80)
Cretomyrma Dlussky 1975	Female (wingless)	Taymyr Peninsula	Santonian (80)
Paleomyrmex Dlussky 1975	Male	Taymyr Peninsula	Santonian (80)
Family Armaniidae Dlussky 1983			
Archaeopone Dlussky 1975	Male	Southern Kazakh S.S.R.	Turonian (90)
Armania Dlussky 1983	Female (winged)	Magadan	Cenomanian (100)
Armaniella Dlussky 1983	Female (winged)	Magadan	Cenomanian (100)
Dolichomyrma Dlussky 1975	Female (wingless)	Southern Kazakh S.S.R.	Turonian (90)
Poneropterus Dlussky 1983	Male	Magadan	Cenomanian (100)
Pseudarmania Dlussky 1983	Female (wingless)	Magadan	Cenomanian (100)
Incertae Sedis (unplaced to family)			
Cretopone Dlussky 1975	Female (wingless?)	Southern Kazakh S.S.R.	Turonian (90)
Petropone Dlussky 1975	Female (wingless)	Southern Kazakh S.S.R.	Turonian (90)

basis of the morphological evidence. The most parsimonious explanation of the data is that the winged females and males from the Soviet deposits are queens and males of eusocial colonies of which the wingless *Sphecomyrma* and wingless Soviet species are the workers.

In proposing this view, I wish to emphasize that the disagreement is not over the facts assembled earlier by Dr. Dlussky and myself. There is only one exception, the supposed divided condition of the hind trochanter in the Armaniidae, to be discussed later. Instead, the disparity is due to a difference of interpretation of the facts with reference to higher classification and phylogeny. This is all to the good. By such contrasts the basic issues can be better clarified and the gathering of new data stimulated.

Materials and Methods

All of the characters known to vary among the Cretaceous species were first broken into character states. Each genus and species was then redescribed character by character in matrix form to allow close comparison of taxa. The holotypes of two of the best preserved and taxonomically most important Soviet species, *Armania robusta* and *Pseudarmania rasnitsyni,* were studied closely in comparison with the descriptions. I also worked with all of the four known specimens of the North American formicoids, placed in *Sphecomyrma* (Wilson 1985a).

In addition, measurements were made of a wide range of contemporary queen and worker ants and the winged and wingless females of other, nonsocial aculeate families deposited in the Museum of Comparative Zoology, in order to detect consistent differences in body proportions between social and nonsocial females. A single representative specimen was taken from the series available in each species, and the species in turn were selected to provide a large amount of phyletic diversity. This information was needed to infer the level of social evolution of the earliest fossils. Further measurements were utilized from the monograph of Miocene ants of North America by Carpenter (1930). All of these fossils belong to extinct species but surviving subfamilies and in some cases surviving genera. They were considered useful because as rock fossils their shapes were likely to have been distorted in the same manner as some of the Soviet specimens, which had been similarly preserved.

TABLE 2. Character-state analysis of Cretaceous formicoid taxa based on the worker or queen castes.

Character	Worker						Queen			
	Sphecomyrma freyi	*Cretomyrma arnoldii*	*Cretomyrma unicornis*	*Petropone petiolata*	*Cretopone magna*	*Dolichomyrma longipes*	*Armania robusta*	*Pseudarmania rasnitsyni*	*Pseudarmania aberrans*	*Armaniella curiosa*
Mandible shape	Slender, 2-toothed	Slender, 2-toothed	?	?	?	?	Slender, 2-toothed	Slender, 2-toothed	?	?
Head shape	Circular	?	?	Circular	?	About 1.5× longer than broad	Circular	Circular	?	?
Clypeus form	Broad, simple	Broad, trapezoidal	?	?	?	Broad, simple	Broad, simple	Broad, simple	?	?
General antennal form	Scape only 0.3× as long as funiculus	?	?	?	?	?	Scape only 0.3× as long as funiculus	Scape only 0.2× as long as funiculus	?	?
Funiculus	Flexible	Flexible	?	?	?	?	?	?	?	?
Compound eyes	Large	?	?	Medium	Large	Large	Large	?	?	?
Ocelli	3	?	?	?	?	?	?	?	?	?
Scutum/scutellum	Distinct, convex	?	?	?	?	?	Not applicable	Not applicable	?	?
Petiole	Node constricted front and rear	Node constricted front and rear	Node constricted front and rear	Node constricted front and rear	?	Node constricted front and rear	Node narrowly constricted in front, weakly constricted in rear	?	Node narrowly constricted in front, weakly constricted in rear	Node narrowly constricted in front, weakly constricted in rear
Gaster	Ovoid, unconstricted	Ovoid, unconstricted	Ovoid, unconstricted	Ovoid, unconstricted	Ovoid, possibly unconstricted	Ovoid, weakly constricted	Ovoid, unconstricted	Ovoid, unconstricted	Ovoid, unconstricted	Ovoid, unconstricted

TABLE 2. Continued.

Character	Worker						Queen			
	Sphecomyrma freyi	*Cretomyrma arnoldii*	*Cretomyrma unicornis*	*Petropone petiolata*	*Cretopone magna*	*Dolichomyrma longipes*	*Armania robusta*	*Pseudarmania rasnitsyni*	*Pseudarmania aberrans*	*Armaniella curiosa*
Sting	Extrusible	Extrusible	?	Extrusible	Extrusible	?	Extrusible	Extrusible	Extrusible	?
Trochanter	1-jointed	1-jointed	?	1-jointed	1-jointed	?	1-jointed	1-jointed	2-jointed?	1-jointed
Tarsal claws	Toothed	Toothed	?	?	?	?	?	?	?	?
Tibial spurs	1, 2, 2	1, 2, 2	?, ?, 2	?	?	?	?, ?, 2	?	?, 2, 0	0, 0, 2
Abdominal segment IV	Freely articulated	Freely articulated	?	?	Freely articulated	Apparently freely articulated	Freely articulated	?	Freely articulated	?
Metapleural gland	Present	Present	?	?	?	?	Apparently present	?	Apparently present	?
General comments	(See text)	(See text)	Protuberance on propodeum[1]	Poorly preserved[2]	Poorly preserved[2]	(See text)	(See text)	(See text)	(See text)	(See text)
Source of data	Wilson et al. (1967)	Dlussky (1975, pers. comm.)	Dlussky (1975)	Dlussky (1975)	Dlussky (1975)	Dlussky (1975)	Dlussky (1983); direct study of holotype	Dlussky (1983); direct study of holotype	Dlussky (1983)	

[1] Horn-like protuberance on propodeum of uncertain nature.
[2] Indeterminate, could be *Sphecomyrma*.

TABLE 3. Character-state analysis of Cretaceous formicid genera based on males.

Character	Species (Male)			
	Paleomyrmex zherichini	*Archaeopone kzylzharica*	*Archaeopone taylori*	*Poneropterus sphecoides*
Ratio, length of 3d antennal segment to 2d segment	2–3	?	5	3
Genitalia covered by terminal abdominal tergite, or not	Not covered	Not covered	Not covered	Covered
Petiole trapezoidal (or nearly cubical) versus tapered anteriorly (delimiting a node)	Trapezoidal	Trapezoidal	Trapezoidal	Tapered
Trochanter	1-jointed	?	?	?
Source of data	Dlussky (1975)	Dlussky (1975)	Dlussky (1983)	Dlussky (1983)

In particular, it was important to take into account the possible increase in length of the abdomen, which tends to occur when this softest of body parts is crushed laterally or dorsoventrally.

Head length was adopted as a reliable index of body size as a whole (see Wilson 1971). The size of the abdomen is critical as an indicator of reproductive as opposed to nonreproductive status in females, because it is the main part of the body containing the ovaries, organs that are proportionately large in fully social hymenopterans. The abdomen in turn was defined in two ways for the present study. First, a "functional" definition marks the abdomen as the posteriormost discrete body part, commonly called the gaster by ant specialists—the ovary-bearing portion behind the one or two segments of the waist. In addition, a strictly homologous definition of the abdomen (or, more precisely, posteriormost major body tagma) was used: all of the true abdominal segments from II posteriad, including the one or two segments of the waist in ants and a few aculeate wasps. The first, functional definition is intuitively the better, because the waist has by definition been reduced to a relatively thin, largely musculated portion that increases the mobility of the gaster. However, both measures were employed in order to evaluate the situation as fully as possible.

The results were evaluated with reference to the following criteria derived from standard systematic practice on modern faunas of insects and other animals. Individual taxa, whether species, genera, or representative of higher taxa, should be distinguished and named only if they differ by character states. The states can be relatively minor in the case of species, but should be more substantial in the case of genera (e.g., in ants they include the number of antennal and palpal segments, presence or absence of clypeal teeth, and presence or absence of antennal scrobes; and they preferably should exist in multiples). In the case of families, traits should be truly major, as for example the presence or absence of the petiole, presence or absence of principal exocrine glands, and the pattern of wing folding—again preferably occurring in multiples. Dlussky (1975, 1983) appears to have used lighter criteria of the kind more commonly accepted in paleontology, in which it is recognized that fossils separated by large geographic distances and stretches of geological time are more likely to belong to different genera or higher categories. Add to this the fact that characters are often obscured due to imperfections in fossilization, and reliance is therefore placed on minor character states.

Because of the importance of the evolutionary issues involved, I believe the criteria employed in assessing the early ant fossils must be the stricter ones used in neontology. That is, it is preferable not to recognize taxa unless the character states separating them can be seen and are of approximately the same magnitude used in recognizing contemporary taxa of the same rank.

Results

The results of the character-state analysis for all of the Cretaceous formicoid genera are summarized in Tables 2 and 3. An inspection shows that no single character state or combination of states can be used to separate a distinct family, the Armaniidae, if neontological standards are applied. Moreover, almost none of the genera can be unambiguously supported from the existing evidence. The only exception is *Cretomyrma,*

TABLE 4. Higher classification of Mesozoic (Cretaceous) ants proposed in the present analysis.

Superfamily Formicoidea Latreille 1802
Family Formicidae Latreille 1802

Formicidae Latreille 1802, *Hist. Nat. Gen. Part., Crust. & Ins.* 3:352. Type genus: *Formica.*
Sphecomyrmidae Dlussky 1983, *Paleontol. Zhurn.* 1983, no. 3, p. 65. Type genus: *Sphecomyrma* Wilson and Brown 1967. NEW SYNONYMY.
Armaniidae Dlussky 1983, *Paleontol. Zhurn.* 1983, no. 3, p. 66. Type genus: *Armania* Dlussky 1983. NEW SYNONYMY (tentative).

Subfamily Sphecomyrminae Wilson and Brown 1967
Sphecomyrma Wilson and Brown 1967

Sphecomyrma Wilson and Brown 1967, *Psyche* 74:8. Type species: *S. freyi* Wilson and Brown 1967.
Dolichomyrma Dlussky 1975, *Trans. Paleontol. Inst.* 147: 121. Type species: *D. longiceps* Dlussky. NEW SYNONYMY (tentative).
Paleomyrmex Dlussky 1975, *Trans. Paleontol. Inst.* 147:118. Type species: *P. zherichini* Dlussky. NEW SYNONYMY (tentative).
Archaeopone Dlussky 1975, *Trans. Paleontol. Inst.* 147:120. Type species: *A. kzylzharica* Dlussky. NEW SYNONYMY (tentative).
Armania Dlussky 1983, *Paleontol. Zhurn.* 1983, no. 3, p. 67. Type species: *A. robusta* Dlussky. NEW SYNONYMY (tentative).
Pseudarmania Dlussky 1983, *Paleontol. Zhurn.* 1983, no. 3, p. 69. Type species: *P. rasnitsyni* Dlussky. NEW SYNONYMY (tentative).
Armaniella Dlussky 1983, *Paleontol. Zhurn.* 1983, no. 3, p. 71. Type species: *A. curiosa* Dlussky. NEW SYNONYMY (tentative).
Poneropterus Dlussky 1983, *Paleontol. Zhurn.* 1983, no. 3, p. 73. Type species: *P. sphecoides* Dlussky. NEW SYNONYMY (tentative).

Cretomyrma Dlussky 1975

Cretomyrma Dlussky 1975, *Trans. Paleontol. Inst.* 147:115. Type species: *C. arnoldii* Dlussky.

Incertae Sedis

Petropone Dlussky 1975, *Trans. Paleontol. Inst.* 147:119. Type species: *P. petiolata* Dlussky.
Cretopone Dlussky 1975, *Trans. Paleontol. Inst.* 147:119. Type species: *C. magna* Dlussky.

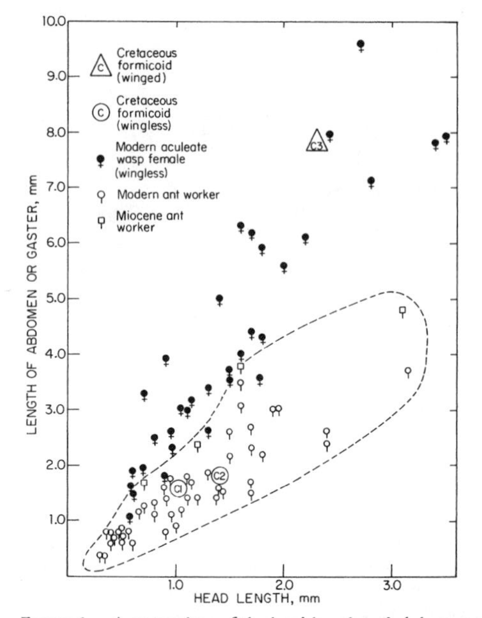

FIGURE 1. A comparison of the head length and abdomen length (functionally defined) in representative specimens of Cretaceous formicoids, modern ant workers, Miocene ant workers, and modern wingless aculeate wasps. Cretaceous formicoids: C1, *Sphecomyrma freyi* holotype worker; C2, *Dolichomyrma longiceps* holotype worker; C3, *Pseudarmania rasnitsyni* holotype winged female. The list of other specimens used is given in the Appendix.

possessing a hornlike protuberance on the propodeum (in *C. unicornis*), more compact gaster, and proportionately longer legs; a second, weaker possibility is *Dolichomyrma,* with an elongate head and slight constriction of the gaster (Dlussky 1975, 1983, and new details provided the author *in litt.*). It cannot be denied that still other genera might be represented by the Soviet fossils; we are only sure that few can be defined by neontological standards with existing data. Hence the synonymy suggested in Table 4 is for the moment the appropriate nomenclatural arrangement.

Figures 1 and 2 show that the measurable, relatively undistorted wingless females among the North American and Soviet fossils, in other words *Sphecomyrma freyi* and *Dolichomyrma longiceps,* have small abdomens in proportion to the remainder of the body, by both the functional and strictly homologous definitions of the abdomen. In this respect they resemble the worker caste of modern ants more than they do the wingless females of modern nonsocial aculeate wasps. Moreover, the winged female type of *Pseudarmania rasnitsyni,* a relatively undistorted specimen, is within the range of the queens of modern ants (as well as the females of nonsocial aculeate wasps), as shown in Fig. 3. In sum, the wingless females among the Cretaceous fossils are best interpreted as worker ants rather than wingless reproductive aculeate wasps belonging to solitary

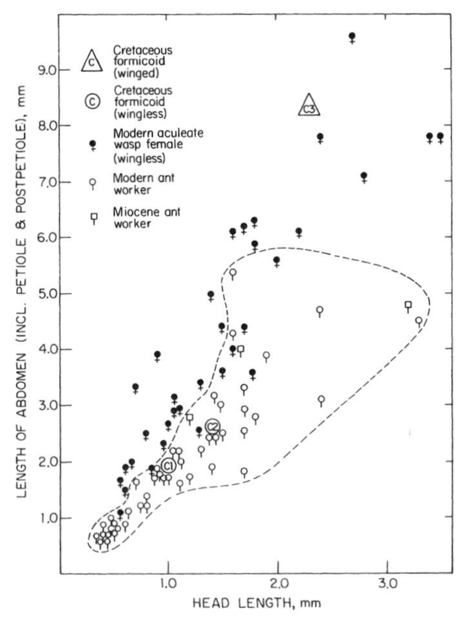

FIGURE 2. A comparison of the head length and abdomen length (defined by strict homology) using the same specimens as in Fig. 1.

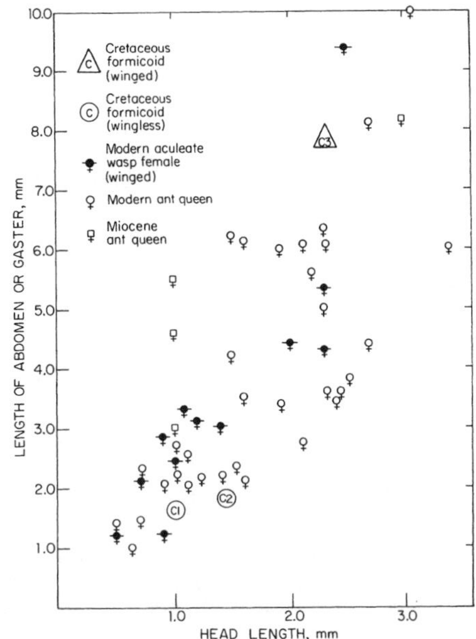

FIGURE 3. A comparison of the head length and abdomen length (functionally defined) in representative specimens of Cretaceous formicoids, modern ant queens, Miocene ant queens, and modern winged aculeate wasps. Cretaceous formicoids: same conventions as in Fig. 1. The list of other specimens is given in the Appendix.

species, while the winged females are probably the queens.

Several additional observations support this conclusion. In his description, Dlussky (1983) states that a diagnostic trait of the Armaniidae is the approximately equal length of the first and third antennal segments, that is, of the scape and second funicular segments. Although this is quite correct, another and more revealing way of putting the matter is to say that the scape is quite short relative to the funiculus, while the second funicular segment is long relative to the first and third funicular segments. It turns out that these are precisely the same distinctive traits used to define the Sphecomyrminae. Hence both the winged and wingless Cretaceous females share the same unusual character state in antennal form, another reason for associating them closely.

Dlussky (1983) gives as another diagnostic trait of the Armaniidae the possession of a second, free trochanter on the middle and hind legs. This condition is weakly indicated in the drawing of *Pseudarmania aberrans,* but it is wholly lacking in the drawings of all of the other armaniids, and by direct examination I confirmed that it is indeed absent in the holotypes of *Armania robusta* (the type genus and species of the family) and *Pseudarmania rasnitsyni.*

In the winged females of the Armaniidae the petioles are more broadly attached posteriorly to the gaster than in the sphecomyrmine wingless females. But this is not a subfamilial or even species-level character. It is a common difference between the queens and workers belonging to the same species among modern ants, and hence it cannot be reliably used as a taxonomic character to separate higher formicoid taxa.

Discussion

The most parsimonious interpretation of the Cretaceous formicoid fossils, neatly joining the facts we know, is that they all belong to the subfamily Sphecomyrminae of the family Formicidae, or true ants. Furthermore, so long as contrary evidence is lacking, the Cretaceous fossils should all be placed provisionally in the genera *Sphecomyrma* and *Cretomyrma*. It is entirely possible that other genera, and even taxa in additional subfamilies or still higher categories, existed in Cenomanian to Santonian times and might be represented by the existing fossils, but until supporting evidence emerges, the conservative taxonomic arrangement suggested here (Table 4) is both more accurate and heuristic.

This interpretation means that the three phases represented among the Cretaceous fossils are most reasonably interpreted to be queen, worker, and male formicoids, respectively, in other words, what we would call ants as opposed to wasps. This hypothesis is more clearly depicted by juxtaposing the best preserved representatives of the three phases as though they are members of the same colony (see Fig. 4). The hypothesis receives considerable support from the size differences between the best-preserved winged fossils and the best-preserved wingless ones, consistent with their being queens and workers. It receives additional support from the fact that the proportionate size of the abdomen in the Cretaceous ants is closer to modern ants than to modern aculeate wasps (Figs. 1, 2).

Dlussky (1983) made two inferences connecting anatomy to behavior inclining him to the hypothesis that the Cretaceous formicoids were not eusocial. The first is that the tips of antennal funiculi are too far removed from the mandibles to allow the precise coordination required for social behavior: "The antennae in these insects did not permit them to control the manipulation of small objects, so that they could not have transported their brood or entered into trophallaxis with their larvae—that is, they could not have been true social insects." This supposition is surely incorrect. The eusocial vespid wasps have similarly proportioned antennae yet experience no difficulty in transporting all prey objects and placing them on the larvae. They also engage in trophallaxis, or liquid food exchange. The flexibility of the funiculi contribute to these skills, and the twists and curves of the Cretaceous ant antennae suggest that their funiculi were likewise flexible. To this may be added the fact that some modern ants, such as the primitive *Amblyopone pallipes* (Traniello 1982) and more advanced *Pogonomyrmex badius* (Wilson 1971) do not engage in trophallaxis. Hence this form of food exchange was not essential for the evolution of eusocial behavior in ants.

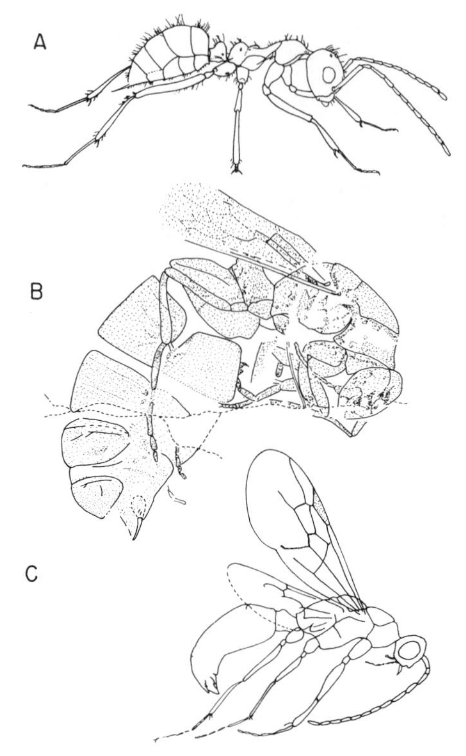

FIGURE 4. The three castes of *Sphecomyrma*, the most primitive known ants, as provisionally associated in the present study. A, Worker: the holotype of *Sphecomyrma freyi*, Cretaceous (Santonian) of New Jersey. B, Winged queen: the holotype of *Armania robusta*, Cretaceous (Cenomanian) near Magadan, northeastern Siberia. C, Male: the holotype of *Paleomyrmex zherichini*, Cretaceous (Santonian) of the Taymyr Peninsula, north-central Siberia.

Dlussky also inferred that the short, wasplike mandibles of the Cretaceous formicoids "indicates that these insects did not build true nests, and could have used only pre-existing hollows." But this overlooks the fact that some primitively eusocial wasps and bees use similar mandibles to build quite elaborate nests, mostly from carton and wax. A few, such as the halictid bees, excavate soil in a very antlike fashion.

To summarize, nothing in the observable anatomy of the Cretaceous formicoids precludes their having possessed a eusocial organization, characterized by brood care, overlap of adult generations, and division of labor between reproductive and nonreproductive castes. Although direct evidence either way is lacking on the matter, these insects could also have constructed nests in the soil, rotting wood, or arboreal cavities.

The low accumulation rate of Cretaceous formicoids to date indicates that they occurred in low densities compared with modern ants. Only two individuals (*Sphecomyrma canadensis*) have been found so far among thousands of insects in amber from Alberta Province, Canada (Carpenter et al. 1939; J. F. McAlpine, pers. comm.). Formicoids constituted just 13 of the 1,200 insect impressions in the Magadan collection and 5 of the 526 impressions in the Kazakhstan collection, in other words about 1% in both cases (Dlussky 1983). These figures contrast sharply with Oligocene and Miocene deposits. In the Florissant and other shales of North America (Carpenter 1930), as well as the Baltic amber of northern Europe (Wheeler 1914) and amber of the Dominican Republic (Wilson 1985b), the ants are among the most abundant insects, making up a large minority of all insect specimens. Thus the adaptive radiation that took place in the late Cretaceous or early Tertiary, yielding at least three of the dominant modern subfamilies (Myrmicinae, Dolichoderinae, Formicinae) by mid-Eocene times (Wilson 1985a), was accompanied by a marked increase in abundance.

The ants are seen to have paralleled the mammals by achieving dramatic increases in diversity and abundance around the close of the Mesozoic Era. How they accomplished this breakthrough and managed to sustain a dominant position in the insect world to the present time is a matter of unusual interest, which additional paleontological studies should help to illuminate.

Acknowledgments

I am grateful to A. P. Rasnitsyn for the loan of the Magadan fossils, and to W. L. Brown, F. M. Carpenter, James Carpenter, and G. M. Dlussky for a critical reading of the manuscript. I am additionally grateful to Dr. Dlussky for supplying previously unpublished details on the anatomy of *Cretomyrma*. My research was supported by National Science Foundation Grant BSR-84-21062.

Literature Cited

CARPENTER, F. M. 1930. The fossil ants of North America. Bull. Mus. Comp. Zool. Harvard. *70*:1–67.

CARPENTER, F. M., J. W. FOLSOM, E. O. ESSIG, A. C. KINSEY, C. T. BRUES, M. W. BOESEL, AND H. E. EWING. 1939. Insects and arachnids from Canadian amber. Univ. Toronto Stud., Geol. Ser. *40*:7–62.

DLUSSKY, G. M. 1975. Formicoidea, Formicidae, Sphecomyrminae. Pp. 114–122. In: Rasnitsyn, A. P., ed., The Higher Hymenoptera of the Mesozoic. Trans. Paleontol. Inst. AN SSR *147* [in Russian].

DLUSSKY, G. M. 1983. A new family of Upper Cretaceous Hymenoptera: an "intermediate link" between the ants and the scolioids. Paleontol. Zhurn. no. *3*:65–78 [in Russian].

EYSINGA, F. W. B. VAN. 1978. Geological time table (published chart). Elsevier; Amsterdam.

TRANIELLO, J. F. A. 1982. Population structure and social organization in the primitive ant *Amblyopone pallipes* (Hymenoptera: Formicidae). Psyche. *89*:65–80.

WHEELER, W. M. 1914. The ants of the Baltic amber. Schrift. Phys.-ökon. Ges. Königsberg. *55*:1–142.

WILSON, E. O. 1971. The Insect Societies. Belknap, Harvard Univ. Press; Cambridge.

WILSON, E. O. 1985a. Ants from the Cretaceous and Eocene amber of North America. Psyche. *92*:205–216.

WILSON, E. O. 1985b. Invasion and extinction in the West Indian ant fauna: evidence from the Dominican amber. Science. *229*: 265–267.

WILSON, E. O., F. M. CARPENTER, AND W. L. BROWN. 1967a. The first Mesozoic ants. Science. *157*:1038–1040.

WILSON, E. O., F. M. CARPENTER, AND W. L. BROWN. 1967b. The first Mesozoic ants, with the description of a new subfamily. Psyche. *74*:1–19.

Appendix

Miocene and modern species used in the measurements of Figs. 1–3 are listed here. All ants, fossil and living, are members of the family Formicidae. The families of the aculeate wasp species are listed individually.

MIOCENE ANT WORKERS. *Aphaenogaster mayri, Archiponera wheeleri, Hypoclinea antiqua, Protazteca quadrata.*

MIOCENE ANT QUEENS. *Formica robusta, Hypoclinea rohweri, Miomyrmex impactus, Pseudomyrmex extinctus.*

MODERN ANT WORKERS. *Acanthoclinea dentata, Amblyopone australis, Aneuretus simoni, Anillidris bruchi, Araucomyrmex tener, Aphaenogaster longiceps, Bothriomyrmex flavus, Brachymyrmex obscurior, Crematogaster parabiotica, Diabolus coniger, Dorymyrmex planidens, Engramma wolfi, Erebomyrma urichi, Forelius andinus, Froggattella kirbyi, Gigantiops destructor, Heteroponera flava, Hypoclinea abrupta, Iridomyrmex sanguinea, Lasius alienus, Leptomyrmex fragilis, Liometopum apiculatum, Monoceratoclinea tricornis, Monomorium cyanea, Myrmecia dichospila, Myrmecia dispar, Myrmica incompleta, Neivamyrmex opaciventris, Neoforelius tucumanus, Nothomyrmecia macrops, Notoncus ectatommoides, Odontomachus opaciventris, Oecophylla smaragdina, Pheidole cephalica, Pogonomyrmex desertorum, Proatta butteli, Pseudomyrmex gracilis, Semonius schultzei, Sericomyrmex* sp., *Solenopsis nitens, Tapinoma melanocephalum, Technomyrmex albipes, Thaumatomyrmex zeteki, Typhlomyrmex rogenhoferi, Zatapinoma* sp.

MODERN ANT QUEENS. *Amblyopone australis, Aenictus binghami, Aneuretus simoni, Aphaenogaster longipes, Brachymyrmex obscurior, Calomyrmex albertisi, Camponotus novaboracensis, Eciton dulcius, Erebomyrma urichi, Formica subpolita, Gigantiops destructor, Gnamptogenys concinna, Heteroponera flava, Iridomyrmex sanguinea, Lasius alienus, Leptomyrmex fragilis, Liometopum apiculatum, Myrmecia dichospila, Myrmecia dispar, Myrmica incompleta, Neivamyrmex humilis, Neivamyrmex opaciventris, Notoncus ectatommoides, Odontomachus opaciventris, Oecophylla smaragdina, Opisthopsis* sp., *Polyrhachis gagates, Pogonomyrmex desertorum, Proatta butteli, Pseudolasius mayri, Pseudomyrmex gracilis, Sericomyrmex* sp., *Solenopsis nitens, Tapinoma melanocephalum, Thaumatomyrmex zeteki, Typhlomyrmex rogenhoferi.*

ACULEATE WASP FEMALES (WINGED). *Alphadryinus bocainanus* (Dryinidae), *Anteon gaullei* (Dryinidae), *Aphelopus varicornis* (Dryinidae), *Ceropales brethesi* (Pompilidae), *Clystopsenella longiventris* (Scolebythidae), *Hypodynerus coarctatus* (Vespidae), *Euodynerus farquharensis* (Vespidae), *Pseudoisobrachium complanatum* (Bethylidae), *Psoropempula mongana* (Pompilidae), *Scolebythus madecassus* (Scolebythidae), *Sphaeropthalma auripilis* (Mutillidae), *Ycaploca evansi* (Scolebythidae).

ACULEATE WASP FEMALES (WINGLESS). *Acrodontochelys cubensis* (Dryinidae), *Aelurus gayi* (Tiphiidae), *Aglyptacros eureka* (Tiphiidae), *Apenesia browni* (Bethylidae), *Ariphron tryphonoides* (Tiphiidae), *Bruesiella formicaria* (Tiphiidae), *Chyphotes attenuatus* (Mutillidae), *Dasymutilla arenivaga* (Mutillidae), *Dromopompilis* sp. (Pompilidae; a brachypterous rather than completely apterous species), *Elaphroptera* sp. (Tiphiidae), *Embolemus nearcticus* (Embolemidae), *Ephuta* sp. (Mutillidae), *Eurycros furtivus* (Tiphiidae), *Glyptacros angustior* (Tiphiidae), *Glyptometopa americana* (Tiphiidae), *Gonatopus frequens* (Dryinidae), *Hemithynnus* sp. (Tiphiidae), *Leucospilomutilla cerbera* (Mutillidae), *Methocha californica* (Tiphiidae), *Myrmosa unicolor* (Tiphiidae), *Myrmosula parvula* (Tiphiidae), *Nealgoa banksii* (Rhopalosomatidae), *Olixon* sp. (Rhopalosomatidae; an extremely brachypterous species), *Photopsis zenobia* (Mutillidae), *Plumarius* sp. (Plumariidae), *Pristocerca cockerelli* (Bethylidae), *Pseudomethoca oceola* (Mutillidae), *Reedomutilla* sp. (Mutillidae), *Rhagigaster laevigatus* (Tiphiidae), *Thynnoides* sp. (Tiphiidae), *Timulla leona* (Mutillidae), *Typhoctes peculiaris* (Mutillidae), *Probethylus* sp. (Sclerogibbidae), unidentified sp. no. 3 (Tiphiidae).

1990

"The dominance of social insects," pp. 1–5, and "Hawaii: a world without social insects," pp. 91–96 in E. O. Wilson, *Success and Dominance in Ecosystems: The Case of the Social Insects* (Ecology Institute, Oldendorf/Luhe, Germany)

In these two brief essays, which appeared as chapters in my book *Success and Dominance in Ecosystems: The Case of the Social Insects,* I describe the extraordinary dominance of social insects, especially ants and termites, in terrestrial environments around the world, as well as the impact of their absence on the fauna and flora of Hawaii for the millions of years prior to the unintended importation of alien species into the islands by humanity. These accounts fortify the principle that the ants and termites are commonly keystone species, whose presence or absence exerts a disproportionate effect on ecosystems. On page 314, an outline of the overall contents of *Success and Dominance in Ecosystems* is also given.

❋ ARTICLE 26

I The Dominance of the Social Insects

Social insects saturate most of the terrestrial environment. In ways that become fully apparent only when we bring our line of sight down to a millimeter of the ground surface, they lay heavily on the rest of the fauna and flora, constraining their evolution.

That fact has struck home to me countless times during my life as a biologist. Recently it came again as I walked through the mixed coniferous and hardwood forests on Finland's Tvärminne Archipelago. My guides were Kari Vepsäläinen and Riitta Savolainen of the University of Helsinki, whose research has meticulously detailed the distribution of ants in the archipelago and the histories of individual colonies belonging to dominant species. We were in a cold climate, less than 800 kilometers from the arctic circle, close to the northern limit of ant distribution. Although it was mid-May, the leaves of most of the deciduous trees were still only partly out. The sky was overcast, a light rain fell, and the temperature at midday was an unpleasant (for me) 12°C. Yet ants were everywhere. Within a few hours, as we walked along trails, climbed huge moss-covered boulders, and pulled open tussocks in bogs, we counted nine species of *Formica* and an additional eight species belonging to other genera, altogether about one-third the known fauna of Finland. Mound-building Formicas dominated the ground surface. The nests of several species, especially *F. aquilonia* and *F. polyctena*, were a meter or more high and contained hundreds of thousands of workers. Ants seethed over the mound surfaces. Columns traveled several tens of meters between adjacent mounds belonging to the same colony. Other columns moved up the trunks of nearby pine trees, where the ants attended groups of aphids and collected their sugary excrement. Swarms of solitary foragers deployed from the columns in search of prey. Some could be seen returning with geometrid caterpillars and other insects. We encountered

2

a group of *F. polyctena* workers digging into the edge of a low mound of *Lasius flavus*. They had already killed several of the smaller ants and were transporting them homeward for food. As we scanned the soil surface, peered under rocks, and broke apart small rotting tree branches, we were hard put to find more than a few square meters anywhere free of ants. In southern Finland they are in fact the premier predators, scavengers, and turners of soil. Exact censuses remain to be made, but it seems likely that ants make up 10% or more of the entire animal biomass of the Tvärminne Archipelago.

Two months earlier, in the company of Bert Hölldobler of the University of Würzburg, F. R. Germany (then at Harvard University, USA), I had walked and crawled on all fours over the floor of tropical forest at La Selva, Costa Rica. The ant fauna was radically different and much more diverse than in Finland. The dominant genus was *Pheidole*, as it is in most tropical localities. Within a 1.5 km^2 area along the Rio Sarapiquí, my students and I have collected 34 species of *Pheidole*, of which 16 are new to science. The total ant fauna in the sample area probably exceeds 150 species. That is a conservative estimate, because Neotropical forests have some of the richest faunas in the world. Manfred Verhaagh (personal communication) collected about 350 species belonging to 71 genera at the Rio Pachitea, Peru. That is the world record at the time of this writing. I identified 43 species belonging to 26 genera from a single leguminous tree at the Tambopata Reserve, Peru (Wilson, 1987a). From my experience in ground collecting in many Neotropical localities, I am sure that an equal number of still different species could have been found on the ground within a radius of a few tens of meters around the base of the tree. In other words, the fauna of the Tambopata Reserve is probably equivalent to that at the Rio Pachitea.

The abundance of ants at Neotropical localities, as opposed to species diversity, is comparable to that on the Tvärminne archipelago, and they occupy a great many more specialized niches as well. In addition to a large arboreal fauna, lacking in Finland*, leaf-cutter ants raise fungi on newly harvested vegetation, *Acanthognathus snare* tiny

* In an interview with the vice-rector of the University of Helsinki, I mentioned the 43 ant species found on one tree in Peru. He asked how many kinds occur on one tree in Finland, whereupon Kari Vepsäläinen answered for me: "If you cut down all the forests of Finland, you will find two species."

collembolans with their long traplike mandibles, *Prionopelta* species hunt campodeid diplurans deep within decaying logs, and so on in seemingly endless detail. Roughly one out of five pieces of rotting wood contains a colony of ants, and others harbor colonies of termites. Ants absolutely dominate in the canopies of the tropical forests. In samples collected by Terry Erwin by insecticidal fogging in Peru, they make up about 70% of all of the insects (personal communication). In Brazilian Terra Firme forest near Manaus, Fittkau and Klinge (1973) found that ants and termites together compose a little less than 30% of the entire animal biomass. These organisms, along with the highly social stingless bees and polybiine wasps, make up an astonishing 80% of the entire insect biomass.

While few quantitative biomass measurements have been made elsewhere, my own strong impression is that social insects dominate the environment to a comparable degree in the great majority of land environments around the world. Very conservatively, they compose more than half the insect biomass. It is clear that social life has been enormously successful in the evolution of insects. When reef organisms and human beings are added, social life is ecologically preeminent among animals in general. This disproportion seems even greater when it is considered that only 13,000 species of highly social insects are known, out of the 750,000 species of the described insect fauna of the world.

In short, 2% of the known insect species of the world compose more than half the insect biomass. It is my impression that in another, still unquantified sense these organisms, and particularly the ants and termites, also occupy center stage in the terrestrial environment. They have pushed out solitary insects from the generally most favorable nest sites. The solitary forms occupy the more distant twigs, the very moist or dry or excessively crumbling pieces of wood, the surface of leaves – in short, the more remote and transient resting places. They are also typically either very small, or fast moving, or cleverly camouflaged, or heavily armored. At the risk of oversimplification, the picture I see is the following: *social insects are at the ecological center, solitary insects at the periphery.*

This then is the circumstance with which the social insects challenge our ingenuity: their attainment of a highly organized mode of colonial existence was rewarded by ecological dominance, leaving what must have been a deep imprint upon the evolution of the remainder of

4

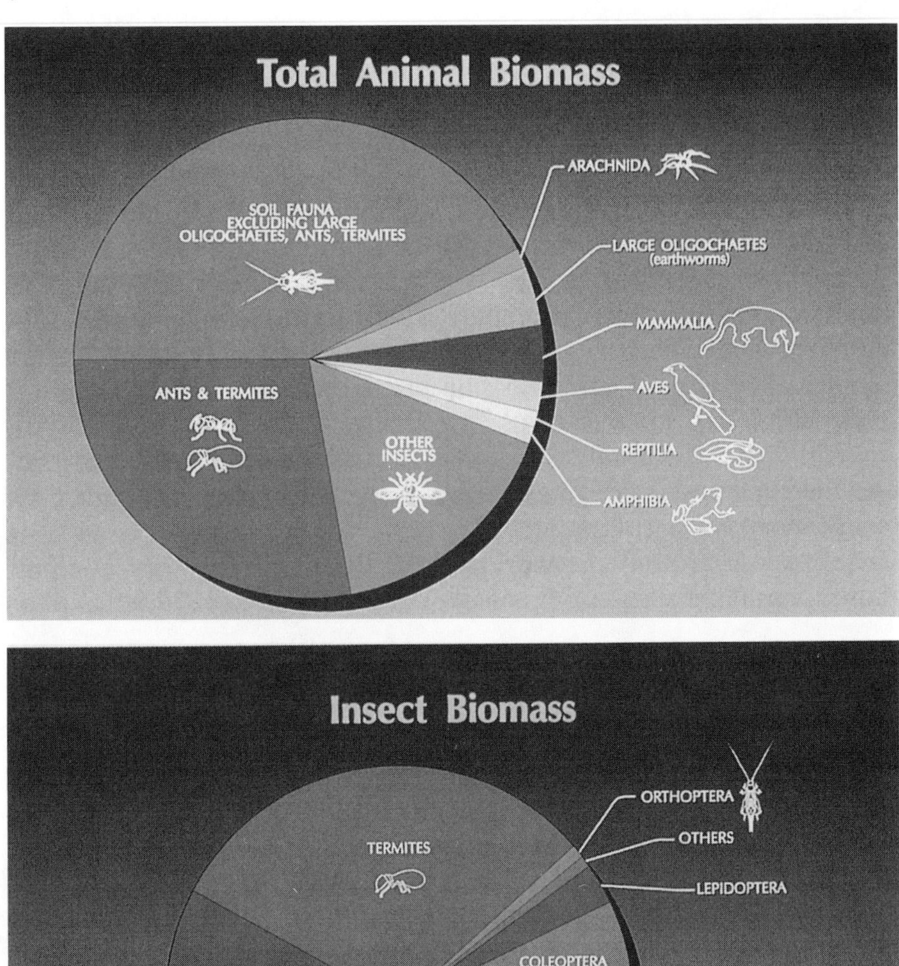

Fig. 1. The apportionment of biomass among groups of animals and insects respectively, in rain forest near Manaus, Brazil. (From Wilson, 1988; based on data from Fittkau and Klinge, 1973, and reproduced with the permission of the National Geographic Society.)

5

Fig. 2. In Brazilian tropical forest, the biomass of ants is approximately four times greater than the biomass of all of the vertebrates (mammals, birds, reptiles, and amphibians) combined. The difference is represented here by the relative sizes of an ant, *Gnamptogenys pleurodon*, and a jaguar. The measures are provided in Fig. 1. (Original drawing by Katherine Brown-Wing.)

terrestrial life. In this book I will explore the themes in ecological research suggested by this achievement. They include:

- In a more explicitly heuristic framework, the general meaning of success and dominance in organisms other than man.
- The general characteristics of social insects and the manner in which they might have contributed to success.
- The nature of social solutions to particular environmental challenges.
- The impact of the social insects on the terrestrial fauna, in other words the actual workings of their dominance.

To a degree unappreciated even by most field biologists, social insects, especially ants and termites, dominate the terrestrial environment. It is tempting to think that colonial life has contributed to this preeminence, and that the social insects might serve as a paradigm for the study of dominance in ecosystems generally.

VIII Hawaii: A World Without Social Insects

In order to assess the impact of a dominant group of organisms, it would be extremely useful to have biotas free of the dominant group that can serve as evolutionary controls. This baseline is not easily found, because dominant groups are also as a rule very geographically widespread. The eusocial insects in particular have almost completely filled the terrestrial world. But there is one place to look. They did not, prior to the coming of man, inhabit the easternmost archipelagoes of the Pacific. In particular, they did not reach Hawaii. This most isolated of all archipelagoes evolved a rich endemic fauna and flora in the absence of termites, ants, and eusocial bees and wasps (Zimmerman, 1948; Wilson and Taylor, 1967; Williamson, 1981).

The massive weight of the social insects was therefore lifted from the plants and animals that departed from their midst and colonized Hawaii. Insects and other arthropods were freed from predation by ants and social wasps. Conversely, predators and scavengers did not have to compete with ants and social wasps; and invertebrate decomposers of wood did not have to contend with termites in the rotting logs and stumps. On the negative side, plants were deprived of the protection of ants and the pollination services of social bees.

How did the Hawaiian biota respond to this release? Unfortunately, we cannot directly read off the results, because there is an additional force working in Hawaii that is easily conflated with the absence of social insects: the disharmonic nature of the biota as a whole. The Hawaiian biota, as expected from its extreme remoteness, has evolved from a limited number of stocks, which have radiated extensively thereafter. By 1980, 6500 endemic insect species had been described, and these are believed to have evolved from about 250 original immigrant species (Williamson, 1981). A typical case is the oecanthine tree crickets, comprising 3 genera and 54 species, or 43% of the entire known

92

oecanthine fauna of the world, all derived from a single species that colonized the islands no more than 2.5 million years ago (Otte, 1989). Disharmony of this kind means that not just social insects but many other major stocks of invertebrates are absent in the native fauna. Their absence as predators, herbivores, and decomposers must also be taken into account when assessing the histories of the sweepstakes winners.

The problem, while not readily soluble, is nevertheless tractable. Given the great ecological importance of social insects and the general significance of dominance in community evolution, the Hawaiian biota deserves a new look with social insects in mind. It is entirely possible that certain traits of the Hawaiian fauna usually ascribed to disharmony and reduced dispersal opportunity, such as extreme local abundance and flightlessness, are due at least in part to the lack of pressure from social insects, especially ants. What I offer now as a first analysis is a set of properties of biotas expected from the absence of social insects, without the attempt (or even the capacity, given the present scarcity of ecological knowledge) of disentangling the effects from those due to the absence of other, ecologically equivalent invertebrate groups as a reflection of disharmony in the fauna.

Scale insects and other honeydew-producing insects protected by ants elsewhere will be scarcer relative to related groups. This prediction is confirmed but vitiated by the disharmonic nature of the fauna. There are no native coccids, fulgorids, or aphids, among the groups most avidly attended by ants elsewhere. Their absence could be ascribed either to the absence of ant protectors or bad luck in the dispersal sweepstakes. The latter hypothesis seems somewhat less likely in view of the fact that aphids are excellent dispersers. There is only one butterfly species belonging to the Lycaenidae, a family whose caterpillars are heavily attended by ants, but the native butterfly fauna of Hawaii is, inexplicably, very small overall. Mealybugs (Pseudococcidae), also much favored by ants, are represented by 3 endemic genera and 14 species, but are heavily outweighed in diversity by the homopterous families Cixiidae, Delphacidae, and Psyllidae, which are not attended by ants (Zimmerman, 1948).

Both herbivores and predaceous insects will occur in denser, less protected populations. This prediction is dramatically confirmed. A very high percentage of the endemic insect species are flightless, and also generally "sluggish," to use Perkins' (1913) term, with populations persisting on the same tree or bush for years. Otte (personal communication) has referred to the conspicuous abundance of endemic crickets and their "lackadaisical" behavior. Many of the species walk about in the open where they can be easily picked up with the fingers, in sharp contrast to the cryptic, fast-moving species that occur in other faunas. Caterpillars such as those of the pyraustid moth genera *Margaronia* and *Omiodes*, the extremely diverse drosophilid flies, and a few other dominant groups are comparably abundant and accessible, or at least were so in the last century in the less disturbed habitats. These are the kinds of insects most vulnerable to ant predation. No fewer than 36 ant species have been introduced by man, including the notorious omnivore and pest species *Pheidole megacephala*. The widespread destruction of native Hawaiian insects by ants is well known. Zimmerman (1948) states that "the introduction of a single species of ant, the voracious *Pheidole megacephala*, alone has accounted for untold slaughter. One can find few endemic insects within that scourge of native insect life. It is almost ubiquitous from the seashore to the beginnings of damp forest." Otte has observed the same displacement in the case of native crickets on Hawaii and the Society Islands, which also lacked ants before they were introduced by human commerce. There are other major causes of extinction of Hawaiian native insects, including habitat destruction and the incursion of alien parasites and diseases. But the important point with reference to the question of social insect dominance is the documented extreme vulnerability of the native insects to introduced ants in both disturbed and undisturbed habitats, which is consistent with observations in other parts of the world where ants are native. (Ants may also have played a key role in the retreat of the rich and abundant Hawaiian land snail fauna, comprising over a thousand species, although I am not aware of studies addressing this possibility.) The converse conclusion is equally important: the local abundance of behaviorally vulnerable, epigaeic insects is consistent with the absence of native ants, whether or not it explains the phenomenon entirely.

The non-formicid predators in the mesofauna (0.2 to 2.0 mm body length range), especially carabid beetles and spiders, should be more diverse and abundant. Also, predators should have evolved in mesofaunal arthropod groups that are not predaceous in other parts of the world. I call attention here especially to carabid beetles and spiders, because it is my experience in many other parts of the world that large numbers of species belonging to these two predatory groups have similar microhab-

94

itat preferences to ants. They occur abundantly in the litter and soil and especially under rocks not yet colonized by ants. In the summit forest of Mt. Mitchell in North Carolina, USA, where ants are very scarce, I found carabids and spiders to be more abundant, or at least more conspicuous, than at lower elevations. Darlington (1971) and Cherix (1980) present evidence that ants generally reduce the abundance of ground-dwelling carabids and spiders in both the tropics and temperate zones, especially those species specialized to live in soil and rotting vegetation. And as expected, carabids and spiders are both very diverse and abundant in the native forests of Hawaii. Other mesofaunal predatory groups that have radiated include the nabid bugs, staphylinid beetles, dolichopodid flies, and muscid flies of the genus *Lispocephala*. Groups that have moved into the ant predator zone include the geometrid moth *Eupithecia*, whose caterpillars ambush insects like praying mantises, and the damselfly *Megalagrion*, whose predator nymphs have left the aquatic environment entirely to hunt on the ground, especially under clumps of ferns. As I have stressed, these adaptive radiations and major ecological shifts may have been favored by the absence of competing predators in addition to ants, due to the general disharmonic nature of the fauna. Yet it is hard to imagine their occurring at all if a well-developed ant fauna had been present.

Non-formicid scavengers should be diverse and prominent relative to those in ecologically otherwise comparable faunas. Ants are strongly dominant as the scavengers of small arthropod corpses in most parts of the world. It is to be expected that this largely empty niche was filled by other groups on Hawaii, perhaps (at a guess) staphylinid and histerid beetles, but I know of no studies addressing the matter.

Wood borers other than termites should be very prominent. In the absence of termites, we should expect to find a greater diversity and abundance of insects that bore through dead wood, especially the softer, rotting "wet" wood favored by so many termite species elsewhere in the world. Again, studies appear not to have been directed specifically to this hypothesis. Candidate groups include beetles of the families Anobiidae, Cerambycidae, Curculionidae, Elateridae, and Eucnemidae, which have in fact radiated extensively on Hawaii.

Solitary wasps and bees should be relatively diverse and very abundant. The solitary eumenid wasp genus *Odynerus* is represented by over 100 endemic species on Hawaii and, until the last century at least, was extremely abundant. The solitary hylaeid bee genus *Hylaeus* contains at least 50 endemic species, all derived from a single ancestor. The relation of these minor evolutionary explosions to the absence of social wasps and bees is an intriguing possibility but has not yet been explored.

Extrafloral nectaries and elaiosomes will be reduced or absent in the native flora. In general, extrafloral nectaries serve to attract ants, which in turn protect the plants from herbivores. Extrafloral nectaries are substantially scarcer in the Hawaiian flora than elsewhere, in agreement with the prediction (Keeler, 1985). Eleven endemic species and 6 indigenous species do have the nectaries, which may be attended by protector arthropods other than ants or simply reflect phylogenetic inertia. Significantly, three other indigenous (but not endemic) species having extrafloral nectaries elsewhere lack them on Hawaii, again supporting the prediction. Elaiosomes are seed appendages attractive to ants that induce the ants to disperse the seeds. No study has been made to my knowledge of their relative abundance on Hawaii.

To conclude, Hawaii has long fascinated biologists for its superb adaptive radiations, and depressed conservationists for the continuing destruction of those same evolutionary wonders. I suggest that the value of the biota is enhanced still further by the realization that it is a natural laboratory, the unique site of an experimental control, for the assessment of the impact of social insects on the environment. These insects are so dominant in almost all other parts of the world that their absence in the original, native Hawaiian fauna and flora provides an exceptional opportunity to study the effects of ecological release on the part of taxa that would otherwise have interacted with them most strongly. At the very least, the absence of social insects should be taken into more explicit consideration in future studies of the biota. What is clearly needed are deeper studies of the life cycles of native Hawaiian taxa in comparison with sister taxa elsewhere, and especially on other islands with and without key elements such as social insects.

Hawaii, the most remote archipelago in the world and home of a rich endemic fauna and flora, was evidently never colonized by social insects before the coming of man. The absence of these dominant elements means that the Hawaiian native biota is a controlled experiment in which we can observe the effects of freedom from social insects, especially ants and termites. It seems probable that the circumstance was a major contributor to some of the tendencies characterizing the Hawaiian biota

96

as a whole, including flightlessness, lack of evasive behavior, increased abundance and diversity of carabid beetles and spiders, adaptive shift to predation in some terrestrial insect groups, and the loss of extrafloral nectaries in flowering plants.

IX General Summary

Success is most heuristically defined as an evolutionary concept: the duration of a clade (a species and all its descendents) through geological time. Dominance on the other hand is an ecological property: the relative abundance of a clade, especially as its numbers cause it to appropriate energy, accumulate biomass, and affect the life and evolution of the remainder of the biota. Species numbers, the breadth of adaptive radiation, geographic range, and the magnitude of evolutionary innovations are enhanced by dominance, and they in turn are likely to increase success, but they nevertheless remain separate properties of clades.

The causes of success and dominance are best elucidated by detailed studies of living organisms that most strikingly display them. This is the rationale here for addressing the subject through the "eusocial" insects, which include the termites, ants, and several clades of bees and wasps. Eusociality is defined as the combination of three traits: care of young, overlap of adult generations in the same colony, and a division of labor between reproductive and nonreproductive castes.

The eusocial insects, which date from Cretaceous times, have already attained moderate longevity in comparison with sister groups. Their tenacity is further indicated by the fact that so far as known no major eusocial clade, at the rank of family or above, has ever gone extinct. Far more important, however, is the overwhelming dominance of the eusocial insects. In some habitats, especially tropical forests, they make up 30% or more of the animal biomass, and comparable figures may hold in grassland, deserts, and even some temperate forests. Ants are the major predators and scavengers in the mesofauna (0.2 to 2.0 mm body length range), while termites are the premier brokers of wood decomposition. Together the two groups turn more soil than other animals. Their impact is evident in the life cycles and behavior of many kinds of organisms. In a most general sense, ants and termites hold the central microhabitats where interference competition allows the control of stable resources, while solitary insects "fill the cracks" of less stable, accessible resources.

98

It is possible to identify the sources of dominance in the eusocial insects with some confidence because these insects, especially the ants and termites, are so conspicuously abundant, their abundance is manifestly tied to competitive superiority, and this superiority clearly arises from unique traits in social organization. The inferences made are strengthened by the comparison of large numbers of species of eusocial insects and nonsocial insects in sister taxa, and the direct establishment of the links between competitive superiority and social organization through field and laboratory studies.

The advantages of colonial life leading to dominance in the social insects are evidently the following: (1) series-parallel operations of multiple workers, insuring a higher percentage of success in the completion of tasks; (2) the ability to engage in more aggressive, even suicidal behavior with positive payoff in terms of inclusive fitness; (3) a superior ability to protect and bequeath resources such as nest sites and large or sustainable food sources; and (4) more efficient homeostasis and control of the nest microenvironment.

The existence of a worker caste, which is largely or entirely nonreproductive, has allowed colonies to evolve such advanced division of labor and high levels of integration as to constitute superorganisms. The small number of studies of efficiency and fitness conducted to date suggest that at least some of the colonial properties have approached evolutionarily local optima. The eusocial insect colony thus faces its competitors as a proportionately huge organism, blanketing a substantial trophophoric field with nearly continuous exploration and defense.

An ideal circumstance in the analysis of dominance by a group of organisms would be the existence of an ecosystem free of the group that could serve as a control. Hawaii is such a place, having never been colonized by eusocial insects, and its biota appears to have evolved in the direction expected in an environment free of social insects. However, this conclusion is vitiated, or at least complicated, by the generally disharmonic nature of the Hawaiian biota, so that important groups ecologically similar to eusocial insects are also missing.

1992

"The effects of complex social life on evolution and biodiversity," *Oikos* 63: 13–18

In this study I reversed the usual cause-and-effect approach to sociobiology and behavioral ecology. Instead of considering how the environment affects the evolution of colonial life, I examined the manner in which colonial existence of even a small number of species can profoundly affect biomass and diversity, not only of their own species, but also, through symbiosis and competition, of many other species through the ecosystem.

OIKOS 63: 13–18. Copenhagen 1992

The effects of complex social life on evolution and biodiversity

Edward O. Wilson

Wilson, E. O. 1992. The effects of complex social life on evolution and biodiversity. – Oikos 63: 13–18.

Social vertebrates display faster chromosomal evolution, faster species turnover, and higher levels of allelic heterozygosity than nonsocial vertebrates. At least some species also display substantially more genetic differentiation among geographically spaced social groups. The explanation for this pattern may be that matrilines stay together over multiple generations to compose such groups, and a relatively small number of males inseminate them. In contrast, higher social insects, comprising the termites, ants, eusocial bees, and eusocial wasps, have slower rates of evolution, similar to that of other, related groups of insects. The explanation appears to be that their population structure is very different from that of vertebrates: from the viewpoint of genetics their colonies are actually individuals (superorganisms), and the population equivalent to the local vertebrate society (= population) is the population of colonies. Moreover, outbreeding is extensive. The four major assemblages of highly social organisms – the colonial invertebrates, eusocial insects, nonhuman social mammals, and man – are characterized for the most part by great biomass, interspecific competitive superiority, and low species diversity. Competitive superiority in colonial invertebrates and eusocial insects over nonsocial animals is achieved substantially by social means. Lower species diversity on the other hand is due substantially to large organismic size, where invertebrate clones and insect colonies are the equivalent of the organisms composing nonsocial species.

E. O. Wilson, Harvard University, Museum of Comparative Zoology, Canbridge, MA 02138–2902, USA.

The usual form of sociobiological analysis addresses the prerequisites in heredity and the environment for the origin and evolution of sociality. In this article I will reverse the procedure and pose the question of how sociality affects evolution, biological diversity, and certain aspects of the environment.

Sociality and rate of evolution

Bush et al. (1977) found that among 225 genera of vertebrates, speciation and chromosomal evolution have proceeded faster in genera whose species are organized into troops or harems, such as those of horses and many primates. They found the same trend in genera with limited vagility, patchy distribution, and strong individual territoriality, a category which includes (for example) many rodents. The criteria used by these authors were: (1) rate of chromosome change inferred from phylogenies of living species; (2) rate of appearance of new species, inferred from phylogenies of living species and fossils; and (3) rate of generic extinctions, hence turnover, in monophyletic clades (species and their descendants).

Marzluff and Dial (1991) obtained what might seem at first a result contrary to that of Bush et al.: hyperdiverse living groups of vertebrates, that is, families with many genera and genera with many species, are not significantly more social than less diverse groups. However, the two sets of results do not conflict. The data reviewed by Marzluff and Dial represent the standing crops of genera and species, while those reviewed by Bush et al. refer to turnover rates. It is possible to have very high turnover – new taxa replacing old ones – and hence rapid evolution overall, but achieve only a small

Accepted 4 November 1990

standing crop base. The turnover is due to the dynamism of the evolutionary process, while the standing diversity is constrained by population size at equilibrium (K) and hence the number of species that can be packed together in the same community.

This brings us to the question of the causes of rapid evolution in social vertebrates. Bush et al. suggested that social groups, by staying clear of one another through behavioral repulsion, separately constitute demes with low effective breeding size. The groups are more prone to allele and chromosome type fixation by genetic drift. Thus, in accordance with Sewall Wright's shifting balance hypothesis, these socially demarcated populations might "experiment" with temporarily less adaptive combinations. If a new, favorable combination were to be struck (in Wright's metaphor, the population crosses the adaptive valley to a new adaptive peak), divergent evolution among the populations would be accelerated. And if the populations were further isolated by a geographic barrier, such as a deep valley or riverine corridor forest, the divergence might lead to full species formation, with intrinsic isolating mechanisms thereafter preventing free genetic exchange with other, sister populations. Finally, in taxonomic groups where only a few species can coexist, sympatry of the newly formed species would tend to result in the extinction of those less well adapted – in other words increase the rate of turnover.

Unfortunately for this neat explanation, Melnick (1987, 1988) noted from his research on rhesus macaques and that of earlier investigators on mammals generally that mammalian societies apparently exchange too many individuals to allow protracted evolution based on sustained genetic drift. For at least the rhesus macaques of Dunga Gali, Pakistan, Melnick also excluded the possibility that divergence among local troops is due to fission and formation of new troops along matrilines, in other words, females and their descendants.

Nonetheless, the data on social vertebrates confirm a high degree of differentiation among social groupings of mammals, just as Bush et al. stressed. Melnick offers the following alternative explanation to inbreeding and genetic drift. Matrilines stay together to compose each troop over multiple generations. Only a small number of males inseminate the females, as a result of dominance interactions, and they soon leave, as do their sons, without inbreeding. The total result is a strong genetic divergence among the troops, even when these groups are stationed only a short distance apart. If geographical isolation occurred, speciation might soon follow. One can imagine that intergroup selection is also an accelerating force, with certain genetic combinations prevailing over others in territorial disputes and takeovers.

In a parallel study, Nevo et al. (1984) found that among 127 vertebrate species surveyed with isozyme electrophoresis, social species possessed higher levels of genic heterozygosity than nonsocial species. Nevo's own work on *Spalax* mole rats of the Middle East indicates that heterozygosity in local populations increases with the level of environmental stress, in particular physical stress associated with arid environments. Nevo has suggested (pers. comm.) that social organization is a form of stress that could promote higher degrees of behavioral polymorphism and hence allelic heterozygosity. But it also seems likely that chance variation arising among geographically proximate groups, as postulated by Bush et al. and Melnick, would result in elevated genic diversity within groups through gene flow, even when gene flow between groups is moderately restricted across a few generations at a time.

When we turn to the second great pinnacle of social evolution, the eusocial insects (Wilson 1975), we encounter a radically different situation from that of the vertebrates. The rate of evolution at the level of the species and above is slower by as much as an order of magnitude. Slow evolution, at least slow macroevolution, is a trait held in common with most other insects. All of the 28 orders of insects alive in the Cretaceous Period are still alive today, with several extending all the way back to the late Paleozoic, whereas only one Cretaceous mammalian order, the Marsupialia, still persists. Within the single insect order Hymenoptera, 25 of the 36 families present in the Cretaceous are alive, whereas this is true for only one mammalian family, the Didelphidae or opossums. In contrast to the fast-evolving mammals, the early Tertiary faunas of social insects have a decidedly modern cast, down to the level of the genus and even the species group. A species of the contemporary stingless bee genus *Trigona* has been found in Late Cretaceous amber (Michener and Grimaldi 1988). No fewer than 56% of the ant genera of the early Oligocene Baltic amber still survive. A modern facies is even more evident in the Dominican amber, which is apparently early Miocene in age. Here 34 genera or 92% of the total 37 are extant. Furthermore, the majority of species belong to living species groups, and a few are difficult to separate from living ants at the species level (Wilson 1985). The eusocial insects (ants, bees, wasps, and termites characterized by a nonreproductive worker caste) appear to be about as conservative in evolution as their closest relatives among the solitary insects (Wilson 1990).

If the rapid evolution and high levels of heterozygosity of social mammals are to be explained as the impact of social organization on population structure, as the evidence seems to indicate, how are we to account for the complete lack of such correspondence in the social insects? The answer is that a colony of social insects, unlike a troop of macaques or herd of elephants, is not ordinarily a population, at least not in the ordinary sense applied in population genetics. Rather, it is the counterpart of a vertebrate organism, with the queen as the reproductive system and the workers as the rest of the body. The queen is inseminated by one to several males at the start of the life of the colony, or in the case of termites, is accompanied by a long-lived consort male. The offspring workers rarely reproduce on their own while the queen still lives, and in the vast majority of species, virgin primary reproductives leave the nest to start their own colonies alone and elsewhere. Soon after departure they mate with individuals from other nests, often in the midst of large panmictic swarms drawn from many other nests. In short, gene flow appears to be as free and random as it is in solitary insects. The equivalent in the social insects of a conventional local breeding population of vertebrates or solitary insects is the population of colonies.

There are exceptions to the population structure of the social insects just characterized. Some ant species in the genera *Formica, Linepithema* (formerly part of *Iridomyrmex*), *Myrmica, Pheidole,* and *Pristomyrmex* are organized into supercolonies or unicolonial populations, in which there are no colonial boundaries and the local population is just one great sprawling colony serviced by large numbers of queens (Hölldobler and Wilson 1990). Much of the mating occurs locally, within the territories of the supercolonies. These aggregations are thus structured somewhat like vertebrate society-populations, and have a similar potential for rapid evolution. Yet they do not present the appearance of much cumulative change. They are typically closely related to species that form ordinary colonies, and in fact may represent evolutionary dead ends that do not evolve much further on their own. The supercolony species are also poor in standing diversity compared with their ordinary relatives. Here the reason seems clear. Supercolony species typically dominate the local environment, aggressively excluding colonies of many of the related species. They are also broad generalists, preying on arthropods, scavenging corpses, and collecting honeydew excreta from aphids and other homopterous insects. It is hard to imagine two recently evolved sister species with supercolonies that overlap in the same habitat.

Sociality, success, and dominance

In considering the status of any group of organisms, biologists intuitively use the concepts of success, dominance, and diversity – and just as often conflate them. For the purposes of evaluating sociality in insects with greater precision, I recently proposed the following definitions (Wilson 1990):

Success. Longevity of the entire clade (a species and all its descendants) through geological time. Paleontologists often use the average duration of genera or families as a measure of a still higher taxon, such as the class Anthozoa or phylum Porifera. What they mean is the clade that starts with the ancestral species first displaying the diagnostic traits of the genus or family, plus all of the species descended from that species. What they exclude is pseudoextinction, the "termination" of a line by evolution of new diagnostic traits without extinction of the populations bearing the ancestral traits. A species evolving radically different traits may be said to give rise to a second species. Thus there are two chronospecies, the ancestor and the descendant, but the line has not gone extinct; the two forms are counted together as part of the same clade. Clade longevity is intuitively the best criterion of success, because it represents the simplest and most direct measure of species-level selection over long periods of time. It is a pure Darwinian criterion of the evolutionary process.

Dominance. Relative abundance, especially as it effects the appropriation of biomass and energy and impacts the populations and evolution of the rest of the biota. Dominance, as I have used the word (Wilson, 1990), is thus not merely population density but the degree to which the species influence the remainder of the community. The influence in general will certainly be correlated with density, and all other things being equal it must be identical to density. But other things are very seldom equal. Influence will rise as organism size increases, hence more biomass is appropriated. It will rise as basal metabolism increases, resulting in the appropriation of more free energy. It will rise with the number of links in the food webs, hence the number of species affected. For parasites it will rise with the degree of virulence. For free-living organisms generally it will rise as more of the physical environment is disturbed, by soil excavation during nest-building, by adventitious destruction of vegetation during trail construction and feeding, and so on. How these varied concomitants of population density and their impact are to be generally measured is a problem not easily resolved, but the usefulness of refining the concept of dominance in particular cases is clear enough.

Speciosity. The number of species, which in turn can be usefully separated into the geographically expanding components of diversity (Magurran 1988): alpha (the number of species occurring together within a given habitat), beta (the rate of turnover in species along a transect in passing from one habitat to another), and gamma (the number for all of the sampled localities taken together). The greater the alpha and beta diversity, the greater the adaptive radiation.

Adaptive radiation. The array of niches occupied by the various species of a clade, especially those that coexist in the same area simultaneously.

Geographic range. The entire area occupied by a clade at a particular time.

Success, which is an evolutionary quality, and dominance, which is an ecological trait, are correlated. All mathematical models to date have confirmed the strong intuitive impression that the risk of extinction decreases with maximum population size (K) and increases with the temporal coefficient of variation in population size (CV). Pimm et al. (1988) confirmed this result in an

Table 1. Evolutionary and ecological traits of the four major groups of organisms with the most complex modes of social organization. +, traits strongly developed; –, weakly developed; ±, with many clades in both categories; ?, indeterminate. Human beings are judged to have undergone adaptive radiation, but by culture rather than speciation.

Group	Traits				
	Success (clade longevity)	Dominance (abundance, impact)	Speciosity (species numbers)	Adaptive radiation	Breadth of geographic range
Colonial invertebrates	±	+	–	–	+
Eusocial insects (termites, ants, eusocial bees, eusocial wasps)	+	+	–	±	+
Nonhuman social mammals	–	±	–	–	+
Man (*Homo*)	?	+	–	+	+

analysis of census data of 100 species of British land birds living in small to moderate sized populations on 16 islands. Stanley (1987) found that among marine bivalve mollusks living along the Pacific rim during Pleistocene times, abundant clades survived the longest. Yet success (longevity) and dominance (abundance and impact on biota) are qualitatively different phenomena and can be uncoupled and opposed in particular cases. A clade can enjoy enormous success by virtue of chronological persistence, but also be so rare as to have negligible influence on the ecosystem in which it lives. Examples include many "living fossils", such as onychophorans and coelacanths, which tend to be local and rare. At the opposite extreme, a clade can be so abundant for a while as to exert a profound and lasting impact on the rest of life, yet be short-lived. The most striking examples of all time may include the extinct clades of social mammals. (We hope and trust that this assemblage will not soon include man.)

A clade is likely to increase its longevity not only if its populations are larger but also if it contains many species, representing an array of adaptive types, and occupies a broad geographical range; because the clade then has "spread its bets" across many independent gene pools, among which one or more may possess traits that insure survival when the remainder of the clade succumbs. This relation is likely to exist as a broad correlation, yet it too can be uncoupled in particular cases. A tardigrade genus such as *Echiniscoides,* for example, might be worldwide and ancient yet monotypic and scarce; a species swarm of cichlid fishes can be middle-aged and extremely diverse, yet limited to a single African lake; and so on through all of the combinations of the population-level traits pertaining to abundance and diversity.

In Table 1 I have used the population-level traits to categorize the four animal (and human) assemblages possessing the most complex social organizations. With the exception of *Homo,* none of these assemblages is monophyletic, which is good for purposes of analysis, since each thus embraces multiple evolutionary experiments. The higher social insects include at least twelve independent lines, and the nonhuman social mammals are at least that polyphyletic. The colonial invertebrates are distributed as independent clades across two phyla, the Cnidaria and Ectoprocta (Bryozoa).

What emerges clearly from this subjective comparison is that highly social organisms tend to be ecologically dominant and geographically widespread, but not speciose. Their record of longevity is decidedly mixed, with eusocial insects showing remarkable persistence since their origins, and colonial invertebrates and nonhuman social mammals less so.

The ecological dominance of social animals is striking. In terra firme rainforest near Manaus, eusocial insects make up about 80% of the insect biomass and a third of the entire animal biomass (Fittkau and Klinge 1973). In some samples from the Peruvian rainforest canopy, ants compose 70% or more of the individual insects (T. L. Erwin and J. E. Tobin, pers. comm.). The prevalence of ants in particular holds equally well in arid habitats. It also extends into the cold temperate zones, where for example ants remain the dominant insects in southern Finland. Wood and Sands (1978), Brian (1983), Hölldobler and Wilson (1990), Wilson (1990) and others have reviewed evidence documenting the impact of ants and termites on soil structure and composition, seed dispersal, plant distribution, and the biology of myriads of species of insects and other animals.

There can be little doubt that this overwhelming presence is due to social organization. The following traits arising from social life have been identified by experiments and field observations as conveying competitive superiority over otherwise similar nonsocial insects (Wilson 1990): (1) series-parallel operations, by which workers change occupations from moment to moment in a way that keeps larval care, nest construction, and other sequential tasks continuously well attended; (2) high levels of altruism, conferring combat advantage over enemies with a reduced cost of inclusive fitness in case of injury or death; (3) ability of the colony as a whole to control nest sites and food sources and bequeath them to later generations; and (4) use of social homeostatic devices, such as nest design, water cooling, and heating from massed bodies, to control microclimate and permit longer occupancy of nest sites near stable resources. With these advantages ants and termites occupy center stage in the terrestrial environment. They have pushed out solitary insects from the generally most favorable nest sites. The solitary forms occupy the most distant twigs, the very moist or dry or excessively crumbling pieces of wood, the surface of leaves, in other words, the more remote and transient microniches. They are also either very small, or fast moving, or cleverly camouflaged, or heavily armored. At the risk of oversimplification, the picture I see is one of social insects at the ecological center, solitary insects at the periphery.

Marine zoologists have independently forged a similar conception of the dominance of colonial invertebrates (Jackson 1985, Vermeij 1987). By Early Cambrian times sponge-like colonial Archaeocyatha formed reefs, to be succeeded in this role by bryozoans, sponges and corals, and then noncolonial worms, foraminiferans, and brachiopods during the Paleozoic; next sponges and algae and then rudist pelecypods in the Mesozoic; and finally corals, sponges, and coralline algae during the Cenozoic. A gradual increase in differentiation of zooids and in integration of the colonies occurred during the Mesozoic and Cenozoic in the corals and cheilostome bryozoans, possibly enhancing the competitive ability of these prominent colonial forms.

Colonial invertebrates displace solitary invertebrates in stable hard-bottom communities around the world. Being larger than solitary animals, they can overgrow them and are less likely to be pulled up by turbulence or eaten by predators. In contrast, solitary immobile animals tend to prevail in less stable environments, including the rocky intertidal zone, unconsolidated bottoms, and small or transient substrata such as shells and seaweeds. Like solitary insects, solitary benthic invertebrates fill in the smaller and more ephemeral niches where their competitively superior social counterparts cannot take hold.

That *Homo sapiens* has also achieved great biomass and competitive superiority by means of social organization is too clear to warrant further discussion.

Remarkably, sociality promotes dominance but not diversity. The numbers of species of the most highly social forms is very small compared to related solitary forms in the current world biota. The number of described species of eusocial insects is as follows: termites, 2200; ants, 8800; eusocial bees, roughly 1000; eusocial wasps, 800. The total number of described insect species, social and solitary combined, is about 750 000. In other words, 2% of the known insect species of the world have appropriated more than half of the biomass. The story is similar for the colonial invertebrates: among 240 000 multicellular invertebrates outside the insects, fewer than 5000 are sponges, 6000 are corals, and 4000 are bryozoans. The highly social nonhuman mammals are even less speciose, comprising the lion, 3 species of canids, 2 elephants, and only about 35 band-forming primates. Being close to the top of the food chain, most of the mammals are also less than dominant. An exception is formed by the elephants, which are capable of building dense populations that destroy extensive areas of woodland and alter the soil through physical disturbance and the deposit of large amounts of excrement.

The apparent reason for lessened biodiversity of the most social animals is size. For the most part these animals are very large. A coral head is essentially one organism, a superorganism. The same is true of an ant or termite colony. Both are orders of magnitude larger than single organisms belonging to solitary species. It is a well-established principle that within a monophyletic assemblage, such as the beetles, birds, and mammals, the numbers of species belonging to different body sizes are highest near (but not precisely at) the lower extreme and taper off gradually toward the upper extreme (Dial and Marzluff 1988, May 1988). The generally accepted explanation, which is also plausible for animal societies (in other words, superorganisms) is that the larger the organism the larger the spatial niche and the fewer the species that can coexist stably.

References

Brian, M. V. 1983. Social insects: ecology and behavioural biology. – Chapman and Hall, New York.

Bush, G. L., Case, S. M., Wilson, A. C. and Patton, J. L. 1977. Rapid speciation and chromosomal evolution in mammals. – Proc. Natl. Acad. Sci. U.S.A. 74: 3942–3946.

Dial, K. P. and Marzluff, J. M. 1988. Are the smallest organisms the most diverse? – Ecology 69: 1620–1624.

Fittkau, E. J. and Klinge, H. 1973. On the biomass and trophic structure of the central Amazonian rain forest. – Biotropica 5: 2–15.

Hölldobler, B. and Wilson, E. O. 1990. The ants. – Belknap Press of Harvard Univ. Press, Cambridge, MA.

Jackson, J. B. C. 1985. Distribution and ecology of clonal and aclonal benthic invertebrates. – In: Jackson, J. B. C., Buss, L. W. and Cook, R. E. (eds), Population biology and evolution of clonal organisms. Yale University Press, New Haven, CT., pp. 297–355.

Magurran, A. E. 1988. Ecological diversity and its measurement. – Princeton Univ. Press, Princeton, NJ.

Marzluff, J. M. and Dial, K. P. 1991. Does social organization influence diversification? – Am. Midl. Nat. 125: 126–134.

May, R. M. 1988. How many species are there on Earth? – Science 241: 1441–1449.

Melnick, D. J. 1987. The genetic consequences of primate social organization: a review of macaques, baboons, and vervet monkeys. – Genetica 73: 117–135.

– 1988. Why are we social? – Ann. Rep., Center for Advanced Study in the Behavioral Sciences, Stanford, CA.

Michener, C. D. and Grimaldi, D. A. 1988. A *Trigona* from Late Cretaceous amber of New Jersey (Hymenoptera: Apidae: Meliponinae). – Am. Mus. Nov. 2917: 1–10.

Nevo, E., Beiles, A. and Ben-Shlomo, R. 1984. The evolutionary significance of genetic diversity: ecological, demographic and life history correlates. – In: Mani, G. S. (ed.), Evolutionary dynamics of genetic diversity. Lecture Notes in Biomathematics 53: 13–213.

Pimm, S. L., Jones, H. L. and Diamond, J. 1988. On the risk of extinction. – Am. Nat. 132: 757–785.

Stanley, S. M. 1987. Periodic mass extinctions of the Earth's species. – Bull. Am. Acad. Arts Sci. 40: 29–48.
Vermeij, G. J. 1987. Evolution and escalation: an ecological history of life. – Princeton Univ. Press, Princeton, NJ.
Wilson, E. O. 1975. Sociobiology: the new synthesis. – Belknap Press of Harvard Univ. Press, Cambridge, MA.
– 1985. Invasion and extinction in the West Indian ant fauna: evidence from the Dominican amber. – Science 229: 265–267.
– 1990. Success and dominance in ecosystems: the case of the social insects. – Ecology Institute, Oldendorf-Luhe, Germany.
Wood, T. G. and Sands, W. A. 1978. The role of termites in ecosystems. – In: Brian, M. V. (ed.), Production ecology of ants and termites. Cambridge Univ. Press, New York, pp. 245–292.

1992

"*Pheidole nasutoides,* a new species of Costa Rican ant that apparently mimics termites," *Psyche* (Cambridge, MA) 99(1): 15–22 (with Bert Hölldobler, first author)

Ants, being dominant insects in biomass and aggressiveness, are among the organisms most mimicked by other organisms. A huge variety of arthropods have evolved to look like ants, smell like them, or both. The visual mimics include beetles, flies, hemipterans, and spiders. Their resemblance protects them from birds and other visual predators reluctant to attack the more formidable kinds of ants that sting or release toxic secretions. A few ant species with small, timid colonies mimic ants of other species that have large, aggressive colonies. This report describes a unique case of an ant species that appears to mimic the soldier caste of large, aggressive termites.

The termitelike ant illustrates another principle, the large number of extremely rare species in the highly biodiverse tropical mainland rainforests. When Bert Hölldobler found the one colony of *Pheidole nasutoides* described here, we at once set out to find others at the same locality, but without success. In later years we continued our quest off and on, and others, including experienced entomologists and students, made efforts of their own. As of this writing, the species has never been found again.

Reprinted from PSYCHE, Vol 99, No. 1, 1992

PHEIDOLE NASUTOIDES, A NEW SPECIES OF COSTA RICAN ANT THAT APPARENTLY MIMICS TERMITES

BY BERT HÖLLDOBLER[1] AND EDWARD O. WILSON[2]

Mimicry is a commonplace among species of ants (Hölldobler and Wilson, 1990). However, to our knowledge no ant or any other social hymenopteran has been recorded that mimics termites, despite the fact that many termite species have formidable defenses against both vertebrates and invertebrates that would seemingly make them ideal models. In 1985, at the La Selva Field Station and Biological Reserve of the Organization for Tropical Studies, in northeastern Costa Rica, we discovered a new species of *Pheidole* whose major workers in life astonishingly resembled those of *Nasutitermes*. These strange-looking ants were nesting in the low arboreal zone at the edge of secondary rain forest, in a habitat especially favored by *Nasutitermes*. Our critical study was limited to a brief observation of the *Pheidole* at the nest site and in the laboratory. On three later visits to this locality during the subsequent five years we searched the vicinity of the original find and widely elsewhere through the La Selva station, in an attempt to locate additional colonies and conduct more detailed studies. We were never successful; the new *Pheidole* remains known only from the original colony. Because of its unique qualities and possible unusual biological significance, we have chosen to describe it now, and with the description the hypothesis that this ant is indeed a rare termite mimic. It is our hope that others will continue the search until the species is rediscovered and can be studied more definitively.

[1] Theodor Boveri Institut für Biowissenschaften (Biozentrum) der Universität, Lehrstuhl für Zoologie II, Am Hubland, 8700 Würzburg, Germany
[2] Museum of Comparative Zoology, Harvard University, 26 Oxford Street, Cambridge, Massachusetts 02138-2902, USA
Manuscript received 1 March 1992

Pheidole nasutoides, new species

(Figures 1, 2)

Diagnosis. A small species closest to *P. defecta* Santschi of Guatemala, and in lesser degree to members of the *P. tepicana* group of the southwestern United States and Mexican Plateau. *P. nasutoides* shares with these species a relatively small size in both castes; a single pair of hypostomal teeth in the major worker; and many aspects of general habitus, including the mesonotal hump reduced or absent in profile in both castes. It is distinct from members of the *tepicana* group in being completely dimorphic instead of trimorphic (i.e., with one rather than two major castes) and in many details of sculpturing. The major worker differs from that of *defecta* in the possession of a small but distinct mesonotal hump, proportionately longer petiolar peduncle, a larger propodeal spiracle, and extensive foveolation on the mesothorax and propodeum, the surfaces of which are thereby opaque. *P. nasutoides* differs from both the *tepicana* group and *defecta* in possessing angulate humeri in the major and minor, and from all other known *Pheidole* in the unique coloration of the major.
Holotype major. Head Width 0.80 mm, Head Length 0.78 mm, Scape Length 0.50 mm, Eye Length 0.12 mm, Pronotal Width 0.34 mm. Body form and pattern of surface sculpture as shown in Figure 1. Color medium yellow, except for the light brown "mask" as depicted in Figure 1 (the intensity and shape of the mask varies considerably among the major workers in the type series).
Paratype minor. (Typical specimen from holotype nest series.) Head Width 0.48 mm, Head Length 0.54 mm, Scape Length 0.54 mm, Eye Length 0.10 mm, Pronotal Width 0.32 mm. Body form and pattern of surface sculpture as shown in Figure 2. Concolorous medium yellow. COSTA RICA: La Selva Field Station and Biological Reserve, near Puerto Viejo de Sarapiqui, Heredia Province, 10°26'N, 84°00'W, 25 March 1985.

Redescription of *Pheidole defecta*

Pheidole defecta Santschi 1923 is a "lost" species, poorly described at the beginning and still known only from the unique major holotype. We have undertaken to redescribe it here, as follows. Head Width 1.10 mm, Head Length 1.10 mm, Scape Length

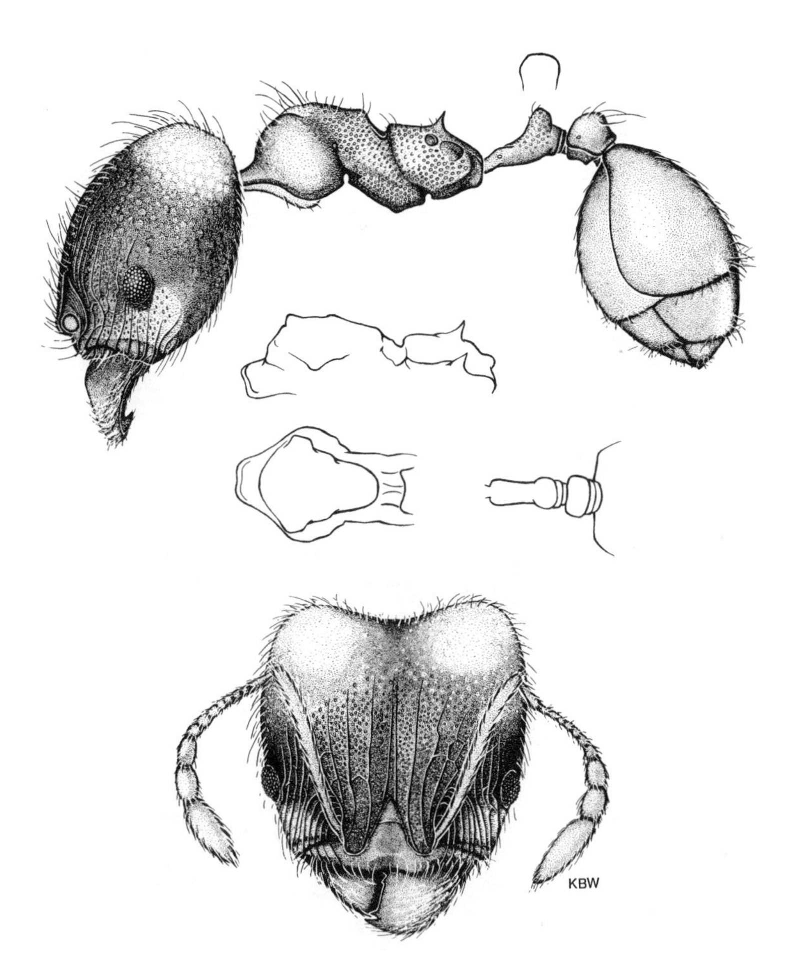

Fig. 1. Holotype major worker of *Pheidole nasutoides*, new species. The deeper stippling on the head represents the extent of the light brown "mask" of the face.

0.64 mm, Eye Length 0.16 mm, Pronotal Width 0.56 mm. Body form as depicted in Figure 3. Especially notable are the short, thick petiolar peduncle and the relatively sparse pilosity. Small patches occur on the mesothorax and sides of the petiole, as well as in the intercarinular spaces between the eye and frontal lobes, that are shallowly foveolate and feebly shining to subopaque. The

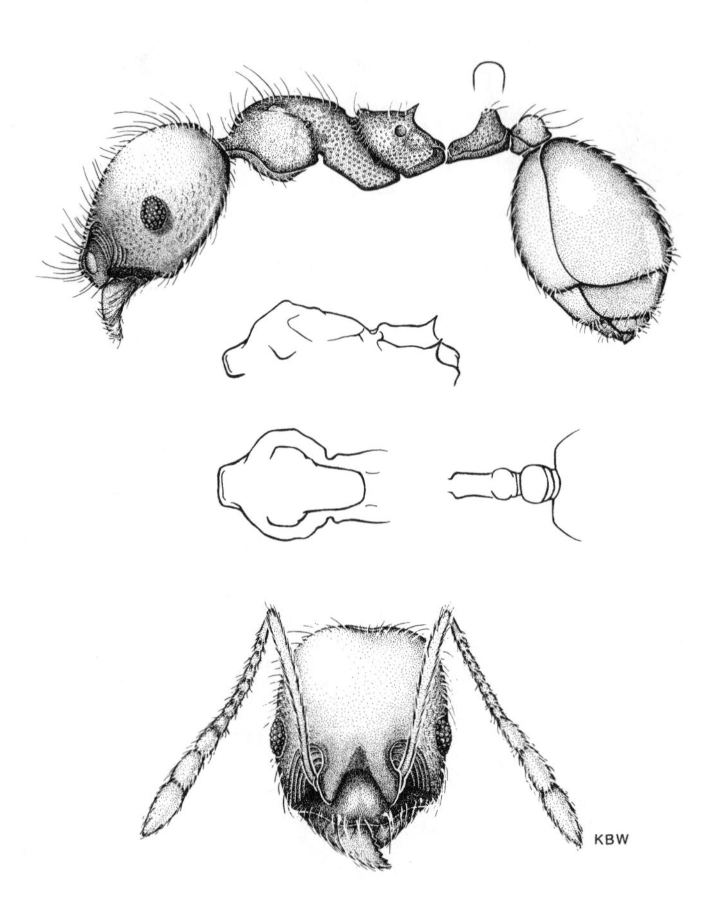

Fig. 2. Paratype minor worker of *Pheidole nasutoides*, new species.

remainder of the body is smooth and shining. Concolorous medium yellow. Type locality: "Guatemala," no further data.

P. defecta is possibly a member of the *tepicana* group, although the lack of other specimens prevents us from judging whether the worker caste is polymorphic. Also, if *defecta* does belong in this group, it is the southernmost known member. If it is fully

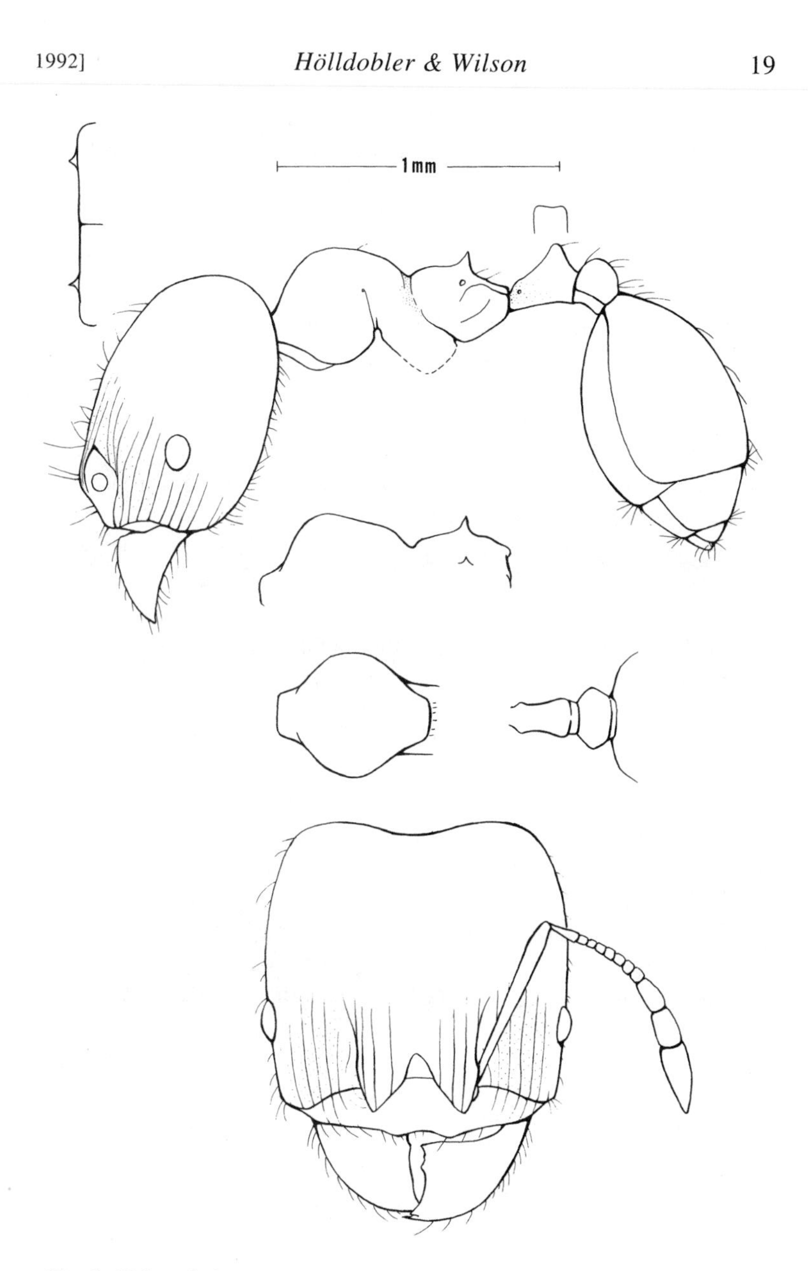

Fig. 3. Unique holotype major worker of *Pheidole defecta* Santschi. The stippling demarcates limited areas of shallow foveolation. The bidentate hypostomal border, similar in appearance to that of *P. nasutoides*, is depicted in the upper left-hand corner.

dimorphic, the usual condition of *Pheidole* and one displayed by *P. nasutoides*, then it may be regarded as phyletically close to *nasutoides* and forming with that species a small, distinctive Central American species group (the "*defecta* group").

Natural History

The type colony was found nesting in a round mass of dried, thatch-like vegetation about 1.5 meters up in the moderately dense foliage of a small tree, which was located at the edge of a second-growth forest bordering the open experimental fields of the La Selva station. When the nest was disturbed, more than a hundred major and minor workers of *P. nasutoides* rushed out and ran in erratic looping patterns to form a spreading wave away from the nest (as shown in Fig. 4). The resemblance of the majors to *Nasutitermes* nasute soldiers under similar circumstances was remarkable. In particular, the mask of the *Pheidole* majors is roughly shaped like that of the nasute termites and contrasts with the light remainder of the body in the same way (Fig. 5). The illusion was heightened when the ants were in motion, creating a *Nasutitermes*-like gestalt. Hölldobler, who discovered the nest, in fact first thought that the ants were *Nasutitermes* and nearly passed them by. During the brief time the colony was observed live in the laboratory, the resemblance remained close. Otherwise, the colony seemed typical for a species of *Pheidole*. Adult males were present, but neither alate nor dealate queens were recovered.

We remain puzzled by our failure to locate other *P. nasutoides* nests despite prolonged effort in the La Selva area. It is possible that the species is simply very rare, existing in extremely sparse populations. Alternatively, it may be normally a dweller of the high canopy, a zone we did not explore. The nest found was at the edge of a disturbed forest patch, and might have fallen from a higher location.

Fig. 4. Reconstruction, with imaginary details in the surrounding vegetation and animals, of the type colony of *Pheidole nasutoides* at La Selva, Costa Rica. The nest of the ants has been disturbed (in this hypothetical scenario) by a *Dendrobates* frog, a species known to feed on ants. Both majors and minors swarm out and run swiftly in erratic looping paths over the vegetation. This motion, together with their unique color pattern, causes the large-headed major workers to resemble nasute soldiers of the termite genus *Nasutitermes*. Several of the termite soldiers, on a foraging expedition, pause on a leaf to the left.

Acknowledgements

The illustrations were prepared by Katherine Brown-Wing. Research was supported in part by a grant from the U.S. National Science Foundation No. DEB-89-15314.

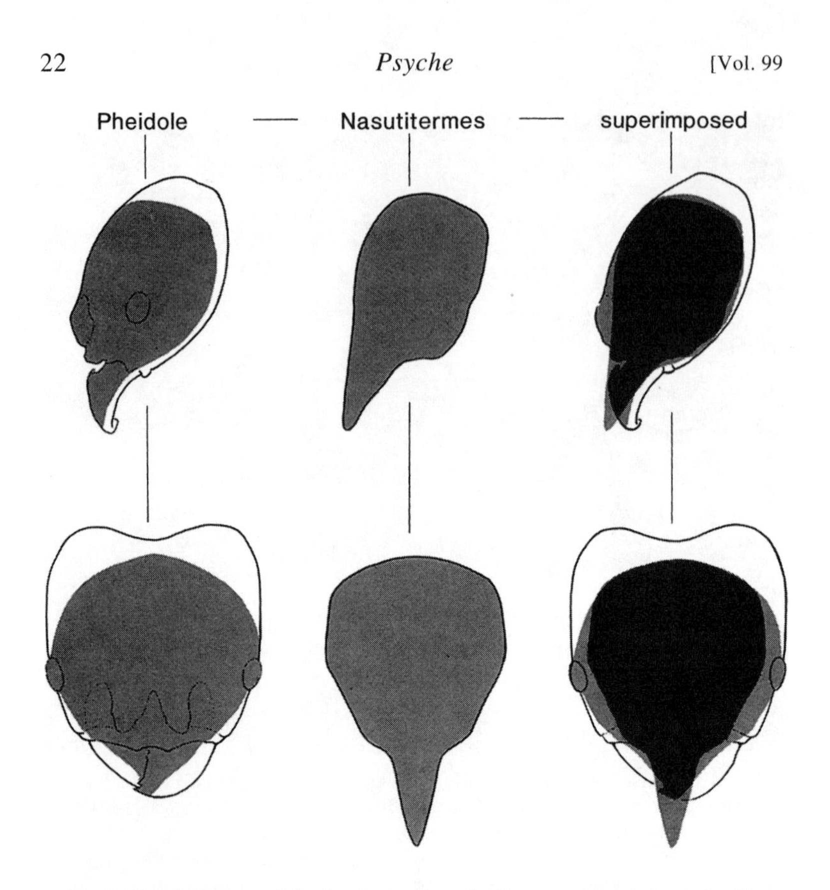

Fig. 5. The dark areas of the head capsules of a *P. nasutoides* major and a *Nasutitermes* nasute soldier, depicted separately and superimposed.

REFERENCES

HÖLLDOBLER, B. AND E. O. WILSON
1990. *The Ants*. Belknap Press of Harvard University Press, Cambridge, Mass. xi + 732 pp.

SANTSCHI, F.
1923. *Pheidole* et quelques autres fourmis néotropiques. *Ann. Ent. Soc. Belg.*, **63**: 45–69 (description of *P. defecta*: pp. 54–55).

2003

"In memory of William Louis Brown (1922–1997)," from E. O. Wilson, *Pheidole in the New World: A Dominant, Hyperdiverse Ant Genus* (Harvard University Press, Cambridge, MA), pp. i–iii

I feel it appropriate to insert here a tribute to Bill Brown, an important figure in the history of myrmecology (the scientific study of ants). He was my mentor in that discipline and, as the first two articles in the next section of this collection illustrate, one of my first collaborators. It is commonplace to say the departed was unique and the mold that made him is now broken. In "Uncle Bill's" case I believe both happen to be true.

In Memory of William Louis Brown (1922-1997)

William L. Brown Jr. at 30,
collecting at Ferntree Gully,
Dandenong Ranges, Australia, 1952.
(Courtesy of Doris E. Brown.)

Bill Brown at 39, Cornell, 1961. (Photo by Thomas Eisner.)

I knew Bill Brown for 50 years, and I've never met anyone else remotely like him. He was unique, and I don't think we'll ever see his like again, not just for the rareness of his character, but for the uniqueness of the time in which he lived and worked on his beloved ants. I've thought a lot about what made Bill different, what caused him to burn with such a pure inner light, and I've come up with this: the devotion to his art. His *art*. He was a scientist to the bone, a hard-core factual investigator, relentless for more information, skeptical in mood, all those things and yet . . . myrmecology, the study of ants, was an art form to him. It was the center of his creative life, and he was a very creative man. The passion he radiated about this subject turned younger people (we all seemed younger than Uncle Bill) into acolytes, into apprentices; and there was no prize the academic world could offer us more than a rare, measured compliment from the master, something like "Yeah, that's pretty good; that's really interesting."

Bill was my mentor. I first met him through correspondence in 1947, when I was 18, and had already taken my vows, so to speak, in ant taxonomy, just as he had been in contact with *his* mentor, William S. (Bill) Creighton, since he was 16. In natural history, addiction occurs early. In the summer of 1950, I rode a Greyhound Bus all the way from Mobile, Alabama, to Boston, and stayed with Bill and his wife, Doris, in their little apartment near Harvard, as they prepared to leave for Australia and momentous field research in that still myrmecologically underexplored continent. We worked in the Museum of Comparative Zoology ant collection together, and he gave me the kind of plain, sincere, egalitarian treatment he was to bestow on dozens of other students in his field in the decades to follow. He welcomed you, he treated you with respect, he stood in awe with you before the intricacy of the subject, he gladly taught and learned, he created a sense that here in this little discipline was something—to borrow from F. Scott Fitzgerald, the kind of writer Bill so admired—something commensurate to man's capacity for wonder. In 1950, he was 28 and I was 21, and the whole world seemed ours to possess.

Off he went to Australia, and in the years to follow just about every other place in the world where interesting ants are to be found, and he was often the first to do serious collecting in each. In time he became one of the most widely traveled naturalists of all time, bar none.

He was a key transitional figure in the history of myrmecology. From 1937, when William Morton Wheeler died, into the 1960s, there were very few researchers working on the classification and ecology of ants, and Bill carried the torch to close the gap. He played a major role in changing ant systematics from a thicket of trinomials and quadrinomials into a consistent binomial system based on the modern biological species concept. He was the first to recognize the major phyletic division of the myrmecioid and poneroid subfamily groups. He made judicious rearrangements of genera and tribes, "sank" innumerable worthless names into synonymy, and crafted clear, precise revisions widely through the Formicidae and most especially in his favorite groups, the Dacetini and Ponerinae.

His open, supportive nature drew young people in, and he played a major role in starting the current boom times in which hundreds are engaged in myrmecology around the world. We have truly followed in his footsteps.

Bill Brown was a working-class guy with a first-class mind and a noble heart. He had steel-hard integrity and his generosity to others knew no bounds. He hated a phony, to use one of his favorite words, could smell one a city block away, and put currency in only two things: solid accomplishment and integrity of purpose and representation. He never looked down on anybody socially and he never looked up to anybody. He was unimpressed by rank and status. He never played academic politics, never sought academic recognition or status, but instead waved aside compliments and put others first in research collaboration; he was a righteous man to the inner core.

Bill at 64,
in the field at La Selva, Costa Rica,
March 1987.

Toward the end of his life we used to talk about his dream field trip: get into a rich rainforest, sit at a table with a pan and collecting materials, and have graduate students scour the surrounding woods for ant nests which they bring to him to sort, study, and preserve. Dig through some of the "juicy red logs" where ant diversity peaks. In March 1987, while he was still sufficiently vigorous, I invited him along on a field trip to La Selva, the Organization for Tropical Studies station in Costa Rica. I sat him down on a field chair with a table in front of him, and *I* was the graduate student, hustling the ant nests for him to go through; and we talked incessantly about the treasures we found. It was a glorious three days.

In his later years Bill filled the same role in myrmecology and among his wider circle of students and admirers as the grizzly bear plays in the conservation movement. You didn't see him very often, and some younger researchers never did, but it made you feel good just to know he was there. It made things seem right. Now that he's gone, a big gap has been opened up in our consciousness that will never be completely closed.

2006

"Ant plagues: a centuries-old mystery solved"

The next article is the unabridged version of a "Brief Communication" published earlier in *Nature* ("Early ant plagues in the New World," *Nature* 433: 32, 6 January 2005). It represents a departure from my earlier research in that it combines historical and biological research, in this instance to solve a mystery almost 500 years old. The project grew out of my larger study on the biogeography and ecology of the ants of the Caribbean Islands, which I have been conducting since the mid-1980s.

Ant plagues

A centuries-old mystery solved

Edward O. Wilson

The identity and origin of the West Indian plague ants of the early sixteenth and late eighteenth centuries, and why they attained such high population levels, have long been mysteries. Here I collate historic accounts with an analysis of the present-day Caribbean ant fauna to suggest that two species were responsible: the tropical fire ant Solenopsis geminata *on Hispaniola and St. John around 1518–19; and either the big-headed ant* Pheidole megacephala, *of African origin or, less likely, the native Jelski's pheidole,* Pheidole jelskii, *in the Lesser Antilles during 1760–70. The most probable causes of the outbreaks were newly formed symbioses of the ants with honeydew-excreting homopteran insects.*

During or soon after the Hispaniolan smallpox epidemic of 1518–19, which killed off a great majority of the surviving indigenous Taino people, a plague of stinging ants irrupted within the fledgling Spanish settlements. According to an eyewitness account by the premier Columbian historian Fray Bartolomé de las Casas (de las Casas, 1875/1997) the insects destroyed crops of oranges, pomegranates, and cassia over a wide portion of the island and invaded dwellings in such numbers that the inhabitants were obliged to set their bedposts in small troughs of water. So desperate was their situation that the colonists turned to divine providence for relief, choosing as intercessor St. Saturnin, to whom they then prayed and held processions through the streets of Santo Domingo. A second such plague occurred on the island of St. John (de las Casas, 1875/1997), and according to tradition a similar formicid onslaught caused the abandonment in 1534 of the village of Sevilla Nueva on Jamaica (Schomburgk, 1848).

More than two centuries later, ant plagues spread through the Lesser Antilles: Barbados in 1760, Martinique in 1763, and Grenada in 1770. On Barbados, the naturalist R. H. Schomburgk later wrote, "Every sugar-plantation between St. George's and St. John's, a space of about twelve miles, was destroyed in succession, and the country was reduced to a state of the most deplorable condition" (Schomburgk, 1848). The ants were so dense, he added with perhaps just a touch of hyperbole, that they covered roads for miles at a stretch. In some places impressions made by the hooves of horses remained visible for only a moment or two, until they were filled up by the ants. No saints were chosen to save the crops of the Lesser Antilles, but large rewards were offered—£20,000 for example on Grenada—to anyone who found a way to stem the tide. No solution was forthcoming, but in these islands, as on Hispaniola earlier, the plagues diminished on their own.

A smaller irruption occurred on Barbados in 1814 (Schomburgk, 1848), and ants were noted to be ubiquitous and serious house pests by several authors during the 1600s and 1700s (Ligon, 1673; Schomburgk, 1848), but what can be regarded as full-blown plagues, causing severe agricultural damage and general economic distress, appear to be limited in the West Indies to those in the early sixteenth and late eighteenth centuries.

What were the plague ants? Why did they multiply to plague proportions? The answers would be significant for three reasons. First, the identification of the agent of the earliest environmental crisis experienced by European colonists of the New World is a matter of no small historical importance. Next, *Formica omnivora,* cited by early authors as the plague species but not identifiable today, remains a classic mystery of ant taxonomy. Finally, the challenge of the West Indian ant plagues provides an opportunity for combined historical and biological research, in which a distinction might be made between the kinds of problems addressed in such hybrid studies that are soluble and those that are inherently insoluble.

Methods

The plague ant problem came to my attention in the course of a study of the biogeography and natural history of the West Indian ants. During fieldwork from 1995 to 2004, while visiting the historic plague islands of Hispaniola (Dominican Republic), Grenada, and Barbados, I paid special attention to species that might have been the plague agents. Among the more relevant places visited were the ruins of the fort and monastery of Concepción de la Vega, built by Columbus in 1496 and abandoned in 1536, one of the locations, along with the old town of Santo Domingo, visited by Fray Bartolomé during the plague. With the aid of scholars, I also combed the contemporary and near-contemporary accounts of the plague ants and fitted together the bits of information concerning their appearance and habits. I matched this information against that of the now reasonably well-known modern ant fauna. By process of elimination, I narrowed the faunal list down to candidates that possess all the available defining traits.

The Identity of Linnaeus's *Formica omnivora*

The plague ants were referred by Schomburgk in 1848 as *F. omnivora*. This entity was one of the seventeen species of ants described by Carolus Linnaeus in the founding document of binomial taxonomy of 1758, all of which he placed in the genus *Formica* (the fourteen recognizable today are scattered through ten genera in four subfamilies) (Linnaeus, 1758). In 1926 William M. Wheeler used the same name in the first and only independent assessment of the West Indian plague ants since that of Schomburgk (de Réaumur, 1936).

Unfortunately, as Wheeler recognized, *F. omnivora* cannot be easily linked to either the plague species or to any species recognized in post-Linnaean ant taxonomy. Linnaeus drew the binomen from a Jamaican house pest ant briefly alluded to by Robert Browne, who used the pre-Linnaean name *Formica domestica omnivora* (the omnivorous house ant) (Browne, 1756). Linnaeus had a specimen before him, for which, in a six-line description, he gave only two useful clues: "thorace bidentato, petiolo binodoso." This much tells us that *F. omnivora* is a member of the subfamily Myrmicinae and that it can also be excluded within the Myrmicinae as either a fire ant or thief ant (genus *Solenopsis*) or a member of the related genus *Monomorium*, both of which genera contain species otherwise credible as plague ants. In 1773, Charles DeGeer, a close associate of Linnaeus, published a crude figure of a specimen that he identified as *F. omnivora*, but its depicted thorax is not spinose (DeGeer, 1773). Moreover, the specimen was not from the West Indies but from Suriname, home of an extremely diverse ant fauna, which, like the West Indian fauna, was otherwise almost entirely unknown to DeGeer. The chances that DeGeer could make a correct match of his specimen with a West Indian species, even if he had the latter in his collection, was very small.

A search in the collection of the Linnean Society of London, the most likely repository, has turned up no specimens authenticated by Linnaeus as *F. omnivora*. But even if one were to be found, yet did not originate from the Lesser Antilles during the plague period, it would be of very doubtful value in pinpointing the plague species. *F. omnivora* will likely stay in the limbo of species *incertae sedis*. A continued attempt to diagnose it holds little or no promise of identifying the plague species.

Narrowing the List of Suspects

The identity of the Hispaniolan plague ant is not difficult to make on the basis of information from the first-hand account of Fray Bartolomé de las Casas (1875/1997). The ant described by him was very aggressive, it stung painfully, it occurred in dense populations in the root systems of shrubs and trees, it did not cut aboveground vegetation yet somehow damaged the root systems, and it was also a pest of houses and gardens. The only species also present in the modern West Indian ant fauna possessing all these qualities is the tropical fire ant *Solenopsis geminata.* This inference is strengthened because a second fire ant species, the red imported fire ant *Solenopsis invicta,* attained plague proportions in the Gulf states in the 1940s following its introduction from the La Platte region of Brazil and northern Argentina (Wilson and Eads, 1949; Lofgren and van der Meer, 1986). Of incidental interest, *S. invicta* reached the West Indies by 1982, where it is now spreading rapidly (Davis et al., 2001). The *S. geminata* is evidently native to the southern United States and very likely to at least part of the circum-Caribbean area; its closest relatives are *Solenopsis aurea* and *S. amblychila* of the southwestern United States (Krieger and Ross, 2002), as opposed to the large complex of *Solenopsis* fire ant species of South America. Because of the ease with which *S. geminata* spreads naturally and by human transport, today flourishing on even the smallest Caribbean islands and in a wide variety of habitats, it can probably also be judged as a pre-Columbian inhabitant of Hispaniola.

The 1760–70 plague ants of the Lesser Antilles possess the same traits as *S. geminata* save one; in the several accounts of these species, including details from an eyewitness on Grenada (Castles, 1804), not a single mention is made of defensive aggressiveness or stinging by the ants. This evidence is negative, but it is a very large negative. Attacks by fire ants are unavoidable around their nests, and once having been experienced by even the most casual observer, would surely have been mentioned. For the fire ant *S. geminata* the sting is the source of the common name: each sting feels like a lighted match held too close, and when a person disturbs a nest and remains in contact, those parts of the body that can be reached by the ants are soon covered by stinging ants.

The Lesser Antillean plague ants were also described as having a two-segmented petiole, placing them within the Myrmicinae, and as varying a great deal in size (Schomburgk, 1848), suggesting the presence of major and minor subcastes (traits shared with *S. geminata*). There is one other possible clue: on Barbados in the mid-1600s there was an ant that was a serious house pest, and whose workers lifted and carried large food items such as cockroaches in coordinated fashion (Ligon, 1673). Fire ants do not use this conspicuous maneuver, but it is a notable trait of *Pheidole.*

All of these characteristics point to the ant genus *Pheidole.* Among the many species known from the West Indies and recently monographed (Wilson, 2003), only two are candidates: *P. jelskii* (Jelski's pheidole), a native species, and *Pheidole megacephala* (the big-headed ant), of African origin. The major workers of *Pheidole jelskii* when disturbed or crushed produce a fetid odor from skatole produced in the hypertrophied poison gland (Law et al., 1965). This trait matches the "sulphureous" smell said to have invariably emanated from crushed plague ants on Grenada (Castles, 1804). It is unique to *P. jelskii* among all the ants of the West Indies, with the possible (and untested) exception of the closely related but far less common *P. fallax*, which is found in the Greater Antilles but unknown from the Lesser Antilles (Wilson, 2003). Also, the Lesser Antillean plague ant was said to be dark brown (Schomburgk, 1848), which fits *P. jelskii* but not the brownish yellow *P. megacephala*.

Yet there is also strong evidence against *P. jelskii*. First, it invariably nests in open ground, making conspicuous crater nests with central, slit-shaped entrances, unmentioned in eighteenth and nineteenth century accounts. It seldom if ever is in closed vegetation with nest galleries extending into the root systems, a common habit of the plague ants and present-day *P. megacephala*. Further, although its workers are today possibly the most abundant ants of the West Indies, they do not aggregate in large local masses. Rather, their colonies are well spaced out. Nor has *P. jelskii* spread to other continents or ever been known to attain plague or near-plague proportions in modern times. Finally, even though extremely abundant, it is not a house pest. *P. jelskii* is thus very different from both *S. geminata* and *P. megacephala*, which possess all these plague-ant traits.

The attribution of a "sulphureous" odor and dark color is therefore puzzling, but it might easily have been due to a conflation of *P. jelskii* and *P. megacephala* if both species had been abundant on Grenada in the 1760s. The two are very similar to unaided vision, and John Castles, from whom the observation originated (Castles, 1804), was a planter, not a naturalist. Still, *P. megacephala* is today rare or absent on Grenada (Bennett and Alam, 1985; Wilson and Cover, unpublished Barbados field notes, 1998) as well as Barbados; and, although this circumstance may not be significant, Schomburgk reported that the plague species had already declined to small numbers by 1848.

It may never be possible to choose between *P. jelskii* and *P. megacephala* as the plague ant of the Lesser Antilles, or even exclude the possibility that both played the role, but the preponderance of evidence favors *megacephala*, a global invasive ant that has caused similar problems elsewhere in tropical and subtropical localities, for example Madeira, Bermuda, Hawaii, and northern Australia (Holway et al., 2002).

The cause of the damage

All but 20 or so of the 11,000 known species of ants of the world are harmless or even beneficial to humanity. Moreover, they are collectively essential for the functioning of many of Earth's terrestrial ecosystems (Hölldobler and Wilson, 1990). Only the attine leafcutter ants of the New World are known to attack vegetation directly as an important source of food, yet entire plantations of Hispaniola during or shortly after 1518–20 were wiped out. "As though fire had fallen from the sky and scorched them," Fray Bartolomé de las Casas (1875/1997) recorded, "they stood scorched and burned out." The sugar cane of the Grenadian fields in the 1770s similarly faded: the juice pressed from the remnant stalks was so meager and of such poor quality that it was not worth harvesting (Schomburgk, 1848).

The only viable hypothesis appears to be symbiosis with other insects that attack plants directly. The two plague ant suspects, *Solenopsis geminata* and *Pheidole megacephala,* heavily attend sapsucking coccids, mealy bugs, and other homopterous insects (Hölldobler and Wilson, 1990). The ants protect these insects—in the case of *S. geminata* with vicious, repeated stinging—in exchange for their abundant excrement, which is rich in sugar and amino acids. It seems very probable, but may be beyond proof, that the plagues followed on the inadvertent introduction in cargo of one or more new homopterous species to the West Indies. These pests, at first unopposed by any parasites or predators natural to them, bloomed into dense populations. The ants, profiting from the increased food supply, similarly flourished. The symbiosis of the two kinds of insects created the plague.

If this hypothesis is correct, the vehicles to Hispaniola, at least consistent in timing, may well have been the first plantains brought from the Canary Islands in 1516. The Spanish, not recognizing the role of the homopterous sapsuckers in the midst of the myriad kinds of insects teeming around their crops, would understandably put the blame on the stinging ants. It was not until the late eighteenth century, on Grenada, that naturalists began to suspect the involvement of homopterans in the West Indian ant plagues (de Réaumur, 1742).

Acknowledgments

This study could not have been completed without the assistance and advice of M. Barnes, B. Bolton, S. Cover, M. Deyrup, R. G. Echevarría, B. D. Farrell, I. Farrell, M. G. Fitten, K. M. Horton, J. Marsden, M. Mejia, and K. Watson. The research was supported in part by the Putnam Expeditionary Fund of Harvard University and the Edward O. Wilson Foundation.

References

Bennett, F. D., and M. M. Alam. *An Annotated Check-list of the Insects and Allied Terrestrial Arthropoda of Barbados* (Bridgetown, Barbados: Caribbean Agricultural Research and Development Institute, 1985).

Browne, R. *The Civil and Natural History of Jamaica,* 1st ed. (London: B. White and Son, 1756), 440.

Castles, J. Observations on the sugar ants. *Philosophical Transactions of the Royal Society of London,* abridged. 16 (1804): 688–694.

Davis, L. R., Jr., R. K. Vandermeer, and S. D. Porter. Red imported fire ants expand their range across the West Indies. *Florida Entomologist* 84 (2001): 735–736.

DeGeer, C. *Memoires pour Servir à l'Histoire des Insectes,* Tome III (Stockholm: Pierre Hesselberg, 1773), 609, pl. 31, figs. 23–24.

de las Casas, F. B. *Historia de las Indias,* Book III, Capítulo CXXXVIII. Madrid: Impr. de M. Ginesta, 1875–1876, 270–271. Translated to English by Sandra Ferdman in *The Oxford Book of Latin American Short Stories,* ed. R. González Echevarría. (New York: Oxford University Press, 1997).

de Réaumur, R. A. F. . *Histoire des Fourmis,* a hitherto unpublished journal of 1742–1743. Translated to English by W. M. Wheeler. (New York: Knopf, 1936), footnote 37 by Wheeler, 232–239.

Hölldobler, B., and E. O. Wilson. *The Ants* (Cambridge: Belknap Press of Harvard University Press, 1990).

Holway, D. A., L. Lach, A. V. Suarez, N. D. Tsutsui, and T. J. Case. The causes and consequences of ant invasions. *Annual Review of Ecological Systems* 33 (2002): 181–233.

Krieger, M. J. B., and K. G. Ross. Identification of a major gene regulating complex social behavior. *Science* 295 (2002): 328–332.

Law, J. H., E. O. Wilson, and J. A. McCloskey. Biochemical polymorphism in ants. *Science* 149 (1965): 544–546.

Ligon, R. *A True and Exact History of the Island of Barbadoes* (London: Peter Parker, 1673).

Linnaeus, C. *Systema Naturae,* Vol. I, 10th ed. rev. (impensis Laurentii Salvii, Holmiae, 1758–59), 579–582.

Lofgren, C. S., and R. K. Vander Meer, eds. *Fire Ants and Leaf-cutting Ants: Biology and Management* (Boulder, CO: Westview Press, 1986).

Schomburgk, R. H. *History of Barbados* (London: Longman, Brown, Green and Longmans, 1848), 640–643.

Wilson, E. O. *Pheidole in the New World: A Dominant, Hyperdiverse Ant Genus* (Cambridge: Harvard University Press, 2003).

Wilson, E. O., and J. H. Eads. A report on the imported fire ant *Solenopsis saevissima* var. *richteri* Forel in Alabama. Special report to the Alabama Department of Conservation. Montgomery, Alabama, July 16, 1949.

PART II

Biodiversity Studies: Systematics and Biogeography

1953

"The subspecies concept and its taxonomic application," *Systematic Zoology* 2(3): 97–111 (with William L. Brown, Jr., second author)

In the early 1950s, while I was still a graduate student at Harvard, the concept of race was a pivotal yet ambiguous and troubling part of the Modern Synthesis of evolutionary theory. In the imagined trajectory of the origin of species, geographical races, or subspecies as they were more formally known, seemed the logical transition stage between a genetically unified parental species and the split of such a species into two or more daughter species. Thus subspecies were conceived as still segments of the parental species yet not isolated within it by reproductive incompatibility. Systematists had a natural impulse to reify subspecies, usually defined as genetically and geographically distinct populations, and having separated them as discrete units, to label them with Latinized trinomens. Thus the mouse species *Peromyscus maniculatus* could be divided into at least three races: *Peromyscus maniculatus maniculatus, Peromyscus maniculatus artemesiae,* and *Peromyscus maniculatus rufinus.* But, as William L. Brown and I concluded in a series of informal lunchtime discussions, there were many weaknesses in such formal recognition. The article reproduced here explores these problems, among which the most serious is discordance in the distribution patterns of genetically distinct traits. For example, size may vary geographically one way, allowing the division of the species into one set of races, whereas skin color varies another way that prescribes a different set of races. Such discordance of hereditary traits is the chief reason why human races are so difficult to define. Indeed, in the 1950s physical anthropologists varied wildly in their definition of human races and hence their count of the number of races existing in the human species.

Our article stirred considerable controversy that lasted during much of the 1950s. It made some of our fellow systematists angry with us, but in the end it played an important role in shifting the analysis of geographic variation away from the description of subspecies as units and toward analysis of individual traits and their diagnostic gene ensembles. It also helped to diminish the misdirected and often hurtful attempts to define human races as objective units.

Looking back now, I believe Brown and I overreached in our recommendations on classification. There are populations that are both isolated, as on islands, and possess enough concordant traits to qualify as objective geographic units, and there are advantages to giving them formal trinomens. Some of these populations are proving to be distinct species, different enough so that even if they were contiguous no interbreeding would occur, but others deserve denotation as objectively definable subspecies.

Reprinted from SYSTEMATIC ZOOLOGY, Volume 2, Number 3, September 1953

The Subspecies Concept and Its Taxonomic Application

E. O. WILSON and W. L. BROWN, JR.

ELEVEN years ago, Mayr summed up and co-ordinated in his *Systematics and the Origin of Species* the taxonomic principles and methods that had gradually come to be recognized as basic and practical by many of the specialists working with the better-known groups of animals. Founded in the genetical precepts of neo-Darwinian evolutionary theory, Mayr's synthesis dealt most importantly with the nature of the species, which he held to be an objective and definable phenomenon, and with the geographical variation shown by populations composing the species. His species criterion was the occurrence in nature of free interbreeding, actual or potential, between members of a population or between populations; different species, he believed, are those populations possessing any factors intrinsic to their member individuals that will act to prevent interbreeding *between* the populations of a degree as free as that *within* each population. The basic reasonableness and operational advantages of Mayr's criterion struck an immediate wide and favorable response among many segments of taxonomic opinion, and his principle has been applied with considerable enthusiasm and with generally improved results to many and varied groups of Recent, sexually reproducing animals.

Along with widespread approval, this version of "population systematics" has aroused some outright opposition, as well as some more tempered criticism of particular phases of Mayr's argument. The outright opposition comes largely from those who either have not read carefully enough the various expositions of population systematics, starting with Mayr's, or who for some reason have failed to understand what we regard as for the most part a clear and simple thesis. Most of those so opposed, like M. W. de Laubenfels (1953) and Ruggles Gates (1951), insist upon regarding Mayr as having postulated that species are basically separated by *sterility barriers*. Starting with this thoroughly mistaken notion, de Laubenfels, Gates, and their school find it easy to bowl over straw men in all directions. De Laubenfels, for instance, is horrified to note that "Some dictionaries, many lesser zoologists, and the one whom many consider to be the greatest living systematist, propose a criterion of complete genetic isolation for species determination. Already they propose that most kinds of wild ducks are all one species. . . . Even many wood warblers are all one species [References?] . . . . By the geneticist's definition tigers are at best a subspecies of the lion, and bison merely a race of domestic cattle." The case of the lion and the tiger especially is so often used in this connection that we feel it would not be superfluous to make an example of it by pointing out the characteristics that prove these two forms species: (1) the breeding ranges of the lion and tiger overlap broadly in southern Asia, and the two species have occurred, at least in the recent historic past, in closely contiguous territories in India; (2) there is no sign that the Indian lion has been genetically affected through interbreeding with the surrounding tiger populations, or vice versa; (3) our principal reference (Burton, 1933) offers no evidence for any hybridization between lions and tigers in nature; (4) differences in breeding behavior, for instance, while not very well studied, seem nevertheless to be of a kind that may well act to prevent genetic interchange between the two species; (5) hybridization, even if it did occur, would not justify reducing the two species to races unless it could be demonstrated to be free and introgressive at the zones of actual contact.

Of course, Mayr, Dobzhansky, Stebbins, and many others have repeatedly observed that hybrid sterility is only one of several possible intrinsic mechanisms that can prevent two populations from interbreeding effectively. In the place of *intrinsic failure to interbreed freely in nature,* which is Mayr's criterion, de Laubenfels and Gates somehow mistakenly and persistently substitute *cross-sterility.* Once the nature of this misinterpretation is fully realized, there is little excuse left for accepting criticism of population systematics aimed in this direction.

Criticism of a more useful sort comes from other sources. First may be mentioned the objections of the paleontologists (latest reference: Simpson, 1952). These workers place emphasis upon the difficulties arising when population systematics is applied to species-evolution as it occurs through geological time. Obviously, species do arise from units that are not distinct as species, so that the intermediate time stages, however brief, destroy the sharpness of the species criterion. More important is the apparent fact that species populations may evolve new characters in time without undergoing any splitting; for instance, one finds a population represented in each of several successive strata by what appear to be slightly and progressively differing species that can only be taken as cross-sections of a continuum. This criticism does not destroy the basis of population systematics, but it leads us to re-emphasize an important qualification that must be made: *species distinctions hold only for the consideration of a single time-transect.*

A greater difficulty is expressed in the disagreement of some observers concerning the interpretations to be placed upon allopatric populations (i.e., populations with geographically separate breeding ranges) that are not very obviously distinct as species. Most authors rely on taxonomic judgement in treating such populations, but some have tried to apply a rule whereby allopatry uniformly marks either all species or all subspecies. The situation with regard to sympatric versus allopatric populations deserves careful attention for the following reasons. If two populations have separate geographical ranges, there may exist between them any degree of interbreeding potential from full to none at all. Thus, theoretically and probably in fact, allopatric populations may show every degree of divergence up to that of full species, and will in this sense blur the fine distinction that characterizes the species as a category. If, however, two allopatric populations extend their ranges unto geographical contiguity, or otherwise become sympatric, their interbreeding reaction theoretically can be expected to establish very quickly whether or not they have diverged to the species level. If they interbreed freely and produce a hybrid population that is in no way reproductively or selectively inferior to the parent populations, then they are clearly to be regarded as conspecific. If they do not interbreed, or if their hybrids are relatively rare and sporadic or otherwise show a reduced ability to form a self-maintaining population as compared with the parent populations, then the latter must be counted as separate species.

Different populations newly arrived at sympatry after having reached an intermediate degree of loss of interbreeding potential will probably go quickly and unequivocally to either the species or subspecies level: any partial intrinsic barriers will be strongly selected either for or against because of the simple fact that it is disadvantageous for the parental populations to maintain the mass production of inferior or sterile hybrids. There is some evidence that this selective process may be among the most important mechanisms involved in species formation (Dobzhansky, 1952, p. 208). Consequently, sympatry is more than an observer's criterion for deciding whether two populations are distinct as species; in any given case it may have been the final and essential factor that actually forced the species separation.

Since no such mechanism is operative in the differentiation of allopatric populations, there will be no clear-cut lower delimitation of species. *These populations must be dealt with arbitrarily* by gauging the genetic divergence through observed characters—morphological, physiological, and behavioral—according to standards based on comparison with the observed divergence of related sympatric species populations.

Therefore, Mayr's interbreeding criterion for the species, if qualified by the restriction of absolutely definable units to a single time-transect and to sympatric situations (the "non-dimensional species"; Mayr, 1949), and extended arbitrarily but with obvious justification through the analogy of character divergence to allopatric populations, seems to provide a natural, consistent, and practicable baseline for systematic theory.

Geographical Variation: The Subspecies Concept

Along with his analysis of the nature of the species, Mayr (1942) gave an extensive review of the evidence on variation within the species. He was mainly concerned with variation of populations as correlated with geography, and particularly with the properties and evolutionary significance of the subspecies, a category generally regarded as synonymous with the geographical race. The subspecies were conceived of as genetically distinct, geographically separate populations belonging to the same species and therefore interbreeding freely at the zones of contact. Many populations previously considered species were found to fit these conditions and were combined as subspecies in a single polytypic species. Mayr also extended the racial category to include closely related but geographically isolated populations, particularly those inhabiting different islands of tropical archipelagoes.

The taxonomic field has not been slow to exploit the opportunities opened up by the general recognition of the geographical race as a formal taxonomic category, expressible nomenclatorially as the trinomial subspecies. At the present time, it is clear that a great part of the total taxonomic effort is directed toward the detection, characterization, and formal nomenclatorial registration of "new" subspecies. This is particularly true in the case of specialist fields dealing with animal groups in which a large proportion of the full species have already been formally described and named, leaving the burden of the unceasing search for novelties to rest upon the subspecific populations.

The past two decades have witnessed an increasing tendency on the part of taxonomists to rely upon the theoretical basis so firmly promulgated by Mayr. With the progressive accumulation of seemingly sound trinomials in relatively well-worked groups such as the birds, there has grown up a complacency in systematics concerning the objectivity and usefulness of the subspecies. Specialists in many less well-worked groups, and especially those where insufficient time and material are available for detailed analysis of geographical variation, have all but forgotten the early claims of subjectivity for the race, and have come to regard it as a concrete geographical population capable of being recognized by one or a few "diagnostic" characters most accessible for study in preserved material. Many massive revisions have of late depended on the authenticity of this notion.

The tacit but very fundamental theoretical assumption most systematists make is that when characters vary geographically, their variation is co-ordinated. In terms of evolutionary genetics, the predominant genome of a given population constitutes a "coadaptive system," an aggregation of genes which are best adapted as a unit to the special environment of the population (Ford, 1945). As a result, the geographical distribution of genes, and with them the resultant phenotypes, will be concordant.

While this concept of character concordance follows evolutionary theory well, the factual background from which it is drawn does not rightfully inspire the confidence taxonomists as a group place in it. Taxonomists seem to have forgotten the great complexities and disparities revealed in racial patterns by some really thorough analyses of geographical variation made in the past. Most of the prominent commentators on the theory of speciation have been careful to emphasize the inherently subjective and even arbitrary nature of racial limits. Here is a vastly unappreciated statement by Mayr (1942):

> We have stated repeatedly that every one of the lower systematic categories grades without a break into the next one; the local population into the subspecies, the subspecies into the monotypic species, the monotypic species into the polytypic species, the polytypic species into the superspecies, the superspecies into the species group. This does not mean that we find the entire graded series within every species group. It simply means that in the absence of definite criteria it is, in many cases, equally justifiable to consider certain isolated forms as subspecies or as species, to consider a variable species monotypic or to subdivide it into two or more geographical races, to consider well-characterized forms as subspecies of a polytypic species or to call them representative species.

From our experience in the literature we are convinced that the subspecies concept is the most critical and disorderly area of modern systematic theory—more so than taxonomists have realized or theorists have admitted. Particular confusion surrounds the drawing of the lower limits of the subspecies category within that spectrum of classes recognized by Mayr as extending from "the local population into the subspecies." The difficulties in this delimitation stem from four outstanding features of geographical variation: (1) the tendency for genetically independent characters to show independent geographical variation; (2) the capacity for characters to recur in more than one geographical area, yielding polytopic races; (3) the common occurrence of the microgeographical race; (4) the necessary arbitrariness of any degree of population divergence chosen as the lowest formal racial level. It is our purpose now to illustrate these four features with the aim of re-evaluating the nature of geographical variation and of throwing new light on the subspecies concept as it is applied in taxonomy.

Independent geographical variation. Abundant examples of this phenomenon can be drawn from most careful analyses of geographical variation in a wide variety of animal groups. In his exceptionally complete work on *"Lymantria" dispar*, Goldschmidt (1940) finds eight characters which vary geographically (excluding chromosome size; cf. Makino and Yosida, 1949), none of which is in exact geographical concordance with any of the others. Several of the characters may be used by themselves to make striking racial divisions by cabinet standards, or they may be used in various combinations to achieve different results. Goldschmidt formally establishes five races by utilizing combinations of characters in size and coloration, while at the same time recognizing that "the number of subspecific types could be greatly increased by going into more and more intricate differences." In fact, Goldschmidt's data affirm that the number of races discernible increases as a function of the number of characters taken into consideration. This classic work is doubly important because it illustrates that physiological characters, such as degree of sexuality, rate of larval development, vary geographically just as do the more obvious adult morphological characters ordinarily used in lepidopteran taxonomy.

Moore (1944) surveys variation of the common leopard frog, *Rana pipiens* Schreber, in eastern North America, giving special attention to the characters stressed by authorities who had formerly divided the species into three geographical groups (species or races). After a very thorough analysis of these taxonomic features plus some others newly introduced, he plots twelve of them in a table of "population formulas" listed geographically. These show clearly that, over broad ranges, there is essentially no consistent maintenance of groupings of characters from one broad region to the next. One can detect gradual clines, step clines, and sudden mid-distribution cline reversals for each character, and the clines obviously do not all have their axes lying along the same compass directions.

One very dramatic character is that concerning the presence or absence of the oviducts in the male. Moore's map (Figure 3) shows oviductless males exclusively inhabiting the Mississippi Basin and the states to the east, from Long Island to northern Florida, while the populations situated peripherally, in the Florida Peninsula, New England, the northern and far western states, and southern Texas have, with scattered exceptions, males with oviducts. It seems reasonable to conclude that the oviductless condition dominates and is spreading outward from some center of origin, gradually displacing the oviduct-present character. The point here is that the distributional pattern of the character shows an obvious lack of correlation with those of the external "taxonomic" features. Is it to be ignored by taxonomists on this account?

Moore later (1946) detailed his findings after conducting interbreeding experiments between frogs from different populations, and found that impairment of embryonic development reached lethal proportions in crosses between parents from the northern and southern extremes of the area sampled, while those populations separated by smaller distances showed intermediate or no hybrid impairment. North-south differences were emphasized in these experiments, but some limited east-west tests gave similar results.

A further extension of his studies led Moore (1949) to consider variation of characters presumably having a much greater adaptive significance than those external ones earlier studied. The new characters included embryonic temperature tolerances and rates of development, which show a north-south difference of a more or less clinal nature; egg size, showing clinal reduction from north to south, but with a striking reversal in Mexico; and form of egg mass, concerning which data were insufficient and show only that variation may possibly run from east to west as well as from north to south. Combining Moore's studies, it is interesting to note that the most promising of the few possible "correlated breaks" in some of the external adult characters comes to the north of New Jersey in the East, whereas by the criteria of egg size and embryonic temperature tolerance, both demonstrated to be adaptively crucial characteristics, the New Jersey populations are not significantly different from the northern populations and belong with the latter instead of with the southern populations.

Moore quite logically rejects the validity of the former broad racial divisions, and points instead to the more uniform concatenation of characters that may be found within each of the many small, allopatric local populations. We agree that his findings accord with his judgement that "there is no generally accepted and easily applied criterion for recognizing subspecies."

LeGare and Hovanitz (1951), in a detailed study of genetically based adult color variation in Californian populations of the butterfly *Melitaea chalcedona* present data suggesting a racial split between the populations of the Little San Bernardino-Mojave Desert mountain area from those to the north and west. However, larval color varies as much as adult color and shows a different geographical deployment. The larvae from several northern populations are yellow, those from the southwest coast are largely deep black, while those from the desert area show replacement of the black by gray. Despite several confused and contradictory statements on the part of LeGare and Hovanitz with regard to the relationship of color and size, it is clear that these are poorly correlated geographically. Thus all three characters studied tend to vary independently, and as in the case of *Lymantria*, several racial divisions can be drawn depending on which characters are used and in what combinations.

A simple case of discordant variation involving a pair of characters has been described by Mayr (1942) for the bird *Paradisaea apoda*. This species is distributed linearly in the lowlands around the coast of New Guinea. Coloration of the back lightens in a cline extending around the eastern tip of the island and terminating in the north at Goodenough Bay, while coloration of the plumes lightens in a cline which commences to the northwest at Cape Ward Hunt and terminates at the Huon Peninsula. Mayr uses the resultant superimposition patterns to demarcate at intervals five races. One wonders what new racial lines could have been drawn had other, less obvious characters been carefully analyzed.

Polytopic races. If races are delimited by a single character, it is easily within the realm of possibility that this character may be selected to predominance in more than one population of the species. Dice (1940) reports the apparent independent origin of populations of the races *Peromyscus maniculatus rufinus* and *P. m. artemisiae* in western North America; these have arisen through the selection of certain coat color alleles best suited to the color of their environmental background. Cazier (in Mayr, Linsley, Usinger, 1953) has found a similar origin for certain color races in the tiger beetle genus *Cicindela*. We have observed the polytopic occurrence of a distinctive racial character involving appendage length in populations of the ant *Lasius niger* (L.) occupying eastern Asia and the eastern Mediterranean and Atlantic Islands region. Mayr (in the work cited) has expressed the opinion that such populations be recognized under a single subspecific name if no other characters vary geographically to form racial patterns. The extreme taxonomic difficulties arising when the distribution becomes more complex are self-evident and need no further comment here.

The microgeographic race. Even when only one or a few characters are employed by the taxonomist, these often vary so elaborately and extensively that nearly every local population is distinguishable from all the others. The best-known examples of this phenomenon are in the snail genera *Achatinella, Partula, Cepaea, Io, Polymidas, Liguus, Europtis, Orion, Chondrothyra*, etc., the first three of which have been discussed so often in general papers on evolution that they need little additional comment here. The most obvious variation is in shell color patterns, but variation in sculpture, size, coiling, etc., also occurs, and the resultant characters can be used in combination to distinguish endless distinct populations even by the most stringent racial standards.

Microgeographic races are especially prominent in snails because of the sedentary habits of these organisms and their tendency to form isolated local colonies. The same phenomenon is evident in more active animals restricted to habitats of a discontinuous or isolated kind, such as bogs, desert streams, and caves. Examples can be drawn from such diverse groups as butterflies (Higgins, 1950), cave beetles (Valentine, 1945), and *Dendroica* warblers (Hellmayr, 1935; Bond, 1950).

The chief disadvantage inherent in formally recognizing microgeographical races is that regardless of how valid the distinctiveness and internal concordance of their characters may prove them, the list of their trinomials must reach stupendous proportions in time. The result is a top-heavy nomenclature helping little of itself to clarify the nature of the geographical variation, but which instead will certainly obscure it as synonymies are recognized and diagnoses shifted. This is apparently the situation being approached in certain rodent groups. In the pocket gophers *Thomomys bottae* and *T. tal-*

poides a total of thirty-five races has already been recognized from Utah alone (Durrant, 1946), and the area has not been so exhaustively worked as to preclude the possibility that many more races remain undetected.

The microgeographical race as conceived in present evolutionary literature is an unusually well differentiated deme, or local communal population. There is no reason to believe that it is an exceptional phenomenon or anything more than the extreme of the tendency prevalent in all geographically variable species to form local populations of a homogeneous and distinctive genetic constitution. If several independent characters enter into the geographical variation, it is reasonable to assume that many demes can be distinguished by racial standards ordinarily applied in taxonomy if enough of the characters are used in combination. This is in fact the condition described in *Rana pipiens* by Moore, and it is reflected by the many references of geneticists and taxonomists to special "strains" typifying geographical localities.

The arbitrary lower limit of the subspecies. Even when the discrepancies arising from discordant geographical variation are eliminated by the use of one or a very few characters, systematists are faced with the fact that there is no real lower limit to the subspecies category. It has been affirmed repeatedly in a variety of animal groups that racial populations show all degrees of divergence from the lowest level of statistical reliability of mean difference to complete differentiation, with no particular tendency to fall either way. Obviously the only way to resolve this situation taxonomically is to establish an arbitrary lower limit above which populations will be formally recognized as subspecies. This subject has been dealt with thoroughly in the recent text on animal systematics by Mayr, Linsley, and Usinger (1953), and there is no need to treat it in any detail here. The point we wish to emphasize is that no arbitrary lower limit will ever be completely satisfactory, for even if only one character is used, there will always be borderline cases of an extremely vexing nature. Samples defined with vague, untrustworthy characters will often fall above a fixed lower limit, while samples usefully distinguished by striking characters will often fall below it. Furthermore, any hard and fast line will unavoidably produce a condition in which some populations are recognized formally as races while others, essentially of the same constitution but of a slightly lower statistical level, are not recognized.

This difficulty concerning the lower limit of the subspecies is well known to most taxonomists who have devoted much serious attention to the problem. Some have compromised the situation by choosing the level of statistical reliability most nearly conforming to their preconceived notion of what should constitute a valid race in the particular group under study. This appears to have been the procedure followed by Austin (1952), for instance, in his study of Pacific petrels: "A subspecific name designating a geographical population is of no practical use unless at least three-quarters or more of the individuals of that population can be correctly assigned by their morphological characters alone." Austin chooses the "84% from 84%" rule of Simpson and Roe, making the illuminating statement that the "97% from 97%" rule would be too stringent, since "Among the petrels it is rare indeed to find the means of any character separated by two standard deviations, allowing a 97% separation." Austin's method is in no way irregular as modern systematic practice goes, a fact that should signal a general re-examination of the relationship between the "taxonomic intuition" and the choice of hard statistical bases of differentiation.

It is apparent that in their application of the subspecies concept most revisionary workers have misinterpreted the nature of geographical variation as revealed by the more careful analyses in the literature. It is also apparent that taxonomic revisions, using as they do relatively small samples and usually only one or two independent diagnostic characters, rarely present any valid general information with regard to the nature of geographical variation in its own right. Most formally named subspecies are in effect little more than special cases deduced from the established concept of subspeciation, and their validity is no stronger than the concept itself. For this reason it is important that we do not stop at disclosing the inconsistencies of the concept; rather, we should attempt to revise it to conform as rigorously as possible to fact. From the data supplied in studies such as those by Goldschmidt, Moore, Welch (1938), Crampton (1932), Vanzolini (1951), Brown and Comstock (1952), and others, it is possible to draw several outstanding conclusions having an important bearing on the taxonomic application of the subspecies concept.

1. Where one character varies geographically, other genetically variable characters can be found to vary also.

2. The geographical variation of independent characters tends to be discordant to some degree. The degree of concordance increases with the degree of isolation of populations, but complete concordance from locality to locality is rarely if ever attained. In fact, complete concordance of several known independent characters in an isolated population may (usually?) be a good indication that the population has attained species level. For example, Goldschmidt's *Lymantria dispar hokkaidoensis* shows concordance of at least three characters, more than any other race of this species, but at the same time it appears to be sufficiently cross-sterile with adjacent races to justify recognition as a distinct species.

3. It follows from (2) that the greater the number of characters, the greater will be the total discordance. As a result, the racial lines first drawn from the most prominent "diagnostic" characters will be increasingly obscured or contradicted by the addition of characters, and the situation will be resolved only by either recognizing additional races marked by different combinations of characters, or by recognizing only the major tendencies in concordance. The first of these two solutions, that of recognizing all racial limits by whatever characters can be used to demarcate them by conventional standards, may be the better one in populations that have differentiated *in situ*, i.e., without initial isolation. When this approach is used, the number of distinguishable races has been found in practice to increase at a slightly more than arithmetical progression with the addition of characters used in combination. The second solution, involving the determination of what might be called *peaks of concordance*, seems the more promising where distinguishable populations are totally isolated or are undergoing secondary intergradation. However, since races are then defined according to character peak concordance, nonconforming characters will of necessity have to be omitted, while the extensiveness of the intergrade zones of the species will increase in proportion to the number of characters included in the peaks. The taxonomist will find himself faced with a dilemma: he must either ignore certain poorly conforming characters or else he must incorporate them in his subspecies diagnosis and thereby broaden the zones of intergradation.

4. It would not be too much of a truism to mention that the greater the geographical area encompassed, the less homogeneous will be the population. Conversely, it appears that in geographically very variable species the only thoroughly homogeneous and concordant units, if any exist at all, are the demes (*sensu* Carter, 1951), which tend to be isolated and completely panmictic within themselves. Where clines occur they are marked between but not within these populations.

As noted previously, most taxonomic recognition of subspecies so far has proceeded on the oversimplified "coadaptive system" concept of the race, which assumes that genetically independent characters will tend to be concordant in their geographical variation. We believe that this assumption has resulted in the establishment of a basic fallacy in the taxonomic method of studying geographical variation. The tendency in this method has been to delimit races on the basis of one or several of the most obvious characters available in preserved material; the remainder of the geographically variable characters are then ignored, or if they are considered at all, they are analyzed only in terms of the subspecific units previously defined. A slight variation of the procedure is to choose several discordant characters, employ them in combinations of two or three to establish racial limits, and then analyze each character individually in terms of these limits.

A case in point is the recent study of the red-eyed towhee *Pipilo erythrophthalmus* by Dickinson (1952). Seven characters are used in various combinations to demarcate four races ranging successively from *alleni* in the south to the typical *erythrophthalmus* in the north. Wing length, plumage and iris color, and tail-spot size vary clinally along the succession of races. Culmen, tarsus, and toe length are greatest in the two intermediate "races" *rileyi* and *canaster*. The total picture of the variation gleaned from this study gives the strong impression that the intermediate forms are nothing more than segments of a broad, partly clinal intergrade zone connecting two extreme terminal populations. This is the conclusion reached by Huntington (1952, *vide infra*) in his analysis of remarkably similar variation found in the eastern purple grackle (*Quiscalus quiscula*); the same kind of characters vary in the same way in both the towhee and grackle, and the intermediate zones in both are geographically very close. The increase in culmen-tarsus-toe lengths in the intermediate towhee populations seems comparable to the increase in culmen-wing lengths in the grackle intergrade zone. In addition, the variation of independent characters in the towhee is obviously quite discordant, as evidenced by the rather poor correspondence of the iris-color distribution as charted by Dickinson with the racial limits previously decided upon.

From Dickinson's data alone it cannot be proved that the geographical pattern in the towhee is the same as in the grackle, and that it may therefore be best expressed by the recognition of two races; yet the fallacy in Dickinson's method of analysis stands out clearly enough. His entire treatment is predicated on the shaky assumption that the races he has defined represent concrete biological units, and this despite his introductory warning: "In ornithological studies in large part the taxonomist is dealing with continuous variates and with variation that appears graphically as a cline. Under such circumstances lines of demarcation must be vague." Having established the four races, Dickinson thereupon uses them as sample groupings from which to analyze each character individually. Only one character, iris color, is plotted geographically as an independent variate. As a result, the true nature of the clinal trends can be inferred only from gross comparisons of the racial diagnoses. Instead of outlining the geographical variation of each character and then synthesizing from it the overall racial pattern, Dickinson has done just the reverse, thereby closing the door to further analysis and interpretation of the data which he has so laboriously gathered and presented.

Because of its closely similar nature and quite different approach, Huntington's analysis of geographical variation in the purple grackle deserves further attention. Much as in the towhee, four races can be demarcated arbitrarily along a southeast-northwest cline, but Huntington chooses to synonymize the intermediate two, *ridgwayi* and *stonei*, as segments of a clinal intergrade zone between the southern nominate race and the northern *versicolor*. Culmen length and wing length vary independently and discordantly with color, this time along a north-south cline. Huntington analyzes these two characters separately to demonstrate that both increase unexpectedly in size (with respect to their over-all clinal trends) at the zone of intergradation of the color characters. This Huntington suggests may be due to a heterotic effect caused by the secondary intergradation of the two terminal races. By deciding upon the racial units *after* the variation of the genetically independent characters has been analyzed separately, Huntington arrives at what appears to be a more natural classification than that proposed by Dickinson for the towhee. But even more important, his data are presented in such a fashion as to allow ready incorporation into future studies of this species.

Insular Races

The more critical reader may have noted by this time a special condition of the foregoing critique of the subspecies concept: the published analyses of geographical variation that have been considered are in nearly every case concerned with Holarctic continental species. Our review of the literature convinces us that really critical analyses of this sort are virtually lacking for insular populations, and herein rests a point. Much of the background of the modern subspecies concept has been drawn from taxonomic studies of insular and montane groups, all of which are essentially the same in their marked fragmentation into completely isolated populations. Special emphasis in this respect has been laid on birds, and it is not too much to say that the development of the entire theory of geographical speciation has been dominated in large part by ornithological leadership. Yet a survey of ornithological taxonomic literature, including the long series of papers by Mayr, Zimmer, Amadon, Lack, and others (cf. Mayr, 1951), has convinced us that the morphological and distributional data on relevant bird populations leave much to be desired, and in fact offer very little definitive information on the two central topics, independent character variation and the subspecies-species evolutionary transition, as they apply to insular populations.

This literature is characterized by two outstanding shortcomings. First, a very limited number of characters is used; taxonomic revisions are typically based on studies of variation in size, external proportions, and color. Even the detailed analyses of (continental) geographical variation, such as those by Dickinson and Huntington just discussed, are based on these same few characters. To these we may add Miller's well-known *Junco* revision, which is the most thorough of all such studies on birds known to us. We have already stressed the weaknesses of any infraspecific classification based on limited numbers of characters. It would be of the utmost interest to see an ornithological revision employing the same number and kinds of characters studied by Goldschmidt in *Lymantria* and Moore in *Rana;* these might include internal features, egg color and size, morphological and physiological nestling characters, microscopic barbule structure, epidermal sculpture, and many others. This sort of work may well be rendered unduly difficult by the limitations of standard ornithological materials and methods, and it would perhaps be presumptuous to suggest a shift of technique. Nevertheless, it is important to emphasize the little-appreciated point that ornithological studies do not remotely approach in morphological detail those published on some other groups of animals.

The second shortcoming of ornithological revisions is the paucity of data on the subspecific versus specific status of insular and other isolated populations. It is true that sharp character discontinuities are often set from isolate to isolate; this allopatric pattern occurs in so many groups as to create a striking faunal picture, especially in tropical archipelagoes. Again we need to point out that few characters have been determined to participate in the discontinuities, and little information has been obtained on concordance of variation, especially as it occurs between islands and island groups. Furthermore, it is a fact that many of these striking racial differences are based on limited samples (occasionally consisting of a single specimen) which may have originated from the same immediate locality or even from a single clutch. There is no way of knowing, on the basis of the mass of published taxonomic work, whether or not the study of additional, less obvious, and possibly discordant characters in larger bird samples might reveal alternative racial divisions among island groups, and finer divisions on single islands, such as occur in snails and probably in other animal groups.

A stronger aspect of this shortcoming is seen in the other direction. Where the several characters utilized show marked concordance, there is always the distinct possibility, previously mentioned, that the allopatric populations have already attained species level. A certain amount of evidence is accumulating to indicate that this may be a very common phenomenon. We have already mentioned the example of Goldschmidt's *Lymantria dispar hokkaidoensis,* which, showing a high degree of character concordance, is partially cross-sterile with adjacent populations. Kinsey (1936) describes four pairs of sympatric species of the *Cynips dugèsi* complex in southern Mexico, none of them more strongly differentiated than are the numerous isolated populations to the north; the extreme paucity of hybrids between the northern allopatric populations may be taken as additional support for the contention that they really represent member species of a superspecies, notwithstanding the high degree of isolation. Most edifying, however, are the numerous cases cited in the literature of pairs of closely related species with contiguous or narrowly overlapping ranges. Taxonomists often consider such pairs to be races of a single species until their true relationship is verified by a careful investigation of their interaction in the zone of contact. The significance of this particular kind of taxonomic clarification has been reviewed for ornithology by Mayr (1951), who uses the expression "pseudo-conspecific pairs of allopatric species" to refer to pertinent cases.

Summing up, we must affirm that present knowledge of insular races, including those of birds, is actually too limited to allow close comparison with the patterns elucidated in studies of continental races. While it is true that striking discontinuities often occur between island or other isolated populations, any interpretation of these discontinuities must carry two serious qualifications. First, insular races, like most continental races, have been defined on the basis of limited numbers of characters, often in assorted combinations, without consideration of the possibly discordant variation of other, more cryptic characters. Second, where some degree of concordance is demonstrated, the excellent possibility that the populations have already attained species level has very seldom been ruled out.

Subspecies: the Taxonomic Application

Because the geographical race has a demonstrably flimsy conceptual basis, it is unfortunate that it has become through the years a deeply rooted taxonomic resort. That the race has become so integral a part of our systematics is due largely to the circumstance that, under the more hierarchical-sounding alias "subspecies," it has established itself gradually but ever more firmly as a unit that could and should be dignified with a Latin name. Caught up in the wave of enthusiasm for the new systematics, the International Commission of Zoological Nomenclature, meeting at Paris in 1948, gave its most recent formal sanction to the named subspecies at the same time that it quite rightly consigned the "variety" and other minor categories without geographical connotation to an inferior rank. In effect, the Commission again officially recognized subspecific names on a level of availability with those given to full species insofar as priority is concerned, and again gave formal recognition to the employment of the neo-Linnaean trinomial. The pool of available trivial names, many of which may never prove assignable to definite species, has been more firmly fixed at discouragingly vast proportions by this action.

If it is now clear that the subspecies trinomial is fast becoming an unquestioned and traditional fixture, it is equally clear, at least to us, that in its assumed function as a formal means of registering geographical variation within the species it tends to be both illusory and superfluous.

Mayr sums up our general philosophy perfectly in his very recent (1953) advice, offered in a different connection to those in attendance at the birth of a struggling taxonomy of viruses:

> The history of all classification, whether dealing with inanimate objects or with organisms, shows that early attempts of classification are based on superficial similarities and very often on single characters, while all improvements of classification are due to ever more penetrating analysis and a broadening of the basis of classification by including more and more characters. The soundest classifications are those built on the greatest possible number of clues. Reciprocally, it can be stated that, in sound classifications, there is usually a fair concordance of the various characters.

The application of this logic to our present knowledge of geographical variation cannot fail to stir a feeling that the trinomial has outlived its usefulness in taxonomy. We are encouraged to note that ornithologists have been among the first to apprehend this circumstance. Lack (1946), after grappling with trinomials in the European robin and finding them based uneasily on convergent polyphyletic characters and complex clinal trends, concludes:

> The use of subspecific names not only implies discontinuity where none may exist, but also unity where there may, in fact, be discontinuity. . . . Certainly, in the case of *Erithacus rubecula*, it is both simpler and more accurate to describe subspecific variation in terms of geographical trends, and to omit altogether the tyranny of subspecific names.

Mayr (1951), in reviewing twelve years of progress in the study of bird speciation, observes, "Instead of expending their energy on the describing and naming of trifling subspecies, bird taxonomists might well devote more attention to the evaluation of trends in variation."

We are inclined to feel even more strongly about the situation. We are convinced that unless our own sampling of the taxonomic literature has badly deceived us, we shall soon begin to observe the withering of the trinomial and its cumbersome appurtenances—the types, the tinted labels, the ponderous subspecies lists gravely entered in a thousand catalogues, the awkward labelling of masses of "intergrade" specimens, and all of the other procedural details that so unnecessarily consume the few effective working hours a modern taxonomist has. We anticipate the time when the taxonomist, if he wants to apply a formal Latinized name to his sample, will have first to produce indications that the population represented has the characteristics of a species. The more irresponsible or naive worker will not then be able, after a weak gesture in the direction of systematic study, to retire to the comfortable, safe nebulosity of a subspecies designation under a name having guaranteed availability against the future contingency that someone will perform the labor necessary to define a good species fitting his type. The study of geographic variation may eventually become just what the term implies, and not merely remain the subspecies mill it so largely is today.

The possibility that some International Congress not too far in the future will see fit to relegate unborn subspecific names to the nomenclatural limbo now occupied by the variety, the natio, the aberration, the forma, etc., inevitably brings up the question of the kind of reference shorthand we shall need to aid in the description of geographical variation. Fortunately, all the reference we require for this purpose is contained in (1) the correct determination to species, and (2) the locality and ecological data that will have to accompany any specimen worth studying. Thus, in publications, we can speak of "*Rana pipiens* Schreber, Montauk Pt., New York"; or "*R. pipiens*, southeastern corner of J. B. Smith farm, 5 miles west of Montauk Pt., in cattail swamp"; or even "*R. pipiens* from Long Island," ". . . from the East Coast," and so on. The precision or breadth of the geographical designation will vary according to the needs of the investigator and with the actual geographical distribution of the character or combination of characters under study. Inevitably, perhaps, repeatedly discussed populations will come to be referred to as "Montauk A," "Reelfoot Lake," "Rock Island," and so forth, but this will no more prove a pitfall than is the geographical vernacular familiarly applied to "strains" of *Drosophila virilis* (Patterson and Stone, 1952), or the locality names by which experienced trappers can often distinguish a series of pelts.

If a character combination of a population remains at all co-ordinate and consistent in its territorial occupancy, there is every reason why we should refer to it merely by mentioning the species concerned and either the locality or full distribution that it occupies. There is no evident advantage in the use of the recommended form "*montaukensis*" over "Long Island race" or "Montauk A." If we find at Lhasa a population of mice of known species that carries a distinctive black cheek stripe, the name "*lhasensis*" conveys this no more readily than does "Lhasa race." If it subsequently be found that the entire Tibetan Plateau is inhabited by mice carrying black cheek stripes, "Lhasa race" is readily expanded, so that we can speak of the "Tibetan race" just as easily as, and interchangeably with, "Lhasa race." The city of Lhasa remains a feature of the Tibetan Plateau, and so do the black-cheeked mice of both places. The very informality and flexibility of a vernacular system are among its most appealing characteristics. A geographical vernacular designation lacks the esoteric authoritarianism surrounding the Latin trinomial, but it is this very quality of trinomials that we consider most misleading, cumbersome, and generally repellent, especially to the uninitiated. The geographical vernacular is more broadly communicable, more frankly expressive, fully as mnemonic, at least as certain in the long run to be precise, and it cuts the taxonomic red tape to practically nothing. Its present unostentatious use in many individual papers in several taxonomic fields reveals no serious operational drawbacks. In short, we feel that the facts we have outlined call for serious, conscious consideration of the desirability of eventual abandonment of the subspecies trinomial and its replacement by a system of reference based on the vernacular employment of relevant geographical names.

Summary

1. Mayr's criterion for the species, that of free interbreeding of populations in nature, when qualified by the conditions of sympatry and synchrony, and extended by morphological analogy to isolated populations, has proved to be objective and practicable for taxonomic work.

2. Roughly, the subspecies has been defined as a genetically distinct geographical fraction of the species. The assumption has been followed, tacitly or otherwise, that when secondary characters vary geographically, this variation tends to follow whatever "diagnostic" characters are chosen to delimit races, and that the subspecies in general can be shown upon further analysis to be a concrete unit. This assumption is demonstrated herein to be contravened by the data available in the literature dealing with geographical variation.

3. Three other prominent features affecting the subspecies concept render it even more subjective and arbitrary in taxonomic practice: the polytopic race, the microgeographic race, and the artificiality of quantitative methods of defining the formal lower limits of the subspecies.

4. Most taxonomic analysis at the intraspecific level has been directed toward the end of naming and characterizing new

subspecies. This tends to be an inefficient and misleading method. It is felt that geographical variation should be analyzed first in terms of genetically independent characters, which would then be employed synthetically to search for possible racial groupings.

5. Although "insular" races (as opposed to contiguous "continental" races) appear at times to be exceptionally clear-cut and have been extensively used in generalizations on raciation, the data in most available analyses are in all respects insufficient to evaluate the intricacies of this process. It is not even certainly known in most such cases whether distinctive isolated populations are races or species.

6. We feel that as the analyses of geographical variation become more complete, the trinomial nomenclatorial system will be revealed as inefficient and superfluous for reference purposes. It is suggested that, for the study of such variation, the use of the simple vernacular locality citation or a brief statement of the range involved is adequate and to be preferred to the formal Latinized trinomial.

REFERENCES

Austin, O. L., Jr. 1952. Notes on some petrels of the North Pacific. *Bull. Mus. Comp. Zool., 107*:391–407. See p. 392.

Bond, J. 1950. Check-list of the birds of the West Indies. Acad. Nat. Sci. Phila. See pp. 135–138.

Brown, F. M., and Comstock, W. P. 1952. Some biometrics of *Heliconius charitonius* (Linnaeus) (Lepidoptera, Nymphalidae). *Amer. Mus. Novit.,* No. 1574, 1–53.

Burton, R. G. 1933. The book of the tiger, with a chapter on the lion in India. Mayflower Press, Plymouth, England.

Carter, G. S. 1951. Animal evolution. Sidgwick and Jackson Ltd., London.

Crampton, H. E. 1932. Studies on the variation, distribution, and evolution of the genus Partula. The species inhabiting Moorea. Publ. Carnegie Inst., No. 410.

Dice, L. R. 1940. Ecologic and genetic variability within species of *Peromyscus*. *Amer. Nat., 74*:212–221.

Dickinson, J. C. 1952. Geographical variation in the red-eyed towhee of the eastern United States. *Bull. Mus. Comp. Zool., 107*:273–352.

Dobzhansky, Th. 1952. Genetics and the origin of species (third ed.). Columbia Univ. Press.

Durrant, S. D. 1946. The pocket gophers (genus *Thomomys*) of Utah. *Univ. Kansas Publ., Mus. Nat. Hist., 1*:1–82.

Ford, E. B. 1945. Butterflies (The New Naturalist series). Collins, London. See p. 283.

Gates, R. R. 1951. The taxonomic units in relation to cytogenetics and gene ecology. *Amer. Nat., 85*:31–50.

Goldschmidt, R. 1940. The material basis of evolution. Yale Univ. Press.

Hellmayr, C. E. 1935. Catalog of the birds of the Americas and adjacent islands. *Field Mus. Nat. Hist., Zool. Ser. (Chicago), 13.* See pp. 362–385.

Higgins, L. G. 1950. A descriptive catalogue of the Palaearctic *Euphydryas* (Lepidoptera: Rhopalocera). *Trans. Roy. Ent. Soc. Lond., 101*:435–499.

Huntington, C. E. 1952. Hybridization in the purple grackle. *Syst. Zool., 1*:149–170.

Kinsey, A. C. 1936. The origin of higher categories in Cynips. Indiana Univ. Publ., Sci. Ser., vol. 4.

Lack, D. 1946. The taxonomy of the robin *Erithacus rubecula* (Linn.). *Bull. Brit. Ornith. Club, 66*:55–65.

de Laubenfels, M. W. 1953. Trivial names. *Syst. Zool., 2*:42–45.

LeGare, M. J., and Hovanitz, W. 1951. Genetic and ecologic analyses of wild populations in Lepidoptera. II. Color pattern variation in *Melitaea chalcedona*. *Wasmann J. Biol., 9*: 257–310.

Makino, S., and Yosida, T. 1949. A critical study of the chromosomes of *Lymantria dispar* L. in relation to the question of racial difference. *Cytologia, 14*:145–157.

Mayr, E. 1942. Systematics and the origin of species. Columbia Univ. Press.

——— 1949. The species concept: semantics vs. semantics. *Evolution, 3*:371–372.

——— 1951. Speciation in birds. Progress report on the years 1938–50. *Proc. Xth Internat. Ornith. Congr.,* pp. 91–131.

——— 1953. Concepts of classification and nomenclature in higher organisms and microorganisms. *Ann. N. Y. Acad. Sci., 56*: 391–397.

Mayr, E., Linsley, E. G., and Usinger, R. L. 1953. Methods and principles of systematic zoology. McGraw-Hill, New York.

Moore, J. A. 1944. Geographic variation in *Rana pipiens* Schreber of eastern North America. *Bull. Am. Mus. Nat. Hist., 82*: 349–369.

——— 1946. Incipient intraspecific isolating mechanisms in *Rana pipiens*. *Genetics, 31*: 304–326.

——— 1949. Geographic variation of adap-

tive characters in *Rana pipiens* Schreber. *Evolution, 3*:1–24.
PATTERSON, J. T., and STONE, W. S. 1952. Evolution in the genus *Drosophila*. Macmillan Co., New York.
SIMPSON, G. G. 1952. The species concept. *Evolution, 5*:285–298.
VALENTINE, J. M. 1945. Speciation and raciation in Pseudanophthalmus (Cavernicolous Carabidae). *Trans. Conn. Acad. Arts Sci., 36*:631–672.
VANZOLINI, P. E. 1951. *Amphisbaena fuliginosa. Bull. Mus. Comp. Zool., 106*:1–67.
WELCH, D'A. A. 1938. Distribution and variation of *Achatinella mustelina* Mighels in the Waianae Mountains, Oahu. B. P. Bishop Mus. Bull., No. 152.

EDWARD O. WILSON is a Junior Fellow of the Society of Fellows of Harvard University, working at the Biological Laboratories, and **WILLIAM L. BROWN, JR.** is Assistant Curator of Insects at the Museum of Comparative Zoology, Harvard University. The authors wish to express their appreciation to J. C. Bequaert, W. J. Clench, P. J. Darlington, W. H. Drury, J. C. Greenway, Jr., Ernst Mayr, B. O. Shreve, and other colleagues for offering references, critical suggestions, and other aid indispensable in the preparation of this paper. Their suggestions in the main have been incorporated, although this does not mean that they all agree fully with the proposed taxonomic application of the authors' theoretical findings.

1956

"Character displacement," *Systematic Zoology* 5 (2): 49–64 (with William L. Brown, Jr., first author)

In the early 1950s, while analyzing the systematics and biogeography of *Lasius,* an ant genus distributed across temperate Eurasia and North America, I discovered a curious phenomenon in the two closely related species *Lasius flavus,* which occurs across North America, and *L. nearcticus,* which occurs only in the East. *Lasius flavus* did not remain the same in the eastern United States as compared with the west, nor did it converge toward *L. nearcticus* as expected from hybridization. Rather, *L. flavus* diverged from *L. nearcticus.* It seemed as though *L. nearcticus* was genetically repulsing *L. flavus.* During our lunchtime discussions, Bill Brown and I sifted through the literature, coming up with additional examples discovered by other authors, including David Lack in his pioneering 1947 book *Darwin's Finches.*

We organized this information, coining the term "character displacement" to cover the known cases. I believe we were also the first to make a clear distinction between displacement due to reinforcement of prezygotic isolating mechanisms (which reduce wasteful production of less viable hybrids, a concept introduced by Theodosius Dobzhansky) and avoidance of competition, which we were able to clarify in our article. Character displacement has since been shown to occur commonly among animal species, although it is far from universal. Our article helped to inspire G. Evelyn Hutchinson's famous "Homage to Santa Rosalia" (or "Why are there so many different kinds of animals?" *The American Naturalist,* 1959, 145–159), which was an early contribution to evolutionary ecology—a field the great Yale biologist helped to create.

Reprinted from SYSTEMATIC ZOOLOGY, Volume 5, Number 2, June 1956

Character Displacement

W. L. BROWN, JR. and E. O. WILSON

IT IS the purpose of the present paper to discuss a seldom-recognized and poorly known speciation phenomenon that we consider to be of potential major significance in animal systematics. This condition, which we have come to call "character displacement," may be roughly described as follows. Two closely related species have overlapping ranges. In the parts of the ranges where one species occurs alone, the populations of that species are similar to the other species and may even be very difficult to distinguish from it. In the area of overlap, where the two species occur together, the populations are more divergent and easily distinguished, i.e., they "displace" one another in one or more characters. The characters involved can be morphological, ecological, behavioral, or physiological; they are assumed to be genetically based.

The same pattern may be stated equally well in the opposite way, as follows. Two closely related species are distinct where they occur together, but where one member of the pair occurs alone it converges toward the second, even to the extent of being nearly identical with it in some characters. Experience has shown that it is from this latter point of view that character displacement is most easily detected in routine taxonomic analysis.

By stating the situation in two ways, we have called attention to the dual nature of the pattern: species populations show displacement where they occur together, and convergence where they do not. Character displacement just might in some cases represent no more than a peculiar and in a limited sense a fortuitous pattern of variation. But in our opinion it is generally much more than this; we believe that it is a common aspect of geographical speciation, arising most often as a product of the genetic and ecological interaction of two (or more) newly evolved, cognate species during their period of first contact. This thesis will be discussed in more detail in a later section.

Character displacement is not a new concept. A number of authors have described it more or less in detail, and a few have commented on its evolutionary significance. We should like in the present paper to bring some of this material together, to illustrate the various aspects the pattern may assume in nature, and to discuss the possible consequences in taxonomic theory and practice which may follow from a wider appreciation of the phenomenon.

Two Illustrations

An example of character displacement outstanding for its simplicity and clarity has been reviewed most recently by Vaurie (1950, 1951). This involves the closely related rock nuthatches *Sitta neumayer* Michahelles and *S. tephronota* Sharpe. *S. neumayer* ranges from the Balkans eastward through the western half of Iran, while *S. tephronota* extends from the Tien Shan in Turkestan westward to Armenia. Thus, the two species come to overlap very broadly in several sectors of Iran (Fig. 1). Outside the zone of overlap, the two species are extremely similar, and at best can be told apart only after careful examination by a taxonomist with some experience in the complex (Vaurie, personal communication). Both species show some geographical variation, and it seems clear from Vaurie's account (1950, Table 5, pp. 25–26) that such races as bear names have been raised for character discordances in various combinations. It therefore appears safe to ignore the subspecies analysis as such and to concentrate on the variation of the independent characters themselves.

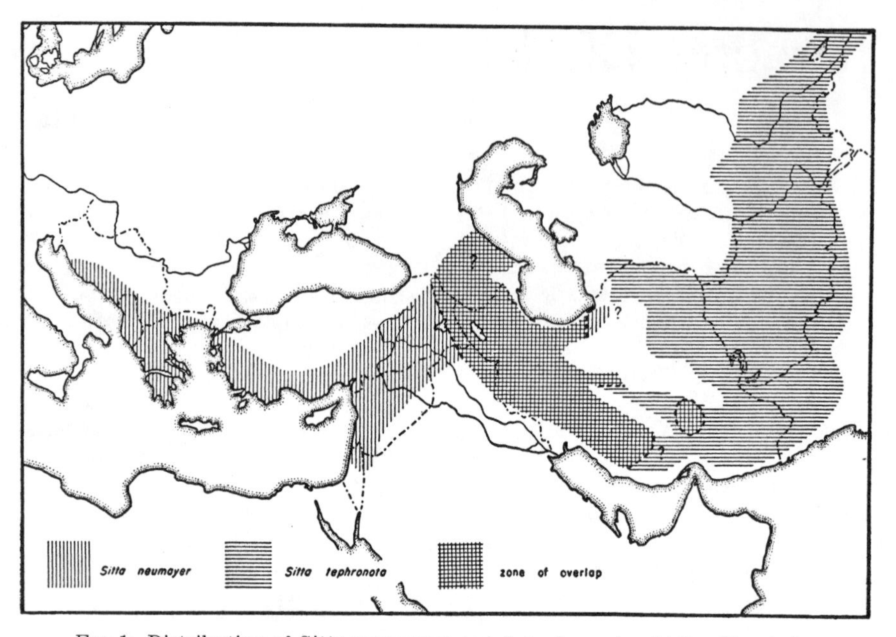

FIG. 1. Distribution of *Sitta neumayer* and *S. tephronota*. (After Vaurie.)

These show quite remarkable displacement phenomena in the Iranian region of overlap between the species, where the two species apparently usually occur in more or less equal numbers (see Fig. 2). In this region, *S. neumayer* shows distinct reductions in overall size and bill length, as well as in width, size, and distinctness of the facial stripe. *S. tephronota,* on the other hand, shows striking positive augmentation of all the same characters in the overlap zone, so that it is distinguishable from sympatric *neumayer* at a glance. Vaurie concludes, we think quite correctly, that the differences within the zone of overlap constitute one basis upon which the two species can avoid competition where they are sympatric. The case of these two nuthatches has already received considerable attention both in the literature and elsewhere, and it bids fair to become the classic illustration of character displacement.

A more complicated case involving multiple character displacement is seen in the ant genus *Lasius* (Wilson, 1955). Where they occur together, in forested eastern North America, the related species *L. flavus* (Fabr.) and *L. nearcticus* Wheeler show differences in the following seven characters: antennal length, ommatidium number, head shape, degree of worker polymorphism, relative lengths of palpal segments, cephalic pubescence, and queen size. In western North America and the Palaearctic Region, where *nearcticus* is absent, *flavus* is convergent to it in all seven characters. In this shift, each character behaves in an independent fashion; e.g., scape length becomes exactly intermediate between that of the two eastern populations, ommatidium number increases in variability and overlaps the range of the two, and queen size changes to that of *nearcticus*. In North Dakota, at the western fringe of the *nearcticus* distribution, the *flavus* population is at an intermediate level of convergence (Fig. 3).

There is some evidence that this dual displacement-convergence pattern is associated with competition and ecological

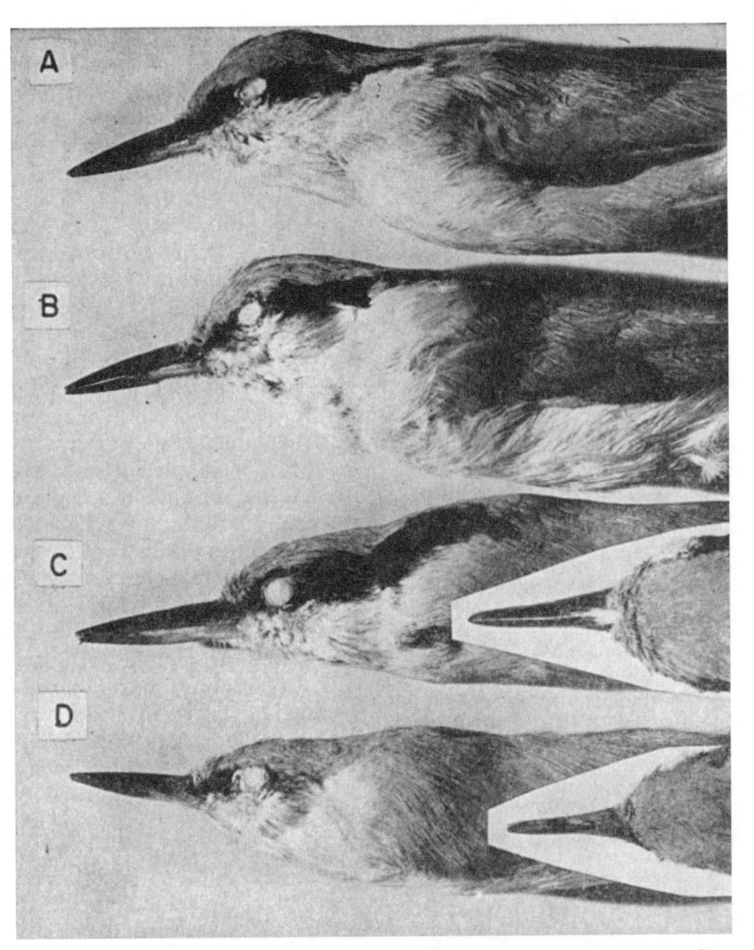

FIG. 2. Size and shape of the bill and facial stripe in *Sitta neumayer* and *S. tephronota*: *A, S. neumayer* from Dalmatia; *B, S. tephronota* from Ferghana; *C, S. tephronota* and *D, S. neumayer,* both from Durud, Luristan, in western Iran. (After Vaurie.)

displacement between the two species. So far as is known, they have similar food requirements. But in eastern North America, where they occur together, *flavus* is mainly limited to open, dry forest with moderate to thin leaf-litter, while *nearcticus* is found primarily in moist, dense forest with thick leaf-litter. There is little information available on the western North American and Asian *flavus* populations, but in northern Europe this species is known to be highly adaptable, preferring open situations, but also occurring commonly in moist forests.

Some Additional Examples

In the following paragraphs we wish to present a number of cases selected from the literature (with two additional unpublished examples) which we have interpreted as showing character displacement. In so doing we are trying to document the thesis that character displacement occurs widely in many groups of animals and in a range of particular patterns. But at the same time we are obliged to give warning, perhaps unnecessarily for the critical reader, that most of these cases in-

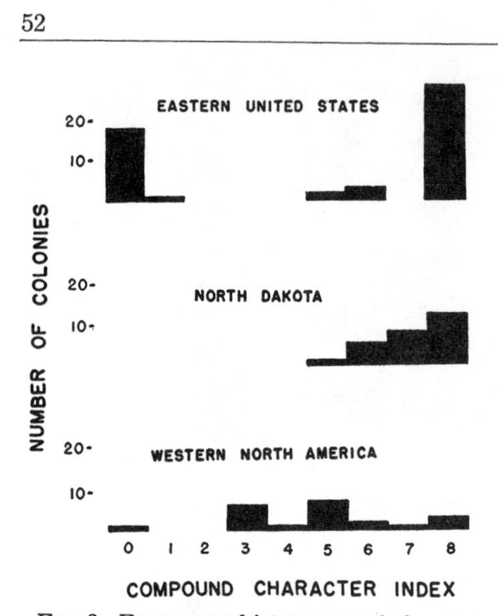

FIG. 3. Frequency histograms of the compound character index of the ants *Lasius nearcticus* (0–1) and *L. flavus* (3–8) in three broad geographic samples. For each colony typical *nearcticus* characters are given a score of 0, typical eastern *flavus* characters a score of 2, and intermediate characters a score of 1. The four characters most clearcut in the eastern United States are used: maxillary palp proportions, antennal scape index, compound eye ommatidium number and head shape. Thus, completely typical *nearcticus* colonies score a total of 0 and completely typical eastern *flavus* 8, with the various ranks of intermediates falling in between (after Wilson, 1955).

volve discontinuously distributed populations, that as a result the species status of these populations with respect to one another has not been ascertained with complete certainty, and that explanations alternative to character displacement are therefore assuredly possible. We ask only that the reader bear through and consider our interpretation in each case.

Birds of the Genus Geospiza. A striking case of character displacement has been described by David Lack in his classic, *Darwin's Finches* (1947). Lack has shown that in the Galapagos certain species of *Geospiza* are often absent on smaller islands, in which case their food niche is filled by other species of the genus. The populations of the latter tend to converge in body size and beak form to the absent species, so much so as to make placement of these populations to species difficult. Lack has demonstrated that body size and beak form are generally important in *Geospiza* in both food getting and species recognition. The dual displacement-convergence pattern we are interested in occurs, at least once, in the following situation. The larger ground-finch *Geospiza fortis* Gould and the smaller *G. fuliginosa* Gould differ from each other principally in size and beak proportion. On most of the islands, where they occur together, the two species can be separated easily by a simple measurement of beak depth, i.e., a random sample of ground-finches (excluding from consideration the largest ground-finch *G. magnirostris* Gould) gives two completely separate distribution curves in this single character. But on the small islands of Daphne and Crossman a sample of ground-finches gives a single unimodal curve exactly intermediate between those of *fortis* and *fuliginosa* from the larger islands. Analysis of beak-wing proportions has shown that the Daphne population is *fortis* and the Crossman population is *fuliginosa;* according to Lack's interpretation each has converged toward the other species, filling the ecological vacuum its absence has created.

Birds of the Genus Myzantha. Among the Australian honey-eaters of the genus *Myzantha,* a light-colored species, *M. flavigula,* occupies the greater part of the arid inland. Toward the wet southwestern corner of the continent, *flavigula* blends gradually into a darker population, usually referred to as "subspecies *obscura.*" In southeastern Australia, in higher-rainfall country, *flavigula* is replaced by two forms—*M. melanocephala,* mostly in the wettest districts, and *M. melanotis* of the subarid Victorian-South Australian mallee district. The southwestern (*obscura*) and one of the southeastern populations (*melanotis*) are extremely similar, differing by what are described as trifling characters of plumage shading, so that some authors consider them conspecific.

The members of an ornithological camp-out in the Victorian mallee, however, have found that *melanotis* there nests sympatrically with both *melanocephala* and *flavigula*, and that at this place the three behave as distinct species without intergradation. Thus we find the two morphologically very similar forms, *obscura* and *melanotis*, flanking the much more widely distributed and differently colored species, *flavigula*, but showing exactly opposite interbreeding reactions with *flavigula*. *Obscura* appears to represent merely the terminus of a cline for melanism produced by *flavigula* in the southwest, where, it may be noted, there is no other competing dark form of the same species group (Fig. 4).

Judging by the findings of the mallee observers, *melanotis* is clearly to be regarded as a species distinct from *flavigula*, including the southwestern *obscura* population. In this we follow Condon (1951), and not Serventy (1953), though the latter has furnished the most comprehensive analysis of the situation.

Serventy's dilemma is keyed by his statement that ". . . it would be unreal to treat *melanotis*, obviously so akin to south-western *obscura*, as a separate species from it. . . ." Here one plainly sees the conflict between two species criteria: one based on morphological similarity, and one on interbreeding reaction in the zone of sympatry.

From the data presented, we interpret the *Myzantha* situation as a case of character displacement. *M. flavigula* tends to produce, in the less arid extremities of its range, populations with darker plumage. In the southwest, it has done just this; presumably, melanism is connected adaptively in some way, directly or indirectly, with increased moisture ("Gloger's Rule"), or plant cover, or both. In the southeastern mallee, however, the melanistic tendencies presumed to be latent or potential in *flavigula* toward the wetter extremes of its range are suppressed in the presence of the darker species *melanotis* (and possibly also *melanocephala*). It would be interesting to know more about

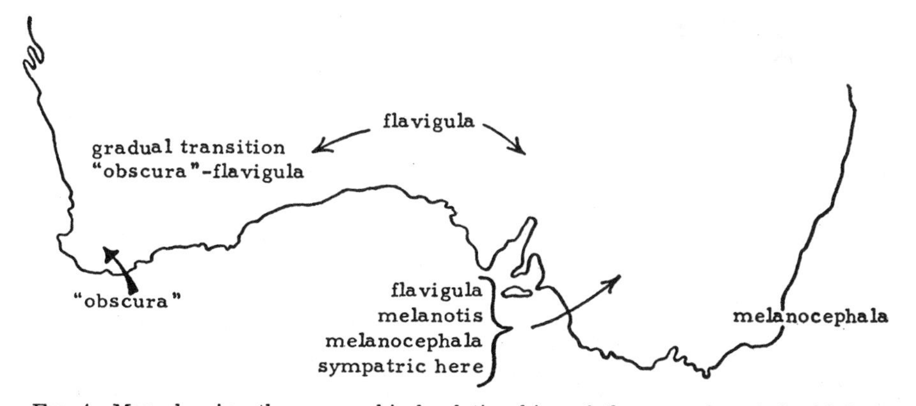

FIG. 4. Map showing the geographical relationships of three species of the bird genus *Myzantha* in southern Australia, based on the discussion of Serventy, 1953. *M. flavigula*, the light-colored bird of arid central Australia, grades into a darker population ("race *obscura*") in southwestern higher-rainfall districts. In southeastern Australia, in the Victorian mallee belt, transitional and mixed ecological conditions allow three non-intergrading species to breed side by side: *M. flavigula; M. melanotis*, a species characteristic of the mallee scrub; and *M. melanocephala*, a southeastern bird of the higher-rainfall districts. *M. melanotis* and the "*obscura*" population are extremely similar, and have been considered synonymous or at least conspecific in the past.

the ecological distribution, food, and habits of the three *Myzantha* species within the region where they occur together.

Parrots of the Genus Platycercus. Serventy (1953) also reviews, among other cases that may involve character displacement, the situation in the rosellas of southeastern Australia (Fig. 5). The crimson rosella (*Platycercus elegans*) is a species of the wooded eastern areas—mostly those with higher rainfall nearest the coast. On Kangaroo Island, off the coast of South Australia, occurs a crimson population that appears to be *elegans* from a strictly morphological viewpoint. Beginning on the mainland opposite Kangaroo Island is a cline connecting the crimson form to an inland, arid-country yellow form (*P. flaveolus*) inhabiting the red gums of the rivers and dry creeks in the Murray-Darling Basins. However, in the Albury district of the upper Murray River and elsewhere up the other rivers, *flaveolus* overlaps or closely approaches the true southeastern *elegans* along a wide front without interbreeding (for a recent detailed account, see Cain, 1955).

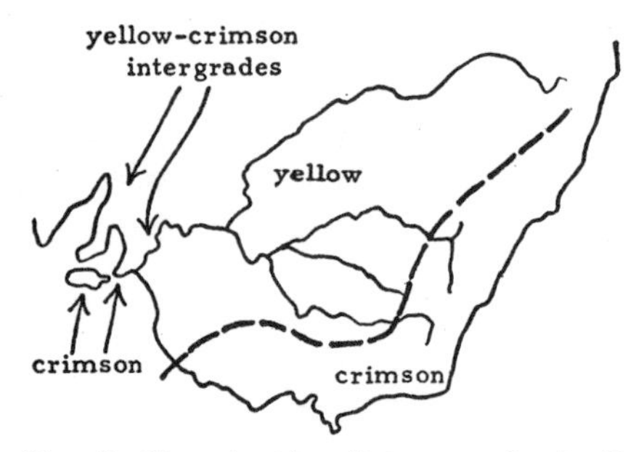

FIG. 5. Map showing the approximate distribution of color forms of the rosellas (parrots) of the *Platycercus elegans* complex in southeastern Australia. The heavy pecked line indicates roughly the inland margin of the southeastern highlands and the higher-rainfall districts, and also the inland limit of the range of the crimson-trimmed *P. elegans*. Inside this line, along the upper reaches of the Murray-Darling river systems, the closely related *P. flaveolus*, a yellow-trimmed form, approaches and may even meet the range of *P. elegans* at some points without producing intergrades. Downstream, *P. flaveolus* grades through a series of intermediately-colored populations culminating in the crimson-trimmed flocks of Kangaroo Island, which are apparently outwardly indistinguishable from those of the true eastern *elegans*. (Adapted from Cain, 1955.)

It is interesting to note that the cline from yellow to crimson in South Australia follows broadly the regional increase in moisture and luxuriance of forest vegetation; both rise to peaks in the ravines at the western end of Kangaroo Island. We suggest that the South Australian clinal population on the mainland, and probably even the crimson populations of Kangaroo Island, are referable to *flaveolus*, which can here produce a wet-adapted crimson form free of displacement pressure from *elegans*.

Birds of the Cape Verde Islands. Bourne (1955) in his review of the birds of the Cape Verde Islands, has presented several cases of character displacement so concisely and pointedly that we can quote him directly:

> The two shearwaters [breeding in the Cape Verde Islands], Cory's shearwater *Procellaria diomedea* and the Little Shearwater *Procellaria baroli*, take similar foods (fish and cephalopods) differing only in size; competition for food between the two species is reduced by the development of different breeding seasons. Elsewhere in its range *Procellaria diomedea* breeds at the same stations as the medium Manx Shearwater *Procellaria puffinus*, which takes similar foods but breeds slightly earlier. There is a dramatic difference in size, and particularly the size of bill, between those races of *Procellaria diomedea* which breed with *Procellaria puffinus* and the form *[P. diomedea] edwardsi* which breeds alone at the Cape Verde Islands, the latter having a bill exactly intermediate in size between that of the northern races and that of *Procellaria puffinus*. It seems likely that *edwardsi* takes the food that is divided between both species elsewhere. It may be remarked that one race of *Procellaria puffinus*, *mauretanicus* of the Balearic Islands, avoids competition with *Procellaria diomedea* by breeding unusually early and leaving the area when the larger species prepares to nest; it is significant that this is the only race of the

species which has a large bill resembling that of *P. d. edwardsi*. It would appear that the bill-size and the breeding seasons of these shearwaters vary with the amount of competition occurring between different species breeding at the same site. . . .

Where the two kites *Milvus milvus* and *Milvus migrans* occur together the latter is the species which commonly feeds over water. The race of *Milvus milvus* found in the Cape Verde Islands closely resembles *Milvus migrans* in the field, and very commonly feeds along the shore and over the sea. It may replace *Milvus migrans*, but it seems likely that with the Raven *Corvus corax*, which also abounds along the shore, it replaces the gulls *Larus* spp. which usually scavenge along the shore elsewhere but have failed to colonize the barren coast of the islands.

Bourne's opinion concerning which species are replaced is a little confusing in this case, since elsewhere *Milvus*, notably *M. migrans* in India, often tends to replace or at least dominate the gulls in scavenger-feeding situations around seaports (Brown, personal observation). The absence of *migrans* seems to us the probable chief reason for the convergence characteristics in the Cape Verde Islands populations of *milvus*.

Bourne cites one additional case:

The Cane Warbler *Acrocephalus brevipennis* [a species precinctive to the Cape Verde Islands] is closely related to large and small sibling species *Acrocephalus rufescens* and *A. gracilirostris* which occur together in the same habitats on the [African] mainland. Where the ranges of these two species overlap they are sharply distinct in size and voice; where they occur apart these distinctions are less marked (Chapin, 1949). *A. brevipennis* is probably related to the larger species, *A. rufescens*, but in the absence of the smaller species it is exactly intermediate in all its characters except the bill, which is large, resembling that of *A. rufescens*. The large bill may be part of the general trend seen on islands, or a consequence of competition for food with the smaller *Sylvia* warblers.

Birds of the Genus Monarcha. Mayr (1955 and personal communication) has described a case of displacement in the monarch flycatchers of the Bismarck Archipelago. *Monarcha alecto* and *M. hebetior eichhorni* occur together through the main chain of the Bismarcks, from New Britain north onto New Hanover, but beyond, on isolated St. Matthias, *M. hebetior hebetior* occurs alone; this last is an ambiguous variant combining several features of *alecto* and *eichhorni*. Mayr suggests the following evolutionary scheme: *hebetior* differentiated from *alecto* as an isolate on St. Matthias and later reinvaded the range of *alecto* on New Britain and New Ireland, where it diverged further under displacement pressure from the latter until it became the present *eichhorni*.

It seems to us that this situation can be more simply explained by assuming that the Bismarcks were first populated by a stock which evolved within the Archipelago and became the species *hebetior*. The later entry of *alecto* into the chain was followed by the displacement of *hebetior* as far as the sympatry extended, leaving the St. Matthias isolate to represent the undisplaced relict of the original *hebetior*.

Fishes of the Genus Micropterus. The two basses *Micropterus punctulatus* and *M. dolomieu* have ranges which include a large part of the eastern United States and are mostly coextensive (Hubbs and Bailey, 1940). Of the two, however, only *punctulatus* is known to occur in Kansas, western Oklahoma, and the Gulf States south of the Tennessee River drainage system. In the Wichita Mountains of western Oklahoma there is a population, described as *M. punctulatus wichitae*, which is intermediate between typical *punctulatus* and *dolomieu*. Its affinity to *punctulatus* is shown by the fact that in a number of characters it grades without a break into *punctulatus*, so that some specimens are indistinguishable from typical *punctulatus*, and in its agreement with *punctulatus* in the critical character of scale-row counts. Hubbs and Bailey seem to favor the theory of a hybrid origin for *wichitae*, but they consider this "no more plausible than the view that the similarities between *wichitae* and *dolomieu* are caused by parallel development, or the view that *wichitae* is a relict of a generally extinct transitional stage between *punctulatus* and *dolomieu*." We, of course, are inclined to favor parallel development, resulting specifically from the absence of the displacing influence of *dolomieu*, as the simplest and most plausible explanation.

Away to the south, many of the Texas populations of *punctulatus* are peculiar in showing converging trends toward *dolomieu*, but less strongly, so that Hubbs and Bailey consider them as possible intermediates between *punctulatus* and *wichitae*. In northern Alabama and Georgia there is a form described as a distinct species (*M. coosae*), which combines some of the characters of *punctulatus* and *dolomieu*, besides showing some peculiar to itself. *Coosae* is completely allopatric to *dolomieu*, and there is some evidence that it may hybridize extensively with the sympatric *punctulatus*. We should like to suggest the possibility here that *coosae* is conspecific with *punctulatus* and represents a section of the *punctulatus* population tending to converge toward *dolomieu* where that species is absent.

In summary, it appears to us likely that *wichitae*, the Texas populations, and possibly even *coosae*, each of which shows intermediate characters, are not products of introgressive hybridization, but may instead represent true *punctulatus* stocks that have tended to converge toward *dolomieu* in the absence of displacing influence from that species.

Frogs of the Genus Microhyla. W. F. Blair (1955) concludes from his study of two North American frogs of the genus *Microhyla:*

> The evidence now available shows that there are geographic gradients in body size in both *Microhyla olivacea* and *M. carolinensis*. The former species shows a west to east decrease in body length, while the latter shows an east to west increase. The clines are such, therefore, that the largest *carolinensis* and the smallest *olivacea*, on the average, occur in the overlap zone of the two species. This pattern of geographic variation in body size parallels the pattern of geographic variation in mating call reported by W. F. Blair (1955) [in press] in which the greatest call differences in frequency and in length occur in the overlap zone. One of these call characteristics, frequency, probably is directly related to body size, for smaller anurans of any given group tend to have a higher pitched call than larger ones of the same group. The other, length of call, appears unrelated to size.
>
> The differences in body size, like those in mating call, belong to a complex of isolation mechanisms (W. F. Blair, 1955) which tends to restrict interspecific mating in the overlap zone of the two species. The existence of the greatest size differences as well as the greatest call differences where the two species are exposed to possible hybridization supports the argument (*op. cit.*) that these potential isolation mechanisms are being reinforced through natural selection.

Frogs of the Genus Crinia. A most interesting case in the Australian genus *Crinia* has recently been called to our attention by A. R. Main (*in litt.*). Where they occur together in Western Australia, as around Perth, the two species *C. glauerti* and *C. insignifera* have markedly different calls. *C. glauerti* has a rattling call resembling "a pea falling into a can and bouncing"; oscilloscope analysis shows this to consist of evenly spaced single impulses at the rate of about 16 per second. *C. insignifera* produces a call "similar to a wet finger being drawn over an inflated rubber balloon . . . we refer to this call as a 'squelch.'" Oscilloscope analysis shows the squelch to have a duration of about 0.25 second and to consist of impulses crowded together. Around Perth and in other localities where it is sympatric with *insignifera*, *glauerti* individuals are occasionally heard to produce the beginnings of the "squelch" by running 12–15 impulses together, but this occurrence is extremely rare. Along the south coast of Western Australia, however, where *glauerti* occurs alone, the call is commonly modified by running 30 or more single impulses together to produce a squelch almost identical to the ear with that of *insignifera*. Thus, in effect, where this species occurs alone it has extended the variability of its call to include the sounds typical of both species. According to Main, the two species show color differences in the breeding males and different ecological preferences; laboratory crosses show reduced F_1 viability. It seems evident to us (Brown and Wilson) that displacement in this case is associated with the reinforcement of reproductive barriers, the breakdown of which would result in inferior hybrids. This aspect will be discussed more fully in a later section.

Ants of the Genus Rhytidoponera. The ants of the Australian *Rhytidoponera metallica* group (revised by Brown, ms.) are widespread and often among the dominant insects of given localities. The common greenhead (*R. metallica*) is the most successful species—a metallescent green or purple ant adapted to a variety of habitats ranging from desert to warm, open woodland, and the only species of the group at all abundant across the dry interior of Australia. In the southeastern and southwestern ("Bassian") corners of the continent, where the rainfall is higher and luxuriant forests occur, *metallica* is replaced by similar species of the same group that nearly or quite completely lack metallic coloration (Fig. 6).

In the east, two such species make the replacement, *R. tasmaniensis* Emery and *R. victoriae* André. *R. tasmaniensis* is the larger of the two, has the fine gastric sculpture of *metallica,* and is usually reddish brown, with bronzy-brown gaster. It is virtually identical with *metallica,* except for color. *R. victoriae* is smaller, more blackish, and has relatively coarser gastric striation. *R. tasmaniensis* is found in a variety of woodland situations, but apparently is excluded from the very wettest forests, which are occupied by *victoriae.* Nevertheless, the two species exist in abundance side by side over large parts of southeastern Australia without a sign of interbreeding. At some points, such as on the moist temperate grasslands west of Melbourne, both species occur together with *metallica,* but maintain their distinctness.

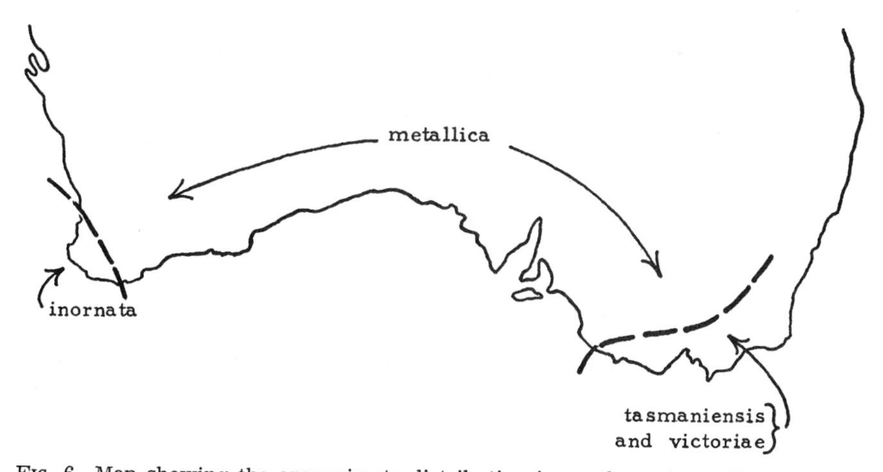

FIG. 6. Map showing the approximate distribution in southern Australia of four closely related common species of ants of the *Rhytidoponera metallica* group. *R. metallica* is nearly or quite the only representative of its group in the more arid central regions, and occurs in open situations in the southeast and southwest as well. In the moister forests of the southeast, *metallica* is replaced by the small, dark *R. victoriae* and the larger, more reddish *R. tasmaniensis,* which frequently occur side by side in the same localities. In the mesic wooded areas of the extreme southwestern corner of Australia, *metallica* is replaced by *R. inornata,* a distinct species which in size and color resembles closely, and broadly overlaps in variation, the two eastern forest species.

In the southwestern corner of Australia, *metallica* is replaced in the wetter parts of the region by non-metallescent *R. inornata* Crawley (though the two species overlap in the Darling Range and undoubtedly elsewhere). The interesting feature here is that *R. inornata* varies in size and color so as to cover the variation in these attributes of both southeastern non-metallic species, *tasmaniensis* and *victoriae*. In fact, one might speak of the two southeastern forms as mutually-displacing equivalents of the southwestern *inornata*, the latter being nearest the generalized type of the group because it has never suffered close competitive pressure and the character displacement that helps to relieve that pressure. This example illustrates the existence of a dual character-displacement pattern where the convergent population is clearly at, or above, the species level.

Slave-making Formica *Ants*. A simpler case in the ants involves the famous Holarctic slavemakers of the *Formica sanguinea* group. In a recent revision (Wilson and Brown, 1955) only three really distinct species are recognized in the group: *F. sanguinea* Latreille, widely distributed through temperate and northern Eurasia, where it is the only species; *F. subnuda* Emery, of boreal and subboreal North America; and *F. subintegra* Emery, ranging through temperate North America and overlapping the range of *subnuda* in the northern United States and along the Rocky Mountain chain.

The two most different forms are *subnuda* and *subintegra*, which can be separated on several external characters. *F. sanguinea* is closely related to *subnuda* in form and habits and is treated as a separate species only arbitrarily, on the basis of slight morphological discontinuities. At the same time, *sanguinea* has pilosity intermediate between that of the two American species, and its clypeal notch, a second important diagnostic character, is more like that of *subintegra* than like that of *subnuda*. We have interpreted this pattern to represent a displacement of *subnuda* away from *subintegra* where these two species meet and interact, while the Palaearctic equivalent of *subnuda* (i.e., *sanguinea*) has tended to converge toward *subintegra* as a consequence of its filling the "adaptive vacuum" which a companion species might otherwise occupy. Of course in this case, as in all others under present consideration, there is no way of determining how much "displacement" has occurred as a process in the sympatric populations as opposed to "convergence" in the unispecific one. The final pattern observed may in fact be the result of one of these two processes alone.

American Scarabaeid Beetles. Howden (1955, p. 207) discusses the status of two geotrupine beetles considered by him to represent subspecies of the species *Eucanthus lazarus* (Fabricius). His *E. l. lazarus* is stated to range widely over the United States, but records from the Gulf States, excepting Florida, are scanty. *E. l. subtropicus* Howden, on the other hand, is restricted to the southeastern states, and is best represented in Florida, Georgia, Alabama and neighboring states.

Howden is puzzled by the apparent fact that "intermediates" between the two forms came from areas "not bordering the Gulf of Mexico," despite the circumstance that it is in this region that the main overlap falls. Intergrades came from areas "on the East Coast," and from Miami, Florida, and, "Occasional northern specimens appear to exhibit most of the characters of *subtropicus*." However, in particular limited localities, presumably near or in the zone of overlap, Howden was able to name the populations one way or the other with little difficulty.

Although the situation in the Florida Peninsula is not clear from Howden's account, the "intermediate" and more typical-appearing *E. l. lazarus* occurring together with *subtropicus* in the Miami area may really represent undisplaced populations of *subtropicus*. If this is the case, then we would favor Howden's alternative interpretation, and consider *lazarus* and *subtropicus* as closely related but distinct species.

Crabs of the Genus Uca. Jocelyn Crane (in Allee *et al.*, 1950, p. 620) notes that in fiddler crabs of the genus *Uca* differentiation in behavior and often in the coloration of the male is greater if the species are found together than if they are found in different habitats or regions.

The Evolution of Character Displacement

Divergence between two species where they occur together, coupled with convergence where they do not, is a pattern that strongly suggests some form of interaction in the evolutionary history of the pair. The usual case may be one in which the members of the pair are cognate (derived from the same immediate parental population) and have recently made secondary contact following the geographical isolation that has mediated their divergence to species level. In such cases, the "terminal" populations, to which overlap does not yet extend, are not affected by the contact and remain closely similar to each other. But where contact has been made, there are two important ways in which the sympatric populations can interact to augment their initial divergence.

The first type of interaction might best be termed *reinforcement* [1] of the reproductive barriers. It may happen that the species continue to interbreed to some extent, and either the resulting inseminations are ineffectual, or the hybrids produced are inviable or sterile, resulting in what geneticists have termed "gamete wastage." Consequently, any further ethological or genetic divergence reducing this wastage will be strongly favored by natural selection (Dobzhansky, 1951; Koopman, 1950; Kawamura, 1953).

Of conceivably equal or greater importance is the process of *ecological displacement*. It seems clear from an *a priori* basis that any further ecological divergence lessening competition between the overlapping populations will be favored by natural selection if it has a genetic basis (Mayr, 1949). That such a process actually occurs is suggested by abundant indirect evidence from ornithology (Lack, 1944), as well as the cases already cited above.

It seems unnecessary to go into a detailed discussion of these previously elaborated concepts, except to point out that secondary divergence of this nature inevitably entails phenotypic "characters" of the type employed in ordinary taxonomic work. Character displacement therefore may be considered as merely the aspects of such divergence that are recognizable to the taxonomist and some other favored organisms. It is interesting to note that the tendency toward displacement of characters is opposed by the pressure for mimicry. One can imagine some elaborate interactions between the two tendencies, particularly in the evolutionarily fertile tropics.

Competition

The concept of competition has been the focus of much important disagreement among ecologists and other biologists, and it deserves close and persistent investigation. However, were it not that Andrewartha and Birch (1954) criticize the use of the concept by Lack and others to explain distribution and variation of birds and other animals, we might well have avoided discussing it here altogether. Andrewartha and Birch (p. 25) seem to consider that competition is an idea of lesser, perhaps even negligible, importance in biology. They think that the tendency for closely related species to inhabit different areas or exploit different ecological niches (as reported, for instance, by Lack) may conceivably have originated from causes "quite different" from competition. They do not offer alternatives that seem to us anything like as satisfactory as Lack's hypotheses.

Andrewartha and Birch make a point

[1] *Reinforcement* is a familiar term in psychology that has been applied to speciation processes (Blair, 1955).

when they ask for more direct evidence for the action of competition, but it is clear that they have failed to appreciate the amount of evidence that does exist in the literature. However, interspecific competition of the direct, conspicuous, unequivocal kind is apparently a relatively evanescent stage in the relationship of animal individuals or species, and therefore it is difficult to catch and record (just as is the often parallel crisis in the rise of reproductive barriers between two newly diverging species). What we usually see is the result of an actually or potentially competitive contact, in which one competitor has been suppressed or is being forced by some form of aggressive behavior to take second choice, or in which an equilibrium has been established when the potential competitors are specialized to split up the exploitable requisites in their environment. A third possible result is the dispersion of potential competitors in space (Lack, 1954). Surely the cases of character displacement we have considered above, especially those for which we have some ecological data, are pertinent examples of correlation between sympatry (with the possibility of competition) and genetic fixation of specializations resulting in the avoidance of competition. The respective convergent unispecific populations outside the sympatric zones are the "controls" for these observations.

The case in which Lack (1944) cites the distribution of the chaffinches (*Fringilla*) in the Canary Islands is held up to special criticism by Andrewartha and Birch. Lack demonstrates that *F. teydea,* endemic to the islands of Gran Canaria and Tenerife, occupies only the coniferous forests at middle altitudes. On the same islands there also occurs a form of the widespread *F. coelebs,* presumably a relatively recent arrival from the Palaearctic mainland, but this bird occurs only in the tree-heath zone above, and in the broadleaf forests below, the coniferous belt. On the island of Palma, however, *F. teydea* is absent, and there a form of *F. coelebs* occupies the coniferous forest as well as the broadleaf zone. Andrewartha and Birch conclude that, "So far as the case is stated, there is no direct evidence that the two species could not live together if they were put together." It is obvious from this that Lack's critics are not going to be satisfied by any ordinary kind of evidence.

What emerges starkly from contemporary discussion of "competition" is the great variation in the meanings with which different authors freight the word. Andrewartha and Birch, while differing with Nicholson (1954) on most important points, do manage to agree with him that the correct kernel of meaning of competition is contained in the expression "together seek." We would adopt the part of their definition that deals with the common striving for some life requisite, such as food, space or shelter, by two or more individuals, populations or species, etc. This seems to us to be close to the definitions preferred by the larger dictionaries we have consulted.

But Andrewartha and Birch, following many other writers, allow their competition concept to include another idea—that expressing direct interference of one animal or species with the life processes of another, as by fighting. On the surface, this inclusion of aggression as an element of competition may seem to some familiar and reasonable, but we wonder whether the concept of competition could not be more useful in biology if it were more strictly limited to "seeking, or endeavoring to gain, what another is endeavoring to gain at the same time," the first meaning given in *Webster's New International Dictionary, Second Edition, Unabridged.* It is noteworthy that competition as defined by this dictionary fails to include the idea of aggression in any direct and unequivocal way.

It may therefore be more logical in the long run to regard the various kinds of aggression between potential competitors (the outcome of which is so often predictable) as another method, parallel with character displacement and dispersion—and genetically conditioned in a similar fashion—by which organisms seek to lessen or avoid competition. Surely it is significant that aggressive behavior often seems most highly developed in cases where a conspecific, or closely related, potential competitor occurs with the aggressor, yet shows little or no displacement in behavior or form. In contrast are the many cases of complete mutual tolerance shown by closely related organisms that live side by side and are differentially specialized in behavior or form.

Character Displacement versus *Hybridization*

Since both divergent and "intermediate" populations are involved in the displacement patterns we have been describing, it is clear that the convergent populations might easily be mistaken as representing products of interspecific hybridization between the two species displacing each other. This is especially true if the convergent populations are small and isolated, or if only a single one is developed. Lack, for instance, in an early paper (1940) interpreted the Daphne and Crossman populations of *Geospiza* as being of hybrid origin, changing his mind only after he had begun to consider more fully the influence of competition on speciation (in *Darwin's Finches,* 1947).

To take another possible example, Miller (1955) describes what he calls a "hybrid" between the woodpeckers *Dendrocopos scalaris* and *D. villosus.* This specimen, a female, was shot in the Sierra del Carmen, Coahuila, Mexico, at about 7000 feet altitude, near the lower limits of the coniferous belt capping the Sierra. Up to, or near, this altitude, Miller found the Sierra to support a population of *scalaris,* but despite intensive collecting, he found no sign of occupancy by the other putative parent species, *villosus. D. scalaris* reaches a higher point in these mountains than it usually does in the neighboring regions of desert scrub and bottomland—its habitat wherever it has been studied—in Mexico, Arizona, New Mexico and parts of Texas. In general, the *villosus* populations of this part of North America are restricted to the higher coniferous belts, but *villosus* and *scalaris* are in contact at some stations where pinyon-oak-juniper meets coniferous forest. Presumably *scalaris* extends farther vertically in the Sierra del Carmen because *villosus* is not present to limit its upward expansion. According to Miller, *villosus* probably does not occur within 200 miles of the Sierra at the present time.

The specimen, thoroughly described and figured by Miller, is indeed intermediate in many respects between the *scalaris* and *villosus* of northern Mexico. However, there seems to be nothing in the information presented to prevent one's interpreting this as a large, unusually dark specimen of *scalaris,* instead of as a hybrid. There is no good reason to deny the possibility that *scalaris* can produce somewhat *villosus*-like variants at the upper limits of its range when *villosus* is absent.

Other examples we have already cited in the present paper show the difficulty in deciding between displacement and hybridization where the species involved are incompletely known. This situation adds considerable complication to the analysis of interspecific hybridization in nature, for it is clear that the alternative explanation of displacement should at least be taken into account.

One thing seems certain; the "hybrid index," better called "compound character index," can by itself be no sound indication that the situation plotted really involves hybridization. This leads us to ask whether even such elaborate and beautifully documented studies of "hybrid" situations as that made by Sibley (1950, 1954) on the towhees of southern Mexico (*Pipilo erythrophthalmus s. lat.* and *P. ocai*) are not really just illustrations of character displacement. In some of the higher mountains of the southeast (Orizaba, Oaxaca), the two very differently colored forms (species) meet but remain distinct. Farther west are found various populations that apparently grade between the extreme *erythrophthalmus* form and the *ocai* form to various degrees of intermediacy, as expressed by Sibley in his "hybrid index."

Some of the *ocai*-form populations at the western end of the range (*P. ocai alticola*) are stated to be distinct from the other races of *ocai* by a characteristic melanization of the head region, which Sibley thinks is due to introgression from *erythrophthalmus* populations found to the north in the Sierra Madre Occidental. Despite this indication of introgression, the western populations at the *ocai* end of the gradients studied are indexed at, or extremely close to, zero, the figure indicating a population of "pure" *ocai*. Aside from what seems to be a variation in "purity" standards for *ocai* here, it is interesting to note that the western populations and those others among the apparent intermediates of the southern Plateau Region can all conceivably, on present evidence, be interpreted as *erythrophthalmus* that have converged toward *ocai* in the absence of the "true" *ocai* form represented by the upland, sympatric southeastern samples.

It seems possible that some strong selective pressure may be acting in the southern Plateau region to produce an *ocai* coloration-type in finchlike birds, and that *erythrophthalmus* may yield to this pressure wherever the true *ocai* is absent in this area. A very *ocai*-like bird of a related genus, *Atlapetes brunneinucha*, reaches the northern limit of its range in the southern Plateau area, and it is possible that the striking similarity marks some adaptive relationship to which both it and the *Pipilo* stock respond. It might even be that mimicry is involved between the sympatric *Atlapetes* and *Pipilo* stocks, although this is nothing more than the sheerest speculation in view of our very incomplete knowledge of the relative distribution of the two forms and other aspects of their biology and their environment, including their predators. At any rate, character displacement must for the time being be considered a reasonable alternative explanation of the variation of southern Mexican *Pipilo* in this group.

It may perhaps be argued that the "hybrid" populations of *Pipilo* are more variable than the presumed parental populations, and that this in itself is a strong indication of hybridization. We do not believe, however, that the case should be decided on this kind of evidence. To start with, tailspot length, the one character used in Sibley's study that has also been analyzed at length in other populations of *P. erythrophthalmus*, shows very considerable variation in areas far removed from the likely influence of *ocai*. According to the data of Dickinson (1953), the Florida population ("race *alleni*") has a coefficient of variation in this character of about 22 in the male; the range of variation is from 6.1 to 27.5 mm. The northeastern (nominate) race shows a corresponding coefficient of about 12, with a range of variation of from 24.0 to 55.0 mm. Furthermore, the chestnut-tinted pileum characteristic of *ocai-erythrophthalmus* "hybrids" occasionally crops up in the eastern North American samples of *erythrophthalmus*. But even if it were true that variation in the direction of *ocai* could be demonstrated only in the *ocai* "area of influence," this could not be taken as proof of hybridization, because an increase in variation is also a common quality of the "convergent" populations in character displacement patterns.

Character Displacement and Taxonomic Judgment of Allopatric Populations

Foremost among the problems of taxonomic theory today is the tantalizing conundrum concerning the status of the allopatric (isolated) population. Few authors hesitate to assign such populations either subspecific or specific rank, and most, it is hoped, appreciate the fact that their decisions are essentially arbitrary. As Mayr (1942) says, "The decision as to whether to call such forms species or subspecies is often entirely arbitrary and subjective. This is only natural, since we cannot accurately measure to what extent reproductive isolation has already evolved." There does not seem to be any definable threshold between polytypic species composed of such subspecific "units" and the superspecies composed of allopatric sister species. However, it is entirely possible that by the time an isolated population attains an ascertainable level of character concordance, it has already passed the species line; i.e., the more sharply defined an isolated subspecific population is by conventional standards, the less likely it is to be infraspecific in reality.

The phenomenon of character displacement should be borne heavily in mind in considering this matter of allopatric populations. If the present conception is correct, related sympatric species will generally show more morphological differences than similarly related allopatric ones. Hence the degree of observed difference between sympatric species cannot be considered a reliable yardstick for measuring the real status of related allopatric populations, nor can the differences among the latter be taken too seriously as indications of their relationships. In fact, the morphological standards set for determining which completely allopatric populations have reached species level may be much too strict in current practice. Despite impressions that might be gained from recent literature, many systematists have realized that in different allopatric populations (of the same species-group or genus), the degree of morphological divergence may be poorly correlated with the amount of reproductive isolation holding between them (Moore, 1954; Kawamura, 1953). In other words, where there is any question whatsoever about the objective species status of two closely related but geographically separated populations, morphology alone cannot be expected to answer it definitely.

Unfortunately, allopatric species or "subspecies" designated as such on a purely morphological basis frequently enter into theoretical discussions as though they were objectively established realities, when in fact they are usually no more than arbitrary units drawn for curatorial convenience.

Summary

Character displacement is the situation in which, when two species of animals overlap geographically, the differences between them are accentuated in the zone of sympatry and weakened or lost entirely in the parts of their ranges outside this zone. The characters involved in this dual divergence-convergence pattern may be morphological, ecological, behavioral, or physiological. Character displacement probably results most commonly from the first post-isolation contact of two newly evolved cognate species. Upon meeting, the two populations interact through genetic reinforcement of species barriers and/or ecological displacement in such a way as to diverge further from one another where they occur together. Examples of the phenomenon, both verified and probable, are cited for diverse animal groups, illustrating the various aspects that may be assumed by the pattern.

Character displacement is easily confused with a different phenomenon: interspecific hybridization. It is likely that many situations thought to involve hybridization are really only character displacement examples, and in cases of suspected hybridization, this alternative should always be considered. Displacement must also be taken into account in judging the status (specific *vs.* infraspecific) of completely allopatric populations. It is clear that, in the case where the species are closely related, sympatric species will tend to be more different from one another than allopatric ones. Thus, degrees of difference among related sympatric populations cannot be used as trustworthy yardsticks to decide the status of apparently close, allopatric populations.

Acknowledgements

We are grateful for information, advice and other aid received from numerous colleagues in the course of preparing this contribution. Especially to be thanked are J. C. Bequaert, W. J. Bock, W. J. Clench, P. J. Darlington, A. Loveridge, A. R. Main, E. Mayr, A. J. Meyerriecks, K. C. Parkes, R. A. Paynter, and E. E. Williams. Dr. C. Vaurie kindly offered the use of his figures to illustrate the *Sitta* case and gave us the benefit of some unpublished observations. Our acknowledgement is not meant to imply that any of those listed necessarily support the arguments we advance.

REFERENCES

ALLEE, W. C., EMERSON, A. E. and others. 1950. Principles of animal ecology. W. B. Saunders Co.

ANDREWARTHA, H. G., and BIRCH, L. C. 1954. The distribution and abundance of animals. Univ. Chicago Press.

BLAIR, W. F. 1955. Size differences as a possible isolating mechanism in *Microhyla*. *Amer. Naturalist, 89*:297–301.

BOURNE, W. R. P. 1955. The birds of the Cape Verde Islands. *Ibis, 97*:508–556, cf. 520–524.

CAIN, A. J. 1955. A revision of *Trichoglossus haematodus* and of the Australian platycercine parrots. *Ibis, 97*:432–479, cf. 457–461, 479.

CONDON, H. T. 1951. Notes on the birds of South Australia: occurrence, distribution and taxonomy. *S. Aust. Ornith., 20*:26–68.

DICKINSON, J. C. 1952. Geographical variation in the red-eyed towhee of the eastern United States. *Bull. Mus. Comp. Zool. Harv., 107*:273–352.

DOBZHANSKY, TH. 1951. Genetics and the origin of species. 3rd Ed. Columbia Univ. Press.

HOWDEN, H. F. 1955. Biology and taxonomy of the North American beetles of the subfamily Geotrupinae . . . *Proc. U. S. Nat. Mus., 104*:159–319, 18 pls.

HUBBS, C. L., and BAILEY, R. M. 1940. A revision of the black basses (*Micropterus* and *Huro*) with descriptions of four new forms. *Misc. Publ. Zool. Univ. Mich.*, No. 48, 51 pp.

KAWAMURA, T. 1953. Studies on hybridization in amphibians. V. Physiological isolation among four *Hynobius* species. *J. Sci. Hiroshima Univ.* (*B, 1*) *14*:73–116.

KOOPMAN, K. F. 1950. Natural selection for reproductive isolation between *Drosophila pseudoobscura* and *Drosophila persimilis*. *Evolution, 4*:135–148.

LACK, D. 1940. Evolution of the Galapagos finches. *Nature, 146*:324–327.

——— 1944. Ecological aspects of species formation in passerine birds. *Ibis, 86*:260–286.

——— 1947. Darwin's finches. Cambridge Univ. Press.

——— 1954. The natural regulation of animal numbers. Oxford Univ. Press.

MAYR, E. 1942. Systematics and the origin of species. Columbia Univ. Press.

——— 1949. Speciation and selection. *Proc. Amer. Phil. Soc., 93*:514–519.

——— 1955. Notes on the birds of northern Melanesia. *Amer. Mus. Novitates,* No. 1707: 1–46, cf. p. 29.

MOORE, J. A. 1954. Geographic and genetic isolation in Australian amphibia. *Amer. Naturalist, 88*:65–74.

MILLER, A. H. 1955. A hybrid woodpecker and its significance in speciation in the genus *Dendrocopos*. *Evolution, 9*:317–321.

NICHOLSON, A. J. 1954. An outline of the dynamics of animal populations. *Australian J. Zool., 2*:9–65.

SERVENTY, D. L. 1953. Some speciation problems in Australian birds . . . *Emu, 53*:131–145, with further references.

SIBLEY, C. G. 1950. Species formation in the red-eyed towhees of Mexico. *Univ. Calif. Publ. Zool., 50*:109–194.

——— 1954. Hybridization in the red-eyed towhees of Mexico. *Evolution, 8*:252–290.

VAURIE, C. 1950. Notes on Asiatic nuthatches and creepers. *Amer. Mus. Novitates,* No. 1472:1–39.

——— 1951. Adaptive differences between two sympatric species of nuthatches. *Proc. Xth Internat. Ornith. Congr., Uppsala, June 1950*:163–166, 3 figs.

WILSON, E. O. 1955. A monographic revision of the ant genus *Lasius*. *Bull. Mus. Comp. Zool. Harv., 113*:1–205, ill.

WILSON, E. O., and BROWN, W. L., JR. 1955. Revisionary notes on the *sanguinea* and *neogagates* groups of the ant genus *Formica*. *Psyche, 62*:108–129.

WILLIAM L. BROWN, JR. is Associate Curator of Insects at the Museum of Comparative Zoology, Harvard University. **EDWARD O. WILSON** is a Junior Fellow of the Society of Fellows of Harvard University.

1958

"Patchy distributions of ant species in New Guinea rain forests," *Psyche* (Cambridge, MA) 65(1): 26–38

Here I gave one of the early descriptions of fine-grained beta diversity in animals. In the 1950s pristine tropical rainforests were thought to be highly diverse at each site but relatively homogeneous from one site to another. But during long treks through relatively undisturbed rainforest along the Huon Peninsula and around the Lae area of present-day Papua New Guinea, I found striking microgeographic patchiness in the occurrence of ant species, resulting in comparable variation in faunal composition over distances of only a few kilometers. While in the field thinking about the pattern I had unearthed, I believed I had made a great discovery. It was important but not entirely original. One day in the late 1960s, the young, great ecologist Robert H. MacArthur (who died at 42 in 1972) called me to report the discovery he had made of the phenomenon in birds. It was a special pleasure to tell him I had done the same in ants. Fifty years later this phenomenon is called beta diversity, and it is recognized to be widespread, if not universal. The causes and ecological consequences of beta diversity, such as those in the New Guinea ants, are now under close study but still poorly understood.

Reprinted from PSYCHE, Vol. 65 March, 1958 No. 1

PATCHY DISTRIBUTIONS OF ANT SPECIES IN NEW GUINEA RAIN FORESTS

BY EDWARD O. WILSON
Biological Laboratories, Harvard University

While recently engaged in field work in New Guinea the author had several excellent opportunities to study local areal distribution of rain forest ants. During one three-week period in April, 1955, a walk was made from Finschhafen, on the eastern tip of the Huon Peninsula, west for a distance of 45 kilometers through the midmountain rain forests of the Dedua-Hube regions to Tumnang and Laulaunung, thence south for thirty kilometers to Butala on the southern coast. In the vicinity of Lae intensive collecting was conducted over a distance of twelve kilometers in recently continuous lowland rain forest within the triangle formed by Didiman Creek, Bubia, and the section between the Busu and Bupu Rivers.

Areal distributions of individual species were found to be almost universally patchy, despite the external appearance of uniformity of the rain forest environment. Furthermore, in the cases of species abundant enough to be studied in some detail, the patchiness seemed to obtain at two levels of distribution, which for purposes of description here will be referred to as "microgeographic" and "geographic".

Microgeographic patchiness. The species common enough to be studied in detail are also relatively adaptable, occurring usually in spots of variable canopy density (see below) and sometimes in more than one major forest type (e.g., *Leptogenys dimunuta* (Fr. Smith), which ranges from medium lowland rain forest to dry, monsoon forest). In this respect, at least, they seem to be no more specialized than the majority of temperate ant species. At the same time, they show definite preferences for certain local environmental con-

---

EXPLANATION OF PLATE 3

Plate 3. Primary medium-aspect rain forest near the lower Busu River, Northeast New Guinea. A bulldozer trail cuts through the lower left hand corner of the picture.

PSYCHE, 1958 VOL. 65, PLATE 3

WILSON — NEW GUINEA RAIN FOREST

ditions. At the Busu River and in other lowland rain forest sites investigated, ant species tended to be segregated into local areas, sometimes a hectare in extent or less, which could be distinguished from adjacent areas by their specific canopy densities. When the total range of possible canopy densities at the Busu River, from the open aspect that fringes savanna areas, to the most closed aspect, ordinarily found covering sloughs, was arbitrarily divided into three divisions (open, medium, dense) and their faunas studied, the following microgeographic segregation of ant species was noted.

"Open rain forest". (Plate 5) Broken canopy; considerable ground insolation; leaf litter 2 to 15 cm. thick; leaf mold present but thin and relatively dry; soil loose, well aerated, and relatively dry; moss scarce on both ground and tree trunks; A-stratum trees generally less than thirty meters high; lianes and plank buttresses much less common than in other two divisions; recumbent vines common on ground; soil and rotting logs generally thoroughly penetrated with dense root and rhizome growth; undergrowth relatively dense; sufficient to make human progress across the forest floor difficult. This is the aspect of old second-growth forest and may be created naturally by the fall of large forest trees or, in mountainous areas, by rockslides. It is also a more or less permanent feature of the fringe of forest, generally one to two hundred meters wide, that borders savanna areas. Occasional spots deep within rain forest approach the open aspect even though an immediate cause, such as a large fallen tree, is not in evidence. Ant species that appear to reach their maximum density in open rain forest at the Busu River included *Platythyrea parallela* (Fr. Smith), *Diacamma rugosum* (Le Guillou), *Odontomachus simillimus* (Fr. Smith), and *Cardiocondyla paradoxa* (Emery). In the canopy of the open forest, species of *Crematogaster,* especially subgenus *Xiphocrema,* and of *Technomyrmex* increased generally, while those of *Iridomyrmex* decreased.

"Medium rain forest". (Plates 3, 4). By far the largest lowland area in the Lae area is covered by forest of the following aspect: closed canopy; ground insolation slight;

leaf litter as in open aspect; underlying leaf mold rich and moist; soil loose, well aerated and drained and relatively moist; moss common on the surface of the ground, on rotting wood lying on the ground and on tree trunks; A-stratum trees average 40 meters or more in height; plank buttresses common; lianes and epiphytes abundant; under growth sparse, making human progress across the forest floor easy. The majority of endemic ant species are concentrated in this division. Examples of genera that reached maximum density (in 1955) on the floor of the Busu forest included *Ponera, Myopias, Ectomomyrmex, Pheidole, Strumigenys, Rhopalothrix, Myrmecina,* and *Pristomyrmex.* In the canopy *Iridomyrmex* heavily predominated.

"Dense rain forest". Closed canopy; little or no ground insolation; leaf litter thin, with one-quarter or more of the ground surface completely bare; leaf mold very poorly developed; soil dense, less well drained and moister than in medium forest; parts of the ground surface occasionally holding shallow pools of water after heavy rains; moss abundant, especially on larger rotting logs; A-stratum trees as tall as in medium forest; plank buttresses common; lianes and epiphytes abundant; undergrowth very sparse, even more so than in medium forest. Ants reaching maximum density in various strata of this division at the Busu River included some species of *Pheidologeton, Tetramorium, Leptomyrmex* and *Iridomyrmex.*

Careful analysis would probably reveal many finer details of microgeographic segregation than those indicated here, for the rain forest is an extraordinarily complex mosaic of local habitats, exhibiting seemingly endless nuances and combinations of erosion states, growth and death of vegetation, composition of leaf mold, and other environmental features. Ant species did not appear to be limited to any of these particular divisions within the rain forest proper. At most, the divisions probably serve as density foci, from which the species are constantly pressing out into adjacent, less favorable habitats,

Geographic patchiness. Ant species apparently show extensive and unpredictable variation in population density over short geographic distances above and beyond that already noted with respect to microgeographic habitat segregation. This phenomenon was first observed in the Hube area of the Huon Peninsula, where, through a few kilometers distance in seemingly uniform mid-mountain forest, dominant species of the genera *Aphaenogaster* (*Planimyrma*), *Meranoplus*, and *Leptomyrmex* showed conspicuously irregular density patterns. The impression was gained that even within the most favored habitats these species showed irregular density patterns. This type of discontinuous distribution is superimposed on the mosaic, habitat-correlated microgeographic patchiness, and the two conditions probably grade into each other. It can be predicted that superimposition of the two levels of patchiness will result in very irregular and complex individual species distributions, which in turn will have a profound effect on the differentiation of local faunas at localities separated by as little as a few kilometers distance. Such an effect was in fact observed in the lowland rain forests of the Lae area, as described below.

Differentiation of local faunas in the Lae area. In 1955 most of the area to the north of Lae and east of the nearby Busu River was covered with a mixture of primary and secondary rain forest, with occasional savanna enclaves. Native villages were not numerous, and native agriculture had not made serious inroads into the forest. In the vicinity of Bubia, to the northwest of Lae, extensive land was under cultivation, partly by the Government Agricultural Experiment Station, but even here the forest was still partly intact, and primary tracts were still accessible[1].

[1] According to both Mr. Henry G. Eckhoff and Mr. Carl M. Jacobson (pers. commun.), who were among the first European settlers of the Lae district, extensive clearing of the forests of this area is a comparatively recent event. Prior to 1925 the only European settlement was the mission station at Malahang, on the coast near Lae. In 1925 a small amount of ground was cleared at Didiman Creek to establish the Goverment Agricultural Experimental Station. Between 1925 and 1930, further clearing proceeded in the vicinity but was still restricted to the present town limits of Lae. Mr. Eckhoff, who arrived in 1928, states that in 1929-30, "My wife and I were the only residents of Lae other than the air freighting companies. There were no other agricultural activities". The next principal development was the establishment of a poultry farm just outside the Lae township by Mr. Jacobson. During the Second World War a road was built from Lae through Bubia to the airfield at Nadzab. Since 1945 clearing for agricultural purposes has proceeded to a limited extent of either side of this road.

Three localities within this forested area were chosen as sites of intensive collecting (see figure 1). The Busu-Bupu forest was the least disturbed of the three; lumbering operations had commenced in the collecting area only the year before, and most of the forest seemed in primary condition. The Didiman Creek site contained a tract of forest, at least partly second-growth in nature, that had been preserved within the Government Agricultural Experimental Station on the northern edge of Lae. At Bubia, extensive forest tracts, primary at least in part, extended to the east of the Jacobson Plantation. The forest tracts at these three localities represent relict segments of what can reasonably be assumed to have been continuous, predominantly primary lowland forest as recently as thirty years ago. Bubia and

TABLE I.

	BUBIA	DIDIMAN CR.	BUSU R.
Cardiocondyla paradoxa Emery	X	XX	XX

Crematogaster (Rhachiocrema) sp. nov.	—	XX	—
Tetramorium validiusculum Emery	XXX	X	XX
Tetramorium ornatum Emery	—	XX	XX
Triglyphothrix fulviceps Emery	—	XX	XX
Aphaenogaster dromedarius Emery	—	—	XX
Meranoplus hirsutus (Fr. Smith)	X	—	XX
Leptomyrmex fragilis (Fr. Smith)	XX	—	XX
Pseudolasius breviceps Emery	XXX	XX	X

Subjective estimates of relative abundance of some dominant ant species at three neighboring localities in New Guinea. A dashed line means absent, or at least never observed; a single X, present but collected only once or twice; double-X moderately abundant; triple-X among the two or three most abundant species at the locality. Since collecting trips were wide-ranging, these estimates reflect most closely the relative abundance of colonies, rather than number of workers or biomass. Further explanation in text.

Busu-Bupu regions were almost connected by continuous forest even as late as 1955. There is no reason to believe

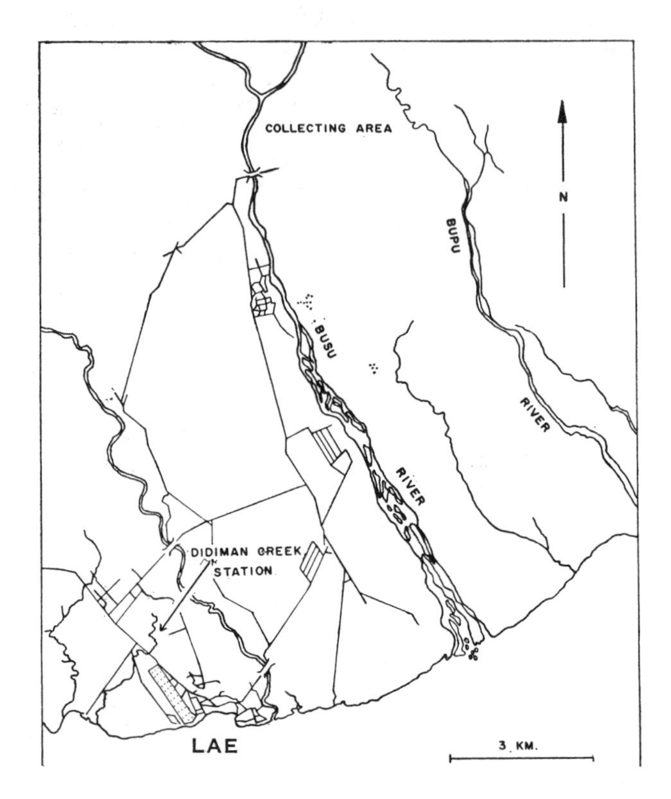

Figure 1. Map of the Lae area in 1955, showing Didiman Creek and the Busu-Bupu area, two of the collecting stations studied with respect to local distribution of species. The third station, Bubia, is located 12.5 kilometers to the northwest of the town of Lae.

Explanation of Plate 4

Plate 4. Floor of primary medium-aspect rain forest near lower Busu River. An overhead tree has just been felled to allow in an unusual amount of sunlight. The exposed portion of the machete is approximately 20 inches, or 50 centimeters, in length. The greatest concentration of species and individual colonies to be found anywhere in New Guinea nest in small pieces of rotting wood in this situation.

PSYCHE, 1958 WILSON — NEW GUINEA RAIN FOREST VOL. 65, PLATE 4

that the forests at the three localities, or the ant faunas in them, had been seriously disturbed by man. All three localities contained rich endemic Papuan faunas, with virtually no infiltration of introduced species.

Subjective impressions of the relative abundance of several of the dominant ant species are presented in Table 1. In each of the three localities, all of the major microgeographic areal divisions were studied. Each locality was visited at least twice during the author's two month stay in the Lae area, and a minimum of four days devoted to intensive collecting. Under these conditions, only the commonest species could be compared, but differences in local abundance of these were so striking that it seems safe to predict that similar patchy distributions are exhibited by other, less dominant members of the fauna.

Discussion: The Evolutionary Implications of Patchiness

In any appraisal of comparative ecology, the New Guinea ant fauna is to be characterized first of all by the exceptional richness of its species and the great size of its biomass. The present study has shown that in addition to sheer size, an additional factor adds greatly to the total faunal complexity. This is the discordant patchy distribution of individual species. The fractioning of species into small subpopulations that are partially isolated from one another probably results in relatively high rates of evolution, whether through random drift or differential selective pressures or both (see for instance Kimura, 1955, and Ford, 1955). Moreover, as a result of discordant patchiness, no two localities harbor exactly the same fauna. Considering that several hundreds of species are thus involved, it is clear that the spatiotemporal structure of the entire New Guinea fauna must present the appearance of a great kaleidoscope. The effects of such a structure on the evolution of individual species of

---

Explanation of Plate 5

Plate 5. Floor of primary open-aspect rain forest near the lower Busu River. The undergrowth at this spot is made up preponderantly of an unidentified speces of *Selaginella*.

Psyche, 1958 Wilson — New Guinea Rain Forest Vol. 65, Plate 5

ants, as well as of other kinds of animals, must be considerable. It very possibly hastens the genetic divergence of local populations and plays an important role in the "exuberance" and amplitude that characterizes evolution in the tropics. Probably as the fauna increases in size, in passing from temperate to tropical areas or from small islands to large ones, the diversifying effects of a kaleidoscopic population structure increase exponentially.

There is abundant evidence that similar features of population structure occur in other groups of organisms in tropical forests. Aubreville (1938), in his "mosaic" or "cyclical" theory of regeneration, has described a kaleidoscope pattern in forest trees of the Ivory Coast. Richards (1952) doubts whether the mosaic theory holds for all rain forest associations, but accepts its validity in special cases where certain conditions have been met.

> "The poor regeneration of the dominant species in African Forests seems in all probability to indicate that the composition of the community is changing. If the forest is in fact 'untouched and primitive', as Aubreville claims . . . . the changes must be cyclical as the Mosaic theory implies. On the other hand, if the community has undergone disturbance in the past, the present combination of species [in a given sample plot] may be a seral stage and the changes part of a normal (not cyclical) process of development toward a stable climax".

Moreau (1948) finds patchiness a common feature in the distribution of rain forest birds in Tanganyika. Where a species is absent from a locality, it is usually replaced by a related species (from the same family), but not always, leaving some inexplicable gaps. The following example is typical:

> "Nearly all the montane forests of eastern Africa from Kenya southward are occupied by one or both of the little barbets, *Pogoniulus bilineatus* and *Viridobucco leucomystax*. On Hanang Mountain, where both these species are missing, *Pogoniulus pusillus*, normally a bird of deciduous trees at lower altitudes, appears in the mountain forests

(Fuggles-Couchman, unpublished). But this does not happen in the neighboring forests of the Mbulu District, where the fruit-eating barbets are not represented at all."

Additional examples from other animal groups and other parts of the tropics (as well as the temperate zones) could be cited to show that patchiness is a widespread phenomenon, on both a very local (microgeographic) and broader (geographic) scale. To all such cases Richards' conditions must be applied, i.e., it must be asked whether patchiness has not arisen exclusively as a result of man-made disturbances. But patchiness as a result of natural disturbances, such as tree falls and stream erosion, is a good possibility also, and should be considered in the future. In the author's present opinion, much of the patchiness observed in New Guinea ant populations has actually arisen through natural disturbances, since enclaves of second-growth vegetation are a normal feature of remote, undisturbed forest. This argument has been taken up in somewhat more detail elsewhere (Wilson, 1959).

Summary

The population structure of individual Papuan ant species is shown to be generally irregular. Patchiness exists at both a local, clearly ecological level, and a broader, "geographic" level not easily correlated with environmental influences. The combined irregularities in the distributions of multiple species result in distinct shifts of faunal composition and relative abundance over distances of only a few kilometers even in relatively continuous, homogeneous rain forest. The theoretical implications of discordant patchiness with respect to rapid evolution are discussed.

Literature Cited

Aubreville, A.

1938. La forêt coloniale: les forêts de l'Afrique occidentale française. Ann. Acad. Sci. Colon., Paris, 9: 1-245.

Ford, E. B.

1955. Rapid evolution and the conditions which make it possible. Symp. Quant. Biol. (Cold Spring Harbor), 20: 230-238.

KIMURA, M.
1955. Stochastic processes and distribution of gene frequencies under natural selection. Symp. Quant. Biol. (Cold Spring Harbor), 20: 33-53.

MOREAU, R. E.
1948. Ecological isolation in a rich tropical avifauna. J. Animal Ecol., 17: 113-126.

RICHARDS, P. W.
1952. *The tropical rain forest*. Cambridge University Press.

WILSON, E. O.
1959. Adaptive shift and dispersal in a tropical ant fauna. Evolution (in press).

1961

"The nature of the taxon cycle in the Melanesian ant fauna," *The American Naturalist* 95: 169–193

The article to follow introduced and named the taxon cycle, which has since been documented in other groups of animals, and set the stage for the formulation of the theory of island biogeography by Robert H. MacArthur and myself two years later.

My writings on the taxon cycle culminated my fieldwork in Melanesia during 1954–55 and systematic research on the Melanesian and Polynesian ant fauna conducted off and on during the subsequent six years at Harvard University. They were, I believe, the first to join ecological data obtained in the field with data on classification and phylogeny obtained in museum studies, to obtain a picture of the effects of ecological adaptation on biogeographic patterns.

Vol. XCV, No. 882 The American Naturalist May–June, 1961

THE NATURE OF THE TAXON CYCLE IN THE MELANESIAN ANT FAUNA*

EDWARD O. WILSON

Biological Laboratories, Harvard University,
Cambridge, Massachusetts

The central contribution of biogeography to general biology is the description of the history of biotas. Aside from its relevance to evolutionary theory, biogeographic history has an immediate significance in population studies: we can expect that the role of individual taxa in ecosystems is influenced both by their geographic origin and by their duration as members of the community. Taxa penetrating from arid source areas will probably fill niches different from those filled by related taxa from moister regions. As a rule, newcomer taxa will undoubtedly affect communities differently from related taxa of long residence. Island biotas derived by radiation of limited stocks show important differences from those derived from more diverse "balanced" stocks, and so on. Of all the major factors that shape community organization, the variables of biogeographic history are probably the least understood. This is due simply to the great complexity of the subject and the tedious nature of its study, which requires revisionary taxonomy as the basic analytical instrument.

The purpose of this paper is to extend an earlier effort (Wilson, 1959a) to synthesize certain information on the zoogeography, speciation patterns and gross ecology of a limited fauna, the ants of Melanesia. In the first report just mentioned, only the subfamily Ponerinae was considered. Faunal sources and expansion patterns of the modern ponerine species were deduced; speciation was shown to be accomplished chiefly by multiple invasions accompanied by major shifts in habitat preferences. In this second study the following groups have been added, following more recent revisionary work by W. L. Brown (1958, 1960, and ms.) and the present author (1957, 1958a, 1959b, c, and ms.): Cerapachyinae, Dolichoderinae, and the myrmicine genera Pheidole, Crematogaster and Strumigenys. These include perhaps 50 per cent of all of the known Melanesian ant species. Not all of the taxa were well enough known to include in all of the analyses; hence, the particular taxa employed are cited with each analysis. Numerical data pertaining to the Asian fauna are based on the catalog by Chapman and Capco (1951), extended and corrected wherever possible by more recent revisionary work. Additional data have been taken from the valuable faunal monographs of Mann (1919, 1921). Emphasis has been shifted somewhat to

*Contribution to a symposium on Modern Aspects of Population Biology. Presented at the meeting of the American Society of Naturalists, cosponsored by the American Society of Zoologists, Ecological Society of America and the Society for the Study of Evolution. American Association for the Advancement of Science, New York, N. Y., December 27, 1960.

a consideration of certain aspects of the formation of individual faunas, the conditions underlying the origin of expanding taxa, and the interaction of expanding and confined taxa.

Certain expressions have special meanings in these analyses and must be defined at the outset:

"Central" tropical Asia: arbitrarily defined as mainland tropical Asia west to and including India and north to and including the "Oriental" portion of southern China, plus Sumatra, Borneo, and Java.

New Guinea: refers to the mainland only.

Expanding taxa: species extending natively over more than a single archipelago, or higher taxa containing such species. Far-ranging species extending beyond certain arbitrary limits are referred to as Stage-I species (see below).

Stage-I species, Asia-based: ranging from tropical Asia, the presumed source area, east to as far as the Moluccas or Micronesia, or beyond.

Stage-I species, New Guinea-based: ranging from mainland New Guinea, the presumed source area, to as far as the Moluccas, Solomon Islands, Micronesia, or any combination of these. Species ranging to Queensland, Aru, Manus, Bismarck Archipelago, or Waigeo but not beyond are arbitrarily not classified as Stage-I.

Stage II and III species: Species of more restricted ranges interpreted as belonging to other phases of the speciation cycle (see Wilson, 1959a).

RELATION OF AREA TO FAUNAL NUMBER

It can be shown that as the area of islands increase, resident faunas of some animal groups logarithmically increase approximately as

$$F = bA^k,$$

where F is the number of resident species and A is the land mass in square miles. In the Ponerinae-Cerapachyinae of Melanesia and the Moluccas, k is approximately 0.7 (figure 1). It is a fact of uncertain significance that k shows considerable variation among different major animal groups and among different faunas. In the Carabidae and herpetofauna of the Greater Antilles and associated smaller islands it is approximately 0.3. In the breeding land and fresh-water birds it is approximately 0.4 in the islands of the Sunda Shelf (Indonesia) but close to 0.5 in the islands of the Sahul Shelf (New Guinea and environs).

The considerable scatter in the area-fauna measurements of the Ponerinae-Cerapachyinae is evidently due to two principal factors: (1) differences in ecology; for example, Kandavu and Vanua Levu are so extensively cultivated as to support smaller indigenous faunas; (2) simply incomplete collecting, which undoubtedly accounts in large part for the seeming paucity of the Halmahera, Bismarcks, and Rennell faunas.

If only those islands are considered which are along the main line of the Sunda-Melanesian arc, which still possess large tracts of native vegetation,

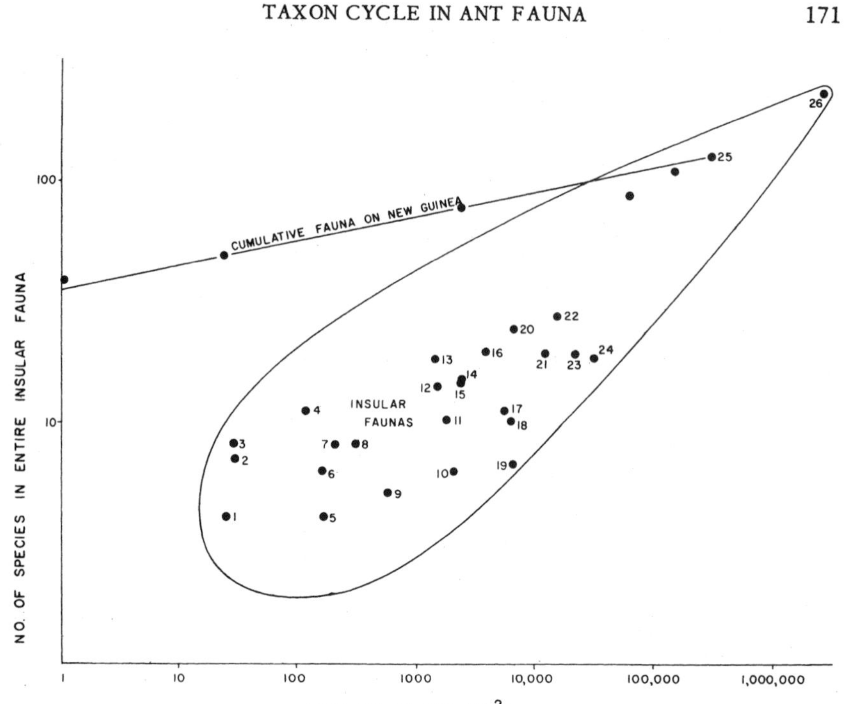

FIGURE 1. The relationship of area to number of ponerine and cerapachyine ant species in the faunas of various Moluccan and Melanesian islands. 1. Ternate; 2. Malapaina; 3. Ugi; 4. Florida; 5. Kandavu; 6. Taviuni; 7. Ndeni; 8. Amboina; 9. Rennell; 10. Vanua Levu; 11. Espiritu Santo; 12. San Cristoval; 13. Santa Isabel; 14. Malaita; 15. Waigeo; 16. Viti Levu; 17. New Hebrides (entire); 18. Ceram; 19. Halmahera; 20. Fiji (entire); 21. New Britain; 22. Solomons (entire); 23. Bismarcks (entire); 24. Moluccas (entire); 25. New Guinea; 26. central tropical Asia. The cumulative New Guinea localities given in the upper curve are as follows: lower Busu River; triangle formed by the lower Busu River, Didiman Creek, and Bubia; all of the Huon Peninsula; northeast New Buinea; northeast New Guinea plus Papua; all of New Guinea.

and which have been reasonably well collected, a much stronger correlation appears, with a slope (k) of about 0.6 and an origin (b) of between two and three. If literally true, this would mean that an island one square mile in area can hold only two or three species, while one under one-tenth of a square mile could hold only one species. These predictions fit very closely the actual faunal size of very small islands in Polynesia, which has been better analyzed (Wilson and Taylor, ms.).

From the details of this analysis, an interesting fact emerges: the size of individual faunas is not correlated with their nearness to the source areas of tropical Asia and New Guinea. Intuitively, one might expect the Moluccas, which are main stepping stones of faunal movement to and from New Guinea, to have a larger fauna than the more remote, peripheral Solomons and Fiji. Yet the reverse is true. The best collected Moluccan Islands (Ternate,

Amboina, Ceram) have somewhat sparser faunas than the best collected parts of the Solomons and Fijis. It is true, on the other hand, that the number of phylogenetic stocks (that is, species groups) declines significantly from the Moluccas and New Guinea eastward. But the total number of species does not deviate from the expected. On Fiji a relatively small number of stocks have diversified to "fill" the Fijian "quota."

The data suggest that individual insular faunas approach upper limits set by the size of the islands. In other words, they are in a saturated or near-saturated condition. It can be inferred that, as a rule, new species can invade an island only if resident species are extinguished to make room for them. Other lines of evidence support this generalization. In New Guinea lowland rain forests, common native ant species show patchy distributions that are poorly correlated with habitat and which result in a limitation of size of local faunas (Wilson, 1958b). In coconut plantations of the Solomon Islands, mixed populations of native and introduced species show clear-cut mosaic distributions determined less by vegetation than by interaction and replacement of competing species (E. S. Brown, 1959). The process of replacement usually involves intercolonial fighting, and it strictly limits the size of faunas of small sample areas. The phenomenon is not unique. Segregation of species by competition in the British ant fauna has already been well demonstrated in the studies of Brian (1952, 1955, 1956a, 1956b).

It is reasonable to expect that the tendency toward mosaic distributions will result in a pattern of fixation-versus-elimination of competing species on very small islands. In Polynesia, where data are now complete enough to allow a reasonably full analysis (Wilson and Taylor, ms.), this proves to be the case. The mosaic pattern is extended to include entire islands and even archipelagoes. On the major islands of Melanesia and Polynesia, and on islands as small as Nuku Hiva and Hiva Oa, members of the same species groups commonly exist together. But among yet smaller islands, such as Rotuma, Raratonga, and Fakaofo, related species tend to exclude one another in an unpredictable manner, forming mosaic patterns. On still smaller islands, intergeneric replacement is evident. The latter phenomenon is consistent with the findings of Brown, who observed intergeneric strife and replacement as a common occurrence in the coconut plantations of the Solomons. As a result of this phenomenon of small-faunal diversification, the summed faunas of entire archipelagoes are larger than could be predicted from knowledge of a limited sample of individual insular faunas. As noted already, a similar process operates at the level of local faunas of large islands to enrich the total insular fauna. The phenomenon can be conveniently referred to as *faunal drift*. This expression is used simply to infer that the composition of small local faunas varies in an unpredictable manner, that is, there is a subjective element of randomness. It remains to be seen to what extent faunal drift is really the result of chance phenomena such as accidents of colonization.

We can next inquire whether faunal drift results in the stabilization of local faunal size, with the result that local faunal size, that is, the number of species occurring in a set sample area, is independent of island size and hence of the total number of species occurring on the island. It turns out that this is not the case. As shown in figure 2, local faunal size increases with the area of the island. Local fauna here is defined as the fauna of an area of approximately one square mile (2.5 km.^{2}) in the lowlands, and encompassing both rain forest and marginal habitats.

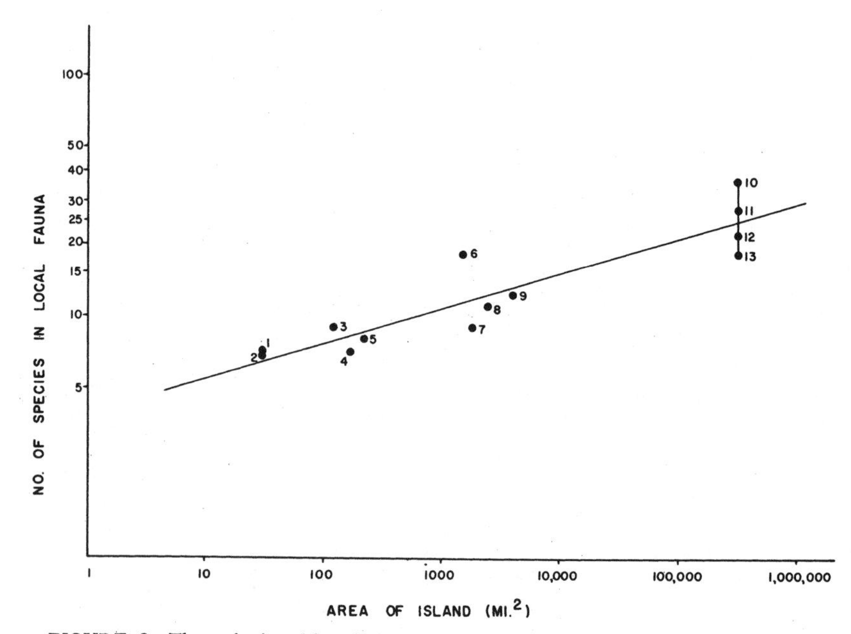

FIGURE 2. The relationship of the area of Melanesian islands to the number of ponerine and cerapachyine species occurring within local faunas on them. The following local faunas are given: 1. Pawa, Ugi; 2, Malapaina; 3. Tulagi, Florida; 4. Somo Somo, Taveuni; 5. Graciosa Bay, Santa Cruz; 6. Fulakora, Santa Isabel; 7. Luganville, Espiritu Santo; 8. Auki, Malaita; 9. Nadarivatu, Viti Levu; 10–12. localities on New Guinea.

With respect to total faunal size in Melanesian ants, the following generalization holds: when the total number of species occurring on an island does not exceed ten species, all of them can be expected to occur in the local fauna, as just defined. When the total fauna includes about 20 species, the local fauna contains from approximately half to all of them. When the total fauna contains over 100 species (New Guinea), local faunas contain only between ten and 30 per cent.

As the sample area on a great island such as New Guinea is progressively decreased, the decrease in local faunal size is less marked than that in decreasing sample areas comprising whole islands (figure 1). Thus the level of "saturation" is higher in local faunas of large islands than in small ones.

THE SOURCES OF THE EXPANDING SPECIES

Two geographical criteria have been used in this study to estimate the origin of a given expanding species: (a) primarily, the center of the present range of the species, and (b) secondarily, the site of maximum diversification and geographical center of the ranges of its closest relatives. In practice the two predicted centers nearly always coincide, at least to the nearest two adjacent archipelagoes. Following are examples of several of the most diverse distribution patterns that have been encountered and the decisions made about them.

(1) *Diacamma rugosum.* Ranges continuously from India to New Guinea. Two distantly related endemic species occur in the Moluccas but the great bulk of the genus occurs in the Oriental Region and is actively speciating and spreading from there. Estimated origin: Oriental Region.

(2) *Odontomachus tyrannicus.* Occurs over most of New Guinea and in the peripheral islands of New Britain, Japan, and Waigeo. The closest related species are limited to New Guinea. Estimated origin: New Guinea.

(3) *Odontomachus saevissimus.* Occurs continuously from the Moluccas to New Britain. A closely related species, *emeryi*, is endemic to the Solomons. Speciation of other members of the *saevissimus* group is active in both the Oriental Region and in New Guinea but primarily in the latter. Estimated origin: New Guinea.

(4) *Odontomachus simillimus.* Occurs continuously from Ceylon through most of Polynesia. The three most closely related species occur in New Guinea; one, *cephalotes*, has spread to offshore islands from New Guinea. Estimated origin: New Guinea.

(5) *Platythyrea parallela.* Ranges more or less continuously from Ceylon to Samoa and occurs (introduced by commerce) in the Society Islands. The bulk of the genus occurs in the Oriental Region. Only one other species, *quadridenta*, is native to Melanesia; this is very distinct from *parallela*, being related to the Oriental *sagei*. Estimated origin: Oriental; questionable.

(6) *Ponera biroi.* Widespread on New Guinea, occurs in addition on New Britain and in the Solomons. A closely related species, *eutrepta*, is limited to Fiji, but the other three members of the *biroi* species groups are endemic to New Guinea. Estimated origin: somewhere in New Guinea plus Bismarcks; New Guinea is the more likely specific source, but both archipelagoes should be included jointly in statistical measures of faunal origins.

The patterns of distributions of several ant subfamilies in Melanesia are summarized in figure 3. Here are presented the limits of distribution of species centered in Melanesia. Those judged to be penetrating from the Oriental Region are omitted. It will be noted that the centers of distribution of the great majority of these Melanesian species are in New Guinea or, at most, New Guinea plus the Bismarck Archipelago. From this center species can be observed in virtually every stage of expansion, some reaching only to immediately adjacent archipelagoes, others outward beyond the Moluccas and Fiji. The ranges tend to be equiformal, that is, extending in both direc-

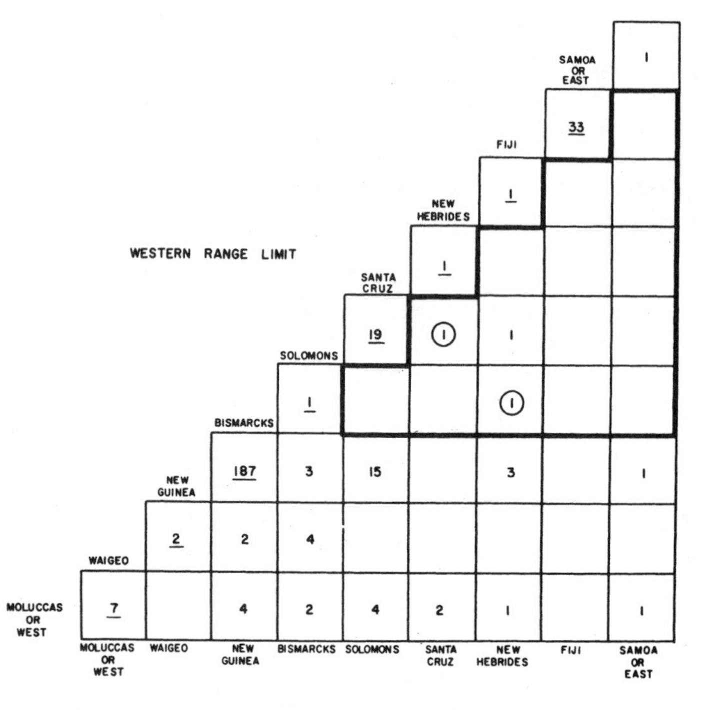

FIGURE 3. Range limits of Melanesian species of Ponerinae, Cerapachyinae, Dolichoderinae, Pheidole, Crematogaster, and Strumigenys. The number of endemic species on each archipelago is underscored.

tions from New Guinea, and in the sum their limits form concentric rings centered on New Guinea. The pattern in the vicinity of Fiji is notably different. Although 33 species are endemic to Fiji, not a single species extends from there to adjacent archipelagoes. There is, in striking contrast to New Guinea, no sign of the movement of species originating on Fiji outward from that archipelago. The Solomons fauna presents a similar pattern to that around Fiji; although being closer to the complex fauna of New Guinea, it is more difficult to analyze. Three species extend from the Solomons to immediately adjacent archipelagoes. Two, however, are evidently relicts: *Anochetus isolatus*, which is a member of a fragmented and evidently receding superspecies (Wilson, 1959b), and an undescribed species of Crematogaster (Orthocrema), which is known only from Espiritu Santo and New Ireland. The third species, *Turneria pacifica*, is classified as originating in the Solomons. The percentage of autochthonous expanding species is still much smaller in the Solomons than in New Guinea, or even New Guinea and the Bismarcks measured jointly.

In table 1 is presented an analysis of the differences in contributions of expanding species of New Guinea, the Solomons, and the Fiji. It will be

TABLE 1

The relation of land area to percentage of expanding species in the resident fauna

Taxa included	Opposed land masses	Number of endemic species	Number of autochthonous expanding species	Difference in frequencies of expanding species	χ^2	P	χ^2_c	Pc	Statistical interpretation of frequency difference
Ponerinae, Cerapachyinae, Dolichoderinae, Myrmicinae	New Guinea	187	35	0.108	1.59	0.20	0.86	0.30	Not significant
	Solomons	19	1						
Ponerinae, Cerapachyinae, Dolichoderinae, Myrmicinae	New Guinea	187	35	0.158	5.86	0.02	4.61	0.03	Significant
	Fiji	33	0						
Ponerinae, Cerapachyinae, Dolichoderinae, Myrmicinae	New Guinea plus Bismarcks	191	32	0.143	5.25	0.02	4.02	0.04	Significant
	Fiji	33	0						
All ant species	New Guinea plus Bismarcks	. . .	. . .	(0.120)	6.25	0.01	5.79	0.02	Significant
	Fiji	58	1						

The frequencies of expanding species originating from Melanesian archipelogoes of different areas are compared. The larger proportion of such species originating from New Guinea is significant at the 95 per cent confidence level with reference to Fiji but not with reference to the Solomons, because of the smaller sample of species from the latter islands. In the last two rows New Guinea and Bismarcks are combined to give the lowest possible estimate of autochthonous expanding species. In the last row the composition of the total Fijian ant fauna is given; that of the total New Guinea-Bismarcks fauna cannot be directly estimated and is assumed to be the same as in the better analyzed taxa, comprising approximately 50 per cent of the species. Both χ^2 and Yates' correction of χ^2 for assumed continuous distribution are given.

noted that whereas approximately 15 per cent of the New Guinea (or New Guinea plus Bismarcks) fauna has expanded to adjacent archipelagoes, none of the Fijian species have, and only one, or five per cent, of the Solomons fauna can be considered to have done so. The difference between the New Guinea and Fijian faunas is significant at the 95 per cent level. That between the New Guinea and Solomons faunas is not significant, but this could be attributed simply to the smaller available sample.

TABLE 2

The relation of land area to percentage of contributed interpenetrating species in the Ponerinae and Cerapachyinae

Competing faunas	Per cent share of land mass	Per cent share of native species	Per cent contribution of interpenetrating Stage-I species	χ^2	χ^2_c	$\frac{P}{P_c}$	Statistical interpretation
Fiji vs. New Guinea	2.28	16.10	$\frac{0}{1} = 0$	...	...	$\frac{\dots}{>0.50}$	Not significant
Fiji vs. central tropical Asia	0.25	9.44	$\frac{0}{2} = 0$	...	...	$\frac{\dots}{>0.50}$	Not significant
Solomons vs. New Guinea	5.22	17.76	$\frac{0}{9} = 0$	...	0.86	$\frac{\dots}{\text{ca. } 0.30}$	Not significant
Solomons vs. central tropical Asia	0.57	10.50	$\frac{0}{10} = 0$	...	0.60	$\frac{\dots}{\text{ca. } 0.40}$	Not significant
New Guinea vs. central tropical Asia	10.93	35.21	$\frac{2(\times 100\%)}{18} = 11.11$	4.40	3.40	$\frac{0.04}{0.06}$	Significant

The ponerine-cerapachyine faunas of opposing pairs of land masses are compared with reference to the exchange between them of interpenetrating species. No significant difference exists between the share of land mass and share of interpenetrating species. The χ^2 and probability values apply to the difference between percentage share in combined native faunas and percentage share in combined interpenetrating species. In the case of New Guinea (versus central tropical Asia) the difference is significant at the 95 per cent confidence level. Samples of species from other Melanesian islands are too small to allow a formal test, but even so it will be noted that the contribution of interpenetrating species by the smaller archipelogoes is consistently nil. (See also figure 4.)

We may now turn to another index of faunal dispersal, the degree to which species of one fauna penetrate other faunas. The *interpenetration* of competing faunas is measured simply as the number of species extending from one archipelago and colonizing another. Here it is possible to compute not only the exchange of faunas within Melanesia but also that between Melanesia and tropical Asia. In table 2 are given the known amounts of interpenetration within the Ponerinae and Cerapachyinae of central tropical Asia, New Guinea, the Solomons, and Fiji. These measures are coupled with land mass and total faunal size and the correlations shown graphically in figure 4. It will be seen that the percentage contribution of interpenetrating species of a given archipelago is nearly the same as its percentage share of the land mass. Note that it is theoretically possible, and interesting, to substitute percentage share of population size for percentage share of land mass. The contribution of interpenetrating species is not linearly correlated with percentage share of total faunal size, as one might intuitively expect. The difference between these two measures is significant at the 95 per cent level in the case of New Guinea versus central tropical Asia. A formal statistical difference cannot be demonstrated in the case of the smaller archipelagoes, due to smallness of sample size. But the relationship is numerically absolute in each case, that is, the contribution of interpenetrating species of the smaller archipelagoes is zero.

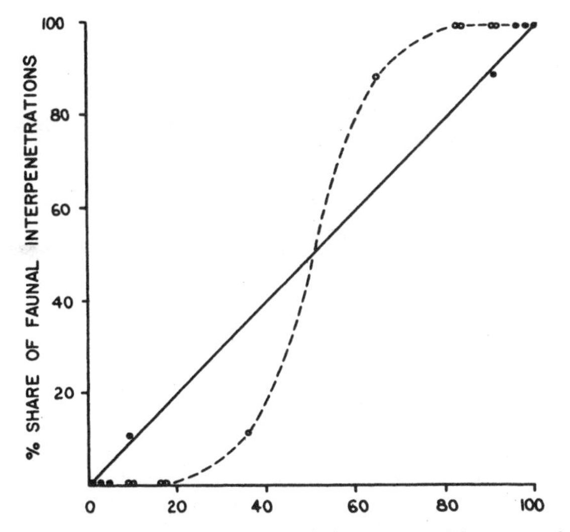

FIGURE 4. The correlation of share of interpenetrating ponerine and cerapachyine species with share of land area (solid circles and line) and with share of total number of species in competing pairs of archipelagoes (open circles and dashed line). See also table 2 and further explanation in the text.

Faunal interpenetration is here defined as a direct indicator of faunal dominance. This interpretation is clarified if we consider what the joint contributions would be to an intermediate island of exactly equal accessibility. The faunas of the two source archipelagoes would "compete" to fill the island's faunal quota. It can be inferred that their percentage share of faunal contributions would be the same as the percentage share of interpenetrating species exchanged between them.

BIOLOGICAL CHARACTERISTICS OF EXPANDING SPECIES

The Melanesian ant fauna appears to stem almost exclusively from species moving out of three source areas: tropical Asia, New Guinea, and Australia. In most of the analyses to follow, only the Ponerinae, Cerapachyinae, and selected myrmicine genera will be considered. All of the expanding species in these taxa are either Asia- or New Guinea-based, simplifying the procedures. It may be noted that Australia-based expanding species in the Dolichoderinae appear to conform to the general biological characterizations to be described in the other groups.

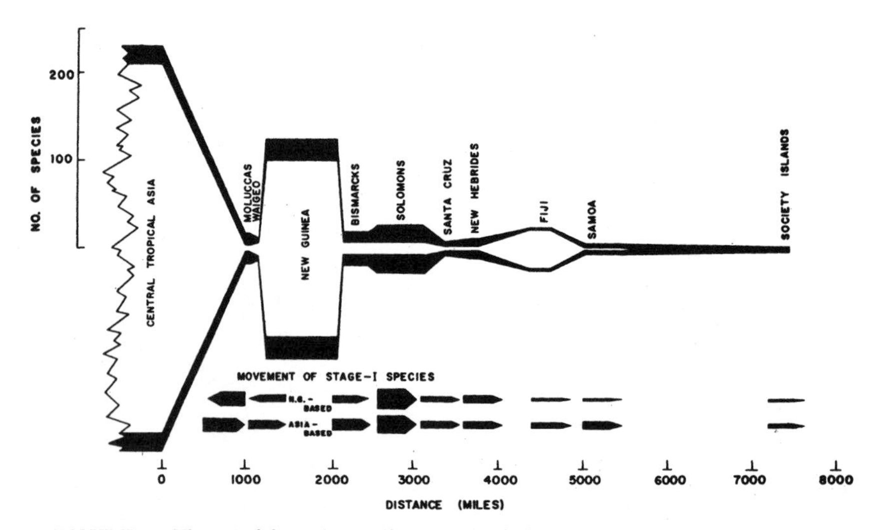

FIGURE 5. The partition of ponerine-cerapachyine species of various archipelagoes into Stage I (shaded) or Stage II and III (blank). The Stage-I species are further distinguished as Asia-based or New Guinea-based.

In figure 5 are given the partitions of some of the faunas of major Oriental-Pacific archipelagoes into speciation stages. A significant characteristic can be seen on inspection: the absolute number of Stage-I (expanding) species does not vary greatly from island to island. What varies markedly is the proportion of Stage II and III species (endemics or near-endemics). As the size of the island decreases, the absolute number of Stage-I species declines only slowly, while that of Stage II and III species declines rapidly. On islands the size of Waigeo and Ndeni (Santa Cruz) in the main Melanesian arc, Stage-I species predominate (figure 6). Fiji shows the same negative correlation between island size and proportion of Stage-I species, but the proportionality remains overall much lower.

As shown elsewhere (Wilson, 1959a) the expanding (Stage-I) species in New Guinea occur preponderantly in marginal habitats. Marginal habitats are defined as those containing the smallest number of ant species. They

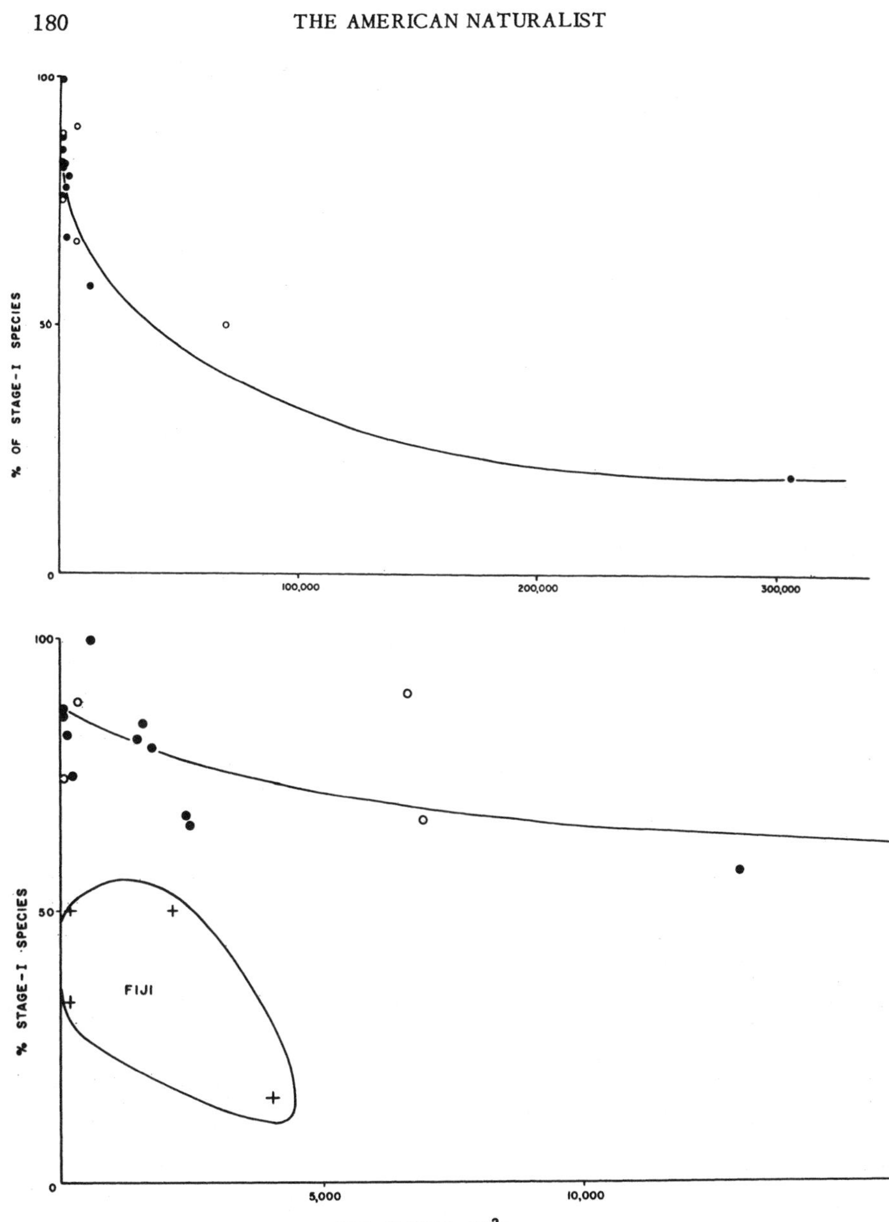

FIGURE 6. The percentage of Stage-I ponerine-cerapachyine species on islands with various areas. Open circles: Moluccas and Celebes. Closed circles: Melanesia exclusive of Fiji. Crosses: Fiji.

include the littoral zone, savannah, monsoon forest, and "open" rain forest. Stage-I species also occur in a significantly wider range of major habitats. In a later section (p. 186) it will be shown that in comparison with members of endemic Melanesian genera and subgenera, the Stage-I species *as a group*

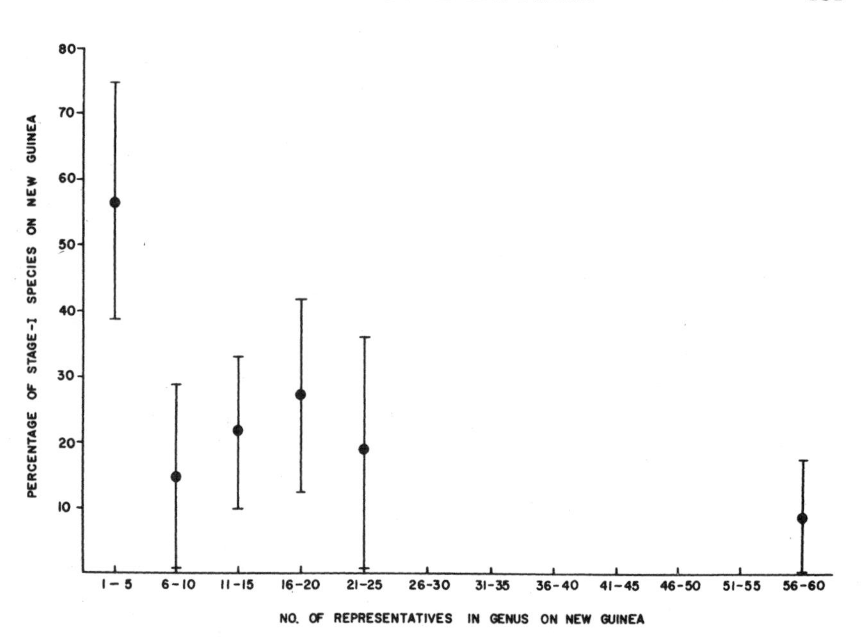

FIGURE 7. The relation of frequency of Stage-I species to size of the genus on New Guinea. The frequency of all genera combined in each size class is given, along with the 95 per cent confidence limits. The subfamilies included are the Ponerinae, Cerapachyinae, and Myrmicinae.

show greater latitude in nest-site choice and colony size. It is further the author's subjective impression, based on too few data to analyze quantitatively, that *individual* Stage-I species show relatively wide latitude in next-site choice, but not in colony size.

It would be of interest now to inquire further into the conditions under which expanding species originate and spread. A valuable new clue is provided by the fact that a negative correlation exists between the size of the genus and the percentage of Stage-I species in the genus. As shown in figure 7, genera containing less than six species have a significantly higher percentage, about 58 per cent, of Stage-I species. Moreover, the number of Stage-I species that a single genus contains at the present time is correlated with the size of the genus but approaches a strict limit. As shown in table 3 and figure 8, even the largest genera (Pheidole, Crematogaster, Strumigenys, Odontomachus) have been able to generate no more than three Asia-based Stage-I species or four New Guinea-based ones. Further, there is a limit to the number of Stage-I species of any genus that coexist on a single island. On the largest of the Melanesian islands, New Guinea, only the dolichoderine genus Iridomyrmex has as many as seven Stage-I representatives. Iridomyrmex is exceptional in being Australia-based and in having Stage-I species conspicuously successful around human settlements. The largest ponerine and myrmicine genera have no more than five Stage-I representatives on New Guinea.

TABLE 3

Some faunal characteristics of ponerine, cerapachyine, myrmicine, and dolichoderine genera occurring in tropical Asia and Melanesia. Further explanation in the text.

Genus	No. of species in central tropical Asia	No. of species originating in central tropical Asia	No. of Stage-I species originating in central tropical Asia	No. of species in New Guinea	No. of species originating in New Guinea	No. of Stage-I species originating in New Guinea	No. of Stage-I species present on New Guinea from all sources	Presence of endemics in other faunal regions
Amblyopone	7	7	0	1	0	0	1	+
Prionopelta	1	1	1	2	2	1	1	+
Myopopone	1	1	1	1	0	0	1	−
Mystrium	1	1	1	1	0	0	1	+
Rhytidoponera	0	0	0	12	12	1	1	+
Gnamptogenys	12	12	0	6	6	0	0	+
Proceratium	2	2	0	1	1	0	0	+
Discothyrea	2	2	0	1	1	1	1	+
Leptogenys	25	25	2	13	12	1	3	+
Anochetus	21	21	1	6	5	1	2	+
Odontomachus	8	8	2	17	14	3	5	+
Platythyrea	7	7	1	2	1	0	1	+
Bothroponera	19	19	1	2	?	?	?	+
Ectomomyrmex	13	13	0	6	6	1	1	−
Centromyrmex	1	1	0	0	0	0	0	+
Cryptopone	2	2	2	4	2	1	3	+
Diacamma	11	11	1	1	0	0	1	−
Emeryopone	1	1	0	0	0	0	0	−
Brachyponera	4	4	1	2	1	1	2	+
Mesoponera	2	2	0	2	2	1	1	+
Trachymesopus	4	4	2	3	1	1	3	+
Ponera	30	30	2	20	18	3	5	+
Pseudoponera	2	2	1	0	0	0	0	−
Myopias	6	6	0	14	14	0	0	−
Harpegnathos	2	2	0	0	0	0	0	−
Odontoponera	1	1	1	0	0	0	0	−
Cerapachys	12	12	0	9	9	1	1	+
Phyracaces	5	5	0	2	2	0	0	+
Lioponera	2	2	0	1	1	0	0	−
Simopone	2	2	0	0	0	0	0	−
Sphinctomyrmex	?	?	?	?	?	?	?	+
Crematogaster	70	70	1	26	25	4	5	+
Pheidole	81	81	2	58	56	4	5	+
Strumigenys	16	16	1	21	20	3	4	+
Hypoclinea	20	20	1	1	0	0	1	+
Monoceratoclinea	0	0	0	2	2	0	0	−
Leptomyrmex	0	0	0	3	3	1	1	+
Technomyrmex	10	10	1	2	1	0	1	+
Turneria	1	?	?	3	3	1	1	−
Iridomyrmex	1	0	0	13	8	2	7	+

In seeking an explanation for this phenomenon, we may look directly to the role of interspecific competition. There is excellent additional evidence to favor the hypothesis that competition is decisive. The Stage-I species on New Guinea include no closely related pairs. Stage-I species in the same

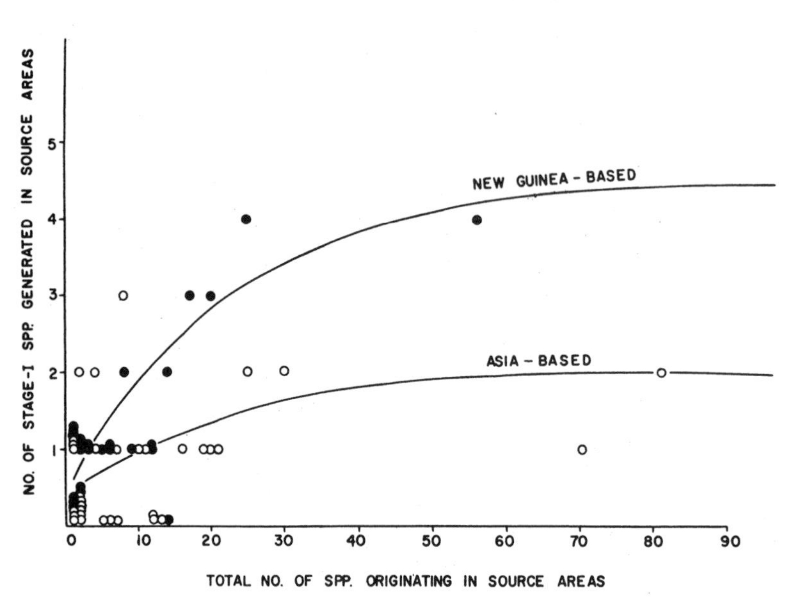

FIGURE 8. The relation of size of genus to number of contained Stage-I species in the Ponerinae, Cerapachyinae, and Myrmicinae in Asia (open circles) and in New Guinea (closed circles). Note that no genus has been able to generate more than four Stage-I species from one source area. See also table 3 and further explanation in the text.

genus tend to be markedly different from one another in morphology, ecology, and behavior. In most cases where the same broad species group is represented by more than one Stage-I member, for example, *Odontomachus simillimus* with *O. cephalotes*, *O. malignus* with *O. saevissimus*, the members occur in different major habitats. Additional supporting evidence is found in the phenomenon of *ecological release*: although the Stage-I species tend to be restricted to marginal habitats on New Guinea and Fiji, which have large endemic faunas, they are not so restricted in the more depauperate Solomon Islands and New Hebrides. On Espiritu Santo, where the author conducted field studies, several Stage-I species, for example, *Odontomachus simillimus*, *Pheidole oceanica*, *P. sexspinosa*, *P. umbonata*, were among the dominant ant species in deep virgin rain forest. At lowland stations on New Guinea, the same species were much sparser and limited to marginal habitats. Other Stage-I species that have marginal distributions on New Guinea, including *Rhytidoponera araneoides*, *Iridomyrmex cordatus*, and *Oecophylla smaragdina*, are dominant in the rain forests of the Solomon Islands (Mann, 1919). There can be no question that interspecific exclusion underlies this phenomenon. Suitable nesting sites and trophophoric fields are virtually saturated with ant colonies in both New Guinea and outer Melanesia. Where large native faunas exist, there is literally no room for significant populations of Stage-I species. Reference has already been made to the role of direct colony conflict in interspecific exclusion of Stage-I ant introduced spe-

TABLE 4

Asia-based genera		Represented in other faunal regions +	−	New Guinea-based genera		Represented in other faunal regions +	−
Producing Stage-I species	+	7	3	Producing Stage-I species	+	10	0
	−	17	28		−	12	10

Classification of the smaller ant genera (1–5 species in the Oriental-Melanesian region) according to whether Stage-I species are produced (+) or not (–) and whether other member species occur in other faunal regions (+) or not (–). All subfamilies are included. Only three of the ant genera limited to the Oriental-Melanesian region produce Stage-I species within it. Further explanation in text.

cies in the Solomon Islands. This may be but one form of competition that produces the displacement-release effect. (Other modes of competition among ant species are discussed by Brian, 1952.) The existing zoogeographic evidence indicates that ecological release is a secondary phenomenon, appearing when initially displaced, expanding species reach depauperate archipelagoes.

Expanding species are able to cross water gaps effectively and colonize empty islands with startling ease. By 1921, less than 40 years after its denudation, the island of Krakatau had been colonized by the four ponerine species, *Odontoponera transversa*, *Ponera confinis*, *Brachyponera luteipes*, and *Odontomachus simillimus* (Wheeler, 1924). *Brachyponera luteipes* had been established by no later than 1909 (Forel, 1909). By 1933 an additional species, *Anochetus graeffei*, was established and a sixth species, *Myopias breviloba*, was known from winged queens (Wheeler, 1937). These six species are among the most widespread of the native Oriental ponerine ants. Together they make a fauna approximately the same size as those found on islands of comparable area in the Moluccas and Melanesia. Similar rapid colonization to near-saturation level occurred on the "empty" islands of Verlaten and Sebesi, near Krakatau.

Finally, the genera and subgenera that have produced Stage-I species are in the great majority represented by distinct species in other faunal regions. This marked characteristic is illustrated in table 4. Expressed another way, taxa now generating Stage-I species in the Oriental-Melanesian region have usually had a record of success in other parts of the world. And, conversely, taxa that are confined to this region are seldom able to generate Stage-I species within it.

BIOLOGICAL CHARACTERISTICS OF AUTOCHTHONOUS AND RETREATING TAXA

Certain ant genera and subgenera are either endemic to Melanesia or clearly contracting within it, that is, centered in Melanesia and represented elsewhere by scattered relict endemics. These include Adelomyrmex s. str.,

Adelomyrmex (Arctomyrmex), Aphaenogaster (Planimyrma), Ancyridris, Dacetinops, Arnoldidris, Archaeomyrmex, Willowsiella, Poecilomyrma, Crematogaster (Rhachiocrema), Crematogaster (Xiphocrema), Pheidole (Electropheidole), Pheidole (Pheidolacanthinus), Monoceratoclinea, Mesoxena, Camponotus (Myrmegonia), Camponotus (Condylomyrma), Polyrhachis (Dolichorhachis). Endemic Melanesian taxa generally contain a relatively small number of species limited to one of the three following archipelagoes: New Guinea, Solomons, and Fiji. It is not possible to determine whether they are relicts or autochthonous taxa at maximum range. (New Caledonia, which has a virtually independent ant fauna derived from Australia, is not considered here.)

In contrast with the above set of taxa are genera and subgenera, headquartered in tropical Asia or tropical Asia plus New Guinea, which are generating Stage-I species and are not notably disjunct. These include Aenictus, Mystrium, Leptogenys, Anochetus, Odontomachus, Platythyrea, Cryptopone, Diacamma, Brachyponera, Trachymesopus, Ponera, Pseudoponera, Odontoponera, Cardiocondyla, Crematogaster (Orthocrema), Strumigenys, Rhopalomastix, Calyptomyrmex, Pheidole s. str., Pheidologeton, Monomorium, Vollenhovia, Tetramorium, Hypoclinea, Technomyrmex, Pseudolasius, Paratrechina (Nylanderia), Oecophylla, Camponotus (Dinomyrmex), Camponotus (Myrmamblys), Camponotus (Colobopsis), Camponotus (Tanaemyrmex), Polyrhachis (Polyrhachis), Polyrhachis (Chariomyrma), Polyrhachis (Myrma), Polyrhachis (Myrmhopla).

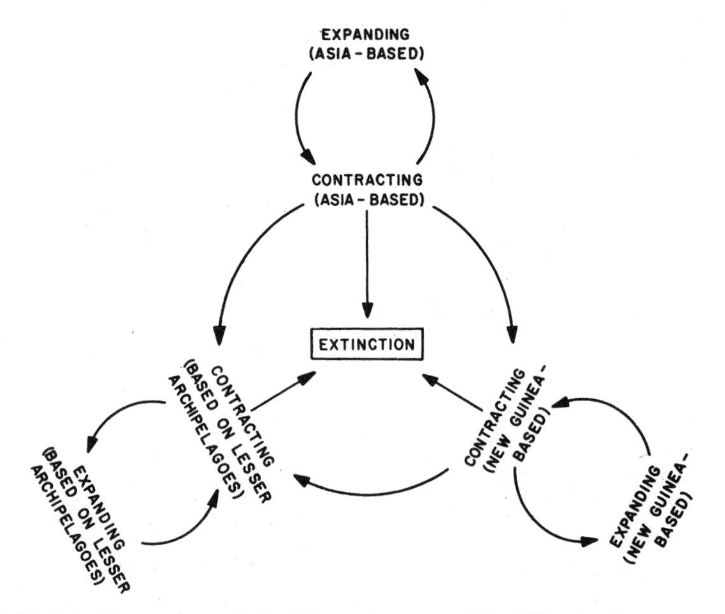

FIGURE 9. The taxon cycle in Melanesian ants. The following features are postulated: the taxon can undergo alternate expansion and contraction, with or without speciation, for an indefinite period of time; it can shift its headquarters from a large land mass to a smaller one but not in the opposite direction.

For convenience, the relative evolutionary positions of the two sets of taxa can be classified as "contracting" and "expanding" in the simple dichotomy represented in figure 9. Taxa endemic to the Solomons or Fiji can be safely regarded as either contracting or at least permanently confined, according to the evidence previously discussed. Taxa endemic to or centered on New Guinea with relicts outside may either be retreating or potent, that is, destined to expand; it is not feasible to speculate further on this

TABLE 5

	Expanding taxa			Receding or autochthonous taxa			χ^2	P	χ_c^2*	P_c*	Statistical interpretation
Characteristic	+++	++	+	+++	++	+					
Spinescence	3	33	1	12	5	1	20.93	< 0.0001	18.04	< 0.0001	Highly significant
Cryptobiosis of non-spinescent forms	5	...	26	3	...	2	3.99	0.04	2.59	0.10	Doubtfully significant
Colony size	3	10	3	0	2	4	6.70	0.01	4.14	0.04	Significant
Terricolous species: small log nest site	5	...	18	4	...	1	6.40	0.01	4.00	0.04	Significant
Prominent use of odor trails	10	...	16	0	...	8	4.36	0.04	2.70	0.10	Doubtfully significant
Species limited chiefly to the mid-mountain rain forests	1	...	31	6	...	6	12.04	< 0.001	9.29	< 0.001	Highly significant
Species limited entirely to inner rain forest habitats	7	...	23	13	...	2	16.25	< 0.001	14.01	< 0.001	Highly significant

* Yates—corrected for continuity.

Some biological differences between expanding genera and confined (retreating or autochthonous) genera in Melanesia. Each qualitative characteristic has been classified as present (+++) or absent (+). Continuously varying characteristics have been roughly classified quantitatively into three classes. Colony size is based on estimates given in Wilson (1959d) and is partitioned into three groups with adult populations of 90 or less (+), 100–800 (++), or 1000 or greater (+++). Not all taxa were well enough known to classify with respect to each characteristic. Only characteristics are shown in which these differences between the two sets of taxa are significant at the 95 per cent confidence level.

distinction. In any case, it is clear that the two sets of taxa belong to different major stages in cyclical evolutionary development in the Oriental-Melanesian fauna (Wilson, 1959a). We may now raise the following question: do taxa undergo parallel biological changes correlated with these zoogeographic episodes? To explore this possibility, the two groups of taxa were scanned for biological differences among diverse morphological, ecological, and behavioral characters. Certain differences were found, as shown in table 5. Background information concerning most of these characters, apart from their evolutionary classification, has already been presented in earlier papers (Wilson 1959a, b).

DISCUSSION: A THEORY OF THE TAXON CYCLE

Enough zoogeographic and ecological data have now been accumulated to justify a preliminary reconstruction of the generalized taxon cycle of the Melanesian ant fauna. Let us start with the actual process of speciation. The evidence suggests that the chief geographic barriers are the water gaps. Speciation probably occurs by internal fragmentation of some populations on the great island of New Guinea, but this appears to be a minor phenomenon, principally involving retreating endemic species. As a rule, semispecies and superspecies, comprising the populations at the threshold of speciation, break at the water gaps. Furthermore, the wider the water gaps, the more frequent the breaks between the allopatric populations. Finally, accumulations of related species on single archipelagoes or islands is chiefly the result of multiple invasions (Wilson, 1959a).

Expanding species in Melanesia originate almost entirely from tropical Asia, New Guinea, and Australia. These are moreover the "potent" species that must from time to time give rise to new taxa. However, the fossil record reveals that ant evolution has been relatively conservative since the early Tertiary (Wheeler, 1914; Carpenter, 1931). The zoogeographic evidence indicates that the origin of new potent taxa of higher rank is a rare event in tropical Asia and Melanesia. Among the 41 smaller ant genera confined to tropical Asia and New Guinea, only three (Myopopone, Odontoponera, Pseudoponera) contain expanding species. Of 51 ant genera of comparable representation in tropical Asia and New Guinea but with representation in other faunal regions (hence, older genera?), 17 contain expanding species. (See table 3.) Other distinctive endemic genera have been evolved in the Solomons, Fiji, and New Caledonia but are strictly limited in size and show no sign of extending their ranges. Thus, the origination of higher taxa is a relatively common event in the Oriental-Melanesian region, but the new products are usually strictly limited to the archipelagoes in which they are born. The combined evidence indicates strongly that the creation of the occasional potent new genera and higher taxa is confined to the large source areas of tropical Asia and New Guinea. Indeed, since Myopopone, Odontoponera, and Pseudoponera are all Asia-based, it is possible that tropical Asia alone serves as a significant source area of potent taxa higher than the species group. Retreating taxa can shift their headquarters from larger to smaller land masses but not in the opposite direction (figure 9).

At lower taxonomic levels, the ants seem to conform to the rule expressed earlier by Darlington (1957, 1959) for vertebrates, that dominant taxa tend to arise in and spread from the largest favorable land masses. It is possible to go a step further and specify that in Oriental-Melanesian ants, the degree of faunal interpenetration is closely correlated with land mass and only secondarily correlated with faunal size, as shown in figure 4.

There is good reason therefore to focus special attention on the ecology of the large land masses generating Stage-I species: tropical Asia and New Guinea. From our somewhat more advanced knowledge of the New Guinea fauna, the following generalization can safely be made. Stage-I species are being produced under conditions of intense and complex competition. As a prelude to expansion out of New Guinea, they became adapted to a wide variety of marginal habitats, containing sparse ant faunas. They are excluded for the most part from the rich inner rain forest habitats by variable numbers of endemic species ecologically similar to them. The endemic competitors are characterized by a high degree of specialization correlated with a more complex partitioning of the environment (Wilson, 1959d) and by smaller individual populations. On other Melanesian islands with depauperate endemic faunas, Stage-I species penetrate the inner forest, mount dense populations there, and fill the available nesting sites.

From the evidence concerning ecological release, the following general prediction can be made: the ecological amplitude of both expanding and endemic species should be negatively correlated with the size of the island on which they occur and hence with the size of the local fauna to which they belong. This prediction cannot be rigorously tested at the present time, but it does seem to be supported by some additional evidence concerning increased variation in morphology of Fijian Ponera (Wilson, 1958a, p. 344) and in feeding habits of Fijian Strumigenys (Brown and Wilson, 1959, p. 289).

Species centered in the marginal habitats of New Guinea have greater opportunities for dispersal. They are poised along the coast and river banks which are the best points of departure for neighboring islands. The smaller corridor islands that can be used as stepping-stones possess the same simplified environments, both physical and biotic, that characterize the marginal habitats on New Guinea.

As the island stepping-stones decrease in size, the percentage of Stage-I species in their faunas increase (figure 6). All of the islands of the Sunda-Melanesian arc, with the possible exception of Fiji, support a strong complement of Stage-I species. On small and medium islands there is therefore a significant constriction of older faunal elements (figure 5). It can be deduced that since the number of Stage-I species does not vary greatly according to islands, the turnover of Stage-I species will probably not vary greatly either. But since the number of older (Stage II and III) species decreases markedly with decrease in island area, we can expect Stage-I species to replace them faster on smaller islands. This prediction seems to be verified by the pattern of distributions of disjunct species groups in Melanesia. As a rule, disjunction involving a hiatus of one or more archipelagoes occurs only after the refugium populations have diverged to species level. This is best interpreted as a result of the fact that speciation ordinarily occurs as the taxon retreats into the inner forest habitats, evidently under displacement pressure from Stage-I species (Wilson, 1959a). Taxa disappear first from the smaller islands and then, apparently progressively, from larger islands. The oldest and most divergent members of disjunct taxa are concentrated as relicts on the largest islands, for example, on New Guinea, the larger Solomons, and Viti Levu. Within the hiatuses the disjunct taxa are commonly represented by ecological vicars from other taxa which are conspicuously abundant and widespread ("ecologically released" species). The following example is unusually clear-cut. Several Iridomyrmex species utilize rubiaceous ant-plants of the genera Myrmecodia and Hydnophytum as their chief nesting sites. On New Guinea these plants are occupied primarily by *I. cordatus* in the marginal habitats and by *I. scrutator* in the inner forest. The ant-plants are saturated by ant colonies and the replacement is therefore virtually absolute. In the Solomon Islands *scrutator* is absent, and *cordatus* is abundant in both marginal habitats and the inner forest. On Fiji *cordatus* is absent, and the ant-plants are occupied by a third species, *nagasau*, which is closely related to *scrutator* and presumably cognate with it.

In summary, expanding (Stage-I) species in Melanesia originate almost entirely from tropical Asia, New Guinea, and Australia. Consequently, these land masses have probably been the ultimate source of all new taxa generated in the Indo-Australian Region and the immediate source of all potent new taxa. The Stage-I species evidently serve an important additional role in displacing, fragmenting, and directing the evolution of older resident species. It can be added that the taxon cycle dates no further back than the early Tertiary, when the radiation of modern ant genera began. Throughout the Cenozoic Melanesia has been broken into numerous islands, at least intermittently (Umbgrove, 1949; Derrick, 1951; Grover, 1955), thus facilitating speciation by multiple invasions. In Miocene times, the Solomon and Fiji Islands were mostly submerged. The evolution of the modern ant fauna of these islands may not date beyond this epoch. The Fijian fauna especially has a modern cast, with no indisputably ancient representatives among its endemic taxa.

A major attribute of evolutionary success in taxa is seen to be the ability to move member species into marginal habitats, at least temporarily. By examining the expanding species we might hope to discover other biological attributes that provide success, in other words, to define new biological rules that apply to the phenomenon of "general adaptation" (Darlington, 1959). Beyond generating marginal-habitat species, however, the expanding taxa appear to be distinguished by only one other common characteristic: great diversity among themselves. In fact, as noted already, one ecologically divergent elements are able to travel through the marginal-habitat channel simultaneously. From this evidence it would seem logical to conclude that general adaptation involves the acquisition of a marked ecological difference. Perhaps the larger this difference the more successful will be the taxon. But of course complete and permanent escape from the faunal equation is impossible and replacement must be inevitable, starting on the smaller islands and in the poorer habitats. Therefore, a second common quality of general adaptation is undoubtedly the ability to replace competitors in the zones of ecological overlap. Perhaps the two qualities are related as follows: the penetration of a new major niche provides the ancestral species with an unmolested population reservoir that allows it, at least for a time, to mount sufficient populations and new adaptations to usurp other niches already occupied by competitor taxa.

The endemic higher taxa, including both confined autochthons and retreating relicts, show more biological uniformity than the expanding taxa, as indicated in table 5. Individually they occur in fewer major habitats (see Wilson, 1959a, p. 137) and occupy a smaller range of nest sites. Collectively they can be characterized by other biological features that are highly variable in expanding taxa. It must be asked whether the collective features are not purely accidental, that is, random correlations derived from the scanning of a great many independent characteristics, or whether they represent a coadaptive complex correlated in some way with the declining position of the taxa in the faunal balance. In fact, most of the features do seem to be associated with the restriction of the endemic taxa to the inner forest habitat. Spinescence may not be part of the coadaptive system; this character is found in terricolous ants that forage above ground and probably serves as a protection against predators (Wilson, 1959d). It may be significant that most of the non-spinescent endemic taxa are notably cryptobiotic, that is, confined in foraging to soil cover and rotting wood. There is a strong tendency for the terricolous species to nest in small rotting logs and branches in the leaf litter; this is the preferred nest site of inner rain forest ant species generally. Colonies are relatively small, a trait closely associated with the restricting nature of the small-log nest site. Finally, odor trails are seldom if ever used, a negative characteristic generally associated with small colony size in ant species. Thus, it appears that endemic taxa are concentrated in and largely restricted to the habitats and nest sites where the greatest number of Melanesian ant species occur. The restriction is reflected in certain other adaptive characters.

If the taxon cycle described here is true for Melanesian ants, it is not necessarily true for other kinds of organisms. Ants are peculiar in several important respects: they are social, highly territorial, and so abundant as to play major roles in the ecosystems in which they live (Branner, 1912; Sernander, 1906; Tevis, 1958). As a result interspecific competition may be more important in their evolution than it is in most other groups of organisms. If the origination of potent new taxa of ants is limited to the larger land masses, this is not necessarily the case in birds (Mayr, 1954), Drosophila (Carson, 1955), or morabine grasshoppers (White, 1959). Indeed, the entire form of the taxon cycle may be altered in groups with markedly different ecology and population structure. It is one of the tasks of comparative zoogeography to determine the extent of this variation in histories.

SUMMARY

Undisturbed ant faunas of islands in the Moluccas-Melanesian arc are for the most part "saturated," that is, approach a size that is correlated closely with the land mass of the island but only weakly with its geographic location (figure 1). In the Ponerinae and Cerapachyinae combined the saturation level can be expressed approximately as $F = 3A^{0.6}$, where F is the number of species in the fauna and A the area of the island in square miles.

Interspecific competition, involving some degree of colonial warfare, plays a major role in the determination of the saturation curve. It deploys the distribution of some ant species into mosaic patterns and increases the diversification of local faunas. Perhaps because of the complex nature of the Melanesian fauna, differences between local faunas appear that give the subjective impression of randomness. Despite the action of species exclusion, the size of local faunas occurring within a set sample area increases with the total size of the island (figure 2).

Water gaps break populations and initiate speciation in Melanesia. Endemic insular faunas build up primarily by the process of multiple invasion.

Expanding species now on Melanesia originated almost exclusively from tropical Asia, New Guinea, and Australia. Faunal dominance, measured by the degree of faunal interpenetration, is a direct function of land area and is less directly related to insular faunal size (figure 4). Taxa originating in Melanesia exclusive of New Guinea are almost all confined to the archipelagoes of their birth.

The following taxon cycle is postulated. A taxon maintains its headquarters in a given land mass indefinitely, expanding and contracting cyclically, or else it declines to extinction. The headquarters can be shifted from a larger to a smaller land mass (for example, from New Guinea to Fiji) but not in the reverse direction (figure 9). Taxa originating in Melanesia exclusive of New Guinea are almost all confined to the archipelagoes of their birth.

On New Guinea, expanding species occur primarily in marginal habitats. In the inner rain forest habitats they are replaced by the large endemic faunas. On archipelagoes with small endemic faunas, the expanding species are ecologically "released," becoming abundant in the inner forest habitats and otherwise increasing their ecological amplitude. As a group they are characterized by great diversification among themselves. No genus among those studied has produced more than three Asia-based or four New Guinea-based Stage-I species (figure 8). No genus studied, including the largest and most successful, has a total of more than seven Stage-I species on New Guinea. Sympatric Stage-I species in the same genus tend to be ecologically and morphologically very dissimilar.

The following rule is predicted: the ecological amplitude of individual species, both expanding and endemic, should be negatively correlated with the size of the local fauna to which they belong and hence the size of the island on which they occur. The prediction is based on the phenomenon of ecological release of Stage-I species and appears to be supported by some fragmentary evidence relating to Fijian endemic species.

Expanding species evidently play a major role in the fragmentation and speciation of older taxa. By dominating the faunas of smaller islands they maintain hiatuses in the ranges of the disjunct taxa. By saturating the marginal habitats they restrict the older taxa to the inner rain forest.

Autochthonous and retreating taxa show certain common biological characteristics coadaptive with restriction to the inner forest habitats. These involve nest site preference, colony size, and foraging behavior (table 5).

Three general attributes of success are recognized in the expanding Melanesian ant taxa: the acquisition of a significant ecological difference, which presumably reduces interspecific competition, the ability to penetrate the marginal habitats, and the ability to disperse across water gaps. It is suggested that the attributes are causally related in the sequence given. Success in the marginal habitats gives expanding species the advantage needed to encompass and progressively replace older resident taxa.

ACKNOWLEDGMENTS

I am indebted to William L. Brown and Philip J. Darlington for a critical reading of the manuscript. Dr. Brown also generously supplied data from his unpublished revisions of Myopias and Strumigenys. This study has been partly supported by a grant from the National Science Foundation.

LITERATURE CITED

Branner, J. C., 1912, Geologic work of ants in tropical America. Pp. 303-333. Smithsonian Institution Report for 1911.

Brian, M. V., 1952, Interaction between ant colonies at an artificial nest-site. Entomol. Monthly Mag. 88: 84-98.

1955, Food collection by a Scottish ant community. J. Animal Ecol. 24: 336-351.

1956, Segregation of species of the ant genus *Myrmica*. J. Animal Ecol. 25: 319-337.

1956 (1958), Interaction between ant populations. Proc. Xth Intern. Congr. Entomol. 2: 781-783.

Brown, E. S., 1959, Immature nutfall of coconuts in the Solomon Islands. II. Changes in ant populations and their relation to vegetation. Bull. Entomol. Research 50: 523-558.

Brown, W. L., 1958, Contributions toward a reclassification of the Formicidae. II. Tribe Ectatommini. Bull. Mus. Comp. Zool. Harvard 118: 175-362.

1960, Contributions.... III. Tribe Amblyoponini. Bull. Mus. Comp. Zool. Harvard 122: 145-230.

Brown, W. L., and E. O. Wilson, 1959, The evolution of the dacetine ants. Quart. Rev. Biol. 34: 278-294.

Carpenter, F. M., 1930, The fossil ants of North America. Bull. Mus. Comp. Zool. Harvard 70: 3-66.

Carson, H. L., 1955, The genetic characteristics of marginal populations of Drosophila. Cold Spring Harbor Symp. Quant. Biol. 20: 276-287.

Chapman, J. W., and S. R. Capco, 1951, Check list of the ants (Hymenoptera: Formicidae) of Asia. Mon. Inst. Sci. Technol. (Manila, P. I.) 1: 1-327.

Darlington, P. J., 1957, Zoogeography. John Wiley & Sons, Inc., New York, N. Y.

1959, Area, climate and evolution. Evolution 13: 488-510.

Derrick, R. A., 1951, The Fiji Islands. 334 pp. Gov. Printing Dept., Suva, Fiji Islands.

Forel, A., 1909, Ameisen aus Java und Krakatau beobachtet und gesammelt von Herrn E. Jacobson. Notes Leyden Mus. 31: 221–232.

Grover, J. C., 1955, Geology, mineral deposits, and prospects of mining development in the British Solomon Islands Protectorate. Mem. Interim Geol. Surv. Brit. Sol. Is. 1: 1–149.

Mann, W. M., 1919, The ants of the British Solomon Islands. Bull. Mus. Comp. Zool. Harvard 63: 273–391.

1921, The ants of the Fiji Islands. Bull. Mus. Comp. Zool. Harvard 64: 401–499.

Mayr, E., 1954, Change of genetic environment and evolution. *In* Evolution as a process, ed. by J. Huxley *et al.* Pp. 157–180. George Allen & Unwin Ltd., London, England.

Sernander, R., 1906, Entwurf einer Monographie der europäischen Myrmekochoren. 410 pp. Uppsala.

Tevis, L., 1958, Interrelations between the Harvester Ant *Veromessor pergandei* (Mayr) and some desert ephemerals. Ecology 39: 695–704.

Umbgove, J. H., 1949, Structural history of the East Indies. 63 pp. Cambridge Univ. Press, Cambridge, England.

Wheeler, W. M., 1914, The ants of the Baltic amber. Schrift. Physik. Öken. Ges. Königsberg 55: 1–142.

1924, Ants of Krakatau and other islands in the Sunda Strait. Treubia 5: 1–20.

1937, Additions to the ant-fauna of Krakatau and Verlaten Island. Treubia 16: 21–24.

White, M. J. D., 1959, Speciation in animals. Austral. J. Sci. 22: 32–39.

Wilson, E. O., 1957, The *tenuis* and *selenophora* groups of the ant genus *Ponera.* Bull. Mus. Comp. Zool. Harvard 116: 355–386.

1958a, Studies on the ant fauna of Melanesia, I-IV. Bull. Mus. Comp. Zool. Harvard 118: 101–153, 303–371.

1958b, Patchy distributions of ant species in New Guinea rain forests. Psyche 65: 26–38.

1959a, Adaptive shift and dispersal in a tropical ant fauna. Evolution 13: 122–144.

1959b, Studies on the ant fauna of Melanesia, V. Bull. Mus. Comp. Zool. Harvard 120: 483–510.

1959c, Studies on the ant fauna of Melanesia, VI. Pacific Insects 1: 39–57.

1959d, Some ecological characteristics of ants in New Guinea rain forests. Ecology 40: 437–447.

Wilson, E. O., and R. W. Taylor, 1961, The ants of Polynesia. (ms.)

1963

"An equilibrium theory of island biogeography," *Evolution* 17(4): 373–387 (with R. H. MacArthur, first author)

Reproduced here is the original presentation of the general theory of island biogeography, which treats the assembly, equilibration, and decline of biodiversity on islands and other discrete ecosystems. The model caused little response among biologists until MacArthur and I formulated it anew, with substantially more documentation and links to a wider range of topics in ecology, in our 1967 book *The Theory of Island Biogeography* (Princeton University Press). The response to this later, full-dress presentation was widespread and almost uniformly positive. It generated many related studies and soon became standard in the literature and practice of ecology, biogeography, and the later discipline of conservation biology.

Reprinted from EVOLUTION, Vol. 17, No. 4, December 24, 1963
Made in United States of America

AN EQUILIBRIUM THEORY OF INSULAR ZOOGEOGRAPHY

ROBERT H. MACARTHUR[1] AND EDWARD O. WILSON[2]

Received March 1, 1963

THE FAUNA–AREA CURVE

As the area of sampling A increases in an ecologically uniform area, the number of plant and animal species s increases in an approximately logarithmic manner, or

$$s = bA^k, \tag{1}$$

where $k < 1$, as shown most recently in in the detailed analysis of Preston (1962). The same relationship holds for islands, where, as one of us has noted (Wilson, 1961), the parameters b and k vary among taxa. Thus, in the ponerine ants of Melanesia and the Moluccas, k (which might be called the *faunal coefficient*) is approximately 0.5 where area is measured in square miles; in the Carabidae and herpetofauna of the Greater Antilles and associated islands, 0.3; in the land and freshwater birds of Indonesia, 0.4; and in the islands of the Sahul Shelf (New Guinea and environs), 0.5.

THE DISTANCE EFFECT IN PACIFIC BIRDS

The relation of number of land and freshwater bird species to area is very orderly in the closely grouped Sunda Islands (fig. 1), but somewhat less so in the islands of Melanesia, Micronesia, and Polynesia taken together (fig. 2). The greater variance of the latter group is attributable primarily to one variable, distance between the islands. In particular, the distance effect can be illustrated by taking the distance from the primary faunal "source area" of Melanesia and relating it to faunal number in the following manner. From fig. 2, take the line connecting New Guinea and the nearby Kei Islands as a "saturation curve" (other lines would be adequate but less suitable to the purpose), calculate the predicted range of "saturation" values among "saturated" islands of varying area from the curve, then take calculated "percentage saturation" as $s_i \times 100/B_i$, where s_i is the real number of species on any island and B_i the saturation number for islands of that area. As shown in fig. 3, the percentage saturation is nicely correlated in an inverse manner with distance from New Guinea. This allows quantification of the rule expressed qualitatively by past authors (see Mayr, 1940) that island faunas become progressively "impoverished" with distance from the nearest land mass.

[1] Division of Biology, University of Pennsylvania, Philadelphia, Pennsylvania.
[2] Biological Laboratories, Harvard University, Cambridge, Massachusetts.

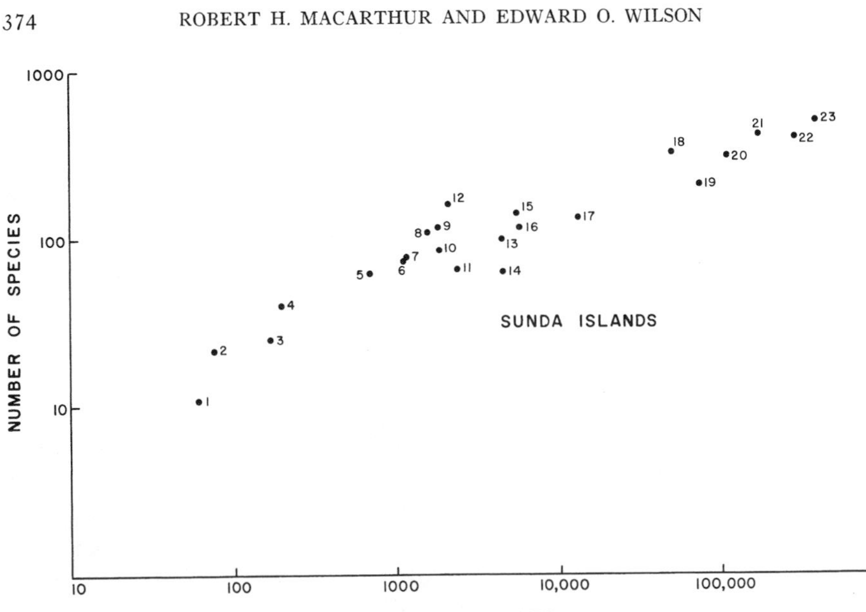

FIG. 1. The numbers of land and freshwater bird species on various islands of the Sunda group, together with the Philippines and New Guinea. The islands are grouped close to one another and to the Asian continent and Greater Sunda group, where most of the species live; and the distance effect is not apparent. (1) Christmas, (2) Bawean, (3) Engano, (4) Savu, (5) Simalur, (6) Alors, (7) Wetar, (8) Nias, (9) Lombok, (10) Biliiton, (11) Mentawei, (12) Bali, (13) Sumba, (14) Bangka, (15) Flores, (16) Sumbawa, (17) Timor, (18) Java, (19) Celebes, (20) Philippines, (21) Sumatra, (22) Borneo, (23) New Guinea. Based on data from Delacour and Mayr (1946), Mayr (1940, 1944), Rensch (1936), and Stresemann (1934, 1939).

AN EQUILIBRIUM MODEL

The impoverishment of the species on remote islands is usually explained, if at all, in terms of the length of time species have been able to colonize and their chances of reaching the remote island in that time. According to this explanation, the number of species on islands grows with time and, given enough time, remote islands will have the same number of species as comparable islands nearer to the source of colonization. The following alternative explanation may often be nearer the truth. Fig. 4 shows how the number of new species entering an island may be balanced by the number of species becoming extinct on that island. The descending curve is the rate at which *new* species enter the island by colonization. This rate does indeed fall as the number of species on the islands increases, because the chance that an immigrant be a new species, not already on the island, falls. Furthermore, the curve falls more steeply at first. This is a consequence of the fact that some species are commoner immigrants than others and that these rapid immigrants are likely, on typical islands, to be the first species present. When there are no species on the island ($N = 0$), the height of the curve represents the number of species arriving per unit of time. Thus the intercept, I, is the rate of immigration of species, new or already present, onto the island. The curve falls to zero at the point $N = P$ where all of the immigrating species are already present so that no new ones are arriving. P is thus the number of species in the "species pool" of immigrants. The shape of the rising curve in the same figure, which represents the

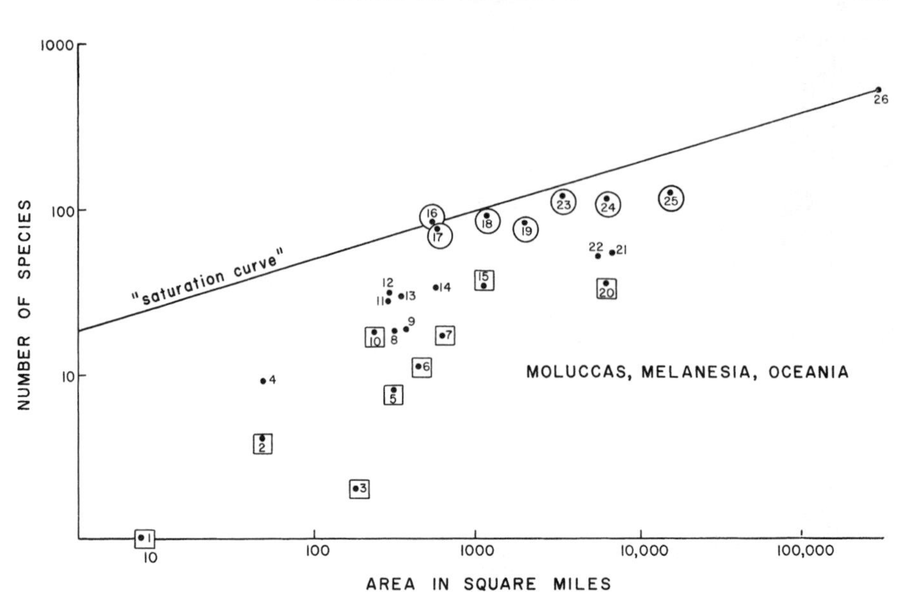

FIG. 2. The numbers of land and freshwater bird species on various islands of the Moluccas, Melanesia, Micronesia, and Polynesia. Here the archiplagoes are widely scattered, and the distance effect is apparent in the greater variance. Hawaii is included even though its fauna is derived mostly from the New World (Mayr, 1943). "Near" islands (less than 500 miles trom New Guinea) are enclosed in circles, "far" islands (greater than 2,000 miles) in squares, and islands at intermediate distances are left unenclosed. The saturation curve is drawn through large and small islands at source of colonization. (1) Wake, (2) Henderson, (3) Line, (4) Kusaie, (5) Tuamotu, (6) Marquesas, (7) Society, (8) Ponape, (9) Marianas, (10) Tonga, (11) Carolines, (12) Palau, (13) Santa Cruz, (14) Rennell, (15) Samoa, (16) Kei, (17) Louisiade, (18) D'Entrecasteaux, (19) Tanimbar, (20) Hawaii, (21) Fiji, (22) New Hebrides, (23) Buru, (24) Ceram, (25) Solomons, (26) New Guinea. Based on data from Mayr (1933, 1940, 1943) and Greenway (1958).

rate at which species are becoming extinct on the island, can also be determined roughly. In case all of the species are equally likely to die out and this probability is independent of the number of other species present, the number of species becoming extinct in a unit of time is proportional to the number of species present, so that the curve would rise linearly with N. More realistically, some species die out more readily than others and the more species there are, the rarer each is, and hence an increased number of species increases the likelihood of any given species dying out. Under normal conditions both of these corrections would tend to increase the slope of the extinction curve for large values of N. (In the rare situation in which the species which enter most often as immigrants are the ones which die out most readily—presumably because the island is atypical so that species which are common elsewhere cannot survive well—the curve of extinction may have a steeper slope for small N.) If N is the number of species present at the start, then $E(N)/N$ is the fraction dying out, which can also be interpreted crudely as the probability that any given species will die out. Since this fraction cannot exceed 1, the extinction curve cannot rise higher than the straight line of a 45° angle rising from the origin of the coordinates.

It is clear that the rising and falling curves must intersect and we will denote by $\hat{s}$ the value of N for which the rate of immigration of new species is balanced by

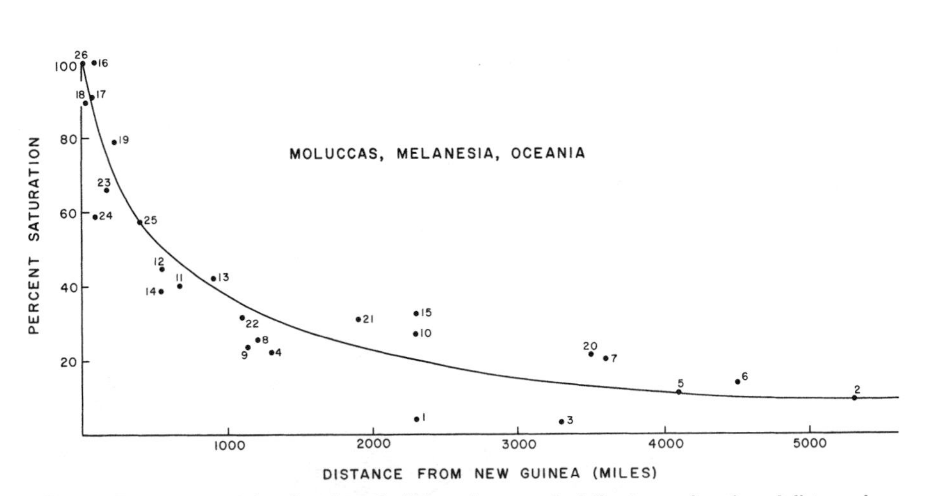

FIG. 3. Per cent saturation, based on the "saturation curve" of fig. 2, as a function of distance from New Guinea. The numbers refer to the same islands identified in the caption of fig. 2. Note that from equation (4) it is an oversimplification to take distances solely from New Guinea. The abscissa should give a more complex function of distances from all the surrounding islands, with the result that far islands would appear less "distant." But this representation expresses the distance effect adequately for the conclusions drawn.

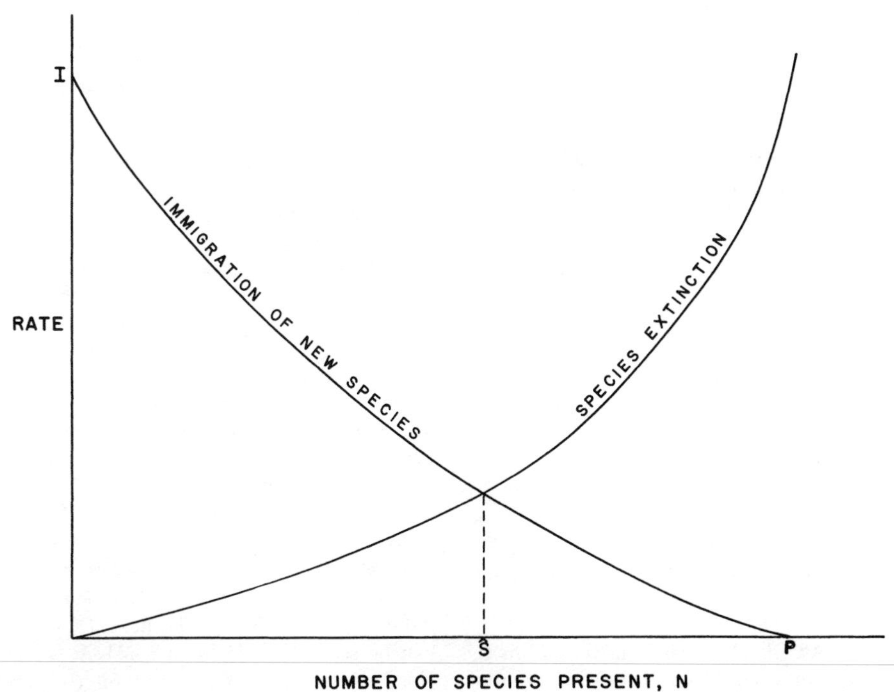

FIG. 4. Equilibrium model of a fauna of a single island. See explanation in the text.

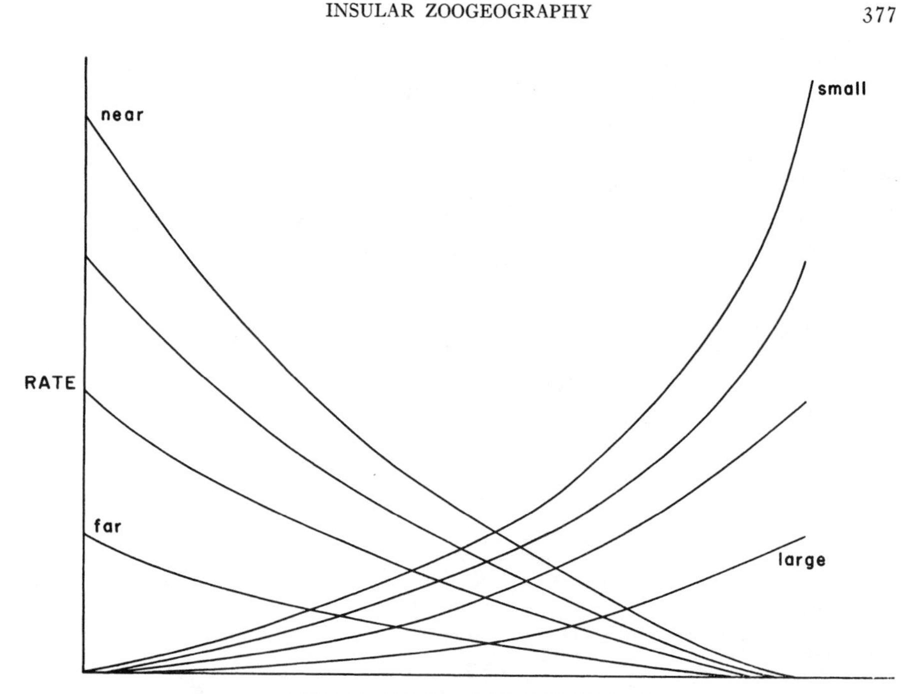

FIG. 5. Equilibrium model of faunas of several islands of varying distances from the source area and varying size. Note that the effect shown by the data of fig. 2, of faunas of far islands increasing with size more rapidly than those of near islands, is predicted by this model. Further explanation in text.

the rate of extinction. The number of species on the island will be stabilized at $\hat{s}$, for a glance at the figure shows that when N is greater than $\hat{s}$, extinction exceeds immigration of new species so that N decreases, and when N is less than $\hat{s}$, immigration of new species exceeds extinction so that N will increase. Therefore, in order to predict the number of species on an island we need only construct these two curves and see where they intersect. We shall make a somewhat oversimplified attempt to do this in later paragraphs. First, however, there are several interesting qualitative predictions which we can make without committing ourselves to any specific shape of the immigration and extinction curves.

A. An island which is farther from the source of colonization (or for any other reason has a smaller value of I) will, other things being equal, have fewer species, because the immigration curve will be lower and hence intersect the mortality curve farther to the left (see fig. 5).

B. Reduction of the "species pool" of immigrants, P, will reduce the number of species on the island (for the same reason as in A).

C. If an island has smaller area, more severe climate (or for any other reason has a greater extinction rate), the mortality curve will rise and the number of species will decrease (see fig. 5).

D. If we have two islands with the same immigration curve but different extinction curves, any given species on the one with the higher extinction curve is more likely to die out, because $E(N)/N$ can be seen to be higher [$E(N)/N$ is the slope of the line joining the intersection point to the origin].

E. The number of species found on islands far from the source of colonization will grow more rapidly with island area than will the number on near islands. More precisely, if the area of the island is denoted by A, and $\hat{s}$ is the equilibrium number of species, then d^2s/dA^2 is greater for far islands than for near ones. This can be verified empirically by plotting points or by noticing that the change in the angle of intersection is greater for far islands.

F. The number of species on large islands decreases with distance from source of colonization faster than does the number of species on small islands. (This is merely another way of writing E and is verified similarly.)

Further, as will be shown later, the variance in $\hat{s}$ (due to randomness in immigrations and extinctions) will be lower than that expected if the "classical" explanation holds. In the classical explanation most of those species will be found which have at any time succeeded in immigrating. At least for distant islands this number would have an approximately Poisson distribution so that the variance would be approximately equal to the mean. Our model predicts a reduced variance, so that if the observed variance is significantly smaller than the mean for distant islands, it is evidence for the equilibrium explanation.

The evidence in fig. 2, relating to the insular bird faunas east of Weber's Line, is consistent with all of these predictions. To see this for the non-obvious prediction E, notice that a greater slope on this log-log plot corresponds to a greater second derivative, since A becomes sufficiently large.

The Form of the Immigration and Extinction Curves

If the equilibrium model we have presented is correct, it should be possible eventually to derive some quantitative generalizations concerning rates of immigration and extinction. In the section to follow we have deduced an equilibrium equation which is adequate as a first approximation, in that it yields the general form of the empirically derived fauna–area curves without contradicting (for the moment) our intuitive ideas of the underlying biological processes. This attempt to produce a formal equation is subject to indefinite future improvements and does not affect the validity of the graphically derived equilibrium theory. We start with the statement that

$$\Delta s = M + G - D, \qquad (2)$$

where s is the number of species on an island, M is the number of species successfully immigrating to the island per year, G is the number of new species being added per year by local speciation (not including immigrant species that merely diverge to species level without multiplying), and D is the number of species dying out per year. At equilibrium,

$$M + G = D.$$

The immigration rate M must be determined by at least two independent values: (1) the rate at which propagules reach the island, which is dependent on the size of the island and its distance from the source of the propagules, as well as the nature of the source area, but not on the condition of the recipient island's fauna; and (2) as noted already, the number of species already resident on the island. Propagules are defined here as the minimum number of individuals of a given species needed to achieve colonization; a more exact explication is given in the Appendix. Consider first the source region. If it is climatically and faunistically similar to other potential source regions, the number of propagules passing beyond its shores per year is likely to be closely related to the size of the population of the taxon living on it, which in turn is approximately a linear function of its area. This notion is supported by the evidence from Indo-Australian ant zoogeography, which indicates that the ratio of faunal exchange is about equal to the ratio of the areas of the source regions (Wilson, 1961). On the other hand, the number of propagules reaching the recipient island probably varies linearly with the angle it subtends with reference to the center of the source region. Only near islands will vary much because of this factor. Finally, the number of propagules reaching the recipient island is most likely to be an exponential function of its distance from the source region. In the simplest case, if the probability that a given propagule ceases its overseas voyage (e.g., it falls into the sea and dies) at any given instant in time remains constant, then the fraction of propagules reaching a given distance fits an exponential holding-time distribution. If these assumptions are correct, the number of propagules reaching an island from a given source region per year can be approximated as

$$\alpha A_i \frac{\text{diam}_i \, e^{-\lambda d_i}}{2\pi d_i}, \qquad (3)$$

where A_i is the area of the source region, d_i is the mean distance between the source region and recipient island, diam_i is the diameter of the recipient island taken at a right angle to the direction of d_i, and α is a coefficient relating area to the number of propagules produced. More generally, where more than one source region is in position, the rate of propagule arrival would be

$$\frac{\alpha}{2\pi} \sum_i \frac{\text{diam}_i}{d_i} A_i e^{-\lambda d_i}, \qquad (4)$$

where the summation is of contributions from each of the *i*th source regions. Again, note that a propagule is defined as the minimum number of individuals required to achieve colonization.

Only a certain fraction of arriving propagules will add a new species to the fauna, however, because except for "empty" islands at least some ecological positions will be filled. As indicated in fig. 4, the rate of immigration (i.e., rate of propagule arrival times the fraction colonizing) declines to zero as the number of resident species (s) approaches the limit P. The curve relating the immigration rate to degree of unsaturation is probably a concave one, as indicated in fig. 4, for two reasons: (1) the more abundant immigrants reach the island earlier, and (2) we would expect otherwise randomly arriving elements to settle into available positions according to a simple occupancy model where one and only one object is allowed to occupy each randomly placed position (Feller, 1958). These circumstances would result in the rate of successful occupation decelerating as positions are filled. While these are interesting subjects in themselves, a reasonable approximation is obtained if it is assumed that the rate of occupation is an inverse linear function of the number of occupied positions, or

$$\left(1 - \frac{s}{P}\right). \qquad (5)$$

Then

$$M = \frac{\alpha(1 - s/P)}{2\pi} \sum_i \frac{\text{diam}_i}{d_i} A_i e^{-\lambda d_i}. \qquad (6)$$

We know the immigration line in fig. 4 is not straight; to take this into account we must modify formula 5 by adding a term in s^2. However, this will not be necessary for our immediate purposes.

Now let us consider G, the rate of new productions on the island by local speciation. Note that this rate does not include the mere divergence of an island endemic to a specific level with reference to the stock species in the source area; that species is still counted as contributing to M, the immigration rate, no matter how far it evolves. Only new species generated from it and in addition to it are counted in G. First, consider an archipelago as a unit and the increase of s by divergence of species on the various islands to the level of allopatric species, i.e., the production of a local archipelagic superspecies. If this is the case, and no exchange of endemics is yet achieved among the islands of the archipelago, the number of species in the archipelago is limited to

$$\sum_{i=1}^{\infty} n_i \hat{s}_i, \qquad (7)$$

where n_i is the number of islands in the archipelago of *i*th area and $\hat{s}_i$ is the number of species occurring at equilibrium on islands of ith area. But the generation of allopatric species in superspecies does not multiply species on single islands or greatly change the fauna of the archipelago as a whole from the value predicted by the fauna–area curve, as can be readily seen in figs. 2 and 3. G, the increase of s by local speciation on single islands and exchange of autochthonous species between islands, probably becomes significant only in the oldest, largest, and most isolated archipelagoes, such as Hawaii and the Galápagos. Where it occurs, the exchange among the islands can be predicted from (6), with individual islands in the archipelago serving as both source regions and recipient islands. However, for most cases it is probably safe to omit G from the model, i.e., consider only source regions outside the archipelago, and hence

$$\Delta s = M - D. \qquad (8)$$

The extinction rate D would seem intuitively to depend in some simple manner on (1) the mean size of the species populations, which in turn is determined by the size of the island and the number of species belonging to the taxon that occur on it; and (2) the yearly mortality rate of the organisms. Let us suppose that the probability of extinction of a species is merely the probability that all the individuals of a given species will die in one year. If the deaths of individuals are unrelated to each other and the population sizes of the species are equal and nonfluctuating,

$$D = sP^{N_r/s}, \qquad (9)$$

where N_r is the total number of individuals in the taxon on the recipient island and P is their annual mortality rate. More realistically, the species of a taxon, such as the birds, vary in abundance in a manner approximating a Barton–Davis distribution (MacArthur, 1957) although the approximation is probably not good for a whole island. In s nonfluctuating species ordered according to their rank (K) in relative rareness,

$$D = \sum_{i=1}^{s} p^{(N_r/s)\sum_{1=i}^{K} 1/(s-i+1)}. \qquad (10)$$

This is still an oversimplification, if for no other reason than the fact that populations do fluctuate, and with increased fluctuation D will increase. However, both models, as well as elaborations of them to account for fluctuation, predict an exponential increase of D with restriction of island area. The increase of D which accompanies an increase in number of resident species is more complicated but is shown in fig. 4.

Model of Immigration and Extinction Process on a Single Island

Let $P_s(t)$ be the probability that, at time t, our island has s species, λ_s be the rate of immigration of new species onto the island, when s are present, μ_s be the rate of extinction of species on the island when s are present; and λ_s and μ_s then represent the intersecting curves in fig. 4. This is a "birth and death process" only slightly different from the kind most familiar to mathematicians (cf. Feller, 1958, last chapter). By the rules of probability

$$P_s(t+h) = P_s(t)(1-\lambda_s h - \mu_s h) + P_{s-1}(t)\lambda_{s-1}h + P_{s+1}(t)\mu_{s+1}h,$$

since to have s at time $t+h$ requires that at a short time preceding one of the following conditions held: (1) there were s and that no immigration or extinction took place, or (2) that there were $s-1$ and one species immigrated, or (3) that there $s+1$ and one species became extinct. We take h to be small enough that probabilities of two or more extinctions and/or immigrations can be ignored. Bringing $P_s(t)$ to the left-hand side, dividing by h, and passing to the limit as $h \rightarrow 0$

$$\frac{dP_s(t)}{dt} = -(\lambda_s + \mu_s)P_s(t) + \lambda_{s-1}P_{s-1}(t) + \mu_{s+1}P_{s+1}(t). \qquad (11)$$

For this formula to be true in the case where $s = 0$, we must require that $\lambda_{-1} = 0$ and $\mu_0 = 0$. In principle we could solve

(11) for $P_s(t)$; for our purposes it is more useful to find the mean, $M(t)$, and the variance, $\mathrm{var}(t)$, of the number of species at time t. These can be estimated in nature by measuring the mean and variance in numbers of species on a series of islands of about the same distance and area and hence of the same λ_s and μ_s. To find the mean, $M(t)$, from (11) we multiply both sides of (11) by s and then sum from $s=0$ to $s=\infty$. Since $\sum_{s=0}^{\infty} sP_s(t) = M(t)$, this gives us

$$\frac{dM(t)}{dt} = -\sum_{s=0}^{\infty} (\lambda_s + \mu_s) sP_s(t) + \sum_{s-1=0}^{\infty} \lambda_{s-1}[(s-1)+1]P_{s-1}(t) + \sum_{s+1=0}^{\infty} \mu_{s+1}[(s+1)-1]P_{s+1}(t).$$

(Here terms $\lambda_{-1}\cdot 0 \cdot P_{-1}(t) = 0$ and $\mu_0 \cdot (-1)P_0(t) = 0$ have been subtracted or added without altering values.) This reduces to

$$\frac{dM(t)}{dt} = \sum_{s=0}^{\infty} \lambda_s P_s(t) - \sum_{s=0}^{\infty} \mu_s P_s(t) = \overline{\lambda_s(t)} - \overline{\mu_s(t)}. \quad (12)$$

But, since λ_s and μ_s are, at least locally, approximately straight, the mean value of λ_s at time t is about equal to $\lambda_{M(t)}$ and similarly $\overline{\mu_s(t)} \sim \mu_{M(t)}$. Hence, approximately

$$\frac{dM(t)}{dt} = \lambda_{M(t)} - \mu_{M(t)}, \quad (13)$$

or the expected number of species in Fig. 4 moves toward $\hat{s}$ at a rate equal to the difference in height of the immigration and extinction curves. In fact, if $d\mu/ds - d\lambda/ds$, evaluated near $s=\hat{s}$ is abbreviated by F, then, approximately $dM(t)/dt = F(\hat{s} - M(t))$ whose solution is $M(t) = \hat{s}(1-e^{-Ft})$. Finally, we can compute the time required to reach 90% (say) of the saturation value $\hat{s}$ so that $M(t)/\hat{s} = 0.9$ or $e^{-Ft} = 0.1$.
Therefore,

$$t = \frac{2.303}{F}. \quad (13a)$$

A similar formula for the variance is obtained by multiplying both sides of (11) by $(s-M(t))^2$ and summing from $s=0$ to $s=\infty$. As before, since $\mathrm{var}(t) = \sum_{s=0}^{\infty} (s-M(t))^2 P_s(t)$, this results in

$$\begin{aligned}\frac{d\,\mathrm{var}(t)}{dt} &= -\sum_{s=0}^{\infty} (\lambda_s + \mu_s)(s-M(t))^2 P_s(t) \\ &\quad + \sum_{s-1=0}^{\infty} \lambda_{s-1}[(s-1-M(t))+1]^2 P_{s+1}(t) \\ &\quad + \sum_{s+1=0}^{\infty} \mu_{s+1}[(s+1-M(t))-1]^2 P_{s+1}(t) \\ &= 2\sum_{s=0}^{\infty} \lambda_s (s-M(t))P_s(t) \\ &\quad - 2\sum_{s=0}^{\infty} \mu_s (s-M(t))P_s(t) \\ &\quad + \sum_{s=0}^{\infty} \lambda_s P_s(t) + \sum_{s=0}^{\infty} \mu_s P_s(t). \end{aligned} \quad (14)$$

Again we can simplify this by noting that the λ_s and μ_s curves are only slowly curving and hence in any local region are approximately straight. Hence, where derivatives are now evaluated near the point $s = M(t)$,

$$\lambda_s = \lambda_{M(t)} + [s-M(t)]\frac{d\lambda}{ds}$$
$$\mu_s = \mu_{M(t)} + [s-M(t)]\frac{d\mu}{ds}. \quad (15)$$

Substituting (15) into (14) we get

$$\begin{aligned}\frac{d\,\mathrm{var}(t)}{dt} &= 2(\lambda_{M(t)} - \mu_{M(t)}) \sum_{s=0}^{\infty} (s-M(t))P_s(t) \\ &\quad + 2\left(\frac{d\lambda}{ds} - \frac{d\mu}{ds}\right) \sum_{s=0}^{\infty} (s-M(t))^2 P_s(t) \\ &\quad + [\lambda_{m(t)} + \mu_{M(t)}] \sum_{s=0}^{\infty} P_s(t) \\ &\quad + \left(\frac{d\lambda}{ds} + \frac{d\mu}{ds}\right) \sum_{s=0}^{\infty} (s-M(t))P_s(t),\end{aligned}$$

which, since $\sum_{s=0}^{\infty} P_s(t) = 1$ and

$\sum (s - M(t)) P_s(t) = M(t) - M(t) = 0$,

becomes,

$$\frac{d\,\mathrm{var}(t)}{dt} = -2\left(\frac{d\mu}{ds} - \frac{d\lambda}{ds}\right)\mathrm{var}(t) + \lambda_{M(t)} + \mu_{M(t)} \, . \qquad (16)$$

This is readily solved for var(t):

$$\mathrm{var}(t) = e^{-2[(d\mu/ds)-(d\lambda/ds)]t} \times \int_0^t (\lambda_{M(t)} + \mu_{M(t)}) e^{2[(d\mu/ds)-(d\lambda/ds)]t}\, dt \, . \qquad (16a)$$

However, it is more instructive to compare mean and variance for the extreme situations of saturation and complete unsaturation, or equivalently of $t =$ near ∞ and t = near zero.

At equilibrium, $\frac{d\,\mathrm{var}(t)}{dt} = 0$, so by (16)

$$\mathrm{var}(t) = \frac{\lambda_{\hat{s}} + \mu_{\hat{s}}}{2\left(\frac{d\mu}{ds} - \frac{d\lambda}{ds}\right)} . \qquad (17)$$

At equilibrium $\lambda_{\hat{s}} = \mu_{\hat{s}} = x$ say and we have already symbolized the difference of the derivatives at $s = \hat{s}$ by F (cf. eq. [13a]). Hence, at equilibrium

$$\mathrm{var} = \frac{X}{F} . \qquad (17a)$$

Now since μ_s has non-decreasing slope $X/s \leqslant d\mu/ds\,|_{s=\hat{s}}$ or $X \leqslant \hat{s}\, d\mu/ds\,|_{\hat{s}}$.

Therefore, variance $\leqslant \frac{\hat{s}\, d\mu/ds}{d\mu/ds - d\lambda/ds}$ or, at equilibrium

$$\frac{\mathrm{variance}}{\mathrm{mean}} \leqslant \frac{d\mu/ds}{d\mu/ds - d\lambda/ds} . \qquad (18)$$

In particular, if the extinction and immigration curves have slopes about equal in absolute value, (variance/mean) $\leqslant$ ½. On the other hand, when t is near zero, equation (16) shows that var$(t) \sim \lambda_0 t$. Similarly, when t is near zero, equations (13) or (14) show that $M(t) \sim \lambda_0 t$. Hence, in a very unsaturated situation, approximately,

$$\frac{\mathrm{variance}}{\mathrm{mean}} = 1 . \qquad (19)$$

Therefore, we would expect the variance/mean to rise from somewhere around ½ to 1, as we proceed from saturated islands to extremely unsaturated islands farthest from the source of colonization.

Finally, if the number of species dying out per year, X (at equilibrium), is known, we can estimate the time required to 90% saturation from equations (13a) and (17a):

$$\frac{2.303}{t} = \frac{X}{\mathrm{variance}}$$

$$t = \frac{2.303\ \mathrm{variance}}{X} \doteq \frac{2.303}{2}\,\frac{\mathrm{mean}}{X} . \qquad (19a)$$

The above model was developed independently from an equilibrium hypothesis just published by Preston (1962). After providing massive documentation of the subject that will be of valuable assistance to future biogeographers, Preston draws the following particular conclusion about continental versus insular biotas: "[The depauperate insular biotas] are not depauperate in any absolute sense. They have the correct number of species for their area, provided that each area is an isolate, but they have far fewer than do equal areas on a mainland, because a mainland area is merely a 'sample' and hence is greatly enriched in the Species/Individuals ratio." To illustrate, "in a sample, such as the breeding birds of a hundred acres, we get many species represented by a single pair. Such species would be marked for extinction with one or two seasons' failure of their nests were it not for the fact that such local extirpation can be made good from outside the 'quadrat,' which is not the case with the isolate." This point of view agrees with our own. However, the author apparently missed the precise distance effect and his model is consequently not predictive in the direction we are attempting. His model is, however, more accurate in its account of

TABLE 1. *Number of species of land and freshwater birds on Krakatau and Verlaten during three collection periods together with losses in the two intervals (from Dammerman, 1948)*

	1908			1919–1921			1932–1934			Number "lost"	
	Non-migrant	Migrant	Total	Non-migrant	Migrant	Total	Non-migrant	Migrant	Total	1908 to 1919–1921	1919–1921 to 1932–1934
Krakatau	13	0	13	27	4	31	27	3	30	2	5
Verlaten	1	0	1	27	2	29	29	5	34	0	2

relative abundance, corresponding to our equation (10).

THE CASE OF THE KRAKATAU FAUNAS

The data on the growth of the bird faunas of the Krakatau Islands, summarized by Dammerman (1948), provide a rare opportunity to test the foregoing model of the immigration and extinction process on a single island. As is well known, the island of Krakatau proper exploded in August, 1883, after a three-month period of repeated eruptions. Half of Krakatau disappeared entirely and the remainder, together with the neighboring islands of Verlaten and Lang, was buried beneath a layer of glowing hot pumice and ash from 30 to 60 meters thick. Almost certainly the entire flora and fauna were destroyed. The repopulation proceeded rapidly thereafter. Collections and sight records of birds, made mostly in 1908, 1919–1921, and 1932–1934, show that the number of species of land and freshwater birds on both Krakatau and Verlaten climbed rapidly between 1908 and 1919–1921 and did not alter significantly by 1932–1934 (see table 1). Further, the number of non-migrant land and freshwater species on both islands in 1919–1921 and 1932–1934, i.e., 27–29, fall very close to the extrapolated fauna–area curve of our fig. 1. Both lines of evidence suggest that the Krakatau faunas had approached equilibrium within only 25 to 36 years after the explosion.

Depending on the exact form of the immigration and extinction curves (see fig. 4), the ratio of variance to mean of numbers of species on similar islands at or near saturation can be expected to vary between about ¼ and ¾. If the slopes of the two curves are equal at the point of intersection, the ratio would be near ½. Then the variance of faunas of Krakatau-like islands (same area and isolation) can be expected to fall between 7 and 21 species. Applying this estimate to equation (19a) and taking t (the time required to reach 90% of the equilibrium number) as 30 years, X, the annual extinction rate, is estimated to lie between 0.5 and 1.6 species per year.

This estimate of annual extinction rate (and hence of the acquisition rate) in an equilibrium fauna is surprisingly high; it is of the magnitude of 2 to 6% of the standing fauna. Yet it seems to be supported by the collection data. On Krakatau proper, 5 non-migrant land and freshwater species recorded in 1919–1921 were not recorded in 1932–1934, but 5 other species were recorded for the first time in 1932–1934. On Verlaten 2 species were "lost" and 4 were "gained." This balance sheet cannot easily be dismissed as an artifact of collecting technique. Dammerman notes that during this period, "The most remarkable thing is that now for the first time true fly catchers, *Muscicapidae,* appeared on the islands, and that there were no less than four species: *Cyornis rufigastra, Gerygone modigliani, Alseonax latirostris* and *Zanthopygia narcissina.* The two last species are migratory and were therefore only accidental visitors, but the sudden appearance of the *Cyornis* species in great numbers is noteworthy. These birds, first observed in May 1929, had already colonized three islands and may now be called common there. Moreover the *Gerygone,* unmistakable from his gentle note and common along the coast and in the mangrove forest, is certainly a new acquisition." Extinctions are less susceptible of proof but the following evidence is suggestive. "On the other hand two species mentioned by Jacobson (1908) were not found in 1921 and have not been observed since, namely the small kingfisher *Alcedo coerulescens* and the familiar bulbul *Pycnonotus aurigaster*." Between 1919–1921 and 1932–1934 the conspicuous *Demiegretta s. sacra* and *Accipter* sp. were "lost," although these species may not have been truly established as breeding populations. But "the well-known grey-backed shrike (*Lanius schach bentet*), a bird conspicuous in the open field, recorded in 1908 and found breeding in 1919, was not seen in 1933. Whether the species had really completely disappeared or only diminished so much in numbers that it was not noticed, the future must show." Future research on the Krakatau fauna would indeed be of great interest, in view of the very dynamic equilibrium suggested by the model we have presented. If the "losses" in the data represent true extinctions, the rate of extinction would be 0.2 to 0.4 species per year, closely approaching the predicted rate of 0.5 to 1.6. This must be regarded as a minimum figure, since it is likely that species could easily be lost and regained all in one 12-year period.

Such might be the situation in the early history of the equilibrium fauna. It is not possible to predict whether the rate of turnover would change through time. As other taxa reached saturation and more species of birds had a chance at colonization, it is conceivable that more "harmonic" species systems would accumulate within which the turnover rate would decline.

Prediction of a "Radiation Zone"

On islands holding equilibrium faunas, the ratio of the number of species arriving from other islands in the same archipelago (G in equation no. 2) to the number arriving from outside the archipelago (M in no. 2) can be expected to increase with distance from the major extra-archipelagic source area. Where the archipelagoes are of approximately similar area and configuration, G/M should increase in an orderly fashion with distance. Note that G provides the best available measure of what is loosely referred to in the literature as adaptive radiation. Specifically, adaptive radiation takes place as species are generated within archipelagoes, disperse between islands, and, most importantly, accumulate on individual islands to form diversified associations of sympatric species. In equilibrium faunas, then, the following prediction is possible: adaptive radiation, measured by G/M, will increase with distance from the major source region and after corrections for area and climate, reach a maximum on archipelagoes and large islands located in a circular zone close to the outermost range of the taxon. This might be referred to as the "radiation zone" of taxa with equilibrium faunas. Many examples possibly conforming to such a rule can be cited: the birds of Hawaii and the Galápagos, the murid rodents of Luzon, the cyprinid fish of Mindanao, the frogs of the Seychelles, the gekkonid lizards of New Caledonia, the Drosophilidae of Hawaii, the ants of Fiji and New Caledonia, and many others (see especially in Darlington, 1957; and Zimmerman, 1948). But there are conspicuous exceptions: the frogs just reach New Zealand but have not radiated there; the same is true of the insectivores of the Greater Antilles, the terrestrial mammals of the Solomons, the snakes of Fiji, and the lizards of Fiji and Samoa. To say that the latter taxa have only recently reached the islands in question, or that they are not in equilibrium, would be a premature if not facile explanation. But it is worth considering as a working hypothesis.

Estimating the Mean Dispersal Distance

A possible application of the equilibrium model in the indirect estimation of the mean dispersal distance, or λ in equation

(3). Note that if similar parameters of dispersal occur within archipelagoes as well as between them,

$$\frac{G}{M} = \frac{A_1 \,\text{diam}_1\, d_2}{A_2 \,\text{diam}_2\, d_1} e^{\lambda(d_2 - d_1)}, \qquad (20)$$

and

$$\lambda = \ln \frac{A_2 \,\text{diam}_2\, d_1 G}{A_1 \,\text{diam}_1\, d_2 M} \Big/ (d_2 - d_1)\,, \qquad (21)$$

where, in a simple case, A_1, diam_1, and d_1 refer to the relation between the recipient island and some single major source island within the same archipelago; and A_2, diam_2, and d_2 refer to the relation between the recipient island and the major source region outside the archipelago.

Consider the case of the Geospizinae of the Galápagos. On the assumption that a single stock colonized the Galápagos (Lack, 1947), G/M for each island can be taken as equal to G, or the number of geospizine species. In particular, the peripherally located Chatham Island, with seven species, is worth evaluating. South America is the source of M and Indefatigable Island can probably be regarded as the principal source of G for Chatham. Given G/M as seven and assuming that the Geospizinae are in equilibrium, λ for the Geospizinae can be calculated from (21) as 0.018 mile. For birds as a whole, where G/M is approximately unity, λ is about 0.014 mile.

But there are at least three major sources of error in making an estimate in this way:

1. Whereas M is based from the start on propagules from an equilibrium fauna in South America, G increased gradually in the early history of the Galápagos through speciation of the Geospizinae on islands other than Chatham. Hence, G/M on Chatham is actually higher than the ratio of species drawn from the Galápagos to those drawn from outside the archipelago, which is our only way of computing G/M directly. Since λ increases with G/M, the estimates of λ given would be too low, if all other parameters were correct.

2. Most species of birds probably do not disperse according to a simple exponential holding-time distribution. Rather, they probably fly a single direction for considerable periods of time and cease flying at distances that can be approximated by the normal distribution. For this reason also, λ as estimated above would probably be too low.

3. We are using $\hat{S}_G/\hat{S}_M$ for G/M, which is only approximate.

These considerations lead us to believe that 0.01 mile can safely be set as the lower limit of λ for birds leaving the eastern South American coast. Using equation no. 12 in another case, we have attempted to calculate λ for birds moving through the Lesser Sunda chain of Indonesia. The Alor group was chosen as being conveniently located for the analysis, with Flores regarded as the principal source of western species and Timor as the principal source of eastern species. From the data of Mayr (1944) on the relationships of the Alor fauna, and assuming arbitrarily an exponential holding-time dispersal, λ can be calculated as approximately 0.3 mile. In this case the first source of error mentioned above with reference to the Galápagos fauna is removed but the second remains. Hence, the estimate is still probably a lower limit.

Of course these estimates are in themselves neither very surprising nor otherwise illuminating. We cite them primarily to show the possibilities of using zoogeographic data to set boundary conditions on population ecological phenomena that would otherwise be very difficult to assess.

Finally, while we believe the evidence favors the hypothesis that Indo-Australian insular bird faunas are at or near equilibrium, we do not intend to extend this conclusion carelessly to other taxa or even other bird faunas. Our purpose has been to deal with general equilibrium criteria, which might be applied to other faunas, together with some of the biological implications of the equilibrium condition.

Summary

A graphical equilibrium model, balancing immigration and extinction rates of species, has been developed which appears fully consistent with the fauna–area curves and the distance effect seen in land and freshwater bird faunas of the Indo-Australian islands. The establishment of the equilibrium condition allows the development of a more precise zoogeographic theory than hitherto possible.

One new and non-obvious prediction can be made from the model which is immediately verifiable from existing data, that the number of species increases with area more rapidly on far islands than on near ones. Similarly, the number of species on large islands decreases with distance faster than does the number of species on small islands.

As groups of islands pass from the unsaturated to saturated conditions, the variance-to-mean ratio should change from unity to about one-half. When the faunal buildup reaches 90% of the equilibrium number, the extinction rate in species/year should equal 2.303 times the variance divided by the time (in years) required to reach the 90% level. The implications of this relation are discussed with reference to the Krakatau faunas, where the buildup rate is known.

A "radiation zone," in which the rate of intra-archipelagic exchange of autochthonous species approaches or exceeds extra-archipelagic immigration toward the outer limits of the taxon's range, is predicted as still another consequence of the equilibrium condition. This condition seems to be fulfilled by conventional information but cannot be rigorously tested with the existing data.

Where faunas are at or near equilibrium, it should be possible to devise indirect estimates of the actual immigration and extinction rates, as well as of the times required to reach equilibrium. It should also be possible to estimate the mean dispersal distance of propagules overseas from the zoogeographic data. Mathematical models have been constructed to these ends and certain applications suggested.

The main purpose of the paper is to express the criteria and implications of the equilibrium condition, without extending them for the present beyond the Indo-Australian bird faunas.

Acknowledgments

We are grateful to Dr. W. H. Bossert, Prof. P. J. Darlington, Prof. E. Mayr, and Prof. G. G. Simpson for material aid and advice during the course of the study. Special acknowledgment must be made to the published works of K. W. Dammerman, E. Mayr, B. Rensch, and E. Stresemann, whose remarkably thorough faunistic data provided both the initial stimulus and the principal working material of our analysis. The work was supported by NSF Grant G-11575.

Literature Cited

Dammerman, K. W. 1948. The fauna of Krakatau 1883–1933. Verh. Kon. Ned. Akad. Wet. (Nat.), (2) **44**: 1–594.

Darlington, P. J. 1957. Zoogeography. The geographical distribution of animals. Wiley.

Delacour, J., and E. Mayr. 1946. Birds of the Philippines. Macmillan.

Feller, W. 1958. An introduction to probability theory and its applications. Vol. 1, 2nd ed. Wiley.

Greenway, J. G. 1958. Extinct and vanishing birds of the world. Amer. Comm. International Wild Life Protection, Special Publ. No. 13.

Lack, D. 1947. Darwin's finches, an essay on the general biological theory of evolution. Cambridge University Press.

MacArthur, R. H. 1957. On the relative abundance of bird species. Proc. Nat. Acad. Sci. [U. S.], **43**: 293–294.

Mayr, E. 1933. Die Vogelwelt Polynesiens. Mitt. Zool. Mus. Berlin, **19**: 306–323.

——. 1940. The origin and history of the bird fauna of Polynesia. Proc. Sixth Pacific Sci. Congr., **4**: 197–216.

——. 1943. The zoogeographic position of the Hawaiian Islands. Condor, **45**: 45–48.

——. 1944. Wallace's Line in the light of recent zoogeographic studies. Quart. Rev. Biol., **19**: 1–14.

Preston, F. W. 1962. The canonical distribution of commonness and rarity: Parts I, II. Ecology, **43**: 185–215, 410–432.

Rensch, B. 1936. Die Geschischte des Sundabogens. Borntraeger, Berlin.

STRESEMANN, E. 1934. "Aves." *In* Handb. Zool., W. Kukenthal, ed. Gruyter, Berlin.
———. 1939. Die Vögel von Celebes. J. für Ornithologie, **87**: 299–425.
WILSON, E. O. 1961. The nature of the taxon cycle in the Melanesian ant fauna. Amer. Nat., **95**: 169–193.
ZIMMERMAN, E. C. 1948. Insects of Hawaii. Vol. 1. Introduction. University of Hawaii Press.

APPENDIX: MEASUREMENT OF A PROPAGULE

A rudimentary account of how many immigrants are required to constitute a propagule may be constructed as follows. Let η be the average number of individuals next generation per individual this generation. Thus, for instance, if $\eta = 1.03$, the population is increasing at 3% interest rate.

Let us now suppose that the number of descendants per individual has a Poisson distribution. If it has not, due to small birth rate, the figures do not change appreciably. Then, due to chance alone, the population descended from immigrants may vanish. This subject is well known in probability theory as "Extinction probabilities in branching processes" (cf. Feller 1958, p. 274). The usual equation for the probability ζ of eventual extinction (Feller's equation 5.2 with $P(\zeta) = e^{-\eta(1-\zeta)}$, for a Poisson distribution), gives

$$\zeta = e^{-\eta(1-\zeta)}.$$

Solving this by trial and error for the probability of eventual extinction ζ, given a variety of values of η, we get the array shown in table 2. From this we can calculate how large a number of simultaneous immigrants would stand probability just one-half of becoming extinct during the initial stages of population growth following the introduction. In fact, if r pairs immigrate simultaneously, the probability that all will eventually be without descendants is ζ^r. Solving $\zeta^R = 0.5$ we find the number, R, of pairs of immigrants necessary to stand half a chance of not becoming extinct as given in table 3. From this it is clear that when η is 1, the propagule has infinite size, but that as η increases, the propagule size decreases rapidly, until, for a species which increases at 38.5% interest rate, one pair is sufficient to stand probability 1.2 of effecting a colonization. With sexual species which hunt for mates, η may be very nearly 1 initially.

TABLE 2. *Relation of replacement rate (η) of immigrants to probability of extinction (ζ)*

η	1	1.01	1.1	1.385
ζ	1	0.98	0.825	0.5

TABLE 3. *Relation of replacement rate (η) to the number of pairs (R) of immigrants required to give the population a 50% chance of survival*

η	1	1.01	1.1	1.385
R	∞	34	3.6	1

1965

"A consistency test for phylogenies based on contemporaneous species," *Systematic Zoology* 14(3): 214–220

In the early 1960s it was widely believed that the only way to reconstruct phylogenies (evolutionary family trees) was by phenetics, the overall similarity of species in many characteristics measured and correlated together. In the study reported here, I showed that another, possibly more rigorous approach is by what is today called cladistics: the character-by-character analysis of species to deduce the pattern of their ancestry and descent. I was unaware at the time of the publication, as were all but a few other systematists, of the much more extensive writings of Willi Hennig in Germany on the same concept. When Hennig's work became more widely know, and key points available in English, it started the revolution in cladistic analysis that is today the foundation of phylogenetic reconstruction. Because I did not try to follow up this train of thought, and in any case likely could not have matched Hennig, I include this piece as a humbling example of failed science.

214 SYSTEMATIC ZOOLOGY

A Consistency Test for Phylogenies Based on Contemporaneous Species

EDWARD O. WILSON

It is commonly stated that phylogenies deduced from data about contemporaneous species cannot be "proved" because, obviously, evolution is a past event recoverable only from fossils. This is true to the extent that no proof exists which has the decisiveness of a witnessed event or the consistency of a physical measurement. Yet scientific proof is rarely direct and is always relative in degree. Evolutionary hypotheses might never be definitive by the standards of experimental biology, but they are valid if they are both falsifiable and heuristic. That is, to be valid they should make concrete predictions that are capable of being negated if the hypothesis is false; and they should point the way to deeper, more meaningful investigations if they are momentarily upheld. Phylogenetic taxonomy has been open to criticism not so much for indirection as for its lack of techniques of formal analysis that render its hypotheses falsifiable and heuristic.

One such procedure that might be employed involves the "weighting" of characters with reference to their phylogenetic significance. Taxonomists intuitively select character states which they postulate to define monophyletic sets of species. The ideal character contains some state that both uniquely defines a set of species and has not been reversed in evolution, so that all existing species which possess this state can be said to have descended from one species in the past that evolved the state. For every such character state that can be identified, a branch in the phylogenetic tree can be added. This extreme form of phylogenetic hypothesis, then, is initiated as a hypothesis about unique, unreversed characters. The formulation perhaps cannot be decisively proven on the basis of contemporaneous species. But can it be disproven? And is anything of biological significance to be gained by the procedure? The following test hopefully gives an affirmative answer to both questions. It is not original in the sense that it offers something very new to taxonomic thinking. Instead, its purpose is to express one common intuitive taxonomic procedure in a new, more rigorous form.

Definitions. Consider a series of m unique, unreversed character states a_1, a_2, a_3, . . . , a_m *each representing a different character*, as yellow dewlap and flattened tail can be said to be states of two separate characters (dewlap color and tail shape) in lizard species. These particular states are interpreted to have appeared during a speciation episode that has resulted in a monophyletic taxon of n contemporaneous species. They now exist in any combination in various of the n species. At one extreme, they may be totally lacking in a given species; at the other extreme, all m character states may occur together in a given species. By *unique* is meant that a given character state a_i appeared in the past only once and in one species. It now exists in one or more descendant species. By *unreversed* is meant that the state has never been lost, i.e., has never reverted to a prior state, in any of the species giving rise to the contemporaneous taxon. The character state itself might have arisen *de novo* as a new structure, it might have appeared as a new state in a series of discrete character states, or it might be arbitrarily recognized as some point and beyond in a continuous morphocline. The taxon therefore is to be treated as a sample space whose points are contemporaneous species; and the character states are events that can occur on the points in the sample space. It is desirable that the character states in the hypothesis be chosen initially for considerations not having to do with the way they jointly define sets of species in the taxon. The properties that can be expected to induce the choice include uniqueness with reference to other taxa, structural complexity, and absence of other states that are clearly annectant or derivative and degenerate in nature.

The hypothesis. The m character states are unique and unreversed.

Testing the hypothesis.[1] Let us label the states such that the possession of a_1 completely defines a set of contemporaneous species A_1, a_2 defines a smaller or equal set A_2, a_3 defines a still smaller or equal set A_3, and so on. Three possible alternative outcomes can now be simply stated (Fig. 1).

I. If the sets of species defined by the character states are non-overlapping, i.e., $A_1 \cap A_2 \cap A_3 \cap \ldots \cap A_m = 0$, the hypothesis cannot be tested.

II. If the sets are overlapping but do not enclose each other (form a series of proper subsets), in the order $a_1, a_2, a_3, \ldots, a_m$, the hypothesis is rejected.

III. If the sets are overlapping and A_2 is wholly enclosed in (is a proper subset of) A_1 (and A_3 is enclosed in A_2, and so on) the arrangement is consistent with the hypothesis but does not definitely prove it. If Situation III holds, the following phylogenetic hypothesis is also consistent: $A_o \cap \tilde{A}_1$, $A_1 \cap \tilde{A}_2, \ldots, A_{m-1} \cap \tilde{A}_m$, A_m are the contemporaneous branches of a phylogenetic tree of the kind illustrated in Figure 2.

Examples of reasonably long sequences of character states that pass the consistency test are probably familiar to most taxonomists. Two such sequences from the ants

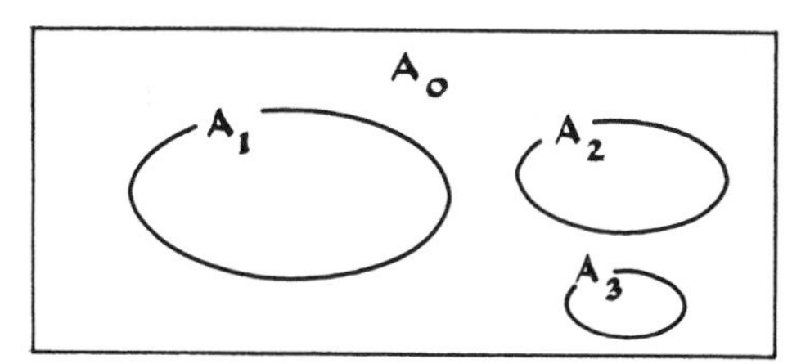

I. Test not applicable

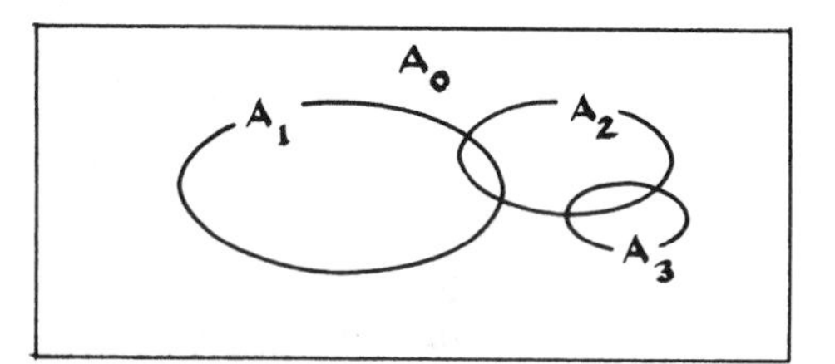

II. Test failed, hypothesis rejected

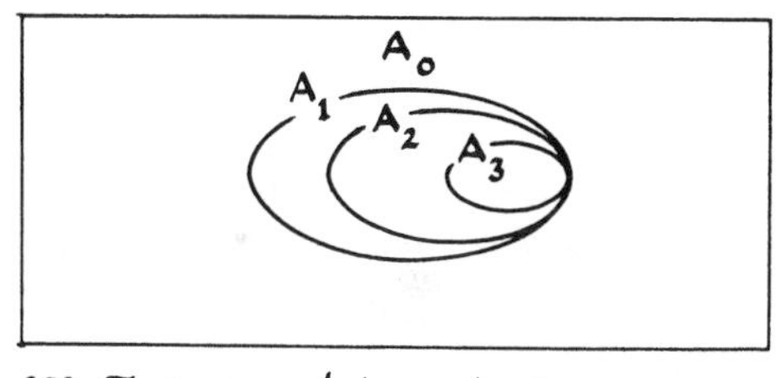

III. Test passed, hypothesis not rejected

FIG. 1. The consistency test. The rectangle encloses the taxon under consideration. It must be reasonably discrete in many characters from all other taxa. Each ellipse encloses a set of species A_i characterized by a character state a_i hypothesized to be unique and unreversed. Only one state per character is considered. A_o is the set of species not bearing any character state hypothecated to be unique and unreversed.

(family Formicidae) are given in Table 1.

Suppose that the m character states are mutually consistent, as in Situation III. Although it is not possible from this fact alone to prove the phylogenetic hypothesis of Figure 2, we might still be able to narrow

[1] The following conventions from set theory are used: A_1 symbolizes the set of all species that possess character state a_1. $\tilde{A}_1$ symbolizes the set of all species that do *not* possess character state a_1. $A_1 \cap A_2$ indicates those species that are in both A_1 and A_2, i.e., possess both character states a_1 and a_2. $A_1 \supset A_2$ means that A_2 is contained wholly within A_1, i.e., all species that possess a_2 also possess a_1.

the permissible alternative explanations somewhat. First, consider the model which is the opposite of the one under consideration, namely that the m states have appeared and disappeared in a random manner with reference to each other during the evolution of the taxon. This hypothesis can be tested in the following manner. Imagine the circumstance, among all possible circumstances, in which there would exist the highest probability of the nested pattern arising by chance combinations alone. This is the simple case illustrated in Figure 3. There are $m + 1$ species evolving separately during the time that the characters are fixed (at random with reference to each other) to produce the nested pattern. Given that at the time m of the species acquired a_1, $m - 1$ acquired a_2, $m - 2$ acquired a_3, *et seq.*, with a single species acquiring a_m, the probability that the resulting sets A_i could be nested by chance alone is

$$P(A_1 \supset A_2 \supset \ldots \supset A_m) = \frac{[m!2!][(m-1)!3!]\ldots[3!(m-1)!][2!m!]}{[(m+1)!]^{m-1}}.$$

All other situations in which the character states are fixed independently to give a nest of sets are equally or less probable. In other words, the equation above, based on the random model illustrated in Figure 3, gives an upper limit for the probability that the m character states were evolved randomly with reference to each other. Applying it for various values of m we find that $P(A_1 \supset A_2 \supset \ldots \supset A_m)$ is 6/125 for four character states, 1/225 for five character states, 16/84,035 for six character states, and 9/153,664 for seven character states. In order for this explicit formulation to be valid it is necessary that the character states be chosen initially without reference to the kind of classification they would engender in the taxon. In practice such selection would come about in the first study of a group of species, before the distribution of various character states with reference to each other are considered.

In sum, if four or more character states hypothecated to be unique and unreversed then pass the consistency test, we are reasonably justified in considering them correlated in some historical manner, regardless of the pathways of speciation taken by the taxon in the past. With much more confidence, this rule can be based on five or more characters.

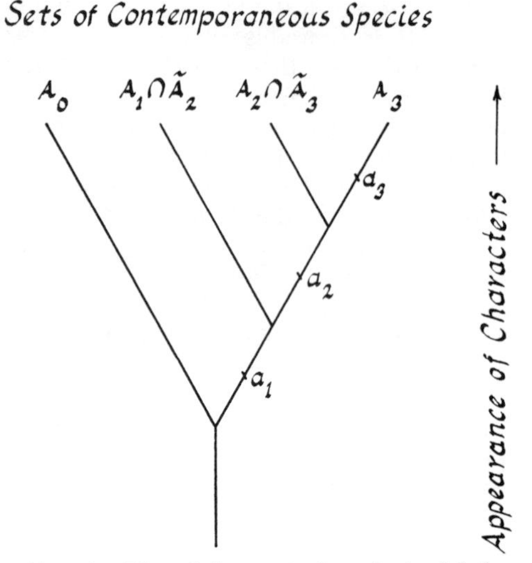

FIG. 2. The phylogenetic hypothesis (cladogram) that is permissible if the character states a_1, a_2, and a_3 pass the consistency test. Each character state represents a *different* character. A_0, $A_1 \cap \tilde{A}_2$, $A_2 \cap \tilde{A}_3$, and A_3 represent sets of contemporaneous species. The nodes labeled a_1, a_2, and a_3 mark the appearance of these character states in time. The ends of the branches are arbitrarily arranged along equal intervals because the consistency test by itself gives no information about over-all similarity of the sets of species.

Suppose the consistency test is thus passed with reasonable confidence. There are five alternative ways in which the character states could be correlated:

1) The m character states were fixed at random. Later, there was differential survival among the species according to their respective combinations of the m states, resulting in the modern consistent pattern. Unless we also postulate genetic drift, this explanation subsumes that the ways that the m states are combined are at one time selectively neutral and later selectively significant. The probability certainly exists

TABLE 1—SEQUENCES OF INTERCONSISTENT CHARACTER STATES IN THE FORMICIDAE.

	Character State	Group Defined within the Taxon
Series No. 1 (Taxon = Aculeate Hymenoptera)	Metapleural gland	Family Formicidae
	Pulvinate poison gland	Subfamily Formicinae
	Sepalous proventriculus	"Section Euformicinae"
	Dense, appressed pilosity in discrete soldier caste	Subgenus *Machaeromyrma* of *Cataglyphis*
Series No. 2 (Taxon = Genus *Lasius*)	"*Niger*-type" male mandible	*Lasius* exclusive of Subgenus *Chthonolasius*
	Metapleural guard hairs lost in female castes	Subgenus *Dendrolasius*
	β-form queen	*L. teranishii* and *L. spathepus*
	Appendages covered with long, silvery pilosity	*L. spathepus*

but seems intuitively relatively small. *Or*,

2) A superordinate character state, e.g., a_1 with reference to a_2, always or with very high frequency appears soon after the subordinate appears; but it also originates in a certain fraction of the species without the subordinate character state. This possibility seems even more remote than (1). *Or*,

3) A subordinate state, e.g., a_2 with reference to a_1, occurs only after the superordinate state is present. But it can still be nonunique and reversible within the species bearing the superordinate state. For example, a_2 could still appear and disappear many times over in species bearing a_1. This is perhaps more likely than (1) and (2); however, if the subordinate character states really could appear and disappear in multiple fashion within the set of species bearing superordinate character states, it would be necessary for each of the subordinate states to have changed in concert to preserve the consistency observed in the contemporaneous taxon. *Or*,

4) The character states could first have appeared together and then been lost in concert to produce the precise pattern. *Or*,

5) The states are unique and unreversed within the taxon. This seems the most likely hypothesis. It is certainly the simplest.

Heuristic value. Consistent phylogenetic schemes, even when based entirely on contemporaneous species, are useful for two reasons: they serve to confirm the identity of the most unusual and stable character states, and they make exact predictions about state combinations in the species yet to be discovered. While remaining explicitly vulnerable, they are a valuable scientific procedure, comprising that part of taxonomic research which has the greatest general interest. This positive aspect of phylogenetic analysis holds whether or not enough characters pass the consistency test to allow the random hypothesis to be rejected. It also holds whether or not the phylogeny deduced is correct in detail and regardless of its effect on formal classification.

An example illustrating the heuristic value of cladistic analysis can be taken from my recent revision of the ant genus *Aenictus* (Wilson, 1964). Two character states, the "Typhlatta spots" of the head and presence of teeth on the anterior margin of the clypeus, were among those initially guessed to be unique and unreversed, but they proved not to be interconsistent. In particular, *Aenictus currax* and *A. huonicus* possessed Typhlatta spots but appeared to lack clypeal teeth. Since these two species are very similar in all characters studied, they were inferred to be closely related. Also, since they are both endemics of New Guinea,

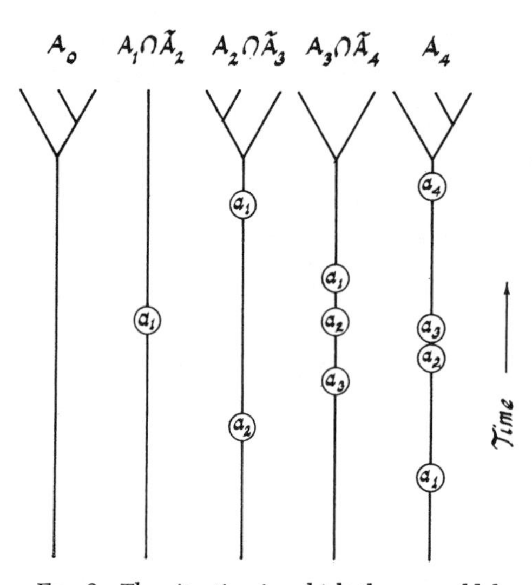

FIG. 3. The situation in which there would be the highest probability of m character states (in this case $m = 4$) passing the consistency test while evolving in a random manner. There are $m + 1$ species during the appearance of the states. The placement of the numbers indicates the appearance of the states in time. They are scattered arbitrarily in this diagram to suggest the condition of randomness. All other situations would give equal or higher probabilities. After the states appear the $m + 1$ species may or may not speciate further to produce the contemporaneous taxon, as exemplified by the irregular branching near the ends of the phyletic lines.

which lies on the periphery of the range of the species group to which they belong, it was guessed that any deviant character states shown by them would be more likely to be derived than original. This second deduction was based on a rule shown by Indo-Australian ant species generally. A closer, second examination of the *Aenictus* species resulted in support for the hypothesis: workers of *A. currax* were found to have hidden, rudimentary teeth. As a consequence, it was concluded that clypeal teeth have been lost secondarily in *A. huonicus*. It is my impression that similar logical sequences are often, even routinely followed in taxonomic revisions. Taxonomists seldom spell their procedures out, however, as I have done in the *Aenictus* revision.

Relation to classification. Formal classifications need not be isomorphic with phylogenetic schemes that are simply cladistic in nature. The sets of species $A_i \cap \tilde{A}_{i+1}$ defined by characters that continue to pass the consistency test may or may not be recognized as taxa. It is conceivable, for example, that a species in A_2 differs from one in $A_1 \cap \tilde{A}_2$ only by the character state a_2 but is different from other species in A_2 by many other characters. In this case it would be valid taxonomic procedure either to lump $A_1 \cap \tilde{A}_2$ and A_2 or to combine the one species from A_2 with $A_1 \cap \tilde{A}_2$; and it would be dubious procedure to split $A_1 \cap \tilde{A}_2$ and A_2 as taxa. This conclusion has been reached by members of both the phylogenetic (Simpson, 1961) and numerical schools (Sokal and Sneath, 1963).

Even so, the present study together with independent and parallel attempts to formalize cladistic analysis (e.g., the articles by Sokal and Camin and by Throckmorton in this issue of *Systematic Zoology*) indicate that we can hope to distinguish with confidence between "constant characters" and "fickle characters." In the aggregate, constant characters reflect phylogenies more accurately than fickle characters, and it would appear that insofar as we wish to transmit evolutionary information in our classifications constant characters should be given greater weight. Such classifications may not be as stable and reproducible as those based on the averaged similarity of unweighted characters, but they have more biological interest. Taxonomy should be more than the blind clustering of taxa according to over-all similarity, as suggested by the "numerical taxonomists." In spite of the attractive simplicity of the latter technique and its undoubted usefulness in special cases, it seems to be of dubious value as a broad philosophy of classification. The main objection is that numerical taxonomy has up to the present offered little hope of yielding new biological information, precisely because it has not been constructed with reference to any real biological questions. Put in another way, taxonomy is a language that can be designed according to any one of many sets of rules. The rules selected should be of maximum heuristic value; beyond that, it is only necessary that they be stated very plainly. While not intending to disparage multivariate statistics or the considerable technical achievements of the numerical taxonomists, I would regard a taxonomy based automatically and *a priori* on unweighted characters as a desirable measure only in cases where phylogenetic hypotheses cannot in any way be tested. Even at its best this procedure should never be accepted as a doctrine. In fact, it seems more likely than ever before that taxonomists will eventually develop standard methods for the combination of phenetic measures and cladistic inferences into truly phylogenetic classifications. To do so would be one of the great achievements of modern evolutionary biology.

Summary

Cladograms of contemporaneous species are most rigorously constructed from a hypothesis which postulates unique, unreversed character states (Fig. 2). Many such phylogenetic schemes, if false, can be quickly discarded by a simple consistency test illustrated in Figure 1. If a set of four or more character states $a_1, a_2, a_3, \ldots, a_m$ found in m different characters in a taxon are selected without reference to the grouping of species within the taxon and then are found to characterize successively smaller sets of species in such a way as to pass the consistency test, it is reasonable to conclude that the character states evolved in the taxon in a non-random manner with respect to each other. The relation of this inference to phylogeny and the heuristic value of phylogenies based solely on contemporaneous species are discussed.

Acknowledgments

I am very grateful to Eli Minkoff and Angelo Serra for critical readings of the manuscript. Several other persons, including A. F. Bartholomay, W. H. Bossert, W. L. Brown, H. E. Evans, R. Inger, E. MacLeod, E. Mayr, G. G. Simpson, R. R. Sokal, and R. W. Taylor, have discussed various aspects of the problem and provided help and encouragement. The consistency test was developed in conjunction with a recent systematic study (Wilson, 1964) of the Indo-Australian doryline ants supported by a grant from the National Science Foundation.

REFERENCES

SIMPSON, G. G. 1961. Principles of animal taxonomy. Columbia University Press, New York, 247 p.

SOKAL, R. R., and P. H. A. SNEATH. 1963. Principles of numerical taxonomy. W. H. Freeman, San Francisco, 359 p.

WILSON, E. O. 1964. The true army ants of the Indo-Australian area (Hymenoptera: Formicidae: Dorylinae). Pacific Insects 6:427–483.

Appendix

The following is a proof of the proposition that the model in which $m \geqslant 3$ character states are fixed in $m+1$ species gives the highest probability, among all possible models, that the m character states could have evolved at random with respect to each other and still have been fixed interconsistently. Let the array of numbers of species in each group A_i vary and any given array be labeled with a number j; in the extreme case $j = 1$, there exists the extreme model just cited in which the number of species in A_0 is $m+1$, the number in A_1 is m, the number in A_2 is $m-1$, and so on. In short, the array $j = 1$ contains the smallest number of species possible. In a second array $j = 2$, A_0 might contain $m+2$ species, A_1 m species, A_2 $m-1$ species, and so on. The probability that a given array occurred as the character states were fixed is p_j and $\sum_j p_j = 1$.

The probability that in an array j the m character states would be fixed in a given pattern with respect to one another can be designated q_{jk}. In particular let us label as $k = \alpha$ the condition in which the character states turn out to be interconsistent. What is desired is the maximum value of $\sum_j p_j q_{j\alpha}$ for all possible arrays j. Now it is intuitively apparent and has been borne out by inspection of many cases (but not formally proved for all possible cases) that where $m \geqslant 3$ the maximum value of $q_{j\alpha}$ is obtained when the array j contains the smallest number of elements, i.e., in the case $j = 1$. The maximum value for $q_{j\alpha}$ is $q_{1\alpha}$, which is a constant when m is chosen. Consider any array j that occurred in evolution. Then given some value of m, $\sum_j p_j q_{j\alpha} \leqslant \sum_j p_j q_{1\alpha}$ for all j and, since $\sum_j p_j q_{1\alpha} = q_{1\alpha} \sum_j p_j$ for the special "alpha case" and $\sum_j p_j = 1$, it follows that $\sum_j p_j q_{j\alpha} \leqslant q_{1\alpha}$.

EDWARD O. WILSON is in the Biological Laboratories, Harvard University, Cambridge, Massachusetts 02138.

1965

"The challenge from related species," *in* H. G. Baker and G. L. Stebbins, eds., *The Genetics of Colonizing Species* (Academic Press, New York), pp. 7–27

This article was intended to address biogeography at two levels. At the first level, it connected my conception of the taxon cycle (1961) to the theory of island biogeography presented two years later by Robert H. MacArthur and myself, as well as to the process of character displacement. The latter I had studied with W. L. Brown, and William H. Bossert had simulated the process with mathematical models. At the second level, it treated the very important but still obscure relation of biogeographic patterns and ecological phenomena with the evolutionary processes that created them. The discussion following the article by several leading evolutionary biologists is especially instructive of the thinking on these and related topics during this period of research.

REPRINTED FROM
THE GENETICS OF COLONIZING SPECIES
© 1965
ACADEMIC PRESS INC., NEW YORK

The Challenge from Related Species

EDWARD O. WILSON
BIOLOGICAL LABORATORIES, HARVARD UNIVERSITY, CAMBRIDGE, MASSACHUSETTS

It is commonly stated that the true test of a theory is its predictive power. By this criterion we must admit to the lack of any theory worthy of the name that treats the subject of the ecology of colonization. The failure is dramatized in the case of biological control in economic entomology, a field where correct prediction is all-important to the success of costly projects. Clausen (1956) reports that by the mid 1950's about 485 species of imported agents had been selected and liberated in the continental United States against 77 pest species. Of these, only 95 became established to give varying degrees of control against 22 of the pest species. On a world-wide basis about one-quarter of such introductions have proven successful (Beirne, 1962). Vertebrate ecology has not done any better. Wodzicki (1950), De Vos *et al.* (1956), Elton (1958), and other recent authors have made it clear that even when the histories of introductions are reviewed with the advantage of hindsight, few useful generalizations are possible.

What is true of this special subject is symptomatic of population ecology generally. There is a temptation to conclude, as many authors have, that the subject will be illuminated as deeper, more detailed studies of individual species are made. But as Watt (1962) has said, "The dead files of ecologists abound with vast sets of unused data which were obtained from studies fundamentally populational in intent." What is certainly also required is better theory. It must relate more directly to empirical information than was the case in the classical formulations of Bailey, Lotka, Thompson, Volterra, and others in the 1920's and 1930's. Yet it must not include more than the design of *ad hoc* models fitted to particular species. Simplifying ideas are demanded that can be heuristic to the study of single cases and yet stand up under relentless generalization.

My purpose in this paper is to discuss some recent contributions to ecology from two related fields that have hitherto been notably lacking in quantitative ideas: zoogeography and speciation theory. Several of my colleagues and I have been concentrating on the interactions of related species during the process of colonization. Our concern has been the laws that express the exchange of species among faunas, the immigration and extinction rates of species, and the genetic changes caused by disoperation among closely related species. We have employed both inductive generalization from zoogeographic data and deductively formulated models.

THE CONCEPT OF FAUNAL DOMINANCE

Within historical times and perhaps earlier, extinction rates of animal species have been higher on islands than on continents. Islands have received more of their species from continents than vice versa, and, where climate does not differ radically, small continents have received a disproportionate share of species in exchanges with larger continents. On the basis of this generalization, evolutionists from Darwin (1859) and Matthew (1915) to Simpson (1953), Darlington (1957, 1959), Brown (1957), and others have speculated on genetic properties engendered within larger, more diversified faunas that are alleged to give species superior competitive ability. A corollary of the concept is the notion of islands as "evolutionary traps," where populations tend to become more homozygous and specialized than comparable populations on the mainland (Mayr, 1942). Darlington (1959) has reasoned as follows:

> "Adaptation involves selection of advantageous mutations, and mutations occur in proportion to number of individuals in populations . . . If, therefore, large area and favorable climate increase the size of populations, that should accelerate general adaptation and the evolution of dominant animals."

More recently, Mayr (1963) has demurred somewhat on the basis that too many exceptional cases are known to make a firm rule. It is apparent that, as is the case with much current evolutionary theory, the idea of a genetic factor in faunal dominance has never been subjected to any serious quantitative testing. Since the concept of faunal dominance applies to the past history of natural populations (as opposed to laboratory populations) the ultimate test must be made on the basis of the statistical treatment of zoogeographic data.

Faunal exchange can be somewhat more rigorously analyzed in the following manner. The simplest biological hypotheses testable by zoogeographic data can be generalized in the equation

$$\frac{s_{1,2}}{s_{2,1}} = l\left(\frac{S_1}{S_2}\right)^m = l\left(\frac{A_1}{A_2}\right)^{\frac{m}{3}}.$$

where A_2 is set arbitrarily as the larger of the two areas, so that $0 < A_1/A_2 \leq 1$, and

$s_{1,2}$ is the number of species at a given moment extending from region 1 to region 2 that originated in region 1.
$s_{2,1}$ is the number of species at the same moment extending from region 2 to region 1 that originated in region 2.
S_1 and S_2 are the number of species in regions 1 and 2, respectively.
A_1 and A_2 are the area in square miles of regions 1 and 2, respectively.
l and m are fitted coefficients.

The relationship between the area of a region in square miles and the number of species living on it has been determined empirically. For most groups of animals studied thus far the relationship can be stated (MacArthur and Wilson, 1963) approximately as

$$S = cA^b$$

where the exponent b, referred to as the "faunal coefficient," is usually in the range 0.3–0.4.* For the purposes of our present study we can safely take the faunal coefficient as equal to $\frac{1}{3}$. For simplicity the reasonable assumption can be made that within a taxon $c_1 = c_2$ for the two regions, so that

$$\frac{S_1}{S_2} = \frac{c_1 A_1^{\frac{1}{3}}}{c_2 A_2^{\frac{1}{3}}} = \left(\frac{A_1}{A_2}\right)^{\frac{1}{3}}.$$

Several alternative conjectures about faunal exchange can now be made and their consequences examined in the empirically determinable ratio $s_{1,2}/s_{2,1}$. In considering them, the reader is asked to keep in mind that they are derived from simple biological hypotheses and do not represent an exhaustive list of empirical curves to which the meager zoogeographic data can be fitted.

Conjecture I. The exchange is based on a random flux of species. The faunal exchange ratio $s_{1,2}/s_{2,1}$ is simply proportional to the ratio of species (S_1/S_2) in the two regions. In other words, $m = 1$ and

$$\frac{s_{1,2}}{s_{2,1}} = l\left(\frac{A_1}{A_2}\right)^{\frac{1}{3}}$$

approximately.

* In the same article I gave the faunal coefficient of Melanesian ants erroneously as 0.5; it is closer to 0.3.

Conjecture II. The exchange is based on a random flux of individual organisms. The faunal exchange ratio $s_{1,2}/s_{2,1}$ is simply equal to the ratio of the areas (A_1/A_2) of the two regions. Here m is equal to the reciprocal of the faunal coefficient, or 3 approximately, and $l = 1$, so that

$$\frac{s_{1,2}}{s_{2,1}} = \frac{A_1}{A_2}.$$

The total number of individuals in a source region is the obvious biological attribute that can be expected to be a rectilinear function of the area. Model II can be applied to numbers of individuals to give it direct biological relevance, but area rather than numbers is the immediately measurable quantity.

Conjecture III. The exchange of species is based on the flux of individuals whose competitive ability increases rectilinearly with the numbers of individuals in the source populations. As in Conjecture II, the exponent is equal to unity, but l is allowed to vary in such a way that the ratio of exchange is altered in a rectilinear fashion.

Conjecture IV. The exchange of species is based on the flux of individuals whose competitive ability increases disproportionately with the numbers of individuals in the source populations. Another conceivable result of a genetic factor intensifying faunal dominance would be $m > 3$, producing a geometric increase in the exchange ratio with increase in area ratio.

Relevant zoogeographic data are very scarce. In an earlier paper (Wilson, 1961, p. 177), I gave data on the distribution of species of Melanesian ponerine ants which can be applied to the models. The exchange ratio of New Guinea to tropical Asia is 0.125, the area ratio is 0.123, and the resident species ratio is 0.543. The difference between $s_{1,2}/s_{2,1}$ and S_1/S_2 is significant in this case at the 95% level (by the Yates-corrected chi-square test) so that Conjecture I has been tentatively eliminated (for this one fauna alone!). The exchange ratios of Fiji–New Guinea, Fiji–Asia, Solomons–New Guinea, and Solomons–Asia are all zero. Therefore Conjecture II cannot be eliminated. At the 99% confidence level, Conjectures III and IV cannot be eliminated except where m is equal to or greater than 6 [in other words $(m/3) \geq 2$]. The simplest equation of faunal exchange satisfying our current limited information can be stated as follows:

$$\frac{s_{1,2}}{s_{2,1}} = \left(\frac{A_1}{A_2}\right)^k, \qquad (\tfrac{1}{3} < k < 2).$$

The extrapolation of these inequalities is shown in Fig. 1. Verbally expressed, the ratio of species exchanged can be said to be a rectilinear or weakly geometric function of the ratio of the areas of the source regions. In the absence of other data, it seems reasonable to take the function as rectilinear for purposes of a first approximation. The results seem to suggest that species from large source regions do not have a higher average intrinsic competitive ability over those from small source regions. In view of the enormous complexity of the situation, the formulation is admittedly weak. It is offered here as a first small

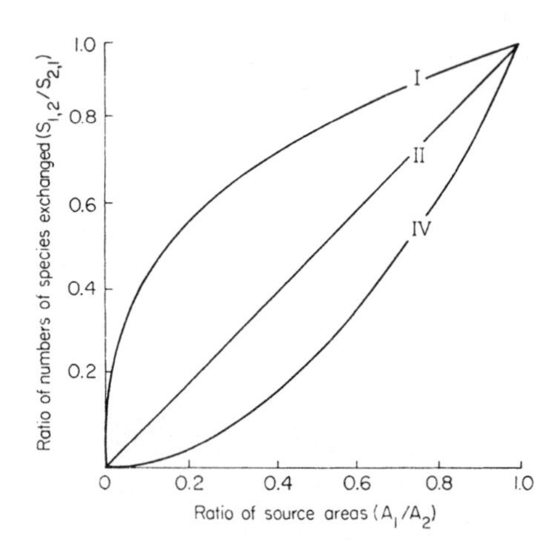

Fig. 1. The rule of faunal dominance. The estimated range within which the curve relating the species-exchange ratio and source-area ratio falls, deduced from the Indo-Australian Ponerinae and Myrmicinae. The curve has a slope lower than that of I (based on Conjecture I) and greater than or equal to that of IV (the lower limit if Conjecture IV with $m = \frac{2}{3}$ is true). The curve based on Conjecture II is also given.

step beyond the practice of qualitative evaluation based on anecdotal evidence.

THE BALANCE OF SPECIES

A crucial factor affecting the success of colonizing species is the degree of saturation of the fauna into which the species is entering. In classical theory, an "impoverished" or "depauperate" fauna of a given taxon offers more opportunity to an invader because of the greater number of "empty niches" thereby made available. Impoverished faunas are understood to be most readily found in remote oceanic islands. Let us examine the meaning of these words more carefully. Like so many reasonable-sounding expressions in biology, they are operationally very hard to define and, in consequence, ambiguously used in the current literature. In a recent paper MacArthur and Wilson (1963) have given reasons for redefining the status of faunas with reference to the equilibrium of species. A summary of some of the primary conditions of the equilibrium condition is given in Fig. 2. This model allows a number of new predictions that can be applied to existing zoogeographic data, e.g., the number of species on islands remote from the principal faunal

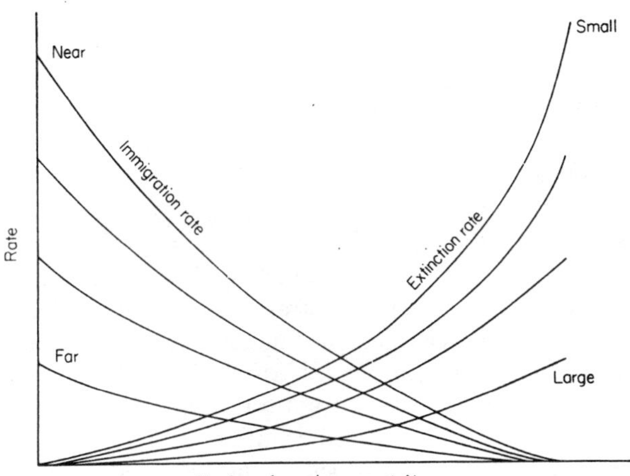

FIG. 2. The attainment of faunal equilibria. In the model, as the number of species increases (moving right on the abscissa), the immigration rate declines because of the decreasing number of available new immigrants, whereas the extinction rate increases because of the decreasing average population size. The number of species at equilibrium is obtained by dropping the intersection point of a given immigration and a given extinction curve to the abscissa. As the immigration rate increases, as on islands nearer to the principal source regions, or the extinction rate decreases, as on larger islands, the equilibrium number increases. Modified from MacArthur and Wilson (1963).

source increases with size more rapidly than does the number of species on near islands. The Pacific bird fauna, which has been faunistically unusually well analyzed, fulfills the equilibrium conditions reasonably well. Other insular faunas have not yet been reconsidered with reference to the model. We expect that its equilibrium criteria can be applied not only to island faunas but also to faunas occupying bodies of water, mountains, and any other isolated or semi-isolated habitats.

By studying Fig. 2, it is possible to reassess the concept of "impoverishment" of an insular fauna. Consider the birds of Hawaii. Before the advent of European settlement, the number of bird species was at or close to equilibrium. In other words, over long periods of time the extinction rate equaled the immigration rate; for every new species that successfully colonized the archipelago, an average of one species became extinct. Perhaps there were very slow secular changes in the total number of species owing to the ever improving adaptation of the fauna as a whole to the Hawaiian environment, but the resulting over-all increment in new species per unit time must have been insignificant in comparison to the constant turnover rate. In this sense, then, the Hawaiian fauna was not "impoverished." With the settlement of Hawaii by Europeans in the nineteenth century, however, and the deliberate introduction of new bird species into the islands by "Acclimatization Societies," the immigration rate was tremendously increased. This had the effect of altering the immigration curve from the equivalent of "far" to the direction of "near" in Fig. 2. If the new inpouring of immigrants were to be held constant, the number of Hawaiian bird species—native plus introduced—would move to a new, much higher equilibrium level. In this special sense, therefore, the old Hawaiian fauna can perhaps be regarded as "impoverished." If, on the other hand, all further importations were strictly forbidden so that the immigration rate returned to the old natural level, the number of species would gradually decline to a third equilibrium not radically different from the pre-European level. In short, the temporary injection of new immigrants does not insure any permanent enrichment of an insular fauna. Only a permanently increased immigration rate can do that.

It is probably true that continents contain equilibrium faunas subject to the same laws as islands; continents are in fact semi-isolated from other continents and, hence, insular with reference to each other. If new immigrant species could be introduced into a continental fauna in numbers high enough so that the percentage increase in immigration rate would equal that seen on islands, there is no reason to doubt that a new equilibrium point would be approached which would show the same percentage increase over the earlier equilibrium as on islands. It is interesting to conjecture that if most of the new immigrants could be taken from the Hawaiian fauna, we would obtain a seemingly clear picture of the "competitive superiority" of Hawaiian species. Also, if we took random samples of the same size from an island and a continent, respectively, with approximately the same physical environments and introduced them to a second island (or continent), we might expect the survival rates in the two samples to be approximately equal. It is true that insular species have been subject to greater alterations within historical times, but the simplest explanation for this difference is that because of the smaller original size of insular faunas the immigration rate has been increased to a relatively much higher level. The introduction of twenty new species per century may increase the immigration rate of a given balanced taxon on Hawaii a thousandfold, but only double it on a continent. The two faunas will seek new equilibrium levels that are very different proportionate to the original faunas. But if the immigration rates were returned to the old, natural levels (as, for instance, by strict quarantine laws) both faunas could be expected to return eventually to approximately their original sizes. As in the preceding section, we must conclude that a closer look at this particular kind of quantitative evidence does not support the notion that continental species are intrinsically superior to insular species. In fact, it does not really support any conclusion other than the obvious fact that continents support more species than islands.

COLONIZATION AS AN ACTION OF EVOLUTIONARY STRATEGY

I agree with the conclusion of Wallace (1959) and Lewontin (1961) that the best measure of fitness of a population, as opposed to that of an individual genotype, is the period of time it survives. Relative reproductive rate, which is used to define the fitness of the individual genotype, is of minor importance in population fitness. Over a long period of time the average rate of growth of any persistent population is approximately zero. A high reproductive potential, together with other attributes that biogeographers tend intuitively to associate with success, such as high density, broad geographic range, and ecological diversity, are best thought of as actions of "strategies" that adapt certain species to certain environments.* Rare, specialized species are by no means *ipso facto* imperiled species; there is no necessary intrinsic reason why their mean survival time should not equal or surpass related species that are widespread and abundant.

Colonization can be an effective strategy under the following conditions. If a species is able to populate many semi-isolated regions, such as a system of oceanic islands, its populations will be tested by a variety

* "A strategy for Player I is a complete enumeration of all actions Player I will take for every contingency that might arise, whether the contingency be one of chance or one created by a move of the opposing player" (Karlin, 1959). In an admirably lucid article Lewontin (1961) has called attention to ways in which mathematical game theory might eventually be brought to bear on evolutionary problems.

of physical and biotic environments. Over a period of time, the greater the number of regions colonized, the higher the probability that the species will survive somewhere, so that endemic foci persist out of which colonization can proceed again. If colonizing power, however, always gave an unsullied advantage, we would expect all species to have maximized their dispersive and reproductive potential. This is obviously not the case. For most evolving species there is a point short of the maximum attainable dispersal power at which fitness is greatest; in some species, such as the flightless birds and insects of islands and mountaintops, this point is drastically short and has been arrived at by reversed evolution. To be a superb colonizer one must not only scatter a large fraction of each generation into lethal habitats; one must also adapt to the comparatively unstable habitats, such as coastal swamp forest, sandy beaches, and atoll scrub, that form the main ports of entry into new regions. Only a minority of species can fill such an adaptive category at any one time.

The above considerations have led me to the following prediction: the class of species picturesquely termed "fugitive species" by Hutchinson (1951) produces the bulk of the colonizers of new regions. This class, recognized by empirical studies of continental faunas in such groups as copepods, birds, and ants (e.g., Brian, 1958; MacArthur, 1958) consists of those species that specialize in the occupation of transient habitats. As the habitats are altered serally, or competitor species begin to infiltrate the area, the fugitive species disappear. They are able to survive by the relatively quick and temporary occupancy of suitable new habitats as these first become available; in contrast, other species specialize in more persistent tenancy of relatively stable habitats. The fugitive species are characterized by superior dispersal and reproductive powers. The fugitive strategy enables a minority of species to survive indefinitely within the kaleidoscopically changing environments of single regions. It also *preadapts* species to colonize neighboring regions, such as offshore islands or adjacent continents. If regions exchange species reciprocally, the strategy of colonization in the sense in which I first mentioned it can be regarded as a simple extension of the fugitive adaptation.

In Fig. 3 are presented some of the features of what I have called the *taxon cycle* of the Melanesian ant fauna. Supporting data have been collected in two previous papers (Wilson, 1959, 1961). A primary feature of the cycle is the superior ability of species adapted to marginal habitats to disperse across water gaps in the Indo-Australian region. The marginal habitats are defined as those with the fewest species. In this part of the wet tropics they include the littoral zone, savanna, monsoon forest, and the border zones of the lowland rain forest. Expanding species, which I interpret as being at least the loose equivalents of continental fugitive species, compete with ecologically similar endemic species. The endemic species as a group are more restricted to the

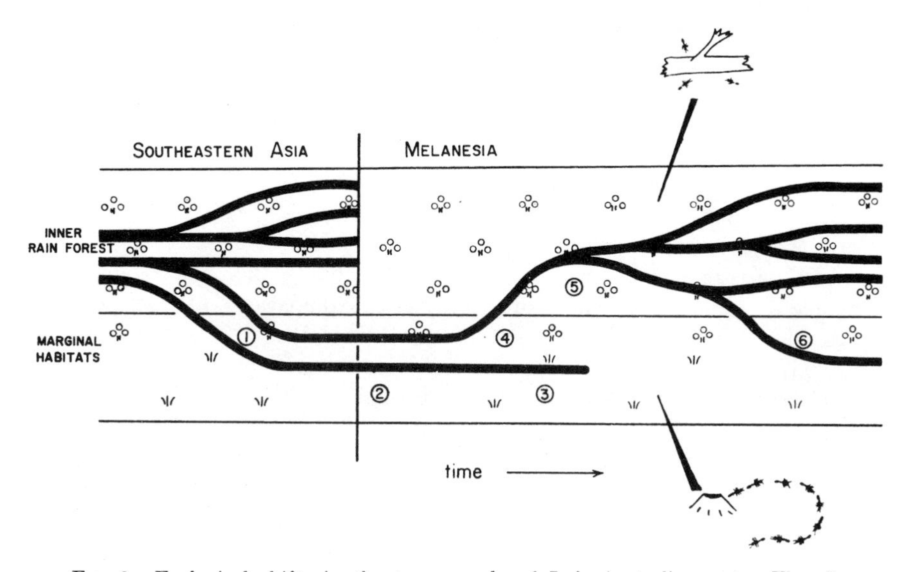

FIG. 3. Ecological shifts in the taxon cycle of Indo-Australian ants. The diagram represents the inferred evolution of ponerine and myrmicine species groups in Melanesia, in this case tracing the history of taxa derived ultimately from Asia. (1) Species or infraspecific populations adapt to marginal habitats in southeastern Asia, then cross the water gaps to New Guinea and colonize marginal habitats there (2). In time these pioneer populations either become extinct (3) or invade the inner rain forest of Melanesia (4). Having successfully adapted to the inner rain forest, they ultimately diverge to species level (5). As diversification continues the source fauna in Asia may be contracting, so that in time the group as a whole becomes Melanesian-centered. A few of the Melanesian species may readapt to the marginal habitats (6) and expand back into Asia. Also indicated diagrammatically are changes in biology correlated with the adaptive shift; in the rain forest the species tend to form smaller colonies that nest in small rotting pieces of wood and fail to employ odor trails. Modified from Wilson (1959).

inner rain forest. Certain biological differences are correlated with this major habitat difference, as also represented in Fig. 3. The zoogeographic evidence indicates that the expanding species play a role in the restriction of the endemic species to the inner rain forest, an effect that lessens the amount of gene flow among the insular populations and promotes speciation. In time the expanding species enter the inner habitats themselves, are constricted by other marginal-habitat competitors, and speciate. The cycle can be (and is) completed by the emergence of some species back into the marginal habitats. Due to the faunal dominance effect discussed in the first section, the cycle in Melanesia is initiated primarily by species entering from Asia. As a result, proliferating species groups centered in Asia occasionally lose ground in Asia, so that their "headquarters" are shifted to New Guinea; but the reverse shift must occur relatively rarely. Similarly, retreating groups may shift their headquarters from New Guinea to Fiji, but probably rarely if ever in the reverse direction. The conception of the center of the taxon cycle shifting eastward in the Indo-Australian region completes the outline of ant speciation as I currently envisage it. It remains to be said that the taxon cycle is merely a formalization of a series of established statistical trends which have been collated in the simplest logical manner. Alternative and more satisfying explanations are no doubt possible; at best the taxon cycle, being statistical in nature, can be expected to be frequently violated by individual species groups.

The added consideration of speciation, as illustrated in the Melanesian ant fauna, complicates our prior definition of population fitness as applied to colonizing species. So long as a species colonizes reciprocally without speciating, all its populations can be considered together as a unit. But if the exchange is not reciprocal, that is, if one population feeds successful propagules into another and receives none in return, the source population will not benefit from its superior colonizing ability. In this case, we can say that colonizing ability per se is not adaptive for the source population, but is preadaptive for any recipient population that benefits. Furthermore, colonizing ability loses its value if complete speciation occurs; for then reciprocal exchange adds nothing to either derivative gene pool and if anything will result in disoperation between the newly sympatric species (see the following section).

Thus, a species that colonizes and also permits speciation is apt to be "hoisted with its own petard" by generating its own competitors. To complete this line of reasoning by a frankly speculative extension, it is conceivable that if a species loses colonizing power to the extent that it begins to speciate, the speciation process becomes autocatalytic as follows: The creation of new species reduces the rate of colonization and the decreased rate of colonization increases, at least for a time, the rate of speciation through over-all decrease in gene flow. If speciation continues, the number of species will approach an asymptote determined either by the maximum number of related species that can coexist, or the cutoff of colonization altogether, or both. At present I know of no way of analyzing speciation to shed light on such rate processes.

THE ANALYSIS OF REPRODUCTIVE CHARACTER DISPLACEMENT

When one species invades the range of a closely related second species, the two may compete or hybridize in such a way as to lower their average genotype fitness. Any new genes whose phenotypes reduce the interaction and, hence, raise average genotype fitness will tend to increase in frequency. Such evolution can be expected invariably to result in divergence of the species, a phenomenon often referred to as character displacement. The usual evidence for displacement cited in a particular case is the pattern of geographic variation in which one species diverges from the other species more or less precisely where the two are sympatric. The process is potentially important in determining not only the success of colonizing species but also the final outcome of geographic speciation itself. In practice the effects of ecological and reproductive displacement are often difficult or impossible to distinguish. Nevertheless, once isolated the latter is more tractable to analytical treatment and is a promising starting point for a quantitative theory of speciation.

When Mendelian populations with imperfect intrinsic isolating mechanisms come together they will, given enough time, either fuse completely as one species or else displace reproductively to become more perfectly isolated. No other equilibrium state seems possible. The semispecies state is unstable, although it does occur commonly enough at any given time in many groups of plants and animals to create taxonomic confusion (Grant, 1957). In a recent study Bossert (1963) has developed a model of semispecies evolution which allows, for the first time, prediction of the outcome of specific cases in which details of the intrinsic isolating mechanisms are known. Several initial restrictions are necessary, each of which can be eventually relaxed without serious damage to the major qualitative results based on them. They are as follows:

(1) The populations are panmictic except with reference to the components under consideration.

(2) The populations are equally abundant.

(3) The characters forming the prezygotic (i.e., premating) barriers are normally distributed.

(4) The prezygotic barriers are based on polygenes that contribute additively to the phenotype.

(5) The prezygotic genotypes produce mean phenotypes whose values range over some finite interval, e.g., 0, 150 units, and the phenotype of

each genotype varies uniformly through some lesser interval, e.g., 10 units.

(6) The postzygotic barrier, if one exists, is created by a single pair of alleles. At the start, species *A* is homozygous for *a*, species *B* is homozygous for *b*, and the initial F_1 hybrids—if any exist—are heterozygotes (*ab*).

(7) Each adult organism is allowed to mate only once.

The model requires the following special redefinition of species with reference to their hybrids. After a hybrid is created, it is then considered to be a member of the species that provides its mate; or, if it mates with another hybrid, it is considered a member of the species into which the majority of its offspring mate. Gene flow into a given species is then defined as the proportion of its matings that are hybrid backcrosses. Error is defined as the frequency of interspecific crosses.

Next to be considered are the effects of the postzygotic barrier, defined in the Bossert model as lowered heterozygote fitness. First, a curious feature of population interaction is to be noted, which holds true whether the postzygotic barriers are unifactorial or polygenic. It is obvious that if the heterozygotic fitness is unity or above, and hybridization occurs at all, the populations will eventually fuse. No displacement in the prezygotic barrier will occur at all, because the hybrids are favored. But what is less apparent is the fact that even if heterozygotic fitness is less than unity (but greater than zero), and if new isolating mechanisms are not permitted to evolve in the interim, the populations will still fuse. The reason is that while hybridizing the populations exchange an increasing number of the postzygotic barrier genes. As a result the two populations come increasingly to resemble both each other and the hybrids, at least insofar as the frequency of the barrier genes is concerned. The relative hybrid fitness consequently rises. Then, as the penalty of committing "error" drops, displacement in the prezygotic barrier slows down. Eventually, displacement stops altogether, and the populations finally fuse.

The final act of speciation can, therefore, be viewed as a race between hybridization and displacement. If there is no error (i.e., the prezygotic devices are perfect) then there is neither hybridization nor displacement. If there is error, but the heterozygote fitness is unity or greater, there will be no displacement, and gene flow will lead to fusion. If there is error and the heterozygote fitness is zero, displacement will take place in the absence of gene flow. Finally, if there is error, and the heterozygote fitness is less than unity but greater than zero, displacement will compete with gene flow, and the outcome will depend on their relative rates. A further complication is that new intrinsic barriers may arise in response to the lowered hybrid fitness. These can check gene flow and eventually reduce it to zero. In a certain intermediate range of values of gene flow, we can for the moment only guess at the outcome.

The graph in Fig. 4 shows the expected average gene flow over the first fifty generations for wide ranges of error and heterozygote fitness values. It was derived from a recursion equation solved repeatedly with the aid of a digital computer. In using the graph to make predictions (for both the unifactorial and polygenic cases) hybrid fitness can be approximated by taking as its equivalent the average fitness of the backcross offspring. The decision as to whether two populations will fuse or separate was made in the following way. The fitness of *aa*, the homozygote of the postzygotic barrier allele in species *A*, was set at slightly greater than 1 and its fate followed as gene flow proceeded by computer simulation into species *B*. It was determined that when average gene flow is set to exceed 0.1 in the first fifty generations, the frequency of *a* in population *B* after fifty generations is equal to or greater than 0.5. In this circumstance, *a* would almost certainly be expected to replace *b* in population *B* even if gene flow were stopped at this point. Hence, the two populations would have erased the postzygotic barrier. If the fitness of *aa* were set at slightly less than 1, gene *b* would then replace *a* in population *A*; and the postzygotic mechanism would still be lost. Conversely, if gene flow averaged less than 0.00316 in the first fifty generations, and gene flow were then stopped at this early period, it was shown that *a* would almost certainly be lost in *B* (and *b* in *A*). Hence, the postzygotic barrier would be preserved.

The complete solution of problems of this kind is sharply limited to cases where a full knowledge of prezygotic isolating mechanisms is obtainable. In cases where habitat differences are complex, or courtship behavior depends on visual or chemical cues difficult to interpret, or other technically refractory problems arise, there is at present no way to apply the theory. The model, however, does allow us to predict species separation with imperfect knowledge of prezygotic devices, providing the known devices result in gene flow below the critical figure.

Besides allowing an entrée to the hitherto seemingly intractable problem of determining the evolutionary status of allopatric populations, the Bossert model has led to certain additional new predictions about reproductive character displacement. The following apply to single-component systems:

(1) Displacement is likely to be both clear-cut and rapid in many ordinary situations. Error reduction (and hence displacement) will proceed rapidly until, depending on the heritability of the character, a

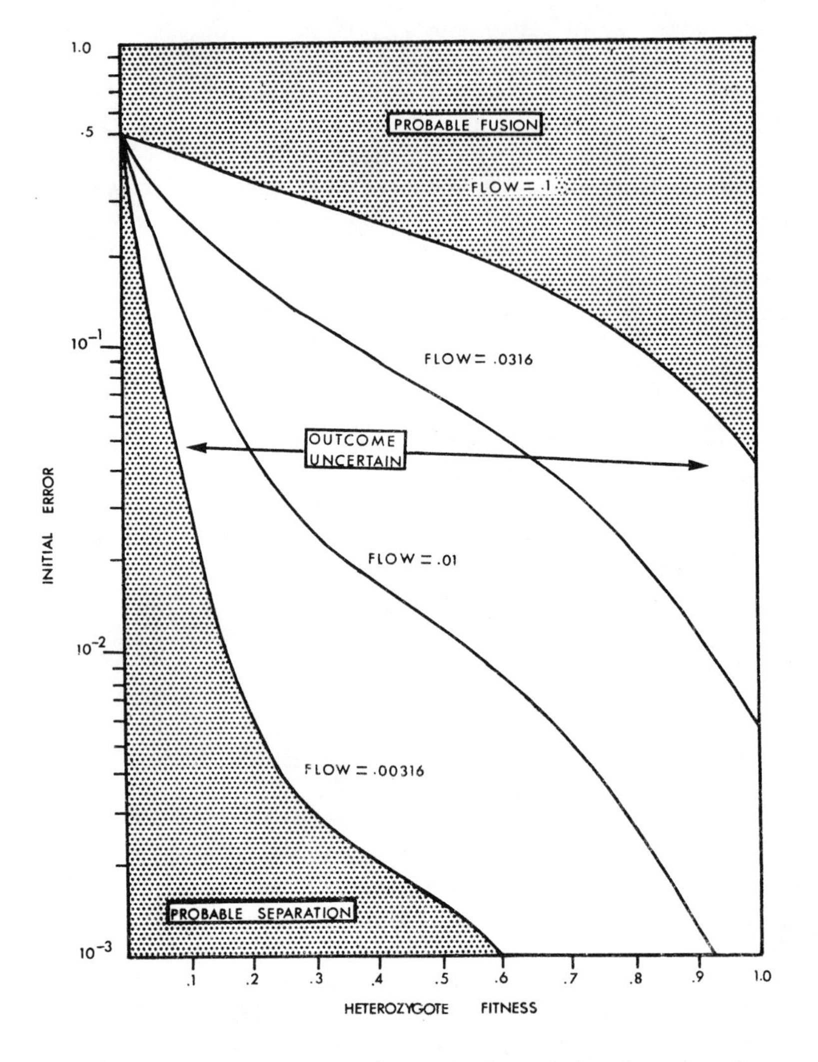

FIG. 4. Prediction of the outcome of contact of populations in various degrees of reproductive isolation. Average gene flow over the first fifty generations of sympatry is given as a function of initial mating error and fitness of the heterozygote of the postzygotic barrier alleles. The two homozygotes of the postzygotic barrier alleles are assumed here to have fitness 1. The zones outside the 0.1 and 0.00316 gene-flow isoclines are indicated as almost certainly yielding, respectively, fusion and separation. Modified from Bossert (1963).

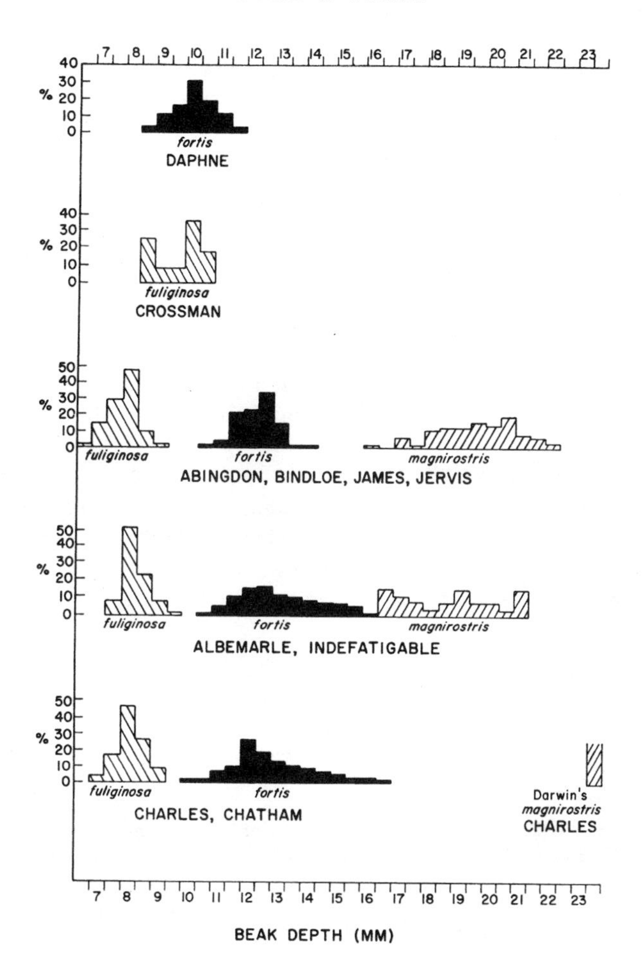

FIG. 5. Patterns of variation in Darwin's finches (genus *Geospiza*) which are interpreted as having been influenced by character displacement. Modified from Lack (1947). The mean differences in beak depth among sympatric species are on the order of 50%. On Charles, Chatham, Albemarle, and Indefatigable, where *fortis* and *fuliginosa* are the only two species of *Geospiza* that exist in significant numbers, the *fortis* distribution is skewed away from that of *fuliginosa*. Where *fortis* is bracketed by significant numbers of both *fortis* and *magnirostris*, on Abingdon, Bindloe, James, and Jervis, the skewness is lacking. (Data on dispersion of *fuliginosa* is insufficient to observe skewness.) These exceptionally complete data conform to two of the principal predictions of the Bossert model.

mean difference of about 3 to 5 standard deviations is reached. If even a slight amount of stabilizing selection also operates to oppose the displacement, the mean difference will stabilize in this range.

(2) The frequency distributions will be skewed away from each other during the displacement process and the degree of skewness can perhaps be used as a measure of the displacement pressure.

(3) The displacement rates of the two species will conform to a "conservation of momentum" law, i.e.,

$$\frac{d\Phi_1}{d\Phi_2} = \frac{n_1}{n_2},$$

where Φ_1 and Φ_2 are the phenotypes of the displacing character in the two species, and n_1 and n_2 are the numbers of individuals in the two breeding populations.

Predictions 1 and 2 seem to me to be fulfilled in the well-known case of displacement in *Geospiza,* as illustrated in Fig. 5. The merit of the Bossert model is that it does lead to new predictions that are testable through further taxonomic research and, eventually and more conclusively, through experiments on natural populations.

References

Beirne, B. P. (1962). Trends in applied biological control of insects. *Ann. Rev. Entomol.* **7,** 387–400.

Bossert, W. H. (1963). Simulation of character displacement in animals. Ph.D. Dissertation, Harvard University, Division of Engineering and Applied Mathematics.

Brian, M. V. (1958). Interaction between ant populations. *Proc. 10th Intern. Congr. Entomol., Montreal, 1956* **2,** 781–783.

Brown, W. L. (1957). Centrifugal speciation. *Quart. Rev. Biol.* **32,** 247–277.

Clausen, C. P. (1956). Biological control of insect pests in the Continental United States. *U. S. Dept. Agr. Bull.* **1139.**

Darlington, P. J. (1957). "Zoogeography: the Geographical Distribution of Animals." Wiley, New York.

Darlington, P. J. (1959). Area, climate, and evolution. *Evolution* **13,** 488–510.

Darwin, C. (1859). "On the Origin of Species." John Murray, London.

De Vos, A., Manville, R. H., and Van Gelder, G. (1956). Introduced mammals and their influence on native biota. *Zoologica* **41,** 163–191.

Elton, C. S. (1958). "The Ecology of Invasions." Methuen, London.

Grant, V. (1957). The plant species in theory and practice. *Publ. Am. Assoc. Advan. Sci.* **50,** 39–80.

Hutchinson, G. E. (1951). Copepodology for the ornithologist. *Ecology* **32,** 571–577.

Karlin, S. (1959). "Mathematical Methods and Theory in Games, Programming, and Economics." Vol. 1, Matrix Games, Programming, and Mathematical Economics." Addison-Wesley, Reading, Massachusetts.

Lack, D. (1947). "Darwin's Finches." Cambridge Univ. Press, London and New York.

Lewontin, R. C. (1961). Evolution and the theory of games. *J. Theoret. Biol.* **1,** 382–403.

MacArthur, R. H. (1958). Population ecology of some warblers of northeastern coniferous forests. *Ecology* **39,** 599–619.

MacArthur, R. H., and Wilson, E. O. (1963). An equilibrium theory of insular zoogeography. *Evolution* **17,** 373–387.

Matthew, W. D. (1915). Climate and evolution. *Ann. N. Y. Acad. Sci.* **24,** 171–318.

Mayr, E. (1942). "Systematics and the Origin of Species." Columbia Univ. Press, New York.

Mayr, E. (1963). "Animal Species and Evolution." Harvard Univ. Press, Cambridge, Massachusetts.

Simpson, G. G. (1953). "The Major Features of Evolution." Columbia Univ. Press, New York.

Wallace, B. (1959). Studies of the relative fitnesses of experimental populations of *Drosophila melanogaster*. *Am. Naturalist* **93,** 295–314.

Watt, K. E. F. (1962). Use of mathematical models in population ecology. *Ann. Rev. Entomol.* **7,** 243–260.

Wilson, E. O. (1959). Adaptive shift and dispersal in a tropical ant fauna. *Evolution* **13,** 122–144.

Wilson, E. O. (1961). The nature of the taxon cycle in the Melanesian ant fauna *Am. Naturalist* **95,** 169–193.

Wodzicki, K. A. (1950). Introduced mammals of New Zealand, an ecological and economic survey. *New Zealand Dept. Sci. Ind. Res. Bull.* **98.**

Discussion of Paper by Dr. E. O. Wilson

Waddington: Colonizing species may be derived from marginal populations—those found at the margin of the species' range. In these marginal parts, is the total fauna usually less than that of the central regions? Are there grounds for supposing that colonies of the species in the marginal zones are less specialized in the sense that they are not confined to a narrow ecological niche, as in the main range of the species?

Mayr: Let us distinguish in our discussion between marginal and peripheral. Marginal is an ecological term meaning living near a pessimal condition. Peripheral means along the periphery of the species range. If we use the two terms interchangeably, we are likely to get confused.

Waddington: Yes, marginal is not fundamentally a geographical term, but an ecological one. The question still stands. In the marginal, not necessarily peripheral, parts of the species distribution, the species is presumably reaching some limit to its adaptation, but are these populations actually competing with as rich or varied a set of competitors as they are in the central parts?

E. Wilson: I define marginal habitats in a special way. They are those habitats with the smallest number of species. It appears that there are certain correlates of marginal habitats, so defined, in Melanesia which are of interest. One of them is that they are less stable, in that they are more apt to be disturbed in a given period of time. I have no evidence to determine whether marginal species are superior generalists in aspects other than habitat choice. The evidence is also insufficient to determine whether they have higher reproductive rates, i.e., higher "*r*." The evidence is sufficient to determine that they have superior dispersal ability.

Waddington: But are they more dispersive—do they cover a larger range of territory?

E. Wilson: Oh, yes. If you accept the way I have classified habitats, species occupying marginal habitats occur in a wider range of these major habitats and they also have wider geographic distribution. Now my reason for suggesting that they are preadapted for colonization was that they are extremely similar, in the kind of habitats that they occupy, to what M. V. Brian, R. H. MacArthur, G. E. Hutchinson, and others have pin-pointed by long-term, local studies as "fugitive" species.

Ehrendorfer: I would like to add another ecological aspect which I think is important to our discussion, namely, the distinction of stable (and "buffered") from labile (and "unbuffered") habitats. Our botanical evidence shows that both fugitive and colonizing species occur in marginal habitats. But such marginal habitats as river sandbanks and open alluvial forests are labile types of habitat and harbor many colonizing and widespread species but hardly any fugitive types. On the other hand, rock fissures and talus slopes, in spite of being marginal, often are stabilized and "buffered" habitats, and contain many fugitive but much fewer widespread types. To what an extent do your data from ants compare with such a concept? I don't think that a classification of habitats into "marginal" and "central" is sufficient relative to the problems under discussion; "stable" versus "labile" may be more important.

E. Wilson: I completely agree. Within the major divisions of what I call marginal habitats, on the basis of the number of species living in them, there are undoubtedly microhabitats which are relatively stable. We simply haven't taken our ecological studies of ants this far.

Dobzhansky: I am particularly interested in your semispecies since I'm working with some semispecies. Maybe "three-quarter" species! The question which arises in this connection is this: very often you have the F_1 hybrid between "three-quarter" species not being inferior—at least in one sex—to the parents and conceivably even heterotic. It is in the backcross that things really become difficult. How does that agree with the Bossert model?

E. Wilson: Bossert foresaw that contingency; his hybrid fitness is, in fact, that of the backcross offspring.

Dobzhansky: On your diagram it was the F_1.

E. Wilson: It was heterozygote fitness. Really the important factor here is the probability that a gene will succeed in getting from one species into another one. It doesn't matter whether it is killed off in the heterozygote or not. It is, if you like, a measure of the success of the introgression of genes.

Stebbins: I do not feel that botanists can accept your decimation of the idea that a semispecies can exist for some time because we have one genus where we happen to possess fossils. This is *Quercus,* in which all the evidence is that modern entities behave as semispecies and that entities exactly comparable to them, both morphologically and ecologically, are to be seen from the Mio-Pliocene. To me, this would suggest a fairly permanent condition. And I think it is true, at least in woody plants, that the semispecies condition can be as permanent as the existence of families of some mammals.

I should also like to ask if Dr. Bossert has constructed models in which the assumption is made that the hybrid in a particular environment is superior to either parent? I think this is the most important assumption so far as plant hybrids are concerned. I have seen over and over again, situations in which two sympatric species will actually form hybrids and these will disappear unless there is disturbance or a new habitat made into which these hybrids migrate and where they are better adapted. This can be seen here on the Monterey Peninsula. Two species of *Ceanothus,* when I first saw them in 1942, were forming a beautiful hybrid swarm simply because a firebreak had been put through the area. Now that firebreak has grown up and all one can see is the parental species. The habitat where the hybrids were superior disappeared. There are other situations, however, where these intermediate habitats have spread and have actually become more common than the original habitats.

E. Wilson: Of course, what you just stated is that the prezygotic devices are very good, but I don't see that it contradicts any conclusion of mine. Now as to the permanence of semispecies in taxonomic groups, I think that our disagreement might be semantic and I wonder if you'd care to define semispecies.

Stebbins: You could consider two entities semispecies if, in some parts of their range, they live sympatrically, retaining their identity and looking like perfectly good species—whereas in other parts of the range, the same two species form extensive hybrid swarms such that no two individuals are alike. This is a very prominent situation in oaks.

E. Wilson: Yes, defined in this way some semispecies can be stable. We need to agree on terminology.

Lewontin: I'm a little bothered by this sort of anti-genetic approach at the beginning; the notion that there isn't really much to say about the genetics of colonizing species because such species can be regarded as a random sample of all species. I think this is probably true at one level, but I think we ought to use a finer screen.

E. Wilson: I was hoping somebody would bring this up. Would you agree with me that we are talking about two levels of understanding of the same problem? I was dealing with what is essentially a stochastic explanation of aggregate species behavior much as one would attempt to explain diffusion of gas by a stochastic model of molecular motion. Our species, however, are more than particles of gas; we can do much better at explaining their individual velocity and direction and so this has to be investigated too.

Lewontin: No, all I'm saying is that sometimes we have processes that look like homogeneous processes at one level, like the diffusion of a gas, but this doesn't change the fact that they are really mixtures of heterogeneous processes which have some intrinsic interest in themselves.

E. Wilson: Why I use a formulation of this sort is simply, I think, to attempt to be predictive about large numbers of species, and this is what interests the zoogeographer. You see, at this stage we are seeking simplifying ideas that deal with large numbers of species. And this is a different level of explanation than the one you're using.

Lewontin: In fact, the colonization process itself will lead to selection within the species. The population that will do the colonization within the species will not necessarily be a random sample of all possible populations of a given species. In other words, at the species level it may be random, but at the population level, it may not. There may be populations within a species that are especially good at colonizing.

1967

"An estimate of the potential evolutionary increase in species density in the Polynesian ant fauna," *Evolution* 21(1): 1–10 (with Robert W. Taylor, second author)

While reflecting on the evolutionary implication of *The Theory of Island Biogeography,* published with R. H. MacArthur in 1967, I became aware that the equilibrium of species numbers on islands and other more-or-less isolated ecosystems is likely to be really a quasi-equilibrium—approximately stable in the short term but gradually shifting over long periods. Later, in 1969 (see "The species equilibrium," to follow in this volume), I envisioned the quasi-equilibrium as comprising four stages in which the final is the evolutionary stage. At the final stage the associated species have had time to coevolve so as to have an overall lower extinction rate. With the degree of isolation and the immigration rate also being held more or less constant, biodiversity rises to its highest species number.

About this time I had just completed with Robert W. Taylor the first comprehensive study of the Polynesian ant fauna. We were able to demonstrate that ants are relatively poor at overseas dispersal, so much so that evidently no native species and certainly no endemics occur in the Pacific archipelagoes east of Tonga and Samoa. These remote eastern islands have been settled instead by "tramp" species transported accidentally by human commerce, and enough have been thus established to create an area-species curve indicative of the earliest stage of quasi-equilibrium. This circumstance provided an unusual opportunity to address an important question in evolutionary ecology: By how much will coevolution increase the number of species from the early-immigrant period of the quasi-equilibrium? And, conversely, how much poorer is a tramp fauna compared with an otherwise equivalent native fauna? By comparing the native species of the western Polynesian archipelagoes, which include many endemics, with the tramp faunas of the eastern Polynesian archipelagoes, it was possible to make a first rough estimate of the amount of increase due to evolution: between 0.5- and 2.0-fold.

AN ESTIMATE OF THE POTENTIAL EVOLUTIONARY INCREASE IN SPECIES DENSITY IN THE POLYNESIAN ANT FAUNA[1]

Edward O. Wilson and Robert W. Taylor[2]
Biological Laboratories, Harvard University, Cambridge, Massachusetts

Received September 20, 1965

An exceptionally high proportion of the ant species of Polynesia are "tramps," that is, species carried about the world by human commerce. A truly native fauna is present as well on the islands of western Polynesia. In the eastern and northern archipelagoes, however, there may be in fact no truly native species at all, the faunulae here consisting of synthetic aggregations of tramp species. Some of these species originated ultimately in the Indo-Australian area, some in Africa, and some in the New World tropics. Most have been carried accidentally to the Pacific islands on European ships, certainly within the last 400 years and probably no earlier than the time of Cook's first voyage (1768–1771). This tramp fauna, despite its heterogeneity and youth, has achieved an orderliness of sorts. In particular, the number of species on an island is well correlated with the island's area, and some related species have acquired complementary (mosaic) inter- and intra-island distributions that can be explained easily only as the outcome of competitive exclusion. In the course of current studies on the Indo-Australian fauna we have come to recognize that the Polynesian ants offer an unusual research opportunity; namely the chance to compare the synthetic, poorly coadapted faunulae of the eastern and northern archipelagoes with the older, native, and better coadapted faunulae of the western archipelagoes. A detailed account of the taxonomy and distribution of individual species is being given elsewhere (Wilson and Taylor, 1967). In the present article we will describe certain particular phenomena which might be of general interest.

Relationships of the Polynesian Fauna

Our survey has included all of the islands from the Ellice group, Rotuma, Samoa, and Tonga, eastward to and including the Marquesas and Easter Island, and north to Hawaii. New Zealand was excluded from the current analysis, since it is faunistically very distinct, most or all of its few native species having been drawn directly from Australia (Brown, 1958). The Wallis Islands and Futuna have been intensively collected only recently by Mr. George Hunt, and their ant faunulae will be the subject of a later separate report by Hunt and Wilson.

[1] Based on research supported by U. S. National Science Foundation grant no. GB1634.

[2] Present address: Entomology Division, C.S.I.R.O., Canberra, A.C.T., Australia.

TABLE 1. *A classification of the Polynesian ant species according to origin.*

1. Endemic to one or more Polynesian archipelagoes:
   Ectomomyrmex insulanus, Ponera loi, P. woodwardi, Strumigenys mailei, Pheidole aana, P. atua, Vollenhovia pacifica, V. samoensis, Rogeria exsulans, Adelomyrmex samoanus, Camponotus navigator, C. rotumanus, C. flavolimbatus, C. conicus, C. nigrifrons, Polyrhachis rotumana.
2. Continuously distributed from Indo-Australian area into Polynesia:
   Prionopelta kraepilini, Platythyrea parallela, Ponera incerta, P. tenuis, Hypoponera confinis, H. punctatissima, Cryptopone testacea, Odontomachus simillimus, Anochetus graeffei, Eurhopalothrix procera, Smithistruma dubia, Strumigenys szalayi, Pheidole fervens, P. oceanica, P. umbonata, P. sexspinosa, Solenopsis papuana, Oligomyrmex atomus, Monomorium talpa, Tetramorium pacificum, T. tonganum, Rogeria sublevinodis, Iridomyrmex anceps, Tapinoma minutum, Technomyrmex albipes, Paratrechina minutula, Camponotus chloroticus.
3. "Tramp species": certainly distributed by recent human commerce:
   Hypoponera opaciceps, H. zwaluwenburgi, Trachymesopus stigma, Leptogenys maxillosa, Syscia typhla, Trichoscapa membranifera, Strumigenys godeffroyi, S. lewisi, S. rogeri, Quadristruma emmae, Pheidole megacephala, Solenopsis geminata, Monomorium destructor, M. latinode, M. floricola, M. fossulatum, M. minutum, M. pharaonis, Triglyphothrix striatidens, Tetramorium caespitum, T. guineense, T. simillimum, Cardiocondyla emeryi, C. nuda, C. wroughtoni, Iridomyrmex humilis, Tapinoma melanocephalum, Anoplolepis longipes, Plagiolepis alluaudi, P. exigua, Paratrechina bourbonica, P. vaga, P. longicornis, Brachymyrmex obscurior, Camponotus variegatus.
4. "Tramp species" intercepted in quarantine at Honolulu but not yet established in Polynesia:
   Brachyponera solitaria, Tetramorium caespitum, Wasmannia auropunctata.
5. Uncertain status:
   Amblyopone zwaluwenburgi, Ponera swezeyi, Smithistruma mumfordi, Oligomyrmex tahitiensis, Chelaner antarcticum.

The Polynesian ant fauna is now relatively well known. Samoa, the archipelago of principal interest to us, has been especially well surveyed. Large collections made in the 1920's by the London School of Hygiene and Tropical Medicine Expedition were studied by Santschi (1928). In the period 1938–1962, five other entomologists, O. H. Swezey, E. C. Zimmerman, T. E. Woodward, G. Ettershank, and R. W. Taylor, made additional collections. A five-week tour by Taylor and his wife in 1962 was devoted wholly to collection and study of the Samoan ants.

Classification of the 83 known Polynesian species according to biogeographic origin is given in Table 1, and the composition of the faunulae of the better known islands is given in Figures 1 and 2. Two conclusions of immediate relevance can be drawn on the basis of these partitions, and on some additional considerations of particular species presented elsewhere by Wilson and Taylor (1967).

First, it is apparent that, prior to the coming of man, few if any native species ranged east of Rotuma, Samoa, and Tonga. The evidence is as follows:

No certain endemics are known to occur in Polynesia beyond these islands. Five species which prior to 1950 were considered to be peculiar to the peripheral areas of the Pacific [namely *Ponera swezeyi* (Wheeler), *Syscia typhla* Roger (=*Cerapachys silvestrii* Wheeler), *Quadristruma emmae* (Emery) (=*Epitritus wheeleri* Donisthorpe), *Chelaner antarcticum* (White) (=*Monomorium rapaense* Wheeler), and *Plagiolepis alluaudi* Forel (=*P. mactavishi* Wheeler = *P. augusti* Emery)] are now known to occur elsewhere. Three others, *Amblyopone zwaluwenburgi* (Williams) of Hawaii, *Smithistruma mumfordi* (Wheeler) of the Marquesas Islands, and *Oligomyrmex tahitiensis* Wheeler of Tahiti, are still unknown elsewhere but belong to poorly collected and taxonomically little-known genera. Moreover, *O. tahitiensis* is known solely from the sexual castes and cannot even be compared with most of the other Indo-Australian members of the genus, which are known only from the worker caste. Thus, the status of these three remaining "endemics" is very dubious. Also, despite a plethora of subspecific and varietal names applied in the literature to populations of species now living in the central and eastern Pacific, we have detected only a single example of true geographic variation in these populations. The case is furthermore a rather trivial one involving color and the thickness of propodeal spines in samples of *Pheidole sexspinosa* Mayr from the Marquesas and Society islands.

About 20 Indo-Australian ant species range to some point east or north of Rotuma, Samoa, and

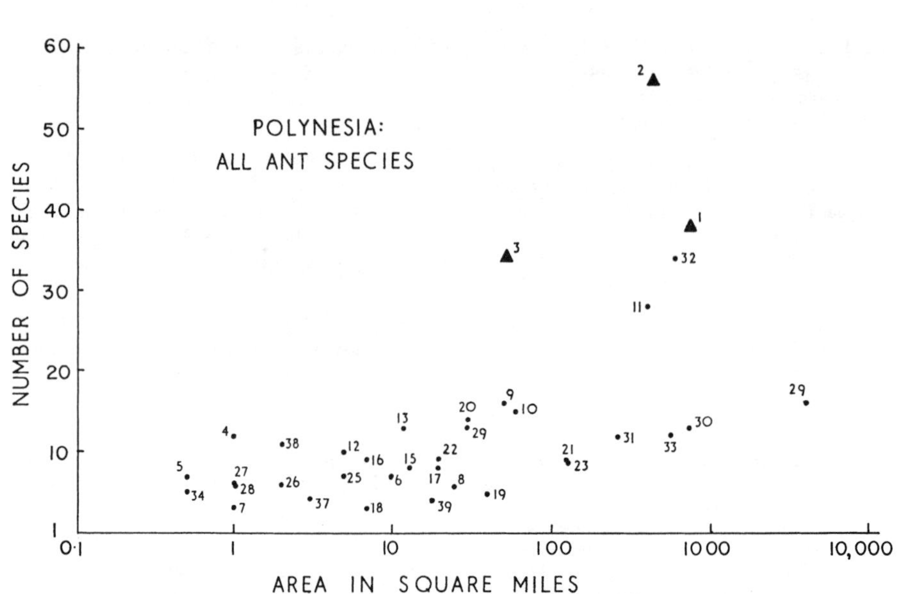

FIG. 1. Total number of ant species plotted against island area for each of the better collected Polynesia islands. Note the high values for the three large Samoan islands (1–3), this is due to the presence of a substantial native fauna in addition to the tramp species. SAMOA: (1) Savai'i, (2) Upolu, (3) Tutuila. TOKELAU: (4) Fakaofo. DANGER ISLANDS: (5) Motu Kotawa. SOCIETY ISLANDS: (6) Bora Bora, (7) Mehetia, (8) Huahine, (9) Moorea, (10) Raiatea, (11) Tahiti. AUSTRAL ISLANDS: (12) Rimatara, (13) Raivavae, (14) Maria Island, Northeast islet, (15) Rapa. GAMBIER ISLANDS: (16) Mangareva. MARQUESAS: (17) Eiao, (18) Hatutu, (19) Ua Pu, (20) Ua Huka, (21) Hiva Oa, (22) Tahuata, (23) Nuku Hiva, (24) Fatu Hiva, (25) Mohotane. PITCAIRN: (26). HENDERSON: (27). FLINT: (28). HAWAII: (29) Hawaii, (30) Maui,(31) Molokai, (32) Oahu, (33) Kauai, (34) Nihoa, (35) French Frigate Shoals, (36) Laysan, (37) Wake, (38) Midway, (39) Kure (Ocean).

Tonga, but these might easily have been transported there by man. For example, *Iridomyrmex anceps* (Roger), one of the most widespread of the Indo-Australian dolichoderines, was unknown until recent years from east of the Solomon Islands. In 1955 it was collected on Aitutaki in the Cook Islands, and in 1956 at Nandi, the international airport community of Viti Levu, Fiji. Intensive collecting has not yet revealed it in the intermediately situated Samoan islands, and the case for its transport to Aitutaki by human commerce is therefore strong. Several others of the Indo-Australian elements in the central and eastern Pacific are certainly known to be tramp species, having become established in the New World as well. Three others—*Odontomachus simillimus* Fr. Smith, *Tetramorium pacificum* Mayr, and *Pheidole fervens* Fr. Smith—have been intercepted at quarantine stations in Hawaii, and the last two have been taken in quarantine in New Zealand.

Second, the native species of western Polynesia are drawn almost exclusively from the Indo-Australian area. Almost all of the endemics have close relatives in Australia or Melanesia, mostly the latter. The single exception is the Samoan *Rogeria exsulans* Wilson and Taylor, which apparently belongs to a group otherwise known only from the Neotropical Region.

STABILIZATION OF THE NEWLY ASSEMBLED FAUNULAE

The central and eastern archipelagoes have thus been populated, in large part or even entirely, by species carried there by human commerce. Some of the Indo-Australian species might have been stowaways on the canoes of the early Polynesian voyagers. But others, originating ultimately in native populations in Africa and the New World, must have come with European ships no earlier than 400 years ago. The

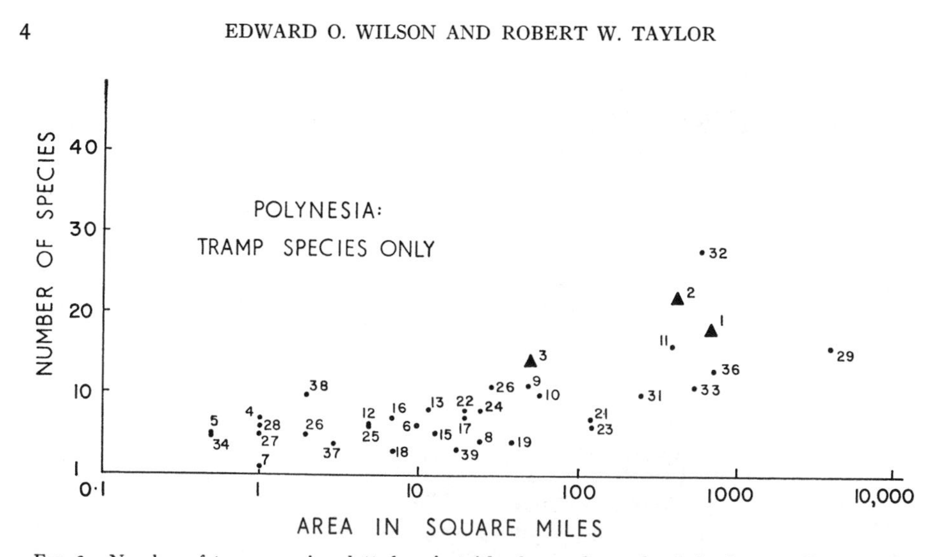

FIG. 2. Number of tramp species plotted against island area for each of the better collected Polynesian islands. The number code is the same as in Figure 1. With the native species removed, the Samoan islands are more consistent with the remainder of Polynesia.

question arises then: Are the islands continuing to fill up as quickly as new species reach them, or are the species densities stabilizing through competitive interactions? Several lines of evidence suggest that some degree of stabilization has occurred:

1) Although the 38 tramp species have been widely and—judging from Hawaiian and New Zealand quarantine records—repeatedly carried through the Pacific, only a fraction inhabit any given island at one time. Twenty-eight occur on Oahu, an island which is both relatively large and which has the most active foreign trade of all the Pacific islands and, with it, probably the highest immigration rate of tramp ant species. Most islands have fewer than nine species, or one-fourth the number in the available pool. As seen in Figure 2, there is a good correlation between the area of each island and the number of tramp species on it. The slope of the log-log plot of the same data is approximately 0.22. This compares favorably with the value of 0.14 obtained for native ant species of the Solomon Islands from the data of Figure 4. It is intermediate between the low (and probably underestimated) value of the Solomon Islands and the values of 0.4–0.5 obtained for land and freshwater birds in the same part of the world (MacArthur and Wilson, 1963). Thus, numbers of species seem to have been limited, and this seems to have occurred in an orderly fashion.

2) One of the more convincing forms of evidence of competitive displacement is a detailed mosaic complementarity of distribution between ecologically similar species. This phenomenon occurs in several sets of the Polynesian tramp species. The clearest example involves the large, aggressive species of *Pheidole*. *P. fervens* Fr. Smith, a widespread Indo-Australian element, is unknown from Samoa at the present time, but it is a dominant ant in the Society Islands. *P. megacephala* (Fabricius), a pantropical species of African origin, well known for its competitive interactions with other ant species, is dominant on Upolu in Samoa, but it is rare or absent in the Society Islands. *P. oceanica*, another Indo-Australian element, replaces *megacephala* on Savai'i, Samoa, and occurs on Upolu only on the western side facing Savai'i; it is relatively uncommon in the Society Islands. Elsewhere in Polynesia the complementarity among the three species is maintained. *Fervens* occurs in Tonga and Pitcairn; it is

only occasional in the Marquesas and is quite unknown in Hawaii. *Megacephala* is unknown from Tonga and Pitcairn, but it is dominant on the Marquesas and in Hawaii.

Similar complementary patterns also occur between *Cardiocondyla emeryi* Forel and *C. nuda* (Mayr) and, less clearly, between *Paratrechina bourbonica* (Forel) and *P. vaga* (Forel). In Hawaii *Solenopsis geminata* (Fabricius) is displaced to drier habitats by *Pheidole megacephala*. On small islands in the Dry Tortugas, Florida, and around Puerto Rico, these two species replace one another almost entirely, so that both are seldom found on the same island. Our data are not yet sufficient to determine whether this is also true in Polynesia, though it is strongly implied by records from smaller islands outside the Hawaiian group. Finally, *Iridomyrmex humilis* (Mayr) is proceeding to eliminate all but a few ant species within its restricted area of distribution in Hawaii, a habit it has exhibited in other parts of the world where it has been introduced.

Not all ant genera have developed complementary distributions or shown other obvious signs of competitive interaction. The four species of *Tetramorium*, for example, are often sympatric, as are the five species of *Ponera* found in Samoa.

3) Some of the tramp species have quite unstable populations, a fact which has been nicely demonstrated by E. S. Brown (1958) in his studies of the coconut grove species of the Solomon Islands. There is some evidence that fluctuations, perhaps great enough in magnitude to produce occasional extinction, have also occurred in Polynesia. In the earlier years of this century, *Tapinoma melanocephalum* (Fabricius) was common enough to be a house pest in Honolulu but apparently had disappeared, at least temporarily, by 1949 (Clagg, 1957). *Trachymesopus stigma* (Fabricius) was apparently common in Samoa up to and including 1940, but has not been encountered by collectors since 1956. *Pheidole fervens* was collected in Samoa in 1926 and 1940, but has not been found since 1956; while

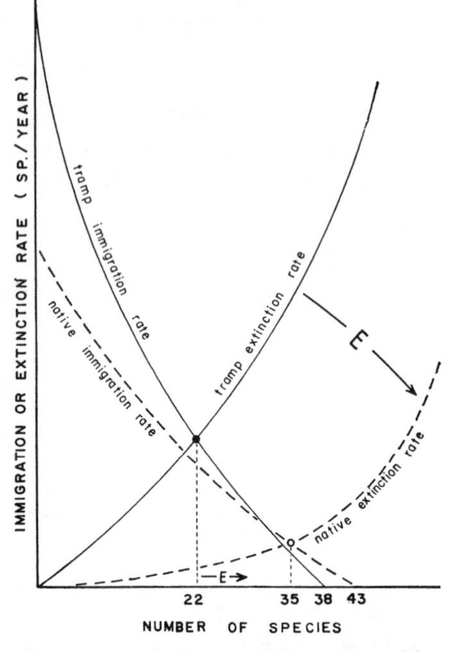

FIG. 3. An equilibrium model (based on MacArthur and Wilson, 1963) of the ant faunula of Upolu, Samoa. An attempt is made to estimate the probable increase in the number of species of tramp origin from the present "quasi-equilibrium" (at 22 species) that would result if the extinction rate were lowered, by means of evolution (E ⟶), toward the native level. The number of known native species on Upolu is 35, and this number is taken as the minimum to which the tramp quasi-equilibrium could move if given enough time.

its apparent nemesis *P. megacephala* has increased in abundance on Upolu since the 1920's.

To summarize, the numbers of tramp species on individual Polynesian islands have remained well below 38, the number available from the total species pool. The stabilization has evidently been effected in part by competitive replacement within at least two or three groups of related species. Conspicuous population fluctuations have occurred in some species in this century, probably resulting in a few extinctions; the role of competition in the fluctuations is suspected but not proven. Detailed histories of most individual species in Poly-

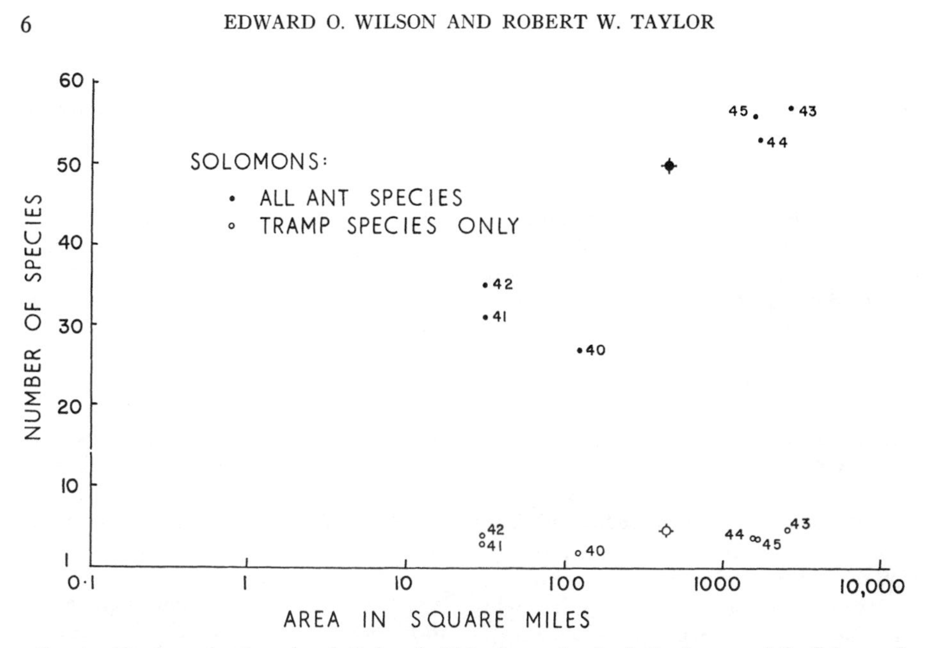

FIG. 4. Numbers of ant species plotted against island area for the better known of the Solomon Islands; based on the data of Mann (1919). The large closed circle represents the estimated total number of species that would occur on an island with the same area as Upolu (430 square miles). The large open circle represents the estimated number of these species of tramp origin on the same island. (40) Florida, (41) Malapaina, (42) Ugi, (43) Malaita, (44) San Cristoval, (45) Santa Ysabel.

nesia are not yet available. The account by Wilson and Taylor (1967) is intended as a first chapter of a record to be carried on in the future.

The Potential Increase of Species Density Through Evolution

The data just presented can be applied to the following abstract problem. Suppose a new faunula were assembled on an island from a pool of species chosen at random from various parts of the world and hence relatively ill-adapted to each other. Immigrants are placed on the island continuously from the beginning of the episode. As the number of species is increased, the rate of extinction will increase until in time it comes to equal the rate of immigration (in species per year). At that point of course the number of species will cease to change. Let us assume that the number on the island reaches the equilibrium at some value less than the number in the species pool.

Now suppose the faunula is allowed to remain at equilibrium long enough for evolution to occur. In evolutionary time the species become better adapted to the local island environment and to each other. As this happens, population stability should increase, since those species unable to achieve it will be preferentially eliminated. As a consequence the species extinction rates will decrease and, providing that immigration rates are not altered, species density will increase.

Hence, in evolutionary time, our imaginary newly assembled fauna is not in a true equilibrium but rather in a *quasi-equilibrium*. Over a period of a few generations, such as might be observed in a human lifetime, the rate of species extinctions would appear to equal the rate of species immigrations. But followed through a much longer period of time, it could be seen to have been almost imperceptibly less. The species density at quasi-equilibrium was gradually increasing in this time by means of evolution.

Presumably the quasi-equilibrium number would go on increasing indefinitely (but at an ever decreasing rate) if the pool of immigrant species were large enough and the environment remained otherwise stable. Indeed, since the pool is finite, the ultimate determinant must be environmental stability. As we expand our time scale from that of evolutionary time, during which evolution is effective, to that of geological time, in which geographic and climatic changes are also effective, environmental stability must become decisive. To digress momentarily, it follows that since tropical regions are more stable than temperate ones, higher quasi-equilibria can be obtained in them. This factor alone might account for most or all of the difference in species diversity between the tropical and temperate zones.

Returning to the concrete case, it would be of general interest to obtain some estimate of the amount of increase in species numbers that can occur through evolution. The Polynesian ant fauna appears to provide at least an entrée into the problem. In Figure 3 is presented the quasi-equilibrium on Upolu, Samoa, as it might be viewed by means of the equilibrium model of MacArthur and Wilson (1963). The following facts and postulates are introduced into the model:

1) There are 38 species of tramp origin known from Polynesia (Table 1, Classes 3–4). These constitute the pool of available immigrants. If all the 38 species were established on Upolu, the immigration rate would of course be zero because no further species would be available as colonists. We have therefore drawn the *tramp immigration rate curve* to intersect the species axis at 38 species.

2) Actually, only 22 tramp species are known to occur on Upolu. For reasons already given we are postulating that this is a short-term "equilibrium" value, or nearly so. In Figure 3 the *immigration rate curve* and *extinction rate curve for tramp species* on Upolu are therefore drawn to intersect the species axis at 22 species.

3) Forty-three native species are known for all of Polynesia. They represent the pool of Melanesia-based species that have successfully crossed the water gaps to Rotuma, Samoa, or Tonga by presumably natural means. Since Upolu is centrally located and close to eastern Melanesia, all of the 43 species that occur in Polynesia are actual or presumably potential colonists on Upolu. The *native immigration rate curve* for the island is therefore drawn to intersect the species axis at 43 species.

4) Actually, only 35 of the 43 native species occur on Upolu. This is taken as the minimum short-term "equilibrium" number for native species and in Figure 3 the *immigration rate curve* and *extinction rate curve for native species* are drawn to intersect the species axis at 35 species.

5) Since most of the tramp species of the world have spread through Polynesia in less than 400 years, the tramp immigration rates can be taken as higher than the native immigration rates, and the curves have been drawn accordingly. The precise differences are unknown but are not necessary for the solution of the problem we have posed.

6) From evidence already given, the extinction rates of tramp species appear to be high. They are almost certainly higher on the average than those of the native species, in which survival has often been prolonged enough to allow geographic differentiation and even full speciation. The two extinction rate curves are drawn accordingly. Again the precise differences are not needed for the solution to the problem.

A fortunate circumstance that permits the model to be applied in this case is that the number of species in the tramp pool, 38, and the number in the native pool, 43, are nearly the same. The immigration curves of the two groups thus descend to nearly the same point on the abscissa. Also, by definition, the two extinction curves ascend from the same point, the origin. By taking advantage of the circumstance it should be possible to make predictions concerning changes that would occur in the equilibrium numbers if the slopes of the

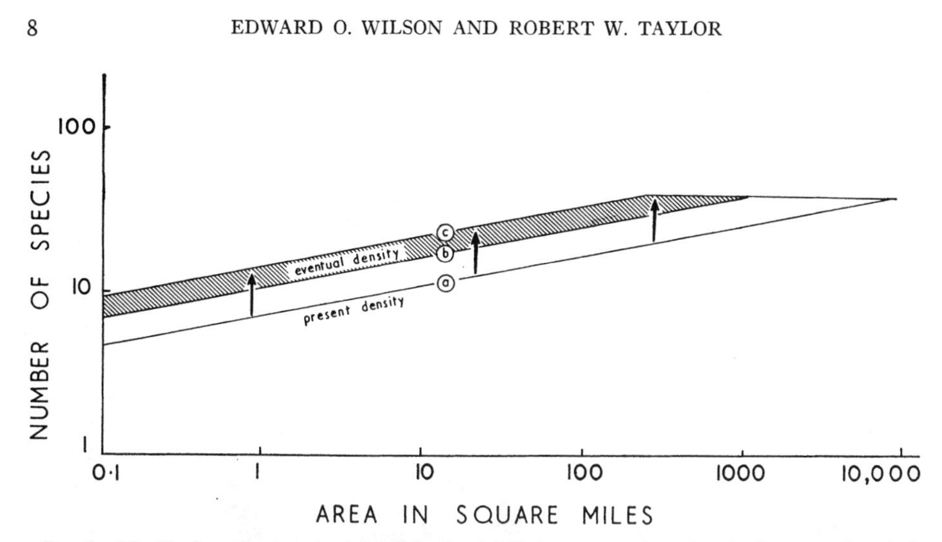

FIG. 5. The Upolu estimate extended to islands of differing area, at least for the Samoa region. It is assumed that the species pool remains at 38 and that the slope of the species-area curve does not change in evolution. (a) The present curve based on the relatively high values for Upolu and Fakaofo. (b) and (c), increases of 1.5 × and 2 × respectively.

immigration and extinction curves were altered.

The particular alteration of interest is the lowering of the tramp extinction curve from its present high level through evolution (indicated in Figure 3 as vector E) toward the lower, native species level. Notice that in a newly assembled faunula, in which imperfectly adapted species were drawn from a much larger but constant pool, immigration rates would tend to remain unchanged, since the recipient island (or continent) is not apt to influence the makeup of the source fauna in any significant way. [In the Indo-Australian ponerine fauna, for example, the ratio of exchange of species in older islands is approximately equal to the ratio of the areas, with the result that a continent or large island will dominate a small island in reciprocal influence (Wilson, 1965).] On the other hand, the extinction rates in the newly assembled faunula can be expected to be sensitive to evolution, since the rates would tend to be lowered by character displacement and other forms of local adaptation.

The question we wish to pose and try to answer is: In the case of an island like Upolu, by how much would the equilibrium species number be increased through evolutionary lowering of the extinction curves? This is a specific case of the more general problem—by how much will species density increase as evolution occurs in a newly assembled fauna? First, it can be affirmed parenthetically that the equilibrium number, or more precisely quasi-equilibrium number, in the newly assembled tramp faunula must be about 22, the number actually occurring there. This is a number quite consistent with the species-area curve presented by the eastern Polynesian faunulae, which are made up more purely of tramp species (Fig. 2). In addition to these tramp species, there are on Upolu 33 native species and one additional species of uncertain classification. These species have stood up to the onslaught of the tramp species by living in the extensive tracts of native vegetation extant on Upolu. Because the great majority of the same species also occur on Savai'i, a larger and even less disturbed island about 11 miles to the west, we conclude that not many extinctions have occurred among the native species of Upolu due to competition from tramp species. The tramp species are concentrated in the cultivated areas and the native species in the native forests, with the two groups overlapping widely in their habitat choices. Thirty-three is the minimum quasi-equilibrium number for native Upolu ants, and we take it to be not far from the true figure. We conclude that if the newly assembled fauna of tramp species were allowed to evolve on an island such as Upolu, without competition from natives and with the immigration rate held constant, the number would increase from about 22 to at least 33, or by a minimum factor of about 1.5.

What is the maximum to which the quasi-equilibrium would increase over a very long time? The answer of course must be no greater than 38, the number in the entire tramp species pool, or by a factor of approximately 2, unless species multiplication occurred. There is another, nontrivial line of evidence that suggests this upper limit to hold even in the presence of a much larger pool of immigrant species and higher immigration rates. In the Solomon Islands, the fauna is much closer to the very rich source fauna of New Guinea and contains correspondingly higher numbers of stocks and species. Even so, the expected number of native ant species on an island the size of Upolu (whose area is 430 square miles) is only about 45, as shown in Figure 4.

To summarize, it is to be expected that species density of a fauna in quasi-equilibrium will increase to some extent as local evolution occurs. In the case of a tropical island the size and position of Upolu, as the faunula changes specifically from a newly assembled, poorly integrated condition to one in which it is adapted enough to produce some species endemic to the island, the number of species can be expected to increase by a factor of between 1.5 and 2. If the newly assembled faunula is better adapted, as it would be if it were drawn entirely from species native to a nearby area, the lower limit would probably be less. In any case, it is hard to imagine a situation in which the species number would be more than doubled, unless the initial assemblage were deliberately chosen for incompatibility among the species.

Estimates can be made for increase in density for the whole range in island areas, at least for the Samoa area, if it is recalled that the species-area curves vary little among different faunas. Such an extended set of estimates is given in Figure 5.

SUMMARY

1. Few if any ant species are native to Polynesia east of Rotuma, Samoa, and Tonga. The central and eastern archipelagoes have been populated in large part by 35 "tramp" species carried there from various parts of the tropics by human commerce. Several other tramp species are known to have been carried to the Pacific but are not yet established in the center and east. The tramp species also occur, along with native ones, in western Polynesia and outside the Pacific area.

2. No one island contains all of the tramp species, and most contain less than one-fourth of them. Several lines of evidence suggest that the species densities have stabilized. Competitive replacement has evidently played a role in the stabilization.

3. The species numbers are considered not to be resting at perfect equilibrium but rather to be at most at a "quasi-equilibrium." If the immigration rates of the 38 tramp species (35 established and 3 others available) were held constant, and after all the species had an opportunity to colonize a given island, the number at quasi-equilibrium could still be expected to increase slowly as local adaptation proceeded through evolution.

4. By comparing the newly assembled tramp faunulae with older native faunulae of Indo-Australian origin in Samoa, an estimate of the potential increase in species numbers through evolution was made. On an island with the size and history of Upolu the factor of increase should be between 1.5 and 2.

5. Estimates can then be made for islands of all areas if the slope of the changing species-area curve holds constant (Fig. 5).

LITERATURE CITED

BROWN, E. S. 1958. Immature nutfall of coconuts in the Solomon Islands. II. Changes in ant populations and their relation to vegetation. Bull. Entomol. Res. **50**: 523–558.

BROWN, W. L. 1958. A review of the ants of New Zealand. Acta Hymenopterologica **1**: 1–50.

CLAGG, C. F. 1957. (Report on Exhibition.) Proc. Hawaiian Ent. Soc. for 1956, p. 197.

MACARTHUR, R. H., AND E. O. WILSON. 1963. An equilibrium theory of insular zoogeography. Evolution **17**: 373–387.

MANN, W. M. 1918. The ants of the British Solomon Islands. Bull. Mus. Comp. Zool. Harvard **63**: 273–391.

SANTSCHI, F. 1928. Formicidae. Insects of Samoa **5**: 41–58.

WILSON, E. O. 1965. The challenge from related species, p. 7–24. *In* Baker, H. G. and G. L. Stebbins [ed.] *The Genetics of Colonizing Species.* Academic Press, New York.

WILSON, E. O., AND R. W. TAYLOR. 1967. The ants of Polynesia. Pacific Insects Monographs (in press).

1969

"The species equilibrium," *in* G. M. Woodwell and H. H. Smith, eds., *Diversity and Stability in Ecological Systems* (Brookhaven Symposia in Biology, no. 22) (Biology Department, Brookhaven National Laboratory, Upton, NY), pp. 38–47

Here I reviewed and connected two of my population-level studies that followed closely upon the publication in 1967 of the theory of island biogeography. The first is the experiment I conducted with Daniel S. Simberloff, then working with me as a Ph.D. student (and later to go on to a distinguished career in ecology) in experimental biogeography. In essence, we arranged to have a series of tiny mangrove islets in Florida Bay fumigated and "defaunated" of all insects and other arthropods, at toxic levels that did not destroy the mangrove vegetation. We then monitored the natural recolonization of the islets by arthropods from the surrounding mangrove forests. Control islets were also monitored after being subjected to all the same procedures except for the final defaunation. We found that equilibrial or at least quasiequilibrial species numbers were reestablished in only 4–6 months. The numbers also reflected the distance effect predicted by island biogeographic theory. This, I believe, was one of the first experiments conducted on entire natural ecosystems.

The second topic provided in the article to follow is my conception of the evolution of quasiequilibria, starting with the arrival of immigrant species and proceeding through two further periods to the final stage of genetic coevolutionary packing.

The Species Equilibrium

EDWARD O. WILSON
Biological Laboratories, Harvard University, Cambridge, Massachusetts 02138

THE THEORY OF SPECIES EQUILIBRIUM

The species equilibrium model proposed by MacArthur and Wilson[1,2] takes one or the other of two basic forms. The *non-interactive equilibrium*, illustrated in Figure 1*A*, is conceived of existing when the population densities of the resident species are so low as to minimize the possibility of extinction by interference. The principal kinds of interference envisaged are competitive exclusion and excessive predation. Non-interactive equilibria can be attained when species-immigration rates onto a newly empty island (or a newly opened habitat of any kind) are so high that the species equilibrium is reached prior to the attainment of population equilibria by individual species. In this circumstance species extinctions still occur – and at a predictably high rate. But they are more likely to be due to density-independent factors operating on the small numbers of propagules and their immediate descendants. Because the extinction rate, measured in numbers of species going extinct per unit time, does not depend on the interaction of species, it can be expected to increase as a linear function of the number of species already present in the community. Hence the extinction curve can be drawn as a straight line, as shown in Figure 1*A*.

By contrast, the *interactive equilibrium*, which is illustrated in Figure 1*B*, does involve exclusions of species by other species. As the number of species in the community is increased, the species extinction rate can be expected to increase in some exponential form. The extinction curve will therefore be concave in an upward direction, as indicated in a very general way in the diagram. Interactive equilibria can be expected to occur in all but the youngest and least stable communities. Simberloff[3] has pointed out that new islands or other vacant habitats that are colonized very rapidly are likely to attain a non-interactive equilibrium first and then shift rather quickly to an interactive equilibrium as the resident populations of individual species approach their saturation levels. The two equilibrial species numbers might be the same; but more likely they would not. The direction of the change cannot be anticipated without further knowledge of the community. If synergistic effects prevail, such as the increase of the survival time of predator species in the presence of more prey, the interactive equilibrium will be higher than the non-interactive one. If the effects of interactive extinctions on the other hand outweigh the synergistic effects, the interactive equilibrium will be lower.

The concept of the non-interactive equilibrium offers the theorist the immediate advantage of great simplicity in design, providing he is willing to make do with a deterministic (as opposed to stochastic) analysis. For if the extinction and immigration curves approximate straight lines, it is possible to state explicitly the relation between extinction rates, immigration rates, and the number of species at equilibrium. Let S be the number of species in the isolated community, $\hat{S}$ the equilibrial number of species, P the number of species in the "pool" (i.e., the number available to immigrate to the community, λ_S and μ_S the immigration and extinction rates in the community in species per unit time, and λ_A and μ_A the average immigration and extinction rates per species. Then from the non-interactive model displayed in Figure 1*A*,

$$dS/dt = \lambda_A(P-S) - \mu_A S\,, \tag{1A}$$

the solution of which is

$$S = \frac{\lambda_A P}{\lambda_A + \mu_A}\left(1 - e^{-(\lambda_A+\mu_A)t}\right). \tag{1B}$$

When t becomes very large, S approaches $\hat{S} = \lambda_A P/(\lambda_A + \mu_A)$. Let us take some lower value, say $S = 0.9\hat{S}$, and label as $t_{0.90}$ the time required to go from zero species to this value of S. Then

$$t_{0.90} = \frac{2.3}{\lambda_A + \mu_A}\,. \tag{2A}$$

In the special case where $\lambda_A = \mu_A$, and recalling that $\mu_A \hat{S} = \mu_S$ by definition,

$$t_{0.90} = \frac{1.2}{\mu_A} = \frac{1.2\hat{S}}{\mu_S}\,. \tag{2B}$$

In the absence of any knowledge of λ_A and μ_A, the ratio $1.2\,\hat{S}/t_{0.90}$ is therefore a good first approximation of the turnover rate, i.e., the extinction rate (which equals the immigration rate), at equilibrium. For small communities, in which the equilibrial number $\hat{S}$ is only a small fraction of the number of species (P) in the species pool, we can expect λ_A to be less than μ_A, since from (1B)

$$\frac{\hat{S}}{P} = \frac{\lambda_A}{\lambda_A + \mu_A}\,. \tag{3}$$

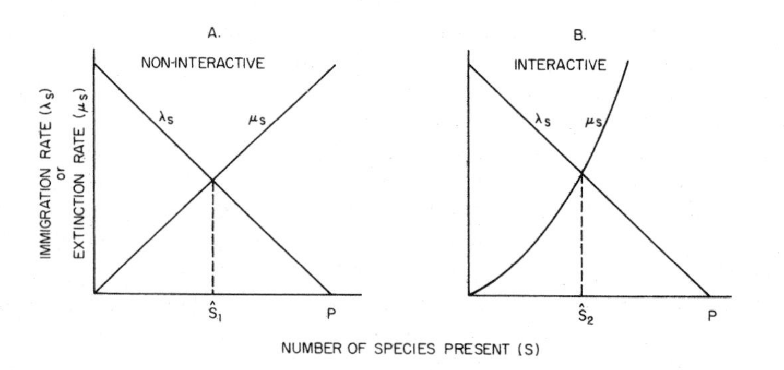

Figure 1. The two forms of the species equilibrium model, specifying the simplest conditions necessary for (*A*) the non-interactive equilibrium and (*B*) the interactive equilibrium.

As the community becomes still smaller, and $\hat{S}$ continues to diminish with respect to P, then $t_{0.90}$ approaches but can never reach $2.3/\mu_A$. This means that the turnover rate at equilibrium, $\mu_{\hat{S}}$, can approach but never reach $2.3\hat{S}/t_{0.90}$. Very few isolated communities close enough to the source areas to attain a non-interactive equilibrium would ever have a value of λ_A that exceeds μ_A and thus be in a position to attain an $\hat{S}$ as great as one-half the entire species pool. It follows that in most conceivable cases

$$\frac{1.2\hat{S}}{t_{0.90}} < \mu_{\hat{S}} < \frac{2.3\hat{S}}{t_{0.90}}.$$

This presents the experimentalist with a favorably narrow range of predictable values with which to check the non-interactive equilibrial hypothesis. Even in cases where the immigration curve is assumed to be concave rather than straight (and the extinction curve remains straight to fulfill the non-interactive condition) the values are unlikely to fall outside this range.

AN EXPERIMENTAL DEMONSTRATION OF A FAUNISTIC EQUILIBRIUM

In order to test some of the predictions of the equilibrium theory and to acquire other kinds of information about the colonization process, Simberloff and Wilson[4-6] devised an experiment in which all the arthropods were removed from very small islands and the subsequent recolonization of the islands was monitored by frequent censuses. The islands selected were red mangrove clumps, 11 to 18 m in diameter, standing isolated in shallow water in the Florida Keys on the western, Florida Bay side. The islands consisted entirely of red mangrove trees (*Rhizophora mangle*) and, with a single exception, were on sand or mud foundations completely covered during high tide. They were located at various distances between 2 and 500 m from the nearest bodies of land.

The Florida Keys are inhabited by an estimated total of 4000 insect species, of which about 500 inhabit mangrove swamps and islands. Of the latter group, ~75 species are commonly found on the small mangrove islands. In addition, some 150 species of spiders, scorpions, pseudoscorpions, centipedes, and arboreal isopods also occur in the Florida Keys; and of these about 20 occur commonly on the small mangrove islands. Thus the total arthropod species pool, which by definition includes species able to emigrate to the islands whether or not they are able to persist and breed there, almost certainly contains on the order of 1000 species. Yet at any given moment only 20 to 40 arthropod species occur on each of the small mangrove islands. Together these comprise tens of thousands of individuals. The actual species lists vary greatly from island to island, and shift markedly over periods of a year or less. These facts alone suggest that the species numbers are at equilibrium. We hoped that by removing all of the species at once, thus perturbing the system in the most direct and drastic manner possible, we would be fortunate enough to be able to document the reattainment of equilibrium in a clear-cut manner, and to measure species immigration and extinction rates during the process.

Six islands were successfully "defaunated" by the following procedure. After a thorough census of the arthropod species had been taken, the entire island was covered by a plastic-impregnated nylon tent and fumigated with methyl bromide at

22 to 25 kg/1000 cubic meters for two hours. A second census was conducted immediately afterward. We found that all arthropods were destroyed by this treatment with the possible exception of a few larvae belonging to two deep-boring beetle species, and several lines of evidence suggest that even these individuals probably died in a few days or weeks from the long-range effects of the poisoning. Variable damage was done to the mangrove vegetation, but in most cases it was minor, and the arthropod microhabitats – now emptied of species – were left intact.

The recolonization of the islands proceeded at a gratifyingly swift pace. Gratifying, I say, not only because Simberloff and I were unprepared to wait out a lifetime to learn the result, but also because the species equilibria were reached while population densities were still very low and competitive exclusion was probably an insignificant factor. In Figure 2 are shown the colonization curves of the four islands that comprise the Sugarloaf Key group in the lower Florida Keys. In every case except E1 (our "distant" island, located 500 m from the nearest source), the number of species rose to a level slightly above the original, pre-defaunation level, and then, about one year following defaunation, slumped back to about the pre-defaunation level. The first high peak was attained when population densities of most of the colonist species were still low. The slump occurred as population densities approached

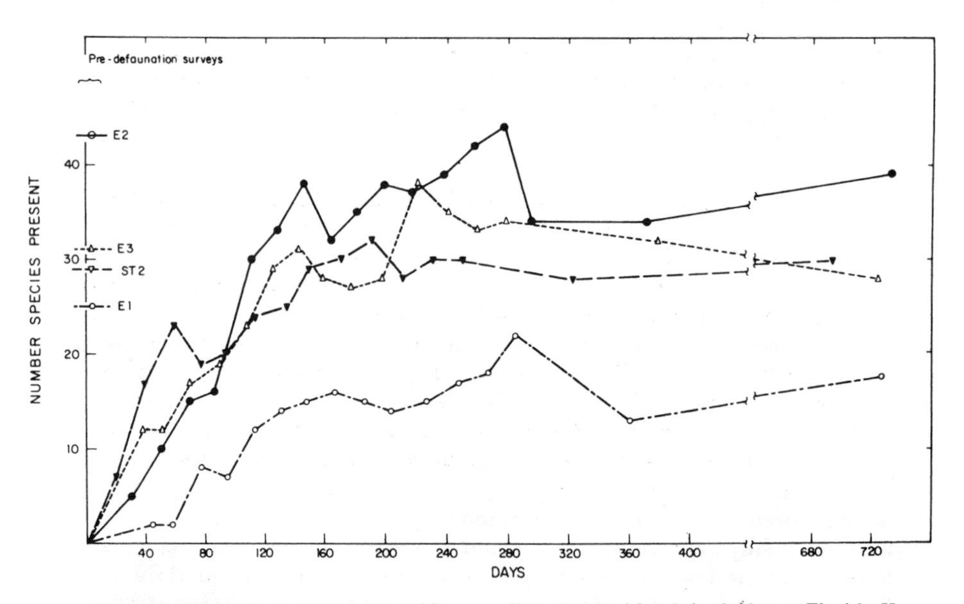

Figure 2. The colonization curves of four small mangrove islands in the lower Florida Keys whose entire faunas, consisting mostly of arthropods, were removed by methyl bromide fumigation. The species numbers just before defaunation and at intervals following it are shown. The number of species is an inverse function of the distance to the nearest source. This effect was evident in the pre-defaunation census and was preserved when the faunas regained equilibrium after defaunation. Thus, the near island E2 has the most species, the distant island E1 the fewest, and the intermediate islands E3 and ST2 intermediate numbers of species. (From Simberloff and Wilson.[6])

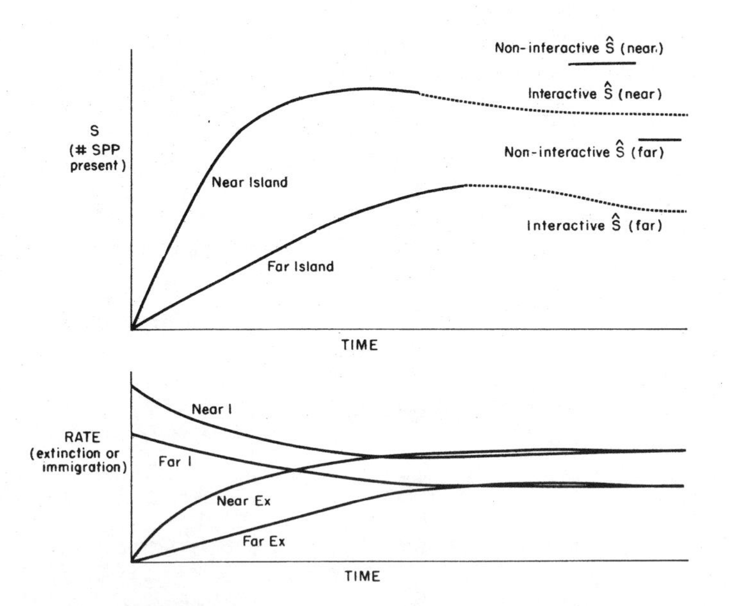

Figure 3. Generalized colonization curves from near and distant islands and a model indicating the origin of the curves as the integrated differences between the immigration time curves (Near I and Far I versus time) and the extinction time curves (Near Ex and Far Ex versus time). (From Simberloff.[3])

the old, pre-defaunation levels. We concluded that the first peak represented the attainment of a non-interactive equilibrium, and that this shortly gave way to an interactive equilibrium which has since persisted for at least two years following defaunation. Figure 3 shows our interpretation of these events in a more graphic form. (The second-year census was taken in March 1969).

It can also be seen that E1, the "distant" island, fulfills a second prediction of the basic equilibrium model. That is, its fauna approached equilibrium more slowly than did those of other, closer islands. This behavior is consistent with the following relation implicit in equation (1B): when distance is increased, the average immigration rate λ_A will decrease, and the term in the outer parentheses in (1B) will approach unity more slowly. Therefore S approaches $\hat{S}$ more slowly.

A detailed history of colonization of the kind obtained for all of the islands is exemplified in Figure 4. These data, together with our notes on the biology of the individual arthropod species, have provided a wealth of new information on the relative dispersal ability of various arthropod taxa, the life stages involved in colonization, and many other facets of the biology of colonization. It is not my intention to review these results here, but rather to single out one additional quantitative estimate that has relevance to the basic equilibrium model. This is the measurement of the species turnover rates, which can be used to test equations (2A) and (2B). In Table 1 are given the values of the turnover rate predicted by the non-interactive form of the model for a "typical" island (E3), one of the experimental group that

THE COLONISTS OF ISLAND E9

PRE 24 45 62 84 101 117 136 153 171 193 210 229 247 266 364

ORTHOPTERA — Gryllidae — Cycloptilum sp.
Cyrtoxipha confusa
Orocharis sp.
DERMAPTERA — Labiduridae — Labidura riparia
COLEOPTERA — Anobiidae — Cryptorama minutum
Tricorynus sp.
Anthicidae — Sapintus fulvipes
Vacusus vicinus
Buprestidae — Actenodes auronotata
Chrysobothris tranquebarica
Cantharidae — Chauliognathus marginatus
Cerambycidae — Styloleptus biustus
Curculionidae — Cryptorhynchus minutissimus
Pseudoacalles sp.
Lathridiidae — Holoparamecus sp.
Oedemeridae — Oxacis sp.
Fam. Unk. — Gen. sp.
THYSANOPTERA — Phlaeothripidae — Haplothrips flavipes
Neurothrips magnafemoralis
Thripidae — Pseudothrips inequalis
CORRODENTIA — Caeciliidae — Caecilius sp. np.
Lachesillidae — Lachesilla n. sp.
Lepidopsocidae — Echmepteryx hageni b
Liposcelidae — Belaphotroctes okalensis
Embidopsocus laticeps
Liposcelis sp. not bostrychophilus
Peripsocidae — Ectopsocus sp. bμ
Peripsocus stagnivagus
Psocidae — Psocidus texanus
Psocidus sp. 1
Trogiomorpha — Gen. sp.
HEMIPTERA — Anthocoridae — Dufouriellus afer
Cixiidae — Oliarus sp.
Miridae — Psallus conspurcatus
Pentatomidae — Oebalus pugnax
Fam. Unk. — Gen. sp.
NEUROPTERA — Chrysopidae — Chrysopa collaris
Chrysopa externa
Chrysopa rufilabris
LEPIDOPTERA — Eucleidae — Alarodia slossoniae
Olethreutidae — Ecdytolopha sp.
Phycitidae — Bema ydda
Psychidae — Oiketicus abbottii
Ptineidae — Nemapogon sp.
Pyralidae — Tholeria reversalis
Saturniidae — Automeris io
Fam. Unk. — Gen. sp.
DIPTERA — Hippoboscidae — Olfersia sordida
Fam. Unk. — Gen. sp.
HYMENOPTERA — Braconidae — Apanteles hemileucae
Apanteles marginiventris
Callihormius bifasciatus
Ecphylus n. sp. nr. chramesi
Iphiaulax epicus
Chalcidae — Gen. sp. 1
Gen. sp. 2
Gen. sp. 3
Gen. sp. 4
Eulophidae — Euderus sp.
Eumenidae — Pachodynerus nasidens
Eupelmidae — Gen. sp.
Formicidae — Brachymyrmex sp.
Camponotus floridanus
Camponotus sp.
Crematogaster ashmeadi
Monomorium floricola
Paracryptocerus varians
Pseudomyrmex elongatus
Pseudomyrmex "flavidula"
Tapinoma littorale
Xenomyrmex floridanus
Gen. sp.
Ichneumonidae — Calliephialtes ferrugineus
Casinaria texana
Pteromalidae — Urolepis rufipes
Sphecidae — Trypoxylon collinum
Vespidae — Polistes sp.
ARANEAE — Araneidae — Argiope argentata
Eriophora sp.
Eustala sp.
Gasteracantha ellipsoides
Metepeira labyrinthea
Nephila clavipes
Clubionidae — Aysha sp.
Dictynidae — Dictyna sp.
Gnaphosidae — Sergiolus sp.
Linyphiidae — Meioneta sp.
Lycosidae — Pirata sp.
Salticidae — Hentzia palmarum
Scytodidae — Scytodes sp.

Figure 4. The history of the colonization of E9, a typical experimental island in the Florida Keys. "Pre" is the pre-defaunation census. Solid entries indicate that a species was seen; shaded, that it was inferred to be present from other evidence; open, that it was not seen and inferred to be absent. (From Simberloff and Wilson.[5])

Table 1

Predicted and Estimated True Turnover Rates on E3, a Typical Experimental Island in the Florida Keys, Where (approximately) $\hat{S}=30$ Species and $t_{0.90}=120$ Days

	Extinction rate at equilibrium (species/day)
Predicted by non-interactive model	
Lower limit ($\mu_{\hat{S}} = 1.2\hat{S}/t_{0.90}$)	0.30
Upper limit ($\mu_{\hat{S}} = 2.3\hat{S}/t_{0.90}$)	0.58
Actual (true value)	
Lower limit, based on observed values	0.08
Estimated upper limit	1.50

was located at an intermediate distance from the source areas, and the estimated limits of the actual values that existed when the equilibrium was first attained. The lower limit is simply the extinction rate of the species actually observed, lacking the upward correction probably necessitated by unseen transients. The upper limit, on the other hand, accounts for the transients in the most generous conceivable manner, by means of a computerized simulation method developed by Simberloff.[3] The maximum estimate of the true number is about 20× the minimum estimate. Both seem intuitively extremely high, entailing 0.3 to 5% on the entire fauna *per day*. Yet this is very close to what the non-interactive model predicts: depending on the relative slopes of the immigration and extinction curves assumed, the turnover rate at equilibrium should fall somewhere in the range 0.30 to 0.58 species/day, or 1 to 2% of the entire fauna per day.

FURTHER SHIFTS IN THE EQUILIBRIUM

There exist universal qualities of biology at the species level which dictate that species equilibria must always be quasi-equilibria. By this I mean that if we observe a well established fauna or flora for short periods of time, we will note that the species extinction rate at least approximately equals the species immigration rate; but if we continue to watch it for a long period of time we will probably note a steady and very slow drift in the average species number, most likely in an upward direction. It seems useful at this point to recognize four kinds of equilibria, which can be expected to appear *seriatim* in each long-lived community. They can be defined as follows (see also Figure 5).

1. *Non-interactive species equilibrium.* As just described, this equilibrium is reached in certain cases prior to the attainment of sufficiently high population densities to make competitive exclusion and perhaps other forms of species interference a major factor in species extinction. The original deterministic form of the non-interactive equilibrium model by MacArthur and Wilson has recently been supplemented by a more precise stochastic form by Simberloff.[3] This aspect of the theory is therefore relatively well worked out, and the process itself has been documented in our Keys experiments.

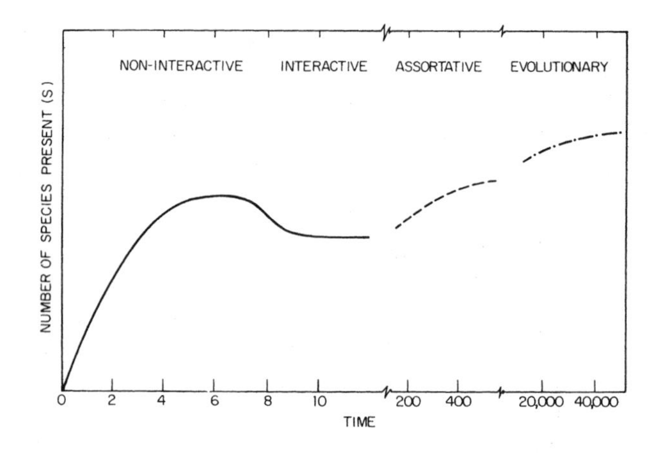

Figure 5. The postulated sequence of equilibria in a community of species through time. The time scale is imaginary, supplied here only to convey the notion of the vastly greater time periods required for shifts to states beyond the initial interactive equilibrium.

2. *Interactive species equilibrium.* When the populations of individual species become dense enough to make competitive exclusion and other forms of species interaction a major factor in species survivorship, or at least a significantly more important factor than when densities are very low, the equilibrium can be said to be interactive. There is no way to predict from theory the magnitude and velocity of change leading from the non-interactive to the interactive state. The causes of extinction in populations are simply too poorly understood. The most that can be said is that in our experimental islands the amount of change was not very great, on the order of 20% or less, while the changeover occurred quite rapidly, within the space of several months.

3. *Assortative species equilibrium.* Examination of Figure 4 confirms that although the number of species equilibrated on E9 (and on the other islands as well) the species *composition* continued to change rapidly. Thus new combinations of species were – and still are – being generated. Inevitably, combinations of longer-lived species must accumulate by this process. Such species persist longer either because they are better adapted to the peculiar physical conditions of the local environment, or else because they are able to coexist longer with the particular set of species among which they find themselves. I propose to refer to the species numbers of such favored combinations as "assortative" equilibria. In certain associations, especially plant communities, the progression is a regular one, a special case that has long been understood by ecologists and referred to as a succession. As a rule, we would expect assortative equilibria to consist of a greater number of species than their ancestral interactive equilibria. But there is no sure way of predicting whether this will always be the case, or, when it occurs, what will be the magnitude of the difference.

4. *Evolutionary species equilibrium.* If the community persists for a sufficiently long period of time, its member species can be expected to adapt genetically to local environmental conditions and to each other. The result should be a lowering of the

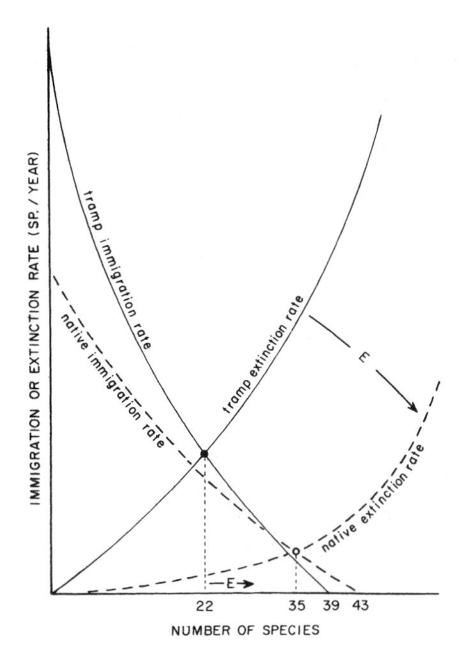

Figure 6. An equilibrium model of the ant fauna of Upolu, Samoa. An attempt is made to estimate the probable increase in the number of species of tramp origin from the present interactive (or assortative) equilibrium, at 22 species, that would result if the extinction rate were lowered by means of evolution (E→) toward the native level. The number of known native species on Upolu is 35, and this number is taken as the minimum to which the tramp equilibrium could move if given enough time. (From Wilson and Taylor.[7])

extinction curve, with little effect on the immigration curve, and consequently a raising of the species equilibrium. Although the process cannot yet be treated in the context of quantitative theory, some idea of the amount of change in species numbers can be gained by comparing newly assembled biotas with older ones containing at least some endemic species. Wilson and Taylor[7] took this approach in an analysis of the Polynesian ant faunas. Most of the islands of Polynesia are inhabited by "tramp" species carried inadvertently by man from many different parts of the world. Most were introduced by modern European commerce during the past 200 years. The total number of such species known to occur in the Pacific area is 38. The number of species found on individual islands increases approximately as the cube root of the area. Islands the size of Upolu, in the Samoan group, contain 15 to 25 tramp species. Upolu itself, for example, contains 22. The islands of extreme western Polynesia also possess old, partly endemic ant faunas. Altogether, 43 undoubted native species occur in this area of which 35 are found on Upolu. Thus the pools of tramp and native species are similar in size (38 versus 43 species), but the equilibrial number of tramp species on Upolu (22) and islands of similar size is much lower than the equilibrial number of native species that occur on Upolu (35). Figure 6 shows how we applied these fragments of information to the basic equilibrium model. The immigration rates of tramp species are known, by historical records, to be much higher than those of native Polynesian species; we have accordingly drawn the immigration curve higher. It follows that the only way that the equilibrial species could be increased in a native fauna is by a lowering of the extinction curve; and this we postulate to have occurred by means of improved adaptation through evolution. We concluded that the transition from the early interactive or assortative equilibria, roughly duplicated at the present time by the synthetic tramp faunas, to the evolutionary equilibria of the native faunas resulted in an approximate doubling of the number of species. And since the area-species relation is a simple power function (MacArthur and Wilson[2]), the same result should apply to other Polynesian islands with different areas.

ADDENDUM

I wish to take this opportunity to make a statement directed at conservationists who, very understandably, will be concerned with our concept of doing ecological experiments. The experimental islands were a very few among hundreds of a similar nature that occur in the Florida Keys. They were selected in consultation with biologists and other officials of the Everglades National Park and Great White Heron National Refuge. Each island consisted of only one or several red mangrove trees, and we made every effort to remove the arthropods without harming the trees themselves or the marine life at the bases of the trees. In fact the only damage done was an accidental browning of part of the foliage on several islands, and the complete killing of the trees on one of the islands. With the exception of the single case in which several trees were killed, the damage was reversible, and such effects can be eliminated altogether by improvement of the defaunation technique in future experiments. Even the arthropods returned to their original levels of diversity within a year of the defaunation. Two years after the experiment was performed, the little islands and their surrounding environments are essentially as we found them. I believe in the principle that such experiments should be designed with extreme care. In no case should ecologists, of all people, be guilty of causing significant irreversible harm to the environment. Particular care should be taken with island ecosystems.

REFERENCES

1. MacArthur, R.H. and Wilson, E.O., *Evolution* **17**, 373 (1963).
2. MacArthur, R.H. and Wilson, E.O., *The Theory of Island Biogeography,* Princeton University Press, 1967.
3. Simberloff, D.S., *Ecology* **50**, 296 (1969).
4. Wilson, E.O. and Simberloff, D.S., *Ecology* **50**, 267 (1969).
5. Simberloff, D.S. and Wilson, E.O., *Ecology* **50**, 278 (1969).
6. Simberloff, D.S. and Wilson, E.O., In preparation.
7. Wilson, E.O. and Taylor, R.W., *Evolution* **21**, 1 (1967).

1971

"The plight of taxonomy," *Ecology* 52(5): 741

This article marks the beginning of a long personal effort, which to others may have seemed quixotic, to help revive taxonomy as a mainstream discipline of biology. By the 1960s taxonomy, or systematics as it is often called when given a more modern emphasis with phylogenetic analysis, had slid to the bottom of the pole of recognition and prestige. Even with the new methodologies in use, it was widely perceived as an antiquated nineteenth-century subject with little to offer except an identification service. This view was obviously wrong. As time wore on, what systematists had always known came to be widely recognized, that most of Earth's biodiversity was (and remains) unexplored. By the early 2000s, major initiatives such as the All-Species Project for exploration, the Tree of Life for phylogenetic reconstruction, and the Encyclopedia of Life for organizing information from all biological disciplines marked the resurgence of taxonomy, the beginning realization of a dream that was kept alive during the lean years.

ECOLOGY

Vol. 52 LATE SUMMER 1971 No. 5

THE PLIGHT OF TAXONOMY

A recent National Science Foundation report on the condition of the principal systematic collections in the United States[1] is an emergency call that ecologists had better heed. Its authors state, soberly and in my opinion with complete justification, that "The long-term under-support now threatening the very existence of this most precious, precarious, irreplaceable resource, if not relieved will inexorably lead to its deterioration, precisely at the time when humanity clamors for solutions for which systematic collections are essential. Help—substantial help, sustained help—is needed desperately and now."

The discipline of systematics is in an equally distressing condition. Taxonomists have been all but excluded from many of the best biology departments in this country, often to be replaced—if at all—by biometricians who occupy themselves more with the philosophy and methodology than with the substance of taxonomy. In the fashion rankings of academic biology, substantive taxonomy long ago settled to the bottom. This must not be permitted to continue. Ecologists, now beginning to savor a windfall of popularity and growing financial support, should recognize their dependence on substantive taxonomy and special responsibility to it. Most of the central problems of ecology can be solved only by reference to details of organic diversity. Even the most cursory ecosystem analyses have to be based on sound taxonomy. And after the first broad measurements of energy flow and geochemical cycling have yielded their important but limited information, what remains of intellectual challenge stems chiefly from details in the biology of particular species. The food nets, the fluctuation of population numbers and biomasses, the diel and seasonal rhythms, the rates and patterns of dispersal, the colonization of empty habitats, microevolution, physiological adaptation, and most other basic topics of ecology, require a deep understanding of the biology of individual taxa. Progress depends not just on the correct identification of species but also on the mastery of larger taxonomic groups of the kind best achieved through deliberate specialization by taxonomists and taxonomically trained ecologists.

There are two clearcut ways in which taxonomy should be strengthened. The first is to give more academic appointments to taxonomists who have mastered the substance, and not just the biometric methodology, of systematics, biologists who are experts on reasonably large groups of organisms on a continental or global basis. They must be allowed to take their place among the theoretical and experimental ecologists now so much in favor in biology departments. The second step is to grant larger amounts of support to the institutions housing the principal systematic collections in this country, as called for by the Steere Committee report. It is to be hoped that ecologists, in their newly acquired influence, will accept that such aid to their intellectual kindred, the taxonomists, is both part of their larger responsibility to science and in their own immediate self interest.

E. O. WILSON
Biological Laboratories
Harvard University
Cambridge, Massachusetts 02138

[1] *The systematic collections of the United States: an essential resource. Part 1. The great collections: their nature, importance, condition, and future.* A report to the National Science Foundation by the Conference of Directors of Systematic Collections; W. C. Steere, Chairman. Published by The New York Botanical Garden, Bronx, New York 10458. xi + 33 pp.

1988

"The biogeography of the West Indian ants (Hymenoptera: Formicidae)," *in* James K. Liebherr, ed., *Zoogeography of Caribbean Insects* (Cornell University Press, Ithaca, NY), pp. 214–230

Ants, thanks to their dominance in the Dominican amber, have the best-known fossil record of any animal group in the West Indies. Using the data on the Dominican amber fauna accumulated principally by Cesare Baroni Urbani and myself in previously published studies, I was able to decipher a major trend: while the contemporary ant fauna has the disharmonic composition typical of oceanic islands, it was a harmonic continental fauna in Miocene times. The reason for the disparity is that when present-day Hispaniola (Dominican Republic plus Haiti) and the other Greater Antillean islands were in the early stages of plate tectonic movement from Central America and therefore close enough to receive a full complement of mainland ant genera, the fauna was rich and continental. With increasing isolation, and the ordinary extinction rate expected to result on more distant islands, the number of genera declined. At the same time a few of the surviving genera experienced adaptive radiations, keeping the overall number of ant species in the larger West Indian islands at a much higher level than otherwise would have been the case.

9 • The Biogeography of the West Indian Ants (Hymenoptera: Formicidae)

Edward O. Wilson

By my count, 89 genera and well-marked subgenera (see footnote in Table 9-1) comprising 383 species of ants have been recorded from the West Indies. Three of the genera (*Codioxenus, Dorisidris, Hypocryptocerus*), constituting 3.4% of the total, are endemic in the special sense of being known so far only from the West Indies; each is also limited to a single island. A total of 176 species, or 46% of the entire fauna, is known solely from the West Indies. Of these, 154, or 40.2% of the entire fauna, have been recorded from a single island—in other words, they might be insular endemics in the strict sense.

Our knowledge of this large and diverse fauna has grown to the point that several interesting generalizations can be made about its origin and dispersal. At the same time, substantial gaps remain, leaving open the possibility for significant discoveries in the future. In particular, most of the genera are in serious need of taxonomic revision, while some of the smaller and medium-sized islands remain largely unexplored. In addition, large collections of ants in the Tertiary (late Oligocene or early Miocene) amber of the Dominican Republic have recently become available, and they are yielding a new and in some cases surprising view of the history of the Greater Antillean fauna.

The purpose of this article is to provide a synopsis of the West Indian ant fauna in terms that can be utilized by biogeographers, while calling attention to some of the principal research opportunities that remain. The islands included in my survey of the living fauna are the Bahamas, the major arc of the Antilles from Cuba to Grenada, and Trinidad and Tobago. I have omitted Margarita, Blanquilla, the Roques archipelago, and the Dutch West Indies (Aruba, Curaçao, Bonaire), partly because they have westward locations well away from the Orinoco delta and hence are less likely to have served as steppingstones to the Lesser Antilles, and partly because their ant

Table 9-1. Diversity and endemicity in the ant faunas of the better-known West Indian islands

		Genera and Subgenera[1]				Species			
Island	Area (km²)	No. of nonendemic genera	No. of endemic genera	Total no. of genera	Percentage generic endemicity	No. of nonendemic species	No. of endemic species	Total no. of species	Percentage species endemicity
Cuba	114,525	45	2	47	4.44	74	72	146	49.66
Bahamas	13,864	28	0	28	0	52	7	59	12.07
Jamaica	10,991	29	0	29	0	53	6	59	10.34
Hispaniola	76,190	36	1	37	2.70	57	31	88	35.63
Mona	52	11	0	11	0	13	0	13	0
Puerto Rico	8,897	31	0	31	0	52	4	56	7.27
Culebra	26	17	0	17	0	26	0	26	0
St. Thomas	83	26	0	26	0	37	2	39	5.13
St. Croix	218	6	0	6	0	7	0	7	0
Antigua	280	7	0	7	0	11	0	11	0
Montserrat	102	6	0	6	0	7	0	7	0
Guadeloupe	1,780	14	0	14	0	15	0	15	0
Dominica	728	22	0	22	0	28	0	28	0
Martinique	1,116	17	0	17	0	20	0	20	0
St. Lucia	616	12	0	12	0	15	1	16	6.67
St. Vincent	389	35	0	35	0	57	6	63	9.68
Barbados	430	18	0	18	0	16	2	18	11.76
Grenada	344	23	0	23	0	38	1	39	2.63
Tobago	300	15	0	15	0	18	0	18	0
Trinidad	4,828	60	0	60	0	126	22	148	14.86

[1]The following relatively well-marked subgeneric distinctions were included in these counts, preliminary to a clearer resolution of their status: *Crematogaster* s.s. and C. (*Orthocrema*); *Leptothorax* (*Macromischa*) and L. (*Nesomyrmex*); *Pachycondyla* s.s. and P. (*Nylanderia*); *Solenopsis* s.s., S. (*Diplorhoptrum*), and S. (*Euophthalma*). All these taxa together count for 12 "well-marked" subgenera equivalent to full genera; the genera to which they belong (e.g., *Crematogaster*) were *not* then added to the counts for the West Indies as a whole.

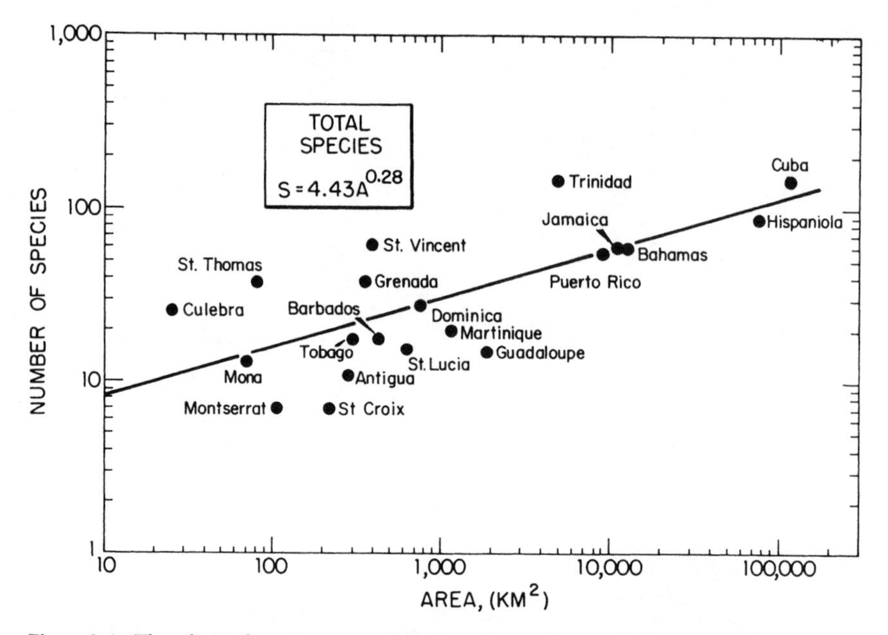

Figure 9-1. The relation between area and total species numbers in the better-known ant faunas of the West Indian islands. Curve estimated by least squares method.

faunas are so poorly known. Finally, I have included a brief account of our rapidly growing knowledge of the ants of Dominican amber.

The Data Base

Locality records of the living fauna were compiled earlier from the literature in catalog form by Kempf (1972). This extraordinarily useful publication has been supplemented by revisions of *Odontomachus* by Brown (1976), the ecitonine army ants by Watkins (1976), *Leptothorax* (*Macromischa*) by Baroni Urbani (1978), and portions of *Pheidole* by Brown (1981), as well as a synopsis of the Cuban ant fauna by Alayo (1974). I have recently provided updated lists of the genera of Haiti and the Dominican Republic (Wilson 1985a–d). Earlier species-level treatments of individual islands include Forel's study (1893) of the unusually thorough collections made by H. H. Smith on St. Vincent, as well as monographs and supplementary studies on Cuba by Wheeler (1913), Hispaniola by Wheeler and Mann (1914) and Wheeler (1936), and Puerto Rico by Smith (1936) and Torres (1984). Levins et al. (1973) made thorough collections of ants on 140 islands and cays of the Puerto Rico bank, plus the nearby islands of St. Croix, Mona, Monito, and Desecheo. They provided a number of interesting ecological

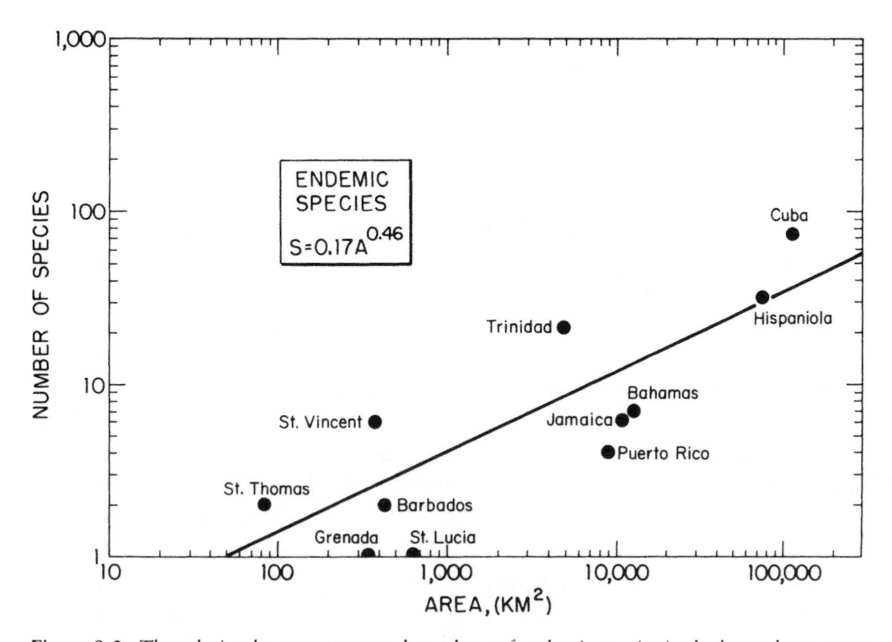

Figure 9-2. The relation between area and numbers of endemic species in the better known ant faunas of the West Indian islands. In some cases, especially the smaller islands, the numbers probably represent overestimates, due to the likelihood that species now known only from a single island (hence counted as "endemic") will eventually be found elsewhere. Curve estimated by least squares method.

and biogeographic generalizations concerning the 52 species encountered, but unfortunately did not publish an island-by-island list (the material is deposited in the Museum of Comparative Zoology, Harvard University).

Additional, unpublished collection data used in the present study, originating especially from Cuba, Hispaniola, and Trinidad, have been obtained by W. L. Brown and me. Current information on the ants of Dominican amber has been summarized in Baroni Urbani and Saunders (1982) and Wilson (1985a–e).

The Relation between Area and Diversity

The relations of area to numbers of genera and species, respectively, on the better-known islands are summarized in Table 9-1 and Figures 9-1 and 9-2. When one takes into account the fact that much of the variance is due to the still very imperfect collecting on most of the islands, the area-diversity curves are seen to be conventional. In particular, the faunistic exponent z (in $S = bA^z$, where S = number of species, A = island area, and b = a taxon area-specific constant) is 0.28, a typical number for closely grouped islands

(MacArthur and Wilson 1967). It is reasonable to suggest that the species numbers are at or near equilibrium, although this cannot be proved entirely from the area-diversity curves. Also, as depicted in Figure 9-2, the slope of the endemic species against area is steeper ($z = 0.46$, with few or no endemics being the rule in islands with areas under 1,000 km^2. Finally, Trinidad is well above the line of the main body of West Indian islands in both its absolute numbers of species and its absolute number of endemic species, but below the lines (see the far right column in Table 9-1) in *percentage* of endemic species. These disparities seem clearly due to Trinidad's origin as a continental island and its connection to the South American mainland during recent geologic times. Its fauna contains no fewer than 17 genera and well-marked subgenera that are widespread in South America but absent in the remainder of the West Indies. (These are *Acanthognathus, Apterostigma, Basiceros, Daceton, Dendromyrmex, Dolichoderus s.s., Eciton, Ectatomma, Lachnomyrmex, Megalomyrmex, Oligomyrmex, Pachycondyla* (*Mesoponera*), *Pachycondyla* (*Neoponera*), *Procryptocerus, Talaridris, Tranopelta*, and *Zacryptocerus s.s.*, i.e. *Z. clypeatus*).

However, while these generalizations appear qualitatively robust, that is, unlikely to be altered in main form by future research, a strong caveat is needed. The West Indian ants are still poorly sampled in comparison with, say, birds, reptiles, and butterflies. Some islands, such as the Caymans, Barbuda, Anguilla, and St. Kitts, are virtually unexplored. Furthermore, while larger islands such as Cuba, Hispaniola, and Trinidad have been better worked, it is safe to predict that visits to remote areas, especially mountain forests, and a generous use of berlese collecting will yield many more species. The resulting general trend will probably be an overall rise in the area-species curve without much change in the slope. On the other hand, the ultimate effect on the curve relating area to the number of endemic species is difficult to predict. Many new endemic species will probably be discovered, especially the rarest and most locally distributed elements, but many other species now known from a single island (hence counted as "endemic") will turn up elsewhere. The processes of discovery are obviously antagonistic—but to a still unknown degree.

With the exception of Trinidad, which is little more than an extension of South America, the Antillean islands have faunas displaying the key traits indicative of a "sweepstakes" origin. In addition to the 17 genera that just reach Trinidad, other dominant elements of the New World fauna extend only to one or a few of the Lesser Antilles, in the great majority of cases to the lower arc of these islands. Examples include *Azteca*, the most abundant and diverse of the Neotropical dolichoderine genera, and other genera that are common and widespread in South and Central America, including *Cephalotes, Hypoclinea, Leptothorax* (*Nesomyrmex*), *Monacis, Myrmicocrypta*, and *Prionopelta*. Moreover, genera and species distributed continuously or near-continuously within the Lesser Antilles show various degrees of penetration northward past Trinidad into the lower arc of islands. Thus *Leptothorax* (*Nesomyrmex*) just reaches Grenada, *Neivamyrmex* reaches Grenada and just beyond, to St. Vincent, and so forth.

A second trait typical of a sweepstakes origin is a disharmonic composition: endemic species clusters occur on the larger islands, which in the case of Cuba and Hispaniola make up a disproportionate share of the native faunas. There are at least four such clusters: the *sphaericus* group or subgenus *Manniella* of *Camponotus* in Cuba (*micrositus, sphaericus, torreï*); the *gilviventris* group or subgenus *Myrmeurynota* of *Camponotus*, which also occurs on the mainland from South America to Mexico, with two species in Cuba (*gilviventris, thysanopus*) and five in Hispaniola (*albistramineus, altivagans, augustei, christophei, toussainti*)[1]; the *Rogeria brunnea* group of Cuba (*brunnea, caraiba, cubensis, scabra*); and *Leptothorax* (*Macromischa*) in all of the Greater Antilles except Jamaica. One group requires qualification: Charles Kugler (pers. comm.), who is currently revising *Rogeria*, considers the Cuban forms still to be of uncertain status even though they are quite likely a cluster of sibling species.

Leptothorax (*Macromischa*) has attained the status of a truly radiated group, as shown in the recent analysis by Baroni Urbani (1978). Cuba has no fewer than 38 species, constituting 26% of the entire known ant fauna. All but one are endemic to the island, and the single exception, *L. androsanus*, is shared only with the nearby Bahamas. As exemplified in Figure 9-3, the morphological diversity of the Cuban species is great enough to encompass what might well be recognized as several distinct genera in other parts of the world. Baroni Urbani believes that this fauna is in fact polyphyletic, with the picture being complicated further by convergence in the key traits of elongation and thickening of the femora and either the enlargement or the loss of the propodeal spines. The elongation of the appendages is correlated with an increase in overall size and thickening of the femora, all of which appear to improve the ant's ability to sting severely from an unusual posture: the gaster is bent all the way beneath the body and the large sting is projected forward. The elongated petiole clearly facilitates this movement, while stronger tibial levator muscles, accommodated by the swollen femora, lend the ants greater stability when they raise their bodies on their legs. Some of the species are metallic blue or green in color (or even gold in certain reflections of light) and can be seen foraging in conspicuous files during daylight hours. Considering the stinging ability of the workers, it is possible that the color is aposematic.

With reference to *Leptothorax* (*Macromischa*) as a whole, Baroni Urbani has concluded that at least two main evolutionary centers exist. One is Cuba,

[1]The "subgenera" of *Camponotus* noted here are of such ambiguous status that they were not included in the genera-plus-subgenera counts of Table 9-1 and elsewhere.

220 · Edward O. Wilson

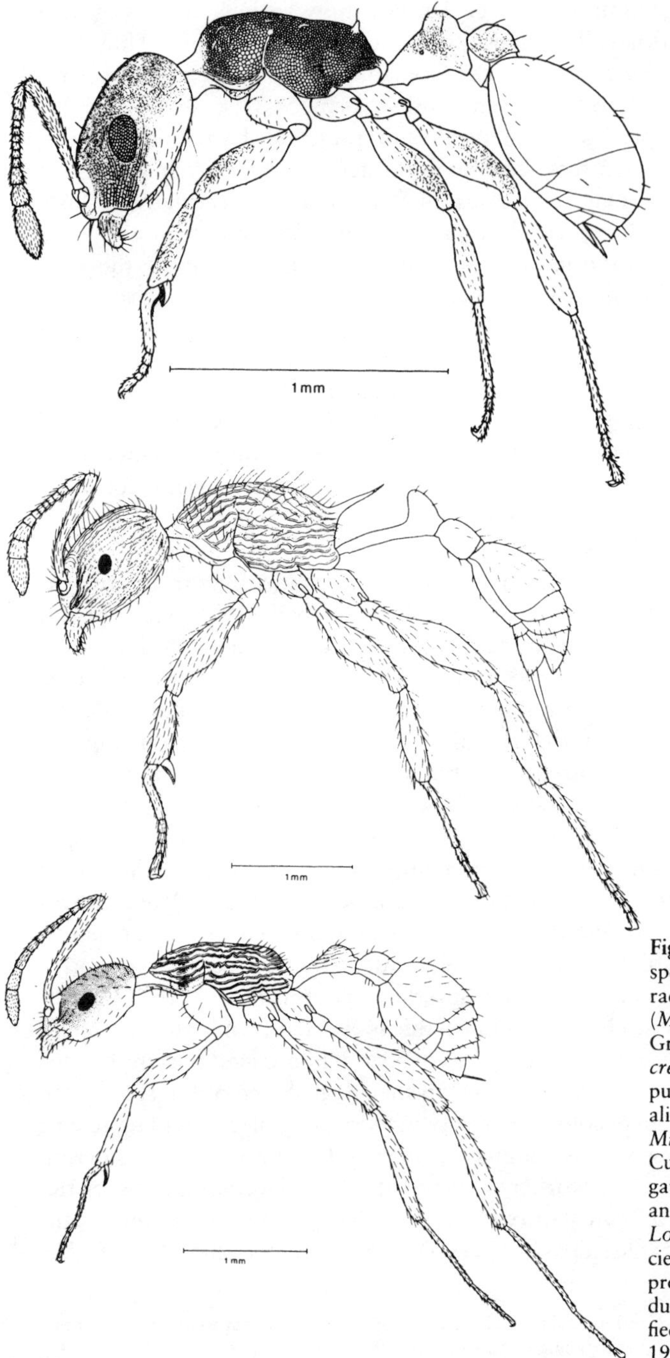

Figure 9-3. Representative species from the adaptive radiation of *Leptothorax* (*Macromischa*) from the Greater Antilles. *Upper*: *L. creolus* of the Dominican republic, a relatively generalized leptothoracine ant. *Middle*: *L. platycnemis*, a Cuban species with elongated legs, propodeal spines and petiolar peduncle. *Lower*: *L. iris*, a Cuban species with elongated legs, lost propodeal spines, and reduced petiolar node. (Modified from Baroni Urbani 1978.)

the species of which are adapted to live in the soil or in crevices of limestone. Another is southern Mexico and Guatemala, which contains arboreal species that often nest in orchids and other epiphytes. The Cuban radiation has spilled over into Hispaniola, with five endemic species, and also into Puerto Rico, which possesses three endemic species and another (*albispina*) shared with nearby Mona. It is surely one of the chief mysteries of West Indian biogeography that not a single species, endemic or otherwise, has been discovered on Jamaica.

The three endemic genera of the West Indies are each limited to a single species. *Codioxenus simulans* and *Dorisidris nitens* are small, inconspicuous ants known only from several collections in Cuba. They belong to the tribe Dacetini, a group of hypogaeic ants specialized (so far as known) as predators on collembolans and other soft-bodied arthropods. *Hypocryptocerus haemorrhoidalis*, a relatively weakly demarcated member of the exclusively arboreal tribe Cephalotini, is common and widespread on Hispaniola. (*Xenometra monilicornis*, previously thought to be an endemic of St. Thomas, is a synonym of *Cardiocondyla emeryi*, an Old World species introduced by human commerce into the West Indies; see Baroni Urbani 1973, and Kugler 1983.)

The Biogeographic Affinities of the Modern Fauna

The Antillean ant fauna can be divided into two sections with reference to geographic origin. The first comprises the species that have spread for varying distances from South America northward through the Lesser Antilles and, in some cases, on into the large islands of the Greater Antilles. Some of these populations are little changed across their ranges, while others have speciated to produce endemic forms on one or more of the islands. Examples include various species complexes within the genera *Acromyrmex*, *Anochetus*, *Atta*, *Azteca*, *Camponotus* (subgenera *Myrmaphaenus*, *Myrmobrachys*, *Myrmosphincta*, *Myrmothrix*, *Pseudocolobopsis*, and *Tanaemyrmex*), *Crematogaster* (subgenus *Orthocrema*), *Cyphomyrmex*, *Hypoclinea*, *Hypoponera*, *Iridomyrmex*, *Leptogenys*, *Monomorium*, *Mycetophylax*, *Mycocepurus*, *Myrmicocrypta*, *Neivamyrmex*, *Nesomyrmex* (usually placed as a subgenus of *Leptothorax*), *Octostruma*, *Odontomachus*, *Pachycondyla*, *Paracryptocerus*, *Pheidole*, *Platythyrea*, *Prionopelta*, *Pseudomyrmex*, *Smithistruma*, *Solenopsis* (subgenus *Diplorhoptrum*), *Strumigenys*, *Tapinoma*, *Trachymyrmex*, and *Wasmannia*.

A smaller group of species and complexes of species are each endemic to one or more of the middle-sized and large islands, especially the four major islands of the Greater Antilles (Cuba, Hispaniola, Jamaica, Puerto Rico). Examples include *Camponotus* (subgenera *Manniella* and *Myrmeurynota*),

Codiomyrmex, Codioxenus, Crematogaster (subgenus *Crematogaster*), *Ephebomyrmex, Gnamptogenys* (*schmitti* group), *Hypocryptocerus, Leptothorax* (subgenus *Macromischa*), *Myrmelachista, Paratrechina* (subgenus *Nylanderia*), *Rogeria* (*brunnea* group), and *Thaumatomyrmex*. Our imperfect knowledge of ant systematics, and of these groups in particular, preclude the identification of the ultimate geographic origin of the endemic Antillean species in the second group.

In addition to the two biogeographic categories just cited, there is at least one species that might have a North American origin: *Prenolepis gibberosa*, an endemic of the mountains of Cuba. Other known *Prenolepis* in the New World are *P. acuminata* from Jalapa on the eastern edge of the Mexican plateau and *P. imparis*, which is one of the most widespread of all ant species in the temperate regions of the United States, Canada, and Mexico. *Prenolepis* is an ancient genus, occurring in the early Oligocene Baltic amber. I have also recently found a specimen in the late Tertiary amber of the Dominican Republic.

The Dominican Amber Fauna

On the basis of the ant fauna, the West Indies today appear to be typical oceanic islands. By this I mean that a limited number of stocks colonized the island in the sweepstakes manner. Also, at least one group, *Leptothorax* (*Macromischa*), has undergone radiation so as to fill some of the vacant adaptive zones. But a strikingly different picture has emerged from the study of late Tertiary amber of the Dominican Republic, the inclusions of which are more characteristic of the contemporary mainland fauna of South and Central America, and particularly of the lowland humid tropical forests.

Earlier research by Baroni Urbani (1980a–d and in Baroni Urbani and Saunders 1982) on material in the Staatliches Museum für Naturkunde, Stuttgart, had disclosed the presence of the genera *Anochetus, Gnamptogenys, Paracryptocerus, Pseudomyrmex*, and *Trachymyrmex*. A poorly preserved set of workers in a single amber piece placed by Baroni Urbani (1980c) in *Leptomyrmex* has now been supplemented by better-preserved workers and a single male (Baroni Urbani and Wilson 1987).

In my own studies, based on 602 pieces of amber containing an estimated 1254 ants and located mostly in the Museum of Comparative Zoology, I have been able to expand this list greatly. The majority of the specimens came from the western amber-bearing region of the Dominican Republic, located in the mountainous La Cumbre region 10 to 20 km northeast of Santiago. A few came from the eastern amber-bearing region south of Sabana de la Mar. Still others, making up fewer than 5% of the specimens, are evidently younger than those from other Dominican localities. Schlee

(1984) placed their age as recently as 200 years, hence ranking them as copal rather than amber, but his evidence was circumstantial: the unusually clear, almost glasslike quality of the amber pieces and their location near the ground surface. In contrast, J. B. Lambert (pers. comm.) found that while the NMR spectra of Cotui material indicate a younger age than that of other Dominican amber, the deposits still seem to be Tertiary rather than Recent. Perhaps the Cotui material spans a substantial period of time, extending even into the Quaternary period. If so, more collecting and analysis might provide us with a unique opportunity to trace the island fauna in a more nearly continuous manner through time.

For purposes of analysis, the fossil and living genera can be conveniently classified into six biogeographic categories based on their presence or absence in the amber and the extent of their retreat since amber times (Table 9-2). Perhaps the single most interesting genus is *Neivamyrmex*, represented by *N. ectopus*, the first fossil army ant ever discovered (Wilson 1985b; see Fig. 9-4 here). No contemporary army ant is known from the northern arc of the Lesser Antilles or any of the Greater Antilles (Watkins 1976). Furthermore, army ants generally have very low colonizing ability across water. In the case of the New World Ecitoninae, *Neivamyrmex nigrescens* has been recorded from the Islas Marias, 100 km off the Mexican Pacific coast, while *N. klugi* occurs on St. Vincent in the lower arc of the Lesser Antilles. No

Table 9-2. The status of ant genera and well-marked subgenera on Hispaniola (Dominican Republic plus Haiti) (From Wilson 1985d)

Present in Dominican Amber

Now extinct worldwide: *Ilemomyrmex, Oxyidris,* new genus near *Rogeria*

Now extinct in Western Hemisphere: *Leptomyrmex*

Now extinct in the Greater Antilles but present elsewhere in the Neotropical region: *Azteca, Dolichoderus, Erebomyrma, Hypoclinea, Leptothorax* (*Nesomyrmex*), *Monacis, Neivamyrmex, Paraponera, Prionopelta*

Now extinct on Hispaniola but present elsewhere in the Greater Antilles: *Cylindromyrmex, Octostruma, Prenolepis*

Still present on Hispaniola: *Anochetus, Aphaenogaster, Camponotus, Crematogaster* (*Acrocoelia*), *Crematogaster* (*Orthocrema*), *Cyphomyrmex, Gnamptogenys, Hypoponera, Iridomyrmex, Leptothorax* (*Macromischa*), *Odontomachus, Pachycondyla* (*Trachymesopus*), *Paratrechina* (*Nylanderia*), *Pheidole, Platythyrea, Pseudomyrmex, Smithistruma, Solenopsis* (*Diplorhoptrum*), *Solenopsis* (*Solenopsis*), *Tapinoma, Trachymyrmex, Zacryptocerus*

New Arrivals

New World genera and well-marked subgenera present in the modern Hispaniolan fauna but unknown in the Dominican amber: *Acropyga, Brachymyrmex, Ephebomyrmex, Eurhopalothrix, Hypocryptocerus, Leptogenys, Monomorium, Mycocepurus, Myrmelachista, Solenopsis* (*Euophthalma*), *Strumigenys, Wasmannia* (possibly introduced by commerce)

Old World genera (or species groups within these genera) introduced within historical times by human commerce: *Cardiocondyla, Paratrechina* (*Paratrechina*), *Tetramorium*

Figure 9-4. The extinct army ant *Neivamyrmex ectopus* from the amber of the Dominican Republic. (From Wilson 1985b.)

ecitonine is known farther away from the mainland. Similarly, the Old World Dorylinae, represented by *Aenictus*, extends only as far east as the Philippine Islands, New Guinea, and Queensland (Wilson 1964). It is wholly unknown from those portions of Micronesia and Polynesia that support a native ant fauna (Wilson and Taylor 1967). The farthest outlier in the western part of the range is a population of *A. fergusoni* on Great Nicobar Island, 160 km from Sumatra. The existence of *Neivamyrmex ectopus* in the Dominican amber is therefore consistent with the common view, based on both geologic and paleobotanical studies (Graham and Jarzen 1969), that the ancestral Antilles were larger and extended closer to the Mexican mainland during the Middle and Late Tertiary than is now the case. Furthermore, the overall closer similarity of *N. ectopus* to contemporary Mexican and United States species, as opposed to South and Central American species (Wilson 1985b), is consistent with a closer approach of the Greater Antilles to Mexico in particular than to the northern coast of South America during the Tertiary period. But this may be the only conclusion that can be drawn. The evidence, in my opinion, does not disclose whether the closer proximity was due to continental drift or merely to the temporary emergence of a larger Antillean land mass that extended westward.

Further evidence of a closer approach of Hispaniola to the mainland, however it occurred, is provided by the presence in the amber of *Dolichoderus, Monacis*, and *Paraponera*, which are today principal elements of moist tropical forests in South and Central America but do not extend north of Trinidad. (*Monacis bispinosus* has been cited from St. Thomas, but I consider this to be either an erroneous record or else based on an introduced population; see Wilson 1985c.) These genera, all comprising large, conspicuous ants, seldom reach offshore islands along the mainland coasts and hence appear to be limited in dispersal ability.

Another important fact revealed by the Dominican fossils is the remarkable retreat of the subfamily Dolichoderinae from the West Indies since Tertiary times. Four genera (*Azteca, Dolichoderus, Hypoclinea, Monacis*) have disappeared entirely from the Greater Antilles and one (*Leptomyrmex*) from the entire Western Hemisphere. Only two (*Iridomyrmex, Tapinoma*) have persisted to present times, while yet another genus, *Conomyrma*, has invaded more recently. The dominant species of the amber fauna was the now-extinct *Azteca alpha*, which, if it was like all modern *Azteca*, was arboricolous. According to W. L. Brown (pers. comm.), who has collected intensively in the Dominican Republic, the dominant arboricolous ants today include members of *Pseudomyrmex, Crematogaster, Zacryptocerus*, and *Camponotus*. In this important respect the West Indian fauna mirrors the general decline of the Dolichoderinae in North and South America, Europe, and Asia, possibly in conjunction with the advance of *Crematogaster* as a competitor of *Iridomyrmex* (Brown 1973).

The generic lists of the amber and modern Hispaniolan faunas can be summarized as follows. Of 38 genera and well-defined subgenera identified in the fossil deposits, 34 have survived somewhere in the New World tropics to the present, although the species studied thus far are extinct. Of the surviving genera and subgenera, 22 persist on Hispaniola. Fifteen genera and subgenera have colonized the island since amber times, restoring the number of genera and well-defined subgenera now present on Hispaniola to 37 (Wilson 1985a–e). It needs to be stressed that while the overall generic diversity has remained the same, the close numerical correspondence (38 and 37 genera, respectively) is surely only a coincidence, likely to be changed with further collecting of living and amber material.

Has the species diversity also stayed constant? The modern Hispaniolan fauna averages 88/37 = 2.35 species per genus. The Hispaniolan amber fauna belonging to the Dolichoderinae and Ecitoninae, the only subfamilies revised so far to the species level, averages 11/8 species = 1.38 per genus. If this ratio were to hold over all 38 genera, the amber fauna would contain 52 species rather than the 88 known from the living fauna. However, sample size represented by the two subfamilies is still too small to be certain that overall ant species diversity was less in amber times than today.

I have also conducted a search for biological traits of living species that were likely to have been shared with the Hispaniolan fossil members of the same genus and might have contributed to the extinction or survival of the genus on the island from Tertiary times (Wilson 1985d). The data were not numerous enough to support effects due to the flightless condition of the queen and hence reduced dispersal power of colonies caused by this trait, despite the clear-cut example of the *Neivamyrmex* army ants. Weak evidence, with confidence just below the 95% level (G test of independence, 2 × 2 tables), exists for an increase in extinction rate associated with large individual or colony size, biological traits that are likely to cause a reduction in the size of populations of colonies. Examples of genera in this condition include *Dolichoderus* and *Paraponera*. A positive relation also exists, in this instance above the 95% confidence level, between extinction and extreme specialization in either prey choice or nest site. Examples of such specialized genera and subgenera that became extinct on Hispaniola (or at least have not yet been found in the modern fauna there) include *Cylindromyrmex, Neivamyrmex, Octostruma*, and *Prionopelta*. Finally, genera and subgenera that occur in both the New World and Old World, and hence show evidence of greater colonizing ability, have also reached the Greater Antilles in greater numbers and become extinct less frequently on Hispaniola since amber times ($P < 0.05$). Examples include *Anochetus, Aphaenogaster, Camponotus* (*Tanaemyrmex*), *Crematogaster s.s., Crematogaster* (*Orthocrema*), *Gnamptogenys, Hypoponera, Iridomyrmex, Odontomachus, Paratrechina* (*Nylanderia*), *Pheidole, Platythyrea, Solenopsis* (*Diplorhoptrum*), *Tapinoma*, and *Trachymesopus*.

Discussion

Several enticing research opportunities have been more clearly revealed by the biogeographic analysis summarized here. First, some of the Antillean islands have never been carefully collected and are either sufficiently large in size, topographically varied enough, or strategically placed to be of more than passing interest. They include the Caymans, Isle of Pines, Barbuda, St. Kitts, Guadeloupe, Martinique, Dominica, Tobago, as well as Margarita and the other islands along the north coast of South America to the west of Trinidad. Jamaica still seems to me to be relatively undercollected among the Greater Antilles. And the more remote forested areas of Cuba and Hispaniola would certainly repay additional berlese sampling of the soil and litter.

Although Baroni Urbani has provided a valuable systematic monograph of *Leptothorax* (*Macromischa*), very little has been learned concerning the biology of this most luxuriantly radiated of the West Indian ant groups. To what extent have the endemic forms, especially the 37 or more on Cuba, diversified in habitat selection, nest site, food habits, defensive behavior, colony size, and social organization? Has *Leptothorax* (*Macromischa*) undergone a typical radiation, in the sense of preferentially occupying niches left vacant by other ant genera that are still absent from the Greater Antilles?

Possibly linked to the *Macromischa* case is the mysterious absence of *Neivamyrmex* and other ecitonine army ants from the Greater Antilles. These insects are for the most part specialized predators on *Camponotus*, *Pheidole*, and other ants. In other islands lacking army ants, for example Fiji, New Caledonia, southern Australia, and Madagascar, cerapachyines such as *Cerapachys* and *Sphinctomyrmex* are disproportionately abundant. But the Cerapachyinae are also absent from the Greater Antilles, or nearly so; the only record to date is *C. biroi* (= *C. seini*), a southeast Asian species introduced into Puerto Rico, and this appears to be relatively scarce. So the question arises: What, if anything, specializes in ant predation in the Greater Antilles? Is it possible that some of the *Leptothorax* (*Macromischa*) species have filled this niche?

I find it also puzzling that the myrmicine genus *Pheidole* has not flourished to a greater extent in the Greater Antilles, despite the fact that it was present as far back as Dominican amber times. Elsewhere it is the dominant ant genus in warm temperate and tropical portions of the New World, with over 500 unchallenged names (and probably at least that many true species) and the greatest numbers of colonies and individual ants in most habitats. However, only 17 described species have been recorded from the Greater Antilles. Of these, a mere 8 are limited to these islands, and none show signs of even an early stage of adaptive radiation. On the other hand, a close study of *Pheidole* might well result in a challenge of my own subjective impression. W. L. Brown (pers. comm.), for example, considers the native *Pheidole* of the Dominican Republic, including several undescribed species he has collected, to be relatively abundant and diverse.

Summary

The known West Indian ant fauna comprises 383 species in 89 genera (including 12 well-marked subgenera; see footnote in Table 9-1). Three of the genera are known only from the West Indies, and in fact only from a single island. In addition, 176 species, or 46% of the total, have so far been recorded only from the West Indies. The highest endemicity is that of Cuba, with 49.7% of its 146 species having that status (see Table 9-1). The great majority of endemic Antillean species are found on islands with areas greater than 1000 km^2 (Fig. 9-2).

The area-species curve appears regular in form, with a coefficient of 0.28, a property that supports but does not prove species equilibria (Fig. 9-1). If Trinidad is excluded because of its status as a continental island close to the South American mainland, the Antillean fauna is disharmonic in the manner typical for oceanic islands. In particular, a number of dominant mainland ant genera, including *Azteca, Cephalotes*, and *Monacis*, are wholly lacking in the Greater Antilles. And one myrmicine group, *Leptothorax* (*Macromischa*), has undergone a dramatic adaptive radiation, producing at least 37 endemic species on Cuba alone.

The modern West Indian ant fauna can be divided into two principal groups with respect to origin. One set comprises species that have spread to varying degrees of penetration from South America northward through the Lesser Antilles and in some cases onto the larger islands of the Greater Antilles. The second set, which is smaller and appears generally older, comprises species that are endemic to one or more of the middle-sized to larger islands, especially the four major islands of the Greater Antilles (Cuba, Hispaniola, Jamaica, Puerto Rico). The ultimate origin of stocks in the second group cannot be determined on the basis of existing knowledge.

In contrast to the contemporary ants, the late Tertiary fauna of the Greater Antilles, as revealed by recent studies of the Dominican amber, is more nearly characteristic of a continental fauna. Several genera were present on Hispaniola, including *Leptomyrmex, Azteca, Dolichoderus, Paraponera*, and the army-ant genus *Neivamyrmex*, that are today abundant elsewhere but wholly absent in the modern Greater Antilles fauna. Because army ants especially are poor colonizers across water, it seems reasonable to conclude that the Greater Antilles were nearer to the mainland (and in particular to Mexico) during late Tertiary times than is the case today.

The Hispaniolan fauna has undergone considerable turnover since Dominican amber times while retaining constancy in the number of genera. Of 38 genera and well-defined subgenera identified in the amber, 34 have survived somewhere in the New World tropics to the present, although the species studied thus far are extinct. (Three genera are apparently entirely extinct: *Ilemomyrmex, Oxyidris*, and an undescribed genus near *Rogeria*.) Of the surviving genera and subgenera, 22 persist today on Hispaniola. Fifteen genera and subgenera have colonized the island since amber times, restoring the number now present on Hispaniola to 37.

A higher extinction rate occurred on Hispaniola in genera and subgenera that are either highly specialized or possess less colonizing ability as evidenced by their restriction to the New World. There is an indication that the same was true for genera with either large individual or colony size, which are biological traits often associated with smaller populations of colonies.

Acknowledgments

I am grateful to William L. Brown and Charles Kugler for advice and additional data provided during the course of this study. The research was supported by National Science Foundation Grant No. BSR–84–21062.

References

Alayo, P. 1974. Introduccion al estudio de los Himenopteros de Cuba. Superfamilia Formicoidea. Acad. Cienc. Cuba, Inst. Zool. 53:1–58.

Baroni Urbani, C. 1973. Die Gattung *Xenometra*, ein objektives Synonym (Hymenoptera, Formicidae). Mitt. Schweiz. Ent. Ges. 46(3–4):199–201.

———. 1978. Materiali per una revisione dei *Leptothorax* neotropicali appartementi al sottogenere *Macromischa* Roger, n. comb. (Hymenoptera: Formicidae). Ent. Basil. 3:395–618.

———. 1980a. First description of fossil gardening ants (Amber Collection Stuttgart and Natural History Museum Basel: Hymenoptera, Formicidae. I: Attini). Stutt. Beitr. Naturk. B 54:1–13.

———. 1980b. *Anochetus corayi* n.sp., the first fossil Odontomachiti ant (Amber Collection Stuttgart: Hymenoptera, Formicidae. II: Odontomachiti). Stutt. Beitr. Naturk. B 55:1–6.

———. 1980c. The first fossil species of the Australian ant genus *Leptomyrmex* in amber from the Dominican Republic (Amber Collection Stuttgart: Hymenoptera, Formicidae. III: Leptomyrmicini). Stutt. Beitr. Naturk. B 62:1–10.

———. 1980d. The ant genus *Gnamptogenys* in Dominican amber (Amber Collection Stuttgart: Hymenoptera, Formicidae. IV: Ectatommini). Stutt. Beitr. Naturk. B 67:1–10.

Baroni Urbani, C., and J.B. Saunders. 1982. The fauna of the Dominican amber: The present status of knowledge. Transactions of the 9th Caribbean Geological Conference (Santo Domingo, Dominican Republic) 1:213–223.

Baroni Urbani, C., and E.O. Wilson. 1987. The fossil members of the ant tribe Leptomyrmecini (Hymenoptera: Formicidae). Psyche 94(1–2):1–8.

Brown, W.L. 1973. A comparison of the Hylaean and Congo–West African rainforest ant faunas. *In* B.J. Meggers, E.S. Ayensu, and W.D. Duckworth, eds., Tropical Forest Systems: A Comparative Review, pp. 161–185. Smithsonian Institution Press, Washington, D.C.

———. 1976. Contributions toward a reclassification of the Formicidae. Part VI. Ponerinae, Tribe Ponerini, Subtribe Odontomachiti. Section A. Introduction, subtribal characters, genus *Odontomachus*. Studia Ent. 19(1–4):67–171.

———. 1981. Preliminary contributions toward a revision of the ant genus *Pheidole* (Hymenoptera: Formicidae), part I. J. Kansas Ent. Soc. 54(3):523–530.

Forel, A. 1893. Formicides de l'Antille St. Vincent, récoltés par Mons. H.H. Smith. Trans. Ent. Soc. Lond. 4:333–418.

Graham, A., and D.M. Jarzen. 1969. Studies in Neotropical paleobotany. I. The Oligocene communities of Puerto Rico. Ann. Missouri Bot. Gardens 56:308–357.

Kempf, W.W. 1972. Catálogo abreviado das formigas da Região Neotropical. Studia Ent. 5(1–4):3–344.

Kugler, J. 1983. The males of *Cardiocondyla* Emery (Hymenoptera: Formicidae) with the description of the winged male of *Cardiocondyla wroughtoni* (Forel). Israel J. Ent. 17:1–21.

Levins, R., M.L. Pressick, and H. Heatwole. 1973. Coexistence patterns in insular ants. Amer. Sci. 61:463–472.

MacArthur, R.H., and E.O. Wilson. 1967. The theory of island biogeography. Monog. Pop. Biol. No. 1. Princeton University Press, Princeton, N.J. 203 pp.

Schlee, D. 1984. Notizen über einige Bernsteine und Kopale aus aller Welt. Stutt. Beitr. Naturk. C 18:29–37.

Smith, M.R. 1936. The ants of Puerto Rico. J. Agric. Univ. Puerto Rico 20(4):819–875.

Torres, J.A. 1984. Niches and coexistence of ant communities in Puerto Rico: Repeated patterns. Biotropica 16(4):284–295.

Watkins, J.F., II. 1976. The Identification and Distribution of New World Army Ants. Baylor University Press, Waco, Texas. x, 102 pp.

Wheeler, W.M. 1913. The ants of Cuba. Bull. Mus. Comp. Zool. Harv. 54(17):477–505.

———. 1936. Ants from Hispaniola and Mona. Bull. Mus. Comp. Zool. Harv. 80(2):195–211.

Wheeler, W.M., and W.M. Mann. 1914. The ants of Haiti. Bull. Amer. Mus. Nat. Hist. 33(1):1–61.

Wilson, E.O. 1964. The true army ants of the Indo-Australian area (Hymenoptera: Formicidae: Dorylinae). Pacific Ins. 6(3):427–483.

———. 1985a. Ants of the Dominican amber (Hymenoptera: Formicidae), 1: Two new myrmicine genera and an aberrant *Pheidole*. Psyche 92(1):1–9.

———. 1985b. Ants of the Dominican amber (Hymenoptera: Formicidae), 2: The first fossil army ants. Psyche 92(1):11–16.

———. 1985c. Ants of the Dominican amber (Hymenoptera: Formicidae), 3: The subfamily Dolichoderinae. Psyche 92(1):17–37.

———. 1985d. Invasion and extinction in the West Indian ant fauna: Evidence from the Dominican amber. Science 229:265–267.

———. 1985e. Ants of the Dominican amber (Hymenoptera: Formicidae), 4: A giant ponerine in the genus *Paraponera*. Israel J. Ent. 19:197–200.

Wilson, E.O., and R.W. Taylor. 1967. The ants of Polynesia. Pacific Ins. Monogr. 14:1–109.

1988

"Editor's Foreword" and "The current state of biological diversity," *in* E. O. Wilson and Frances M. Peter, eds., *BioDiversity* (National Academy Press, Washington, DC), pp. v–vii, 3–18

In 1986, a meeting was held that called the attention of a scientific and lay public to the magnitude and imperilment of biological diversity. It was the National Forum on BioDiversity held in Washington, D.C., during September 21–24, under the auspices of the National Academy of Sciences and the Smithsonian Institution. The book *BioDiversity* originated from the Forum and soon became an international bestseller, at least a bestseller by academic standards. It introduced the term "biodiversity," short for biological diversity, which quickly spread far and wide and is today enshrined in English-language dictionaries.

Because of my work in systematics and conservation, I was asked to serve as editor of the volume. The conception of the Forum is described in my editor's Foreword, and a primer of current knowledge of biodiversity is presented in the opening article.

EDITOR'S FOREWORD

The diversity of life forms, so numerous that we have yet to identify most of them, is the greatest wonder of this planet. The biosphere is an intricate tapestry of interwoven life forms. Even the seemingly desolate arctic tundra is sustained by a complex interaction of many species of plants and animals, including the rich arrays of symbiotic lichens. The book before you offers an overall view of this biological diversity and carries the urgent warning that we are rapidly altering and destroying the environments that have fostered the diversity of life forms for more than a billion years.

The source of the book is the National Forum on BioDiversity, held in Washington, D.C., on September 21–24, 1986, under the auspices of the National Academy of Sciences and Smithsonian Institution. The forum was notable for its large size and immediately perceived impact on the public. It featured more than 60 leading biologists, economists, agricultural experts, philosophers, representatives of assistance and lending agencies, and other professionals. The lectures and panels were regularly attended by hundreds of people, many of whom participated in the discussions, and various aspects of the forum were reported widely in the press. On the final evening, a panel of six of the participants conducted a teleconference downlinked to an estimated audience of 5,000 to 10,000 at over 100 sites, most of them hosted by Sigma Xi chapters at universities and colleges in the United States and Canada.

The forum coincided with a noticeable rise in interest, among scientists and portions of the public, in matters related to biodiversity and the problems of international conservation. I believe that this increased attention, which was evident by 1980 and had steadily picked up momentum by the time of the forum, can be ascribed to two more or less independent developments. The first was the accumulation of enough data on deforestation, species extinction, and tropical biology to bring global problems into sharper focus and warrant broader public exposure. It is no coincidence that 1986 was also the year that the Society for Conservation Biology was founded. The second development was the growing awareness of the close linkage between the conservation of biodiversity and economic development. In the United States and other industrial countries, the two are often seen in opposition, with environmentalists and developers struggling for compromise in a zero-sum game. But in the developing nations, the opposite is true. Destruction of the natural environment is usually accompanied by short-term profits and then rapid local economic decline. In addition, the immense richness of tropical biodiversity is a largely untapped reservoir of new foods, pharmaceuticals, fibers, petroleum substitutes, and other products.

Because of this set of historical circumstances, this book, which contains papers from the forum, should prove widely useful. It provides an updating of many of the principal issues in conservation biology and resource management. It also documents a new alliance between scientific, governmental, and commercial forces—one that can be expected to reshape the international conservation movement for decades to come.

The National Forum on BioDiversity and thence this volume were made possible by the cooperative efforts of many people. The forum was conceived by Walter G. Rosen, Senior Program Officer in the Board on Basic Biology—a unit of the Commission on Life Sciences, National Research Council/National Academy of Sciences (NRC/NAS). Dr. Rosen represented the NRC/NAS throughout the planning stages of the project. Furthermore, he introduced the term *biodiversity*, which aptly represents, as well as any term can, the vast array of topics and perspectives covered during the Washington forum. Edward W. Bastian, Smithsonian Institution, mobilized and orchestrated the diverse resources of the Smithsonian in the effort. Drs. Rosen and Bastian were codirectors of the forum. Michael H. Robinson (Director of the National Zoological Park) served as chairman of the Program Committee, organized one of the forum panels, and served as general master of ceremonies. The remainder of the Program Committee consisted of William Jordan III, Thomas E. Lovejoy III, Harold A. Mooney, Stanwyn Shetler, and Michael E. Soulé.

The various panels of the forum were organized and chaired by F. William Burley, William Conway, Paul R. Ehrlich, Michael Hanemann, William Jordan III, Thomas E. Lovejoy III, Harold A. Mooney, James D. Nations, Peter H. Raven, Michael H. Robinson, Ira Rubinoff, and Michael E. Soulé. David Johnson at the New York Botanical Garden was very helpful in verifying some of the botanical terms used in this book. Helen Taylor and Kathy Marshall of the NRC staff and Anne Peret of the Smithsonian Institution assisted with the wide variety of arrangements necessary to the successful conduct of the forum. Linda Miller Poore, also of the NRC staff, entered this entire document on a word processer and was responsible for formatting and checking the many references. Richard E. Morris of the National Academy Press guided this book through production.

The National Forum on BioDiversity was supported by the National Research Council Fund and the Smithsonian Institution, with supplemental support from the Town Creek Foundation, the Armand G. Erpf Fund, and the World Wildlife Fund. The National Research Council Fund is a pool of private, discretionary, nonfederal funds that is used to support a program of Academy-initiated studies of national issues in which science and technology figure significantly. The NRC Fund consists of contributions from a consortium of private foundations including the Carnegie Corporation of New York, the Charles E. Culpeper Foundation, the William and Flora Hewlett Foundation, the John D. and Catherine T. MacArthur Foundation, the Andrew W. Mellon Foundation, the Rockefeller Foundation, and the Alfred P. Sloan Foundation; the Academy Industry Program, which seeks annual contributions from companies that are concerned with the health of U.S. science and technology and with public policy issues with technological content; and the National Academy of Sciences and the National Academy of Engineering endowments. The publication of this volume was supported by the National Research Council Dissemination Fund, with supplemental support from the World Wildlife Fund. We are deeply grateful to all these organizations for making this project possible.

Finally, and far from least, Frances M. Peter marshalled the diverse contributions in the present volume and was essential to every step of the manuscript editing process. The cover for *Biodiversity* was derived from a forum poster designed by artist Robert Goldstrom.

E. O. WILSON

CHAPTER

1

THE CURRENT STATE OF BIOLOGICAL DIVERSITY

E. O. WILSON
Frank B. Baird, Jr. Professor of Science, Harvard University,
Museum of Comparative Zoology, Cambridge, Massachusetts

Biological diversity must be treated more seriously as a global resource, to be indexed, used, and above all, preserved. Three circumstances conspire to give this matter an unprecedented urgency. First, exploding human populations are degrading the environment at an accelerating rate, especially in tropical countries. Second, science is discovering new uses for biological diversity in ways that can relieve both human suffering and environmental destruction. Third, much of the diversity is being irreversibly lost through extinction caused by the destruction of natural habitats, again especially in the tropics. Overall, we are locked into a race. We must hurry to acquire the knowledge on which a wise policy of conservation and development can be based for centuries to come.

To summarize the problem in this chapter, I review some current information on the magnitude of global diversity and the rate at which we are losing it. I concentrate on the tropical moist forests, because of all the major habitats, they are richest in species and because they are in greatest danger.

THE AMOUNT OF BIOLOGICAL DIVERSITY

Many recently published sources, especially the multiauthor volume *Synopsis and Classification of Living Organisms*, indicate that about 1.4 million living species of all kinds of organisms have been described (Parker, 1982; see also the numerical breakdown according to major taxonomic category of the world insect fauna prepared by Arnett, 1985). Approximately 750,000 are insects, 41,000 are vertebrates, and 250,000 are plants (that is, vascular plants and bryophytes). The remainder consists of a complex array of invertebrates, fungi, algae, and microorganisms (see Table 1-1). Most systematists agree that this picture is still very incomplete except

4 / BIODIVERSITY

TABLE 1-1 Numbers of Described Species of Living Organisms[a]

Kingdom and Major Subdivision	Common Name	No. of Described Species	Totals
Virus			
	Viruses	1,000 (order of magnitude only)	1,000
Monera			
Bacteria	Bacteria	3,000	
Myxoplasma	Bacteria	60	
Cyanophycota	Blue-green algae	1,700	4,760
Fungi			
Zygomycota	Zygomycete fungi	665	
Ascomycota (including 18,000 lichen fungi)	Cup fungi	28,650	
Basidiomycota	Basidiomycete fungi	16,000	
Oomycota	Water molds	580	
Chytridiomycota	Chytrids	575	
Acrasiomycota	Cellular slime molds	13	
Myxomycota	Plasmodial slime molds	500	46,983
Algae			
Chlorophyta	Green algae	7,000	
Phaeophyta	Brown algae	1,500	
Rhodophyta	Red algae	4,000	
Chrysophyta	Chrysophyte algae	12,500	
Pyrrophyta	Dinoflagellates	1,100	
Euglenophyta	Euglenoids	800	26,900
Plantae			
Bryophyta	Mosses, liverworts, hornworts	16,600	
Psilophyta	Psilopsids	9	
Lycopodiophyta	Lycophytes	1,275	
Equisetophyta	Horsetails	15	
Filicophyta	Ferns	10,000	
Gymnosperma	Gymnosperms	529	
Dicotolydonae	Dicots	170,000	
Monocotolydonae	Monocots	50,000	248,428
Protozoa			
	Protozoans: Sarcomastigophorans, ciliates, and smaller groups	30,800	30,800
Animalia			
Porifera	Sponges	5,000	
Cnidaria, Ctenophora	Jellyfish, corals, comb jellies	9,000	
Platyhelminthes	Flatworms	12,200	
Nematoda	Nematodes (roundworms)	12,000	
Annelida	Annelids (earthworms and relatives)	12,000	

TABLE 1-1 Continued

Kingdom and Major Subdivision	Common Name	No. of Described Species	Totals
Mollusca	Mollusks	50,000	
Echinodermata	Echinoderms (starfish and relatives)	6,100	
Arthropoda	Arthropods		
Insecta	Insects	751,000	
Other arthropods		123,161	
Minor invertebrate phyla		9,300	989,761
Chordata			
Tunicata	Tunicates	1,250	
Cephalochordata	Acorn worms	23	
Vertebrata	Vertebrates		
Agnatha	Lampreys and other jawless fishes	63	
Chrondrichthyes	Sharks and other cartilaginous fishes	843	
Osteichthyes	Bony fishes	18,150	
Amphibia	Amphibians	4,184	
Reptilia	Reptiles	6,300	
Aves	Birds	9,040	
Mammalia	Mammals	4,000	43,853
TOTAL, all organisms			1,392,485

[a]Compiled from multiple sources.

in a few well-studied groups such as the vertebrates and flowering plants. If insects, the most species-rich of all major groups, are included, I believe that the absolute number is likely to exceed 5 million. Recent intensive collections made by Terry L. Erwin and his associates in the canopy of the Peruvian Amazon rain forest have moved the plausible upper limit much higher. Previously unknown insects proved to be so numerous in these samples that when estimates of local diversity were extrapolated to include all rain forests in the world, a figure of 30 million species was obtained (Erwin, 1983). In an even earlier stage is research on the epiphytic plants, lichens, fungi, roundworms, mites, protozoans, bacteria, and other mostly small organisms that abound in the treetops. Other major habitats that remain poorly explored include the coral reefs, the floor of the deep sea, and the soil of tropical forests and savannas. Thus, remarkably, we do not know the true number of species on Earth, even to the nearest order of magnitude (Wilson, 1985a). My own guess, based on the described fauna and flora and many discussions with entomologists and other specialists, is that the absolute number falls somewhere between 5 and 30 million.

A brief word is needed on the meaning of species as a category of classification. In modern biology, species are regarded conceptually as a population or series of populations within which free gene flow occurs under natural conditions. This means that all the normal, physiologically competent individuals at a given time are capable of breeding with all the other individuals of the opposite sex belonging to the same species or at least that they are capable of being linked genetically to them through chains of other breeding individuals. By definition they do not breed freely with members of other species.

This biological concept of species is the best ever devised, but it remains less than ideal. It works very well for most animals and some kinds of plants, but for some plant and a few animal populations in which intermediate amounts of hybridization occur, or ordinary sexual reproduction has been replaced by self-fertilization or parthenogenesis, it must be replaced with arbitrary divisions.

New species are usually created in one or the other of two ways. A large minority of plant species came into existence in essentially one step, through the process of polyploidy. This is a simple multiplication in the number of gene-bearing chromosomes—sometimes within a preexisting species and sometimes in hybrids between two species. Polyploids are typically not able to form fertile hybrids with the parent species. A second major process is geographic speciation and takes much longer. It starts when a single population (or series of populations) is divided by some barrier extrinsic to the organisms, such as a river, a mountain range, or an arm of the sea. The isolated populations then diverge from each other in evolution because of the inevitable differences of the environments in which they find themselves. Since all populations evolve when given enough time, divergence between all extrinsically isolated populations must eventually occur. By this process alone the populations can acquire enough differences to reduce interbreeding between them should the extrinsic barrier between them be removed and the populations again come into contact. If sufficient differences have accumulated, the populations can coexist as newly formed species. If those differences have not yet occurred, the populations will resume the exchange of genes when the contact is renewed.

Species diversity has been maintained at an approximately even level or at most a slowly increasing rate, although punctuated by brief periods of accelerated extinction every few tens of millions of years. The more similar the species under consideration, the more consistent the balance. Thus within clusters of islands, the numbers of species of birds (or reptiles, or ants, or other equivalent groups) found on each island in turn increases approximately as the fourth root of the area of the island. In other words, the number of species can be predicted as a constant X (island area)$^{0.25}$, where the exponent can deviate according to circumstances, but in most cases it falls between 0.15 and 0.35. According to this theory of island biogeography, in a typical case (where the exponent is at or near 0.25) the rule of thumb is that a 10-fold increase in area results in a doubling of a number of species (MacArthur and Wilson, 1967).

In a recent study of the ants of Hispaniola, I found fossils of 37 genera (clusters of species related to each other but distinct from other such clusters) in amber from the Miocene age—about 20 million years old. Exactly 37 genera exist on the island today. However, 15 of the original 37 have become extinct, while 15 others not present in the Miocene deposits have invaded to replace them, thus sustaining the original diversity (Wilson, 1985b).

On a grander scale, families—clusters of genera—have also maintained a balance within the faunas of entire continents. For example, a reciprocal and apparently symmetrical exchange of land mammals between North and South America began

3 million years ago, after the rise of the Panamanian land bridge. The number of families in South America first rose from 32 to 39 and then subsided to the 35 that exist there today. A comparable adjustment occurred in North America. At the generic level, North American elements dominated those from South America: 24 genera invaded to the south whereas only 12 invaded to the north. Hence, although equilibrium was roughly preserved, it resulted in a major shift in the composition of the previously isolated South American fauna (Marshall et al., 1982).

Each species is the repository of an immense amount of genetic information. The number of genes range from about 1,000 in bacteria and 10,000 in some fungi to 400,000 or more in many flowering plants and a few animals (Hinegardner, 1976). A typical mammal such as the house mouse (*Mus musculus*) has about 100,000 genes. This full complement is found in each of its myriad cells, organized from four strings of DNA, each of which comprises about a billion nucleotide pairs (George D. Snell, Jackson Laboratory, Maine, personal communication, 1987). (Human beings have genetic information closer in quantity to the mouse than to the more abundantly endowed salamanders and flowering plants; the difference, of course, lies in what is encoded.) If stretched out fully, the DNA would be roughly 1-meter long. But this molecule is invisible to the naked eye because it is only 20 angstroms in diameter. If we magnified it until its width equalled that of wrapping string, the fully extended molecule would be 960 kilometers long. As we traveled along its length, we would encounter some 20 nucleotide pairs or "letters" of genetic code per inch, or about 50 per centimeter. The full information contained therein, if translated into ordinary-size letters of printed text, would just about fill all 15 editions of the *Encyclopaedia Britannica* published since 1768 (Wilson, 1985a).

The number of species and the amount of genetic information in a representative organism constitute only part of the biological diversity on Earth. Each species is made up of many organisms. For example, the 10,000 or so ant species have been estimated to comprise 10^{15} living individuals at each moment of time (Wilson, 1971). Except for cases of parthenogenesis and identical twinning, virtually no two members of the same species are genetically identical, due to the high levels of genetic polymorphism across many of the gene loci (Selander, 1976). At still another level, wide-ranging species consist of multiple breeding populations that display complex patterns of geographic variation in genetic polymorphism. Thus, even if an endangered species is saved from extinction, it will probably have lost much of its internal diversity. When the populations are allowed to expand again, they will be more nearly genetically uniform than the ancestral populations. The bison herds of today are biologically not quite the same—not so interesting—as the bison herds of the early nineteenth century.

THE NATURAL LONGEVITY OF SPECIES

Within particular higher groups of organisms, such as ammonites or fishes, species have a remarkably consistent longevity. As a result, the probability that a given species will become extinct in a given interval of time after it splits off from other species can be approximated as a constant, so that the frequency of species surviving through time falls off as an exponential decay function; in other words, the percentage (but not the absolute number) of species going extinct in each period of time stays the same (Van Valen, 1973).[1] These regularities, such as they are, have been interrupted during the past 250 million years by major episodes of extinction that have been recently estimated to occur regularly at intervals of 26 million years (Raup and Sepkoski, 1984).

Because of the relative richness of fossils in shallow marine deposits, the longevity of fish and invertebrate species living there can often be determined with a modest degree of confidence. During Paleozoic and Mesozoic times, the average persistence of most fell between 1 and 10 million years: that is, 6 million for echinoderms, 1.9 million for graptolites, 1.2 to 2 million for ammonites, and so on (Raup, 1981, 1984).

These estimates are extremely interesting and useful but, as paleontologists have generally been careful to point out, they also suffer from some important limitations. First, terrestrial organisms are far less well known, few estimates have been attempted, and thus different survivorship patterns might have occurred (although Cenozoic flowering plants, at least, appear to fall within the 1- to 10-million-year range). More importantly, a great many organisms on islands and other restricted habitats, such as lakes, streams, and mountain crests, are so rare or local that they could appear and vanish within a short time without leaving any fossils. An equally great difficulty is the existence of sibling species—populations that are reproductively isolated but so similar to closely related species as to be difficult or impossible to distinguish through conventional anatomical traits. Such entities could rarely be diagnosed in fossil form. Together, all these considerations suggest that estimates of the longevity of natural species should be extended only with great caution to groups for which there is a poor fossil record.

RAIN FORESTS AS CENTERS OF DIVERSITY

In recent years, evolutionary biologists and conservationists have focused increasing attention on tropical rain forests, for two principal reasons. First, although these habitats cover only 7% of the Earth's land surface, they contain more than half the species in the entire world biota. Second, the forests are being destroyed so rapidly that they will mostly disappear within the next century, taking with them hundreds of thousands of species into extinction. Other species-rich biomes are in danger, most notably the tropical coral reefs, geologically ancient lakes, and coastal wetlands. Each deserves special attention on its own, but for the moment the rain forests serve as the ideal paradigm of the larger global crisis.

Tropical rain forests, or more precisely closed tropical forests, are defined as habitats with a relatively tight canopy of mostly broad-leaved evergreen trees

---

[1] Van Valen's original formulation, whose difficulties and implications are revealed by more recent research, has been discussed by Raup (1975) and by Lewin (1985). These studies deal with the clade, or set of populations descending through time after having split off as a distinct species from other such populations. They do not refer to the chronospecies, which is just a set of generations of the same species that is subjectively different from sets of generations.

sustained by 100 centimeters or more of annual rainfall. Typically two or more other layers of trees and shrubs occur beneath the upper canopy. Because relatively little sunlight reaches the forest floor, the undergrowth is sparse and human beings can walk through it with relative ease.

The species diversity of rain forests borders on the legendary. Every tropical biologist has a favorite example to offer. From a single leguminous tree in the Tambopata Reserve of Peru, I recently recovered 43 species of ants belonging to 26 genera, about equal to the entire ant fauna of the British Isles (Wilson, 1987). Peter Ashton found 700 species of trees in 10 selected 1-hectare plots in Borneo, the same as in all of North America (Ashton, Arnold Arboretum, personal communication, 1987). It is not unusual for a square kilometer of forest in Central or South America to contain several hundred species of birds and many thousands of species of butterflies, beetles, and other insects.

Despite their extraordinary richness, tropical rain forests are among the most fragile of all habitats. They grow on so-called wet deserts—an unpromising soil base washed by heavy rains. Two-thirds of the area of the forest surface consists of tropical red and yellow earths, which are typically acidic and poor in nutrients. High concentrations of iron and aluminum form insoluble compounds with phosphorus, thereby decreasing the availability of phosphorus to plants. Calcium and potassium are leached from the soil soon after their compounds are dissolved from the rain. As little as 0.1% of the nutrients filter deeper than 5 centimeters beneath the soil surface (NRC, 1982). An excellent popular account of rain forest ecology is given by Forsyth and Miyata (1984).

During the 150 million years since its origin, the principally dicotyledonous flora has nevertheless evolved to grow thick and tall. At any given time, most of the nonatmospheric carbon and vital nutrients are locked up in the tissue of the vegetation. As a consequence, the litter and humus on the ground are thin compared to the thick mats of northern temperate forests. Here and there, patches of bare earth show through. At every turn one can see evidence of rapid decomposition by dense populations of termites and fungi. When the forest is cut and burned, the ash and decomposing vegetation release a flush of nutrients adequate to support new herbaceous and shrubby growth for 2 or 3 years. Then these materials decline to levels lower than those needed to support a healthy growth of agricultural crops without artificial supplements.

The regeneration of rain forests is also limited by the fragility of the seeds of the constituent woody species. The seeds of most species begin to germinate within a few days or weeks, severely limiting their ability to disperse across the stripped land into sites favorable for growth. As a result, most sprout and die in the hot, sterile soil of the clearings (Gomez-Pompa et al., 1972). The monitoring of logged sites indicates that regeneration of a mature forest might take centuries. The forest at Angkor (to cite an anecdotal example) dates back to the abandonment of the Khmer capital in 1431, yet is still structurally different from a climax forest today, 556 years later. The process of rain forest regeneration is in fact so generally slow that few extrapolations have been possible; in some zones of greatest combined damage and sterility, restoration might never occur naturally (Caufield, 1985; Gomez-Pompa et al., 1972).

10 / BIODIVERSITY

Approximately 40% of the land that can support tropical closed forest now lacks it, primarily because of human action. By the late 1970s, according to estimates from the Food and Agricultural Organization and United Nations Environmental Programme, 7.6 million hectares or nearly 1% of the total cover is being permanently cleared or converted into the shifting-cultivation cycle. The absolute amount is 76,000 square kilometers (27,000 square miles) a year, greater than the area of West Virginia or the entire country of Costa Rica. In effect, most of this land is being permanently cleared, that is, reduced to a state in which natural reforestation will be very difficult if not impossible to achieve (Mellilo et al., 1985). This estimated loss of forest cover is close to that advanced by the tropical biologist Norman Myers in the mid-1970s, an assessment that was often challenged by scientists and conservationists as exaggerated and alarmist. The vindication of this early view should serve as a reminder always to take such doomsday scenarios seriously, even when they are based on incomplete information.

A straight-line extrapolation from the first of these figures, with identically absolute annual increments of forest-cover removal, leads to 2135 A.D. as the year in which all the remaining rain forest will be either clear-cut or seriously disturbed, mostly the former. By coincidence, this is close to the date (2150) that the World Bank has estimated the human population will plateau at 11 billion people (The World Bank, 1984). In fact, the continuing rise in human population indicates that a straight line estimate is much too conservative. Population pressures in the Third World will certainly continue to accelerate deforestation during the coming decades unless heroic measures are taken in conservation and resource management.

There is another reason to believe that the figures for forest cover removal present too sanguine a picture of the threat to biological diversity. In many local areas with high levels of endemicity, deforestation has proceeded very much faster than the overall average. Madagascar, possessor of one of the most distinctive floras and faunas in the world, has already lost 93% of its forest cover. The Atlantic coastal forest of Brazil, which so enchanted the young Darwin upon his arrival in 1832 ("wonder, astonishment & sublime devotion, fill & elevate the mind"), is 99% gone. In still poorer condition—in fact, essentially lost—are the forests of many of the smaller islands of Polynesia and the Caribbean.

HOW MUCH DIVERSITY IS BEING LOST?

No precise estimate can be made of the numbers of species being extinguished in the rain forests or in other major habitats, for the simple reason that we do not know the numbers of species originally present. However, there can be no doubt that extinction is proceeding far faster than it did prior to 1800. The basis for this statement is not the direct observation of extinction. To witness the death of the last member of a parrot or orchid species is a near impossibility. With the exception of the showiest birds, mammals, or flowering plants, biologists are reluctant to say with finality when a species has finally come to an end. There is always the chance (and hope) that a few more individuals will turn up in some remote forest remnant or other. But the vast majority of species are not monitored at all. Like the dead of Gray's "Elegy Written in a Country Churchyard," they pass from the Earth without notice.

Instead, extinction rates are usually estimated indirectly from principles of biogeography. As I mentioned above, the number of species of a particular group of organisms in island systems increases approximately as the fourth root of the land area. This has been found to hold true not just on real islands but also on habitat islands, such as lakes in a "sea" of land, alpine meadows or mountaintops surrounded by evergreen forests, and even in clumps of trees in the midst of a grassland (MacArthur and Wilson, 1967).

Using the area-species relationship, Simberloff (1984) has projected ultimate losses due to the destruction of rain forests in the New World tropical mainland. If present levels of forest removal continue, the stage will be set within a century for the inevitable loss of 12% of the 704 bird species in the Amazon basin and 15% of the 92,000 plant species in South and Central America.

As severe as these regional losses may be, they are far from the worst, because the Amazon and Orinoco basins contain the largest continuous rain forest tracts in the world. Less extensive habitats are far more threatened. An extreme example is the western forest of Ecuador. This habitat was largely undisturbed until after 1960, when a newly constructed road network led to the swift incursion of settlers and clear-cutting of most of the area. Now only patches remain, such as the 0.8-square-kilometer tract at the Rio Palenque Biological Station. This tiny reserve contains 1,033 plant species, perhaps one-quarter of which are known only to occur in coastal Ecuador. Many are known at the present time only from a single living individual (Gentry, 1982).

In general, the tropical world is clearly headed toward an extreme reduction and fragmentation of tropical forests, which will be accompanied by a massive extinction of species. At the present time, less than 5% of the forests are protected within parks and reserves, and even these are vulnerable to political and economic pressures. For example, 4% of the forests are protected in Africa, 2% in Latin America, and 6% in Asia (Brown, 1985). Thus in a simple system as envisioned by the basic models of island biogeography, the number of species of all kinds of organisms can be expected to be reduced by at least one-half—in other words, by hundreds of thousands or even (if the insects are as diverse as the canopy studies suggest) by millions of species. In fact, the island-biogeographic projections appear to be conservative for two reasons. First, tropical species are far more localized than those in the temperate zones. Consequently, a reduction of 90% of a tropical forest does not just reduce all the species living therein to 10% of their original population sizes, rendering them more vulnerable to future extinction. That happens in a few cases, but in many others, entire species are eliminated because they happened to be restricted to the portion of the forest that was cut over. Second, even when a portion of the species survives, it will probably have suffered significant reduction in genetic variation among its members due to the loss of genes that existed only in the outer portions.

The current reduction of diversity seems destined to approach that of the great natural catastrophes at the end of the Paleozoic and Mesozoic eras—in other words, the most extreme in the past 65 million years. In at least one important respect, the modern episode exceeds anything in the geological past. In the earlier mass extinctions, which some scientists believe were caused by large meteorite strikes, most of the plants survived even though animal diversity was severely reduced. Now, for the first time, plant diversity is declining sharply (Knoll, 1984).

HOW FAST IS DIVERSITY DECLINING?

The area-species curves of island systems, that is, the quantitative relationship between the area of islands and the number of species that can persist on the islands, provide minimal estimates of the reduction of species diversity that will eventually occur in the rain forests. But how long is "eventually"? This is a difficult question that biogeographers have attacked with considerable ingenuity. When a forest is reduced from, say, 100 square kilometers to 10 square kilometers by clearing, some immediate extinction is likely. However, the new equilibrium will not be reached all at once. Some species will hang on for a while in dangerously reduced populations. Elementary mathematical models of the process predict that the number of species in the 10-square-kilometer plot will decline at a steadily decelerating rate, i.e., they will decay exponentially to the lower level.

Studies by Jared Diamond and John Terborgh have led to the estimation of the decay constants for the bird faunas on naturally occurring islands (Diamond, 1972, 1984; Terborgh, 1974). These investigators took advantage of the fact that rising sea levels 10,000 years ago cut off small land masses that had previously been connected to South America, New Guinea, and the main islands of Indonesia. For example, Tobago, Margarita, Coiba, and Trinidad were originally part of the South American mainland and shared the rich bird fauna of that continent. Thus they are called land-bridge islands. In a similar manner, Yapen, Aru, and Misol were connected to New Guinea. In the study of the South American land-bridge islands, Terborgh found that the smaller the island, the higher the estimated decay constant and hence extinction rate. Terborgh then turned to Barro Colorado Island, which was isolated for the first time by the rise of Gatun Lake during the construction of the Panama Canal. Applying the natural land-bridge extinction curve to an island of this size (17 square kilometers) and fitting the derived decay constant to the actual period of isolation (50 years), Terborgh predicted an extinction of 17 bird species. The actual number known to have vanished as a probable result of insularization is 13, or 12% of the 108 breeding species originally present. The extinction rates of bird species on Barro Colorado Island were based on careful studies by E. O. Willis and J. R. Karr and have been recently reviewed by Diamond (1984).

Several other studies of recently created islands of both tropical and temperate-zone woodland have produced similar results, which can be crudely summarized as follows: when the islands range from 1 to 25 square kilometers—the size of many smaller parks and reserves—the rate of extinction of bird species during the first 100 years is 10 to 50%. Also as predicted, the extinction rate is highest in the smaller patches, and it rises steeply when the area drops below 1 square kilometer. To take one example provided by Willis (1979), three patches of subtropical forest isolated (by agricultural clearing) in Brazil for about a hundred years varied from 0.2 to 14 square kilometers, and, in reverse order, their resident bird species suffered 14 to 62% extinction rates.

What do these first measurements tell us about the rate at which diversity is being reduced? No precise estimate can be made for three reasons. First, the number of species of organisms is not known, even to the nearest order of magnitude. Second, because even in a simple island-biogeographic system, diversity reduction depends on the size of the island fragments and their distance from each other—factors that vary enormously from one country to the next. Third, the ranges of even the known species have not been worked out in most cases, so that we cannot say which ones will be eliminated when the tropical forests are partially cleared.

However, scenarios of reduction can be constructed to give at least first approximations if certain courses of action are followed. Let us suppose, for example, that half the species in tropical forests are very localized in distribution, so that the rate at which species are being eliminated immediately is approximately this fraction multiplied by the rate-percentage of the forests being destroyed. Let us conservatively estimate that 5 million species of organisms are confined to the tropical rain forests, a figure well justified by the recent upward adjustment of insect diversity alone. The annual rate of reduction would then be $0.5 \times 5 \times 10^6 \times 0.007$ species, or 17,500 species per year. Given 10 million species in the fauna and flora of all the habitats of the world, the loss is roughly one out of every thousand species per year. How does this compare with extinction rates prior to human intervention? The estimates of extinction rates in Paleozoic and Mesozoic marine faunas cited earlier (Raup, 1981, 1984; Raup and Sepkoski, 1984; Van Valen, 1973) ranged according to taxonomic group (e.g., echinoderms versus cephalopods) from one out of every million to one out of every 10 million per year. Let us assume that on the order of 10 million species existed then, in view of the evidence that diversity has not fluctuated through most of the Phanerozoic time by a factor of more than three (Raup and Sepkoski, 1984). It follows that both the per-species rate and absolute loss in number of species due to the current destruction of rain forests (setting aside for the moment extinction due to the disturbance of other habitats) would be about 1,000 to 10,000 times that before human intervention.

I have constructed other simple models incorporating the quick loss of local species and the slower loss of widespread species due to the insularization effect, and these all lead to comparable or higher extinction rates. It seems difficult if not impossible to combine what is known empirically of the extinction process with the ongoing deforestation process without arriving at extremely high rates of species loss in the near future. Curiously, however, the study of extinction remains one of the most neglected in ecology. There is a pressing need for a more sophisticated body of theories and carefully planned field studies based on it than now exist.

WHAT CAN BE DONE?

The biological diversity most threatened is also the least explored, and there is no prospect at the moment that the scientific task will be completed before a large fraction of the species vanish. Probably no more than 1,500 professional systematists in the world are competent to deal with the millions of species found in the humid tropic forests. Their number may be dropping, due to decreased professional opportunities, reduced funding for research, and the assignment of a higher priority to other disciplines. Data concerning the number of taxonomists, as well as detailed arguments for the need to improve research in tropical countries, are given by NRC (1980). The decline has been accompanied by a more than 50% decrease in the number of publications in tropical ecology from 1979 to 1983 (Cole, 1984).

The problem of tropical conservation is thus exacerbated by the lack of knowledge and the paucity of ongoing research. In order to make precise assessments and recommendations, it is necessary to know which species are present (recall that the great majority have not even received a scientific name) as well as their geographical ranges, biological properties, and possible vulnerability to environmental change.

It would be a great advantage, in my opinion, to seek such knowledge for the entire biota of the world. Each species is unique and intrinsically valuable. We cannot expect to answer the important questions of ecology and other branches of evolutionary biology, much less preserve diversity with any efficiency, by studying only a subset of the extant species.

I will go further: the magnitude and control of biological diversity is not just a central problem of evolutionary biology; it is one of the key problems of science as a whole. At present, there is no way of knowing whether there are 5, 10, or 30 million species on Earth. There is no theory that can predict what this number might turn out to be. With reference to conservation and practical applications, it also matters why a certain subset of species exists in each region of the Earth, and what is happening to each one year by year. Unless an effort is made to understand all of diversity, we will fall far short of understanding life in these important respects, and due to the accelerating extinction of species, much of our opportunity will slip away forever.

Lest this exploration be viewed as an expensive Manhattan Project unattainable in today's political climate, let me cite estimates I recently made of the maximum investment required for a full taxonomic accounting of all species: 25,000 professional lifetimes (4,000 systematists are at work full or part time in North America today); their final catalog would fill 60 meters of library shelving for each million species (Wilson, 1985a). Computer-aided techniques could be expected to cut the effort and cost substantially. In fact, systematics has one of the lowest cost-to-benefit ratios of all scientific disciplines.

It is equally true that knowledge of biological diversity will mean little to the vast bulk of humanity unless the motivation exists to use it. Fortunately, both scientists and environmental policy makers have established a solid linkage between economic development and conservation. The problems of human beings in the tropics are primarily biological in origin: overpopulation, habitat destruction, soil deterioration, malnutrition, disease, and even, for hundreds of millions, the uncertainty of food and shelter from one day to the next. These problems can be solved in part by making biological diversity a source of economic wealth. Wild species are in fact both one of the Earth's most important resources and the least utilized. We have come to depend completely on less than 1% of living species for our existence, the remainder waiting untested and fallow. In the course of history, according to estimates made by Myers (1984), people have utilized about 7,000 kinds of plants for food; predominant among these are wheat, rye, maize, and about a dozen other highly domesticated species. Yet there are at least 75,000 edible plants in existence, and many of these are superior to the crop plants in widest use. Others are potential sources of new pharmaceuticals, fibers, and petroleum substitutes. In addition, among the insects are large numbers of species that are potentially superior as crop pollinators, control agents for weeds, and parasites and predators of insect pests. Bacteria, yeasts, and other microorganisms are likely to continue yielding new medicines, food, and procedures of soil restoration. Biologists have begun to fill volumes with concrete proposals for the further exploration and better use of diversity, with increasing emphasis on the still unexplored portions of the tropical biota. Some of the most recent and useful works on this subject include those by Myers (1984), NRC (1975), Office of Technology Assessment (1984), Oldfield (1984), and the U.S. Department of State (1982). In addition, an excellent series of specialized publications on practical uses of wild species have been produced during the past 10 years by authors and panels commissioned by the Board on Science and Technology for International Development (BOSTID) of the National Research Council.

In response to the crisis of tropical deforestation and its special threat to biological diversity, proposals are regularly being advanced at the levels of policy and research. For example, Nicholas Guppy (1984), noting the resemblance of the lumbering of rain forests to petroleum extraction as the mining of a nonrenewable resource for short-term profit, has recommended the creation of a cartel, the Organization of Timber-Exporting Countries (OTEC). By controlling production and prices of lumber, the organization could slow production while encouraging member states to "protect the forest environment in general and gene stocks and special habitats in particular, create plantations to supply industrial and fuel wood, benefit indigenous tribal forest peoples, settle encroachers, and much else." In another approach, Thomas Lovejoy (1984) has recommended that debtor nations with forest resources and other valuable habitats be given discounts or credits for undertaking conservation programs. Even a small amount of forgiveness would elevate the sustained value of the natural habitats while providing hard currency for alternatives to their exploitation.

Another opportunity for innovation lies in altering somewhat the mode of direct economic assistance to developing countries. A large part of the damage to tropical forests, especially in the New World, has resulted from the poor planning of road systems and dams. For example, the recent settlement of the state of Rondonia and construction of the Tucurui Dam, both in Brazil, are now widely perceived by ecologists and economists alike as ill-conceived (Caufield, 1985). Much of the responsibility of minimizing environmental damage falls upon the international agencies that have the power to approve or disapprove particular projects.

The U.S. Congress addressed this problem with amendments to the Foreign Assistance Act in 1980, 1983, and 1986, which call for the development of a strategy for conserving biological diversity. They also mandate that programs funded through the U.S. Agency for International Development (USAID) include an assessment of environmental impact. In implementing this new policy, USAID has recognized that "the destruction of humid tropical forests is one of the most important environmental issues for the remainder of this century and, perhaps, well into the next," in part because they are "essential to the survival of vast numbers of species of plants and animals" (U.S. Department of State, 1985). In another sphere, The World Bank and other multinational lending agencies have come under increasing pressure to take a more active role in assessing the environmental impact of the large-scale projects they underwrite (Anonymous, 1984).

In addition to recommendations for international policy initiatives, there has recently been a spate of publications on the linkage of conservation and economic use of tropical forests. Notable among them are *Research Priorities in Tropical Biology* (NRC, 1980), based on a study of the National Research Council; *Technologies to Sustain Tropical Forest Resources* (OTA, 1984), prepared by the Office of Technology Assessment for the U.S. Congress; and the *U.S. Strategy on the Conservation of Biological Diversity* (USAID, 1985), a report to Congress by an interagency task force. Most comprehensive of all—and in my opinion the most encouraging in its implications—is the three-part series *Tropical Forests: A Call for Action*, released by the World Resources Institute, The World Bank, and the United Nations Development Programme (1985). The report makes an assessment of the problem worldwide and reviews case histories in which conservation or restoration have contributed to economic development. It examines the needs of every tropical country with important forest reserves. The estimated cost to make an impact on tropical deforestation over the next 5 years would be U.S. $8 billion—a large sum but surely the most cost-effective investment available to the world at the present time.

In the end, I suspect it will all come down to a decision of ethics—how we value the natural worlds in which we evolved and now, increasingly, how we regard our status as individuals. We are fundamentally mammals and free spirits who reached this high a level of rationality by the perpetual creation of new options. Natural philosophy and science have brought into clear relief what might be the essential paradox of human existence. The drive toward perpetual expansion—or personal freedom—is basic to the human spirit. But to sustain it we need the most delicate, knowing stewardship of the living world that can be devised. Expansion and stewardship may appear at first to be conflicting goals, but the opposite is true. The depth of the conservation ethic will be measured by the extent to which each of the two approaches to nature is used to reshape and reinforce the other. The paradox can be resolved by changing its premises into forms more suited to ultimate survival, including protection of the human spirit. I recently wrote in synecdochic form about one place in South America to give these feelings more exact expression:

> To the south stretches Surinam eternal, Surinam serene, a living treasure awaiting assay. I hope that it will be kept intact, that at least enough of its million-year history will be saved for the reading. By today's ethic its value may seem limited, well beneath the pressing concerns of daily life. But I suggest that as biological knowledge grows the ethic will shift fundamentally so that everywhere, for reasons that have to do with the very fiber of the brain, the fauna and flora of a country will be thought part of the national

E. O. WILSON / 17

heritage as important as its art, its language, and that astonishing blend of achievement and farce that has always defined our species (Wilson, 1984).

REFERENCES

Anonymous. 1984. Critics fault World Bank for ecological neglect. Conserv. Found. News. Nov.-Dec.:1–7.

Arnett, R. H. 1985. General considerations. Pp. 3–9 in American Insects: A Handbook of the Insects of America North of Mexico. Van Nostrand Reinhold, New York.

Brown, R. L., ed. 1985. State of the World 1985: A Worldwatch Institute Report on Progress Toward a Sustainable Society. W. W. Norton, New York. 301 pp.

Caufield, C. 1985. In the Rainforest. A. A. Knopf, New York. 283 pp.

Cole, N. H. A. 1984. Tropical ecology research. Nature 309:204.

Diamond, J. M. 1972. Biogeographic kinetics: Estimation of relaxation times for avifaunas of Southwest Pacific islands. Proc. Natl. Acad. Sci. USA 69:3199–3203.

Diamond, J. M. 1984. "Normal" extinctions of isolated populations. Pp. 191–246 in M. H. Nitecki, ed. Extinctions. University of Chicago Press, Chicago.

Erwin, T. L. 1983. Beetles and other insects of tropical forest canopies at Manaus, Brazil, sampled by insecticidal fogging. Pp. 59–75 in S. L. Sutton, T. C. Whitmore, and A. C. Chadwick, eds. Tropical Rain Forest: Ecology and Management. Blackwell, Edinburgh.

Forsyth, A., and K. Miyata. 1984. Tropical Nature: Life & Death in the Rain Forests of Central & South America. Scribner's, New York. 272 pp.

Frankel, O. H., and M. E. Soulé. 1981. Conservation and Evolution. Cambridge University Press, Cambridge, Mass. 327 pp.

Gentry, A. H. 1982. Patterns of Neotropical plant-species diversity. Evol. Biol. 15:1–85.

Gomez-Pompa, A., C. Vazquez-Yanes, and S. Guevara. 1972. The tropical rain forest: A nonrenewable resource. Science 177:762–765.

Guppy, N. 1984. Tropical deforestation: A global view. Foreign Affairs 62:928–965.

Hinegardner, R. 1976. Evolution of genome size. Pp. 179–199 in F. J. Ayala, ed. Molecular Evolution. Sinauer Associates, Sunderland, Mass.

Knoll, A. H. 1984. Patterns of extinction in the fossil record of vascular plants. Pp. 21–68 in M. H. Nitecki, ed. Extinctions. University of Chicago Press, Chicago.

Lewin, R. 1985. Red Queen runs into trouble? Science 227:399–400.

Lovejoy, T. E. 1984. Aid debtor nations' ecology. New York Times, October 4.

MacArthur, R. H., and E. O. Wilson. 1967. The Theory of Island Biogeography. Princeton University Press, Princeton, N.J. 203 pp.

Marshall, L. G., S. D. Webb, J. J. Sepkoski, Jr., and D. M. Raup. 1982. Mammalian evolution and the great American interchange. Science 215:1351–1357.

Melillo, J. M., C. A. Palm, R. A. Houghton, G. M. Woodwell, and N. Myers. 1985. A comparison of two recent estimates of disturbance in tropical forests. Environ. Conserv. 12:37–40.

Myers, N. 1983. A Wealth of Wild Species: Storehouse for Human Welfare. Westview Press, Boulder, Colo. 300 pp.

Myers, N. 1984. The Primary Source: Tropical Forests and Our Future. W. W. Norton, New York. 399 pp.

NRC (National Research Council). 1975. Underexploited Tropical Plants with Promising Economic Value. Board on Science and Technology for International Development Report 16. National Academy of Sciences, Washington, D.C. 187 pp.

NRC (National Research Council). 1979. Tropical Legumes: Resources for the Future. Board on Science and Technology for International Development Report 25. National Academy of Sciences, Washington, D.C. 331 pp.

NRC (National Research Council). 1980. Research Priorities in Tropical Biology. National Academy of Sciences, Washington, D.C. 116 pp.

NRC (National Research Council). 1982. Ecological Aspects of Development in the Humid Tropics. National Academy Press, Washington, D.C. 297 pp.

Oldfield, M. L. 1984. The Value of Conserving Genetic Resources. National Park Service, U.S. Department of the Interior, Washington, D.C. 360 pp.

OTA (Office of Technology Assessment). 1984. Technologies to Sustain Tropical Forest Resources.

18 / BIODIVERSITY

Congress of the United States, Office of Technology Assessment, Washington, D.C. 344 pp.

Parker, S. P., ed. 1982. Synopsis and Classification of Living Organisms. McGraw-Hill, New York. 2 vols.

Raup, D. M. 1975. Taxonomic survivorship curves and Van Valen's Law. Paleobiology 1:82–86.

Raup, D. M. 1981. Extinction: Bad genes or bad luck? Acta Geol. Hisp. 16:25–33.

Raup, D. M. 1984. Evolutionary radiations and extinction. Pp. 5–14 in H. D. Holland and A. F. Trandell, eds. Patterns of Change in Evolution. Dahlem Konferenzen, Abakon Verlagsgesellschaft, Berlin.

Raup, D. M., and J. J. Sepkoski, Jr. 1984. Periodicity of extinctions in the geologic past. Proc. Natl. Acad. Sci. USA 81:801–805.

Selander, R. K. 1976. Genic variation in natural populations. Pp. 21–45 in F. J. Ayala, ed. Molecular Evolution. Sinauer Associates, Sunderland, Mass.

Simberloff, D. S. 1984. Mass extinction and the destruction of moist tropical forests. Zh. Obshch. Biol. 45:767–778.

Terborgh, J. 1974. Preservation of natural diversity: The problem of extinction-prone species. BioScience 24:715–722.

USAID (U.S. Agency for International Development). 1985. U.S. Strategy on the Conservation of Biological Diversity. An Interagency Task Force Report to Congress. U.S. Agency for International Development, Washington, D.C. 52 pp.

U.S. Department of State. 1982. Proceedings of the U.S. Strategy Conference on Biological Diversity. November 16–18, 1981, Washington, D.C. Publication No. 9262. U.S. Department of State, Washington, D.C. 126 pp.

U.S. Department of State. 1985. Humid Tropical Forests: AID Policy and Guidance. U.S. Department of State Memorandum. Government Printing Office, Washington, D.C. 3 pp.

Van Valen, L. 1973. A new evolutionary law. Evol. Ther. 1:1–30.

Willis, E. O. 1979. The composition of avian communities in remanescent woodlots in southern Brazil. Papeis Avulsos Zool. 33:1–25.

Wilson, E. O. 1971. The Insect Societies. Belknap Press of Harvard University Press, Cambridge, Mass. 548 pp.

Wilson, E. O. 1984. Biophilia. Harvard University Press, Cambridge, Mass. 176 pp.

Wilson, E. O. 1985a. The biological diversity crisis: A challenge to science. Issues Sci. Technol. 2:20–29.

Wilson, E. O. 1985b. Invasion and extinction in the West Indian ant fauna: Evidence from the Dominican amber. Science 229:265–267.

Wilson, E. O. 1987. The arboreal ant fauna of Peruvian Amazon forests: A first assessment. Biotropica 2:245–251.

World Bank. 1984. World Development Report 1984. Oxford University Press, New York. 286 pp.

World Resources Institute, The World Bank, and United Nations Development Programme. 1985. Tropical Forests: A Call for Action. World Resources Institute, Washington, D.C. 3 vols.

1989

"Threats to biodiversity," *Scientific American,* 261(3) (September): 108–116

This article, written for the scientifically interested public, summarizes much of what was understood of global biodiversity at the close of the 1980s. It was composed after the publication of *BioDiversity,* which I edited, and contains some of the key concepts subsequently expanded in my 1992 comprehensive work *The Diversity of Life.* The generalizations and examples contained here remain relevant at the time of writing.

The article marks another important juncture in science and human affairs. In 1989, when it was published, conservation biology was a fledgling discipline. Founded in 1985, the Society for Conservation Biology became for a time the fastest growing biological organization, and within 20 years acquired 8,000 members from around the world. Its journal *Conservation Biology,* begun in 1987, has proved critically important in collecting and reviewing the important discoveries of basic and applied research. The impact of the discipline has helped to establish the following broader principle: providing their research is thoroughly documented and peer reviewed, and the subject they address is urgent, scientists will not lose their credibility if they enter the arena of public policy.

Threats to Biodiversity

Habitat destruction, mostly in the tropics, is driving thousands of species each year to extinction. The consequences will be dire—unless the trend is reversed

by Edward O. Wilson

The human species came into being at the time of greatest biological diversity in the history of the earth. Today as human populations expand and alter the natural environment, they are reducing biological diversity to its lowest level since the end of the Mesozoic era, 65 million years ago. The ultimate consequences of this biological collision are beyond calculation and certain to be harmful. That, in essence, is the biodiversity crisis.

In one sense the loss of diversity is the most important process of environmental change. I say this because it is the only process that is wholly irreversible. Its consequences are also the least predictable, because the value of the earth's biota (the fauna and flora collectively) remains largely unstudied and unappreciated. Every country can be said to have three forms of wealth: material, cultural and biological. The first two we understand very well, because they are the substance of our everyday lives. Biological wealth is taken much less seriously. This is a serious strategic error, one that will be increasingly regretted as time passes. The biota is on the one hand part of a country's heritage, the product of millions of years of evolution centered on that place and hence as much a reason for national concern as the particularities of language and culture. On the other hand, it is a potential source for immense untapped material wealth in the form of food, medicine and other commercially important substances.

EDWARD O. WILSON was one of the first to call attention to the global decline in biological diversity and to sound the alarm on the consequences of its loss. His interest in living organisms, especially ants, stems back to his childhood, which was devoted to the exploration and collection of living things, and to his undergraduate studies in evolutionary biology at the University of Alabama. He received his Ph.D. in biology from Harvard University, where he is now Frank B. Baird, Jr., Professor of Science and Curator in Entomology. Wilson has made major contributions to a number of fields, including the behavior and evolution of social insects, chemical communication and the evolution of social behavior. He has been awarded the National Medal of Science, the Pulitzer Prize in general nonfiction for his book *On Human Nature* and the Tyler Prize for environmental achievement.

It is a remarkable fact, given the interdependence of human beings and the other species that inhabit the planet, that the task of studying biodiversity is still in an early stage. Although systematics is one of the two oldest formal disciplines of biology (the other is anatomy), we do not even know to the nearest order of magnitude the number of species of organisms on the earth. With the help of other specialists, I have estimated the number of species that have been formally described (given a Latinized scientific name) to be about 1.4 million. Even conservative guesses place the actual number of species at four million or greater, more than twice the number described to date.

Terry L. Erwin of the Smithsonian's National Museum of Natural History believes the number of species to be even greater. With the help of co-workers, he applied an insecticidal fog to the forest canopy at localities in Brazil and Peru in order to obtain an estimate of the total number of insect and other arthropod species in this rich but still relatively unexplored habitat. By extrapolating his findings to moist tropical forests around the world and by including a rough estimate of the number of ground-dwelling species in his calculations, Erwin arrived at a global total of 30 million species. Even if this number proves to be a considerable overestimate, the amount of biodiversity in the world is certain to be projected sharply upward in other, compensatory ways.

Groups such as the mites and fungi, for example, are extremely rich and also very underexplored, and habitats such as the floors of the deep sea are thought to harbor hundreds of thousands of species, most of which remain undescribed. Even the number of bacterial species on the earth is expected to be many times greater than the 3,000 that have been characterized to date. To take one example, an entirely new flora of bacteria has recently been discovered living at depths of 350 meters or more beneath the ground near Hilton Head, South Carolina. Even new species of birds continue to turn up at an average rate of two per year.

Systematists are in wide agreement that whatever the absolute numbers, more than half of the species on the earth live in moist tropical forests, popularly referred to as rain forests. Occupying only 6 percent of the land surface, these ecosystems are found in warm areas where the rainfall is 200 centimeters or more per year, which allows broad-leaved evergreen trees to flourish. The trees typically sort into three or more horizontal layers, the canopy of the tallest being 30 meters (about 100 feet) or more from the ground. Together the tree crowns of the several layers admit little sunlight to the forest floor, inhibiting the development of undergrowth and leaving large spaces through which it is relatively easy to walk.

The belief that a majority of the

TROPICAL RAIN FORESTS, such as this one in northern Costa Rica, are among the most species-rich habitats on the earth. The enormous biological diversity found in these forests can be explained by the fact that the most species-rich groups on the planet, the invertebrates and flowering plants, are concentrated there. The vegetation, much of it broad-leaved evergreens, is extremely lush; the tallest trees tower as much as 30 meters (100 feet) above the rain-forest floor.

DEFORESTATION IS OCCURRING at a rapid rate around the world. In Costa Rica (*top*), as well as in parts of South America, rain forest is often cut and the land fenced in and converted to pasture. Unlike temperate forests, where nutrients accumulate in the soil, tropical forests typically have poor-quality soil. Consequently, within two or three years after being cleared, soil that once supported dense vegetation becomes too nutrient-poor to provide much grass for grazing cattle. In the U.S. (*bottom*), the impact of large-scale logging operations can be clearly seen in this photograph of a mountain range in the state of Washington. The scattered logs in the foreground are trees that have been cut and stripped of their branches and are waiting to be collected.

planet's species live in tropical rainforest habitats is not based on an exact and comprehensive census but on the fact that the two overwhelmingly species-rich groups, the arthropods (especially insects) and the flowering plants, are concentrated there. Other extremely species-rich environments exist, including the coral reefs and abyssal plains of the oceans and the heathlands of South Africa and southwestern Australia, but these appear to be outranked substantially by the rain forests.

Every tropical biologist has stories of the prodigious variety in this one habitat type. From a single leguminous tree in Peru, I once retrieved 43 ant species belonging to 26 genera, approximately equal to the ant diversity of all of the British Isles. In 10 selected one-hectare plots in Kalimantan in Indonesia, Peter S. Ashton of Harvard University found more than 700 tree species, about equal to the number of tree species native to all North America. The current world record at this writing (certain to be broken) was established in 1988 by Alwyn H. Gentry of the Missouri Botanical Garden, who identified approximately 300 tree species in each of two one-hectare plots near Iquitos, Peru.

Why has life multiplied so prodigiously in a few limited places such as tropical forests and coral reefs? It was once widely believed that when large numbers of species coexist, their life cycles and food webs lock together in a way that makes the ecosystem more robust. This diversity-stability hypothesis has given way during the past 20 years to a reversed cause-and-effect scenario that might be called the stability-diversity hypothesis: fragile superstructures of species build up when the environment remains stable enough to support their evolution during long periods of time. Biologists now know that biotas, like houses of cards, can be brought tumbling down by relatively small perturbations in the physical environment. They are not robust at all.

The history of global diversity is reflected in the standing diversity of marine animals, the group best represented in the fossil record. The trajectory can be summarized as follows: after the initial "experimental" flowering of multicellular animals, there was a swift rise in species number in early Paleozoic times (some 600 million years ago), then plateaulike stagnation for the remaining 200 million years of the Paleozoic era and finally a slow but steady climb through the Mesozoic and Cenozoic eras to diversity's present all-time high [*see illustration at right*].

The overall impression gained from examining these and comparable sets of data for other groups of organisms is that biological diversity was hard won and a long time in coming. Furthermore, the procession of life was set back by five massive extinction episodes during the Ordovician, Devonian, Permian, Triassic and Cretaceous periods. The last of these is by far the most famous, because it ended the age of dinosaurs, conferred hegemony on the mammals and ultimately, for better or worse, made possible the origin of our own species. But the Cretaceous crisis was minor compared with the great Permian crash some 240 million years ago, which extinguished between 77 and 96 percent of all marine animal species. As David M. Raup of the University of Chicago has observed, "If these estimates are even reasonably accurate, global biology (for higher organisms, at least) had an extremely close call." It took five million years, well into Mesozoic times, for species diversity to begin a significant recovery.

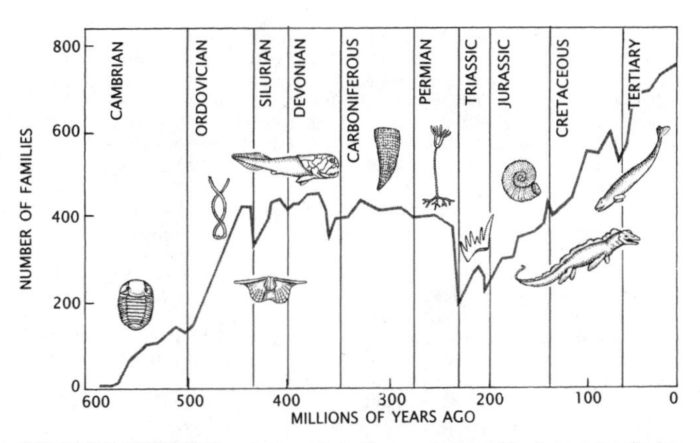

BIOLOGICAL DIVERSITY has increased slowly over time, set back occasionally by mass extinctions. There have been five mass-extinction events so far: at the close of the Ordovician, Devonian, Permian, Triassic and Cretaceous periods, when the number of families of marine organisms declined by 12, 14, 52, 12 and 11 percent, respectively. The extinction event at the end of the Permian was by far the most severe; since then diversity has slowly increased to its present all-time high. It is now declining at an unprecedented rate, however, as a result of human activity.

What lessons can be drawn from these extinction episodes of the past? It is clear that recovery, given sufficient time, is sometimes possible. It is also true that in some cases new species can be created rapidly. A large minority of flowering-plant species have originated in a single generation by polyploidy—a multiplication of chromosome sets, either within a single individual or following the hybridization of two previously distinct species. Even geographic speciation, in which populations diverge genetically after being separated by a barrier such as a strait or desert, can in extreme cases lead to the evolution of new species in as few as from 10 to 100 generations. Hence, it might be argued that when a mass extinction occurs the deficit can be made up in a relatively short time. But under such circumstances pure *numbers* of species mean little. What matters more, in terms of the spread of genetic codes and the multiple ways of life they prescribe, is diversity at the higher taxonomic levels: the number of genera, families and so on.

A species is most interesting when its traits are sufficiently unique to warrant its placement in a distinct genus or even a higher-level taxon, such as a family. A concrete example helps to illustrate my point. In western China a new species of muntjac deer was recently discovered, which appears to differ from the typical muntjac of Asia only in chromosome number and in a few relatively minor anatomical traits. Human beings intuitively value this slightly differentiated species, of course, but not nearly so much as they value the giant panda, which is so distinctive as to be placed in its own genus (*Ailuropoda*) and family (Ailuropodidae).

Within the past 10,000 years biological diversity has entered a wholly new era in the turbulent history of life on the earth. Human activity has had a devastating effect on species diversity, and the rate of human-induced extinctions is accelerating. The heaviest pressure has hitherto been exerted on islands, lakes and other isolated and strongly circumscribed environments. Fully one half of the bird species of Polynesia have been eliminated through hunting and the destruction of native forests. In the 1800's most of the unique flora of trees and shrubs on St. Helena, a tiny island in the South Atlantic, was lost forever when the island was completely deforested. Hundreds of fish species that are endemic to Lake Victoria, formerly of great commercial value as food and aquarium fish, are now threatened with extinction as the result of the careless introduction of one species of fish, the Nile perch. The list of such biogeographic disasters is extensive.

Serious as the episodes of pinpoint destruction are, they are minor compared with the species hecatomb caused by the clearing and burning of tropical rain forests. Already the forest has been reduced to approximately 55 percent of its original cover (as inferred from soil and climate profiles of the land surface), and it is being further reduced at a rate in excess of 100,000 square kilometers a year. This amount is 1 percent of the total cover, or more than the area of Switzerland and the Netherlands combined.

What is the effect of such habitat reduction on species diversity? In archipelago systems such as the West Indies and Polynesia, the number of species found on an individual island corresponds roughly to the island's area: the number of species usually increases with the size of the island, by somewhere between the fifth and the third root of the area. Many fall close to the central value of the fourth root. The same relation holds for "habitat islands," such as patches of forest surrounded by a sea of grassland. As a rough rule of thumb, a tenfold increase in area results in a doubling of the number of species. Put the other way, if the island area is diminished tenfold, the number of species will be cut in half.

The theory of island biogeography, which has been substantiated at least in broad outline by experimental alterations of island biotas and other field studies, holds that species number usually fluctuates around an equilibrium. The number remains more or less constant over time because the rate of immigration of new species to the island balances the extinction rate of species already there, and so diversity remains fairly constant. The relation between the theory of island biogeography and global diversity is an important one: if the area of a particular habitat, such as a patch of rain forest, is reduced by a given amount, the number of species living in it will subside to a new, lower equilibrium. The rich forest along the Atlantic coast of Brazil, for example, has been cleared to less than 1 percent of its original cover; even in the unlikely event that no more trees are cut, the forest biota can be expected to decline by perhaps 75 percent, or to one quarter of its original number of species.

I have conservatively estimated that on a worldwide basis the ultimate loss attributable to rain-forest clearing alone (at the present 1 percent rate) is from .2 to .3 percent of all species in the forests per year. Taking a very conservative figure of two million species confined to the forests, the global loss that results from deforestation could be as much as from 4,000 to 6,000 species a year. That in turn is on the order of 10,000 times greater than the naturally occurring background extinction rate that existed prior to the appearance of human beings.

Although the impact of habitat destruction is most severely felt in tropical rain forests, where species diversity is so high, it is also felt in other regions of the planet, particularly where extensive forest clearing is taking place. In the U.S. alone, some 60,000 acres of ancient forests are being cut per year, mostly for lumber that is then exported to Japan and other countries in the Pacific rim. Most severely affected are the national forests of the Pacific Northwest, from which some 5.5 billion board-feet of timber were harvested in 1987, and Alaska's Tongass National Forest, where as much as 50 percent of the most productive forestland has been logged since 1950. Although reforestation in these areas is possible, the process of regrowth may last 100 years or more.

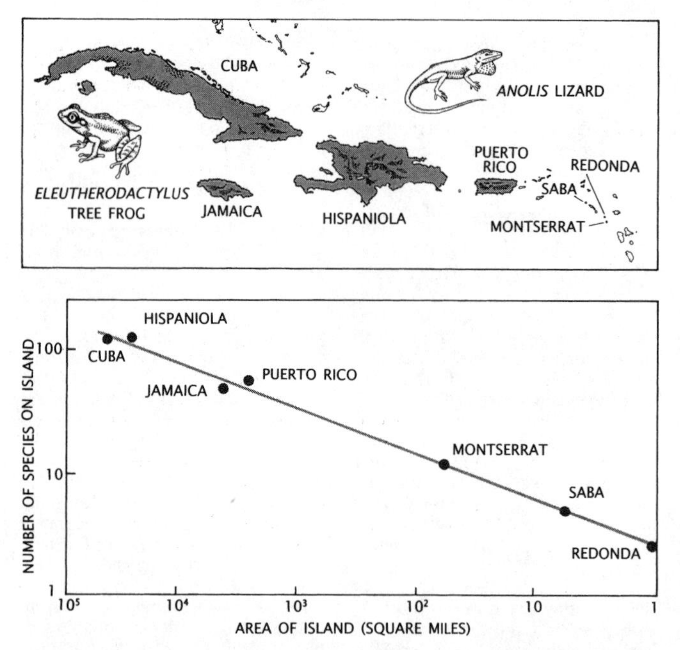

NUMBER OF SPECIES on an island corresponds to its size. As a general rule, when the area of an island increases tenfold, the number of species on it doubles. This is easily demonstrated for an island archipelago, such as the West Indies (*top*), where there are numerous islands of different sizes. The numbers of species of reptiles and amphibians on five islands, including *Anolis* lizards and *Eleutherodactylus* tree frogs, were counted and the combined total plotted against the area of each island. As the curve shows (*bottom*), a large island, such as Cuba, has more than twice as many species as, say, the smaller island of Saba. These findings have important implications for conservation biology because the data can be used to predict species loss from habitat destruction and to determine the optimal size of wildlife preserves.

How long does it take, once a habitat is reduced or destroyed, for the species that live in it to actually become extinct? The rate of extinction depends on the size of the habitat patch left undisturbed and the group of organisms concerned. In one ingenious study, Jared M. Diamond of the University of California at Los Angeles and John W. Terborgh of Duke University counted the number of bird species on several continental-shelf islands, which until about 10,000 years ago had been part of the mainland but then became isolated when the sea level rose. By comparing the number of species per island with the number of species on the adjacent mainland, Diamond and Terborgh were able to estimate the number of species each island had lost and to correlate the rate of species loss with island size.

Their model has been reasonably well confirmed by empirical studies of local bird faunas, and the results are sobering: in patches of between one and 20 square kilometers, a common size for reserves and parks in the tropics and elsewhere, 20 percent or more of the species disappear within 50 years. Some of the birds vanish quickly. Others linger for a while as the "living dead." In regions where the natural habitat is highly fragmented, the rate of species loss is even greater.

These extinction rates are probably underestimates, because they are based on the assumption that the species are distributed more or less evenly throughout the forests being cut. But biological surveys indicate that large numbers of species are confined to very limited ranges; if the small fraction of the forest habitat occupied by a species is destroyed, the species is eliminated immediately. When a single ridge top in Peru was cleared recently, more than 90 plant species known only from that locality were lost forever.

Ecologists have begun to identify "hot spots" around the world—habitats that are rich in species and also in imminent danger of destruction. Norman Myers, an environmental consultant with wide experience in the tropics, has compiled a list of threatened rain-forest habitats from 10 places: the Chocó of western Colombia, the uplands of western Amazonia, the

SATELLITE IMAGE of the northern tip of Prince of Wales Island in Tongass National Forest, Alaska, shows the extent to which the region has been clear cut. Areas that have recently been cleared and are barren of tree cover are indicated in pink; those that have been cut but have started to recover are light green; areas where the forest has not yet been disturbed are dark green. The image covers about 400 square miles.

Atlantic coast of Brazil, Madagascar, the eastern Himalayas, the Philippines, Malaysia, northwestern Borneo, Queensland and New Caledonia. Other biologists have similarly classified certain temperate forest patches, heathlands, coral reefs, drainage systems and ancient lakes. One of the more surprising examples is Lake Baikal in Siberia, where large numbers of endemic crustaceans and other invertebrates are endangered by rising levels of pollution.

The world biota is trapped as though in a vise. On one side it is being swiftly reduced by deforestation. On the other it is threatened by climatic warming brought on by the greenhouse effect. Whereas habitat loss is most destructive to tropical biotas, climatic warming is expected to have a greater impact on the biotas of the cold temperate regions and polar regions. A poleward shift of climate at the rate of 100 kilometers or more per century, which is considered at least a possibility, would leave wildlife preserves and entire species ranges behind, and many kinds of plants and animals could not migrate fast enough to keep up.

The problem would be particularly acute for plants, which are relatively immobile and do not disperse as readily as animals. The Engelmann spruce, for example, has an estimated natural dispersal capacity of from one to 20 kilometers per century, so that massive new plantings would be required to sustain the size of the range it currently occupies. Margaret Davis and Catherine Zabinski of the University of Minnesota predict that in response to global warming four North American trees—yellow birch, sugar maple, beech and hemlock—will be displaced northward by from 500 to 1,000 kilometers. Hundreds of thousands of species are likely to be similarly displaced; how many will adapt to the changing climate, not having migrated, and how many will become extinct is, of course, unknown.

Virtually all ecologists, and I include myself among them, would argue that every species extinction diminishes humanity. Every microorganism, animal and plant contains on the order of from one million to 10 billion bits of information in its genetic code, hammered into existence by an astronomical number of mutations and episodes of natural selection over the course of thousands or millions of years of evolution. Biologists may eventually come to read the entire genetic codes of some individual strains of a few of the vanishing species, but I doubt that they can hope to measure, let alone replace, the natural species and the great array of genetic strains composing them. The power of evolution by natural selection may be too great even to conceive, let alone duplicate. Without diversity there can be no selection (either natural or artificial) for organisms adapted to a particular habitat that then undergoes change. Species diversity—the world's available gene pool—is one of our planet's most important and irreplaceable resources. No artificially selected genetic strain has, to my knowledge, ever outcompeted wild variants of the same species in the natural environment.

It would be naive to think that humanity need only wait while natural speciation refills the diversity void created by mass extinctions. Following the great Cretaceous extinction (the latest such episode), from five to 10 million years passed before diversity was restored to its original levels. As species are exterminated, largely as a result of habitat destruction, the capacity for natural genetic regeneration is greatly reduced. In Norman Myers's phrase, we are causing the death of birth.

Wild species in tropical forests and other natural habitats are among the most important resources available to humankind, and so far they are the least utilized. At present, less than one tenth of 1 percent of naturally occurring species are exploited by human beings, while the rest remain untested and fallow. In the course of history people have utilized about 7,000 plant species for food, but today they rely heavily on about 20 species, such as wheat, rye, millet and rice—plants for the most part that Neolithic man encountered haphazardly at the dawn of agriculture. Yet at least 75,000 plant species have edible parts, and at least some of them are demonstrably superior to crop species in prevalent use. For example, the winged bean, *Psophocarpus tetragonolobus*, which grows in New Guinea, has been called a one-species supermarket: the entire plant—roots, seeds, leaves, stems and flowers—is edible, and a coffeelike beverage can be made from its juice. It grows rapidly, reaching a height of 15 feet in a few weeks, and has a nutritional value equal to that of soybeans.

Wild plant and animal species also represent vast reservoirs of such potentially valuable products as fibers and petroleum substitutes. One example is the babassú palm, *Orbignya phalerata*, from the Amazon basin; a stand of 500 trees produces about 125 barrels of oil a year. Another striking example is the rosy periwinkle, *Catharanthus roseus*, an inconspicuous little plant that originated in Madagascar. It yields two alkaloids, vinblastine and vincristine, that are extremely

PLANTS FROM TROPICAL RAIN FORESTS are the source of food, medicine and other commercially valuable products. The rosy periwinkle, *Catharanthus roseus* (*left*), contains substances that are effective against some cancers, and the babassú palm, *Orbignya phalerata* (*right*), produces bunches of fruit (each one weighing about 200 pounds), from which oil (for cooking and other purposes) can be extracted.

INSECT DIVERSITY is extraordinarily high in tropical rain forests, where millions of species, including this ant from the island of Sulawesi in Indonesia, have yet to be inventoried. The ant, which is unusual for its large eyes and robotlike movements, belongs to the genus *Opisthopsis* but has not yet been given a species name.

effective against Hodgkin's disease and acute lymphocytic leukemia. The income from these two substances alone exceeds $100 million a year. Five other species of *Catharanthus* occur on Madagascar, none of which have been carefully studied. At this moment one of the five is close to extinction due to habitat destruction.

Biological diversity is eroding at a swift pace, and massive losses can be expected if present rates continue. Can steps be taken to slow the extinction process and eventually bring it to a halt? The answer is a guarded "yes." Both developed and developing (mostly tropical) countries need to expand their taxonomic inventories and reference libraries in order to map the world's species and identify hot spots for priority in conservation. At the same time, conservation must be closely coupled with economic development, especially in countries where poverty and high population densities threaten the last of the retreating wildlands. Biologists and economic planners now understand that merely setting aside reserves, without regard for the needs of the local population, is but a short-term solution to the biodiversity crisis.

Recent studies indicate that even with a limited knowledge of wild species and only a modest effort, more income can often be extracted from sustained harvesting of natural forest products than from clear-cutting for timber and agriculture. The irony of cutting down tropical forests in order to grow crops or graze cattle is that after two or three years the nutrient-poor topsoil can no longer support the agricultural activity for which it was cleared in the first place.

Thomas Eisner of Cornell University has suggested that in addition to the compilation of biological inventories, programs should be established to promote chemical prospecting around the world as part of the search for new products. The U.S. National Cancer Institute has begun to do just that: their natural products branch is currently screening some 10,000 substances a year for activity against cancer cells and the AIDS virus.

It has become equally clear that biological research must be tied to zoning and regional land-use planning designed not only to conserve and promote the use of wild species but also to make more efficient use of land previously converted to agriculture and monoculture timber. More efficient land use includes choosing commercial species well suited to local climatic and soil conditions, planting mixtures of species with yields higher than those of monocultures and rotating crops on a regular basis. These methods relieve pressure on natural lands without reducing their overall productivity. No less important are social studies and educational programs that focus directly on the needs of the people who live on the land.

I have enough faith in human nature to believe that when people are both economically secure and aware of the value of biological wealth they will take the necessary measures to protect their environment. Out of that commitment will grow new knowledge and an enrichment of the human spirit beyond our present imagination.

FURTHER READING

HOW MANY SPECIES ARE THERE ON EARTH? Robert M. May in *Science*, Vol. 241, No. 4872, pages 1441-1449; September 16, 1988.

BIODIVERSITY. Edited by E. O. Wilson and Frances M. Peter. National Academy Press, 1988.

CONSERVATION BIOLOGY: THE SCIENCE OF SCARCITY AND DIVERSITY. Edited by Michael E. Soulé. Sinauer Associates, Inc., 1986.

THE PRIMARY SOURCE: TROPICAL FORESTS AND OUR FUTURE. Norman Myers. W. W. Norton & Company, 1984.

MASS EXTINCTIONS IN THE MARINE FOSSIL RECORD. David M. Raup and J. John Sepkoski, Jr., in *Science*, Vol. 215, No. 4539, pages 1501-1503; March 19, 1982.

1991

"The high frontier," *National Geographic,* 180(6) (December): 78–107

It was a special pleasure to write this article on one of the most biodiverse yet least known habitats on Earth. It is superbly illustrated by my former student Mark W. Moffett in collaboration with members of the *National Geographic* staff.

Rain Forest Canopy
The High Frontier

A new breed of scientist risks life and limb to probe the great unexplored world at the top of tropical rain forests. On a steep slope in Costa Rica, ecologist Pierre O. Berner paints rings to monitor the growth of an oak branch.

By EDWARD O. WILSON

Photographs by MARK W. MOFFETT

THE TROPICAL RAIN FOREST I had entered was a shadowed world broken by beams and nuances of greenish sunlight. I had come home to my favorite habitat, the one before which naturalists stand in awe. I was on Barro Colorado, an island in Gatún Lake halfway along the Panama Canal. My visit rekindled a simile that had come to mind in other places and other times: Seen on foot, a rain forest is like the nave of a cathedral, a thing of reverential beauty yet with much of its splendor out of reach in the towers and illuminated clerestories high above.

There was no lack of life around me on the ground. It teemed in the patchwork of light and dark. My attention was pulled to eye level and downward by the closeness of plants and animals in the soil and undergrowth. But I remained aware of a wholly different world a hundred feet above, where brilliant sunlight drenched sprays of vegetation and Babylonian gardens, an errant wind soughed throughout the day, and legions of birds, insects, and other animals specialized for high arboreal life flew and leaped back and forth. This high layer is the powerhouse of the forest, where more than 90 percent of photosynthesis takes place and, in the fullest sense, life begins.

Given a few grains of soil, baby bromeliads spring from seeds on a leaf of the mother plant (above). Bromeliads and other epiphytes—plants that grow atop other plants—thrive in the rain forest canopy, often taking root on mats of wind-deposited soil and decaying vegetation. In Costa Rica botanist Nalini Nadkarni peels a thick mat from a branch (facing page) to study how fast the layers accumulate.

The crown foliage of most tree species grows year-round, to be consumed by animals or to die and rot. The dead material thus produced rains steadily to the ground, bearing the remnants of energy to sustain the kinds of plants and animals among which I now stood. It brings nitrogen, phosphorus, and other nutrients back to the earth, to be sucked up by tree roots and returned to the canopy to restart the cycle of life.

The rain forest canopy, an undiscovered continent as naturalist William Beebe called it, is achingly close to the earthbound observer (map, page 84). But it is almost inaccessible and has remained largely unexplored. During 35 years of visits to tropical forests, I have made repeated attempts to study insects in the canopy. I once followed a logging operation in Papua New Guinea, climbing into the upper branches almost as soon as the trees fell to capture ants, beetles, and other specimens and to take notes. I worked the margin of a forest on the South Pacific island of Espiritu Santo, where the canopy bent down to the shore. On the edge of a ravine in the Brazilian Amazon, I peered for days through binoculars into tree crowns a few yards away. I learned little from these efforts. I was forced to stay with the ground and undergrowth.

Now I walked through the forest of Barro Colorado Island to a 138-foot tower, maintained by the Smithsonian Tropical Research Institute to assist long-term studies of the canopy. Climbing to the top, I could look out over the crowns of all but the highest trees, and peer at foliage close enough to touch.

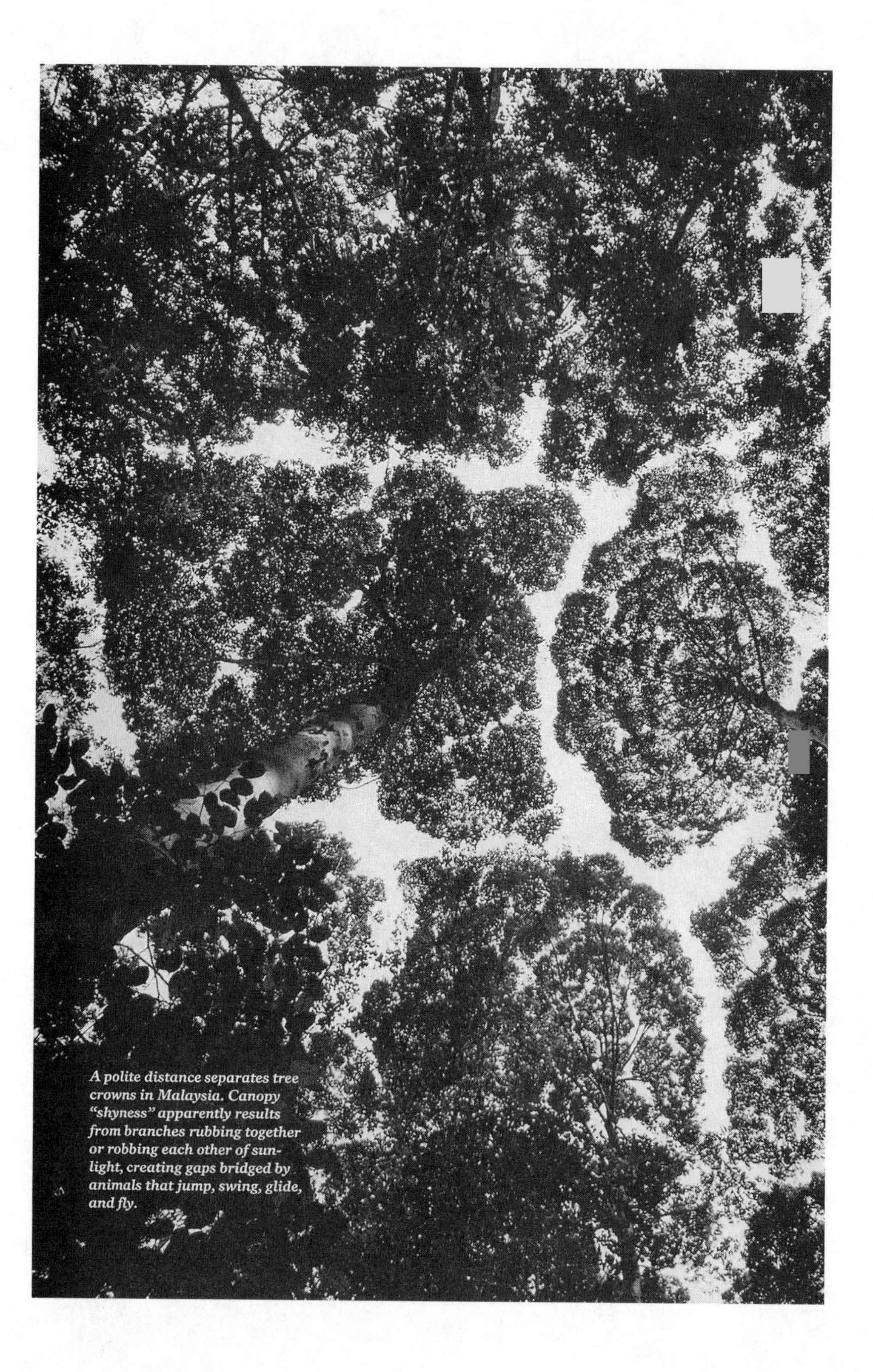

A polite distance separates tree crowns in Malaysia. Canopy "shyness" apparently results from branches rubbing together or robbing each other of sunlight, creating gaps bridged by animals that jump, swing, glide, and fly.

At my fingertips, literally, as I reached out and pulled a tree branch closer, were squadrons of ants gathered around treehoppers, thorn-shaped insects busily sucking the juices of the tender leaf shoots. The ants were not attacking these strange creatures. They were protecting them from spiders, wasps, and other enemies. In exchange, I knew, the treehoppers deposited sugar-laced excrement for the ants to eat. Such are the bonds of symbiosis that hold the rain forest community together.

Another image soon replaced the cathedral as I looked out and away: It now seemed I was floating atop a life-filled sea. All around me the bright green tree crowns of the upper story billowed like waves in a gentle breeze. Arboreal dragonflies soared and darted over the surface in search of insect prey, just as other dragonflies patrol the surfaces of ponds and lakes. Beautiful brown-and-blue charaxine butterflies swirled around one another in territorial dogfights. A pair of toucans glided into a nearby emergent tree, calling noisily. Ants scurried everywhere, hunting for food.

I could look straight down, as though peering into a crystal-clear

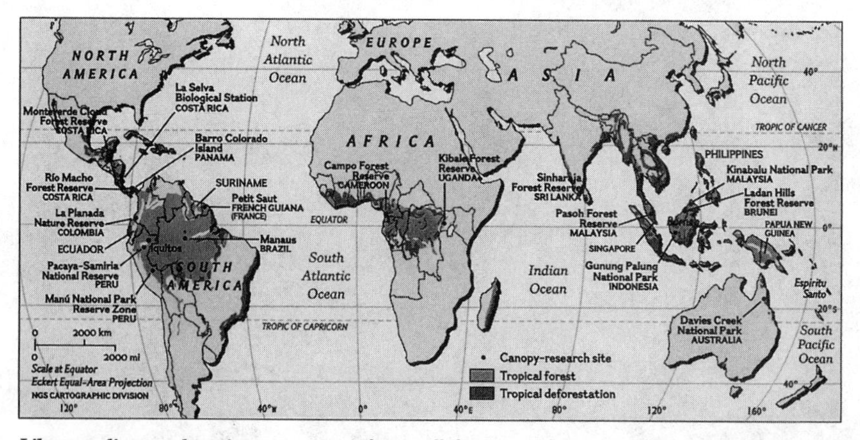

Like an undiscovered continent encircling the globe, tropical rain forests shelter an astonishing abundance of organisms—probably more than half the earth's plant and animal species. Heart of the forest is the canopy, the thick upper foliage where more than 90 percent of photosynthesis takes place. At canopy study sites around the world, scientists race to discover new organisms. With tropical forests being cut at a rate of 55,000 square miles a year, untold numbers of species perish before they can be identified, much less studied.

water column, all the way to the ground. Thirty feet below, hundreds of small flies danced in a midair mating swarm alongside tree crowns of the lower story. Below them giant morpho butterflies sailed by, flashing brilliant blue points of light as they opened their wings, then almost disappearing as they flicked their wings back up, the alternation sending a signal in metronomic rhythm. On the ground, far below and hard to see in the deep shade, logs and fallen tree branches lay scattered on a thin blanket of dead leaves.

Well, I made it to the treetops, I thought, and here I am, born a few years too early, a slightly creaky field biologist now cast in the role of spectator instead of participant, but happy to be that much. A new generation of scientists have begun a serious assault on the

---

EDWARD O. WILSON, winner of the National Medal of Science and two Pulitzer Prizes, is the Frank B. Baird, Jr., Professor of Science and the Curator in Entomology at Harvard University. Zoologist and photographer Mark W. Moffett, a frequent contributor to the magazine, is one of Wilson's former graduate students. He is also one of a number of scientists studying the rain forest canopy with support from your Society.

mysteries of the canopy, and it will be a pleasure to travel with them vicariously. In addition to towers, they use a wide range of imaginative and daring methods to reach and study this part of the forest. Various teams of hard-muscled young men and women around the world lean booms into the upper branches, travel out in gondolas suspended from building cranes, ascend on ropes, lower supporting nets from dirigibles, nail ladders onto tree trunks, and travel along walkways suspended across the crowns of trees.*

Month by month, at an accelerating rate, their efforts have begun to disclose the remarkable and unique traits of the canopy and its inhabitants.

THEY ARE REVEALING unimagined worlds in the foliage of the rain forest, where chunk-headed snakes with catlike eyes feast on frogs and lizards; where an ant known as *Daceton armigerum,* armed with jaws like a bear trap, rotates its head vertically to snatch flies from the air; and where earthworms wriggle through foot-thick soil on tree branches—ten stories in the air. How do the worms and soil get there? That's one of the questions scientists are exploring. In the process, they are turning up thousands of new species, as yet undescribed by science. They've found a poisonous caterpillar in Peru that looks like a miniature dust mop, and in Papua New Guinea giant weevils that carry miniature gardens of mosses and lichens on their backs. So many new species are being found that it is hardly news any more.

The big news is, quite simply, that life is more diverse and more plentiful than anyone had previously known. Of the roughly 1.4 million species of organisms given a scientific name to the present time and those remaining to be studied, many biologists believe the majority are to be found in tropical rain forests.

In just one 25-acre tract in Malaysia, Peter Ashton of Harvard University found 750 species of trees. In another record-breaking survey, Alwyn Gentry of the Missouri Botanical Garden identified 283 tree species in only 2.5 acres near Iquitos, Peru. By contrast, about 700 species make up the entire native tree flora of the United States and Canada.

Animal diversity is equally mind-boggling. From a single tree in Amazonia, I identified 43 ant species, approximately the same number as occur in all the British Isles. At some places in the upper Amazon basin, 1,200 kinds of butterflies occur, about 7 percent of the world's known species.

Yet even these figures pale in comparison with recent estimates of the total diversity of insects, *(Continued on page 92)*

*See "A Raft Atop the Rain Forest," by Francis Hallé, in the October 1990 NATIONAL GEOGRAPHIC.

Wincing in the face of gale-force winds, ecologist Ken Clark mans his weather station in Costa Rica's Monteverde Cloud Forest Reserve. As moist air whips in from the Atlantic Ocean, water condenses on Teflon filaments. Clark later measures nitrogen and other waterborne nutrients to determine how much the atmosphere adds to canopy soil mats and the rest of the ecosystem.

In an endless cycle, rain forest trees shed leaves that fall to the ground, thus enriching the soil tapped by the trees as they produce more leaves.

Stairways to the roof of the forest

How to get to the top? Enterprising scientists always find a way, simple or otherwise. Nalini Nadkarni (below) uses climbing rope and stamina as she gives her son Gus a look at the view in Costa Rica. From a

tower on Panama's Barro Colorado Island (far left), the author reaches out to collect ants, a lifelong passion. Erected by the Smithsonian Institution, the 138-foot-tall structure provides access to a high but narrow study area.

Near Panama City, a construction crane (top) enables researchers in a gondola to explore a 118-foot radius anywhere from ground level to a hundred feet high. After reaching the canopy in Sri Lanka by a succession of ladders (left), scientists built a bamboo platform for pollination studies.

Some Asian tree communities stand barren for several years, then burst into fruit over a large area, an unpredictable phenomenon called "masting." When it happens, scientists need to get into the canopy fast, with equipment such as a lightweight aluminum boom used in Malaysia (second from top).

Canopy animals

Stripped of its foliage, the branch reveals residents and visitors: a snake in naked pursuit of a tree frog, birds and small mammals feasting on worms and insects.

Hovering near their nest, shown hanging at center in the bottom view, paper wasps add their buzz to birdsongs and the whir of katydids—sounds drowned out by the screech of a white-faced monkey alarmed by a rival troop. Meanwhile, the canopy's silent armies, the ants, scurry to meet their mysterious imperatives.

Vascular plants

In a riot of leaves and roots, bromeliads, ferns, and other epiphytes add to the canopy's biomass. Some plants, or their relatives, are also found in temperate climates, either as natives or exotic imports: orchids, philodendrons, ericads (of the same family as blueberries and azaleas), columneas (of the African violet family), and peperomias, kin to the plant that yields black pepper.

Nonvascular plants
Among the first plants to colonize the limb are mosses,
lichens, and other lower epiphytes that spread like carpets, gaining a foothold
on areas with the least soil.
These canopy dwellers
show distinct preferences.
Some species grow only on
branches, some only on
dead surfaces, like the
stubs of broken-off branches.
Others, known as epiphylls,
grow only on leaves.
Soil and detritus
Gravity lays life's foundation
on the topside of branches.
Wind-borne dust mingles
with fallen leaves, animal
droppings, and plant and animal remains. This mixture, a
kind of bog, is especially
deep at the fork, where a
bromeliad flourishes, second
view from top. From the lower branch hangs a nest
fashioned by paper wasps.
At the end of the branches,
Azteca ants maintain football-
shaped gardens, where they
literally grow plants for food.

Layers of life at the top

The profusion of life in the canopy may seem chaotic, but beneath it all lies an ordered pattern of relationships based on the struggle to survive. Plants live on other plants; animals eat other animals. And animals both exploit the plants for food and cover, and aid them by pollinating them and dispersing their seeds.

Four views of the same limb of a fig tree in Costa Rica (above) isolate major plant groupings and an array of animals, revealing the complex inner workings of canopy life.

PAINTINGS BY JOHN D. DAWSON

The bromeliad: a microhabitat

A small universe of interdependent creatures live in and around a large bromeliad, a member of the pineapple family, and the branch that supports it.

A nectar- and fruit-eating bat, *Anoura geoffroyi* **1**, homes in to feed, pollinating the bromeliad blossoms in the process. Going to and from their nest, *Myrmelachista* ants **2**, smaller than grains of rice, use a tunnel in the branch's core, where they swarm over the new, winged daughter queens **3**.

Known as a tank bromeliad for its water-collecting capacity, the plant is also home to creatures that utilize the water: tree frogs **4**, salamanders **5**, beetles **6**, flatworms **7**, flies **8**, katydids **9**, and spiders **10**.

A coin-shaped close-up shows the bromeliad's reservoir to be alive with mosquito larvae **11** and the larva **12** and pupa **13** of a wood gnat.

Orchids **14** and a stagtongue fern **15** share the nutrients of the soil mat. Mistletoe **16**, a parasite, sinks roots into the branch itself.

(Continued from page 85) spiders, and other arthropods living in the canopy. The discovery of the superabundance has been made by enveloping tree crowns with fogs of rapidly acting biodegradable insecticides and collecting the arthropods that fall dying, in the tens of thousands, to the ground.

Yearning for sunlight, earthbound vines called lianas (below) ride piggyback to reach the canopy. One reaches out with a stem that spirals like a corkscrew. Another grabs bark with three-pronged tendrils, while a third hangs on with stout spines. But the bully of the forest, the strangler fig (facing page), isn't content to coexist. After sprouting in the canopy from seeds dropped by birds and bats, it sends roots to the ground that envelop the host tree, which dies and rots away.

How many kinds of arthropods live in the canopy? In a celebrated study published in 1982, Terry L. Erwin of the Smithsonian's National Museum of Natural History arrived at a figure of 20 million for the world as a whole, using the following procedure. He first identified 1,200 species of beetles from canopies of the tree *Luehea seemannii* in Panamanian rain forest. Of these beetle species, 163 were believed to be limited to this tree species. There are about 50,000 tropical tree species worldwide, so that if *L. seemannii* is typical, the total number of canopy-dwelling tropical beetle species is 8.15 million. These beetles represent 40 percent of the tropical canopy species of all arthropods, which, therefore, come to about 20 million. The rain forest canopy contains about twice as many arthropod species as the ground, so that the total number of species—in the canopy and on the ground combined—might well be 30 million. Nigel E. Stork of the Museum of Natural History in London independently evaluated his own counts from the forests of Borneo to produce a possible range of five million to ten million tropical forest arthropods.

Why this huge variation in the estimates of rain forest diversity? A great deal depends on the degree to which insect and other arthropod species are limited to one or at most a very few kinds of trees. Because research in the canopy has been so sparse to date, this key factor remains largely unknown. If the arthropod species turn out to be very restricted in the kinds of trees on which they live, their true numbers may approach 30 million or more. On the other hand, if they are able to exist on a wide range of species, the number will prove to be closer to five million.

Whatever the exact amount of diversity in the rain forest treetops, it must run into the millions of species. The few biologists who can identify rain forest organisms are swamped with new species now pouring in from collections in the canopy and on the ground.

MY OWN EXPERIENCE is typical. Every time I enter a previously unstudied stretch of rain forest, I find a new species of ant within a day or two, sometimes during the first hour. I search the ground and low vegetation, dig into rotting logs and stumps with a gardener's trowel, break open dead twigs and branches lying on the ground, and pull at ferns and other epiphytes growing on tree trunks and newly fallen tree limbs. On a typical day, the first 40 or 50 colonies encountered might be species already known to science, some very familiar to me, some rare and requiring later study under a microscope at higher magnification back home. Then a new species. Then another 20 colonies of established species, and one more new species, and so on in a continuing adventure for many days in a row.

One day Stefan Cover, a curatorial assistant at Harvard University's Museum of Comparative Zoology, returned from the rain forests of northeastern Costa Rica and presented me with a large ant of the genus *Pheidole* strikingly different from anything I'd seen before. I had to have more specimens! Cover drew me a map

A fog of biodegradable insecticide sprayed by Smithsonian's Terry Erwin in Peru will soon bring a rain of spiders, insects, and other invertebrates. Gathered on sheets, such collections show a superabundance of species in the canopy—far more than previously believed.

Mining a trove of invertebrates, Michael Pogue of the Smithsonian sorts specimens collected by fogging in Peru. A single Amazonian tree typically yields more than three pounds of specimens comprising 1,700 species, mostly ants and beetles. From such humble origins come medical and other scientific breakthroughs.

Woe to the novice field scientist who picks up this fluffy moth caterpillar (above). When grabbed, hidden spines break off and release an irritating venom. In the insect equivalent of a wash-and-wax job, one Daceton *ant grooms another (center, right), cleaning it while possibly spreading protective antibiotic substances.*

Colors send signals, say canopy scientists. The orange and black of a young unidentified grasshopper (center, left) probably mimics stinging or foul-tasting insects; its vivid blue trim may also indicate that it tastes bad.

showing the exact spot where he had found the species, near the crossing of two trails and just to the side of an adjacent fallen tree. I went there soon afterward and searched the area carefully and without success. But close by, nesting in the clay soil of one of the trails, I discovered two *more* new species of large, striking *Pheidole* ants, an unexpected gift for a biologist who prospects for ants as others dig for diamonds or gold.

These biological treasures seem endless. I am currently laboring on the classification of more than 300 new species of rain forest ants in the Harvard collection. These represent only a fraction of those already collected and awaiting study.

THE BEST PLACE TO SEE the complete profile of a tropical rain forest and to put the canopy in perspective is, I am sorry to report, where it is being cleared and destroyed. As the forest is sliced along its side and peeled back, and the intervening second growth bulldozed and burned, the trees can be seen to vary greatly in height. A nearly continuous upper canopy is the dominant feature, composed mostly of trees with flattened crowns, nearly horizontal branches, and trunks free of branches for the first 60 feet or so. The term canopy, meaning an overarching cover, is entirely appropriate. Yet rising even above this layer are a few scattered emergents, giants that tower to heights of 200 feet or more. Lower down are trees with narrower trunks and vertically elongated crowns. Some are young individuals of upper-story species struggling their way to the top. Others are mature trees of species specialized for this intermediate, twilight region. Still lower, mostly at the height of a human being or less, are saplings and herbaceous plants.

All this exuberant, multilayered growth is supported by rain, lots of it. Rain forests grow in areas with more than 80 inches of rainfall annually, which allows the growth of broad-leaved evergreen trees. The leaves are typically smooth to the point of slickness and in many species are strongly narrowed at the end. Both features, smoothness of surface and tapered form, promote the rapid runoff of water during torrential rains and help prevent the leaves and branches from breaking because of waterlogging.

Two continents of life do indeed exist as layers in the tropical rain forest. The ground is a dark factory of decomposition, where bacteria, fungi, millipedes, and termites and other wood-feeding insects degrade the fallen plant debris into nutrient molecules. These substances are quickly absorbed by the omnipresent rootlets of the trees, so that little material is present on the ground. The air is still and humid, saturated with the odors of healthy decay.

The canopy is a brilliantly lit, noisy, three-dimensional world. Wind rakes the tree crowns, evaporating moisture away at a rate comparable to that in grasslands, drying the vegetation at times to almost desert-like conditions. Relative humidity ranges from as high as 100 percent at night to less than 30 percent at midday. Sunlight bakes the vegetation, occasionally raising the temperature of the ambient air to more than 90°F, a full 10° higher than at ground level. Frequent rainstorms pound the branches and leaves, breaking away the weak ones. The rain, after filtering through the tree crowns, descends to the forest floor. It arrives as a delayed shower and rivulets that stream down trunks and elevated roots.

The drab coat of the three-toed sloth (top) is a "living bug carpet"—home to beetles, ticks, fleas, and a steady companion, a moth (above) that lays eggs in the sloth's dung. After hatching, the next generation of moths will fly, seek out sloths, and begin the cycle anew.

The exact structure of tropical rain forests varies from continent to continent, but one feature common to most is an abundance of vines. Most conspicuous are the lianas, which are thick, woody climbers. They sprout on the forest floor and then send out long shoots that grasp vegetation as the vines grow up into the crowns. Often they cross to other crowns, binding the trees together. Lianas have great tensile strength. They are composed of both hard and soft tissues, making them flexible and difficult to break or sever.

A real-life Tarzan could not have swung from tree to tree with lianas, which are attached to the ground. (And unlike biologists, he would not have been likely to endure the stinging wasps, biting ants, and spines and saw-toothed edges of the canopy vegetation. Tarzan would have stayed on the ground.)

You can categorize tropical forest plants by the way they respond to light. While lianas take root in the ground and grow upward, strangler figs do the exact opposite: They sprout in the canopy from seeds dropped by birds and bats, then send roots to the ground.

The stem climber *Monstera tenuis* of Central America, a member of the arum family so favored as houseplants, responds to light in yet another, radically different way. Immediately after sprouting from a seed on the forest floor, the *Monstera* grows toward the dark. This orientation, the exact reverse of that used by almost all other kinds of plants, leads its shoots to the deeper shadows around the base of a tree. Once on the tree trunk the *Monstera* apparently switches orientation and grows upward, toward the light.

Lured by lush vegetation, large mammals find forage in the canopy. A chimpanzee snacks on figs in Uganda's Kibale Forest Reserve. In Colombia, a spectacled bear looks for a meal of bromeliads. Largely because its forest habitat is being destroyed, the bear has been declared an endangered species.

ALTHOUGH THE TREE CROWNS and vines alone are enough to create an abounding and unique habitat, the canopy is vastly enriched by yet another dimension. Epiphytes, plants that use the trees as support but do not draw nutrients or water from their tissues, grow in luxuriant masses along the trunks and branches. They are extremely diverse worldwide, comprising 29,000 species of vascular plants in 83 families, more than 10 percent of all higher plants. In addition to orchids, which have the most species, there is a profusion of ferns, bromeliads, gesneriads, figs, arums, and members of the pepper family. There are also many nonvascular epiphytes, such as mosses and lichens.

At the Monteverde Cloud Forest Reserve of Costa Rica, Nalini Nadkarni and her fellow canopy researchers from the Marie Selby Botanical Gardens in Sarasota, Florida, have found one of the most complex assemblages of organisms on earth. There are worlds within worlds: Here and there small trees sprout from the epiphyte root masses, so that trees actually grow on the branches of other trees. Lichens and other epiphylls grow on the leaves of the smaller trees, small insects browse among the epiphylls, and protozoans and bacteria live within the insects.

The epiphytes add immensely to the productivity of the forest, filtering atmospheric nutrients and capturing solar energy that

would otherwise bounce and scatter away from the naked tree limbs. For the ecosystem as a whole, they act as supplemental foliage in the rain forest trees.

The most elaborate form of symbiosis in epiphytes, one that epitomizes the complexity of the rain forest canopy, is the ant garden. I have encountered many gardens in Brazil and Suriname at the edge of rain forests, where a bit of the canopy dips close to the ground. Spherical in shape and somewhat smaller than a soccer ball, they were made of soil and masses of vegetable fibers chewed and shaped by the ants. They bristled with small, succulent epiphytic plants that sprouted from their surfaces in all directions.

When I touched one—in fact, when I just stood with my face one or two feet away—large *Camponotus* ants swarmed out onto the surface and sprayed clouds of formic acid in my direction. The defenders seemed frantic to get to me, and I was thankful they could not fly like wasps or bees. In a lifetime of studying ants, I have never seen a species so suicidally aggressive and intimidating.

Protected and fertilized by the ants, the garden epiphytes flourish. In return, they provide their residents with food and shelter. The plants belong to at least 16 genera of philodendrons, bromeliads, figs, and cactuses. They are specialized for this strange existence, limited almost entirely to the gardens. Similarly, the most abundant ant species in the gardens appear to depend on their epiphyte partners.

The canopy serenade begins at dusk for Cathy Langtimm, sitting 70 feet up in a Costa Rican forest. Her shotgun microphone picks up the cries of tiny tree-dwelling mice, often mistaken for bird or insect sounds.

"The call is surprisingly loud, like a two-note whistle," says Langtimm, of the University of Florida. "The mice probably use sound to mark territory or locate mates. The canopy is a tangle of branches that makes it hard to attract mates by leaving pheromone trails." The calls are short—only a second or so—to evade predators.

How do the gardens get started? There seems to be no other way to explain it than to say that the ants plant gardens and harvest crops. They are initially attracted to fleshy appendages on seeds, called elaiosomes, that serve as food. The seeds may also smell like the ants' own larvae. After the ants have cleaned off these nutritious fragments, they drop the still intact seeds into the recesses of their nest, allowing the epiphytes to sprout and grow out through the nest walls. As the plants mature, the ants harvest food from them in the form of fruit, nectar, and elaiosomes.

TROPICAL RAIN FORESTS began to take form roughly 140 million years ago, near the beginning of the Cretaceous period in the age of dinosaurs. Early in that time, when most of the world's climate was tropical to subtropical, flowering plants originated and later spread as dominant elements around the world. Many of the species became partners with insects, which derived food from them while serving as pollinators and transporters of seeds. These insects, including a medley of wasps, beetles, bees, flies, butterflies, and moths, also rose to dominance. Some insect groups increased vastly in numbers and diversity simply by preying on flowering plants—feeding on them without pollinating them or giving other services in return. Still others consumed decaying vegetation or else simply used the plants for shelter.

Biologists who struggle with the task of collecting and classifying the immense cornucopia of biological diversity ponder the important ecological question it raises: Why are there so many species? Most experts on biological diversity agree that more than half the kinds of plants and animals in the world live in rain forests, yet this ecosystem covers less than 6 percent of the land surface. Of these species, more than half either live in the canopy or, in the case of the trees and vines, create it with their upper foliage.

The mystery of why so much of life is invested in rain forests is still far from solved, but clues abound. One is the greater climatic stability of the tropical zones. The temperate zones experience wide seasonal swings in temperature each year. The tropics as a whole, and rain forest areas in particular, enjoy nearly constant temperatures and have never been glaciated. Plants and animals in colder parts of the world are for the most part adapted to survive in a variable environment, and as a result they range widely. A plant species found in New York, for example, is also likely to be found as far away as Tennessee and Michigan. A tropical species, in contrast, is more likely to have evolved to fit a narrow niche in a constant environment. Plant species in South America and elsewhere are often limited to a single valley or ridgetop. So when you add up all the tropical species, there are many more of them.

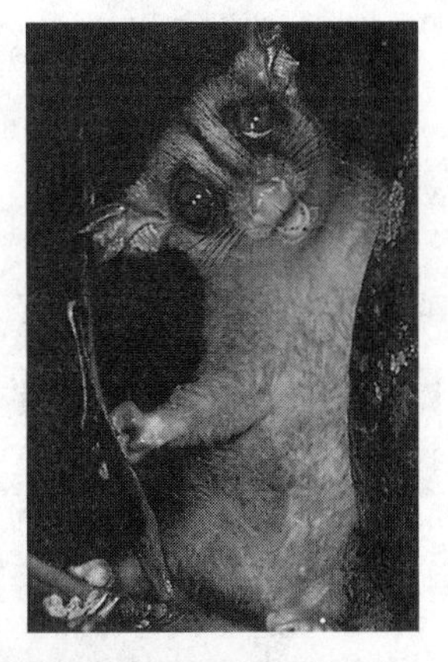

A deft grasp—aided by five digits and an opposable large toe on each foot—serves a small opossum well as it forages in a Brazilian canopy for fruit, insects, lizards, and eggs. The vision of this nocturnal creature is enhanced by huge eyes with widely dilating pupils.

Still more diversity is piled on by physical disturbances that create gaps in the forest. When the canopy is broken open, sunlight falls more abundantly on the ground, and a new burst of vegetation springs up. The species of trees and smaller plants in this assemblage are mostly different from those in the surrounding mature forest. So are many of the insects and animals that live on these gap specialists.

Gaps are continually created at random spots throughout the forest. As a storm passes, sporadic winds are likely to break a few large tree limbs that have grown weak and vulnerable from heavy growths of epiphytes. The rain fills up the axil sheaths of the epiphytes and saturates the humus and clotted dust around their roots. Occasionally a lightning bolt strikes and kills a tree.

Elsewhere a large tree sways in a gust of wind above the rain-soaked soil. Its shallow roots cannot hold, and the entire tree keels over. Its trunk and crown shear through smaller trees to open a hundred-foot path. As the sky clears and sunlight floods the newly opened gaps, the surface temperature rises and humidity falls. The soil and leaf litter dry out and grow warmer, creating a new environment for the plant seeds resting there. In the months to follow, pioneer trees take root. They are very different from the young shade-tolerant saplings and understory plants of the deep forest. Fast growing and short-lived, they form a single canopy far below that of the major forest. Their tissue is relatively soft and vulnerable to attack by herbivores.

One of the dominant gap-dwelling groups of Central and South America, the broad-leaved trees of the genus *Cecropia,* swarm with fierce ants that live in hollowed internodes of the trunk. The ants, a species of *Azteca,* protect the trees from all predators except three-toed sloths and a few other animals specialized to feed on *Cecropia.* The slightest disturbance brings them out by the

hundreds or thousands, biting with their mandibles and emitting noxious defensive secretions. Because of their much smaller size, they are less formidable than the treetop garden ants, but only slightly so. On several occasions in Costa Rica I have tried to dissect good-size *Cecropia* trees to study the inhabitants, and each time found it a harrowing experience.

In short order the little defenders were under my clothing, in my hair, running over my eyeglasses. I had to stop frequently to clean myself off, and I finally gave up. But that, of course, is the point. I was Gulliver tied down by the Lilliputians in a successful defense of their land. Although I was 80 million times heavier than each *Azteca* ant, the tribe prevailed.

When the pioneer vegetation thickens enough to shade the ground, the cycle begins to draw to a close. Conditions of light and temperature improve for mature-forest species, and their seedlings take root and grow. Within a hundred years the gap specialists are gone, and the multiple-layer high forest has returned.

The pioneer species are the sprinters; the slower growing species of the mature forest are the long-distance runners. Together they create a mosaic of vegetation types throughout the forest that is forever changing, a dazzling kaleidoscope of biological diversity. When you walk for a mile or two through mature rain forest, you cut through many of the successional phases from gap to mature stands. Life is continually enriched by the passage of storms and the fall of forest giants.

LIFE IS THUS PILED UPON LIFE in the tropical rain forests. Long periods of uninterrupted evolution have pushed diversity to extremes along several dimensions: epiphyte gardens, intense specialization, delicate symbioses, and a constant turnover of plants and animals that fill the forest gaps. But this great edifice is all a house of cards. Most of the millions of species are so highly specialized that they can be quickly driven to extinction by the disturbance of their forest homes.

Unfortunately examples of such vulnerable species are easy to find. Spix's macaw *(Cyanopsitta spixii),* a beautiful parrot of northeastern Brazil, is on the brink of extinction due to human interference. The forests in which it can live have been largely destroyed during the past century. In addition, the Africanized honeybee, the "killer bee" accidentally introduced into Brazil in the 1950s, has occupied many of the nest sites the macaw needs to reproduce. Finally, local fancy-bird dealers, able to get $18,000 or more for each bird, have reduced the remnant population until only one individual was known to remain in the wild in late 1990.

The fragility of the rain forests extends not just to single species but also to entire local ecosystems. In the early 1980s Alwyn Gentry and Cal Dodson surveyed the plants of Centinela, an isolated mountain ridge on the Pacific side of the Andes in Ecuador. They encountered almost a hundred plant species found nowhere else in the world. A few years later, before the Centinela flora and fauna could be studied further, farmers from surrounding valleys completely cleared the ridge. The unique plant species and almost certainly a host of animal species associated with them were gone.

The tropical rain forests are being reduced in this manner almost everywhere in the world at an accelerating rate. The original cover,

A smorgasbord of fruits plucked from the canopy in Borneo (left) owes its abundance to bats, monkeys, and birds that pollinate flowers and scatter seeds. A fig tree in Singapore (far left) aids its own cause by bearing fruit on its trunk, an irresistible treat to bats.

Tied to a tree in Borneo, Tim Laman aims to loop fishing line over another branch, to haul up climbing rope for his study of strangler figs. Despite the priority given safety, dangers remain for rain forest scientists, drawn to this high frontier by forms of life yet to be discovered.

before human populations had much impact, was about six million square miles. Now it is only three million square miles, less than 6 percent of the world's land surface, roughly the same as the area of the contiguous United States. In 1979, according to surveys conducted by British scientist Norman Myers and the Food and Agriculture Organization of the United Nations, the rain forest and similar, much less extensive monsoon forest were being destroyed at the rate of 29,000 square miles a year. By 1990 the figure had almost doubled, to 55,000 square miles a year, an area larger than the state of Florida.

As the area of a habitat such as a rain forest is decreased, the number of species of plants and animals it can support also declines. The relation between these two qualities of the natural environment, area and diversity, is consistent. A reduction of the habitat to one-tenth its original area means an eventual loss of about half its species. In other words, if a forest of 10,000 square miles and a hundred resident bird species is cut back to 1,000 square miles, it will eventually lose about 50 of the bird species.

This amount of rain forest reduction has already occurred in several of the biologically richest parts of the world, including the Philippines, parts of West Africa, and the Atlantic coast of Brazil. The current rate of deforestation worldwide translates to an eventual annual loss of species of at least half a percent a year. Even that figure is probably a considerable underestimate, because it does not take into account the near-instantaneous destruction of entire communities of endemic species such as that on Ecuador's Centinela. If the rain forests of the world hold ten million or more species, a figure many tropical biologists consider likely, the rate of extinction worldwide may well have already reached 50,000 species a year.

A forlorn island of trees stands in Brazil amid an area cleared for ranching, which along with slash-and-burn farming accounts for 75 percent of all rain forest destruction worldwide.

Aided only by a strap called a peconha between his feet, Jay Malcolm shinnies up to explore a Brazilian forest "fragment" beside a clear-cut. Above him, a trap baited with bananas and peanut butter collects rodents and marsupials for a census. Researchers study fragments to determine how animal species react to such reduced habitats.

CAN HUMANITY AFFORD to lose so much of its natural heritage? The tragedy, as biologists see it, is that large blocks of diversity are being lost before they can be studied scientifically. The great majority of the vanishing species have never even received a scientific name.

Biologists and conservationists watch in dismay as the forests disappear, just as the extraordinarily rich two continents of life, the ground and canopy, are being effectively explored and compared for the first time. It is as though the stars began to vanish at the moment astronomers focused their telescopes.

A large part of the world's greatest biological treasure-house is being leveled for farming, ranching, and logging. The loss is compounded by the fact that if managed properly, the forests can yield a higher rate of income in perpetuity. Enough harvesting

A hundred feet high, daredevil entomologist Jack Longino pops up above the canopy in Costa Rica. He and others wonder how long such vistas will remain unbroken. At present rates of destruction, earth's rain forests could be gone by the middle of the next century, a chilling prospect.

Says a veteran botanist: "Rain forest is the very core of the biology of this planet."

techniques already exist to make this dream a reality. When timber is removed from a succession of narrow strips following the contours of the land, native trees grow back rapidly.

Other materials can be drawn from the forest with even less disturbance. A wide range of natural products were either directly extracted from or at least were discovered in rain forests and transplanted to plantations elsewhere. Their names include both the familiar and unfamiliar: rubber, copal, dammar, chicle, balata, quinine, vanilla, cocoa, coffee, Brazil nuts, avocado, rattan, and a large percentage of our most favored species of houseplants.

The fruits of only about a dozen species of temperate zone plants dominate the commercial market. At least 3,000 may be available in the tropical forests, and of these, 200 are in actual use. Some, like cherimoya, papaya, and mango, have only recently joined bananas in the northern markets.

A fundamental difference exists between the two major forest types of the world, a difference that should concern economists. Whereas the commercial value of the temperate forests comes

almost entirely from timber, the potential value of tropical forests lies mostly in the large array of other products available from the diversity of their species.

An important property of the tropical rain forest products we now use is that they originate in the lower levels, from understory shrubs and smaller trees and the trunks of the larger trees. The high canopy, so little understood biologically, is almost entirely unexplored commercially. No one knows what new foods, pharmaceuticals, fibers, vegetable oils, and other materials await discovery a hundred feet above the ground.

From a beetle without a name atop an orchid in a distant threatened forest may come a cure for cancer.

But such practical concerns are not what I ordinarily think about in more somber moments as I walk the trails of Barro Colorado Island and other remnants of the tropical forests. What comes to mind is what I like to call the ultimate irony of organic evolution: that life, in the moment of achieving self-understanding through the mind of man, has doomed its most beautiful creations. □

2003

"The origins of hyperdiversity," *in* E. O. Wilson, *Pheidole in the New World: A Dominant, Hyperdiverse Ant Genus* (Harvard University Press, Cambridge, MA), pp. 13–18

I undertook the classification and natural history of *Pheidole* ants in part for the unusual challenge they represented: this genus was so difficult that nobody else wanted it. One of the largest of all animal genera, *Pheidole* was, before its revision, the Mt. Everest of ant taxonomy. Museum drawers were filled with new and otherwise indeterminate species. It is also one of the most abundant insects in many tropical and subtropical localities. Ecologists working in such areas almost always encountered *Pheidole* if they dealt with ants at all. But with its taxonomy in chaos, they were forced to sort out morphospecies listed simply as *Pheidole* sp. 1, *Pheidole* sp. 2, and so on. There was no hope of cross-referencing most of these listings except by locating and comparing voucher specimens deposited in museums by other researchers.

I started revising the New World *Pheidole* in 1985 and finished with the publication of the monograph in 2003, working at odd hours in my home laboratory. It was a hobby, a form of relaxation during which I listened to classical and soft rock music. After looking at tens of thousands of specimens from New England to Argentina and making over 5,000 drawings, I had the New World *Pheidole* more or less in hand: 624 species, of which 337 were new to science, altogether comprising 19 percent of all the ant species known from the Western Hemisphere. In the final year, Piotr Naskrecki and Sarah Ashworth, colleagues at Harvard who had become experts in the new techniques of high-resolution digital photography and automontage focusing, added photographs of the types of all the new species, which were then included on a CD-ROM at the back of each copy of the book. Recognizing that the *Pheidole* monograph is a hybrid created in what is likely to be a brief transitional period between print and electronic publishing in taxonomy, I like to call the 794-page *Pheidole* monograph the "last of the great sailing ships."

Pheidole brought me in close contact with the poorly understood phenomenon of hyperdiversity. With the details of an actual case fresh in my mind, I undertook a general analysis of its possible causes, which I have reproduced here. It may be argued that because genera, while representing phylogenetically united clusters, are otherwise arbitrary in the limits put around them, *Pheidole* might be split into two or more genera someday, and then further split until the fragments are no longer hyperdiverse. Perhaps that will happen, but I doubt it. The species are so densely clustered in anatomical traits, and the traits so discordantly distributed among the species, as to make cleavage impracticable or at best misleading at the time of this writing.

❋ Article 46

The Origins of Hyperdiversity

Strong variation in species richness among genera, families, and still higher taxa is a universal but still poorly understood biological phenomenon. Its signature is the hollow curve distribution, first constructed in 1922 by the botanist J. C. Willis and refined by many investigators subsequently.

The hollow curve distribution is drawn from taxonomic data through the following steps. Take an assemblage of higher taxa sharing a common phylogenetic origin, such as a group of genera or a group of families. Count the number of units composing each taxon, such as the number of species in each genus or the number of genera (or species) in each family. For example, within the family Formicidae, comprising all the ants of the world, Bolton (1995a) recognized 296 genera comprising 9,538 species. When the genera are rank ordered according to the numbers of species each contains (i.e., *Camponotus* with 931 species, then *Pheidole* with 525, *Polyrhachis* with 477, *Crematogaster* with 427, *Tetramorium* with 415, and so on downward), and these numbers are represented in a histogram, the result is an approximation of a hollow curve. The hollow curve distribution for the New World ant genera, a subset of the world fauna, is provided in Figure 8.

When a genus or family at or near the top also contains exceptionally large numbers of species with reference to plant and animal diversity as a whole, say into the hundreds of species, it can reasonably be called "extremely dominant" or "overdominant" or, to use the more descriptively precise term I have employed in this monograph, hyperdiverse. Hyperdiverse genera among plants are *Rhododendron* with 1,200 species and *Erica* with 700; among beetles, the weevil genus *Apion* with 1,100, the staphylinid beetle genus *Stenus* with 1,400, and the scarab genus *Onthophagus* (possibly the largest animal genus of all) with over 1500; and among Hymenoptera, the ant genera *Camponotus* and *Pheidole,* each containing well over 1,000 species.

Hollow curve distributions, including those with taxa rich enough to be called hyperdiverse, are not artifacts of particular taxonomic schemes. Nor are they the outcome of the random birth and extinction of species through geologic time. Rather, they are truly biological and thus far more promising for scientific study. This conclusion was confirmed in an important analysis by Dial and Marzluff (1989). Of 85 real (not modeled) taxonomic assemblages they reviewed that had been published by previous authors, 53 had been classified by traditional schemes, in which species with common "diagnostic" traits are clustered into genera or families on the basis of presumed common ancestry and the amount of evolution that has occurred in the various lines thus distinguished. The remaining 32 assemblages had been classified by strict phylogenetic schemes, in which the genera or families represent clades, or groups inferred by minimal path analysis or quantitative phenetic similarity to have a common ultimate origin from a single branching point. The distinction between the two sets of assemblages was drawn from the evaluation of systematic practice by Raikow (1985).

Dial and Marzluff next derived hollow curve distributions from the data of the 85 empirical assemblages and matched them against distributions based on the following four independent models of random process: (1) The Poisson distribution, the first model, assumes that diversification within the group, hence the rank-ordered curve of comparative species or genus numbers, is simply a random property of the group. (2) A second model assumes a rank-ordered distribution originating when species formation and extinction occurred randomly throughout the evolutionary history of the group. (3) The MacArthur broken-stick hypothesis, generating the third distribution, assumes that each genus (or family) garners resources *simultaneously* and randomly, and that the number of species in a genus, or species and genera in a family, corresponds to the resources each such unit controls. (4) Finally, the canonical log-normal model assumes that the resources, hence the number of species or genera based on them, are *sequentially* subdivided among the evolving groups.

In each of the 85 real groups, the most diverse genus or family was significantly more so than predicted by any of the four models. Moreover, the dominant unit—the most diverse genus or family within the assemblage—contained on average 37 percent of the subunits, i.e., species or genera, in the traditionally constructed higher classifications, and they contained 39 percent of the subunits in the phylogenetic classifications. Hence the two approaches used by taxonomists were virtually identical in estimating the size of the dominant unit. Dial and Marzluff concluded that taxonomic dominance by one unit is not only a general but also a nonrandom feature of standing biodiversity.

Hyperdiversity can be recognized as a useful additional distinction when the genus or other higher category typified by *Pheidole* is extremely large with reference to all plants and animals. It is also heuristic in its extreme expression of the

THE ORIGINS OF HYPERDIVERSITY

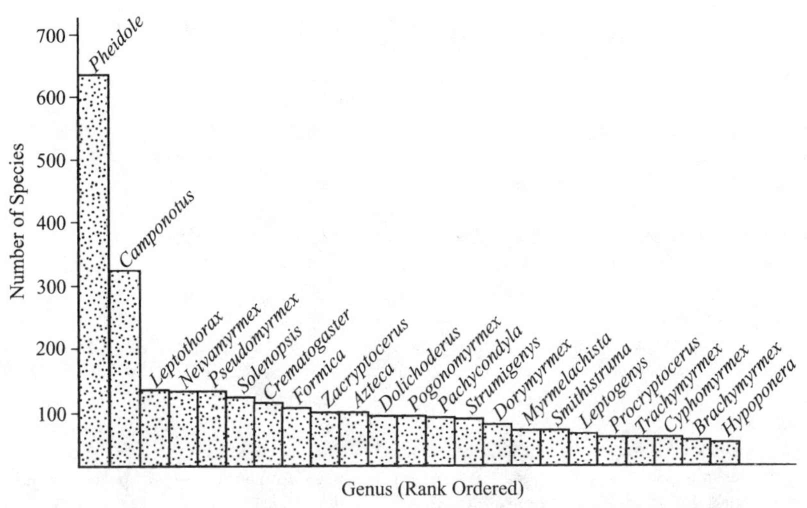

Figure 8 The hollow curve distribution of the 23 most speciose living New World ant genera, based on the census by Bolton (1995a) with the addition of the 337 new species of *Pheidole* described in the present monograph. The 157 smaller genera are not included. In addition to being the most speciose ant genera both in the New World and worldwide, *Pheidole* and *Camponotus* are also large enough to be called hyperdiverse with reference to all the other living genera of plants and animals.

underlying biological causes, making them more accessible to analysis and interpretation. Such is the approach I have taken in the study of *Pheidole*.

In a separate body of evidence favoring the idiosyncratic biological nature of hollow curve distributions, Stanley et al. (1981) and Flessa and Thomas (1985) found that fossil lineages of marine invertebrates tend to proliferate in nonrandom and often synchronous patterns through time. Similarly, Cracraft (1984) has pointed out that sister clades of birds often differ in patterns of diversification, despite their broad sharing of plesiomorphic traits.

What general causes might be adduced to explain high levels of biodiversity? The best understood and documented causes can be summarized as follows: *dominance in biodiversity, including the extreme species richness ranked as hyperdiversity, is attained by a fortunate combination of small size, right demographic factors, preemption during colonization and subsequent incumbency, and a suite of key adaptations potent in opening new niches or excluding competitors*. Further, the most realistic models that can be constructed to describe the origin of the hollow curve distribution are those that assume a sequential origin of groups within a given geographic domain or their sequential invasion into the domain from outside, followed by a peaking and subsidence of biodiversity through evolutionary time.

Among the biasing factors of biodiversity, consider first size of the organism. One of the most solidly established rules of biodiversity, documented many times within most of the major groups of organisms, is that genera or higher taxa with small organisms have the largest number of species. Evolutionary biologists have generally agreed that the smaller the organisms, the more finely their species can divide the environment. Thus a dozen species of epiphytic flowering plants and a thousand kinds of insects can live and breed on a single tree, but at most only one to several arboreal lizard or frog species find the same tree adequate.

When species of a genus or family are divided into classes based on the size of the organisms, and the number of species in each class is plotted as a function of those classes, a unimodal distribution is usually produced. The curve is highest toward the lower size class, but not at the lowest of all. It is tempting to ascribe the drop-off toward the lower limit to the relative difficulty of sampling smaller organisms, but this cannot be the case for birds and mammals, which have been thoroughly sampled, and it is unlikely to be true generally. Dial and Marzluff (1988) have proposed the following alternative hypothesis to explain the result. First, species of small organisms within a group adapt better to variable environments because they reproduce faster and can survive better in restricted areas when environmental change reduces

THE ORIGINS OF HYPERDIVERSITY

the habitable area. On the other hand, species of large organisms do better in stable environments, an evolutionary trend that may be due to their superior ability to control resources. Alternation between variable and stable environments through evolutionary time, Dial and Marzluff argue, tends to cull species of both extreme sizes. In addition, a combination of high speciation rates and swifter colonization of fine-grained environments permits higher standing levels of diversity in the small to medium (but not in the smallest) size classes, pushing the mode toward the lower end of the curve.

High speciation rates, it should be kept in mind, cannot by themselves create or produce large amounts of standing diversity. A group can speciate explosively and over long periods of evolutionary time, but if its overall extinction rate is even greater, the standing diversity will not grow but instead decline toward zero.

In general, a successful clade should be expected to speciate logistically in a manner similar to that of a growing population. The first species with properties favorable to multiplication and radiation will most likely divide to produce one to a small number of species. These young species in turn, by adapting to new niches, then spread the ecological and geographical range of the clade as a whole. At first the multiplication into daughter species will be faster than the rate of species extinction, causing the clade to grow exponentially. But as evolutionary time passes, the still unoccupied niches and the available increment of geographical range must decrease, and the rate of speciation will fall to the rate of species extinction. As more time passes, millions of years in the case of the more persistent clades, and for various combinations of reasons both secular and episodic, the extinction rate surpasses the rate of species formation, and the standing diversity as a whole descends. The life span of clades, from the birth of a species as it cleaves from its sister species to its extinction or the extinction of the last of its descendant species, varies greatly among higher categories. For example, it is about 6 million years for starfish and other echinoderms, 1.9 million years for graptolites (colonial animals distantly related to the vertebrates), 1.2 to 2 million years for ammonites (shelled mollusks resembling modern cephalopods), and 0.5 to 5 million years for mammals.

What other properties promote high levels of diversity? In a far-ranging correlative analysis of land vertebrates and plants, Marzluff and Dial (1991a) found that the more diverse taxa constituting clades, whether the clades are classified as genera, families, or orders, tend to have short generation times, based on an early age at first reproduction and a brief average life span. The most diverse taxa also have superior dispersal ability and high resource availability.

From these generalizations, it may be less surprising to learn that one of every 30 species of organism known in the world is a member of the weevil superfamily Curculionoidea. Containing about 60,000 described species in 5,200 genera, these small insects are ancient, dating to the late Jurassic. They are strong fliers with short generation times, and were favored during their evolutionary history by the origin of the angiosperms, which became their principal hosts and gave them access to a vast food supply to colonize and preempt. One family of curculionoids, the Curculionidae, which are especially efficient at ovipositing deeply into plant tissues, comprises 50,000 of the described species all on its own (Farrell 1998; Oberprieler 1999).

As has generally been believed by evolutionary biologists since the time of Sewall Wright's original theory (1931), and as well is documented for example by Bush et al. (1977), species divided into demes, in other words small, partially isolated populations, evolve and speciate more rapidly than large, continuous populations. Such demes, for example, are often formed in social species fragmented into partially closed packs, troops, clans, or other behaviorally delimited groups. The conventional explanation of the effect, due to Sewall Wright, is that small population size increases inbreeding, which leads to significant amounts of genetic drift, defined as the shift of gene frequencies by chance alone. Genetic drift can also occur by the founder effect, the origin of chance differences in the genomes of small sets of individuals that emigrate to start new demes. An example of the founder effect has been described by Melnick (1988) in the rhesus macaques of Pakistan. In this highly social species, matrilines stay together in well-integrated troops over many generations. As a result, they constitute demes. During genetic and demographic monitoring over 40 months, Melnick and his co-workers discovered that strong differences existing between the rhesus troops are due not to inbreeding, but to a high rate of insemination of females in each troop by a proportionately small number of males.

It is important to keep in mind that swift deme divergence and high speciation rates do not automatically translate into high diversity, because any speciation rate can be balanced by an equal extinction rate. The balance can be reached while standing diversity is low or high. Aware of this disjunction between rate and equilibrium, Marzluff and Dial (1991b) conducted a correlative test of the relationship between degree of sociality and taxonomic diversity in birds, mammals, and insects, and found no consistent relationship.

It is useful to distinguish between two hypotheses concerning the effect of social evolution on taxonomic diversity. The first, the *plesiomorphic driver hypothesis,* suggested for example by Bush et al. (1977) and one I have also intuitively favored (Wilson 1975a), holds that the many demes of socially organized species possess a common (plesiomorphic) trait that drives up the level of taxonomic diversity. The second explanation is the *apomorphic adaptation hypothesis,* favored by Marzluff and Dial on the basis of their correlative evidence. It holds that sociality is not a general plesiomorphic adaptation forcing diversification but rather an apomorphic (derived) trait that adapts organisms to certain environmental challenges; species-level diversity is then determined primarily by the environmental challenges. In fact, since both plesiomorphic and apomorphic traits are likely to be adaptive, the difference between the two explanations may not be as great as first appears.

What then, if not the proneness to deme formation, constitutes the forces that press clades toward high levels of biodiversity? Preemption and incumbency must be well up on the list. There are two ways by which species achieve them. The first is superior long-distance dispersal power, allowing early invasion of newly opened environments such as emerging oceanic archipelagos, freshly formed great lakes, and grasslands and other biomes born during climatic change. The second mode is a breakthrough adaptation that either allows the creation of a major new niche within an already established ecosystem or else confers an edge over rivals and predators in niches already filled. These two serendipities grade one into the other, on the one hand because to develop superior powers of long-distance dispersal is itself a breakthrough adaptation, and on the other hand because a breakthrough adaptation is also likely to serve well in recently created environments. The principal difference between newly assembled and mature ecosystems is that in newly assembled ecosystems the niches are more open and the enemies it contains less formidable.

The advantages to species that colonize a newly evolving ecosystem, or a major niche in an old one, are much the same as those described for new industries by Alfred Chandler in his 1977 classic *The Visible Hand: The Managerial Revolution in American Business.* Like Standard Oil and Ford Motor Company, the 350–400 colonizing species that gave rise to the estimated 10,000 contemporary endemic insect species of Hawaii (Howarth and Mull 1992; Otte 1994) enjoyed considerable unclaimed natural resources and room for evolutionary expansion. For a while they could experiment and still thrive with adaptive half-measures. While their competitors were increasing by subsequent invasions, they still had time to evolve to higher levels of efficiency and competitiveness. As a consequence, many such biotic clades, like pioneer corporations, have enjoyed exceptionally long life.

There is another reason why success begets success. The ability to disperse and diversify is not an accident. Large adaptive radiations are more likely than small adaptive radiations to generate *r*-strategy species biologically specialized for dispersal or species that prefer river banks, littoral zones, and other staging areas where long-distance dispersal is more likely to occur. In addition, a clade prone to produce many kinds of specialized species is more likely to generate at least one specialist that hits upon a breakthrough innovation.

Finally, species-level selection favors the diverse clade. An ensemble of cognate species that is geographically widespread and differentiated into both generalist and specialized species is more likely than its more restricted sister clade to survive environmental change. Even if a large number of its species and demes perish, a few will pull through somewhere, perhaps to renew major diversification.

All of the hypotheses, correlative analyses, and speculations I have reviewed so far have been very broad, even abstract in nature. Constructed by many investigators in steps, they represent the beginnings of a general theory of the origin of biological diversification at the species level. I believe that evolutionary biologists are now at the cusp of a new, much more detailed empirical shift in their investigations. The next step will be to address more explicitly a simple question: *What are the adaptations that promote ecological dominance within a clade?* Two formidable obstacles stand in the way of any easy answer to this question. The first is the extreme particularity of the phenomena to be investigated. Even if we identified the leveraging adaptations we seek clade by clade, the final result might prove to be pure natural history, each account unconnected to the rest in details of biology and environmental circumstance. And then, even if general patterns in the natural history could be spelled out, how could it be proved that the adaptations cited were truly the decisive ones? Clades typically have multiple plesiomorphic traits. It may prove very difficult to choose convincingly among them.

No evolutionary biologist has an easy solution to these problems. But a hard-won answer conceivably exists, and might be obtained by repeated use of the following procedure. *Once the phylogeny of a group of clades has been inferred, choose those that have radiated extensively and repeatedly in different geographic theaters of opportunity. Then find the plesiomorphic traits that are both diagnostic for the most successful clades and rarely if ever reversed within them.*

One of many taxa amenable to this approach is the Trigoniidae, which expanded dramatically in the Triassic to become the dominant family of shallow-burrowing bivalve mollusks of near-shore marine habitats and persisted until their near extinction at the end of the Cretaceous (Stanley 1977). Today only *Neotrigonia* of Australia and New Zealand survives. Chiefly inhabitants of coarse, shifting substrate, the trigoniids were essentially the cockles of the Mesozoic. They possessed a powerful foot that was highly efficient for burrowing and, judging from *Neotrigonia,* probably even conferred the ability to jump. Other, coadapted features contributed to the success of the group. They included enormous, complex hinge teeth that kept the valves aligned at the wide angles required by the oversized foot, and in many species, unusual arrays of knobs or ridges on the shell surface that gripped the substratum during burrowing.

Contemporary leaf beetles (superfamily Chrysomeloidea) and weevils (superfamily Curculionoidea), which compose the single clade Phytophaga, contain about 80 percent of all living herbivorous beetle species, and perhaps half of all herbivorous insect species. The hyperdiversity and ecological dominance of beetles generally was due first to a shift, in the Mesozoic, from feeding mostly on decaying organic material (saprophagy) to feeding mostly on plants (herbivory). The next shift occurred in the Chrysomeloidea and Curculionoidea, during the late Cretaceous and early Tertiary, from specialization on conifers and cycads to concentration on the newly dominant angiosperms (Anderson 1995, 1999; Farrell 1998). The unmistakable key adaptation of the weevils is endophytophagy, the deep penetration of plant tissues to feed on them directly. The most diverse weevil family of all, the Curculionidae, was reached by a series of stepping-stone adaptations, described vividly by Oberprieler (1999):

> From their cucujoid ancestor the weevils inherited the successful coleopteran body form with protective elytra, as well as an incipient phytophagy, probably pollen feeding. Enclosure of the pollen-sacs of their apparent original hosts, Jurassic conifers, into strobili seemingly prompted the elongation of the weevil face into a rostrum, initially for adult feeding as preserved by extant Nemonychidae. In the belid lineage, the elongated rostrum took on the function of a sclerotized ovipositor, representing the first key weevil adaptation and enabling a truly endophytophagous way of life of the larva in nearly all plant tissues. The rise of the angiosperms in the Cretaceous provided the next adaptive zone for weevils, traced by the origin of the Brentidae and a drastic rise in weevil diversity. However, the brentid rostrum remained an ineffective tool for penetrating deep-lying plant tissues, being obstructed by the antennal insertions. Extreme developments of this type of rostrum carry serious disadvantages that can only be sustained in exceptional cases, as exemplified by female *Antliarhinus zamiae*. The development of a more efficient rostrum, in which the elongated basal antennal segments fold back into grooves on the rostrum, characterises the most diverse weevil family, the Curculionidae. It allows deeper penetration of plant tissues with a shorter and stronger rostrum and is evidently the most crucial of all weevil adaptations, opening up the full structural and taxonomic diversity of not only angiosperms but also earlier plant groups to the weevils and resulting in the staggering number of more than 50,000 described and nearly 300,000 estimated species in this family.

Many other such key advances presented by systematists in their respective groups of expertise have been ably summarized and categorized by Heard and Hauser (1995).

Now I will risk a hypothesis that explains the ecological dominance and taxonomic hyperdiversity of *Pheidole*. The one trait that distinguished all known species of *Pheidole* is their unusual large-headed major subcaste. Except for a handful of social parasites scattered through four of the species groups in different parts of the world, there is no evidence of the evolutionary loss of this subcaste in any clade. Even in these instances the parasites live with host colonies that themselves possess majors. And in the one parasitic species (*inquilina*) known to possess a rudimentary worker caste, the major subcaste is still present. The major subcaste is associated in all known *Pheidole* species with another trait that appears coadapted with it. This is the reduction of the sting in both the major and minor to a vestigial condition without, in at least the vast majority of species, a compensatory increase of defensive exocrine secretions, an enhancement that characterizes, for example, the large subfamilies Dolichoderinae and Formicinae.

This suite of changes has allowed *Pheidole* to achieve an extreme division of labor in which an unusually heavy reliance for defense is placed on the major caste. The majors, as the surrogate for the sting and most of the toxic chemical armament of the colonies, are a highly mobile strike force. They have also differentiated dramatically from one clade to the next within *Pheidole* as part of the exceptional adaptive radiation achieved by the genus. In most species, the minors summon them from the nest to attack enemies or help subdue and transport large prey. In a minority of species, for example those composing the *distorta* and *lamia* groups as well as many members of the *tristis* group, the majors are anatomically and behaviorally specialized to serve as house guards, remaining in the nest to fight or block the way of intruders. Many, as in the *pilifera* group and probably most of the *flavens* and *tristis* groups as well, further serve to mill seeds collected by the minors. Some may also function as semi-repletes for the long-term storage of liquid food, although the evidence for this role is scant. Finally, as evidenced by *bicarinata* in the *pilifera* group and *dentata* in the *fallax* group, in which the phenomenon has been experimentally studied, the majors function as an emergency standby caste. They are able to take over the labor roles of the minors, albeit clumsily, when the ranks of the smaller caste have been depleted.

Another specialized trait of *Pheidole,* although not yet well enough studied to be called universal, is the absence of ovaries in both the minors and majors. Combined with the lightness of their chemical armament and sting apparatus, the minors in particular have the appearance of a "throwaway"

THE ORIGINS OF HYPERDIVERSITY

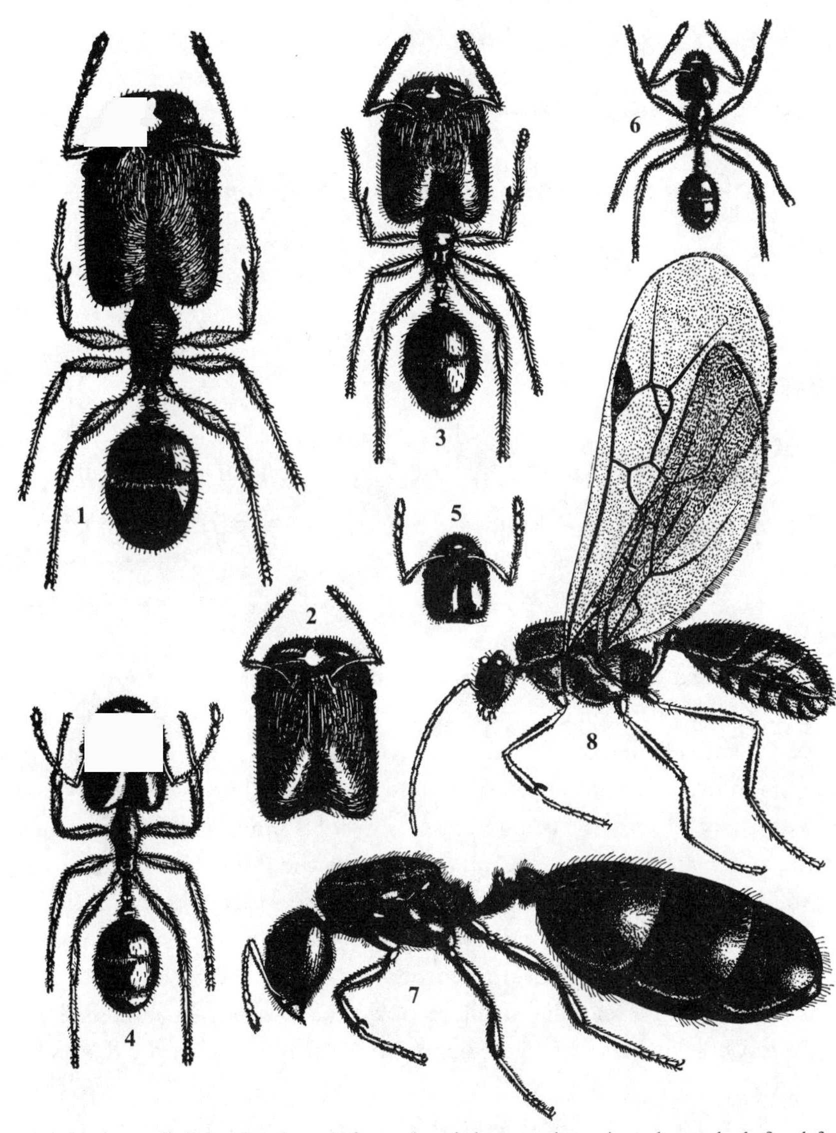

Figure 9 William Morton Wheeler's 1910 illustration of *Pheidole tepicana (= P. instabilis),* which is reproduced here, is the classic depiction of castes in ants. *P. tepicana* is one of several members of the genus that possess continuous polymorphism in the worker caste, culminating at the upper size range in a supermajor: supermajor *(1, 2)*; major-supermajor intermediate *(3)*; "typical" major *(4)*; major-minor intermediate *(5)*; minor *(6)*; queen *(7)*; male *(8)*. (Modified from Wheeler 1910b: 89.)

caste, that is, small, light, cheaply manufactured, and short-lived. These traits join the relatively small size of the ants and their short colony reproductive time to make possible the remarkable ecological success and hyperdiversity of *Pheidole*.

In sum, as I have come to interpret the genus, the typical *Pheidole* colony is an unusually resilient superorganism able to expend and replace minor workers readily while utilizing the major subcaste both for defense and as an emergency labor force in the event of severe depletion of the minors. The majors are especially efficient as a defense force, able to mobilize quickly and thus reduce the need for stings or elaborate, energetically expensive chemical systems. These traits join the relatively small size of the ants and their short colony reproductive time to make possible the remarkable ecological success and hyperdiversity of *Pheidole.*

2000–2004

"A global biodiversity map," *Science* 289: 2279 (2000)

"On the future of conservation biology," *Conservation Biology* 14(1): 1–3 (2000)

"The encyclopedia of life," *Trends in Ecology and Evolution* 18(2): 77–80 (2003)

"Taxonomy as a fundamental discipline," *Philosophical Transactions of the Royal Society of London, Ser. B Biological Sciences* 359: 739 (2004)

The next four arguments are presented on behalf of the completion of the Linnaean enterprise to discover all species on Earth, accompanied by the construction of an electronic Encyclopedia of Life in which all information on each species in turn is kept and updated. Nearly 250 years have passed since the inauguration by Carolus Linnaeus of the binomial system of nomenclature (e.g., *Canis familiaris* for the domestic dog) and the hierarchical classification by degree of similarity (species *familiaris* and similar species in the genus *Canus, Canus* and similar genera in the family Canidae, Canidae and similar families in the order Carnivora, and so on upward to phyla and domains). Yet only a small fraction of Earth's species, perhaps fewer than a tenth, have been discovered and given Latinized scientific names by biologists.

The first essay, "A global biodiversity map," suggests the need and feasibility of discovery of the unexplored biosphere. The second, on the application of systematics to conservation biology, points out the advantage of knowing fully what it is we are trying to save. In this article I describe how the new technologies of high-resolution digital photography and electronic publication can and inevitably will revolutionize the mapping of biodiversity. In the course of the initiative, basic taxonomic research will be extended from the small number of museums with major reference collections to smaller institutions. The capacity will be developed to conduct it even in field camps where collection of specimens occurs. The third essay describes what I believe is the inevitable creation of an electronic encyclopedia, a repository of enormous potential for the future of biology. The final essay, working off the first three, makes the case for reinvigorating taxonomy as a mainstream discipline of biology.

A Global Biodiversity Map

Edward O. Wilson

As genomics and biomedicine are to human health, so ecology and conservation biology are to the planet's health. Unfortunately, compared with their sister disciplines, ecology and conservation biology are still disadvantaged. Their growth is hampered by a seldom-acknowledged deficiency: our ignorance of most of the world's biodiversity, particularly at the level of individual species, where knowledge is foundational to all other studies of diversity and hence of the whole living environment. The number of species given a scientific name is believed to fall between 1.5 million and 1.8 million. The true, full number, including those still undescribed, can only be guessed at to the nearest order of magnitude, with the opinion of many experts gravitating toward the vicinity of 10 million. Reliable biodiversity assessments are limited to a few relatively well-known groups, including the vascular plants, vertebrates, and a small number of invertebrates such as corals and butterflies. If the true number of species is about 10 million, these focal groups add up to fewer than 5% of the total. Bacteria, mites, nematode worms, fungi, beetles, and other major environmental players are necessarily ignored, or at best given "morphospecies" code numbers. Even among the small minority of all species diagnosed and named, fewer than 1% have been subject to the kind of careful biological studies needed to undergird ecology and conservation biology.

To describe and classify all of the surviving species of the world deserves to be one of the great scientific goals of the new century. In applied science, this completion of the Linnaean enterprise is needed for effective conservation practices, for bioprospecting (the search for new natural products in wild species), and for impact studies of environmental change. In basic science, it is a key element in the maturing of ecology, including the grasp of ecosystem functioning and of evolutionary biology. It also offers an unsurpassable adventure: the exploration of a little-known planet.

"What is... needed now is a concerted effort..."

Pieces of a worldwide biodiversity project are being put in place. In 1999, the Megascience Forum of the Organization for Economic Cooperation and Development created the Global Biodiversity Information Facility to coordinate and bring online all electronic databases for various groups of organisms. CD-ROMs of individual groups for different parts of the world proliferate, augmenting a continued flow of traditional print monographs. New electronic technology, increasing exponentially in power, is trimming the cost and time required for taxonomic description and data analysis. It promises to speed traditional systematics by 2 orders of magnitude. What is lacking and needed now is a concerted effort, comparable to the Human Genome Project (HGP), to complete a global biodiversity survey—pole to pole, whales to bacteria, and in a reasonably short period of time.

If treated as a near-horizon goal instead of an eventual destination, the survey will multiply benefits in basic and applied science. The key choke point will also be quickly revealed. It is not the needed tools of informatics, most of which are already at hand. Nor is it a persuasive rationale, which can be readily expressed to scientists and the public alike. Rather, it is the severely limited capacity of museums and other collections-oriented facilities to collect, prepare, and analyze specimens, and the shortage of expert taxonomists to do the job.

According to a recent survey by the Association of Systematic Collections, in North America only 3000 Ph.D.-level researchers are active in the exploration and description of the world's fauna and flora. At a rough guess, another 3000 are engaged elsewhere in the world. Museums, universities, and government agencies in the United States collectively spend between $150 million and $200 million each year on systematics research. These levels are incommensurate with the magnitude of the task and of the benefit it offers humanity.

What will it cost to complete such a map? Suppose there are in fact about 10 million living species. The cost per species, using newly available informatics technology, might be $500, for a total of $5 billion spread over 10 to 20 years, hence roughly comparable to the HGP. As in that enterprise, per-unit cost can be expected to drop as technology is improved, while scientific and practical benefits from the accumulating knowledge grow exponentially.

Edward O. Wilson is Pellegrino University Research Professor at the Museum of Comparative Zoology, Harvard University, 26 Oxford Street, Cambridge, MA 02138–2902, USA. E-mail: ewilson@oeb.harvard.edu.

Editorial

On the Future of Conservation Biology

Conservation biology has been aptly described as a discipline with a deadline, but for those who work in this intensive-care ward of ecology it is more precisely a never-ending avalanche of deadlines. The conservation biologist knows that each imperiled species is a masterpiece of evolution, potentially immortal except for rare chance or human choice, and its loss a disaster. You and I will be entirely forgotten in a thousand years, but, live or die, the black-footed ferret, barndoor skate, Lefevre's riffle shell, Florida torreya, and the thousands of other species now on the brink of extinction will not be forgotten, not while there is a civilization. Our conservation successes, the only truly enduring part of us, will live in their survival.

Conservation biologists are crisis managers who ply the full array of biological organization from gene to ecosystem. Their scientific work is both basic and practical. It is also one of the most eclectic of intellectual endeavors. Consider the following example from recent media headlines: survival of the red-cockaded woodpecker, a bird (an *American* bird no less) turns upon our knowledge of its distribution and natural history, survival of the mature pine woodland in which it lives, the economic and political forces that erode its nest sites, the legislation that protects it, and, not least, the moral precepts that support the very idea of ecosystem and species conservation.

No real basis exists—as some writers have imagined—for conflict between ecosystem studies and single-species studies in conservation biology. Each is vital and intellectually dependent upon the other. Within the broader framework of ecosystem studies, community ecology in particular is about to emerge as one of the most significant intellectual frontiers of the twenty-first century. Although it still has only a mouse's share of science funding, it stands intellectually in the front rank with astrophysics, genomics, and neuroscience. Community ecologists face the daunting challenge of explaining how biotas are assembled and sustained. Most of their effort today is in description and analysis, with closest attention paid to one species or to several species as modules. As time passes, more resources will be put into the mathematical modeling and experimental manipulation of entire assemblages, from the bottom up, species to communities. Biotas, like cells and brains, are prime targets for the emerging field of general complexity theory. They have already been singled out as paradigms of complex adaptive systems and are certain to attract the attention not just of ecologists but also of physicists, molecular biologists, and others who are running short of virgin fields of inquiry.

Like the rest of science, community ecology advances by repeated cycles of reduction and synthesis, in which bottom-up analysis of the working parts explains the complex whole and, in reciprocity, an evolving theory of the complex whole guides further exploration of the working parts. The relevance of this perpetual process to conservation biology is as follows. The more or less independently evolved key working parts are the species. In the future, solid advances in community ecology will depend increasingly on a detailed knowledge of species and their natural history, which feeds and drives theory.

It follows that community ecology and conservation biology are in desperate need of a renaissance of systematics and natural history. By systematics I mean much more than just the phylogenetic analysis of already known species. Phylogenetic reconstruction, currently the dominating focus of systematics, obviously is worth doing, but more scientifically important and far more urgent for human welfare is the description and mapping of the world biota. They are scientifically important because descriptive systematics is the foundation for community ecology. And they are urgent because the development of a mature, accessible knowledge of global biodiversity is necessary for conservation theory and practice.

Few biologists other than systematists appreciate how little is known of Earth's biodiversity. Estimates of the total number of species still vacillate wildly: 3,600,000 at the low end and 111,700,000 at the high end (*Global Biodiversity Assessment*, 1995). The estimated number of species described and given scientific names ranges between 1.5 and 1.8 million. Here also the true number is only a matter of speculation. Even figures for the relatively well-studied vertebrates are spongy. Estimates for the extant fish species of the world, including both described and undescribed, range from 15,000 to 40,000. That figure becomes a veritable black hole in the case of the bacteria and archaea, whose species could with

Conservation Biology, Pages 1-3
Volume 14, No. 1, February 2000

equal ease number either in the thousands or in the tens of millions.

Natural history is still further behind. Even among the named species—never mind those still undiscovered—only a minute fraction, less than 1%, have been studied beyond the essentials of habitat preference and diagnostic anatomy. In general, ecologists and conservation biologists appear not to fully appreciate how thin the ice is on which they skate.

The full exploration of the living part of this planet will be an adventure of megascience, summoning the energy and imagination of our best minds. Its relevance to human welfare was spelled out in the Convention on Biological Diversity of the 1992 Earth Summit, and much of its methodology and possible organizational flowchart by Systematics Agenda 2000. Funding is still limited given the task at hand but is rising under the auspices of organizations such as the Global Environmental Facility and special programs of the U.S. National Science Foundation.

If conservation biology is to mature into an effective science, pure systematics must be accompanied by a massive growth of natural history. For each species, for the higher taxa to which it belongs, and for the populations it comprises, there is value in every scrap of information. Serendipity and pattern recognition are the fruit of encyclopedic knowledge gathered for its own sake. For example, all that can be learned about an endangered conifer on New Caledonia, about the rest of the conifers of New Caledonia, and about every other member of the entire world conifer flora, deserves dedicated pursuit. Periodic summaries of the information are rightfully placed into *Nature, Science, Proceedings of the National Academy of Sciences*, and other mainstream journals. Just being there, they help recruit the media to the good cause. For in order to care deeply about something important it is first necessary to know about it. So let us resume old-fashioned expeditions at a quickened pace, solicit money for permanent field stations, and expand the support of young scientists—call them "naturalists" with pride—who by inclination and the impress of early experience commit themselves to deep knowledge of particular groups of organisms.

Naturalists at heart, conservation biologists in ultimate purpose, they are in every sense of the word modern scientists. Their purview comprises systematics, ecology, and conservation biology, increasingly empowered by methodology for the accumulation and analysis of electronic databases. Their technology expands according to Moore's Law: a doubling of microchip capacity every 18 months. In 1999 a new initiative, Species 2000, set out, at last, to catalog all named species of organisms and thus provide an instantly accessible census of known global biodiversity. In 1999 the Megascience Forum of the Organization for Economic Cooperation and Development (OECD) authorized the creation of the Global Biodiversity Information Facility (GBIF), whose charge is to coordinate and bring on-line all the rapidly accumulating electronic databases for various groups of organisms. The effort will be aided by the growth of regional institutions such as the East Asian Network for Taxonomy and Biodiversity Conservation, headquartered in Seoul, and the Biodiversity Foundation for Africa, based in Bulawayo.

By 2020 or earlier the combined methodology might work as follows. Imagine an arachnologist making a first study of the spider fauna of an isolated Ecuadorian rain forest. He (or she, recognizing with admiration the powerful and growing influence of women scientists in this discipline) sits in camp sorting newly collected specimens with the aid of a portable, internally illuminated microscope. After quickly sorting the material to family or genus, he enters the electronic keys that list character states for, say, 20 characters and pulls out the most probable names for each specimen in turn. Now the arachnologist consults monographs of the families or genera available on the World Wide Web, studying the illustrations, pondering the distribution maps and natural history recorded to date. If monographs are not yet available, he calls up digitized photographs from the GBIF files of the most likely type specimens taken wherever they are—London, Vienna, Sao Paulo, anywhere photographic or electron micrographs have been made—and compares them with the fresh specimens by panning, rotating, magnifying, and pulling back again for complete views. Does this specimen belong to a new species? He records its existence (noting the exact location from his global positioning system receiver), habitat, web form, and other relevant information into the GBIF, and he states where the voucher specimens will be placed—perhaps later to become type specimens. Informatics has thus allowed the type specimens of Ecuadorian spiders in a sense to be repatriated to Ecuador, and new data on its spider fauna to be made immediately and globally available.

The arachnologist has accomplished in a few hours what previously consumed weeks or months of library and museum research. He understands that biodiversity studies advance along two orthogonal axes. First are monographs, which treat all of the species across their entire ranges, and second are local biodiversity studies, which describe in detail the species occurring in a single locality, habitat, or region. When expanded to include more and more groups, local biodiversity studies may eventually cover all local plants, animals, and microorganisms, creating an all-taxa biotic inventory (ATBI), a truly solid base for community ecology in its full complexity.

These cross-cutting databases open new avenues of useful analysis for the conservation biologist. When information on elevation, slope, vegetation cover, soil type, rainfall, and other biotic and abiotic properties of the study site are digitized, overlaid with one another, and matched with similar overlays from the surrounding region, the range of new and rare species can be predicted. At least a good guess can be made about where each in turn is most likely to occur. To single-species searches and mapping can be added the already well-developed technique of gap analysis, in which the overlays include cropland, human habitation, transportation routes, ground and runoff water reserves, and current reserves. With such information available in easily accessible form, regional conservation becomes not only scientifically sound but a great deal easier to achieve in the political arena.

Systematics and natural history also form the requisite empirical base for population viability analyses (PVAs), which are key instruments for predicting the future of species at risk and devising means for pulling them back to safety. Furthermore, PVAs will in time allow the prognosis of exotic species most likely to become invasive, that is, destined to grow from harmless beachhead populations to levels that are economically and environmentally destructive. At the present time we notoriously lack the capacity to identify potential pests such as the zebra mussel, red imported fire ant, green crab, brown tree snake, and miconia before they are irreversibly established. The general public will be unanimously on the side of conservationists in this effort. The zebra mussel alone, while exterminating native mussel populations, also shuts down electrical utilities by clogging water intake pipes. The resulting losses will accumulate, according to the U.S. Fish and Wildlife Service, to 5 billion dollars by the year 2002. This example by itself should have enough weight on the balance sheet to justify major financial support for ecology and conservation biology.

To build encyclopedic hypertexts of systematics and natural history is simultaneously to promote ecotourism, which the governments of many developing countries now see as a principal source of foreign-exchange income. In Costa Rica, for example, tourism with a strong natural-history slant, yielding upwards of a billion dollars a year, has now passed banana and coffee production as the chief source of external income.

Systematics and natural history databases also are obviously necessary for bioprospecting, the search for new pharmaceuticals, agricultural crops, fibers, and other natural products that can be harvested from wild species. The same is true for genes to be used in interspecific transfers, one of the driving forces of the new and future giant industry of genetic engineering.

When large arrays of species are studied for their intrinsic interest, the result is a heuristic surge in basic and applied research in other domains of biology. New phenomena are discovered and research agendas suggested never dreamed of by those with the opposite research strategy, which is to choose a problem within the ambit of existing knowledge and then to search for a species—any species—useful for its solution. Thus, conservation biologists of the coming century will, so long as they draw strength from the groundwork of biodiversity exploration, serve science handsomely and lead humanity toward one of its noblest goals.

Edward O. Wilson

Museum of Comparative Zoology, Harvard University, Cambridge, MA 02138, U.S.A.

Opinion TRENDS in Ecology and Evolution Vol.18 No.2 February 2003 77

The encyclopedia of life

Edward O. Wilson

Museum of Comparative Zoology, Harvard University, 26 Oxford Street, Cambridge, MA 02138-2902, USA

Comparative biology, crossing the digital divide, has begun a still largely unheralded revolution: the exploration and analysis of biodiversity at a vastly accelerated pace. Its momentum will return systematics from its long sojourn at the margin and back into the mainstream of science. Its principal achievement will be a single-portal electronic encyclopedia of life.

Imagine an electronic page for each species of organism on Earth, available everywhere by single access on command. The page contains the scientific name of the species, a pictorial or genomic presentation of the primary type specimen on which its name is based, and a summary of its diagnostic traits. The page opens out directly or by linking to other data bases, such as ARKive, Ecoport, GenBank and MORPHOBANK. It comprises a summary of everything known about the species' genome, proteome, geographical distribution, phylogenetic position, habitat, ecological relationships and, not least, its practical importance for humanity.

The page is indefinitely expansible. Its contents are continuously peer reviewed and updated with new information. All the pages together form an encyclopedia, the content of which is the totality of comparative biology.

The rationale

There are compelling reasons to build such an all-species encyclopedia. Not least is the heuristic power for biology as a whole. As the census of species on Earth comes ever closer to completion, and as their individual pages fill out to address all levels of biological organization from gene to ecosystem, new classes of phenomena will come to light at an accelerating rate. Their importance cannot be imagined from our present meagre knowledge about the biosphere and the species comprising it. Who can guess what the mycoplasmas, collembolans, tardigrades and other diverse and still largely unknown groups will teach us? As the species coverage grows, gaps in our biological knowledge will stand out like blank spaces on maps. They will become destinations toward which researchers will gravitate.

For the first time, the biotas of entire ecosystems can be censused in full. Unknown microorganisms and the smallest invertebrates, which still comprise most species yet lack even a name, will be revealed. Only with such encyclopedic knowledge can ecology mature as a science and acquire predictive power species by species, and from those, ecosystem by ecosystem.

As one result, the human impact on the living environment could be assessed in far more reliable detail than is now possible. Today, for example, we base estimates of species extinction on data from a scattering of taxonomically best known groups, including the flowering plants, land and freshwater vertebrates, and a few invertebrates, such as butterflies and mollusks. These taxa contain only about a quarter of the known species on Earth, and almost certainly a much smaller fraction of those still unknown. Tomorrow, other invertebrates, including insects and nematodes, as well as fungi and nearly all microorganisms, together comprising most species on Earth, as well as essential pathways of the energy and materials cycles, can also be assessed.

The all-species encyclopedia will serve human welfare in more immediately practical ways. The discovery of wild plant species adaptable for agriculture, new genes for enhancement of crop productivity, and new classes of pharmaceuticals can be accelerated. The outbreak of pathogens and harmful plant and animal invasives will be better anticipated and halted. Never again, with fuller knowledge of such extent, need we overlook so many golden opportunities in the living world around us, or be so often surprised by the sudden appearance of destructive aliens that spring from it.

An all-species encyclopedia of life is logically inevitable if for no other reason that the consolidation of biological knowledge is urgently overdue. In its earliest stages, already emerging, it forms a matrix within which comparative studies are rapidly organized. The process will accelerate as traditional taxonomic procedures, still mostly dependent on repeated examinations of type specimens and print literature, are replaced by high-resolution digital photography, nucleic acid sequencing and internet publication. With further documentation organized into the species pages, new lines of research will open at a quickening pace. Model species for laboratory and field research can be more easily found – obedient to the principle that for every problem in biology, there exists a species ideal for its solution.

A growing, single-access species-structured encyclopedia will ease navigation through the immense biological data bases. Aided by computer search engines, patterns can be summoned whose detection would otherwise demand impracticable amounts of effort and time. Principles and theory can be built, deconstructed and rebuilt with an unprecedented power and transparency.

Ultimately, and at a deeper level, the all-species encyclopedia will, I believe, transform the very nature of biology, because biology is primarily a descriptive science. Although it depends upon a solid base of physics and chemistry for its functional explanations, and the theory of natural selection for its evolutionary explanations, it is

Corresponding author: Edward O. Wilson (ewilson@oeb.harvard.edu).

http://tree.trends.com 0169-5347/02/$ - see front matter PII: S0169-5347(02)00040-X

defined uniquely by the particularity of its elements. Each species is a small universe in itself, from its genetic code to its anatomy, behavior, life cycle and environmental role, a self-perpetuating system created during an almost unimaginably complicated evolutionary history. Each species merits careers of scientific study and celebration by historians and poets. Nothing of the kind can be said (at the risk of stating the obvious) for each proton or inorganic molecule.

The taxonomic foundation

Taxonomy, the scientific study and practice of classification, is the foundation to the all-species encyclopedia. However, it is still one of the most underfunded and weakly developed biological disciplines. Worldwide, as few as 6000 biologists work within it. Most people are surprised to learn that most of biodiversity is still entirely unknown. They assume that taxonomy all but wound down generations ago, so that today each new species discovered is a newsworthy event. The truth is that we do not know how many species of organisms exist on Earth even to the nearest order of magnitude. Those formally diagnosed and given latinized scientific names are thought to number somewhere between 1.5 and 1.8 million, with no exact accounting having yet been made from the taxonomic literature. Estimates of the full number, known plus unknown, vacillate wildly according to method. As summarized in the *Global Biodiversity Assessment* [1], they range from an improbable 3.6 million at the low end to an equally improbable 100 million or more at the high end. The commonest order-of-magnitude guess is ten million.

The smaller the organisms, the more poorly known the group to which it belongs. About 69 000 species of fungi have been distinguished and named, but as many as 1.6 million are thought to exist. Of the nematode worms, making up to four of every five animals on Earth (and, it is said, so abundant that if all solid matter on the surface of the planet were to disappear, its ghostly outline could still be seen in nematodes), ~15 000 species are known but millions more might await discovery. Nematodes in turn are dwarfed in diversity by the bacteria and archaeans, the black hole of biological systematics. Although only ~6000 have been formally recognized, approximately that many, almost all new to science, can be found in only a few grams of rich forest soil. Our ignorance of these microorganisms is epitomized by bacteria of the genus *Prochlorococcus*, arguably the most abundant organisms on the planet and responsible for a large part of the organic production of the ocean, yet unknown until 1988. *Prochlorococcus* cells float passively in open water at 70 000–200 000 ml^{-1}, multiplying with energy captured by sunlight. They eluded recognition so long because of their extremely small size. Representing a special group called picoplankton, they are much smaller than conventional bacteria and barely visible at the highest optical magnification.

Even the largest organisms await a full accounting. The global number of amphibian species has grown in the past 15 years by more than a third, from 4000 to 5400. The flowering plants, for centuries among the favorite targets of naturalists, could rise from the present 272 000 to over 300 000: each year ~2000 new species are added to the standard world list of the *International Plant Names Index* (http://www.ipni.org).

The biodiversity agenda

How best might the taxonomic foundation be laid? From 13 to 15 October, 2001, a 'summit' was held at Harvard University by leaders of organizations devoted to comprehensive taxonomic surveys on a global or continental scale. Their aim was to find a way to complete a world census in a foreseeable period of time. Included were the Africa Biodiversity Foundation (headquartered in Bulawayo, Zimbabwe), Census of Marine Life (New York, USA), the Global Biodiversity Information Facility (Copenhagen, Denmark), the Global Taxonomy Initiative of the Convention on Biological Diversity (New York), the Integrated Taxonomic Information System (Washington, DC, USA), and NatureServe (Arlington, USA). Also present were scientist representatives from major collections in North and Latin America, as well as experts in bioinformatics technology. The summit was hosted by the All Species Foundation, newly formed as a facilitator of the overall effort. Its aim is to provide a clearing-house for the frontline initiatives, to assist them in their funding initiatives and development of bioinformatics, to initiate new projects, and to monitor and report progress in the overall enterprise on a continuing basis.

The attendees of the all-species summit agreed that a complete or, more realistically, a nearly complete global biodiversity census is technically feasible within 25 years. The magnitude of the task can be visualized as follows: whereas 10% of species on Earth out of, say (at an educated guess) 10 million–20 million, have been diagnosed during the first 250 years, beginning with Carolus Linnaeus' *Systema Naturae* in the mid-1700s, it is proposed to complete the remaining 90% in one-tenth that time.

The idea of a complete global biodiversity census with a timeline and coordinated initiatives had first been proposed in 1992 [2]. By the mid-1990s, the importance of the new technologies of bioinformatics in descriptive biology had also become apparent [3]. In 2000, explicit proposals were put forth for a census timeline and practical bioinformatics in systematics research [4–8]. By 2002, the implications of the new initiatives were being explored by biologists in several disciplines [9–11], and it could be said quite fairly that a 'biodiversity commons' [12] had come into being within the 'bioinformatics nation' [13].

The full agenda of biodiversity exploration is now unfolding in three overlapping phases. The first is the Catalog of Life, aimed at the organization of information about existing species into an electronic global framework [11]. The Catalog was born of the collaborative efforts of Species 2000, a federation of data bases begun in 1994 by the International Union of Biological Sciences, and headquartered at the University of Reading, UK; the Integrated Taxonomic Information System, begun in 1995

Opinion TRENDS in Ecology and Evolution Vol.18 No.2 February 2003 79

through a partnership among interested agencies of the US Federal Government; the Global Taxonomy Initiative of the Convention on Biological Diversity, a worldwide effort spun from the 1992 Rio Earth Summit; and the Global Biodiversity Information Facility, begun by the Organization for Economic Cooperation and Development in 1996 and now headquartered as an independent operation in Copenhagen.

The second phase of the full biodiversity agenda is the accelerated discovery of life forms still unknown. This achievement, the anticipated moon shot of systematic biology, is envisioned as a future goal by the organizations loosely grouped under what Bisby *et al.* [11] have called the 'Catalog of Life' initiative, and as an immediate goal with a timeline by the All Species Foundation, headquartered in San Francisco, USA [6–8,10].

The final enterprise, the electronic Encyclopedia of Life, which is already being pressed here and there, will expand upon the growing base provided by the taxonomic Catalog of Life. Covering all biological levels, from genome to ecology, it will serve as the ultimate guide to biodiversity.

New technologies

Faith in a sprint to the finish of the global census is engendered by the more advanced revolutions ongoing in bioinformatics and genomics, which together offer the means to transform the traditional methods of taxonomy. The old methods, which still prevail, have been enormously labor intensive and time consuming. To complete a taxonomic analysis of a genus or higher order taxon requires examination of the primary types of each species, subspecies and variety, which are typically scattered among museums in North America and Europe, and often in other continents. The systematist must conduct lengthy tours to examine all these specimens, or else have them sent through by hand or mail, a risky step that not all curators are willing to take. The systematist must also have access to a wide array of books and journals, many of which are old and rare. As a result, the tradition of systematics since Linnaeus has been that of arcane expertise practiced by groups of specialists working on groups of organisms to which they have devoted their professional lives.

With the new technology, the 19th century culture of taxonomy has begun to be replaced. For the first time, type specimens can be illustrated by swiftly made high-resolution digital photographs, the anatomical detail and depth of field of which are beyond those seen in specimens viewed by light microscopy. The photographs can be published on the Internet. When all the primary types of a particular group, say weevils of the family Curculionidae or grasses of the family Gramineae, are digitally photographed and online, they can be accessed immediately by anyone anywhere. When the original diagnoses from print literature are added, experts can proceed with revisions at a speed and an economy vastly greater than enjoyed in the predigital era. In one step, the practice of taxonomy is globalized and democratized and, in a sense, the type specimens are repatriated to their country of origin.

One such program already completed is the 'virtual herbarium' of the New York Botanical Garden. Almost its entire collection of type specimens of some vascular plants, representing 90 000 species, is now finished. Similar initiatives are underway in the insect collections of the Academy of Natural Sciences in Philadelphia, USA and Harvard University's Museum of Comparative Zoology. With more such projects completed, collection by collection around the world, the global iconography will come together like pieces fitted into a mosaic. The result will be the requisite foundation for a swift exploration of biodiversity on Earth and the accompanying growth of the all-species encyclopedia.

Key challenges

Construction of the complete taxonomic base will not, however, be just a smooth compilation of species. The magnitude of biodiversity and the tangle of evolutionary processes that generated it still present formidable problems. First in line is the difficulty of classifying microorganisms and many of the smallest, soft-bodied invertebrates, most of the species of which can be reliably separated only by molecular diagnosis. The difficulty has put all-species inventories out of reach in the past. However, its solution appears close at hand, thanks to the rapid advances occurring in genomics. Already, for example, tens of thousands of species from the major domains of organisms have been at least partially sequenced for small subunit rRNA genes. By April 2002, the last date for which I have seen an accounting, the genomes of no fewer than 61 species of bacteria had been completely sequenced. As the process accelerates, and the cost per base pair continues to drop, genomic data will become standard for taxonomy, as well as for phylogenetic reconstruction, across all groups of organisms.

A second barrier to the all-species inventory is the incongruence of the species concept between major groups. The classic definition of the species in sexually reproducing organisms is a closed gene pool – a population of individuals that are capable of freely interbreeding under natural conditions. This criterion works reasonably well for most animals and plants, but creates difficulties in some plant groups in which hybridization is extensive but short of total. And it fails logically, of course, in the many populations that lack sexual reproduction. The value of the classic definition of reproductive isolation is still unknown in the bulk of microorganisms, where species might have to be delineated arbitrarily by a cutoff percentage of base pairs shared by populations or some other genetic criterion.

The species problem cannot be settled in advance by any formula or legislation. It will probably be broken only as the all-species initiative evolves, illuminating the particularities of species-level variation from one phylogenetic group of organisms to another. As this knowledge grows, the difficulty of defining species will metamorphose into deeper studies of how species-level diversity arises, group by group. Meanwhile, the process of censusing can and should proceed with the best tools and species concepts at hand. Resolution of the species problem will be one of its most important results.

The problems inherent in bioinformatics are also formidable. As electronic search engines are developed, they must be made interoperable within and between phylogenetic groups. They must have quality control, exercised most probably by publication committees comparable to boards of editors of journals. They need to be created, as in the case of GenBank, to provide free public access. In joining the bioinformatics nation, taxonomists and encyclopedists need to address and overcome the growing problem of information overload already bedeviling those managing DNA microarray analyses, airline schedules and bank accounts. And finally, with current floppy disks starting to lose data within a decade and even optical disks in less than a century, improvement in longevity and format transfer methods will be a priority in the technologies adopted.

These obstacles are daunting, but they are of a technical nature eminently vulnerable to human ingenuity. To overcome them, and thereby complete the great Linnaean enterprise, creating the base of the all-species encyclopedia, will secure the rightful place of comparative biology within mainstream science.

References

1 Heywood, V.H. and Watson, R.T. (1995) *Global Biodiversity Assessment*, Cambridge University Press
2 Raven, P.H. and Wilson, E.O. (1992) A fifty-year plan for biodiversity surveys. *Science* 258, 1099–1100
3 Edwards, M. and Morse, D.R. (1995) The potential for computer-aided identification in biodiversity research. *Trends Ecol. Evol.* 10, 153–158
4 Wilson, E.O. (2000) A global biodiversity map. *Science* 289, 2279
5 Wilson, E.O. (2000) On the future of conservation biology. *Conserv. Biol.* 14, 1–3
6 Kelly, K. (2000) All species inventory: a call for the discovery of all life-forms on Earth. *Whole Earth* Fall, 4–9
7 Warshall, P. (2000) Bioinformatics: the master list and virtual museum. *Whole Earth* Fall, 50
8 Lawler, A. (2001) Up for the count? *Science* 294, 769–770
9 Godfray, C.J. (2002) Challenges for taxonomy. *Nature* 417, 17–19
10 Gerwin, V. (2002) All living things, online. *Nature* 418, 362–363
11 Bisby, F.A. *et al.* (2002) Taxonomy, at the click of a mouse. *Nature* 418, 367
12 Moritz, T. (2002) Building the biodiversity commons. *D-Lib Magazine* 8 http://www.dlib.org/dlib/june02/moritz/06moritz.html
13 Stein, L. (2002) Creating a bioinformatics nation. *Nature* 417, 119–120

Published online 18 March 200

Taxonomy as a fundamental discipline

Descriptive taxonomy is not just a service agency for the rest of biology. Its product is far more than a stock inventory of Earth's biodiversity. Rather, given the extreme particularity of species and how little we know about them as a whole, taxonomy can justly be called the pioneering exploration of life on a little known planet. Among its cascade of derivative functions, taxonomy lays the foundations for the phylogenetic tree of life, it provides a requisite database for ecology and conservation science, and, not least, it makes accessible the vast and still largely unused benefits offered by biodiversity to humanity.

Where the rest of biology is largely vertical in orientation, ranging across the levels of biological organization within a very few species, taxonomy is lateral, encompassing the full sweep of life. Only when both dimensions are sufficiently filled out and joined can we expect to achieve a truly unified science of biology.

The exploration of life on planet Earth is still in an early stage. The number of formally described species, each diagnosed and given a Latinized name, is generally thought to lie between 1.5 and 1.8 million. The total number, described and undescribed, has been estimated by various methods and investigators to fall between 3.6 and 100+ million. Most analysts favour 10 million as the nearest order of magnitude. The greatest number of unknown species (or taxa equivalent to the species of multicellular organisms) probably occurs in the prokaryotes, which remain for biodiversity studies the equivalent of the cosmologist's dark matter. About 6000 species have been described worldwide (as of mid-2003), but at least that many occur in a gram of ordinary soil; and some four million, virtually all unknown to science, have been estimated to live in a ton of soil (Curtis *et al.* 2003).

A rough picture of the magnitude of the task facing taxonomists can be sketched as follows. Let us suppose that *ca.* 10% of the species (or comparable taxa) of Earth's animals, plants and micro-organisms have been named and classified. It has taken biologists the 250 years since Linnaeus to accomplish that much. In October 2001 a 'summit conference' of many of the leaders of organizations devoted to continental and global surveys, together with informatics experts, was held at Harvard University to consider the challenge of finishing the Linnaean enterprise. It was agreed that the technology now exists needed to complete, or nearly complete, the discovery of the remaining 90% in one-tenth the duration of the post-Linnaean era, or 25 years. The technologies that make this possible are three in number: (i) high-resolution digital photography of type and other specimens, yielding perfect depth of field by computer-assembled photo-montages; (ii) genomic maps of adequate resolution for species separation; and (iii) Internet publication.

Having conducted taxonomic monographs off and on for five decades, culminating in a revision of one-fifth of the known New World ant species aided towards the end by the new technology (Wilson 2003), I have been impressed by both the need and the feasibility of such an accelerated global effort. The initial wave should be the Internet and CD-ROM publication of e-types, each consisting of multiple-view high-resolution images of holotypes and primary type specimens. By this means a very large majority of the already described species of plants and animals, and even a substantial number of micro-organisms, probably totalling over one million, can be reliably characterized without the laborious and time-consuming processes of visits to and specimen loans from far flung museums. Already, 10% of that number of species have been or will soon be e-typed: 90 000 plant species in the 'virtual herbarium' of the New York Botanical Garden and 10 000 or more insect species in the Entomology Department of the Museum of Comparative Zoology (including, for example, the largest world collection of ants). With more and more e-types thus available anytime and anywhere, the production of comprehensive monographs of higher taxa such as genera and families will accelerate, and with them local biotic studies abetted by fresh rounds of fieldwork and specimen preparation. The global biodiversity map will emerge like a mosaic by a coalescing of such complementary studies, with biotic surveys growing orthogonally to monographs.

Who will conduct these studies? There are at present, at rough estimate, *ca.* 6000 taxonomists at work worldwide on all organisms combined. That is a tiny slice of the biological community as a whole, and their discipline remains one of the weakest and most underfunded (Wilson 2002). From my own museum experience, I believe that twice this number, with several technical assistants each and the aforementioned new technologies, could move the global biodiversity survey to near completion within a single human generation.

Edward O. Wilson
Harvard University Museum of Comparative Zoology, 26 Oxford Street, Cambridge, MA 02138-2902, USA

REFERENCES

Curtis, T. P., Sloan, W. T. & Scannell, J. W. 2003 Estimating prokaryotic diversity and its limits. *Proc. Natl Acad. Sci. USA* **99**, 10 494–10 499.

Wilson, E. O. 2002 The encyclopedia of life. *Trends Ecol. Evol.* **18**, 77–80.

Wilson, E. O. 2003 Pheidole *in the New World: a dominant, hyperdiverse ant genus*. Cambridge, MA: Harvard University Press.

One contribution of 18 to a Theme Issue 'Taxonomy for the twenty-first century'.

PART III

Conservation and the Human Condition

1974

"The conservation of life," *Harvard Magazine* 76(5) (January): 28–31, 35–37

Except in a few remote sites, the rainforests and other natural environments I visited during fieldwork in the tropics were fragmented and degraded to some degree. By the time I wrote the article reproduced here, this depressing experience and a great deal of newly published evidence had convinced me that drastic action is needed to save what remains of natural ecosystems and the species in them. However, like many other scientists, I held back at first from political engagement. The role of the scientist, I thought, was to provide data and ideas, and let government and conservation organizations such as the World Wildlife Fund solve the problem. It took the courage of Peter Raven, a distinguished scientist and Director of the Missouri Botanical Garden, to inspire me to take a more public role. We were in the midst of a worldwide crisis, he argued, and scientists who understood the situation could no longer afford to stand by as advisers. Peter and I followed in the footsteps of Rachel Carson and other prominent scientist-activists to publicize a problem we felt qualified to define.

The article reproduced here was my first venture into conservation activism. It was also the first explicit application to conservation of the theory of island biogeography. This form of reasoning, while primitive here, was to become one of the stanchions of the discipline of conservation biology.

THE FUTURE OF THE GLOBE: WILDLIFE

The conservation of life

We are a long way from understanding all the economic, health, and aesthetic advantages of species diversity. Like latter-day Noahs, we had better work to insure the variety of earth's creatures.

by Edward O. Wilson

In a world of shrinking faith and uncertain trumpets, very few moral precepts are any longer accepted as absolute. We can nevertheless hope that one of them will be the ethic of organic diversity, which goes like this: Man must conduct himself in such a way that he adds as little as possible to the extinction rate of species on earth. Wherever he can, without seriously threatening his own welfare, he should actively reduce the extinction rate, thereby increasing the number of species that can survive in equilibrium on the globe.

Of course there have to be exceptions to this dictum. If the genus *Plasmodium* disappeared from the face of the earth, and took with it all of the agony it causes human beings and wildlife species, few people would mourn. The genus includes the parasites that cause malaria, and we are not likely to delay its extinction. In general, however, we will do well to recognize that man is the steward of the world's natural resources, the self-appointed but still profoundly ignorant steward; that the living part of the environment is still mostly unknown to him; and that he has therefore scarcely begun to conceive of the possible benefits that the world's organisms will ultimately bring in economic welfare, health, and aesthetic pleasure.

To sense the depth of man's ignorance in these matters, consider that biologists do not even know to the nearest order of magnitude how many species exist. Ten years ago the popularly accepted figure for animals was the British ecologist C. B. Williams's estimate of three million, based on extrapolations of species-abundance curves. Now some authors use the figure ten million, an order-of-magnitude conjecture advanced in the manner of physics. The reason for the upward revision is twofold. First, habitats previously thought to be barren or sparsely populated, such as the deep sea floor, have been found to contain a rich variety of organisms. Whole faunas, such as the marine annelids, abyssal benthos, and many insect taxa, are still in the earliest stages of Linnaean exploration. Second, we have discovered that a great many species exist that are very hard to distinguish, that large complexes of poorly defined sibling species are common even in the better known animal and plant groups.

All this lack of information must be balanced by an equal amount of caution. Our best strategy is a holding operation, by which diversity is preserved through any reasonable means until systematics, ecology, and evolutionary theory work their way up from the Stone Age toward some degree of mastery of the essential subject matter.

As an example of the worst thing that biologists might let slip by them, consider the possibility that the animal and plant life of the Atlantic and Pacific could be mingled by migration through a new Panamanian sea-level canal proposed for construction in the 1980's. The present Panama Canal is based on a series of fresh-water locks, which, by lucky circumstance, have prevented the free migration of organisms from the Atlantic to the Pacific.

Three to five million years ago the emergence of the Panama Isthmus cut the straits that connected the Pacific Ocean and Caribbean Sea, isolating the marine populations on either side. The existing ecological differences between the inshore habitats are substantial. The Atlantic coast has moderate tides, sandy beaches, mangrove swamps, and rich coral reefs. The Pacific side is characterized by strong tides, more silty water, periodic upwellings of cold nutrient-rich water, rocky shores created by extensive lava flows, and limited, depauperate coral reefs. Accelerated no doubt by such differences in the physical environment, evolution has proceeded mostly to the species level and beyond. Of the roughly 20,000 species of marine animals and plants that occur on both sides of the Panama Isthmus, perhaps no more than 10 percent are held in common. In the extreme case of the

Edward O. Wilson (opposite) is professor of zoology and curator in entomology at the Museum of Comparative Zoology at Harvard. His chief interests are biogeography and species diversity, and the social behavior of insects, particularly ants. He is the author, with R. H. MacArthur, of The Theory of Island Biogeography *(1967), the original work on that subject, and of* The Insect Societies *(1971). An enthusiastic teacher, he gives Harvard's elementary biology course, as well as a seminar for graduate students on the feasibility of a general science of sociobiology.*

fishes and mollusks, fewer than 1 percent are held in common. What would happen if free exchange of these faunas were permitted through a sea-level canal? On this point biologists have fallen into total disagreement. The following diverse opinions have been expressed in various articles, seminars, and government hearings during the past eight years:

1. There would be only a limited exchange of species, mostly from the Pacific to the Atlantic. Life in the two oceans would not be seriously disturbed.

2. The Atlantic marine biota—the ecological entity made up of all the region's animal and plant life—is richer in species and hence possesses superior competitive ability. If allowed to invade through a sea-level canal, it would cause widespread extinction in the Pacific biota. The combined extinction rates of the Pacific and Atlantic elements might reach 5,000 species.

3. The argument in 2 is based on the postulate that the greater the number of species, the greater their individual competitive ability. An alternate hypothesis, which cannot be excluded on the basis of existing knowledge, is that the greater fluctuation of the Pacific inshore environment induces the evolution of a higher proportion of opportunistic species, capable of wedging their way into existing biotas, especially within areas disturbed to some extent by man's activities. If this model is correct, and the conjecture in 2 is wrong, the flow of organisms would be predominantly from the Pacific to the Atlantic. In either case, the total impact on the two oceans cannot be predicted.

4. An exchange of biotas would be generally unpredictable and dangerous. Species could extinguish each other by excessive amounts of competition or loss of fitness through uncontrolled hybridization.

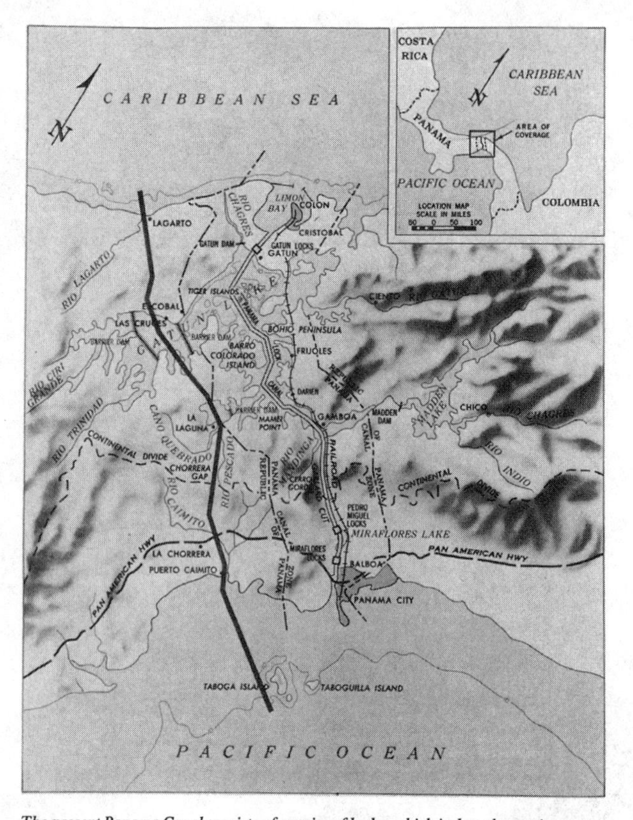

The present Panama Canal consists of a series of locks, which isolate the marine populations of the Pacific Ocean and the Caribbean Sea. A new Panamanian sea-level canal, proposed for construction in the 1980's, by allowing the migration of animals and plants between the two oceans, might cause widespread species extinction. The new canal's route, recommended by the Atlantic-Pacific Interoceanic Canal Study Commission in its special report to the President, is shown here as the heavy black line running ten miles to the west of the existing canal.

In fact, biogeographers—scientists engaged in the biological study of the geographical distribution of plants and animals—have neither the theory nor the previous experience to predict the outcome of an unimpeded exchange of faunas across a sea-level Panamanian canal. This incapacity has become increasingly clear to concerned scientists who have tried to evaluate the evidence dispassionately. Nevertheless, a strongly cautious approach seems mandatory. It is necessitated not just by the very real possibility of widespread species extinction. The introduction of only one wrong species, such as the yellow-bellied sea snake from the eastern Pacific into the

Atlantic Ocean, could inflict enough economic or ecological damage to justify the attempt to prevent any migration at all.

Previous experience with the careless mixing of aquatic biotas, for example in the Great Lakes via the Erie and Welland Canals, indicates that to permit the mixing of the rich tropical Pacific and Atlantic biotas would be playing ecological roulette with all cylinders loaded. Moreover, a unique biogeographic experiment of global proportions would thereby have been performed, without adequate preparation. The natural setting for the experiment took millions of years to develop and cannot be repeated. Biology should be fully prepared before allowing it to proceed even piecemeal. For these reasons two groups of biologists—a University of Miami team supported by the Battelle Memorial Institute, and the Committee of Ecological Research for the Interoceanic Canal (CERIC) of the National Academy of Sciences—have independently recommended that some kind of biological barrier be constructed across the canal before it is opened. The barrier can take any one or a combination of several forms: bubble curtains, ultrasonic screens, intrusions of heated or fresh water, and others. The details will be a straightforward exercise in engineering, infinitely simpler than the one biologists and the rest of humanity would face if the mixing is allowed to proceed.

Biogeographers cannot predict the outcome of mixing the Pacific and Atlantic biotas, except to say that it would be dangerous, because such a prediction requires a solution to one of the most complex problems they can ever conceivably face. Similarly, molecular biologists cannot say how the tissues of man and other higher organisms develop, and behavioral biologists are unable to explain conscious thought, because these problems are also the Mount Everests of their respective disciplines. Like the rest of biology, however, biogeography is far from helpless when dealing with smaller, better circumscribed units.

A quantitative theory called island biogeography can be very helpful to us in our efforts to encourage organic diversity in the world. The most straightforward application of the theory is in the design of natural preserves. Natural habitats have always been fragmented into island-like enclaves. With certain exceptions, such as the arctic tundra, man has intensified this process, reducing the fragments in size and increasing their degree of isolation. The number of species belonging to a single group, such as birds, ants, or flowering plants, that will exist in equilibrium on a given island is a function of the area and the degree of isolation of the island. When the distance to the principal source area is held constant, whether that area is a continent, a set of islands, or just a similar habitat nearby, the number of species (S) increases approximately as a simple power function of the area, as follows: $\log S = a + z\log A$, where A is the area and a and z are fitted constants. When the independent parameter of isolation is increased, a rises at a rate characteristic of each taxon and the part of the world in which the relation is observed. In most cases z falls somewhere between 0.2 and 0.4. A very rough rule of thumb is that a tenfold increase in area results in a doubling of the number of species at equilibrium.

When a nature preserve is set aside, it is destined to become an island in a sea of habitats modified by man. The species number will shift from its original equilibrium due to the area and distance effects just cited. As years pass the diversity will decline, eventually reaching a new, lower steady state. An estimate of the loss can be made by comparing the reserve with the area-species curves of older systems, providing appropriate systems exist under comparable conditions of isolation. Jared Diamond of U.C.L.A. has developed an elegant technique to estimate the relaxation rate and secondary equilibrium values in the case of island birds. He made use of land-bridge islands that were disconnected from New Guinea at known times in the recent geologic past. His results have been confirmed and extended in parallel studies in the West Indies and Central America. Researchers have discovered that significant drops in the number of species in newly disconnected islands take place over a period of decades in the smallest islands, which are comparable in area to small natural reserves on continents, and during centuries for islands comparable in size to our largest national parks. Barro Colorado Island in Panama provides an alarming example of the high potential decrement rate on small islands. B.C.I. is actually a forested hilltop that was surrounded by water fifty to sixty years ago when Lake Gatun was formed in connection with the construction of the Panama Canal. It has been a nature preserve almost since the time of its isolation. Inserting the area of the island (6.5 mi.²) and its known period of

(*continued on page 35*)

Previous experience with the careless mixing of aquatic biotas, for example in the Great Lakes via the Erie and Welland Canals, indicates that to permit the mixing of the rich tropical Pacific and Atlantic biotas would be playing ecological roulette with all cylinders loaded.

THE FUTURE OF THE GLOBE: WILDLIFE·

(*continued from page 31*)

isolation into an extinction model based on the West Indian studies. John Terborgh of Princeton University estimated that the number of resident bird species should have declined from the original 205 observed to a current fauna of 188 species, a loss of sixteen or seventeen species. This is in close agreement with the decline actually observed.

The new information from island biogeography shows that planners and managers of national parks and other natural preserves will be prudent to take the natural extinction rate into account (in addition to the man-induced extinction rate) and to choose appropriate measures to minimize it. The following basic procedures should be included:

1. Individual preserves must be made as large as possible. Since the areas of preserves will always be fixed by political compromise, estimates should be made of the extinction rates, as a function of time and area, of the most vulnerable taxa such as the birds and mammals. Then the minimal areas demanded should be the ones at which the initial and consequently highest extinction rates will be reasonably low. The projected rates should be such that only large increments of reserved land will lower them significantly further. In other words, land acquisition must reach the point of diminishing return with respect to the most extinction-prone groups.

2. Unique habitats and biotas are best contained in multiple preserves,and these isolates should be located as closely together as possible. The reason is that extinction has a strong random component. Species seldom become extinct in every part of their range simultaneously. They tend to persist because ecologically suitable localities that lose them can be recolonized from other localities that are still occupied. Reciprocal intercolonization of preserves can proceed indefinitely through time and, if aided by deliberate transplantations, might extend the life of species well beyond what it would be under natural circumstances.

3. Because biogeographers have discovered that peninsulas have fewer species than central portions of continents, preserves of a fixed area should be as round in shape and continuous as possible. (This principle and those embodied in the first two recommendations are illustrated in the diagram at right.)

4. Extinction models should not be

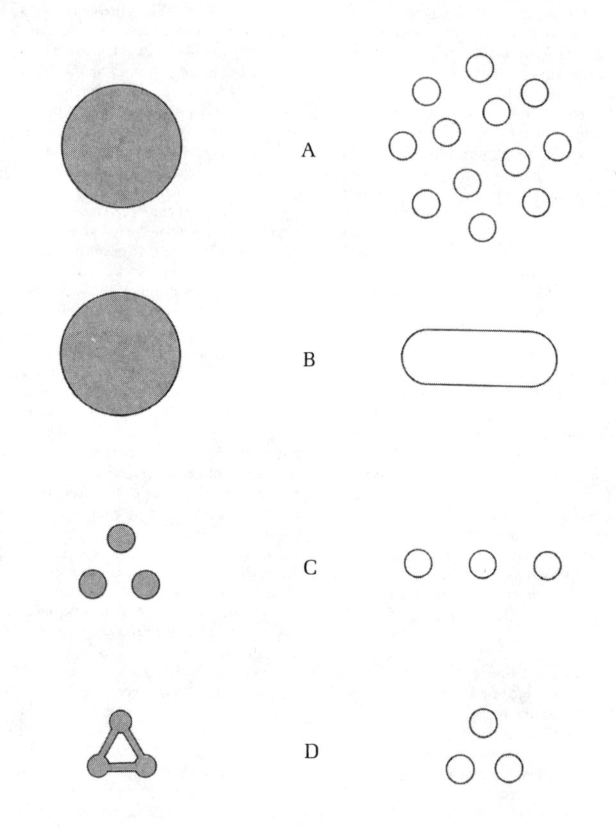

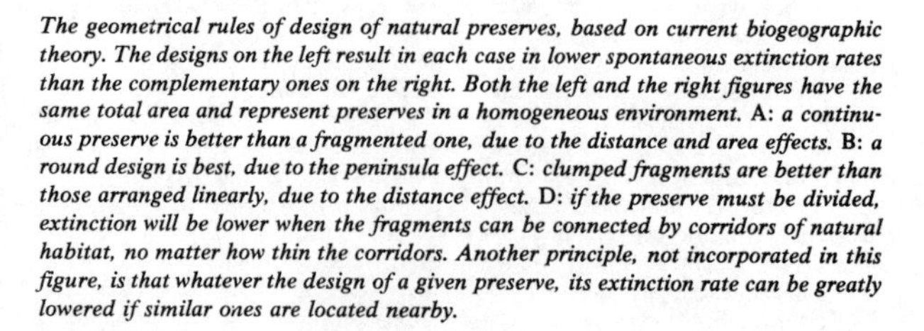
The geometrical rules of design of natural preserves, based on current biogeographic theory. The designs on the left result in each case in lower spontaneous extinction rates than the complementary ones on the right. Both the left and the right figures have the same total area and represent preserves in a homogeneous environment. A: *a continuous preserve is better than a fragmented one, due to the distance and area effects.* B: *a round design is best, due to the peninsula effect.* C: *clumped fragments are better than those arranged linearly, due to the distance effect.* D: *if the preserve must be divided, extinction will be lower when the fragments can be connected by corridors of natural habitat, no matter how thin the corridors. Another principle, not incorporated in this figure, is that whatever the design of a given preserve, its extinction rate can be greatly lowered if similar ones are located nearby.*

restricted to the most conspicuous or vulnerable organisms but should eventually be developed for all taxa. Those displaying the highest degrees of endemicity and vulnerability (the two phenomena are generally correlated) deserve first attention. No group, not even the humblest and most obscure among invertebrates and microorganisms, should be ignored.

It is within the power of science not merely to hold down the rate of species extinction but to reverse it. Among the principal aspects of the ecology of communities now under intensive study is what's called the "species packing problem." In essence, the problem is the identification of those traits that allow certain sets of species, but not others, to be fitted together in the same ecosystem without markedly increasing the species extinction rate. In other words, how tightly can species be packed? During colonization by undisturbed biotas, congenial sets of species are gradually assembled by chance alone, raising the steady-state species number to what has been called the assortative equilibrium. Theoretically, assortative equilibria can be planned that exceed any occurring in nature. Species might even be drawn from different parts of the world—not willy-nilly, as in the careless importations of the past, but after careful ecological analysis has identified them as candidates for insertion into new faunas. Some of the first and most important introductions would surely be "orphan species," those on the brink of extinction in their native range but capable of being fitted into certain alien communities elsewhere. I do not suggest that the state of the art is advanced enough for us to proceed with planned biotic mixing, only that species packing is one of the techniques of applied biogeography that seems likely to become practicable within the next several decades on the basis of current and projected research.

We can create wholly new biological communities where little existed before. An artificial reef in California's Santa Monica Bay, although composed of the humblest materials dumped onto the muddy bottom, greatly increased the diversity of marine life in the area.

Optimism is further justified by the favorable outcome of a few biotic mixtures that have already occurred haphazardly, indicating a degree of flexibility on the part of species that will provide biogeographers with some margin for error. The Kaingaroa Forest of New Zealand, for example, contains 250,000 acres of exotic conifers, including *Pinus radiata*, *P. ponderosa*, *P. contorta*, and *Pseudotsuga taxifolia* from North America, and *Pinus nigra* from southern Europe. Introduced native birds mingle with endemic New Zealand species in this synthetic environment. Ecological differentiation is well marked; no two species have the same feeding habit, and the insectivorous birds exploit all of the major feeding niches except that of woodpeckers. The really surprising fact, however, is that some of the native species are now as abundant in the Kaingaroa Forest as in almost any native forest, and some are more abundant than in most of the remainder of their range. Furthermore, the invertebrate fauna of the forest consists mostly of native species.

Two circumstances are special in the case of the birds. First, the number of species is still small, largely because the New Zealand fauna was poorly developed to start with, and the mixed community has probably not yet met many of the difficulties in packing that would be routine in large continental faunas. Second, forest birds are differentiated to a large degree by foliage height and profile rather than by the species of trees in which they live. Certain kinds of insects that feed on only one or a few kinds of plants, particularly those specializing on hardwoods, would in most instances find it impossible to penetrate the Kaingaroa conifers. Yet the lesson is clear: what works in part by accident can be brought closer to perfection through design.

Ultimately, design might also include the artificial selection of strains, or even the creation of new species, for the purposes of biotic enrichment. If theory and experiment indicate that an orphaned species cannot be fitted into any existing communities, strains might be selected

A reading list on biological diversity and its preservation

The following books and articles are suggested by Edward Wilson for a more extended introduction to the basic principles of ecology, especially those that relate to conservation and the preservation of diversity.

Books

Edward O. Wilson *et al.*, *Life on Earth*, Sinauer Associates, Stamford, Conn., 1973. This elementary biology textbook contains most of the fundamental ideas of modern ecology and biogeography and can serve as a stepping stone to all but the most technical reports on these subjects.

Robert H. MacArthur, *Geographical Ecology: Patterns in the Distribution of Species*, Harper and Row, New York, 1972. A somewhat advanced but definitive textbook that explains the current status of the theory of species diversity.

David W. Ehrenfeld, *Biological Conservation*, Holt, Rinehart, and Winston, New York, 1970. A provocative and clearly written introduction to the practical aspects of the subject.

Articles

W. I. Aron and S. H. Smith, "Ship canals and aquatic ecosystems," Science, Volume 174, pages 13-20, 1971. A balanced review of the perils of mixing animals and plants from different oceans.

Jared M. Diamond, "Distributional ecology of New Guinea birds," Science, Volume 179, pages 759-769, 1973. Describes the method by which the natural extinction rate on islands and other isolated preserves can be estimated.

within captive populations of that species and eventually inserted into one or more existing communities. I do not seriously suggest that such a procedure will be followed in the foreseeable future for any but a very few of the organisms most valued by man. Furthermore, the genetic molding of communities is a technology that cannot be seriously contemplated until the inchoate discipline of population ecology has moved closer to a full solution of the species packing problem.

Many of the earth's major habitats are biological deserts: the open sea, the ice caps, some of the trace-element barrens, and the real deserts, the extremes of which are virtually lifeless. Quite by coincidence, technology is at this moment driving toward two major goals that could transform these areas: an unlimited or at least vastly greater source of energy, and, as one of the principal benefits of the first, the cheap desalinization of sea water. With the achievement of these goals, men will move increasingly onto the land deserts, carrying communities of organisms with them. One may hope that we will not be satisfied with limiting ourselves to a baggage of domestic animals, houseplants, pests, and commensals. It lies easily within our power to create wholly new parks and reserves where nothing existed before in historical times. But what will go into these *de novo* communities? Thought about this subject sharpens one's vision of the future of applied biogeography.

In fact, the deliberate creation of new biological communities has already begun. Large areas of desert-like barrens in Australia have been transformed into agricultural land by the simple addition of zinc, copper, and molybdenum, "trace" elements required for life that were previously present in abnormally low quantities. Marine biologists have discovered that artificial reefs, with rich complements of reef organisms, can be created just by dumping concrete rubble, abandoned automobiles, used automobile tires, and similar inert refuse onto the mud or sand floors of shallow marine waters. Successful experiments of this nature have been conducted off the shores of Florida and California.

What these efforts engender are in effect habitat islands, the biotas of which grow and equilibrate according to the same laws of biogeography governing wholly natural islands. The communities are not likely to be as diverse as those that have evolved for millions of years in the natural islands, yet the process of enrichment can be speeded by the deliberate importation of compatible species to reach new and higher assortative equilibria, to create intricate, fascinating new communities. This is another aspect of biogeographic technology that ongoing basic research might render practicable during the next few decades.

The greatest misfortune that awaits the human intellect is to be no longer faced with something commensurate with its capacity for wonder. If the golden age of science really ends, and research shrinks to a few remote and arcane frontiers accessible only to specialists, the wonder will indeed be gone. By that time even the pre-scientific myths that sustained our ancestors, and intrigue us still, would have largely evaporated—having been accounted for in full, perhaps by the right kind of neurophysiological analysis of the limbic system and hypothalamus.

But this exhaustion of the wonderful will not occur during the lifetime of anyone now living. The ultimate complexity, offering an unexplored terrain of virtually infinite extent, lies in biology. Even after the cell has been torn down and put together again, and the labyrinthine mysteries of tissue development followed to their ends, there lie ahead the much more extensive challenges of ecology and biogeography. The full exploration of organic diversity is a prospect that suits the biocentric human brain, especially those emotive centers that evolved to make us superior hunters and agriculturists. The same instincts that motivate the bird watcher, the butterfly collector, and the backyard gardener can indefinitely sustain the scientifically curious segment of a more sophisticated human population in the pursuits of ecology and biogeography.

The most interesting part of the universe is right here. We have a biotic planet, and the chances of finding another one within scores of light years are remote indeed. A biotic planet, with millions of species on it, is an infinitely more interesting puzzle than any number of lifeless planets. And the possibilities of ecosystems manipulation confront us with mysteries that only generations more of study and creative work can solve. □

1975

"Applied biogeography," *in* Martin L. Cody and Jared M. Diamond, eds., *Ecology and Evolution of Communities* (Belknap Press of Harvard University Press), pp. 522–534 (with Edwin O. Willis, second author)

After proposing the use of principles of biogeography in conservation planning (see "The conservation of life," the preceding article), I found that Edwin O. Willis had also been thinking along the same lines. So the two EOWs collaborated in the more comprehensive, albeit still very elementary, treatment that is reproduced here.

18 Applied Biogeography

Edward O. Wilson and Edwin O. Willis

The Diversity Ethic

In a world of shrinking faith and uncertain trumpets, very few precepts are any longer accepted as absolute. We can nevertheless hope that one of them will be the ethic of organic diversity—that for an indefinite period of time man must add as little as possible to the rate of worldwide species extinction and where possible he should lower it. This precept, which is based wholly on rational considerations, can also be the guiding principle of applied biogeography. It emerges from a recognition that man is the self-appointed but still profoundly ignorant steward of the world's natural resources, that the living part of the environment is still mostly unknown to him, and that he has therefore scarcely begun to conceive of the possible benefits that other organisms will bring in economic welfare, health, and esthetic pleasure. To sense the depth of that ignorance, consider that biologists do not even know to the nearest order of magnitude how many species exist. Ten years ago the popularly accepted figure for animals was C. B. Williams' estimate of three million, based on extrapolations of species abundance curves. Now some authors use the figure ten million, an order-of-magnitude conjecture advanced in the manner of physics. The reason for the upward revision is twofold: the discovery that whole faunas, such as the marine annelids, abyssal benthos, and many insect taxa, are still in the earliest stages of Linnaean exploration; and the growing realization that large complexes of poorly defined sibling species are common even in the better-known animal and plant groups.

All this lack of information must be balanced by an equal amount of caution. Our best strategy is a holding operation, by which diversity is preserved through any reasonable means until systematics, ecology, and evolutionary theory work their way up from the stone age toward some degree of mastery of the essential subject matter. As an example of the worst thing that biologists might let slip by them, consider the possibility that the Atlantic and Pacific biotas could be mingled by migration through the new Panamanian sea-level canal proposed for construction in the 1980s. Three to five million years ago the emergence of the Panama Isthmus cut the straits that connected the Pacific Ocean and the Caribbean Sea, isolating the marine populations on either side. The existing ecological differences between the inshore habitats are substantial. The Atlantic coast has moderate tides, sandy beaches, mangrove swamps, and rich coral reefs. The Pacific side is characterized by strong tides, more silty water, periodic upwellings of cold nutrient-rich water, rocky shores created by extensive lava flows, and limited, depauperate coral reefs. Accelerated no doubt by such differences in the physical environment, evolution has proceeded mostly to the species level and beyond. Of the roughly 20,000 species of marine animals and plants that occur on both sides of the Panama Isthmus, perhaps no more than ten percent are held in common (Newman, 1972). In the extreme case of the fishes and mollusks, fewer than one percent are held in common. What would happen if free exchange of these faunas were permitted through a sea-level canal? On this point biologists have fallen into total disagreement. The following diversity of opinions has been expressed in various articles, seminars, and government hearings during the past ten years:

1. There would be only limited exchange of species, mostly from the Pacific to the Atlantic. The ecosystems would not be seriously disturbed (Topp, 1969; Voss, 1972).

2. The Atlantic marine biota is richer in species and hence possesses superior competitive ability. If allowed to invade through a sea-level canal, it would cause widespread extinction in the Pacific biota. The combined extinction rates of the Pacific and Atlantic elements might reach 5000 species (Briggs, 1969).

3. The Briggs argument (just cited) is based on the postulate that the greater the number of species, the greater their individual competitive ability. An alternative hypothesis that cannot be excluded with existing knowledge is that the greater fluctuation of the Pacific inshore environment induces the evolution of a higher proportion of opportunistic species, capable of wedging their way into existing biotas, especially within areas disturbed to some extent by man's activities. If this model is correct, and Briggs' conjecture wrong, the biotic flow would be predominantly from the Pacific to the Atlantic. In either case, the total impact on the two ecosystems cannot be predicted.

4. An exchange of biotas would be generally unpredictable and dangerous. Species could be removed not only by competitive replacement but also by overwhelming degrees of hybridization with imperfectly isolated geminate forms on the other side of the Isthmus (Rubinoff, 1965).

In fact, biogeography has neither the theory nor the previous experience to predict the outcome of an unimpeded exchange of faunas across the sea-level canal. This incapacity has become increasingly clear to concerned scientists who have tried to evaluate the evidence dispassionately, including Aron and Smith (1971). Therefore, a strongly cautious approach seems mandatory. It is necessitated not just by the very real possibility of widespread species extinction. The introduction of only one wrong species, such as the yellow-bellied sea snake from the eastern Pacific into the Atlantic Ocean (see Graham, Rubinoff, and Hecht, 1971), could inflict enough direct economic or ecological damage to justify the attempt to prevent any migration at all. Furthermore, changes in just a few species in a tightly integrated community could have widespread indirect effects by destabilizing the community (Levins, Chapter 1). Previous experience with the careless mixing of aquatic biotas, for example in the Great Lakes via the Erie and Welland canals, indicates that to permit the mixing of the rich tropical Pacific and Atlantic biotas would be playing ecological roulette with all cylinders loaded. Moreover, a unique biogeographic experiment of global proportions would thereby have been performed, without adequate preparation and in the wrong century. The natural setting for the experiment took millions of years to develop and cannot be repeated. Biology should be fully prepared before allowing it to proceed even piecemeal. For these reasons two groups of biologists, the University of Miami team supported by the Battelle Memorial Institute and the Committee of Ecological Research for the Interoceanic Canal (CERIC) of the National Academy of Sciences, have independently recommended that some kind of biological barrier be constructed across the canal before it is opened (Voss, 1972; Newman, 1972). The barrier can take any one or a combination of several forms: bubble curtains, ultrasonic screens, intrusions of heated or fresh water, and others. The details will be a straightforward exercise in engineering, infinitely simpler than the one biologists and the rest of humanity would face if the mixing is allowed to proceed.

The Design of Nature Preserves

Biogeographers cannot predict the outcome of mixing the Pacific and Atlantic biotas, except to say that it is dangerous, for the reason that this is one of the most complex problems they can ever conceivably face. Similarly, molecular biologists do not understand how metazoan tissues develop, and behavioral biologists cannot explain conscious thought, because these problems are also the Mount Everests of their respective disciplines. Like the rest of biology, however, biogeography is far from helpless when dealing with smaller, better-circumscribed units. The quantitative theory of island biogeography in which Robert MacArthur was so involved can be brought to bear on several kinds of problems of diversity maintenance. Preston (1962), Willis (1971), Wallace (1972), Diamond (1975), and Terborgh (1974a) have pointed out that the most straightforward application is in the design of natural preserves. Natural habitats have always been fragmented into island-like enclaves. With certain exceptions, such as the forests of New England, man has intensified this process, reducing the fragments in size and increasing their degree of isolation. The number of species belonging to a single taxon such as birds, ants, or flowering plants, equilibrates on a given island at a level that is a function of the area and the degree of isolation of the island (MacArthur and Wilson, 1967). Similar effects are seen on "habitat islands" within continents (Cody, Chapter 10). When the distance to the principal source area is held constant, whether that area is a continent, a set of islands, or just a similar habitat nearby, the number of species S increases approximately as a simple power function of the area, as follows: $\log S = a + z \log A$, where A is the area and *a* and *z* are fitted constants. When the independent parameter of isolation is increased, *z* rises at a rate characteristic of each taxon and the part of the world in which the relation is observed. In most cases *z* falls somewhere between 0.2 and 0.4 (cf. May, Chapter 4, Table 5; Diamond, Chapter 14, Figures 2 and 3). A very rough rule of thumb is that a tenfold increase in area results in a doubling of the number of species at equilibrium.

When a nature preserve is set aside, it is destined to become an island in a sea of habitats modified by man. The species number will shift from its original equilibrium because of the area and distance effects just cited. As years pass the diversity will decline, eventually reaching a new, lower steady state. An estimate of the loss can be made by comparing the reserve with the area-species curve of older systems, providing that appropriate systems exist under comparable conditions of isolation. Diamond (1972, 1973, and Chapter 14) has developed an elegant technique to estimate the relaxation rate and secondary equilibrium values in the case of island birds. He made use of land-bridge islands that were disconnected from New Guinea at known times in the recent geologic past. His results have been confirmed and extended by Terborgh (1974b) in parallel studies in the West Indies and Central America. Both Diamond and Terborgh discovered that significant drops in the number of species in newly disconnected islands take place over a period of decades in the smallest islands, which are comparable in area to small natural reserves on continents, and during centuries in islands comparable in size to our largest national parks.

Barro Colorado Island in Panama provides both a test of the theory and an alarming example of the high potential decrement rate on small islands. Barro Colorado actually consists of a hilltop of 15.7 km^2 of lowland tropical forest, which was isolated from surrounding forests about 1914 when Gatún Lake rose around it as part of the formation of the central part of the Panama Canal. Since 1923 the island has been a protected biological reserve, and its forests have been growing to maturity. Inserting the area of the island and its known period of isolation into an extinction model based on the West Indian studies, Terborgh (1974a) estimated that the number of resident bird species should have declined by nearly 10 percent. This is in close agreement with the decline actually observed. Let us examine the history of extinctions in some detail.

The birds of Barro Colorado, fortunately, have been well studied over the years. Chapman (1938) and others worked there during the 1920s and 1930s, Eisenmann (1952) and others visited from 1947 to date, and Willis studied there two to eleven months per year from 1960 to 1970. Of 208 species of birds breeding on the island in Chapman's time, 45 had disappeared by 1970. Several other species were down to one or a few individuals. A grebe and a gallinule have colonized the lake, and three species of the forest edge (two tanagers and a wren) are currently attempting colonization. Other species of the forest edge have attempted colonization but failed. No forest species has reached the island, although points on its edge are only about half a kilometer from sites on the mainland where the species occur.

For two reasons the record of extirpations must be interpreted with care. Early workers probably missed several species, including several tiny flycatchers and an elusive forest dove now present. Several sight records by early workers have to be doubted as possible vagrants or misidentifications. If some of these forest species, notably difficult to detect, were actually breeding earlier, there may have been a higher original avifauna and more extirpations than we think. On the other hand, if other species such as the tiny flycatchers have been successful colonists, the original avifauna may have been lower than we think. We regard a higher figure for the original avifauna to be the more likely, but neither possibility materially affects conclusions of this chapter.

The second reason for careful interpretation presents more of a problem. Barro Colorado was not just a tract of mature forest that became an island of mature forest. Rather, it was a mixture of mature forest and patches of second-growth forest, the latter now growing to maturity. Some 32 of the lost species of birds, or perhaps a few less, are birds of second-growth or forest edge (Willis, 1974). These species lost their habitats as the forest grew. They are "weed" species, abundant on the mainland and easily capable of colonizing patches of secondary forest there. Such opportunistic forms are in little danger of extinction even where cutting of the forests has reduced forests to fragments. Thirteen lost species (Table 1), or perhaps a few more, regularly occur in extensive tracts of tall forest elsewhere, sometimes at lower density than in less mature forests. They can be expected to disappear as the forests are cut.

The particular, idiosyncratic causes of extinction are nearly impossible to pinpoint and are probably varied (Willis, 1974). Several of the lost species nest or feed on the ground. Perhaps leaf litter is reduced as monolayer trees take over the forest (see Horn, 1971), and very likely the sparser ground cover in mature forests provides relatively poor protection from predators. However, ground-living birds are also the ones that are least able to emigrate to the island. High densities of certain mammals, in part due to losses of large predatory mammals, could lead to destructive levels of predation on the nests and of some birds. Finally, because the range of habitats is limited, refugia do not exist during exceptional wet or dry years (cf. discussion of "hot spots" by Diamond, Chapter 14).

Two lost species belong to a group studied most intensively, the birds that follow army ants to feed on the arthropods flushed by these insect predators. Of the original seven species in this guild, the largest (*Neomorphus geoffroyi,* a ground cuckoo) was gone before Willis arrived in 1960. The second largest (*Dendrocolaptes certhia,* a woodcreeper) was down to two pairs in 1960 and disappeared by 1970. The third largest (*Phaenostictus mcleannani,* an antbird) decreased from 15 pairs to one female and a few males by 1970. The fifth species (*Gymnopithys bi-

Table 1. Forest birds extirpated from Barro Colorado Island

Species	Large for guild	Ground nester	Ground forager	Low density in tall forests	Immigration
Harpy eagle (*Harpia harpyja*)	++			*a*	*f*
Barred forest-falcon (*Micrastur ruficollis*)*	+			*b*	
Red-throated caracara (*Daptrius americanus*)*	+			*c*	*f*
Great curassow (*Crax rubra*)	++		+	*a*	*e*
Marbled wood-quail (*Odontophorus gujanensis*)	+	+	+	*b*	*e*
Rufous-vented ground-cuckoo (*Neomorphus geoffroyi*)	++		+	*a*	*e*
Barred woodcreeper (*Dendrocolaptes certhia*)*	+		+	*b*	*e*
Buff-throated Automolus (*Automolus ochrolaemus*)	+	+		*b*	*e*
Black-faced antthrush (*Formicarius analis*)	+		+	*b*	*e*
Sulphur-rumped flycatcher (*Myiobius sulphureipygeus*)				*d*	*e*
White-breasted wood-wren (*Henicorhina leucosticta*)		+	+	*b*	*e*
Nightingale wren (*Microcerculus marginatus*)			+	*b*	*e*
Song wren (*Leucolepis phaeocephalus*)*			+	*b*	*e*

Species marked with an asterisk disappeared during the 1964–1973 decade. Probably other species disappeared during 1971–1973. *a*, low density because large for the ecological guild. *b*, higher densities reached in less mature or dry forests. *c*, wasp-eating, wanders widely. *d*, nests over streams, which are uncommon on Barro Colorado. Several other flycatchers that nest over or near streams in second-growth have also been extirpated. *e*, immigration to Barro Colorado from the mainland unlikely. *f*, birds not now present on mainland, but could immigrate.

color, an antbird) decreased from some 50 to 20 pairs. Only the fourth, sixth, and seventh species, medium to small birds that often forage apart from the ants as well as near them, maintained substantial populations. Even over a ten-year period, reasons for declines were not evident; it can only be said that high predation on nests kept replacements below fairly high losses of adults despite frequent renestings.

Several other losses fit a pattern of early loss of large or specialized species, a pattern to be discussed later as "ecological truncation." The harpy eagle was the largest local raptor of Barro Colorado, the great curassow the largest local frugivore, the barred forest-falcon the largest insect-eater, the caracara the only wasp-eater, and the black-faced antthrush the largest litter insectivore. Some small and generalized birds are doing better than in similar mainland forests nearby, an example of the usual island pattern of "density compensation" (MacArthur, Diamond, and Karr, 1972) by which a large number of individuals of a few small species replace missing species (cf. Cody, Chapter 10; Brown, Chapter 13). The most abundant species on the island has been studied closely (Oniki, in preparation) and has high nest and adult losses but a wide range of foraging behaviors. Maturation of the forest has led to success for several large fruit-eaters and arthropod-eaters; average bird weights are higher in the half of the island covered by tall forest than in the remaining medium forest. Since losses of large species and density compensation go against this trend, small size of the island rather than forest growth is likely to be the main reason for losses. However, losses of small wrens may have followed from reduction of the isolated populations due to maturation of the forest.

Loss rates on Barro Colorado have remained steady or increased slightly, at about ten species per decade. Of these, about three have been forest inhabitants. The latter estimate is close to the one that Diamond (1971) made for losses on Karkar Island off New Guinea and Santa Cruz Island off California. Coiba, an island connected to Panama by a Pleistocene land bridge about 10,000 years ago, is much larger than Barro Colorado but has retained many fewer of the mainland species near it than has Barro Colorado (Wetmore, 1957). One can therefore predict that many more species will ultimately be lost from Barro Colorado.

All of this new information from island biogeography shows that planners and managers of national parks and other natural preserves will be prudent to take the spontaneous extinction rate into account and to choose appropriate measures to minimize it. The following basic procedures should be included:

1. Individual preserves must be made as large as possible. Since the areas of preserves will always be fixed by political compromise, estimates should be made of the extinction rates, as a function of time and area, of the most vulnerable taxa such as the birds and mammals. Then the minimal areas demanded should be the ones at which the initial and consequently highest extinction rates will be reasonably low. The projected rates should be such that only large increments of reserved land will lower them significantly further. In other words, land acquisition must reach the point of diminishing returns with respect to the most extinction-prone groups.

2. Unique habitats and biotas are best contained in multiple preserves, and these isolates should be located as closely together as possible. The reason is that extinction has a strong random component. Species seldom become extinct in every part of their range simultaneously. They tend to persist because ecologically suitable localities that lose them can be recolonized from other localities that are still occupied. Reciprocal intercolonization of preserves can proceed indefinitely through time and, if aided by deliberate transplantations, might extend the life of species well beyond what it would be under natural circumstances.

3. Because of the peninsula effect discovered by biogeographers, preserves of a fixed area should be as round in shape and as continuous as possible. (This principle and those embodied in the first two recommendations are illustrated in Figure 1.)

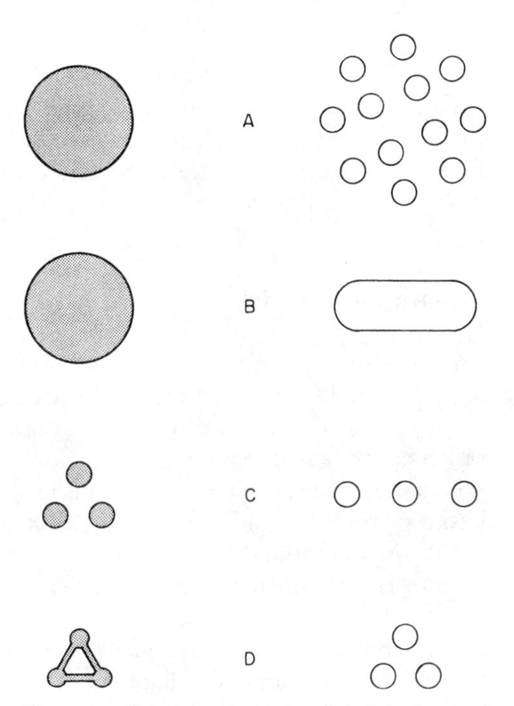

Figure 1 The geometrical rules of design of natural preserves, based on current biogeographic theory. The design on the left results in each case in a lower spontaneous extinction rate than the complementary one on the right. Both the left and the right figures have the same total area and represent preserves in a homogeneous environment. A: a continuous preserve is better than a fragmented one, because of the distance and area effects. B: a round design is best, because of the peninsula effect (cf. MacArthur and Wilson, 1967, pp. 115–116 and Figure 37). C: clumped fragments are better than those arranged linearly, because of the distance effect. D: if the preserve must be divided, extinction will be lower when the fragments can be connected by corridors of natural habitat, no matter how thin the corridors (Willis, 1974). Another principle, not incorporated in this figure, is that whatever the design of a given preserve, its extinction rate can be greatly lowered if similar preserves are located nearby.

4. Extinction models of the kind invented by Diamond and Terborgh should not be restricted to the most conspicuous or vulnerable organisms but should eventually be developed for all taxa. Those displaying the highest degrees of endemicity and vulnerability (the two phenomena are generally correlated) deserve first attention. No group, not even the humblest and most obscure among invertebrates and microorganisms, should be ignored. Thus, in addition to the full taxonomic surveys that are just getting underway in natural preserves, biologists should begin studies of species dynamics and species equilibria.

The Special Problem of Ecological Truncation

Truncation of ecological guilds, the well-known but seldom-emphasized early loss of specialists and of large species, is probably mainly due to the fact that most such species occur at very low densities, require large areas for sustenance, or both (see discussion of so-called incidence functions, the dependence of an island's species composition on its species number or area, in Chapter 14 by Diamond). "Density compensation" for truncation may lead to increased numbers of individuals of generalized small species (MacArthur, Diamond, and Karr, 1972). However, species diversities and biomasses will still be lower.

The ant-following birds of Barro Colorado show ecological truncation clearly, as shown earlier. Diamond (Chapter 14) found a large eagle (*Harpyopsis novaguineae*) and a distinctive kingfisher (*Clytoceyx rex*) gone from all the formerly connected islands off New Guinea. An even better example of truncation is presented by the disastrous extinctions of Hawaiian birds during the past century. These losses were evidently due at least in part to human cultivation of the lowlands and to the introduction of disease-vector mosquitoes (Warner, 1968). The first species to go were such large or specialized species as the big Kioea (*Chaetoptila angustipluma*) and both long-beaked Mamos (*Drepanis* spp.). Many of the species now approaching the end are characterized by intermediate size, distinctive adaptations, or both—the half-beaked *Hemignathus*, for example, the parrotbill *Pseudonestor*, and others. Few members of the endemic Hawaiian family Drepanididae other than small warbler-like species such as the Amakihi (*Loxops virens*) are doing well.

Truncation creates a particular difficulty for the planning of natural reserves, because multiple refuges of a given habitat tend to lose the same specialized species. Diamond (1972) found that the largest (3000 square miles) of the formerly connected islands off New Guinea has retained only 45 out of the 134 land-bridge bird species. Even if parks in the New Guinea of the future should be huge like these offshore islands, many species would still be lost.

The island effect and truncation, taken to their extremes, result in the domination of small parks by rats, cockroaches, sparrows, and similar invaders from nearby human areas. Practically no birds other than pigeons and starlings winter in wooded Tappan Square in Oberlin, Ohio (Margaret F. Smith, personal communication). A thousand such Squares would be far less valuable for the maintenance of biotas than one refuge of a thousand Square units.

Planned Biotic Enrichment

It is within the power of science not merely to hold down the rate of species extinction but to reverse it. Among the principal topics of community ecology now under intensive study is the species-packing problem (MacArthur and Wilson, 1967; MacArthur and Levins, 1967; MacArthur, 1972; Schoener, 1970; May, 1973). One of the more sophisticated recent developments is the specification of "assembly rules" by Diamond (Chapter 14). A central goal of this research is the identification of those traits that allow certain sets of species, but not others, to be fitted together in the same ecosystem without markedly increasing the species extinction rate. During colonization by undisturbed biotas, such congenial sets are gradually assembled by chance alone, raising the steady-state species number to what has been called the assortative equilibrium (Wilson, 1969). Theoretically, assortative equilibria can be planned that exceed any occurring in nature. Species might even be drawn from different parts of the world—not willy-nilly, as in the careless importations of the past, but after careful ecological analysis has identified them as candidates for insertion into new faunas. Some of the first and most important introductions would surely be "orphan species," those on the brink of extinction in their native range but capable of being fitted into certain alien communities elsewhere. We do not suggest that the state of the art is advanced enough for us to proceed with planned biotic mixing, only that species packing is one of the techniques of applied biogeography that seems likely to become practicable within the next several decades, on the basis of current and projected research.

Optimism is further justified by the favorable outcome of a few biotic mixtures that have already occurred haphazardly, indicating a degree of flexibility on the part of species that will provide biogeographers with some margin for error. The Kaingaroa Forest of New Zealand, for example, contains 250,000 acres of exotic conifers, including *Pinus radiata, P. ponderosa, P. contorta,* and *Pseudotsuga taxifolia* from North America, and *Pinus nigra* from southern Europe. Introduced birds mingle with endemic New Zealand species in this synthetic environment. Ecological differentiation is well marked; no two species have the same feeding habit, and the insectivorous birds exploit all of the major feeding niches except that of woodpeckers. The really surprising fact, however, is that some of the native species are now as abundant in the Kaingaroa Forest as in almost any native forest, and some are more abundant than in most of the remainder of their range (Gibb, 1961; personal communication, 1973). Furthermore, the invertebrate fauna of the forest consists mostly of native species (Rawlings, 1961). Two circumstances are special in the case of the birds. First, the number of species is still small, largely because the New Zealand fauna was depauperate to start with, and the mixed community has probably not yet encountered many of the difficulties in packing that would be routine in large continental faunas. As discussed by Cody (Chapter 10), species of depauperate faunas are better able to colonize exotic habitats than are species of rich faunas. Second, forest birds are differentiated to a large degree by foliage height and profile rather than by the species of trees in which they live. Monophagous and oligophagous insects, particularly those specializing on hardwoods, would in most instances find it impossible to penetrate the Kaingaroa conifers. Yet the lesson is clear: what works in part by accident can be brought closer to perfection through design.

Ultimately, design might also include the artificial selection of strains, or even the creation of new species, for the purposes of biotic enrichment. If theory and experiment indicate that an orphaned species cannot be fitted into any existing communities, strains might be selected within captive populations of the species that could eventually be inserted into one or more communities. We do not seriously suggest that such a procedure will be followed in the foreseeable future for any but a very few of the organisms most valued by man. Furthermore, the genetic molding of communities is a technology that cannot be seriously contemplated until the inchoate discipline of population ecology has moved closer to a full solution of the species-packing problem.

Edward O. Wilson and Edwin O. Willis **532**

The Creation of New Communities

Many of the earth's major habitats are biological deserts: the open sea, the ice caps, some of the trace-element barrens, and the real deserts, the extremes of which are virtually abiotic. Quite by coincidence, technology is at this moment striving toward two major goals that could transform these areas: an unlimited or at least vastly greater source of energy, and, as one of the principal benefits of the first, the cheap desalinization of sea water. With the achievement of these goals, men will move increasingly onto the land deserts, carrying communities of organisms with them. We will not be satisfied, it is hoped, with limiting ourselves to a baggage of domestic animals, houseplants, pests, and commensals. It lies easily within our power to create wholly new parks and reserves where nothing existed before in historical times. But what will go into these de novo communities? Thought about this subject sharpens one's vision of the future of applied biogeography.

In fact, the deliberate creation of new biological communities has already begun. Large areas of desert-like barrens in Australia have been transformed into agricultural land by the simple addition of zinc, copper, or molybdenum, "trace" elements required for life that were previously present in abnormally low quantities (Anderson and Underwood, 1959). Marine biologists have discovered that artificial reefs, with rich complements of reef organisms, can be created just by dumping concrete rubble, abandoned automobiles, used automobile tires, and similar inert refuse onto the mud or sand floors of shallow marine waters. Successful experiments of this nature have been conducted off the shores of Florida and California (see Turner, Ebert, and Given, 1969). What these efforts engender are in effect habitat islands, the biotas of which grow and equilibrate according to the same laws of biogeography governing wholly natural islands. The communities are not likely to be as diverse as those that have evolved for millions of years in the natural islands, yet the process of enrichment can be speeded by the deliberate importation of compatible species to reach new and higher assortative equilibria. This is another aspect of biogeographic technology that ongoing basic research might render practicable during the next few decades.

Ecosystem Manipulation: The Ultimate Game

The greatest misfortune that awaits the human intellect is to be no longer faced with something commensurate with its capacity for wonder. If the golden age of science really ends, and research shrinks to a few remote and arcane frontiers accessible only to specialists, the wonder will indeed be gone. By that time even the prescientific myths that sustained our ancestors, and intrigue us still, would have largely evaporated—having been accounted for in full, perhaps by the right kind of neurophysiological analysis of the limbic system and hypothalamus. But this possibility will not materialize during the lifetime of anyone now living. The ultimate complexity, offering an unexplored terrain of virtually infinite extent, lies in biology. Even after the cell has been torn down and put together again, and the labyrinthine mysteries of metazoan development followed to their ends, there lie ahead much more extensive challenges of ecology and biogeography. The full exploration of organic diversity is a prospect that suits the biocentric human brain, especially those emotive centers that evolved to make us superior hunters and agriculturists. The same instincts that motivate the birdwatcher, the butterfly collector, and the backyard gardener can indefinitely sustain the scientifically curious segment of a more sophisticated human population in the pursuits of ecology and biogeography. The very size of the world's biota, comprising millions of species, is itself a challenge that only generations more of study will encompass. The possibilities for ecosystems manipulation, outlined in this essay on applied biogeography, offer creative work that is orders of magnitude even more extensive.

References

Anderson, A. J., and E. J. Underwood. 1959. Trace-element deserts. *Sci. American,* January 1959, pp. 97–106.

Aron, W. I., and S. H. Smith. 1971. Ship canals and aquatic ecosystems. *Science* 174:13–20.

Briggs, J. C. 1969. The sea-level Panama Canal: potential biological catastrophe. *BioScience* 19:44–47.

Chapman, F. M. 1938. *Life in an Air Castle; Nature Studies in the Tropics.* D. Appleton-Century, New York.

Diamond, J. M. 1971. Comparison of faunal equilibrium turnover rates on a tropical island and a temperate island. *Proc. Nat. Acad. Sci. U.S.A.* 68:2742–2745.

Diamond, J. M. 1972. Biogeographic kinetics: estimation of relaxation times for avifaunas of southwest Pacific islands. *Proc. Nat. Acad. Sci. U.S.A.* 69:3199–3203.

Diamond, J. M. 1973. Distributional ecology of New Guinea birds. *Science* 179:759–769.

Diamond, J. M. 1975. The island dilemma: lessons of modern biogeographic studies for the design of natural reserves. *Biological Conservation,* in press.

Eisenmann, E. 1952. Annotated list of birds of Barro Colorado Island, Panama Canal Zone. *Smithsonian Miscellaneous Collections* 117(5):1–62.

Gibb, J. A. 1961. Ecology of the birds of Kaingaroa Forest. *Proc. New Zealand Ecol. Soc.* 8:29–38.

Graham, J. B., I. Rubinoff, and M. K. Hecht. 1971. Temperature physiology of the sea snake *Pelamis platurus:* an index of its colonization potential in the Atlantic Ocean. *Proc. Nat. Acad. Sci. U.S.A.* 68:1360–1363.

Horn, H. S. 1971. *The Adaptive Geometry of Trees.* Princeton University Press, Princeton.

MacArthur, R. H. 1972. *Geographical Ecology. Patterns in the Distribution of Species.* Harper and Row, New York.

MacArthur, R. H., J. M. Diamond, and J. R. Karr. 1972. Density compensation in island faunas. *Ecology* 53:330–342.

MacArthur, R. H., and R. Levins. 1967. The limiting similarity, convergence, and divergence of coexisting species. *Amer. Natur.* 101:377–385.

MacArthur, R. H., and E. O. Wilson. 1967. *The Theory of Island Biogeography.* Princeton University Press, Princeton.

May, R. M. 1973. *Stability and Complexity in Model Ecosystems.* Princeton University Press, Princeton.

Newman, W. A. 1972. The National Academy of Science Committee on the ecology of the interoceanic canal. *Bull. Biol. Soc. Washington* 2:247–259.

Preston, F. W. 1962. The canonical distribution of commonness and rarity: Part II. *Ecology* 43:410–432.

Rawlings, G. B. 1961. Entomological and other factors in the ecology of a *Pinus radiata* plantation. *Proc. New Zealand Ecol. Soc.* 8:47–51.

Rubinoff, I. 1965. Mixing oceans and species. *Natural History* 74:69–72.

Schoener, T. W. 1970. Nonsynchronous spatial overlap of lizards in patchy habitats. *Ecology* 51:408–418.

Terborgh, J. 1974a. Faunal equilibria and the design of wildlife preserves. *In* F. Golley and E. Medina, eds., *Tropical Ecological Systems: Trends in Terrestrial and Aquatic Research.* Springer Verlag, New York.

Terborgh, J. W. 1974b. Preservation of natural diversity: the problem of extinction prone species. *BioScience* 24:715–722.

Topp, R. W. 1969. Interoceanic sea-level canal: effects on the fish faunas. *Science* 165:1324–1327.

Turner, C. H., E. E. Ebert, and R. R. Given. 1969. Man-made reef ecology. *Fish Bull., Dept. Fish and Game, State of Calif.* No. 146, pp. 1–221.

Voss, G. L. 1972. Biological results of the University of Miami deep-sea expeditions. No. 93. Comments concerning the University of Miami's marine biological survey related to the Panamanian sea-level canal. *Bull. Biol. Soc. Washington* 2:49–58.

Wallace, B. 1972. *Essays in Social Biology. Volume 1. People, Their Needs, Environment, Ecology.* Prentice-Hall, Englewood Cliffs, N.J.

Warner, R. E. 1968. The role of introduced diseases in the extinction of the endemic Hawaiian avifauna. *Condor* 70:101–120.

Wetmore, A. 1957. The birds of Isla Coiba, Panamá. *Smithsonian Miscellaneous Collections* 134(9):1–105.

Willis, E. O. 1971. The loss of birds from a tropical forest reserve. Paper presented at Dauphin Island meeting, Wilson Ornithological Society.

Willis, E. O. 1974. Populations and local extinctions of birds on Barro Colorado Island, Panamá. *Ecol. Monogr.* 44:153–169.

Wilson, E. O. 1969. The species equilibrium. *Brookhaven Symp. Biol.* 22:38–47.

1980

"Resolutions for the 80s," *Harvard Magazine* 82(3) (January–February): 21

In late 1979, the editors of *Harvard Magazine* asked seven Harvard professors, "What is the most important problem facing this nation or the world at the start of the decade, and what resolutions should we be making to deal with it? How well will we have coped with the problem by the end of the Eighties?" A psychologist said hunger, an environmental engineer said mass poverty, a professor of divinity said unequal distribution of wealth, a conservative philosopher said too much government, an economist said the systemic problems of capitalism, and a sociologist said a too-aggressive U.S. foreign policy.

I said species extinction. Our responses, at least my own, might well have been soon forgotten, except for the two following sentences I included: "The one process ongoing in the 1980s that will take millions of years to correct is the loss of genetic and species diversity by the destruction of natural habitats. This is the folly our descendents are least likely to forgive us." This expression of the problem of global conservation became one of the most widely quoted in the environmental movement during the 1980s and beyond, and is the reason I include the entire essay here.

sively to define ourselves, the range of our determinant concerns, by lines drawn around neighborhoods, towns, and, in the military clutch, nations? The issue is not a utopia of "one world." The issue is not a sacrifice of our own needs and responsibilities to vague or insidious demands from distant shores. The issue is the ability of some of us, right now, to make effective inroads on a problem—and, as yet, our failure to do so. The soreness and pain belong to others, the scandal (to use a word of Saint Paul's) is ours, and greater with each decade of this century. I can't help but feel that we in America will be getting closer, during the 1980s, to a realization of how much it means to us, morally and spiritually—never mind to boys and girls by the hundreds of millions in other parts of the world—that the "problem" of hunger be ended, once and for all, on this earth. □

E.O. WILSON

Frank B. Baird Jr. Professor of Science

"Species extinction is now accelerating and will reach ruinous proportions. By the late 1980s the extinction rate could easily rise to ten thousand species a year (one species per hour) . . ."

Permit me to rephrase the question as follows: What event likely to occur in the 1980s will our descendants most regret, even those living a thousand years from now? My opinion is not conventional, although I wish it were. The worst thing that can happen—*will* happen—is not energy depletion, economic collapse, limited nuclear war, or conquest by a totalitarian government. As terrible as these catastrophes would be for us, they can be repaired within a few generations. The one process ongoing in the 1980s that will take millions of years to correct is the loss of genetic and species diversity by the destruction of natural habitats. This is the folly our descendants are least likely to forgive us.

Species extinction is now accelerating and will reach ruinous proportions during the next twenty years. No one is sure of the number of living species of plants and animals, including such smaller forms as mosses, insects, and minnows, but estimates range between five and ten million. A conservative estimate of the current extinction rate is one thousand species a year, mostly due to the accelerating destruction of tropical forests and other key habitats. By the late 1980s the figure could easily rise to ten thousand species a year (one species per hour) and it is expected to accelerate further through the 1990s. During the next thirty years, fully one million species could be erased. The current rate is already by far the greatest in recent geological history; it is vastly higher than the rate of production of new species by natural evolution. Furthermore, many unique forms that emerged slowly over millions of years will disappear. In our own lifetime humanity will suffer an incomparable loss in aesthetic value, practical benefits from biological research, and world-wide environmental stability. Deep mines of biological diversity will have been dug out and discarded carelessly and incidentally in the course of environmental exploitation, without our even knowing fully what they contained.

This impoverishment cannot be halted during the 1980s, but it can be slowed. We need to shift the emphasis of conservation from the temperate zone to the tropics, from the preservation of isolated "star" species, such as the harpy eagle and Indian white rhinoceros, to the entire ecosystems in which they live. A more powerful, global conservation ethic should be cultivated. The endemic plants and animals of each nation should be treated by its citizens as part of their heritage, as precious as their art and history. When national leaders such as former president Daniel Oduber Quirós of Costa Rica have the courage to advance the preservation of ecosystems within their domains, they should be accorded international honors up to and including the Nobel Peace Prize, in recognition of the very great contributions they make, not just to their own generation but to generations as far into the future as it is possible to imagine. □

GEORGE RUPP

John Lord O'Brian Professor of Divinity, Dean of the Faculty of Divinity

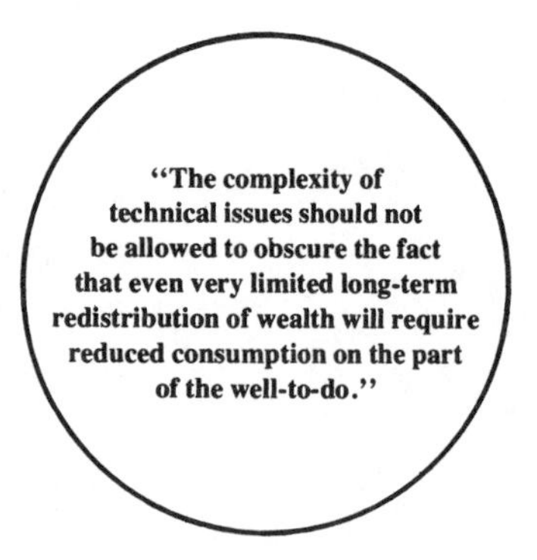

The most important problem facing this nation and the world is an enormously uneven distribution of wealth in the context of increasingly salient limitations in available resources.

The fact that available resources are limited profoundly accentuates the problem of uneven distribution. The easy answers of the recent past become less and less plausible: proportionately the same slices of an ever larger socioeconomic pie may have some attractions even to those with small pieces; but if the pie cannot increase in size

1985

"The biological diversity crisis: a challenge to science," *Issues in Science and Technology* 2(1): 20–29

By the early 1980s I had become fully engaged in the effort to bring the biological diversity crisis more directly into the public domain. This article restates my key theme of Earth as a poorly explored planet that is losing its biological diversity as a consequence of human activity. But I also stressed the need for governmental engagement and the support of research in developing countries, where both the magnitude of biodiversity and its loss are greatest. "This being the only living world we are ever likely to know, let us join to make the most of it." The manuscript was rejected by *Foreign Affairs* as not relevant to that journal's concerns, but accepted by the more science-oriented *Issues in Science and Technology*.

THE BIOLOGICAL DIVERSITY CRISIS: A Challenge to Science

Edward O. Wilson

PROLOGUE: *The worldwide deterioration of natural environments, especially severe in the tropics, is causing the extinction of species at a rate considered by many ecologists to be without precedent in the history of the Earth. Yet the extent of biological diversity, and hence the magnitude of its current decline, has never been precisely measured. Although 1.7 million species of plants, animals, and microorganisms have been formally identified and classified, the total number of species may exceed 30 million, with the great majority living in tropical forests. Each of these species is a unique product of thousands or millions of years of evolutionary history; most contain on the order of a billion bits of genetic information.*

The pool of species diversity is more than a continuing challenge to basic science. It comprises a vast reservoir of potential new crops, pharmaceuticals, and other natural products, as well as plant species capable of restoring depleted soils. Thus, for practical reasons as well as esthetic ones, an increasing number of countries have begun to treat native faunas and floras as part of their national heritage. Conservation and development are now widely understood to be closely linked.

In this article biologist Edward O. Wilson makes a case for a strengthened program to analyze organic diversity. He suggests that a complete catalog of species can be compiled at a relatively low cost, especially if already available computer-aided techniques are used. An attempt to measure global diversity is "a mission worthy of the best effort of science" that would more than pay for itself in economic and environmental benefits.

Edward O. Wilson received his Ph.D. in 1955 from Harvard University, where he now serves as Baird Professor of Science and Curator in Entomology at the Museum of Comparative Zoology. He is the author or coauthor of more than 250 articles and 12 books, including The Insect Societies, Sociobiology: The New Synthesis, *and* On Human Nature, *the last of which received the 1979 Pulitzer Prize in general nonfiction. His scientific awards include the National Medal of Science (1977) and the Tyler Prize for Environmental Achievement (1984).*

THE BIOLOGICAL DIVERSITY CRISIS

Certain measurements are crucial to our ordinary understanding of the universe. What, for example, is the mean diameter of the Earth? It is 12,742 kilometers. How many stars are there in the Milky Way? Approximately 10^{11}. How many genes are there in a small virus particle? There are 10 (in ϕX174 phage). What is the mass of an electron? It is 9.1 x 10^{-28} grams. And how many species of organisms are there on Earth? We do not know, not even to the nearest order of magnitude.

Of course, the number of described species is so impressive that it may appear to be complete. The corollary would be that systematics, defined broadly as the description and analysis of biological diversity, is an old-fashioned science concerned mostly with routine tasks. In fact, about 1.7 million species have been formally named since Linnaeus inaugurated the binomial system of species identification in 1753 (two familiar examples are the white pine, *Pinus strobus*, and the tiger, *Panthera tigris*). Some 440,000 species are plants, including algae and fungi; 47,000 are vertebrates; and according to one meticulous estimate published in 1985,[1] 751,012 are insects. The remainder consists of assorted invertebrates and microorganisms.

But these figures grossly underestimate the diversity of life on Earth, and its true magnitude is still a mystery. In 1964 British ecologist Carrington B. Williams, employing a combination of intensive local sampling and mathematical extrapolation, projected the number of insect species at 3 million.[2] During the next 20 years, systematists described several new complex faunas in relatively unexplored habitats, such as the floor of the deep sea. They also began to employ protein analysis and ecological studies routinely, enabling them to detect many more "sibling species"—populations that are reproductively isolated from other populations but difficult to distinguish on the basis of museum specimens alone. A few writers began to put the world's total as high as 10 million species.

In 1982 the ante was raised threefold again by Terry L. Erwin of the National Museum of Natural History. He and other entomologists developed a technique that for the first time allowed intensive sampling of the canopy of tropical rain forests. This layer of leaves and branches conducts the vast bulk of the photosynthesis for the forest as a whole and is clearly rich in species. It had been largely inaccessible, however, because of its height (a hundred feet or more), the slick surfaces of tree trunks, and the dangers from swarms of stinging ants and wasps at all levels. To overcome these difficulties, the entomologists first fired a projectile with an attached line over one of the upper branches. They then raised a canister containing an insecticide and swift-acting knockdown agent up into the canopy and released the contents as a fog by radio command. As insects and other arthropods fell out of the trees (the chemicals did not harm vertebrates), the researchers collected them in sheets laid on the ground. The number of species proved to be far greater than previously suspected, because of unusually restricted geographical ranges and high levels of specialization on different parts of the trees. Erwin extrapolated a possible total of 30 million insect species, mostly confined to the rain forest canopy.[3] Research is in an even earlier stage on epiphytic plants (such as orchids), roundworms, mites, fungi, protozoans, bacteria, and other mostly small organisms that abound in great diversity in the tree tops.

If astronomers were to discover a new planet beyond Pluto, the news would make front pages around the world. Not so for the discovery that the living world is richer than earlier suspected, a fact of much greater import to humanity. Organic diversity has remained obscure among scientific problems for reasons having to do with both geography and the natural human interest in big organisms. The great majority of organisms in the world are tropical and inconspicuous invertebrates, such as insects, crustaceans, mites, and nematode worms. The mammals, birds, trees, shrubs, and smaller flowering plants of the North Temperate Zone, the subjects of most natural history research and popular writing, comprise relatively few species. In one area of 25 acres of rain forest in Borneo, for example, about 700 species of trees were identified; there are no more than 700 tree species in all of North America.[4] Familiarity with organisms close to home gives the false impression that the Linnaean period of formal taxonomic description has indeed ended. But a brief look almost anywhere else (for example, at the Australian fauna illustrated in Figure 1) shows that the opposite is true.[5]

It may be argued that to know one kind of beetle is to know them all. But a species is not like a molecule in a cloud of molecules--it is a unique population.

Why does this lack of balance in knowledge matter? It may still be argued that to know one kind of beetle is to know them all, or at least enough to get by. But a species is not like a molecule in a cloud of molecules. It is a unique population of organisms, the terminus of a lineage that split off from the most closely related species thousands or even millions of years ago. It has been hammered and shaped into its present form by mutations and natural selection, during which certain genetic combinations survived and reproduced differentially out of an almost inconceivably large possible total.

In a purely technical sense, each species of higher organism—beetle, moss, and so forth—is richer in information than a Caravaggio painting, a Mozart symphony, or any other great work of art. Consider the house mouse, *Mus musculus*. Each of its cells contains four strings of DNA, each of which comprises about a billion nucleotide pairs organized into 100,000 structural genes. If stretched out fully, the DNA would be roughly 1 meter long. But this molecule is invisible to the naked eye because it is only 20 angstrom units in diameter. If we magnified it until its width equaled that of wrapping string, the fully extended molecule would be 600 miles long. As we traveled along its length, we would encounter some 20 nucleotide pairs or "letters" of genetic code per inch. The full information contained therein, if translated into ordinary-sized letters of printed text, would just about fill all fifteen editions of the *Encyclopaedia Britannica* published since 1768.

II

Perhaps because organic diversity is so much larger than previously imagined, it has proved difficult to express as a coherent subject of scientific inquiry. To use the common phrase of experimental science, what is the central problem of systematics? Its practitioners, who by necessity limit themselves to small slices of the diversity, understand but seldom articulate a mission of the kind that enspirits particle physics or molecular genetics. For reasons that transcend the mere health of the discipline, the time has come to focus on such a mission. It can be said that if other disciplines are considered that depend directly upon systematics, including ecology, biogeography, and behavioral biology, an entire hierarchy of important problems is presented. But one stands out, in the sense that progress toward its solution is needed to put the other disciplines on a solid basis. In the case of descriptive systematics (in other words, "taxonomy"), the key question is the number of living species. How many exist in each major group, from bacteria to mammals? I believe that we should aim at nothing less than a full count, a complete catalog of life on Earth. To attempt an absolute measure of diversity is a mission worthy of the best effort of science.

The magnitude and cause of biological diversity is not just the central problem of systematics; it is one of the key problems of science as a whole. It can be said that for a problem to be so ranked, its solution must promise to

Figure 1. Estimated sizes of Australian insect, British insect, and Australian terrestrial vertebrate faunas and levels of taxonomic knowledge of them.

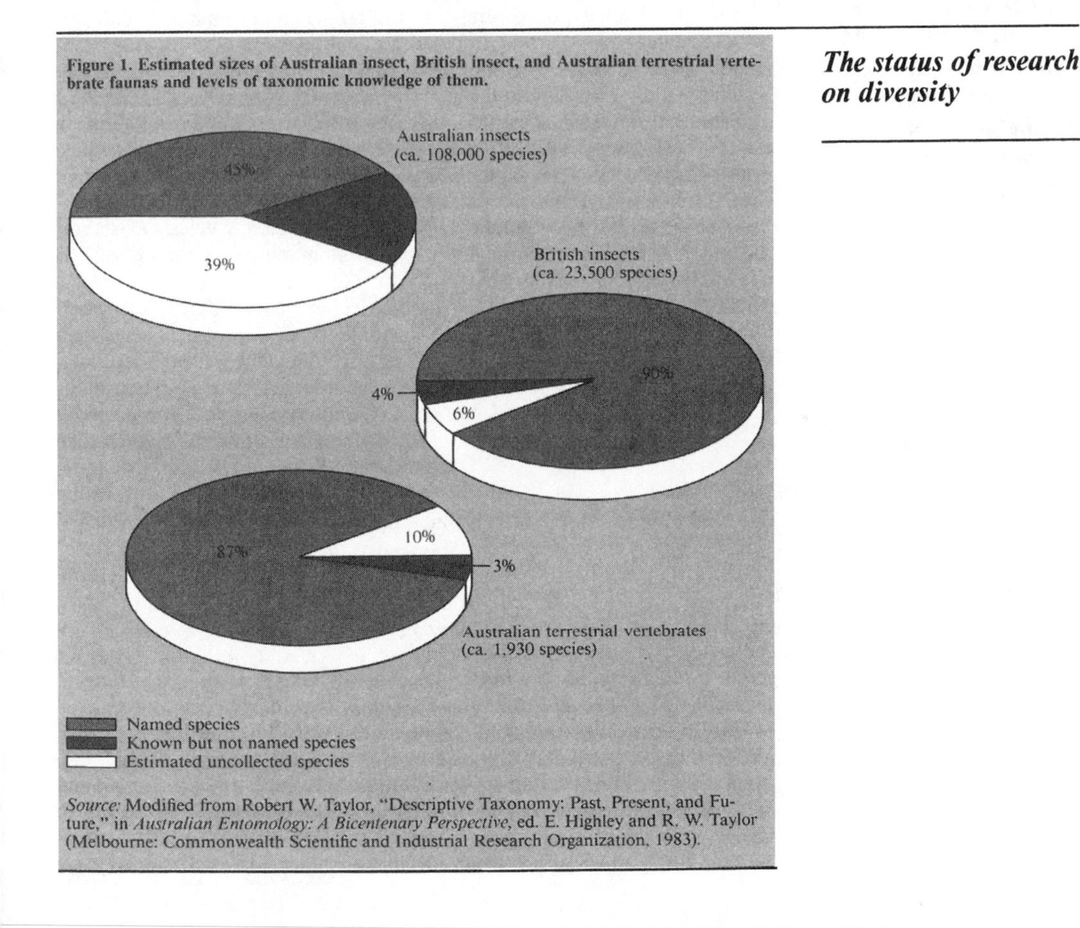

Source: Modified from Robert W. Taylor, "Descriptive Taxonomy: Past, Present, and Future," in *Australian Entomology: A Bicentenary Perspective*, ed. E. Highley and R. W. Taylor (Melbourne: Commonwealth Scientific and Industrial Research Organization, 1983).

The status of research on diversity

yield unexpected results, some of which are revolutionary in the sense that they resolve conflicts in current theory while opening productive new areas of research. In addition, the answers should influence a variety of related disciplines. They should affect our view of man's place in the order of things and open opportunities for the development of new technology of social importance. These several criteria are very difficult to satisfy, of course, but I believe that the diversity problem meets them all.[6]

To this end the problem can be restated as follows: If there are indeed 30 million species, why did there not evolve 40 million, or 2,000, or 1 billion? Many ramifications spring from this ultimate Linnaean question. We would like to know whether something peculiar about the conformation of the planet or the mechanics of evolution itself has led to the precise number that does exist. At the next level down, why is there an overwhelming preponderance of insect species on land but virtually none in the sea? "Hot spots" of disproportionately high diversity of plants and animals occur within larger rain forests, and we need to know their contents and limits, as well as the peculiarities of their evolution at the species level. Would it be possible to increase the diversity of natural systems artificially without destabilizing them? The greater such diversity, the more likely we are to discover new species and genetic varieties for use in agriculture, forestry, and medicine. Only taxonomic analysis can guide research on these and related topics.

Our understanding of the causes of biological diversity is still crude. The science addressing it can be generously put at about the level of physics in the late nineteenth century.

The relation of systematics research to other disciplines of biology becomes clearer by considering the way diversity is created. A local community of plants and animals, of the kind occupying a pond or offshore island, is dynamic. New colonists arrive as old residents die off. If enough time passes, the more persistent populations evolve into local endemic species. On islands as large as Cuba or Oahu, the endemics often split into two or more species living side by side. The total play of these evolutionary forces (immigration, extinction, and species multiplication) determines diversity. To understand each of the forces is automatically to address the principal concerns of ecology, biogeography, and population genetics. Our current understanding of the causes of diversity is still crude. The science addressing it can be generously put at about the level of physics in the late nineteenth century.

III

There is, in addition, a compelling practical argument for attempting a complete survey of diversity. It is well known that only a tiny fraction of the species with potential economic importance has been utilized.[7] Tens of thousands of plants, and millions of animals, have never been studied well enough to assess their potential. Throughout history, for example, a total of 7,000 kinds of plants have been grown or collected as food. Of these, 20 species supply 90 percent of the world's food, and just 3—wheat, maize, and rice—supply more than half. In most parts of the world this thin reservoir of diversity is sown in monocultures particularly sensitive to insect attacks and disease. Yet waiting in the wings are tens of thousands of species that are edible, and many are demonstrably superior to those already in use. In addition, the vast insect faunas contain large numbers of species that are potentially superior as crop pollinators, control agents for weeds, and parasites and predators of insect pests. Bacteria, yeasts, and other microorganisms, which are also poorly known, are likely to yield new medicinals, food, and procedures of soil restoration. Biologists have begun to fill volumes with concrete proposals to explore and make better use of diversity.

The case of natural sweeteners serves as a parable of the potential of untapped resources among wild species. A plant has been found in West Africa, the katemfe (*Thaumatococcus daniellii*), that produces proteins 1,600 times sweeter than sucrose. A second West African plant, the serendipity berry (*Dioscoreophyllum cumminsii*), produces a substance 3,000 times sweeter. The parable is the following: Where in the wild universe does the progression end? To cite a more clearly humanitarian example, one in ten plant species contains anticancer substances of variable potency, but relatively few have been bioassayed. Economists use the expression "opportunity costs" for losses incurred because certain choices were made rather than others. In the case of systematics—or more precisely the neglect of systematics and the biological research dependent upon it—the opportunity costs are very high.

Biological diversity is declining. Destruction of natural environments, a worldwide phenomenon, is reducing the numbers of species and the amount of genetic variation within individual species. The loss is most intense in the tropical rain forests. In prehistoric times these most species-rich of all terrestrial habitats covered an estimated 5 million square miles. Today they occupy 3.5 million square miles and are being cut down at an annual rate of 0.7 percent—that is, 24,500 square miles, about the size of West Virginia. The effect of this deforestation on diversity can be approximated by the following rule of thumb in biogeography. When the area of a habitat is reduced to 10 percent of its original size, the number of species that can persist in it indefinitely will eventually decline to 50 percent. That much habitat reduction has already occurred in many parts of the tropics. The forests of Madagascar now occupy less than 10 percent of their original area, while the once-teeming Brazilian Atlantic forests are down to less than 1 percent of their original size. Even great wilderness areas such as the Amazon and Oronoco basins, equatorial Africa, and Borneo are giving way. If present levels of deforestation continue, the stage will be set within a century for the inevitable loss of about 12 percent of the 700 bird species in the Amazon basin and 15 percent of the plant species in South and Central America.[8]

No comfort should be drawn from the spurious belief that because extinction is a natural process, man is merely another Darwinian agent.

No comfort should be drawn from the spurious belief that because extinction is a natural process, man is merely another Darwinian agent. The rate of extinction is now about 400 times that recorded through recent geological time and is accelerating rapidly. If we continue on this path, the reduction of diversity seems destined to approach that of the great natural catastrophes at the end of the Paleozoic and Mesozoic eras—in other words, the most extreme in 65 million years. And in at least one respect this man-made hecatomb is worse than anything that happened in the geological past. In the earlier mass extinctions, which some scientists believe were possibly caused by large meteorite strikes, most of the plants survived even though animal diversity was severely reduced. Now, for the first time ever, plant diversity too is declining sharply.[9]

IV

A complete survey of life on Earth may appear to be a daunting task. But compared with what has been dared and achieved in high-energy physics, molecular genetics, and other branches of "big science," it is in the second or third rank. To handle 10 million species even with the least efficient old-fashioned methods is an attainable goal. If one specialist proceeded at the cautious pace of an average of ten species per year, including collecting, curatorial work, taxonomic analysis, and publication, about 1 million person-years of work would be required. Given 40 years of productive life per scientist, the effort would consume 25,000 professional lifetimes. That is not an excessive investment on a global scale. The number of systematists worldwide would still represent less than 10 percent of the current population of scientists working in the United States alone and fall short of the standing armed forces of Mongolia and the population of retirees in Jacksonville, Florida. Nor does information storage present an overwhelming problem, even when left wholly to conventional libraries. If each species were given a single, double-columned page for the diagnostic taxonomic description, a figure, and brief biological characterization, and if the pages were bound into ordinary 1,000-page, 6-centimeter-wide hardcover volumes, the 10,000 or so final volumes of this ultimate catalog would fill 600 meters of library shelving. That is far below the capacity of some existing libraries of evolutionary biology. The library of Harvard's Museum of Comparative Zoology, for example, contains 4,850 meters of shelving.

A complete survey of life on Earth may appear to be a daunting task.

But I have given the worst scenario imaginable in order to establish the plausibility of the project. Systematic work could be speeded up many times over by new procedures now coming into general use. The Statistical Analysis System (SAS), a set of computer programs currently running in more than 4,000 institutions worldwide,[10] permits the recording of taxonomic identifications and localities of individual specimens and the automatic integration of data into catalogs and biogeographic maps. Other computer-aided techniques rapidly compare species across large numbers of traits, applying unbiased measures of overall similarity, in a procedure known as phenetics. Still others assist in sorting out the most likely patterns of the phylogeny by which species split apart to create diversity, the method called cladistics. Scanning electron microscopy has speeded up the illustration of insects and other small specimens and has rendered descriptions more accurate. The DELTA system, developed and used at Australia's Commonwealth Scientific and Industrial Research Organization in Canberra, codes data for the automatic identification of specimens.[5, 11] Elsewhere, research is being conducted that may lead to computerized image scanning for purposes of automatic description and data recording.

In North America about 4,000 systematists work on 3,900 systematics collections. But a large fraction of these specialists, perhaps a majority, are engaged only part-time in taxonomic research. More to the point, few can identify organisms from the tropics, where the great majority of species exist and where extinction is proceeding most rapidly. Probably no more than 1,500 professional systematists in the world are competent to deal with tropical organisms, and their number may be declining, owing to decreased professional opportunities, reduced funding for research, and assignment of higher priority to other disciplines.[12] To take one especially striking example, ants and termites make up about one-third of the animal biomass in tropical forests. They cycle a large part of the energy in all terrestrial habitats and include the foremost pests of agriculture, causing billions of dollars in damage yearly. Yet there are exactly eight entomologists worldwide with the general competence to identify tropical ants and termites, and only five of these are able to work at their specialty full-time.

V

It is not surprising to find that neglect of species diversity retards other forms of biological research. Every ecologist can tell of studies delayed or blocked by lack of taxonomic expertise. In one recent and typical case, William G. Eberhard, an entomologist at the University of Costa Rica, consulted most of the small number of available (and overworked) authorities to identify South and Central American spiders used in a study of behavior. He was able to place only 87 of the 213 species included, and then only after considerable delay. He notes that "there are some families (e.g., Pholcidae, Linyphiidae, Anyphaenidae) in which identifications even to genus of Neotropical species are often not possible, and apparently will not be until major taxonomic revisions are done. On a personal level, this has meant that I have refrained from working on some spiders (e.g., Pholcidae—one of the dominant groups of web spiders in a variety of forest habitats, at least in terms of numbers of individuals) because I can't get them satisfactorily identified."[13]

There are exactly eight entomologists worldwide with the general competence to identify tropical ants and termites.

If systematics is an indispensable handmaiden of other branches of research, it is also a fountainhead of discoveries and new ideas, providing a remedy for what biologist and philosopher William Morton Wheeler once called the "dry rot" of academic biology. Systematics has never been given enough credit for this second, vital role. If a biologist can identify only a limited number of species, he is likely to gravitate toward them and end up on well-trodden ground; the remainder of the species remain a confusing jumble. But if he is well-trained in the classification of the organisms encountered, his opportunities multiply. The known facts of natural history become an open book, patterns of adaptation fall into place, and previously unknown phenomena offer themselves conspicuously. By proceeding in this opportunistic fashion, the biologist may discover a new form of animal communication, a previously unsuspected mode of root symbiosis, or a relation between certain species that permits a definitive test of competition theory. The irony of the situation is that such successful research then gets labeled as ecology, physiology, or almost anything else but its *fons et origo*, the study of diversity.

Systematics is linked not only to the remainder of biology but to the fortunes of the international conservation movement, which is now focusing its attention on the threatened environments of the tropics. Plans for systems of ecological reserves have been laid by the International Union for the Conservation of Nature and Natural Resources (IUCN), by UNESCO, and by a growing number of national governments ranging from Australia and Sri

Lanka to Brazil and Costa Rica. The aim is to hold on to the greatest amount of species within the limits imposed by population pressures and the costs of land purchases. The long-term effects of this enterprise can only be crudely predicted until systematics surveys are completed on a country-by-country basis. In the United States a proposal for a National Biological Survey (NABIS) has been presented to Congress. The program would (1) establish a survey to describe all of the plants and animals, (2) fund basic taxonomic studies to this end, and (3) produce identification manuals, catalogs, and other practical aids.[14] If multiplied across many countries, such efforts could bring the full assessment of biological diversity within reach.

Systematics surveys have been a relatively small part of the national research effort in biology. In fiscal year 1985 the National Museum of Natural History, the largest organization of its kind in the United States, spent $12.8 million to support the activities of 85 scientists engaged partly or wholly in taxonomic studies. In the same year the Program in Systematic Biology of the National Science Foundation (NSF) granted $12 million for basic taxonomic research, while other programs of the NSF and Department of Interior provided $13.8 million for support of museum services, studies of endangered species, and other activities related to systematics. Worldwide support for basic tropical biology, including systematics and ecology, is approximately $50 million per year.

Biological diversity is unique in its importance to both developed and developing countries.

Congress addressed the problem of biological conservation in the tropics through a 1980 amendment to the Foreign Assistance Act, which mandates that programs funded through the Agency for International Development (AID) include an assessment of environmental impact. In implementing this policy, AID recognizes that "the destruction of humid tropical forests is one of the most important environmental issues for the remainder of this century and, perhaps, well into the next," in part because they are "essential to the survival of vast numbers of species of plants and animals."[15]

Moving further, AID set up an interagency task force in 1985 to consider biological diversity as a comprehensive issue. In its report to Congress the task force evaluated the current activities of the dozen federal agencies that have been concerned with diversity, including the Smithsonian Institution, the Environmental Protection Agency, and AID itself.[16] The most important recommendations made by the group, in my opinion, are those that call for the primary inventory and assessment of native faunas and floras. In fact, not much else can be accomplished without this detailed information.

AID also supports research programs in which nationals of the recipient countries are the principal investigators and United States citizens serve as collaborators. This arrangement is a proven way to build science and technology in the third world, and it is particularly well suited to tropical biology. Studies of diversity are best conducted at the sites with the maximum amount of diversity. Such studies are labor intensive and require less expensive instrumentation than most kinds of research. Perhaps most importantly, their relevance to national identity and welfare are immediately obvious.

To put the matter as concisely as possible, biological diversity is unique in its importance to both developed and developing countries and in the cost-effectiveness of its study. The United States would do well to seek a formal international agreement among countries, possibly in the form of an "International Decade for the Study of Life on Earth," to improve financial support and access to study sites. To spread technical capability where it is most needed, arrangements could be made to retain specimens within the countries of their origin while nationals are trained to assume leadership in systematics and the related scientific disciplines.

In *Physics and Philosophy*, Werner Heisenberg suggested that science is the best way to establish links with other cultures because it is concerned not with ideology but with nature and man's relation to nature. If that promise can ever be met, it will surely be in an international effort to understand and save biological diversity. This being the only living world we are ever likely to know, let us join to make the most of it. ■

NOTES:

1. Ross H. Arnett, Jr., *American Insects: A Handbook of the Insects of America North of Mexico* (New York: Van Nostrand Reinhold, 1985).
2. Carrington B. Williams, *Patterns in the Balance of Nature* (New York: Academic Press, 1964).
3. The fullest account of the forest canopy work is by Terry L. Erwin, "Beetles and Other Insects of Tropical Forest Canopies at Manaus, Brazil, Sampled by Insecticidal Fogging," in *Tropical Rain Forest: Ecology and Management*, ed. Stephen L. Sutton, Timothy C. Whitmore, and A. C. Chadwick (Edinburgh: Blackwell, 1983), 59–75.
4. Peter S. Ashton, personal communication (May 1985).
5. Robert W. Taylor, "Descriptive Taxonomy: Past, Present, and Future," in *Australian Entomology: A Bicentenary Perspective*, ed. E. Highley and R. W. Taylor (Melbourne: Commonwealth Scientific and Industrial Research Organization, 1983).
6. I am indebted to John Daly of the Agency for International Development for an invaluable discussion of the criteria of key scientific problems.
7. See, for example, *Underexploited Tropical Plants with Promising Economic Value*, 2d ed. (Washington, D.C.: National Academy of Sciences, 1975), known informally as the "Green Book"; also Norman Myers, *A Wealth of Wild Species: Storehouse for Human Welfare* (Boulder, Colo.: Westview Press, 1983), and Margery L. Oldfield, *The Value of Conserving Genetic Resources* (Washington, D.C.: U.S. Department of the Interior, 1984).
8. Daniel S. Simberloff, "Mass Extinction and the Destruction of Moist Tropical Forests," *Zhurnal Obshchei Biol.* 45, no. 6 (1984).
9. Andrew H. Knoll, "Patterns of Extinction in the Fossil Record of Vascular Plants," in *Extinctions*, ed. Matthew H. Nitecki (Chicago: University of Chicago Press, 1984), 21–68.
10. John C. La Duke, David Lank, and Tim Sirek, "Utilization of Statistical Analysis System (SAS) as a Revisionary Tool and Cataloguing Program," *ASC Newsletter* (Association of Systematics Collections) 12, no. 2 (April 1984).
11. Michael J. Dallwitz, "A General System for Coding Taxonomic Descriptions," *Taxon* 29, no. 1 (1980).
12. Data concerning the number of taxonomists, as well as detailed arguments for the need to improve research in tropical countries, are given by Peter H. Raven et al. in *Research Priorities in Tropical Biology* (Washington, D.C.: National Academy of Sciences, 1980). A partial updating has been provided by Stephen R. Edwards, "The Systematics Community: Priorities for the Next Decade," *ASC Newsletter* (Association of Systematics Collections) 12, no. 5 (October 1984), and personal communication (May 1985). Tropical ecology research as a whole is also declining, as documented by N. H. Ayodele Cole in *Nature* 309 (May 17, 1984).
13. William G. Eberhard, personal communication (April 1985).
14. Michael Kosztarab, "A Biological Survey of the United States," *Science* 223, no. 4635 (February 3, 1984).
15. "Humid Tropical Forests: AID Policy and Program Guidance," Department of State memorandum, (Washington, D.C.: 1985).
16. U.S. Agency for International Development, *U.S. Strategy on the Conservation of Biological Diversity: An Interagency Task Force Report to Congress* (Washington, D.C.: February 20, 1985).

1985

"Outcry from a world of wounds," *Discover* 6(6): 64–66

This article was another piece in my attempt in the mid-1980s to help bring the global biodiversity crisis to as broad an audience as possible. It was written in parallel with the preceding "The biological diversity crisis: a challenge to science," which was addressed primarily to policymakers.

VIEWPOINT E. O. WILSON

Outcry from a World of Wounds

lpha and omega. That chiliastic phrase occurred to me recently at the World Wildlife Fund research station near Manaus, Brazil, as I stood simultaneously in the beginning and the end of a tropical environment. I was at the edge of virgin rain forest being cut back and burned by its owners. The transition was so sharp that during a playful moment at the end of a long day of field work, I placed my left foot on the root mat of the still-extant forest and by stretching my legs a little got my right foot down on the bare ground of the newly created pasture.

What this whimsical act symbolized for me should be disturbing to anyone concerned about the future of the global environment. The great Amazonian forest, now being rolled back toward the Guyana border with efficient ferocity, is a treasure house of unimaginable richness. In a single one-hectare (2.5-acre) section of the World Wildlife Fund study area, the Brazilian botanist Leopoldo Teixeira has identified 52 species of the sapodilla family *(Sapotaceae)* among just 93 trees present—one kind of sapotaceous tree per 1.8 individuals. At about the same time, Keith Brown, an American entomologist, spotted 258 kinds of butterflies within a slightly larger area and projected the actual number to exceed 500. I had come to study some of the 200 or so species of ants at this single locality, and within hours I found two species new to science. But this was no great achievement. In the Amazon wilderness the biologist can make more discoveries per day, with less effort, than almost anywhere else in the world.

No one knows the true extent of the diversity of life in tropical forests. According to a recent rough estimate by the Smithsonian Institution's Terry Erwin, there may be tens of millions of kinds of insects alone, most of them limited to the upper branches and leaf sprays where they remain hidden and undiscovered. On the other hand, it's quite easy to calculate the number of survivors in the newly created "pastures." For many groups of organisms you can count the species on the fingers of one hand. The contrast is why I like to say, with admittedly a great deal of personal bias, that no biologist has truly lived who hasn't looked into the heart of a virgin rain forest and witnessed the ultimate in evolution.

We are witnessing the relentless destruction of a treasure house of unimaginable riches, where biologists can make more discoveries in a single day, with less effort, than almost anywhere else on earth.

Tropical environments differ from those in temperate zones in other ways just as consequential. For reasons rooted in both geology and heredity, they're among the most fragile of habitats. We're accustomed to think of tropical forests—"jungles"—as among the most fertile places on earth, ready to be cleared and converted into equally rich farmland. But this is one of the dangerous illusions of our time. Two-thirds of the forest area is underlain by "tropical red earths," which are ancient in geological origin, with few nutrients, and hence easily converted into wasteland. There are also high concentrations of iron and aluminum (good for mining, bad for agriculture), which form insoluble compounds with phosphorus, reducing the amount of that vital substance available to plants. Equally important, calcium and potassium are carried out of the soil and into the drainage systems almost as soon as their compounds are dissolved by the rain.

Nevertheless, during a period of more than 80 million years the rain forest has become adapted to grow tall in this desertic environment. The trees achieve their multiple layers and great mass because they have evolved extraordinary devices to hold on to the scarce nutrients. They absorb these substances from the soil with shallow root mats containing symbiotic fungi specialized for the same purpose. The mats are thin and extensive, often growing partway up the tree trunks. You can sometimes pull them free with your hands, like a loosely nailed carpet. Most of the time the nutrients are locked up in the mass of the living vegetation. This peculiar sequestering of resources means that when tropical red earths are cleared of trees and exposed to the sun, they rapidly lose their ability to support most kinds of vegetation, including the great majority of crops. That's why it's possible to go from a magnificent forest to something approaching a desert in one short, brutal step.

Recent research has shown that seeds of the rain forests are fragile in the extreme. Because of the lush conditions under which they evolved, most kinds begin to germinate within a few days or weeks, which severely limits their ability to disperse across the stripped land. As a result, most sprout and die in the ashes of the newly created clearings. Regeneration of a full-blown forest evidently takes centuries. The process is in fact so slow that few extrapolations have been possible; in some places the return of the rain forest might never occur naturally.

Edward O. Wilson, Baird Professor of Science and the curator in entomology at Harvard's Museum of Comparative Zoology, is best known for his pioneering work in sociobiology.

Discover, 6(6): 64-66 (June 1985)

PHILADELPHIA MUSEUM OF ART

We're accustomed to think of tropical forests as among the most fertile places on earth, ready to be cleared and converted into equally rich farmland. But this is one of the dangerous illusions of our time.

Like many ecologists (accustomed to living in what Aldo Leopold called a world of wounds), I worry a great deal about the rain forests for a special reason. Although this habitat covers less than ten per cent of the land surface of the world, it's home to well over half of the species. And even though only a tiny fraction of the diversity has been examined, we know that it's a cornucopia of potentially useful new plants and animals. My favorite example, out of many that have come to light, is the Amazon babassú palm (*Orbignya martiana*), known locally as the "vegetable cow." The palm's coconut-like fruits occur in bunches of up to 600 with a collective mass of 200 pounds. About 70 per cent of the kernel is composed of a colorless oil that can be converted into margarine, shortening, fatty acids, toilet soap, and detergents. A stand of 500 trees on one hectare can produce 125 barrels of oil a year. After the oil has been extracted, the remaining seedcake, which is about one-fourth protein, serves as excellent animal fodder. Other forest plant species have begun to yield pharmaceuticals, almost every conceivable kind of lumber and vegetable fiber, and efficient (as well as infinitely renewable) petroleum substitutes. In a fashion even less appreciated, the tropical forests are also places of dazzling beauty. Hence, it's all the more tragic that the forests are the most endangered of all major habitats on earth. They're being destroyed or

VIEWPOINT

While we dilate and look in the wrong places, the environment deteriorates. We ought to worry less about ideology and more about science and technology in foreign policy.

seriously degraded at the worldwide rate of about one per cent yearly, which amounts to an area greater than that of Switzerland. Ecologists aren't sure of the exact rate of loss of diversity caused by this destruction, but most agree that thousands of species are being extinguished each year before they've even received a scientific name.

An idea of what is happening can be obtained with this rule of thumb used by biogeographers: when the area of a habitat is reduced to ten per cent of the original, the number of species that can persist in it indefinitely is reduced to one-half. The rain forests of Madagascar, for example, have already shrunk to one-tenth their original size and are being cut back still further. Think of how the Malagasy people will feel in the year 2020, when they've attained a more comfortable life and a higher level of education and finally realize that a large part of their fauna and flora was carelessly thrown away—not thousands of years ago by primitive ancestors but within living memory.

This juggernaut of tropical destruction is being carried along by a demographic momentum unique in history. The world's population has doubled since 1950 (to 4.8 billion); and although the rate of growth has recently begun to slow on all continents except Africa, net population growth continues to shoot upward (because there is still a disproportionate number of young women in the world having children). Eighty-five million people were added during the past year, two-thirds of them in the tropics. By 2050 more than 80 per cent will be living in what is now called the Third World, most of which lies within the tropical and subtropical zones. The heaviest concentrations are in countries already too crowded to sustain a decent standard of living with the technology available to them. Haitians floating ashore at Miami and teenage Salvadorians cradling semiautomatic rifles aren't just the victims of bad government. They inhabit the most overcrowded countries in the Western Hemisphere, exceeded in this respect only by Grenada (another trouble spot) and three other small Caribbean island-nations (Barbados, St. Vincent and the Grenadines, and Trinidad and Tobago). Their plight originates in a shrinking per capita resource base and is leading to an understandably deepening despair.

During at least the next two or three generations, adequate resources won't be conferred on such starveling nations by industrialization or the redistribution of wealth; they must come from crops sown on the land. The U.N. Food and Agriculture Organization estimates that a 60 per cent increase in world food production will be needed by the year 2000 to feed the entire population. This raises a key issue: research and development. We have to assume—indeed, if ever there was a moral mandate, we have to *believe*—that the conservation of biological resources in tropical countries can be achieved at the same time that the standard of living is raised. Even if humanitarian concerns could be dismissed, the economies of the industrial and developing nations are now inseverably linked. During the 1960s and 1970s, exports from the U.S. to Third World countries rose exponentially, and they now account for 36 per cent of the U.S. total. Until recently, Third World countries were the fastest growing market abroad for our goods and services. This is the real reason that political unrest in the tropical zone ripples out to touch our lives on a daily basis.

Let me return a final time to the rain forests to illustrate what can be done to reconcile conservation and development. An ecologist, Carl Jordan of the University of Georgia, has pointed out that nutrients can be preserved and a moderately high timber yield sustained by an appropriate technique of "strip harvesting." A long corridor is logged along the contour of a slope, wide enough to be profitable but narrow enough to keep the nutrients from completely leaching away. After the desired logs are hauled out, the area is left undisturbed until seeds infiltrate from the surrounding forest and sprout. Then the loggers clear-cut another strip a little higher up the slope, so that nutrients washing down are caught by the emerging growth of the first strip. Within a few years the technique can more than pay for itself, leaving the forest largely intact.

There are a lot of schemes like this on the drawing boards, talked about at conferences, and occasionally tried in field stations. But all the while, as we dilate and look in the wrong places, the environment continues to deteriorate.

The roots of most social and economic problems in the tropics are the various forms of biological excess: overpopulation, disease, famine, humus erosion, and the loss of species diversity. Until that cardinal fact is acted upon, military and political solutions are likely to prove no more than short-term expedients. The Congress has responded in a limited fashion by amending the Foreign Assistance Act to make environmental assessment of projects paid for by the Agency for International Development mandatory. This ensures that there's more money for badly needed research and development. It is to be hoped that the World Bank and other multilateral development organizations will assume greater responsibility in kind, and that the trend will grow.

Meanwhile, the industrialized countries should worry a lot less about ideology and more about the role of science and technology in foreign policy. For the most part we just don't have enough knowledge to advise the tropical countries properly and, through their hoped-for success, enhance our own security. The amount of money spent on basic research in tropical biology is inadequate, to say the least. It's downright self-destructive: roughly $50 million worldwide, less than one per cent of the amount spent on health-related research in this country, or the interest that would accrue on the interest on the interest on the interest of the annual U.S. defense budget. □

1987

"The little things that run the world (the importance and conservation of invertebrates)," *Conservation Biology* 1(4): 344–346

Conservation science and practice have traditionally favored big organisms, in other words flowering plants and vertebrates. Except for a few groups, such as butterflies and mollusks, invertebrate animals are generally overlooked, in good part, simply because they are relatively difficult to observe and classify. Here I argue why we should redress the balance.

The Little Things That Run the World*
(The Importance and Conservation of Invertebrates)

On the occasion of the opening of the remarkable new invertebrate exhibit of the National Zoological Park, let me say a word on behalf of these little things that run the world. To start, there are vastly more kinds of invertebrates than of vertebrates. At the present time, on the basis of the tabulation that I have just completed (from the literature and with the help of specialists), I estimate that a total of 42,580 vertebrate species have been described, of which 6,300 are reptiles, 9,040 are birds, and 4,000 are mammals. In contrast, 990,000 species of invertebrates have been described, of which 290,000 alone are beetles—seven times the number of all the vertebrates together. Recent estimates have placed the number of invertebrates on the earth as high as 30 million, again mostly beetles—although many other taxonomically comparable groups of insects and other invertebrates also greatly outnumber vertebrates.

We don't know with certainty why invertebrates are so diverse, but a commonly held opinion is that the key trait is their small size. Their niches are correspondingly small, and they can therefore divide up the environment into many more little domains where specialists can coexist. One of my favorite examples of such specialists living in microniches are the mites that live on the bodies of army ants: one kind is found only on the mandibles of the soldier caste, where it sits and feeds from the mouth of its host; another kind is found only on the hind foot of the soldier caste, where it sucks blood for a living; and so on through various bizarre configurations.

Another possible cause of invertebrate diversity is the greater antiquity of these little animals, giving them more time to explore and fill the environment. The first invertebrates appeared well back into Precambrian times, at least 600 million years ago. Most invertebrate phyla were flourishing before the vertebrates arrived on the scene, some 500 million years ago.

Invertebrates also rule the earth by virtue of sheer body mass. For example, in tropical rain forest near Manaus, in the Brazilian Amazon, each hectare (or 2.5 acres) contains a few dozen birds and mammals but well over one billion invertebrates, of which the vast majority are not beetles this time but mites and springtails. There are about 200 kilograms dry weight of animal tissue in a hectare, of which 93 percent consists of invertebrates. The ants and termites alone compose one-third of this biomass. So when you walk through a tropical forest, or most other terrestrial habitats for that matter, or snorkel above a coral reef or some other marine or aquatic environment, vertebrates may catch your eye most of the time—biologists would say that your search image is for large animals—but you are visiting a primarily invertebrate world.

It is a common misconception that vertebrates are the movers and shakers of the world, tearing the vegetation down, cutting paths through the forest, and consuming most of the energy. That may be true in a few ecosystems such as the grasslands of Africa with their great herds of herbivorous mammals. It has certainly become true in the last few centuries in the case of our own species, which now appropriates in one form or other as much as 40 percent of the solar energy captured by plants. That circumstance is what makes us so dangerous to the fragile environment of the world. But it is otherwise more nearly true in most parts of the world of the invertebrates rather than the nonhuman vertebrates. The leafcutter ants, for example, rather than deer, or rodents, or birds, are the principal consumers of vegetation in Central and South America. A single colony contains over two million workers. It sends out columns of foragers a hundred meters or more in all directions to cut forest leaves, flower parts, and succulent stems. Each day a typical mature colony collects about 50 kilograms of this fresh vegetation, more than the average cow. Inside the nest, the ants shape the material into intricate sponge-like bodies on which they grow a symbiotic fungus. The fungus thrives as it breaks down and consumes the cellulose, while the ants thrive by eating the fungus.

The leafcutting ants excavate vertical galleries and living chambers as deep as 5 meters into the soil. They and other kinds of ants, as well as bacteria, fungi, termites, and mites, process most of the dead vegetation and return its nutrients to the plants to keep the great tropical forests alive.

* *Address given at the opening of the invertebrate exhibit, National Zoological Park, Washington, D.C., on May 7, 1987.*

Conservation Biology
Volume 1, No. 4, December 1987

Much the same situation exists in other parts of the world. The coral reefs are built out of the bodies of coelenterates. The most abundant animals of the open sea are copepods, tiny crustaceans forming part of the plankton. The mud of the deep sea is home to a vast array of mollusks, crustaceans, and other small creatures that subsist on the fragments of wood and dead animals that drift down from the lighted areas above, and on each other.

The truth is that we need invertebrates but they don't need us. If human beings were to disappear tomorrow, the world would go on with little change. Gaia, the totality of life on Earth, would set about healing itself and return to the rich environmental states of a few thousand years ago. But if invertebrates were to disappear, I doubt that the human species could last more than a few months. Most of the fishes, amphibians, birds, and mammals would crash to extinction about the same time. Next would go the bulk of the flowering plants and with them the physical structure of the majority of the forests and other terrestrial habitats of the world. The earth would rot. As dead vegetation piled up and dried out, narrowing and closing the channels of the nutrient cycles, other complex forms of vegetation would die off, and with them the last remnants of the vertebrates. The remaining fungi, after enjoying a population explosion of stupendous proportions, would also perish. Within a few decades the world would return to the state of a billion years ago, composed primarily of bacteria, algae, and a few other very simple multicellular plants.

If humanity depends so completely on these little creatures that run the earth, they also provide us with an endless source of scientific exploration and naturalistic wonder. When you scoop up a double handful of earth almost anywhere except the most barren deserts, you will find thousands of invertebrate animals, ranging in size from clearly visible to microscopic, from ants and springtails to tardigrades and rotifers. The biology of most of the species you hold is unknown: we have only the vaguest idea of what they eat, what eats them, and the details of their life cycle, and probably nothing at all about their biochemistry and genetics. Some of the species might even lack scientific names. We have little concept of how important any of them are to our existence. Their study would certainly teach us new principles of science to the benefit of humanity. Each one is fascinating in its own right. If human beings were not so impressed by size alone, they would consider an ant more wonderful than a rhinoceros.

New emphasis should be placed on the conservation of invertebrates. Their staggering abundance and diversity should not lead us to think that they are indestructible. On the contrary, their species are just as subject to extinction due to human interference as are those of birds and mammals. When a valley in Peru or an island in the Pacific is stripped of the last of its native vegetation, the result is likely to be the extinction of several kinds of birds and some dozen of plant species. Of that tragedy we are painfully aware, but what is not perceived is that hundreds of invertebrate species will also vanish.

The conservation movement is at last beginning to take recognition of the potential loss of invertebrate diversity. The International Union for the Conservation of Nature has an ongoing invertebrate program that has already published a Red Data Book of threatened and endangered species—although this catalog is obviously still woefully incomplete. The Xerces Society, named after an extinct California butterfly, was created in 1971 to further the protection of butterflies and other invertebrates. These two programs are designed to complement the much larger organized efforts of other organizations on behalf of vertebrates and plants. They will help to expand programs to encompass entire ecosystems instead of just selected star species. The new invertebrate exhibition of the National Zoological Park is one of the most promising means for raising public appreciation of invertebrates, and I hope such exhibits will come routinely to include rare and endangered species identified prominently as such.

Several themes can be profitably pursued in the new field of invertebrate conservation:

- It needs to be repeatedly stressed that invertebrates as a whole are even more important in the maintenance of ecosystems than are vertebrates.
- Reserves for invertebrate conservation are practicable and relatively inexpensive. Many species can be maintained in large, breeding populations in areas too small to sustain viable populations of vertebrates. A 10-ha plot is likely to be enough to sustain a butterfly or crustacean species indefinitely. The same is true for at least some plant species. Consequently, even if just a tiny remnant of natural habitat exists, and its native vertebrates have vanished, it is still worth setting aside for the plants and invertebrates it will save.
- The *ex situ* preservation of invertebrate species is also very cost-effective. A single pair of rare mammals typically costs hundreds or thousands of dollars yearly to maintain in a zoo (and worth every penny!). At the same time, large numbers of beautiful tree snails, butterflies, and other endangered invertebrates can be cultured in the laboratory, often in conjunction with public exhibits and educational programs, for the same price.
- It will be useful to concentrate biological research and public education on star species when these are available in threatened habitats, in the manner that has proved so successful in vertebrate conservation. Examples of such species include the tree snails of Moorea, Hawaii, and the Florida Keys; the Prairie

sphinx moth of the Central States; the birdwing butterflies of New Guinea; and the metallic blue and golden ants of Cuba.

- We need to launch a major effort to measure biodiversity, to create a complete inventory of all the species of organisms on Earth, and to assess their importance for the environment and humanity. Our museums, zoological parks, and arboreta deserve far more support than they are getting—for the future of our children.

A hundred years ago few people thought of saving any kind of animal or plant. The circle of concern has expanded steadily since, and it is just now beginning to encompass the invertebrates. For reasons that have to do with almost every facet of human welfare, we should welcome this new development.

EDWARD O. WILSON

Museum of Comparative Zoology
Harvard Uniyersity
Cambridge, Massachusetts, 02138–2902

1989

"The coming pluralization of biology and the stewardship of systematics," *BioScience* 39(4): 242–245

In 1989, when the accompanying essay was published, biology was strikingly lopsided. Most attention and funding were directed to molecular and cellular studies of a small number of species, including a fungus, a nematode worm, and of course *Homo sapiens.* This focus was and remains extremely productive, but one negative result was a neglect of biological diversity. Most of the living world remained unexplored, especially the millions of species of microorganisms, fungi, and invertebrates that form the base of Earth's ecosystems and thence human survival.

As I write today, a shift has begun and we are becoming broader in our view. I identify three reasons for the change. First of all, the urgency of the revitalization of systematics and a broader base of biodiversity conservation has become obvious. The second reason is that systematics, along with its sister disciplines of ecology, biogeography, and conservation biology, has grown much more capable and sophisticated than in the 1980s. Now technology driven, using such procedures as DNA sequencing and computer-based pattern searching, experts on birds, fungi, flowering plants, and all other taxonomic groups are able not only to explore and classify the species composing these groups, but also to more easily address them at any level of biological organization.

Third, the unification of biology is emerging in an increasingly compelling manner. In their bottom-up approach molecular biologists and others who focused on processes at the level of cell and tissue have also embraced evolution. Many now routinely use genomic analysis to reconstruct both phylogenies of the major groups of organisms and the evolution of the genome and proteome. As a result collaborations between molecular and evolutionary biologists have become commonplace.

As part of unification, biology has entered the Age of Synthesis. Reduction of complex systems to their elements and basic processes remains as the cutting edge of biology. But we have also reached a sufficiency of understanding to proceed more deliberately to the grand program of synthesis. Contrary to a common opinion that science is primarily a reductionist enterprise, and therefore simplistic and even dehumanizing, all science grows to maturity by repetitive cycles of reduction and synthesis. The best biology is accomplished by addressing a complex system such as a protein molecule or ecosystem, then locating a point of entry and cleaving it into its elements and processes, and finally, either literally or by mathematical simulation, putting it back together again. Synthesis is much harder than reduction and much slower. The era of biology in which synthesis will come into its own has only begun. It will last a very long time—as long, in fact, as biology remains a creative, growing discipline.

As this essay argues, a necessary theme of biology is that whatever principles emerge in the future must encompass the particularity of species. Each species consists of a unique configuration of genes that have been assembled by a very long history of adaptation to certain other species, and within a particular ecosystem. As the molecular and cell biologists have discovered, each species can consume the attention of many scientific careers. Only from such pluralization vastly extended and added to the central goals of biology will true unity be achieved.

Reprinted from *BioScience*

Roundtable

The coming pluralization of biology and the stewardship of systematics

Certain metaphysical constructions that Gerald Holton[1] has called the "themata" of science hold a grip more powerful—and less vulnerable—than ordinary theories. Isaac Newton's idea of a book of Nature authored by God, Charles Darwin's vision of the grandeur of natural selection, and Friedrich Engels' description of dialectical synthesis are perhaps the most familiar examples. These metaphysical themes have shaped the direction of theory while influencing how scientists think about the totality of their life's work. It is my impression that a thematic shift in biology has begun to occur.

The shift, if I have interpreted it correctly, will eventually move biology toward an earlier, more robust view of its *raison d'etre.* Until the 1950s, biologists emphasized taxonomic groups of organisms rather than levels of organization, as in molecular, cell, organismal, and ecosystem biology. During the 1950s a shift occurred, which was enormously beneficial, to accommodate the beginnings of molecular and cell biology. The new theme was the belief that biological laws or principles must be revealed by intensive level-of-organization analysis. The rising dominance of that worldview, however, was of necessity only temporary.

In the near future, whereas some biologists will continue to think solely in terms of levels of organization and search for the broadest possible generalizations, a greater number will commit themselves to the study of particular groups of organisms across all levels of organization. The theme that impels this shift is the fundamental and unalterable value of each group of organisms complete unto themselves. The principal division of labor will change from the present philosophical stress on levels of biological organization to more emphasis on taxonomic groups of organisms. The coming shift can be visualized as a tilt from a nearly horizontal to a more vertical orientation—rotating not the full 90 degrees but, say, 45 degrees.

More biologists will commit themselves to the study of particular groups of organisms across all levels of organization

The result will be a pluralization of biology and the return of the expert naturalist to a position of leadership in biological research. By pluralization I mean the increased esteem and growth of studies of particular groups of organisms for their own sake. Put another way, taxon-oriented disciplines such as herpetology and nematology will regain ground lost to level-oriented disciplines such as cell biology and ecology. The word *fundamental* will be applied not just to broad generalizations but also to important discoveries about individual taxa, even if the information cannot be readily applied to other taxa.

This shift won't return biology to old-fashioned, purely descriptive natural history. The technical reach of the new naturalist extends from the molecular to the populational levels, as evolutionists learn molecular techniques and molecular biologists interest themselves in the evolution of the organisms they study. As biologists increasingly commit themselves to particular groups of organisms, they seem destined to converge toward a common language and methodology. Herpetologists and nematologists, and the molecular biologists who collaborate with them, have already begun to speak effectively to each other in new, shared tongues.

Why pluralization is likely

The first trend suggesting such an intellectual reorganization is the growing recognition that few if any universal principles exist in biology that are both precise and widely applicable. The vast majority of investigations in molecular, cellular, and other level-oriented disciplines typically reveal truths which, although jointly rooted in the physical sciences, concern only particular species or at most limited groups of species. Consider for example three textbook examples of basic discoveries: endocytosis by neutrophils, the action of juvenile hormones in holometabolous insects, and the density-dependent control of rodent populations. None of these findings applies beyond the taxonomic group in which it was discovered. The chief value is heuristic: the discoveries stimulate the search for parallel or analogous phenomena in broader groups of organisms. They are cited as phenomena to look for elsewhere, representatives of categories that might be abstracted into broader spheres of generality.

New general principles, the grail of biology, are becoming ever more elusive. Density dependence, for example, turns out to be present in some species and absent in others, and when present to take forms that are understandable only with a knowledge of the life cycle of each particular species and the ecosystem in which it lives. The same is true of immuno-

[1]G. Holton, 1977. *The Scientific Imagination: Case Studies.* Cambridge University Press, Cambridge, UK.

by E. O. Wilson

chemistry, and chemoreception, and kin selection, and so on and on. It seems to me a remarkable feature of biology that while factual knowledge grows exponentially, with a doubling time of perhaps 10 to 20 years, the number of broadly applicable discoveries made per investigator per year is declining steeply. A large part of the reason is the historicity of biological phenomena, which generates special cases and crumbles generality in direct proportion to the depth of understanding.

The swift advance in knowledge achieved during the levels-directed revolution also carries within it the seeds of the decline of research confined to one or two layers of organization. Soon after new methods are invented, they are modularized—transformed into streamlined, partially automated packages and made available to all. Subjects such as electron microscopy, amino acid sequencing, and multivariate analysis, once part of a frontier, have been captured in operator's instructions for commercially available instruments. Symbiosis among the disciplines is the natural result. Systematists now routinely compare proteins, and molecular biologists construct phylogenetic trees.

At the same time, biologists are placing new emphasis on the uniqueness of each species, as something more than just a collection of related organisms and a great deal more than an interchangeable unit at the population level of organization. When you have seen one species of chrysomelid beetle, you have emphatically not seen them all. In fact you still know very little about the family Chrysomelidae. In its genes each species contains on the order of a million to a billion bits of information, assembled by an almost inconceivable number of events in mutation, recombination, and natural selection during an average life span, according to taxonomic group, of one to ten million years. And the better each species is understood, the greater is the esteem for research conducted on it.

What does this particularity mean for the understanding of life as a whole? No one knows the number of species of living organisms, including animals, plants, and microorganisms, but there are probably at least 5 million, and the number could be as high as 30 million. Whatever the number, it is believed to represent less than one percent of all the species that ever existed in geological time. We have only begun even a superficial exploration of life on Earth.

To the extent that this panorama is encompassed by biology, pure history will become more important. Because most biological phenomena occur in only a small portion of phyletic lines, their pattern of origin assumes an independent importance. Phylogeny (the branching patterns) and evolutionary grades (the levels of adaptation attained) form the core of biology as surely as those occasional organizational rules that hitherto have so tenuously bound biology into a unified discipline.

New general principles, the grail of biology, are becoming ever more elusive

It also seems to follow that the more fully diversity is explored, the more quickly will the true unifying principles be discovered. The laws of biology are written in the language of diversity. Researchers often speak half humorously about the rule attributed to the physiologist August Krogh: for every biological problem there exists an organism ideal for its solution. What can be called the inverse Krogh rule applies with equal force: for every organism there exists a problem for the solution of which it is ideally suited and others for which it is useless. Colon bacteria are wonderful for genetic mapping but not for meiosis. Langurs and lions have given us the key to understanding infanticide but would have been a wretched first choice for genetic mapping. Every kind of organism has a special place in the epistemological sun.

In short, the future of basic biological research lies substantially in the exploration of diversity. The surest path to discovery will be systematics of a new kind, in which deep knowledge of particular groups of organisms is promoted by research that moves back and forth across all the levels of biological organization. To be a world authority on nematodes or diatoms or palms will take on a new meaning and new expertise—and also new responsibilities.

The case of neurobiology

Neurobiology and behavior illustrate the direction in which most of biological research is moving. The most productive strategy has proved to be the selection and detailed analysis of paradigmatic species to illuminate two or more adjacent levels of organization in the tortuous route from genes to behavior. During the past 30 years a number of such key species have risen to prominence. They range from the simplest to the most complex (including humanity), with each species being used to study some phenomenon for which it provides a relatively easy or even unique access. Neurobiologists thus have demonstrated the utility of Krogh's rule and its inverse.

At the most elementary level is the demonstration of the control of locomotion in the bacterium *Escherichia coli*. The individual bacterium moves by rotating its flagellum like the propeller of a ship. It alters course by changing the direction of the spin of the flagellum, which causes it to tumble and points it in a random new direction. By continuous trial and error it is able to move toward nutrients and away from toxic substances. Partly because of the simplicity of the system, biologists have made remarkable progress in identifying the proteins that recognize chemical stimuli, as well as the central processing proteins. The genes prescribing the key proteins have been located. The extreme simplicity of the behavioral system has thus permitted biologists to characterize behavior all the way down to the level of the genes, but the behavioral patterns can be analogized with only a minute part of any pattern in more complex organisms.

Moving then to a more complicated organism, biologists are making rapid progress in the genetic dissection of *Drosophila* fruit flies, particularly *Drosophila melanogaster,* because these insects are relatively easy to manipulate genetically. Researchers can even create individuals that are mosaics of male and female cells.

These gynandromorphs, to cite one ingenious example from the rapidly growing field of *Drosophila* neurogenetics, are used to locate the sensory and nervous tissues that mediate certain forms of reproductive behavior. Investigators have been able to correlate the sex of the various tissues and the behaviors of the individual fly and thence locate genes for the processing of sensory information and efferent commands through the nervous system. Other research has led to the identification of many genes that control mating and orientation, as well as a portion of the molecular pathways that lead to the behavioral phenotype.

Sophisticated neurophysiological recordings have been the hallmark of research on the sea snail *Aplysia californica.* The roles and patterns of discharge of individual neurons have been traced in considerable detail. The cellular basis (and increasingly the molecular mechanisms) of elementary forms of learning have begun to yield to this approach, which takes advantage of the relative accessibility and anatomical simplicity of the *Aplysia* nervous system.

At a still higher level of organization, the social insects provide instructive paradigms. In the ants, bees, wasps, and termites, most behavior has no meaning except when fitted into a total pattern of different responses on the part of other colony members. One of the most fully understood species is the African weaver ant, *Oecophylla longinoda.* The colony consists of a single queen and a half million or more workers, all of which live in treetop nests constructed of leaves bound together by larval silk. This hexapod empire is held together by powerful attractants and ovary-suppressing pheromones produced by the queen.

The workers use no fewer than five procedures in chemical communication to recruit nestmates in different contexts. Various combinations of chemical and tactile signals are employed to attract nestmates to food discoveries, new territory, new nest sites, territorial invaders at long range, and territorial invaders at close range. Special identification substances, which differ from colony to colony, are deposited in anal droplets to mark the territory. The workers build new nests by forming chains and rows with their own bodies and pulling the edges of leaves in coordinated movements. When the correct arrangement of the leaves is attained, other workers bring out larvae and use them as living shuttles. Holding these grub-like forms in their mandibles, they move them back and forth while the larvae expel silk and bind the leaf edges together.

The laws of biology are written in the language of diversity

Thus although the brain of an individual ant is composed of at most a million nerve cells, compared with the 100 billion or more neurons forming the human brain, and although each ant performs fewer than fifty behavioral acts, the caste system and division of labor permit a complicated and effective repertory at the level of the colony. The insect society can be taken apart, analyzed, and reassembled in much the same manner as the bacterium and *Drosophila,* and with considerably greater ease, to illustrate what may eventually prove to be some of the more general features of biological organization.

Neurobiology and behavioral biology should continue to progress by the skillful use of the comparative method carried across all levels of organization. Again, the dominant theme is pluralism. Particular species are chosen for the levels of organization to which they provide easiest access. The whole mosaic can be fitted together only when the idiosyncrasies of each species are interpreted against the backdrop of its evolutionary history and place in the ecosystem. Enduring biological principles will emerge when enough of this information has accumulated. How much information is needed can only be dimly guessed.

The stewardship of systematics

I have argued that deep expertise in particular groups of organisms, if combined with free-wheeling opportunism in the choice of problems, is the wave of the future for biology. Individual investigators will range more easily from molecule to population, piecing together the clearest images taken from different species of organisms to create a synthesis that will, like an increasingly detailed mosaic, constitute modern biology. It seems equally likely that the progress of the science must depend on the spread of expertise across more and more species, with a full picture of the world biota as the ultimate goal. To intensify pluralism in this fashion means the rebirth of systematics as a principal orienting mode of biology.

The trend toward pluralism places a special obligation on systematists, including the core subset of researchers called taxonomists. Let me define these two occupational terms. As an expert on a group of species, a *systematist* is primarily interested in diversity, including classification, but ranges freely into other aspects of the biology of the favored group. A *taxonomist* is a systematist who is responsible for so many species that there is time only for their classification.

The resources of museums and other institutions housing major collections are already strained in the service of pure taxonomy, quite apart from the wider adventure of systematics. Aside from a few highly exceptional taxa such as mammals and birds, the identification of newly discovered organisms is often delayed for months or years. For many groups of organisms there are no experts at all. Taxonomists, and hence the broader class of systematists, cannot do the job expected of them. As biodiversity studies develop and practical needs also multiply, especially in the tropics, this shortfall could become critical within a very few years.

Systematists themselves are currently in the thrall of what some critics perceive as largely sterile arguments over the methodology of phylogenetic reconstruction. I see the methodological phase through which systematics has recently passed as invigorating and productive, even though it is certain to be greatly augmented or even replaced in some cases by direct reading of the genetic codes. The arguments have produced sound techniques for inferring degrees of similarity and branching during species formation. Even more important, they have resulted in a remarkable improvement in taxonomic procedures, amounting to the standardization of techniques by which results can be replicated and independently tested. But most of this activity is after all just methodology. The time has come for systematics to move on, to meet its destiny. Otherwise, why was all the work done?

For that matter, why does a biologist do research at all? To discover, of course. Alfred North Whitehead said that a scientist does not discover in order to know, he knows in order to discover. But in biology there is much more to the discovery impulse. The uniqueness of phylogenetic lineages means that history counts, and history in turn generates a sense of the sacredness of place, of life—not general and abstract but singular, a group of organisms on a particular piece of land across a measured span of time. Thus biology satisfies the two great expansive drives of the human mind: exploration and intellectual enrichment. The theme of pluralism ensures that biology will never, within any time imaginable, exhaust these two drives.

Systematists who are experts on particular groups of organisms, as opposed to those who study method alone, must overcome a certain reticence in their relationship to the rest of science. Too often I have heard other biologists say that a taxonomist with a grant goes off somewhere and writes a specialized monograph, and that is the end of it. And they say that systematists have not formulated a set of central questions in science that they alone are peculiarly qualified to answer.

If systematics were really an exhausted relic from the premolecular age, we should not try to keep it from entering the long sleep of senescence. But the exact opposite is the case. As it was during a glorious past and will be again, systematics in the broad sense is the key to the future of biology.

The responsible expert is the steward of a chosen taxonomic group in the service of science. He or she knows best which organisms exist and where, which are most endangered, which offer new kinds of problems to be solved, and which are most likely to benefit humankind. The systematist's best strategy is to explain such matters to as broad an audience as possible while inviting the collaboration of other biologists. No one but the systematist can reveal the particular and extraordinary value of the alcyonacean corals, chytrid fungi, anthribid weevils, sclerogibbid wasps, melostomes, ricinuleids, elephant fish, and so on down the long and enchanted roster.

Acknowledgments

I am indebted to the following persons for their critical reading of this article: Bert Hölldobler, Gerald Holton, Ernst Mayr, David M. Raup, and Carroll M. Williams. □

E. O. Wilson is a professor in the Museum of Comparative Zoology at Harvard University, Cambridge, MA 02138.

BioScience is the monthly magazine for biologists in all fields. It includes articles on research, policy, techniques, and education; news features on developments in biology; book reviews; announcements; meetings calendar; and professional opportunities. Published by the American Institute of Biological Sciences, 730 11th Street, NW, Washington, DC 20001-4584; 202/628-1500. 1989 membership dues, including *BioScience* subscription: Individual $42.00/year; Student $23.00/year. 1989 Institutional subscription rates: $93.00/year (domestic), $99.00/year (foreign).

1993

"Biophilia and the conservation ethic," *in* S. Kellert and E. O. Wilson, eds., *The Biophilia Hypothesis* (Island Press, Washington, DC), pp. 31–41

In my earlier work, *Biophilia* (Harvard University Press, Cambridge, MA, 1984), I argued that human nature, our instinctual essence, is inseverably linked to the remainder of the living world. Our natural affinity for life, which I labeled "biophilia," may be key to part of the evolution of the human mind. If this proposition is correct, and there was plenty of evidence for it in 1984 with more to follow, it has profound implications for our understanding and management of the environment. The main argument, with special reference to the conservation ethic, is expanded in the present essay.

CHAPTER 1

Biophilia and the Conservation Ethic

Edward O. Wilson

BIOPHILIA, IF IT exists, and I believe it exists, is the innately emotional affiliation of human beings to other living organisms. Innate means hereditary and hence part of ultimate human nature. Biophilia, like other patterns of complex behavior, is likely to be mediated by rules of prepared and counterprepared learning—the tendency to learn or to resist learning certain responses as opposed to others. From the scant evidence concerning its nature, biophilia is not a single instinct but a complex of learning rules that can be teased apart and analyzed individually. The feelings molded by the learning rules fall along several emotional spectra: from attraction to aversion, from awe to indifference, from peacefulness to fear-driven anxiety.

The biophilia hypothesis goes on to hold that the multiple strands of emotional response are woven into symbols composing a large part of culture. It suggests that when human beings remove themselves from the natural environment, the biophilic learning rules are not replaced by modern

32

versions equally well adapted to artifacts. Instead, they persist from generation to generation, atrophied and fitfully manifested in the artificial new environments into which technology has catapulted humanity. For the indefinite future more children and adults will continue, as they do now, to visit zoos than attend all major professional sports combined (at least this is so in the United States and Canada), the wealthy will continue to seek dwellings on prominences above water amidst parkland, and urban dwellers will go on dreaming of snakes for reasons they cannot explain.

Were there no evidence of biophilia at all, the hypothesis of its existence would still be compelled by pure evolutionary logic. The reason is that human history did not begin eight or ten thousand years ago with the invention of agriculture and villages. It began hundreds of thousands or millions of years ago with the origin of the genus *Homo*. For more than 99 percent of human history people have lived in hunter-gatherer bands totally and intimately involved with other organisms. During this period of deep history, and still farther back, into paleohominid times, they depended on an exact learned knowledge of crucial aspects of natural history. That much is true even of chimpanzees today, who use primitive tools and have a practical knowledge of plants and animals. As language and culture expanded, humans also used living organisms of diverse kinds as a principal source of metaphor and myth. In short, the brain evolved in a biocentric world, not a machine-regulated world. It would be therefore quite extraordinary to find that all learning rules related to that world have been erased in a few thousand years, even in the tiny minority of peoples who have existed for more than one or two generations in wholly urban environments.

The significance of biophilia in human biology is potentially profound, even if it exists solely as weak learning rules. It is relevant to our thinking about nature, about the landscape, the arts, and mythopoeia, and it invites us to take a new look at environmental ethics.

How could biophilia have evolved? The likely answer is biocultural evolution, during which culture was elaborated under the influence of hereditary learning propensities while the genes prescribing the propensities were spread by natural selection in a cultural context. The learning rules can be inaugurated and fine-tuned variously by an adjustment of sensory thresholds, by a quickening or blockage of learning, and by modification of emotional responses. Charles Lumsden and I (1981, 1983, 1985) have envisioned biocultural evolution to be of a particular kind, gene-culture coevolution, which traces a spiral trajectory through time: a certain genotype makes a behavioral response more likely, the response enhances survival and reproductive fitness, the genotype consequently spreads through the population, and the behavioral response grows more frequent. Add to this the strong general tendency of human beings to translate emotional feelings into myriad dreams and narratives, and the necessary conditions are in place to cut the historical channels of art and religious belief.

Gene-culture coevolution is a plausible explanation for the origin of biophilia. The hypothesis can be made explicit by the human relation to snakes. The sequence I envision, drawn principally from elements established by the art historian and biologist Balaji Mundkur, is this:

1. Poisonous snakes cause sickness and death in primates and other mammals throughout the world.
2. Old World monkeys and apes generally combine a strong natural fear of snakes with fascination for these animals and the use of vocal communication, the latter including specialized sounds in a few species, all drawing attention of the group to the presence of snakes in the near vicinity. Thus alerted, the group follows the intruders until they leave.
3. Human beings are genetically averse to snakes. They are quick to develop fear and even full-blown phobias with very little negative reinforcement. (Other phobic elements in the natural environment include dogs, spiders, closed spaces, running water, and heights. Few modern artifacts are as effective—even those most dangerous, such as guns, knives, automobiles, and electric wires.)
4. In a manner true to their status as Old World primates, human beings too are fascinated by snakes. They pay admission to see captive specimens in zoos. They employ snakes profusely as metaphors and weave them into stories, myth, and religious symbolism. The serpent gods of cultures they have conceived all around the world are furthermore typically ambivalent. Often semihuman in form, they are poised to inflict vengeful death but also to bestow knowledge and power.

34

5. People in diverse cultures dream more about serpents than any other kind of animal, conjuring as they do so a rich medley of dread and magical power. When shamans and religious prophets report such images, they invest them with mystery and symbolic authority. In what seems to be a logical consequence, serpents are also prominent agents in mythology and religion in a majority of cultures.

Here then is the ophidian version of the biophilia hypothesis expressed in briefest form: constant exposure through evolutionary time to the malign influence of snakes, the repeated experience encoded by natural selection as a hereditary aversion and fascination, which in turn is manifested in the dreams and stories of evolving cultures. I would expect that other biophilic responses have originated more or less independently by the same means but under different selection pressures and with the involvement of different gene ensembles and brain circuitry.

This formulation is fair enough as a working hypothesis, of course, but we must also ask how such elements can be distinguished and how the general biophilia hypothesis might be tested. One mode of analysis, reported by Jared Diamond in this volume, is the correlative analysis of knowledge and attitude of peoples in diverse cultures, a research strategy designed to search for common denominators in the total human pattern of response. Another, advanced by Roger Ulrich and other psychologists, is also reported here: the precisely replicated measurement of human subjects to both attractive and aversive natural phenomena. This direct psychological approach can be made increasingly persuasive, whether for or against a biological bias, when two elements are added. The first is the measurement of heritability in the intensity of the responses to the psychological tests used. The second element is the tracing of cognitive development in children to identify key stimuli that evoke the responses, along with the ages of maximum sensitivity and learning propensity. The slithering motion of an elongate form appears to be the key stimulus producing snake aversion, for example, and preadolescence may be the most sensitive period for acquiring the aversion.

Given that humanity's relation to the natural environment is as much a part of deep history as social behavior itself, cognitive psychologists have

35

been strangely slow to address its mental consequences. Our ignorance could be regarded as just one more blank space on the map of academic science, awaiting genius and initiative, except for one important circumstance: the natural environment is disappearing. Psychologists and other scholars are obligated to consider biophilia in more urgent terms. What, they should ask, will happen to the human psyche when such a defining part of the human evolutionary experience is diminished or erased?

There is no question in my mind that the most harmful part of ongoing environmental despoliation is the loss of biodiversity. The reason is that the variety of organisms, from alleles (differing gene forms) to species, once lost, cannot be regained. If diversity is sustained in wild ecosystems, the biosphere can be recovered and used by future generations to any degree desired and with benefits literally beyond measure. To the extent it is diminished, humanity will be poorer for all generations to come. How much poorer? The following estimates give a rough idea:

- Consider first the question of the *amount* of biodiversity. The number of species of organisms on earth is unknown to the nearest order of magnitude. About 1.4 million species have been given names to date, but the actual number is likely to lie somewhere between 10 and 100 million. Among the least-known groups are the fungi, with 69,000 known species but 1.6 million thought to exist. Also poorly explored are at least 8 million and possibly tens of millions of species of arthropods in the tropical rain forests, as well as millions of invertebrate species on the vast floor of the deep sea. The true black hole of systematics, however, may be bacteria. Although roughly 4,000 species have been formally recognized, recent studies in Norway indicate the presence of 4,000 to 5,000 species among the 10 billion individual organisms found on average in each gram of forest soil, almost all new to science, and another 4,000 to 5,000 species, different from the first set and also mostly new, in an average gram of nearby marine sediments.
- Fossil records of marine invertebrates, African ungulates, and flowering plants indicate that on average each clade—a species and its descendants—lasts half a million to 10 million years under natural conditions. The longevity is measured from the time the ancestral form

36

splits off from its sister species to the time of the extinction of the last descendant. It varies according to the group of organisms. Mammals, for example, are shorter-lived than invertebrates.

- Bacteria contain on the order of a million nucleotide pairs in their genetic code, and more complex (eukaryotic) organisms from algae to flowering plants and mammals contain 1 to 10 billion nucleotide pairs. None has yet been completely decoded.
- Because of their great age and genetic complexity, species are exquisitely adapted to the ecosystems in which they live.
- The number of species on earth is being reduced by a rate 1,000 to 10,000 times higher than existed in prehuman times. The current removal rate of tropical rain forest, about 1.8 percent of cover each year, translates to approximately 0.5 percent of the species extirpated immediately or at least doomed to much earlier extinction than would otherwise have been the case. Most systematists with global experience believe that more than half the species of organisms on earth live in the tropical rain forests. If there are 10 million species in these habitats, a conservative estimate, the rate of loss may exceed 50,000 a year, 137 a day, 6 an hour. This rate, while horrendous, is actually the minimal estimate, based on the species / area relation alone. It does not take into account extinction due to pollution, disturbance short of clear-cutting, and the introduction of exotic species.

Other species-rich habitats, including coral reefs, river systems, lakes, and Mediterranean-type heathland, are under similar assault. When the final remnants of such habitats are destroyed in a region—the last of the ridges on a mountainside cleared, for example, or the last riffles flooded by a downstream dam—species are wiped out en masse. The first 90 percent reduction in area of a habitat lowers the species number by one-half. The final 10 percent eliminates the second half.

It is a guess, subjective but very defensible, that if the current rate of habitat alteration continues unchecked, 20 percent or more of the earth's species will disappear or be consigned to early extinction during the next thirty years. From prehistory to the present time humanity has probably already eliminated 10 or even 20 percent of the species. The number of bird species, for example, is down by an estimated 25 percent, from 12,000 to 9,000, with

37

a disproportionate share of the losses occurring on islands. Most of the megafaunas—the largest mammals and birds—appear to have been destroyed in more remote parts of the world by the first wave of hunter-gatherers and agriculturists centuries ago. The diminution of plants and invertebrates is likely to have been much less, but studies of archaeological and other subfossil deposits are too few to make even a crude estimate. The human impact, from prehistory to the present time and projected into the next several decades, threatens to be the greatest extinction spasm since the end of the Mesozoic era 65 million years ago.

Assume, for the sake of argument, that 10 percent of the world's species that existed just before the advent of humanity are already gone and that another 20 percent are destined to vanish quickly unless drastic action is taken. The fraction lost—and it will be a great deal no matter what action is taken—cannot be replaced by evolution in any period that has meaning for the human mind. The five previous major spasms of the past 550 million years, including the end-Mesozoic, each required about 10 million years of natural evolution to restore. What humanity is doing now in a single lifetime will impoverish our descendants for all time to come. Yet critics often respond, "So what? If only half the species survive, that is still a lot of biodiversity—is it not?"

The answer most frequently urged right now by conservationists, I among them, is that the vast material wealth offered by biodiversity is at risk. Wild species are an untapped source of new pharmaceuticals, crops, fibers, pulp, petroleum substitutes, and agents for the restoration of soil and water. This argument is demonstrably true—and it certainly tends to stop anticonservation libertarians in their tracks—but it contains a dangerous practical flaw when relied upon exclusively. If species are to be judged by their potential material value, they can be priced, traded off against other sources of wealth, and—when the price is right—discarded. Yet who can judge the *ultimate* value of any particular species to humanity? Whether the species offers immediate advantage or not, no means exist to measure what benefits it will offer during future centuries of study, what scientific knowledge, or what service to the human spirit.

At last I have come to the word so hard to express: spirit. With reference to the spirit we arrive at the connection between biophilia and the environ-

38

mental ethic. The great philosophical divide in moral reasoning about the remainder of life is whether or not other species have an innate right to exist. That decision rests in turn on the most fundamental question of all: whether moral values exist apart from humanity, in the same manner as mathematical laws, or whether they are idiosyncratic constructs that evolved in the human mind through natural selection. Had a species other than humans attained high intelligence and culture, it would likely have fashioned different moral values. Civilized termites, for example, would support cannibalism of the sick and injured, eschew personal reproduction, and make a sacrament of the exchange and consumption of feces. The termite spirit, in short, would have been immensely different from the human spirit—horrifying to us in fact. The constructs of moral reasoning, in this evolutionary view, are the learning rules, the propensities to acquire or to resist certain emotions and kinds of knowledge. They have evolved genetically because they confer survival and reproduction on human beings.

The first of the two alternative propositions—that species have universal and independent rights regardless of how else human beings feel about the matter—may be true. To the extent the proposition is accepted, it will certainly steel the determination of environmentalists to preserve the remainder of life. But the species-right argument alone, like the materialistic argument alone, is a dangerous play of the cards on which to risk biodiversity. The independent-rights argument, for all its directness and power, remains intuitive, aprioristic, and lacking in objective evidence. Who but humanity, it can be immediately asked, gives such rights? Where is the enabling canon written? And such rights, even if granted, are always subject to rank-ordering and relaxation. A simplistic adjuration for the right of a species to live can be answered by a simplistic call for the right of people to live. If a last section of forest needs to be cut to continue the survival of a local economy, the rights of the myriad species in the forest may be cheerfully recognized but given a lower and fatal priority.

Without attempting to resolve the issue of the innate rights of species, I will argue the necessity of a robust and richly textured anthropocentric ethic apart from the issue of rights—one based on the hereditary needs of our own species. In addition to the well-documented utilitarian potential of wild species, the diversity of life has immense aesthetic and spiritual

39

value. The terms now to be listed will be familiar, yet the evolutionary logic is still relatively new and poorly explored. And therein lies the challenge to scientists and other scholars.

Biodiversity is the Creation. Ten million or more species are still alive, defined totally by some 10^{17} nucleotide pairs and an even more astronomical number of possible genetic recombinants, which creates the field on which evolution continues to play. Despite the fact that living organisms compose a mere ten-billionth part of the mass of earth, biodiversity is the most information-rich part of the known universe. More organization and complexity exist in a handful of soil than on the surfaces of all the other planets combined. If humanity is to have a satisfying creation myth consistent with scientific knowledge—a myth that itself seems to be an essential part of the human spirit—the narrative will draw to its conclusion in the origin of the diversity of life.

Other species are our kin. This perception is literally true in evolutionary time. All higher eukaryotic organisms, from flowering plants to insects and humanity itself, are thought to have descended from a single ancestral population that lived about 1.8 billion years ago. Single-celled eukaryotes and bacteria are linked by still more remote ancestors. All this distant kinship is stamped by a common genetic code and elementary features of cell structure. Humanity did not soft-land into the teeming biosphere like an alien from another planet. We arose from other organisms already here, whose great diversity, conducting experiment upon experiment in the production of new life-forms, eventually hit upon the human species.

The biodiversity of a country is part of its national heritage. Each country in turn possesses its own unique assemblages of plants and animals including, in almost all cases, species and races found nowhere else. These assemblages are the product of the deep history of the national territory, extending back long before the coming of man.

Biodiversity is the frontier of the future. Humanity needs a vision of an expanding and unending future. This spiritual craving cannot be satisfied by the colonization of space. The other planets are inhospitable and immensely expensive to reach. The nearest stars are so far away that voyagers would need thousands of years just to report back. The true frontier for humanity is life on earth—its exploration and the transport of knowledge

40

about it into science, art, and practical affairs. Again, the qualities of life that validate the proposition are: 90 percent or more of the species of plants, animals, and microorganisms lack even so much as a scientific name; each of the species is immensely old by human standards and has been wonderfully molded to its environment; life around us exceeds in complexity and beauty anything else humanity is ever likely to encounter.

The manifold ways by which human beings are tied to the remainder of life are very poorly understood, crying for new scientific inquiry and a boldness of aesthetic interpretation. The portmanteau expressions "biophilia" and "biophilia hypothesis" will serve well if they do no more than call attention to psychological phenomena that rose from deep human history, that stemmed from interaction with the natural environment, and that are now quite likely resident in the genes themselves. The search is rendered more urgent by the rapid disappearance of the living part of that environment, creating a need not only for a better understanding of human nature but for a more powerful and intellectually convincing environmental ethic based upon it.

REFERENCES

I first used the expression "biophilia" in 1984 in a book entitled by the name (*Biophilia*, Harvard University Press). In that extended essay I attempted to apply ideas of sociobiology to the environmental ethic.

The mechanism of gene-culture coevolution was proposed by Charles J. Lumsden and myself in *Genes, Mind, and Culture* (Harvard University Press, 1981), *Promethean Fire* (Harvard University Press, 1983), and "The Relation Between Biological and Cultural Evolution," *Journal of Social and Biological Structure* 8(4) (October 1985):343–359. It represents an extension of theoretical population genetics in an effort to include the principles of cognition and social psychology.

Balaji Mundkur traced the role of snakes and mythic serpents in *The Cult of the Serpent: An Interdisciplinary Survey of Its Manifestations and Origins* (State University of New York Press, 1983).

Jared Diamond's study of Melanesian attitudes toward other forms of life and Roger S. Ulrich's review of psychological research on biophilia are presented elsewhere in this volume.

I have reviewed the measures of global biodiversity and extinction rates in greater detail in *The Diversity of Life* (Harvard University Press, 1992).

Clarifying the Concept

41 In evaluating the environmental ethic I have been aided greatly by the writings of several philosophers, including most notably Bryan Norton (*Why Preserve Natural Diversity?*, Princeton University Press, 1987), Max Oelschlaeger (*The Idea of Wilderness: From Prehistory to the Age of Ecology*, Yale University Press, 1991), Holmes Rolston III (*Environmental Ethics: Duties to and Values in the Natural World*, Temple University Press, 1988), and Peter Singer (*The Expanding Circle: Ethics and Sociobiology*, Farrar, Straus & Giroux, 1981).

1993

"Is humanity suicidal?," *New York Times Magazine,* May 30, pp. 24–29

When the editors of the *New York Times Magazine* asked me to address the question in the title of this essay, my answer was no, we aren't suicidal, but we are death for much of the rest of life and, hence, in ultimate prospect, unwittingly dangerous to ourselves. This version of the essay was published in *Biosystems* the same year.

ELSEVIER SCIENTIFIC PUBLISHERS IRELAND

BioSystems 31 (1993) 235–242

Bio Systems

Is humanity suicidal?†

Edward O. Wilson

Museum of Comparative Zoology, Harvard University, Cambridge, MA 02138, USA

Abstract

The world's fauna and flora has entered a crisis unparalleled since the end of the Mesozoic Era, with the extinction rate of species now elevated to more than a thousand times that existing before the coming of humanity. Scientists and policy makers are ill-prepared to moderate this hemorrhaging, because so little is known of the biology of the Earth's millions of species and because so little effort has been directed toward conservation thus far. With the vanished species will go great potential wealth in scientific knowledge, new products, ecosystems services, and part of the natural world in which the human species originated. The need for new research and improved management is thus urgent. If it is not met, humanity will likely survive, but in a world biologically impoverished for all time.

Key words: Biodiversity; Conservation; Evolution; Extinction; Species; Pharmaceuticals; Ecosystems; Rain forests

Imagine that on an icy moon of Jupiter — say, Ganymede — is concealed the space station of an alien civilization. For millions of years its scientists have closely watched the earth. Because their law prevents settlement on a living planet, they have tracked the surface by means of satellites equipped with sophisticated sensors, mapping the spread of large assemblages of organisms, from forests, grasslands, and tundras to coral reefs and the vast planktonic meadows of the sea. They have recorded millennial cycles in the climate, interrupted by the advance and retreat of glaciers and a scattershot of volcanic eruptions.

The watchers have been waiting for what might be called the Moment. When it comes, occupying only a few centuries and thus a mere tick in geological time, the forests shrink back to less than half their original cover. Atmospheric carbon dioxide rises to the highest level in more than a hundred thousand years. The ozone layer of the stratosphere thins, and holes open at the poles. Plumes of nitrous oxide and other toxins rise from fires in South America and Africa, settle in the upper troposphere and drift eastward across the oceans. At night the land surface brightens with millions of pinpoints of light, which coalesce into blazing swaths across Europe, Japan, and eastern North America. A semicircle of fire spreads from gas flares around the Persian Gulf.

It was all but inevitable, the watchers might tell us if we met them, that from the great diversity of large animals and given enough time, one species or another would gain intelligent control of Earth. That role has fallen to *Homo sapiens*, a primate risen in Africa from a lineage that split away from the chimpanzee line five to eight million years ago. Unlike any creature that lived before, we have become a geophysical force, swiftly changing the atmosphere and climate as well as the composition of the remainder of life. Now in the midst of a pop-

†Shorter version first published in the New York Times Sunday Magazine, May 30, 1993, Reprinted here with permission from Prof Wilson and Mr James Atlas.

SSDI 0303-2647(93)01423-Q

ulation explosion, the human species has doubled in population during the past 50 years to 5.5 billion. It is scheduled to double again in the next 50 years. No other single species in evolutionary history has even remotely approached the sheer mass in protoplasm generated by humanity.

Darwin's dice have rolled badly for Earth. It was a misfortune for the living world in particular, many scientists believe, that a carnivorous primate and not some more benign form of animal made the breakthrough. Our species retains hereditary traits that add greatly to our destructive impact. We are tribal and aggressively territorial, intent on private space beyond minimal requirements, and oriented by selfish sexual and reproductive drives. Cooperation beyond the family and tribal levels comes hard.

Worse, our partial carnivory causes us to use the sun's energy at low efficiency. It is a general rule of ecology that (very roughly) only 10% of the energy of production is saved across each step in the food chain. In a way that varies greatly from one habitat to another, about 10% of the sun's energy captured by photosynthesis to produce plant tissue is converted into energy in the tissue of herbivores, the animals that eat the plants. Of that amount, 10% reaches the tissue of the carnivores feeding on the herbivores. Similarly, only 10% is transferred to carnivores that eat carnivores. And so on for another step or two. In a wetlands chain that runs from marsh grass to grasshopper to warbler to hawk, the energy captured during green production shrinks a thousand-fold. In other words, it takes a great deal of grass to support a hawk. Human beings, like hawks, are top carnivores, at the end of the food chain whenever they eat meat, two or more links removed from the plants; if chicken, for example, two links, and if tuna, four links. Even with most societies confined today to a mostly vegetarian diet, humanity is gobbling up a large part of the rest of the living world. We appropriate between 20 and 40% of the sun's energy that would otherwise be fixed into the tissue of natural vegetation, principally by our consumption of crops and timber, construction of buildings and roadways, and the creation of wastelands. In the relentless search for more food we have reduced animal life in lakes, rivers, and now, increasingly, the open ocean. And everywhere we pollute the air and water, lower water tables, and extinguish species.

The human species is, in a word, an environmental abnormality. It is possible that intelligence in the wrong kind of species was foreordained to be a fatal combination for the biosphere. Perhaps one of the laws of evolution across inhabited planets in the universe, documented by the watchers on Ganymede, is that intelligence usually extinguishes itself.

This admittedly dour scenario is based on what can be termed the juggernaut theory of human nature. The conception holds that people are programmed by their genetic heritage to be so selfish that a sense of global responsibility will come too late. Individuals place themselves first, family second, tribe third, and the rest of the world a distant fourth. Their genes also predispose them to plan ahead for one or two generations at most. They fret over the petty problems and conflicts of their daily lives, and respond swiftly and often ferociously to slight challenges to their status and tribal security. But oddly, as psychologists have discovered, people also tend to underestimate both the likelihood and impact of such disasters as major earthquakes and great storms.

The reason for the myopic fog, many evolutionary biologists also contend, is that it was actually advantageous during all but the last few millennia of the two million years of existence of the genus *Homo*. The brain evolved into its present form during this long stretch of evolutionary time, during which people existed in small, preliterate hunter-gatherer bands. Life was precarious and short. A premium was placed on close attention to the near future and early reproduction, and little else. Disasters of a magnitude that occur only once every few centuries were forgotten or transmuted into myth. So today the mind still works comfortably backward and forward for only a few years, spanning a period not exceeding one or two generations. Those in past ages whose genes inclined them to think on the short term lived longer and had more children than those who did not. Prophets never enjoyed a Darwinian edge, this view holds, and it follows that today cosmopolitanism is in short supply.

The rules have recently changed, however. Global crises are rising within the life span of the generation now coming of age, a foreshortening that may explain why young people express more concern about the environment than do their elders. The time scale has contracted because of the exponential growth in both the human population and technologies impacting the environment. Exponential growth, let me add in reminder, is basically the same as the increase of wealth by compound interest. The larger the population, the faster the growth; the faster the growth, the sooner the pulation becomes still larger. Nigeria, to illust ·te by reference to one of our more fecund nations, is 3.3%. The population is expected to double from its 1988 level to 216 million by year 2010. If the same rate of growth could be continued without abatement for yet another hundred years, to 2110, the population will exceed that of the rest of humanity currently in existence. If continued for one thousand years, Nigerians will weigh as much as the planet Earth.

And with people everywhere seeking a better quality of life, the search for resources is expanding even faster than the population. The demand is being met by an increase in scientific knowledge, which doubles every 10–15 years. It is accelerated further by a parallel rise in environment-devouring technology. Because Earth is finite in many resources that determine the quality of life — including arable soil, nutrients, fresh water, and space for natural ecosystems — doubling of consumption at constant time intervals can bring disaster with shocking suddenness. Even when a non-renewable resource has been only half used, it is still only one interval away from the end. Ecologists like to make this point with the French riddle of the lily pond. At first there is only one lily pad in the pond, but the next day it doubles and thereafter each of its descendants double. The pond completely fills with lily pads in 30 days. When is the pond exactly half full? Answer: on the twenty-ninth day.

Yet, mathematical exercises aside, who can safely take the measure of human power to overcome the perceived limits of Earth? The question of central interest is this: are we racing to the brink of an abyss, or are we just gathering speed for a takeoff to a wonderful future? The crystal ball is clouded; the human condition baffles all the more because it is both unprecedented and bizarre, almost beyond understanding.

In the midst of uncertainty, opinions on the human prospect have tended to fall loosely into two schools. The first, exemptionalism, holds that since humankind is transcendent in intelligence and spirit, so must also our species have been released from the iron laws of ecology that bind the remainder of life. No matter how serious the problem, civilized human beings, by ingenuity, force of will, and — who knows — maybe divine dispensation, will find a solution. Population growth? Good for the economy, claim some of the exemptionalists, and in any case a basic human right, so let it run. Land shortages? Try fusion energy to power the desalinization of seawater, then reclaim the world's deserts. (The process might be assisted by towing icebergs to coastal pipelines.) Species going extinct? Not to worry. That is nature's way, or did we interpret Darwin wrongly? Think of mankind as only the latest in a long line of exterminating agents in geological time. In any case, because our species has pulled free of old-style, mindless Nature, we have begun a different order of life. Evolution should now be allowed to proceed along this new trajectory. Finally, resources? The planet has more than enough resources to last indefinitely, if human genius is allowed to address each new problem in turn, without alarmist and unreasonable restrictions imposed on economic development. So hold the course, and touch the brakes lightly.

The opposing idea of reality is environmentalism. When expressed as a full-blown philosophy as opposed to mere cautionary sentiment, it sees humanity as a biological species tightly dependent on the natural world. As formidable as our intellect may be and as fierce our spirit, the argument goes, they are not enough to free us from the constraints of the natural environment in which the human ancestors evolved. We cannot draw confidence from successful solutions to the smaller problems of the past. Many of Earth's vital resources are about to be exhausted, her atmospheric chemistry is deteriorating, and human populations have already grown dangerously large. Natural ecosystems, the wellsprings of a healthful environment, are being irreversibly degraded.

At the heart of the environmentalist world view is the conviction that human physical and spiritual health depend on sustaining the planet in a relatively unaltered state. Earth is our home in the full, genetic sense, where humanity and its ancestors existed for all the millions of years of their evolution. Natural ecosystems — forests, coral reefs, marine blue waters — maintain the world exactly as we would wish it to be maintained. Our body and our mind evolved precisely to live in this particular planetary environment and no other. When we debase the global environment and extingish the variety of life, we are dismantling a support system that is too complex to understand, let alone replace, in the foreseeable future. Space scientists theorize the existence of a virtually unlimited array of other planetary environments, almost all of which are uncongenial to human life. Our own Mother Earth, lately called Gaia, is a specialized conglomerate of organisms and the physical environment they create on a day-to-day basis, which can be destabilized and turned lethal by careless activity. We run the risk, conclude the environmentalists, of beaching ourselves upon alien shores like a great confused pod of pilot whales.

If I have not done so enough already by tone of voice, I will now place myself solidly in the environmentalist school. Not so radical as to wish a turning back of the clock, not given to driving spikes into Douglas firs to prevent logging, and distinctly uneasy with such hybrid movements as ecofeminism, which holds (as summarized by the historian-philosopher Max Oelschlaeger) that

> Mother Earth is a nurturing home for all life and should be revered and loved as in premodern (Paleolithic and archaic) societies. Ecosystematic malaise and abuse is rooted in androcentric concepts, values, and institutions. The many problems of human relations, and relations between the human and non-human worlds, will not be resolved until androcentric institutions, values, and ideology are eradicated.

Still, even though soaked in androcentric culture, I am radical enough to take seriously the question heard with increasing frequency: Is humanity suicidal? Is the drive to environmental conquest and self-propagation set so deeply in our genes as to be unstoppable prior to a collapse of the biosphere?

My short answer — opinion if you wish — is that humanity is not suicidal, at least not in the sense just stated. We are smart enough and have time to avoid an environmental catastrophe of civilization-threatening dimensions. But the technical problems are sufficiently formidable to require a redirection of much of science and technology, and the ethical issues are so basic as to force a reconsideration of our self-image as a species.

Let me first cite some reasons for optimism, then redefine the global problem in an attempt to take its measure. People are awakening to the environment worldwide, and they see in it both danger and opportunity. There are reasons for believing that this time the movement is for real, that we have entered what might someday be generously called the Century of the Environment. Among the signposts are the following:

- The United Nations Conference on Environment and Development (UNCED), held in Rio de Janeiro in June 1992, attracted more than 120 heads of government, the largest number ever assembled in history. They were joined by representatives of conservation organizations, environmental scientists, journalists, and spectators from around the world. The importance of the Rio Conference was not so much in the agreements signed, which are flexible, tenuous, and often heavily compromised, as in formal recognition at the highest levels that the problems exist, and in the attention given to the event by the media. Rio helped move environmental issues closer to the political center stage.
- On November 18, over 1500 senior scientists from 69 countries issued a 'Warning to Humanity,' stating that overpopulation and environmental deterioration put the very future of life at risk. Unprecedented in its representation, and cutting across political ideologies, the group included most of the living Nobel laureates, a majority of

the members of the Pontifical Academy of Sciences, and leaders of many national and international scientific organizations.

• The greening of religion is yet another global trend, with theologians and religious leaders addressing environmental problems as a moral issue. In May 1992 leaders of most of the major American denominations met with scientists as guests of members of the U.S. Senate to formulate a 'Joint Appeal by Religion and Science for the Environment.'

• Polls show that the American public considers environmental quality both to be compatible with economic growth and deserving of high priority. In 1992, 77% of Americans polled thought it possible to balance economic growth and environmental quality; 17% stated willingness to sacrifice economic growth to protect the environment; and only 4% agreed with the reverse. Eighty-five percent favored more spending on research to protect the environment in ways consistent with economic growth, including the improvement of energy efficiency and conservation.

• Conservation of biodiversity is increasingly seen by both national governments and major landowners as important to their country's future. Indonesia, home to a large part of the native Asian plant and animal species, has begun to shift to land management practices that conserve and sustainably develop the remaining rain forests. Costa Rica has created a National Institute of Biodiversity to the same end. Mexico and Taiwan have followed suit with comparable institutes. A pan-African institute for biodiversity research and management has recently been founded with headquarters in Zimbabwe.

• There are favorable demographic signs. The rate of population increase is declining on all continents, although it is still well above zero almost everywhere and remains especially high in Sub-Saharan Africa. Despite entrenched traditions and religious beliefs, the desire to use contraceptives in family planning is spreading around the world. Almost half the women in developing countries employ one method or another, while another 17% state they would use them but still lack access. Demographers estimate that if the demand were fully met, this action alone would reduce the eventual stabilized population by over two billion.

• The environmental problems created by overpopulation are vastly amplified by the acceleration in demand for energy. At the present time the industrialized nations, with only 20% of the world's population, consume 70% of its disposable energy. It is hard to imagine how the developing countries can approach the same quality of life without unacceptable increases in the use of fossil fuels or the proliferation of nuclear reactors. Yet even here, with determination and luck, the problem can be solved. According to John Holdren of the University of California, it might be accomplished with only a tripling of global energy consumption, if both the poor and rich countries use every energy-saving method available and the population does not rise above 10 billion.

In summary, at least some of the gathering environmental problems can be overcome with existing or feasible technology. The will is also potentially there, if for no other reason that the schedule of crises has now been shortened to the extent that they stare us in the face. Yet the awful truth remains that a large part of humanity will suffer no matter what is done. The number of people living in absolute poverty, defined by Robert S. McNamara as life so characterized by malnutrition, illiteracy, and disease as to fall below any reasonable criterion of dignity, has risen during the past 20 years to nearly one billion. It is expected to increase another 100 million by the end of the decade. Whatever progress has been made in the developing countries, and that includes an overall improvement in the average standard of living, is threatened by a continuance of rapid population growth and the deterioration of forests and arable soil.

Our hopes must be chastened further still, and this is in my opinion the central issue, by a key and seldom recognized distinction between the non-living and living environments. Science and the political process can be adapted to manage the non-living, physical environment. The human hand is now upon the physical homeostat. The ozone layer can be fully restored to the upper atmosphere by elimination of CFCs, with these substances peaking at six times the present level and then subsiding during the next half century. Also, with procedures that will prove far more difficult and initially expensive, carbon dioxide and other greenhouse gases can be pulled back to concentrations that slow global warming.

The human hand, however, is not upon the biological homeostat. There is no way in sight to micromanage the natural ecosystems and the millions of species they contain. That feat might be accomplished by generations to come, but then it will be too late for the ecosystems — and perhaps for us. The faith of exemptionalists that humanity can plunge ahead, working out problems seriatim as we go along, is for this reason alone ruinous. Consequently, as I'll now argue, science and public policy should be directed toward preserving natural ecosystems until such time, if ever, our descendants wish to make the great leap and make themselves dependent upon a closely cultivated or even artificial biosphere.

The point of no return, in which both the living world and humanity depend for survival on our day-to-day intervention, has fortunately not been reached. Despite the changes we have made, we still live in a largely natural environment. Humanity is heir to a variety of life large enough to have resisted instant eradication. The amount of this biodiversity — the number of genes, the number of species, and so on up through the levels of biological organization to the number of ecosystems — is close to the greatest that ever existed in geological history.

In making that statement, I grant that it is very difficult to put an exact measure on global biodiversity. Biologists have discovered and given scientific names to about 1.4 million species, but that is only a beginning. The true number has been estimated by various recent authors to lie between 10 and 100 million, with no solid consensus on which is the closer. Even the most familiar groups of animals and plants are imperfectly known. An average of two or three new species of birds turn up somewhere in the world every year. A new previously unknown beaked whale was described from the eastern Pacific in 1991, and another has been sighted from ships in the same region but not yet captured. Novel frogs and fishes from the tropics are a glut in museums. Smaller organisms are a mind-boggling cascade, with unclassified insects, fungi, and bacteria numbering in the millions.

Despite the seemingly bottomless nature of the Creation, mankind has been chipping away, and Earth is destined to become an impoverished planet within a century if present trends continue. Mass extinctions are being reported with increasing frequency in every part of the world. They include half the freshwater fishes of peninsular Malaysia, 10 birds native to Cebu in the Philippines, half the 41 tree snails of Oahu, 44 of the 68 shallow-water mussels of the Tennessee River shoals, as many as 90 plant species growing on Ecuador's Centinela Ridge, and in the United States as a whole, about 200 plant species, with another 680 species and races now classified as in danger of extinction. The main cause of the decline is the destruction of natural habitats, especially tropical forests. Close behind, especially on the Hawaiian archipelago and other islands, is the introduction of rats, pigs, beard grass, lantana, and other exotic organisms that outbreed and extirpate native species.

The few thousand biologists worldwide who specialize on diversity are aware that they can witness and report no more than a tiny percentage of the extinctions actually occurring. The reason is that they have facilities to keep track of only a tiny fraction of the millions of species and a sliver of the planet's surface on a yearly basis. They have devised a rule of thumb to characterize the situation, that whenever careful studies are made of habitats before and after disturbance, extinctions almost always come to light. The corollary: the great majority of extinctions are never observed. Vast numbers of species are apparently vanishing before they can be discovered and named.

There is a way, nonetheless, to estimate the rate of loss indirectly. Independent studies around the world and in fresh and marine waters have revealed a robust connection between the size of a habitat and the amount of biodiversity it contains. Even a small loss in area reduces the number of species. The relation is such that when the area of the habitat is cut to a tenth of its original cover, the number of species eventually drops by roughly one half. Tropical rain forests, thought to harbor a majority of Earth's species (the reason conservationists get so exercised about rain forests) are being reduced by nearly that magnitude. At the present time they occupy about the same areas as that of the 48 coterminous United States, representing a little less than half their original, prehistoric cover; and they are shrinking each year by about 2%, an amount equal to the state of Florida. If the typical value (that is, 90% percent area loss causes 50% eventual extinction) is applied, the projected loss of species due to rain forest destruction worldwide is half a percent, across the board for all kinds of plants, animals, and microorganisms.

Such estimates of extinction based on area are still on the low side because they do not include the additional hemorrhaging due to partial disturbance of the forests, as opposed to clear-cutting, and the introduction of exotic organisms. Nor do they include the whopping sudden losses that occur when the last remnants of local forest are felled, an event occurring with increasing frequency in many tropical countries.

When all of the extinction agents are considered together, it is reasonable to project a reduction by twenty percent or more of the rain forest species by the year 2020, climbing to 50% or more by mid-century, if nothing is done to change current practice. Comparable erosion is likely in other environments now under assault, including many coral reefs and Mediterranean-type heathlands of Western Australia, South Africa, and California.

The ongoing loss will not be replaced by evolution in any period of time that has meaning for humanity. Extinction is now proceeding thousands of times faster than the production of new species. The average life span of a species and its descendants in past geological eras varied according to group (such as mollusks or echinoderms or flowering plants) from about one to ten million years. During the past 500 million years, there have been five great extinction spasms comparable to the one now being inaugurated by human expansion. The latest, evidently due to the strike of an asteroid, ended the Age of Reptiles 66 million years ago. In each case it took over ten million years for evolution to completely replenish the biodiversity lost. And that was in an otherwise undisturbed natural environment. Humanity is now destroying most of the habitats where evolution can occur.

The surviving biosphere remains the great unknown of Earth in many respects. On the practical side, it is hard even to imagine what other species have to offer in the way of new pharmaceuticals, crops, fibers, petroleum substitutes, and other products. We have only a poor grasp of the ecosystems services by which other organisms cleanse the water, turn soil into a fertile living cover, and manufacture the very air we breathe. We sense but do not fully understand what the highly diverse natural world means to our aesthetic pleasure and mental well-being.

Scientists are unprepared to manage a declining biosphere. To illustrate, consider the following mission they might be given. The last remnant of a rain forest is about to be cut over. Environmentalists are stymied. The contracts have been signed, and local landowners and politicians are intransigent. In a final desperate move, a team of biologists is scrambled in an attempt to preserve the biodiversity by extraordinary means. Their assignment is the following: collect samples of all the species of organisms quickly, before the cutting starts; maintain the species in zoos, gardens, and laboratory cultures, or else deep-freeze samples of the tissues in liquid nitrogen; and finally, establish the procedure by which the entire community can be reassembled on empty ground at a later date, when social and economic conditions have improved.

The biologists cannot accomplish this task, not if thousands of them came with a billion-dollar budget. They cannot even imagine how to do it. In the forest patch live legions of species: perhaps 300 birds, 500 butterflies, 200 ants, 50 000 beetles, 1000 trees, 5000 fungi, tens of thousands of bacteria, and so on down a long roster of major groups. Each species occupies a precise niche, demanding a certain place, an exact microclimate, particular nutrients, and temperature and humidity cycles with specified timing to trigger phases of the life cycle. Many, perhaps most, of the species are locked in symbioses with other species; they cannot survive and reproduce unless arrayed with their partners in the correct idiosyncratic configuration.

Even if the biologists pulled off the taxonomic equivalent of the Manhattan Project, sorting and preserving cultures of all the species, they could not then put the community back together again. It would be like unscrambling an egg with a pair of spoons. The biology of the microorganisms needed to reanimate the soil would be mostly unknown. The pollinators of most of the flowers and the correct timing of their appearance could only be guessed. The 'assembly rules,' the sequence in which species must be allowed to colonize in order to coexist indefinitely, would remain in the realm of theory.

The dreams of ecology are like those of astrophysics a hundred years ago, but with the difference that the species are being extinguished before predictive power can be gained. Even more forlorn is the hope that genetic engineers might create new life forms to replace those lost in the dwindling ecosystems. The best that our finest technology can accomplish now or in the foreseeable future is the alteration of a small fraction of genes plus the mixing of natural sets by hybridization. Even these mutant varieties, which retain a large part of the heredity of still-existing species, are unlikely to survive except in environments sufficiently disturbed and impoverished by human activity. I know of no case of a domestic strain that competed successfully with unmodified members of the same species in the native environment.

In its neglect of the rest of life exemptionalism fails definitively. To move ahead as though scientific and entrepreneurial genius will solve each crisis that arises implies that the declining biosphere can be similarly manipulated. But the world is too complicated to be turned into a garden. There is no biological homeostat that can be worked by humanity; to believe otherwise is to risk reducing a large part of Earth to a wasteland.

The environmentalist vision, prudential and less exuberant than exemptionalism, is closer to reality. It sees humanity entering a bottleneck unique in history, constricted by population and economic pressures. In order to pass through to the other side, within perhaps 50–100 years, more science and entrepreneurship will have to be devoted to stabilizing the global environment. That can be accomplished, according to expert consensus, only by halting population growth and devising a wiser use of resources than has been accomplished to date. And wise use for the living world in particular means preserving the surviving natural ecosystems, micromanaging them only enough to save the biodiversity they contain, until such time they can be understood and employed in the fullest sense for human benefit.

References

Kellert, S.R. and Wilson, E.O., 1993, Hypothesis. Island Press (Washington, DC).

Wilson, E.O. 1992, Diversity of Life. Harvard University Press (Cambridge, MA).

1998

"Consilience among the great branches of learning," *Daedalus* 127(1): 131–149

In *Consilience: The Unity of Knowledge* (Knopf, New York, 1988), I reasoned that the natural sciences, especially biology, have matured to the extent that new life has been granted to the original Enlightenment dream of the seventeenth and eighteenth centuries. And not just the natural sciences among themselves, but also the social sciences and humanities, can be united in one skein of cause-and-effect explanations. This essay summarizes some of the key points of that argument. It says that consilience, literally the "jumping together" of explanations from different scholarly fields, is inherently natural to the human mind and accurately reflects the real world.

Edward O. Wilson

Consilience Among the Great Branches of Learning[1]

THE CENTRAL THEME OF THE ENLIGHTENMENT, enhanced across three centuries by the natural sciences, is that all phenomena tangible to the human mind can be rationally explained by cause and effect. Thus humanity can—all on its own—know; and by knowing, understand; and by understanding, choose wisely.

The idea is amplified by what Gerald Holton has called the Ionian Enchantment, the conviction that all tangible phenomena share a common material base and are reducible to the same general laws of nature.[2] The roots of the Enchantment reach to the beginnings of Western science in the sixth century B.C., when Thales of Miletus, in Ionia, considered by Aristotle to be the founder of the physical sciences, proposed that all substances are composed ultimately of water. Although the hypothesis was spectacularly wrong, the ambition it expressed—to attain the broadest possible generalization in cause-and-effect explanations—was destined to become the driving force of Western science.

The success of the scientific revolution may make this perception now appear trivially obvious. Surely, it will seem to many, coherent cause-and-effect explanation is an inevitable consequence of logical thought. But to see otherwise it is only necessary to examine the history of Chinese science. From the first through the thirteenth centuries, as Europe passed from late antiquity through the Dark Ages, science in China flourished. It kept pace with Arab science, even though geographic isolation deprived Chinese scholars of the

Edward O. Wilson is Research Professor and Honorary Curator in Entomology at Harvard University.

ready-made base that Greek culture provided their Western counterparts. The Chinese made brilliant advances in subjects such as descriptive astronomy, mathematics, and chemistry. But they never acquired the habit of reductive analysis in search of general laws that served Western science so well from the seventeenth century on. They consequently failed to expand their conception of space and time beyond what was attainable by direct observation with the unaided senses. The reason, according to Joseph Needham, the principal Western chronicler of the subject, was their emphasis on the holistic properties and harmonious relationships of observable entities, from stars to trees to grains of sand.[3] Unlike Western scientists, they had no inclination to search for abstract codified law in nature. Their reluctance was stimulated to some degree by the historic rejection of the Legalists, who attempted to impose rigid, quantified law during the transition from feudalism to bureaucracy in the fourth century B.C. But of probably greater importance was the fact that the Chinese steered away from the idea of a supreme being who created and supervises a rational, law-governed universe. If there is such a ruler in charge, it makes sense—Western sense at least—to read a divine plan and code of laws into physical existence. If, on the other hand, no such ruler exists, it seems more appropriate to search for separate rules and harmonious relations among the diverse entities composing the material universe. In summary, it can be said that Western scholars but not their Chinese counterparts hit upon the more fortunate metaphysics among the two most available to address the physical universe.

Western scientists also succeeded because they believed that the abstract laws of the various disciplines in some manner interlock. A useful term to capture this idea is *consilience*. The expression is more serviceable than coherence or interconnectedness because the rarity of its usage has preserved its original meaning, whereas coherence and interconnectedness have acquired many meanings scattered among a plethora of contexts. William Whewell, in his 1840 synthesis *The Philosophy of the Inductive Sciences*, introduced consilience as literally a "jumping together" of facts and theory to form a common network of explanation across the scientific disciplines. He said, "The Consilience of Inductions takes place when an Induction, obtained from one class of facts, coincides with an Induction, obtained from another different class. This Consilience is a test of the truth of the Theory in which it occurs."

Consilience proved to be the light and way of the natural sciences. Physics, with its astonishing congruity to mathematics, came to undergird chemistry, which in turn proved foundational for biology. The successful union was not just a broad theoretical consistency, as articulated by Whewell, but an exact folding of principles pertaining to more complex and particular systems into the principles for simpler and more general systems. Organisms, it came to pass, can be reduced to molecules whose properties are entirely conformable to the laws of chemistry, and the elements to which the molecules are composed are in turn conformable to the laws of quantum physics.

To place the organization of modern science in clearer perspective, the disciplines can be tied to the position that their entities occupy in the scale of space and time, while noting that each class of entities represents a level of organization determined by the ensemble of other entities composing them and located lower on the space-time scale.

The consilient view of the natural world is illustrated by the use of the space-time scale to define the disciplines of biology:

Evolutionary space-time. Over many generations entire populations of organisms undergo evolution, which at the most elemental level is a change in the frequencies of the genes in the organisms that compose the populations. The foremost cause of evolution is natural selection, the differential survival and reproduction of the competing genes—or, put more precisely, the differential survival and reproduction of the organisms whose traits are determined by the genes. Natural selection occurs when populations interact with their environment. The subdiscipline broadly covering the phenomena in this segment of space-time is evolutionary biology.

Ecological space-time. Evolution by changes in gene frequency is coarse grained: It becomes apparent only when the history of an entire population is watched across generations. The process of natural selection driving it is finer grained, comprising particular events that affect the birth, reproduction, and death of individual organisms. These are events that can be observed only in a more constricted space and during shorter periods of time, usually the span of a season or less, than is the case for genetic evolution. They are addressed by the discipline of ecology. (Ecology is often put under the rubric of evolutionary biology, when that subject is broadly defined.)

Organismic space-time. Natural selection acts on the anatomy, physiology, and behavior of organisms whose programs of development are prescribed by genes. These properties usually occupy millimeters to meters in space and seconds to hours in time. The subdiscipline treating them is organismic biology.

Cellular space-time. The anatomy, physiology, and behavior of organisms are aggregated phenomena of cells and tissues. Covering micrometers to centimeters, and milliseconds to full generations, they are the province of cellular and developmental biology.

Biochemical space-time. The development and function of cells and tissues are themselves the aggregate products of highly organized systems of molecules. At this latter level, space ranges from nanometers to millimeters, and time usually from nanoseconds to minutes. The responsible discipline is molecular biology.

Two superordinate ideas unite and drive the biological sciences at each of these space-time segments. The first is that all living phenomena are ultimately obedient to the laws of physics and chemistry, with higher levels of organization arising by aggregate behavior at lower levels. The second is that all biological phenomena are products of evolution, and principally evolution by natural selection. The two ideas are expressions of consilience in the following way: Cells and thence organisms, being organized ensembles of molecules, are physicochemical entities, which were assembled not at random but by natural selection. Looked at this way, consilience in biology is the full sweep through the space-time scale, from near-instantaneous molecular process to the transgenerational shifts of gene frequency that compose evolution.

To many critics, especially in the social sciences and humanities, such an extreme expression of reductionism will seem fundamentally wrong-headed. Surely, they will say, we cannot explain something as complex as a brain or an ecosystem by molecular biology. To which most biologists are likely to respond, yes, we can, or we will be able to do so within a few years. The critics in turn call that impossible; such complex systems are distinguished by holistic, emergent properties not explicable by molecular biology, let alone atomic physics. The only fair response to this is yes, put that way, you are right.

Thus arises the paradox of emergence: Complex biological phenomena are reducible but cannot be predicted from a knowledge of molecular biology, at least not contemporary molecular biology. Each higher level of organization requires its own principles, including precisely definable entities, processes, spatial relationships, interactive forces, and sensitivity to external influences, which permit an accurate characterization and perhaps a stab at prediction from knowledge of its elements. Still, the principles, if sound, can be reduced from the top down and stepwise to those formulated at lower levels of organization. An ecosystem, to take the most complicated of all levels, can be broken into the species composing its biota. The species in turn can be analyzed according to the demography of the organisms composing them (population size and growth, birth and death schedules, age structure), along with their interactions with other kinds of organisms and with the physical environment. As part of this study, the organisms can be divided into organ systems, the organ systems into tissues and cells, and so on. The ecosystem, like other biological systems, is not truly hierarchical but heterarchical. It is constrained by the nature of its elements, and the behavior of the elements is determined at least in part by the sequences and proportions in which they are combined. By and large, however, the entities of each level can be reduced; and the principles used to describe the level, if apposite and correct, can be telescoped into those of lower levels and, especially, the next level down. That in essence is the process of reduction, or top-down consilience, which has been intellectually responsible for the enormous success of the natural sciences.

To proceed in the opposite direction, bottom-up, by synthesis—simple to more complex, general to more specific—is far more difficult. Physical scientists have succeeded splendidly at the task. They have interwoven principles of quantum theory, statistical mechanics, and reagent chemistry into stepwise syntheses from subatomic particles to atoms to chemical compounds. Advances in biology, if we measure their success by predictive power, have been much slower. Scanning the space-time scale along which biological complexity increases, we can see progress decelerate to a near stall at the level of protein synthesis. This is a critical juncture in the life sciences. About one hundred thousand kinds of protein molecules are found in the body of a vertebrate animal. Along with the nucleic acids that encode them, they are the essential materials of life. In particular, proteins form most of the basic structure of the body while running its machinery through catalysis of organic chemical reactions. Thanks to advances in technology, biochemists find it relatively easy to sequence the amino acids composing at least the smaller protein molecules, and to map the three-dimensional configuration in which these units are arrayed. It is another matter entirely, however, to predict how amino acids will fold together to create the configuration.[4] Three-dimensional form is all-important in the case of enzymes, which are the protein catalysts, because it determines which substrate molecules the enzyme molecule captures and which reaction it then catalyzes. When procedures are worked out to predict the exact shapes that arise from particular amino acid sequences, the result is likely to be a revolution in biology and medicine. It will permit the design of artificial enzymes and other proteins with desirable properties in biochemical reactions—perhaps superior to those occurring naturally. The difficulty is technical rather than conceptual: Prediction requires the integration of binding forces among all the amino acids simultaneously, an enormous computational problem; and in order to proceed that far it must also measure the forces with a precision beyond the capability of present-day biochemistry.

Even greater challenges are presented by the conceptual reconstruction of cells and tissues from a knowledge of the constituent molecules and chemical processes obtained through reductive analysis. In 1994 the editors of *Science* asked a hundred cellular and developmental biologists to identify the most important unsolved problems in their field of research. Their responses focused prominently on the mechanisms of synthesis.[5] In rank order, the problems most often cited were the following: 1) the molecular mechanisms of tissue and organ development; 2) the connection between development and genetic evolution; 3) the steps by which cells become committed to a particular fate during development; 4) the role of cell-to-cell signaling in tissue development; 5) the self-assembly of tissue patterns during development of the early embryo; and 6) the manner in which nerve cells establish their specific connections to create the nerve cord and brain. Although these problems are formidably difficult, the researchers reported that considerable progress has already been achieved and that the solution of several may be reached within a few years.

To summarize to this point, the consilience of material cause-and-effect explanations is approaching continuity throughout the natural sciences, binding them together across the full span of space and time. Of the two complementary processes of consilience, reduction and synthesis, the more successful has been reduction, because it is both conceptually and technically easier to master. Synthesis good enough to be quantitatively predictive has progressed much more slowly, but it is now inching its way within biology to the level of cell and tissue.

Yet despite the progress of the natural sciences in understanding the natural world, they have remained sequestered from the other great branches of learning. The social sciences and humanities are generally thought to be too grounded in ineffable phenomena of mind and culture, too complex and holistic, and too dependent on historical circumstance to be consilient with the natural sciences.

That venerable perception, I believe, is about to change. The reason is that the natural sciences, doubling in information content every two decades or less, have now expanded to touch the material processes that generate mental and cultural phenomena. Two disciplines—the brain sciences and evolutionary biology—are now filling the ancient gap between dual epistemologies to serve as bridges between the great branches of learning.

The brain sciences are a conglomerate of research activities by neuroscientists, cognitive psychologists, and philosophers ("neurophilosophers") bound together by their conviction that the mind is the brain at work and, as such, can be understood entirely as a biological phenomenon. For their part, evolutionary biologists address the origin of the mental process, which is also considered a biological process. In particular, they focus on the instinct-like emotional responses and learning biases that affect individual development and the evolution of culture.

The key and largely unsolved problem of the brain sciences is the neuron circuitry and neurotransmitter fluxes composing conscious thought. The most important entrée to the problem is brain imaging, the monitoring of brain activity by the direct mapping of its metabolic patterns. The current method of choice in brain imaging is positron emission tomography (PET) scanning, which measures activity in different parts of the brain by the amount of their blood flow—hence the oxygen and energy being delivered to them. The patient is first injected with a small amount of rapidly decaying isotope of oxygen or another harmless radioactive material that emits elementary particles called positrons. The positrons interact with electrons in tissue reached by the isotope, resulting in radiation that can be picked up by a camera. As the patient experiences a sensation, or reflects upon a subject, or feels an emotion, blood flow increases within a tenth of a second in the activated part of his brain, and the corresponding change is detected by the scanner.

An alternative method of brain imaging is functional magnetic resonance imaging (fMRI). Its precursor recording method is static magnetic resonance imaging (MRI), which is based on the response of molecules in body tissues to radio waves after the molecules have been forced into a certain orientation by a powerful magnet. The magnitude of the response rises according to the water content of the tissues, which in turn increases while blood (half of which is water) flows into the active areas. Researchers convert MRI into fMRI, which enables them to use it to monitor brain activity, by recording multiple images through time. The images are then viewed in rapid succession to create moving images in the manner of conventional cinematography. The fMRI method is more efficient in this respect than PET scanning, having been improved to record hundreds of images per minute.

As in all biological research, the overall evolution of brain scanning is toward ever deeper, finer, and faster probes of activity. Other methods directed toward these goals, based on different physical phenomena from those employed in PET and fMRI, have recently opened a new chapter in imaging technology. One method, still limited currently to experimental use in animals, is the application of voltage-sensitive dyes to the surface of the living brain. The electrical conduction of the nerve fibers literally light up the dyes in patterns that can be tracked by photodiode cameras. Images have been recorded in excess of a thousand per second, allowing more nearly continuous monitoring than PET and fMRI scanning.

As the twenty-first century opens, we can expect to witness the invention of even more sophisticated methods of brain imaging, as well as refinement of those already in use. With luck, scientists will eventually reach their ultimate goal of monitoring the activity of intact brains continuously and at the level of individual nerve fibers. In short, the mind as brain-at-work can be made visible.

Brain imaging and experimental brain surgery, together with analyses of localized brain trauma and endocrine and neurotransmitter mediation, have permitted a breakout from age-old subjective conceptions of mental activity. Researchers now speak confidently of a coming solution to the brain-mind problem.

Some students of the subject, however (including a few of the brain scientists themselves), consider that forecast overly optimistic. In their view, technical progress has been largely correlative and has contributed little to a deeper understanding of the conscious mind. They consider it the equivalent of mapping the communicative networks of a city, correlating its activity with ongoing social events, and then declaring the material basis of culture solved. Even if brain activity is mapped completely, they ask, where does that leave consciousness, and especially subjective experience? How to express joy in a summer rainbow with neurobiology? Perhaps these phenomena rise from undiscovered physicochemical phenomena or exist at a level of organization still beyond our comprehension. Or maybe, as a cosmic principle, the conscious mind is just too complicated and subtle ever to understand itself.

This view of the mind as *mysterium tremendum* is, in the opinion of most brain researchers, unjustifiably defeatist. It is the residue of mind-body dualism, the impulse to posit a master integrator—whether corporeal or ethereal—located somewhere in the brain and charged with integrating information from the neural circuits and making decisions. The perception weighs too lightly the alternative and more parsimonious hypothesis: That activity of the neural circuits *is* the mind, and as a consequence nothing more of fundamental aspect is needed to account for mental phenomena at the highest levels. In this view, the hundred million or so neurons, each with an average of thousands of connections to other neurons, are enough to symbolize the thick stream of finely graded information and emotional coloring we introspectively recognize as composing the conscious mind.

To envision the immense amount of information that can be encoded, consider the following hypothetical example supplied by neurobiologists. Suppose that the chemoreceptive brain were programmed to sort and retrieve information by vector coding. Suppose further that combined activities of nerve cells imposing the codes classify individual tastes into combinations of sweetness, saltiness, and sourness. The brain need only distinguish 10 degrees in each of these taste dimensions to discriminate 10×10×10 or 1,000 substances.

A large part, if not the totality, of mental activity comprises scenarios built with such symbolic information. The scenarios are usually reconstructions of the here and now, during which the brain is flooded with fresh sensory information. Many others recreate the past as it is summoned from long-term memory banks. Still others construct alternative possible futures, or pure fantasy.

According to the parsimonious theory of mind, emotions are the modifications of neural activity that animate and focus the scenarios. An act of decision is the prevalence of certain future scenarios over others; those that prevail are most likely to be the ones most conformable to instinct and reinforcement from prior experience. What we think of as meaning is the linkage among neural networks. Learning is the spreading activation that enlarges imagery and engages emotion. The self (to continue the parsimonious theory) is the key dramatic character of the scenarios. It must exist, because the brain is located within the body, and the body is the constant intense focus of real-time sensory experience and decision making.[6]

The primary environment in which the mind develops is culture. This highest level of human activity was defined in 1952 by Alfred Kroeber and Clyde Kluckhohn, out of a review of 164 prior definitions, as follows: "Culture is a product; is historical; includes ideas, patterns, and values; is relative; is learned; is based upon symbols; and is an abstraction from behavior and the products of behavior."[7] It comprises the life of a society, the totality of its religion, myths, art, technology, sports, and all the other systematic knowledge transmitted across generations.

Throughout this century scholars in all the branches of learning have treated culture as an entity apart, comprehensible only on its own terms and not those of the natural sciences. By this conception culture stands apart even if the mind has a reducible, material basis; it must do so first because the fine details of the cultures of individual societies are historically determined, and second because cultures comprise phenomena too complicated, too flickering through time, and too subtle to be subject to natural scientific analysis.

A fixed belief in the independent nature of culture has contributed to the isolation of the social sciences and humanities from the natural sciences throughout modern history. It is the basis of the discontinuity famously cited by C. P. Snow in 1959 as separating the scientific culture from the literary culture. Now there is reason to believe that the difference is not a true epistemological discontinuity, not a divide between two kinds of reality, but something far less forbidding and yet much more interesting. The boundary between the two cultures is instead a vast, unexplored terrain of phenomena awaiting entry from both sides.

The terrain is the interaction between genetic evolution and cultural evolution. We know that culture is learned. At the same time, evidence is mounting that learning is genetically biased; it is becoming increasingly accepted that culture is influenced by human nature. But what exactly is human nature? It is not the genes, which prescribe it, or the cultural universals, which are its most obvious products. It is the epigenetic rules, the hereditary biases that guide the development of individual behavior. There are several examples of epigenetic rules that can be cited in this early stage of investigation.

142 *Edward O. Wilson*

The facial expressions denoting the elementary emotions of fear, loathing, anger, surprise, and happiness are human universals and evidently inherited. They are adjusted by cultural evolution within individual societies to project particular nuances of meaning. The smile, one of the basic elements of emotive communication, appears at two to four months in infants everywhere, virtually independent of environment. It occurs on schedule in deaf-blind infants and even in thalidomide-crippled children who cannot touch their own faces.[8] The tendency to fear snakes is another human universal. It is furthermore widespread, if not universal, in all other Old World primate species. Snakes are among the few stimuli that easily evoke true phobias in people—the deep and intractable visceral reactions acquired with only one or two frightening experiences. They share their power with heights, closed spaces, running water, spiders, and other ancient perils of humanity; a similar degree of sensitivity does not exist for knives, guns, electric sockets, automobiles, and other modern sources of risk. The cultural consequences of the response to snakes, combining fear and intense curiosity, are manifold. Snakes are among the animals most commonly experienced in dreams, even among urbanites who have never seen one in life. They play prominent mythic roles in cultures around the world, taking new forms variously as demons, dragons, seducers, magical healers, and gods.[9]

Automatic incest avoidance is universal in primate species studied to date, including *Homo sapiens*. The generally accepted adaptive explanation is the heightened risk that inbreeding poses of producing defective offspring, and that evolutionary inference is well supported by the evidence. The closer the genetic relationship of parents, the more likely they will bring together matching recessive genes that are deleterious in a double dose. Children of full siblings and of fathers and daughters, for example, have twice the early mortality rate of outbred children. Among those that survive, ten times more suffer genetic defects such as heart deformities, deaf-mutism, mental retardation, and dwarfism. The epigenetic rules, or hereditary developmental biases that prevent incest, are two-layered in apes, monkeys, and other non-human primates. First, all species so far studied for the trait (nineteen worldwide) practice the equivalent of human exogamy: Young individuals leave the parent group and join another before they attain full maturity. Second, all species examined for the possible existence of the Westermarck effect also display that phenomenon. This means that individuals are sexually desensitized to individuals with whom they have been closely associated while very young, normally their parents and siblings. The critical period for the effect in human beings is the first thirty months of life. Out of the Westermarck effect have apparently risen incest taboos with all their supporting arsenal of legends and myths. The effect is enhanced in some but not all societies by a third barrier: the direct observation and correct rational understanding of the ill effects of incest.[10]

Similar examples of epigenetic rules have multiplied in the literature of biology and the behavioral sciences during the past several decades. They have been found in virtually all categories of human behavior, including sexual and parental bonding, the acquisition of language, and even the cardinal role of trust during contract formation. They leave little doubt that a true hereditary human nature exists, and that it includes social behaviors held in common with nonhuman primate species and others that are diagnostically human.

Such is the interdisciplinary subject awaiting study by all the great branches of learning, and I can think of no more important intellectual undertaking. The relation between biological evolution and cultural evolution is, in my opinion, both the central problem of the social sciences and humanities and one of the great remaining problems of the natural sciences.

The process by which genetic evolution and cultural evolution appear to be linked is usually called gene-culture coevolution. The theory of gene-culture coevolution incorporates the two levels of approach I cited earlier as the core of modern biology.[11] Put as briefly as possible, they are that living processes are physicochemical and also self-assembled by natural selection. The first level is composed of proximate explanations, which describe the structures and processes by which an organism responds. The question of interest in any proximate explanation is, How does the phenomenon occur? The second level is composed of ultimate, or evolutionary, explanations, which account for the origin of the structures and processes, usually by the adaptive advantage they confer on organisms. The question of interest at this level is, *Why* does the phenomenon occur? In the case of hereditarily based incest avoidance, the proximate causes are emigration and the Westermarck effect. The ultimate cause is the deleterious effects of inbreeding, which by natural selection has driven the species toward emigration and the Westermarck effect.

The theory of gene-culture coevolution is still spotty and largely untested. Nevertheless, I believe that most researchers on the subject would agree with the following outline of the present form of the theory: People survive and leave offspring to the degree that they learn and adapt to the culture of their society, and the societies themselves flourish or decline in proportion to the effectiveness of their adaptation to their environment and surrounding societies. For hundreds of millennia certain aptitudes and cultural norms have arisen that are consistently adaptive in this Darwinian sense. They include language facility, cooperativeness within the group, exogamy and incest avoidance, rites of passage, territoriality, male polygyny, and parent-offspring bonding. Hereditary epigenetic rules have evolved that pull individual preference, and hence cultural evolution, toward these norms. They comprise the elements of what we subjectively call human nature. The genes prescribing them also increase in frequency as a result of the same process. Spreading through the population, maintained by the edge they give most of the time in survival and reproduction, they have secured the stability of human nature across societies and generations.

To conclude my synopsis of the theory, cultural evolution is much faster than genetic evolution. One result is nongenetic cultural diversity, which scatters particular cultural variants around each central, genetic trend to a degree determined by the strength of the epigenetic rules affecting them. The products of cultural evolution, multiplying rapidly through the population, can improve the fitness of individuals and societies, or they can reduce them. But only if the advantage or disadvantage is sustained for many generations—population genetics theory would suggest at least ten—can the epigenetic rules and the genes prescribing them be replaced. That is why human nature today remains Paleolithic even in the midst of accelerating technological advance. Thus corporate CEOs impelled by stone-age emotions work international deals with cellular telephones at thirty thousand feet.

If it is granted that the human condition is subject to consilient explanation from genes to mind to culture, even as a working hypothesis, the consequences to follow will be considerable. The first is support across the great branches of learning for what can appropriately be called "gap analysis" as a research strategy.[12] Already a mainstay of the natural sciences, gap analysis is the systematic attempt to identify domains of phenomena in which important discoveries are most likely to be made. Its most productive method is reduction, the search for novel phenomena, or at least the search for novel explanations of phenomena already known, by examination of the next level of organization down. Successful reduction confirms the existence of elements in the lower level that interact to create the higher level. In this manner, molecular biology was created de novo from the basic chemistry of macromolecules, and the study of cells and tissues was revolutionized by molecular biology.

The social sciences, I believe, will advance more rapidly if they adopt a consilient worldview and the gap analysis suggested by it leading to reductionist analysis. They have failed to give this approach a try, except in a few sectors such as biological anthropology, largely because of their aversion to biology. The reasons for the aversion are complex, stemming partly from the effort of the social science disciplines—anthropology, economics, political science, and sociology—to maintain intellectual independence, partly from the daunting complexity of the subject, and partly from fear of the misuse of biology to support racist ideology.

Still, biology is the logical foundational discipline of the social sciences. I mean by this assessment biology as broadly defined, including much of contemporary psychology, especially cognitive psychology, which is in the process of being subsumed by neurobiology and the brain sciences. A great majority of social scientists, including the most influential theoreticians in economics, build their models as if this information does not exist. Their conceptions of human behavior come either from folk psychology—intuitive notions that seem right but are often factually wrong—or from notions of the mind as an optimizing device for rational choice. They ignore contrary signs from genetics, neurobiology, cognitive psychology, and the many quirky properties of human nature. For them history began a few thousand years ago with the rise of complex societies, overlooking the fact that it began hundreds of thousands of years ago with the evolutionary origins of human nature in hunter-gatherer bands.

In summary, it is hard to imagine how the social sciences can unite and achieve general, predictive theory without taking a reductionist approach to the phenomena of human nature, both their proximate causes in the machinery of the brain and their ultimate causes in deep, evolutionary history.

The theory and criticism of the arts can also benefit in the same fashion. Let me cite several examples already in hand. We now know, from neurobiology and the brain sciences, how the brain breaks down and classifies the continuously varying wavelength of visible light into four basic colors, namely, blue, green, yellow, and red. The process has been tracked in segments from the base sequences in the DNA that prescribe the cone pigments of the photosensitive retinal cells to the nerve-cell sequences that lead from the retina to the primary visual cortex at the extreme rear of the brain. From anthropological and linguistic studies we know that people in societies around the world fix their color terms toward the centers of the primary colors in the spectrum and away from the intermediate and hence ambiguous wavelengths. Finally, we know that as societies increase their color vocabularies, in the course of cultural evolution, they tend to employ up to eleven basic terms, usually accumulating them in the following sequence: Languages with only two basic color terms use them to distinguish black and white; languages with only three terms identify black, white, and red; languages with only four terms have words for black, white, red, and either green or yellow; languages with only five terms have words for black, white, red, green, *and* yellow; and so on until all eleven terms are included, as exemplified in the English language. The sequence cannot be due to chance alone. If the terms were combined at random, there would be 2,036 possible combinations. But for the most part they are drawn from only 22. Surely this is the kind of information needed to produce a coherent theory of aesthetics in the visual arts.[13]

In another domain relevant to visual aesthetics, neurobiological measurements have shown that the brain is most aroused by abstract designs in which there is about 20 percent repetition of elements. That is the amount of redundancy found in a simple maze, two turns of a logarithmic spiral, or an asymmetrical cross. It seems hardly a coincidence that roughly the same property is shared by a great deal of the art in friezes, grillwork, colophons, and flag designs. Or that it crops up again in the glyphs of ancient Egypt and Mesoamerica as well as the pictographs of Japanese, Chinese, Thai, Bengali, and other Asian languages. The response appears to be innate: Newborn infants gaze longest at figures with about the same amount of redundancy.[14]

In yet another topic of aesthetics, ideal female facial beauty as judged in at least two cultures, European and Japanese, has recently been found to follow some surprising principles. Using blended and artificially altered photographs, psychologists have discovered that the most admired facial features are near the anatomical average of the population but with heightened cheekbones, reduced chin size, enlarged eyes, and shortened distance between the nose and chin.[15] The cause of this effect, if upheld as inborn by further cross-cultural and developmental studies, is unknown. It could represent an innate recognition of the signs of youthfulness and hence greater reproductive potential.

The creative arts themselves, in literature, the visual arts, drama, music, and dance, may not be affected significantly by such knowledge from the natural sciences. The purpose of the arts is to transmit personal experience and emotion directly from mind to mind while avoiding explanation of the logic behind the creative work; thus, *ars est celare artem*, it is art to conceal art. But theory and criticism of the arts, which does attempt this mode of explanation, cannot help but be strengthened by the new information. If the greatest art is indeed that which touches all humanity, as commonly said, it follows that consilient cause-and-effect accounts of human nature will become increasingly foundational to sound theory and criticism.

148 *Edward O. Wilson*

ENDNOTES

[1]This essay presents in much abbreviated form some of the arguments in my book-length exposition of the same general subject, *Consilience: The Unity of Knowledge* (New York: Knopf, 1998).

[2]The Ionian enchantment is discussed by Gerald Holton in *Einstein, History, and Other Passions* (Woodbury, N.Y.: American Institute of Physics Press, 1995).

[3]*The Shorter Science and Civilisation in China: An Abridgment of Joseph Needham's Original Text*, Vol. I, prepared by Colin A. Ronan (New York: Cambridge University Press, 1978).

[4]In characterizing the prediction of three-dimensional protein structure, I benefited greatly from an unpublished paper presented by S. J. Singer at the American Academy of Arts and Sciences in December 1993; he has also kindly reviewed my account.

[5]On the opinions of cell and developmental biologists concerning the frontiers of their field, see "Looking to Development's Future," by Marcia Barinaga, *Science* 266 (1994): 561–564.

[6]Among the many recent works I have used to interpret the consensus of students of the mind-body problem are Patricia S. Churchland, *Neurophilosophy: Toward a Unified Science of the Mind-Brain* (Cambridge, Mass.: MIT Press, 1986); Paul M. Churchland, *The Engine of Reason, the Seat of the Soul* (Cambridge, Mass.: MIT Press, 1995); Antonio R. Damasio, *Descartes' Error: Emotion, Reason, and the Human Brain* (New York: G. P. Putnam, 1994); Daniel C. Dennett, *Consciousness Explained* (Boston: Little, Brown, 1991); J. Allan Hobson, *The Chemistry of Conscious States: How the Brain Changes Its Mind* (Boston: Little, Brown, 1994); and Stephen M. Kosslyn and Oliver Koenig, *Wet Mind: The New Cognitive Neuroscience* (New York: Free Press, 1992).

[7]Alfred Kroeber and Clyde K. M. Kluckhohn, "Culture: A Critical Review of Concepts and Definitions," *Papers of the Peabody Museum of American Archaeology and Ethnology, Harvard University*, vol. 47 (1952), no. 12, 643–644.

[8]On basic facial expressions: the literature, including smiling, is reviewed by Charles J. Lumsden and Edward O. Wilson in *Genes, Mind, and Culture* (Cambridge, Mass.: Harvard University Press, 1981) and by the pioneer behavioral biologist Irenäus Eibl-Eibesfeldt in *Human Ethology* (Hawthorne, N.Y.: Aldine de Gruyter, 1989).

[9]On the fear of snakes and the origin of the serpent myth: Balaji Mundkur, *The Cult of the Serpent: An Interdisciplinary Survey of Its Manifestations and Origins* (Albany, N.Y.: State University of New York Press, 1983) and Edward O. Wilson, *Biophilia* (Cambridge, Mass.: Harvard University Press, 1984).

[10]On incest and its avoidance in human beings and other primates: Arthur P. Wolf, *Sexual Attraction and Childhood Association: A Chinese Brief for Ed-*

ward Westermarck (Stanford, Calif.: Stanford University Press, 1995) and William H. Durham, *Coevolution: Genes, Culture, and Human Diversity* (Stanford, Calif.: Stanford University Press, 1991).

[11]The expression gene-culture coevolution and a first general theory pertaining to it, in the sense of combining models from genetics, psychology, and anthropology, were provided by Lumsden and Wilson in *Genes, Mind, and Culture.* A review and update of the subject are given in my more general book *Consilience.*

[12]"Gap analysis" is a term I have borrowed from conservation biology. It means the method of mapping known ranges of threatened plant and animal species and using the information to select the best sites to set aside as reserves. See J. Michael Scott and Blair Csuti, "Gap Analysis for Biodiversity Surveys and Maintenance," in Marjorie L. Reaka-Kudla et al., eds., *Biodiversity II: Understanding and Protecting Our Biological Resources* (Washington, D.C.: Joseph Henry Press, 1997), 321–340.

[13]A full account of the biological and cultural origins of color perception and vocabulary is given by multiple authors in Trevor Lamb and Janine Bourriau, eds., *Colour: Art & Science* (New York: Cambridge University Press, 1995).

[14]On the optimum amount of redundancy in design: see the review by Charles J. Lumsden and Edward O. Wilson, *Promethean Fire* (Cambridge, Mass.: Harvard University Press, 1983).

[15]On female facial beauty: "Facial Shape and Judgements of Female Attractiveness," by D. I. Perrett et al., *Nature* 368 (1994): 239–242. Other aspects of ideal physical characteristics are discussed by David M. Buss in *The Evolution of Desire* (New York: BasicBooks, 1994).

1998

"Integrated science and the coming century of the environment," *Science* 279: 2048–2049

In this invited essay, on the 150th anniversary of the American Association for the Advancement of Science, and with the new millennium close at hand, I discussed the possible integration of biology with the social sciences and humanities, with emphasis on the promise of applying such interdisciplinary approaches to problems of the environment.

AMERICAN ASSOCIATION FOR THE ADVANCEMENT OF SCIENCE

150 YEARS • 1848-1998

INTEGRATED SCIENCE AND THE COMING CENTURY OF THE ENVIRONMENT

EDWARD O. WILSON, *Pellegrino University Research Professor and Honorary Curator in Entomology at Harvard University, is the author of 18 books, 2 of which have received the Pulitzer Prize; an ardent defender of the liberal arts; and a promoter of global conservation of species and natural ecosystems.*

ILLUSTRATION: ALLAN M. BURCH

The sesquicentennial of the American Association for the Advancement of Science is a good time to acknowledge that science is no longer the specialized activity of a professional elite. Nor is it a philosophy, or a belief system, or, as some postmodernist thinkers would have it, just one world view out of a vast number of possible views. It is rather a combination of mental operations, a culture of illuminations born during the Enlightenment four centuries ago and enriched at a near-geometric rate to establish science as the most effective way of learning about the material world ever devised. The sword that humanity finally pulled, it has become part of the permanent world culture and available to all.

"Science, to put its warrant as concisely as possible, is the organized systematic enterprise that gathers knowledge about the world and condenses the knowledge into testable laws and principles."[*] Its defining traits are first, the confirmation of discoveries and support of hypotheses through repetition by independent investigators, preferably with different tests and analyses; second, mensuration, the quantitative description of the phenomena on universally accepted scales; third, economy, by which the largest amount of information is abstracted into a simple and precise form, which can be unpacked to re-create detail; fourth, heuristics, the opening of avenues to new discovery and interpretation.

And fifth, and finally, is consilience, the interlocking of causal explanations across disciplines. "This consilience," said William Whewell when he introduced the term in his 1840 synthesis *The Philosophy of the Inductive Sciences*, "is a test of the truth of the theory in which it occurs."[†] And so it has proved within the natural sciences, where the webwork of established cause and effect, while still gossamer frail in many places, is almost continuous from quantum physics to biogeography. This webwork traverses vast scales of space, time, and complexity to unite what in Whewell's time appeared to be radically different classes of phenomena. Thus, chemistry has been rendered consilient with physics, both undergird molecular biology, and molecular biology is solidly connected to cellular, organismic, and evolutionary biology.

The scales of space, time, and complexity in the explanatory webwork have been widened to bracket some 40 orders of magnitude. Consider, for example, the webwork's reach from quantum electrodynamics to the birth of galaxies; or the great breadth it has attained in the biological sciences, which are not only united with physics and chemistry but now touch the borders of the social sciences and humanities.

This last augmentation, while still controversial, deserves special attention because of its implications for the human condition. For most of the last two centuries following the decline of the Enlightenment, scholars have traditionally drawn sharp distinctions between the great branches of learning, and particularly between the natural sciences as opposed to the social sciences and humanities. The latter dividing line, roughly demarcating the scientific and literary cultures, has been considered an epistemological discontinuity, a permanent difference in ways of knowing. But now growing evidence exists that the boundary is not a line at all, but a broad, mostly unexplored domain of causally linked phenomena awaiting cooperative exploration from both sides.

Researchers from four disciplines of the natural sciences have entered the borderland:

- Cognitive neuroscientists, outriders of the once but no longer "quiet" revolution, are using an arsenal of new techniques to map the physical basis of mental events. They have shifted the frame of discourse concerning the mind from semantic and introspective analysis to nerve cells, neurotransmitters, hormones, and recurrent neural networks. Working on a parallel track, students of artificial intelligence, with an eye on the future possibility of artificial emotion, search with neuroscientists for a general theory of cognition.
- Combining molecular genetics with traditional psychological tests, behavioral geneticists have started to characterize and even pinpoint genes that affect mental activity, from drug addiction to mood and cognitive operations. They are also tracing the epigenesis of the activity, the complex molecular and cellular pathways of mental development that lead from prescription to phenotype, in the quest for a fuller and much-needed understanding of the interaction between genes and environment.
- Evolutionary biologists, especially sociobiologists (also known within the social sciences as evolutionary psychologists and evolutionary anthropologists), are reconstructing the origins of human social behavior with special reference to evolution by natural selection.
- Environmental scientists in diverse specialties, including human ecology, are more precisely defining the arena in which our species arose, and those parts that must be sustained for human survival.

The very idea of a borderland of causal connections between the great branches of learning is typically dismissed by social theorists and philosophers as reductionistic. This diagnosis is of course quite correct. But consider this: Reduction and the consilience it implies are the key to the success of the natural sciences. Why should the same not be true of other kinds

The author is at the Museum of Comparative Zoology, Harvard University, 26 Oxford Street, Cambridge, MA 02138, USA.

[*]E. O. Wilson, *Consilience: The Unity of Knowledge* (Knopf, New York, 1998), p. 53. [†]W. Whewell, *The Philosophy of Inductive Sciences* (Parker, London, 1840), p. 230. [‡]F. Bacon, *Advancement of Learning* (Tomes, London, 1605). [§]E. O. Wilson, *Consilience*, p.280.

of knowledge? Because mind and culture are material processes, there is every reason to suppose, and none compelling enough to deny, that the social sciences and humanities will be strengthened by assimilation of the borderland disciplines. For however tortuous the unfolding of the causal links among genes, mind, and culture, and however sensitive they are to the caprice of historical circumstance, the links form an unbreakable webwork, and human understanding will be better off to the extent that these links are explored. Francis Bacon, at the dawn of the Enlightenment in 1605, prefigured this principle of integrative science (by which he meant a large part of all the branches of learning) with an image I especially like: "No perfect discovery can be made upon a flat or a level: neither is it possible to discover the more remote or deeper parts of any science, if you stand but upon the level of the same science and ascend not to a higher science."‡

The unavoidable complement of reduction is synthesis, the step that completes consilience from one discipline to the next. Synthesis is far more difficult to achieve than reduction, and that is why reductionistic studies dominate the cutting edge of investigation. To reduce an enzyme molecule to its constituent amino acids and describe its three-dimensional structure is far easier, for example, than to predict the structure of an enzyme molecule from the sequence of its amino acids alone. As the century closes, however, the balance between reduction and synthesis appears to be changing. Attention within the natural sciences has begun to shift away from the search for elemental units and fundamental laws and toward highly organized systems. Researchers are devoting proportionately more time to the self-assembly of macromolecules, cells, organisms, planets, universes—and mind and culture.

"WE ARE DANGEROUSLY BAFFLED BY THE MEANING OF [OUR] EXISTENCE...[WE NEED] AN EXPLANATORY INTEGRATION NOT JUST OF THE NATURAL SCIENCES BUT ALSO OF THE SOCIAL SCIENCES AND HUMANITIES..."

If this view of universal consilience is correct, the central question of the social sciences is, in my opinion, the nature of the linkage between genetic evolution and cultural evolution. It is also one of the great remaining problems of the natural sciences. This part of the overlap of the two great branches of learning can be summarized as follows. We know that all culture is learned, yet its form and the manner in which it is transmitted are shaped by biology. Conversely, the genes prescribing much of human behavioral biology evolved in a cultural environment, which itself was evolving. A great deal has been learned about these two modes of evolution viewed as separate processes. What we do not understand very well is how they are linked.

The surest entry to the linkage, or gene-culture coevolution as it is usually called, is (again in my opinion) to view human nature in a new and more heuristic manner. Human nature is not the genes, which prescribe it, or the universals of culture, which are its products. It is rather the epigenetic rules of cognition, the inherited regularities of cognitive development that predispose individuals to perceive reality in certain ways and to create and learn some cultural variants in preference to competing variants.

Epigenetic rules have been documented in a diversity of cultural categories, from syntax acquisition and paralinguistic communication to incest avoidance, color vocabularies, cheater detection, and others. The continuing quest for such inborn biasing effects promises to be the most effective means to understand gene-culture coevolution and hence to link biology and the social sciences causally. It also offers a way, I believe, to build a secure theoretical foundation for the humanities, by addressing, for example, the biological origins of ethical precepts and aesthetic properties of the arts.

The naturalistic world view, by encouraging the search for consilience across the great branches of learning, is far more than just another exercise for philosophers and social theorists. To understand the physical basis of human nature, down to its evolutionary roots and genetic biases, is to provide needed tools for the diagnosis and management of some of the worst crises afflicting humanity.

Arguably the foremost of global problems grounded in the idiosyncrasies of human nature is overpopulation and the destruction of the environment. The crisis is not long-term but here and now; it is upon us. Like it or not, we are entering the century of the environment, when science and polities will give the highest priority to settling humanity down before we wreck the planet.

Here in brief is the problem—or better, complex of interlocking problems—as researchers see it. In their consensus, "[t]he global population is precariously large, will grow another third by 2020, and climb still more before peaking sometime after 2050. Humanity is improving per capita production, health, and longevity. But it is doing so by eating up the planet's capital, including irreplaceable natural resources. Humankind is approaching the limit of its food and water supply. As many as a billion people, moreover, remain in absolute poverty, with inadequate food from one day to the next and little or no medical care. Unlike any species that lived before, *Homo sapiens* is also changing the world's atmosphere and climate, lowering and polluting water tables, shrinking forests, and spreading deserts. It is extinguishing a large fraction of plant and animal species, an irreplaceable loss that will be viewed as catastrophic by future generations. Most of the stress originates directly or indirectly from a handful of industrialized countries. Their proven formulas are being eagerly adopted by the rest of the world. The emulation cannot be sustained, not with the same levels of consumption and waste. Even if the industrialization of developing countries is only partly successful, the environmental aftershock will dwarf the population explosion that preceded it."§ Recent studies indicate that to raise the rest of the world to the level of the United States using present technology would require the natural resources of two more planet Earths.

The time has come to look at ourselves closely as a biological as well as cultural species, using all of the intellectual tools we can muster. We are brilliant catarrhine primates, whose success is eroding the environment to which a billion years of evolutionary history exquisitely adapted us. We are dangerously baffled by the meaning of this existence, remaining instinct-driven, reckless, and conflicted. Wisdom for the long-term eludes us. There is ample practical reason—should no other kind prove persuasive—to aim for an explanatory integration not just of the natural sciences but also of the social sciences and humanities, in order to cope with issues of urgency and complexity that may otherwise be too great to manage.

ILLUSTRATION: TERESE WINSLOW

Appendix: The Published Works of Edward O. Wilson

1949

Richteri, the fire ant. *Alabama Conservation* 20(12): 8–9.

A report on the imported fire ant *Solenopsis saevissima* var. *richteri* Forel in Alabama. Special report to the Alabama Department of Conservation, mimeographed, 58 pp. (with J. H. Eads; EOW first author)

Records of the order Zoraptera from Alabama. *Entomological News* 60(7): 180–181. (with B. D. Valentine; EOW second author)

1950

Notes on the food habits of *Strumigenys louisianae* Roger (Hymenoptera: Formicidae). *Bulletin of the Brooklyn Entomological Society* 45(3): 85–86.

A new *Leptothorax* from Alabama (Hymenoptera: Formicidae). *Psyche* 57(4): 128–130.

1951

Variation and adaptation in the imported fire ant. *Evolution* 5(1): 68–79.

1952

Notes on *Leptothorax bradleyi* Wheeler and *L. wheeleri* M. R. Smith (Hymenoptera: Formicidae). *Entomological News* 63(3): 67–71.

The *Solenopsis saevissima* complex in South America. *Mémorias do Instituto Oswaldo Cruz* 50: 60–68. (Spanish translation, *ibid.*, 49–59.)

The morphology of the proventriculus of a formicine ant. *Psyche* 59(2): 47–60. (with T. Eisner; EOW second author)

1953

On Flander's hypothesis of caste determination in ants. *Psyche* 60(1): 15–20.

The origin and evolution of polymorphism in ants. *Quarterly Review of Biology* 28(2): 136–156.

Origin of the variation in the imported fire ant. *Evolution* 7(3): 262–263.

The subspecies concept and its taxonomic application. *Systematic Zoology* 2(3): 97–111. (with W. L. Brown; EOW first author)

The ecology of some North American dacetine ants. *Annals of the Entomological Society of America* 46(4): 479–495.

1954

A new interpretation of the frequency curves associated with ant polymorphism. *Insectes Sociaux* 1(1): 75–80.

The case against the trinomen. *Systematic Zoology* 3(4): 174–176. (with W. L. Brown; EOW second author)

The beetle genus *Paralimulodes* Bruch in North America, with notes on morphology and behavior (Coleoptera: Limulodidae). *Psyche* 61(4): 154–161. (with T. Eisner and B. D. Valentine; EOW first author)

1955

A monographic revision of the ant genus *Lasius*. *Bulletin of the Museum of Comparative Zoology, Harvard* 113(1): 1–201.

Ecology and behavior of the ant *Belonopelta deletrix* Mann (Hymenoptera: Formicidae). *Psyche* 62(2): 82–87.

Revisionary notes on the *sanguinea* and *neogagates* groups of the ant genus *Formica*. *Psyche* 62(3): 108–129. (with W. L. Brown; EOW first author)

Division of labor in a nest of the slave-making ant *Formica wheeleri* Creighton. *Psyche* 62(3): 130–133.

The status of the ant genus *Microbolbos* Donisthorpe. *Psyche* 62(3): 136.

1956

Feeding behavior in the ant *Rhopalothrix biroi* Szabó. *Psyche* 63(1): 21–23.

Character displacement. *Systematic Zoology* 5(2): 49–64. (with W. L. Brown; EOW second author)

Aneuretus simoni Emery, a major link in ant evolution. *Bulletin of the Museum of Comparative Zoology, Harvard* 115(3): 81–99. (with T. Eisner, G. C. Wheeler, and J. Wheeler; EOW first author)

New parasitic ants of the genus *Kyidris* with notes on ecology and behavior. *Insectes Sociaux* 3(3): 439–454. (with W. L. Brown; EOW first author)

1957

The discovery of cerapachyine ants on New Caledonia, with the description of new species of *Phyracaces* and *Sphinctomyrmex*. *Breviora* no. 74, 9 pp.

The *tenuis* and *selenophora* groups of the ant genus *Ponera* (Hymenoptera: Formicidae). *Bulletin of the Museum of Comparative Zoology, Harvard* 116(6): 355–386.

Dacetinops, a new ant genus from New Guinea. *Breviora* no. 77, 7 pp. (with W. L. Brown; EOW second author)

Quantitative studies of liquid food transmission in ants. *Insectes Sociaux* 4(2): 157–166. (with T. Eisner; EOW first author)

The organization of a nuptial flight of the ant *Pheidole sitarches* Wheeler. *Psyche* 64(2): 46–50.

Sympatry of the ants *Conomyrma bicolor* (Wheeler) and *C. pyramica* (Roger). *Psyche* 64(2): 76.

Behavior of the Cuban lizard *Chamaeleolis chamaeleontides* (Duméril and Bibron) in captivity. *Copeia* no. 2, p. 145.

A new parasitic ant of the genus *Monomorium* from Alabama, with a consideration of the status of genus *Epixenus* Emery. *Entomological News* 68(9): 239–246. (with W. L. Brown; EOW second author)

1958

Character displacement and species criteria. *Proceedings of the Tenth International Congress of Entomology* (Montreal, 1956) 1: 125–128.

Radioactive tracer studies of food transmission in ants. *Proceedings of the Tenth International Congress of Entomology* (Montreal, 1956) 2: 509–513. (with T. Eisner; EOW second author)

The worker caste of the parasitic ant *Monomorium metoecus* Brown and Wilson, with notes on behavior. *Entomological News* 69(2): 33–38. (with W. L. Brown; EOW first author)

The fire ant. *Scientific American* 198(3) (March): 36–41.

The beginnings of nomadic and group-predatory behavior in the ponerine ants. *Evolution* 12(1): 24–31.

Observations on the behavior of the cerapachyine ants. *Insectes Sociaux* 5(1): 129–140.

Patchy distributions of ant species in New Guinea rain forests. *Psyche* 65(1): 26–38.

Studies on the ant fauna of Melanesia, I: The tribe Leptogenyini; II: The tribes Amblyoponini and Platythyreini. *Bulletin of the Museum of Comparative Zoology, Harvard* 118(3): 101–153.

Recent changes in the introduced population of the fire ant *Solenopsis saevissima* (Fr. Smith). *Evolution* 12(2): 211–218. (with W. L. Brown; EOW first author)

Studies on the ant fauna of Melanesia, III: *Rhytidoponera* in western Melanesia and the Moluccas; IV: The tribe Ponerini. *Bulletin of the Museum of Comparative Zoology, Harvard* 119(4): 303–371.

A chemical releaser of alarm and digging behavior in the ant *Pogonomyrmex badius* (Latreille). *Psyche* 65(2–3): 41–51.

Chemical releasers of necrophoric behavior in ants. *Psyche* 65(4): 108–114. (with N. I. Durlach and L. M. Roth; EOW first author)

1959

Source and possible nature of the odor trail of fire ants. *Science* 129(3349): 643–644.
Adaptive shift and dispersal in a tropical ant fauna. *Evolution* 13(1): 122–144.
Invader of the south. *Natural History* 68(5): 276–281.
Studies on the ant fauna of Melanesia, V: The tribe Odontomachini. *Bulletin of the Museum of Comparative Zoology, Harvard* 120(5): 483–510.
Studies on the ant fauna of Melanesia, VI: The tribe Cerapachyini. *Pacific Insects* 1(1): 39–57.
Some ecological characteristics of ants in New Guinea rain forests. *Ecology* 40(3): 437–447.
Pheromones in the organization of ant societies. *Anatomical Record* 134(3): 653.
The search for *Nothomyrmecia. Western Australian Naturalist* 7(2): 25–30. (with W. L. Brown; EOW second author)
Communication by tandem running in the ant genus *Cardiocondyla. Psyche* 66(3): 29–34.
The evolution of the dacetine ants. *Quarterly Review of Biology* 34(4): 278–294. (with W. L. Brown; EOW second author)
William M. Mann. *Psyche* 66(4): 55–59.
Glandular sources and specificity of some chemical releasers of social behavior in dolichoderine ants. *Psyche* 66(4): 70–76. (with M. Pavan; EOW first author)

1961

The nature of the taxon cycle in the Melanesian ant fauna. *American Naturalist* 95(882): 169–193.
Ants from three remote oceanic islands. *Psyche* 68(4): 137–144. (with R. W. Taylor; EOW second author)
Biocommunication. *In* P. Gray, ed., *The Encyclopedia of the Biological Sciences,* pp. 107–109. Reinhold Publishing Corp., New York.

1962

The ants of Rennell and Bellona Islands. *Natural History of Rennell Island, British Solcmon Islands* 4: 13–23.
Chemical communication among workers of the fire ant *Solenopsis saevissima* (Fr. Smith), 1: The organization of mass-foraging. *Animal Behaviour* 10(1–2): 134–147.
Chemical communication among workers of the fire ant *Solenopsis saevissima* (Fr. Smith), 2: An information analysis of the odour trail. *Animal Behaviour* 10(1–2): 148–158.

Chemical communication among workers of the fire ant *Solenopsis saevissima* (Fr. Smith), 3: The experimental induction of social responses. *Animal Behaviour* 10(1–2): 159–164.

The Trinidad cave ant *Erebomyrma* (=*Spelaeomyrmex*) *urichi* (Wheeler), with a comment on cavernicolous ants in general. *Psyche* 69(2): 62–72.

Behavior of *Daceton armigerum* (Latreille), with a classification of self-grooming movements in ants. *Bulletin of the Museum of Comparative Zoology, Harvard* 127(7): 403–421.

1963

The analysis of olfactory communication among animals. *Journal of Theoretical Biology* 5: 443–469. (with William H. Bossert; EOW second author)

The social biology of ants. *Annual Review of Entomology* 8: 345–368.

Chemical communication among animals. *Recent Progress in Hormone Research* 19: 673–716. (with William H. Bossert; EOW first author)

Pheromones. *Scientific American* 208(5) (May): 100–114.

Social modifications related to rareness in ant species. *Evolution* 17(2): 249–253.

An equilibrium theory of insular zoogeography. *Evolution* 17(4): 373–387. (with Robert H. MacArthur; EOW second author)

1964

The anatomical source of trail substances in formicine ants. *Psyche* 71(1): 28–31. (with Murray S. Blum; EOW second author)

A fossil ant colony: new evidence of social antiquity. *Psyche* 71(2): 93–103. (with Robert W. Taylor; EOW first author)

The true army ants of the Indo-Australian area (Hymenoptera: Formicidae: Dorylinae). *Pacific Insects* 6(3): 427–483.

The ants of the Florida Keys. *Breviora* no. 210, 14 pp.

1965

Trail sharing in ants. *Psyche* 72(1): 2–7.

Purification of the fire ant trail substances. *Nature* 207(4994): 320–321. (with C. T. Walsh and J. H. Law; EOW third author)

Biochemical polymorphism in ants. *Science* 149(3683): 544–546. (with J. H. Law and J. A. McCloskey; EOW second author)

Chemical communication in the social insects. *Science* 149(3688): 1064–1071.

A consistency test for phylogenies based on contemporaneous species. *Systematic Zoology* 14(3): 214–220.

The challenge from related species. *In* H. G. Baker and G. Ledyard Stebbins, eds., *The Genetics of Colonizing Species,* pp. 7–27. Academic Press, New York.

1966

Habitat selection by queens of two field-dwelling species of ants. *Ecology* 47(3): 485–487. (with George L. Hunt; EOW first author)
Behaviour of social insects. *In* P. T. Haskell, ed., *Insect Behaviour* (Symposium of the Royal Entomological Society, no. 3), pp. 81–96. Royal Entomological Society, London.

1967

The first Mesozoic ants, with the description of a new subfamily. *Psyche* 74(1): 1–19. (with F. M. Carpenter and W. L. Brown; EOW first author)
The validity of the "Consistency Test" for phylogenetic hypotheses. *Systematic Zoology* 16(1): 104.
An estimate of the potential evolutionary increase in species density in the Polynesian ant fauna. *Evolution* 21(1): 1–10. (with R. W. Taylor; EOW first author)
The first Mesozoic ants. *Science* 157(3792): 1038–1040. (with F. M. Carpenter and W. L. Brown; EOW first author)
The Theory of Island Biogeography. Princeton University Press, Princeton, NJ. 203 pp. (with R. H. MacArthur; EOW second author)
Biogeographic theory simplified. Mimeographed for Biology 17, Fall 1967, 13 pp.
The superorganism concept and beyond. *In* R. Chauvin and Ch. Noirot, eds., *L'Effet de Groupe Chez les Animaux* (Colloque Internationaux du Centre National de la Recherche Scientifique, no. 173, Paris, 1967), pp. 27–39. Centre National de la Recherche Scientifique, Paris.
Ant fauna of Futuna and Wallis Islands, stepping stones to Polynesia. *Pacific Insects* 9(4): 563–584. (with George L. Hunt; EOW first author)
The ants of Polynesia (Hymenoptera: Formicidae). *Pacific Insects Monograph* no. 14, 109 pp. (with Robert W. Taylor; EOW first author)

1968

The ergonomics of caste in the social insects. *American Naturalist* 102(923): 41–66.
A suggested revision of nomenclatural procedure in animal taxonomy. *Systematic Zoology* 17(2): 188–191. (with Henry F. Howden and Howard E. Evans; EOW third author)
The alarm-defence system of the ant *Acanthomyops claviger. Journal of Insect Physiology* 14(7): 955–970. (with Frederick E. Regnier; EOW second author)
Lignumvitae—relict island. *Natural History* 77(8): 52–57. (with Thomas Eisner; EOW first author)
Recent advances in systematics. *BioScience* 18(12): 1113–1117.
Chemical systems. *In* T. A. Sebeok, ed., *Animal Communication, Techniques of Study and Results of Research,* pp. 75–102. Indiana University Press, Bloomington, IN.

1969

A general method for estimating threshold concentrations of odorant molecules. *Journal of Insect Physiology* 15(4): 597–610. (with W. H. Bossert and F. E. Regnier; EOW first author)

The alarm-defence system of the ant *Lasius alienus*. *Journal of Insect Physiology* 15(5): 893–898. (with F. E. Regnier; EOW second author)

Summary of the conference. *In* Charles G. Sibley, ed., *Systematic Biology* (Proceedings of an International Conference, University of Michigan, Ann Arbor, June 1967), pp. 615–627. Publication No. 1692 of the National Academy of Sciences, Washington, DC.

Experimental zoogeography of islands: defaunation and monitoring techniques. *Ecology* 50(2): 267–278. (with Daniel S. Simberloff; EOW first author)

Experimental zoogeography of islands: the colonization of empty islands. *Ecology* 50(2): 278–296. (with Daniel S. Simberloff; EOW second author)

The species equilibrium. *In* G. M. Woodwell and H. H. Smith, eds., *Diversity and Stability in Ecological Systems* (Brookhaven Symposia in Biology, no. 22), pp. 38–47. Biology Department, Brookhaven National Laboratory, Upton, NY.

1970

Orientation to nest material by the ant, *Pogonomyrmex badius* (Latreille). *Animal Behaviour* 18(2): 331–334. (with Walter Hangartner and Jackson M. Reichson; EOW third author)

Defensive liquid discharge in Florida tree snails (*Liguus fasciatus*). *Nautilus* 84(1): 14–15. (with T. Eisner; EOW second author)

Experimental zoogeography of islands: a two-year record of colonization. *Ecology* 51(5): 934–937. (with Daniel S. Simberloff; EOW second author)

Chemical communication within animal species. *In* E. Sondheimer and J. B. Simeone, eds., *Chemical Ecology*, pp. 133–155. Academic Press, New York.

Introduction. *In* J. W. Johnston, Jr., D. G. Moulton, and A. Turk, eds., *Advances in Chemoreception*, vol. 1: *Communication by Chemical Signals*, pp. 1–3. Appleton-Century-Crofts, New York.

1971

Competitive and aggressive behavior. *In* J. F. Eisenberg and W. Dillon, eds., *Man and Beast: Comparative Social Behavior*, pp. 183–217. Smithsonian Institution Press, Washington, DC.

A Primer of Population Biology. Sinauer Associates, Sunderland, MA. 192 pp. (with William H. Bossert; EOW first author)

The Insect Societies. Belknap Press of Harvard University Press, Cambridge, MA. 548 pp.

Chemical communication and "propaganda" in slave-maker ants. *Science* 172(3980): 267–269. (with F. E. Regnier; EOW second author)

Social insects. *Science* 172(3981): 406.

The evolution of the alarm-defense system in the formicine ants. *American Naturalist* 105(943): 279–289 (with F. E. Regnier; EOW first author).

The prospects for a unified sociobiology. *American Scientist* 59(4): 400–403.

Recruitment trails in the harvester ant *Pogonomyrmex badius. Psyche* 77(4): 385–399. (with Bert Hölldobler; EOW second author)

The plight of taxonomy. *Ecology* 52(5): 741.

Caste evolution as a function of mature colony size in social bees and wasps. In *Entomological Essays to Commemorate the Retirement of Professor K. Yasumatsu,* pp. 215–217. Hokuryukan Publishing Co., Tokyo.

1972

The new science of sociobiology. *Intellectual Digest* 2(5): 38–40.

Animal communication. *Scientific American* 227(3) (September): 52–60.

1973

The queerness of social evolution (The 1972 Founders' Memorial Award Lecture). *Bulletin of the Entomological Society of America* 19(1): 20–22.

Life on Earth. Sinauer Associates, Sunderland, MA. 1033 pp. (with T. Eisner, W. R. Briggs, R. E. Dickerson, R. L. Metzenberg, R. D. O'Brien, M. Susman, W. E. Boggs; EOW first author)

On the queerness of social evolution. *Social Research* 40(1): 144–152. (Reprinted from *Bulletin of the Entomological Society of America* [March 1973].)

Predatory behaviour in the ant-like wasp *Methocha stygia* (Say) (Hymenoptera: Tiphiidae). *Animal Behaviour* 21(2): 292–295. (with Donald J. Farish; EOW first author)

Eminent ecologist, 1973, Robert Helmer MacArthur. *Bulletin of the Ecological Society of America* 54(3): 11–12.

The ants of Easter Island and Juan Fernandez. *Pacific Insects* 15(2): 285–287.

Group selection and its significance for ecology. *BioScience* 23(11): 631–638.

1974

Ecology, Evolution, and Population Biology [Editor] (Readings from *Scientific American*). W. H. Freeman and Co., San Francisco. 319 pp.

The soldier of the ant *Camponotus (Colobopsis) fraxinicola* as a trophic caste. *Psyche* 81(1): 182–188.

Ants of the *Formica fusca* group in Florida. *Florida Entomologist* 57(2): 115–116. (with André Francoeur; EOW first author)

On the estimation of total behavioral repertories in ants. *Journal of the New York Entomological Society* 83(2): 106–112. (with Robert M. Fagen; EOW first author)

The conservation of life. *Harvard Magazine* 76(5): 28–37.

Aversive behavior and competition within colonies of the ant *Leptothorax curvispinosus. Annals of the Entomological Society of America* 67(5): 777–780.

The population consequences of polygyny in the ant *Leptothorax curvispinosus. Annals of the Entomological Society of America* 67(5): 781–786.

1975

The origins of human social behavior. *Harvard Magazine* 77(8): 21–26.

Leptothorax duloticus and the beginnings of slavery in ants. *Evolution* 29(1): 108–119.

Animal Behavior (Readings from *Scientific American*). W. H. Freeman and Co., San Francisco. 339 pp. (coeditor with T. Eisner; EOW second editor)

Sociobiology: The New Synthesis. Belknap Press of Harvard University Press, Cambridge, MA. 697 pp.

Slavery in ants. *Scientific American* 232(6) (June): 32–36.

Evolutionary significance of the social insects. *In* D. Pimentel, ed., *Insects, Science, and Society* (Cornell Centennial Celebration of Entomology), pp. 25–31. Academic Press, New York.

Human decency is animal. *New York Times Magazine,* October 12, pp. 38–50. (Reprinted in *Dialogue* 9[4]: 85–94 [1976].)

Applied biogeography. *In* M. L. Cody and J. Diamond, eds., *Ecology and Evolution of Communities,* pp. 522–534. Belknap Press of Harvard University Press, Cambridge, MA. (with Edwin O. Willis; EOW first author)

Enemy specification in the alarm-recruitment system of an ant. *Science* 190(4216): 798–800.

Some central problems of sociobiology. *Social Science Information* 14(6): 5–18.

1976

The organization of colony defense in the ant *Pheidole dentata* Mayr (Hymenoptera: Formicidae). *Behavioral Ecology and Sociobiology* 1(1): 63–81.

Earth, getting back to nature—our hope for the future (A provocative interview with sociobiologist Edward O. Wilson, by Irving Penn). *House & Garden* February, pp. 64–65.

Academic vigilantism and the political significance of sociobiology. *BioScience* 26(3): 183, 187–190.

Presentation of the Paleontological Society Medal to Frank Morton Carpenter. *Journal of Paleontology* 50(3): 549–550.

Communication by trail laying in ants. *In* Biological Sciences Curriculum Study,

Research Problems in Biology: Investigations for Students, series 2, 2nd ed., pp. 20–22. Oxford University Press, New York.

A social ethogram of the Neotropical arboreal ant *Zacryptocerus varians* (Fr. Smith). *Animal Behaviour* 24(2): 354–363.

The first workerless parasite in the ant genus *Formica* (Hymenoptera: Formicidae). *Psyche* 83(3–4): 277–281.

Sociobiology: a new approach to understanding the basis of human nature. *New Scientist* 13 May, pp. 342–345.

The social instinct. *Bulletin of the American Academy of Arts and Sciences* 30(1): 11–25.

Behavioral discretization and the number of castes in an ant species. *Behavioral Ecology and Sociobiology* 1(2): 141–154.

Comment on D. T. Campbell's "On the conflicts between biological and social evolution and between psychology and moral tradition," which appeared in *American Psychologist* 30: 1103–1126 (1975). *American Psychologist* 31(5): 370–371.

Author's Précis and Author's Reply in "Multiple Review of Wilson's Sociobiology." *Animal Behaviour* 24(3): 698–699, 716–718.

Which are the most prevalent ant genera? *Studia Entomologica* 19(1–4): 187–200.

Sergei Kovalev: A colleague in trouble. *Science* 194: 133–134. (with Thomas Eisner; EOW second author)

1977

The number of queens: an important trait in ant evolution. *Naturwissenschaften* 64: 8–15. (with Bert Hölldobler; EOW second author)

Weaver ants: social establishment and maintenance of territory. *Science* 195(4281): 900–902. (with Bert Hölldobler; EOW second author)

Colony-specific territorial pheromone in the African weaver ant *Oecophylla longinoda* (Latreille). *Proceedings of the National Academy of Sciences of the United States of America* 74(5): 2072–2075. (with Bert Hölldobler; EOW second author)

Biology and the social sciences. *Daedalus* 106(4) (Fall): 127–140. (Reprinted in *Zygon,* 25[3]: 245–262 [1990].)

The Insects (Readings from *Scientific American*). W. H. Freeman and Co., San Francisco. 334 pp. (coeditor with Thomas Eisner; EOW second Editor)

Evolutionary biology seeks the meaning of life itself. *The New York Times: Week in Review,* Sunday, Nov. 27, p. 16E. (Translated by Prof. A. Brito da Cunha and published in *O Estado de S. Paulo,* 27/7/78, "A biologia evolutiva e o significado da vida.")

Weaver ants. *Scientific American* 273(6) (December): 146–154. (with Bert Hölldobler, EOW second author)

Animal and human sociobiology. *In* Clyde E. Goulden, ed., *Changing Scenes in Natural Sciences, 1776–1976,* pp. 273–281. Special Publication no. 12, Academy of Natural Sciences, Philadelphia.

1978

The multiple recruitment systems of the African weaver ant *Oecophylla longinoda* (Latreille) (Hymenoptera: Formicidae). *Behavioral Ecology and Sociobiology* 3(1): 19–60. (with Bert Hölldobler; EOW second author)

The attempt to suppress human behavioral genetics. *Journal of General Education* 29(4): 277–287.

The genetic evolution of altruism. *In* Lauren Wispé, ed., *Altruism, Sympathy and Helping,* pp. 11–37. Academic Press, New York.

Encounter: Marvin Harris and E. O. Wilson debate the claims of sociobiology—the envelope and the twig. *The Sciences* 18(8): 10–15, 27.

On Human Nature. Harvard University Press, Cambridge, MA. 260 pp.

Division of labor in fire ants based on physical castes (Hymenoptera: Formicidae: *Solenopsis*). *Journal of the Kansas Entomological Society* 51(4): 615–636. (dedicated to Charles Duncan Michener on his sixtieth birthday 22 September 1978.)

What is sociobiology? *Society* (Transaction, Rutgers) 15(6):10–14. (First published in Michael S. Gregory, Anita Silvers & Diane Sutch, eds., *Sociobiology and Human Nature* Jossey-Bass Publishers, San Francisco, 1978, pp. 1–12.)

Caste and Ecology in the Social Insects. Princeton University Press, Princeton, NJ. 352 pp. (with George F. Oster; EOW second author).

Altruism. *Harvard Magazine* 81(2): 23–28.

Biophilia. *In* "The Column, Harvard University Press." *New York Times Book Review,* Jan. 14, p. 43.

1979

Of ants and men. *Nieman Reports* 33(1): 4–5.

The evolution of caste systems in social insects. *Proceedings of the American Philosophical Society* 123(4): 204–210.

Reorienting social theory. *The Wilson Quarterly* 3(4): 96–97.

1980

Resolutions for the 80s. *Harvard Magazine* 82(3): 21.

Sex differences in cooperative silk-spinning by weaver ant larvae. *Proceedings of the National Academy of Sciences of the United States of America* 77(4): 2343–2347. (with Bert Hölldobler; EOW first author)

A consideration of the genetic foundation of human social behavior. *In* George W. Barlow and James Silverberg, eds., *Sociobiology: Beyond Nature/Nurture?* (AAAS Selected Symposium no. 35), pp. 295–306. Westview Press, Boulder, CO.

Comparative social theory. *In* Sterling M. McMurrin, ed., *The Tanner Lectures on Human Values,* vol. I, pp. 49–73. University of Utah Press, Salt Lake City.

Caste and division of labor in leaf-cutter ants (Hymenoptera: Formicidae: *Atta*), I: The overall pattern in *A. sexdens. Behavioral Ecology and Sociobiology* 7(2): 143–156.

Caste and division of labor in leaf-cutter ants (Hymenoptera: Formicidae: *Atta*), II: The ergonomic optimization of leaf cutting. *Behavioral Ecology and Sociobiology* 7(2): 157–165.

Translation of epigenetic rules of individual behavior into ethnographic patterns. *Proceedings of the National Academy of Sciences of the United States of America* 77(7): 4382–4386. (Reprinted 1981 in *Revista International* [Lisbon] 1[2]: 59–63.) (with Charles J. Lumsden; EOW second author)

Gene-culture translation in the avoidance of sibling incest. *Proceedings of the National Academy of Sciences of the United States of America* 77(10): 6248–6250. (with Charles J. Lumsden; EOW second author)

The ethical implications of human sociobiology. *The Hastings Center Report* 10(6): 27–29.

1981

Genes, Mind, and Culture. Harvard University Press, Cambridge, MA. 428 pp. (with Charles J. Lumsden; EOW second author)

Epigenesis and the evolution of social systems (The Wilhemine E. Key 1980 Invitational Lecture of the American Genetic Association). *Journal of Heredity* 72(2): 70–77.

The relation of science to theology. *Zygon* 15(4): 425–434.

Who is Nabi? *Nature* 290(5808) (23 April): 623.

Communal silk-spinning by larvae of *Dendromyrmex* tree-ants (Hymenoptera: Formicidae). *Insectes Sociaux* 28(2): 182–190.

Genes, mind, and ideology. *The Sciences* 21(9): 6–8. (with Charles J. Lumsden; EOW second author)

Conservation of tropical forests. *Science* 213(4514): 1314. (with Thomas Eisner, Hans Eisner, Jerold Meinwald, Carl Sagan, Charles Walcott, Ernst Mayr, Peter Raven, Anne Ehrlich, Paul R. Ehrlich, Archie Carr, Eugene P. Odum, and Carl Gans; EOW seventh author)

1982

Alfred Edwards Emerson. *In Biographical Memoirs,* vol. 53, pp. 159–177. National Academy of Sciences, Washington, DC. (with Charles D. Michener; EOW first author)

Précis of *Genes, Mind, and Culture,* with commentary by 23 peers and authors' responses. *The Behavioral and Brain Sciences* 5(1): 1–37. (with Charles J. Lumsden; EOW second author)

The importance of biological field stations. *BioScience* 32(5): 320.

The significance of field research in biology (editorial). *BIOS* 53(1): 2–3.

Of insects and man. *In* M. D. Breed, C. D. Michener, and H. E. Evans, eds., *The Biol-*

ogy of Social Insects (Proceedings of the Ninth Congress of the International Union for the Study of Social Insects, Boulder, Colorado, 1982), pp. 1–3. Westview Press, Boulder, CO.

Toward a humanist biology. *The Humanist* 42(5): 38–41, 56.

Sociobiology, individuality, and ethics: a response. *Perspectives in Biology and Medicine* 26(1): 19–29.

La coevoluzione di geni e cultura. *Laboratorio di scienze dell'uomo* 1(3–4): 277–295. (with Charles J. Lumsden; EOW second author)

1983

Queen control in colonies of weaver ants (Hymenoptera: Formicidae). *Annals of the Entomological Society of America* 76(2): 235–238. (with Bert Hölldobler; EOW second author)

Promethean Fire: Reflections on the Origin of Mind. Harvard University Press, Cambridge, MA. 216 pp. (with Charles J. Lumsden; EOW second author)

Sociobiology and human beings. *The Psychohistory Review* 11(2–3): 5–14.

Foreword. *In* Joseph Shepher, *Incest: A Biosocial View,* pp. xi–xii. Academic Press, New York.

Sociobiology and the Darwinian approach to mind and culture. *In* D. S. Bendall, ed., *Evolution from Molecules to Men,* pp. 545–553. Cambridge University Press, Cambridge.

The evolution of communal nest-weaving in ants. *American Scientist* 71(5): 490–499. (with Bert Hölldobler; EOW second author)

Sociobiology: From Darwin to the present. *In* Charles L. Hamrum, ed., *Darwin's Legacy* (Nobel Conference XVIII, Gustavus Adolphus College, St. Peter, Minnesota), pp. 53–75. Harper & Row, San Francisco.

Caste and division of labor in leaf-cutter ants (Hymenoptera: Formicidae: *Atta*), III: Ergonomic resiliency in foraging by *A. cephalotes. Behavioral Ecology and Sociobiology* 14(1): 47–54.

Caste and division of labor in leaf-cutter ants (Hymenoptera: Formicidae: *Atta*), IV: Colony ontogeny of *A. cephalotes. Behavioral Ecology and Sociobiology* 14(1): 55–60.

Sociobiology and the idea of progress (The Katzir-Katchalsky Lecture, Weizmann Institute of Science, Rehovot, Israel, 14 May 1980). *In* M. Balaban, ed., *Biological Foundations and Human Nature,* pp. 199–216. Academic Press, New York.

1984

Human sociobiology: a preface. *Journal of Human Evolution* 13: 1–2.

Million-year histories: species diversity as an ethical goal. *Wilderness* 48(165) (Summer): 12–17.

Biophilia. Harvard University Press, Cambridge, MA. 159 pp.

The drive to discovery. *The American Scholar* 53(4) (August): 447–464.

Clockwork lives of the Amazonian leafcutter army. *Smithsonian* 15(7): 92–101.

Snakes and psyche: explorations. *OMNI* 7(3) (December): 38–40.

Discussion: *Genes, Mind and Culture. Zygon* 19(2) (June): 213–232.

The relation between caste ratios and division of labor in the ant genus *Pheidole* (Hymenoptera: Formicidae). *Behavioral Ecology and Sociobiology* 16(1): 89–98.

New approaches to the analysis of social systems. *In* N. Keyfitz, ed., *Population and Biology*, pp. 41–51. Ordina, Liège, Belgium. (Abridged from 1980 Wilhelmine Key Lecture of the American Genetic Association).

Tropical social parasites in the ant genus *Pheidole,* with an analysis of the anatomical parasitic syndrome (Hymenoptera: Formicidae). *Insectes Sociaux* 31(3): 316–334.

On incest and mathematical modeling (authors' response). *The Behavioral and Brain Sciences* 7(4): 742–744. (with Charles J. Lumsden; EOW second author)

Behavior of the cryptobiotic predaceous ant *Eurhopalothrix heliscata,* n. sp. (Hymenoptera: Formicidae: Basicerotini). *Insectes Sociaux* 31(4): 408–428. (with W. L. Brown, Jr.; EOW first author)

1985

Correlates of variation in the major/minor ratio of the ant, *Pheidole dentata* (Hymenoptera: Formicidae). *Annals of the Entomological Society of America* 78(1): 8–11. (with Ardis B. Johnston; EOW second author)

The principles of caste evolution. *In* B. Hölldobler and M. Lindauer, eds., *Experimental Behavioral Ecology* (Fortschritte der Zoologie, no. 31), pp. 307–324. Sinauer Associates, Sunderland, MA.

A case for competition (letter). *The Sciences* 25(1) (January/February): 10, 12.

Conservation and development are linked. *Harvard Gazette* February 1, pp. 6, 11.

An interview with E. O. Wilson on sociobiology and religion, by Jeffrey Saver. *Free Inquiry* 5(2) (Spring): 15–22.

In the queendom of the ants: a brief autobiography. *In* Donald A. Dewsbury, ed., *Leaders in the Study of Animal Behavior: Autobiographical Perspectives,* pp. 464–484. Bucknell University Press, Cranbury, NJ.

Outcry from a world of wounds: viewpoint. *Discover* 6(6) (June): 64–66.

In praise of sharks. *Discover* 6(7) (July): 40–42, 48, 50–53.

The sociogenesis of insect colonies. *Science* 228(4707): 1489–1495.

Between-caste aversion as a basis for division of labor in the ant *Pheidole pubiventris* (Hymenoptera: Formicidae). *Behavioral Ecology and Sociobiology* 17(1): 35–37.

Invasion and extinction in the West Indian ant fauna: evidence from the Dominican amber. *Science* 229(4710): 265–267.

Caste-specific techniques of defense in the polymorphic ant *Pheidole embolopyx* (Hymenoptera: Formicidae). *Insectes Sociaux* 32(1): 3–22. (with Bert Hölldobler; EOW first author)

Altruism and ants. *Discover* 6(8) (August): 46–51.

Ants of the Dominican amber (Hymenoptera: Formicidae), 1: Two new myrmicine genera and an aberrant *Pheidole. Psyche* 92(1): 1–9.

Ants of the Dominican amber (Hymenoptera: Formicidae), 2: The first fossil army ants. *Psyche* 92(1): 11–16.

Ants of the Dominican amber (Hymenoptera: Formicidae), 3: The subfamily Dolichoderinae. *Psyche* 92(1): 17–37.

The biological diversity crisis: a challenge to science. *Issues in Science and Technology* 2(1) (Fall): 20–29. (Reprinted in *BioScience* 35[11]: 700–706.)

In Phillip L. Berman, ed., *The Courage of Conviction,* pp. 204–212. Dodd, Mead & Company, Inc., New York. (Autobiographical sketch)

The evolution of ethics. *New Scientist* 108(1478) (17 October): 50–52. (with Michael Ruse; EOW second author)

The search for faunal dominance. *In* George E. Ball, ed., *Taxonomy, Phylogeny, and Zoogeography of Beetles and Ants: A volume dedicated to the memory of Philip Jackson Darlington, Jr. (1904–1983)* (Series Entomologica, vol. 33), pp. 389–493. Dr. W. Junk Publishers, Dordrecht, Boston (Member of Kluwer Academic Pub. group).

Ants from the Cretaceous and Eocene amber of North America. *Psyche* 92(2–3): 205–216.

Time to revive systematics (editorial). *Science* 230(4731): 1227.

From the heart of a rain forest. *Focus* (WWF-US), 7(6) (November/December): 6.

The relation between biological and cultural evolution. *Journal of Social and Biological Structure* 8(4) (October): 343–359. (with Charles J. Lumsden; EOW second author)

Ants of the Dominican amber (Hymenoptera: Formicidae), 4: A giant ponerine in the genus *Paraponera. Israel Journal of Entomology* 19: 197–200.

1986

Nest area exploration and recognition in leafcutter ants (*Atta cephalotes*). *Journal of Insect Physiology* 32(2): 143–150. (with Bert Hölldobler; EOW second author)

Soil-binding pilosity and camouflage in ants of the tribes Basicerotini and Stegomyrmecini (Hymenoptera: Formicidae). *Zoomorphology* 106(1): 12–20. (with Bert Hölldobler; EOW second author)

Foreword. *In* K. C. Kim and L. Knutson, eds., *Foundations for a National Biological Survey,* p. vii. Association of Systematics Collections, c/o Museum of Natural History, University of Kansas, Lawrence, KS 66045.

Moral philosophy as applied science. *Philosophy* 61(236) (April): 173–192. (with Michael Ruse; EOW second author)

Sociobiology and sociology converging: an evaluation of Lopreato's *Human Nature and Biocultural Evolution. Revue Européenne des Sciences Sociales* 24(73): 5–8.

Ecology and behavior of the primitive cryptobiotic ant *Prionopelta amabilis* (Hymenoptera: Formicidae). *Insectes Sociaux* 33(1): 45–58. (with Bert Hölldobler; EOW second author)

Caste and division of labor in *Erebomyrma,* a genus of dimorphic ants (Hymenoptera: Formicidae: Myrmicinae). *Insectes Sociaux* 33(1): 59–69.

Ecology and behavior of the Neotropical cryptobiotic ant *Basiceros manni* (Hymenoptera: Formicidae: Basicerotini). *Insectes Sociaux* 33(1): 70–84. (with Bert Hölldobler; EOW first author)

A harvest of decay. *Natural History* 95(8): 54–57. (Photographs by Thomas Eisner)

On genetic determinism and morality. *Chronicles* 10(8) (August): 12–14.

The defining traits of fire ants and leaf-cutting ants. *In* Clifford S. Lofgren and Robert K. Vander Meer, eds., *Fire Ants and Leaf-Cutting Ants: Biology and Management,* Ch. 1, pp. 1–9. Westview Press, Boulder, CO.

Storm over the Amazon. *In* Daniel Halpern, ed., *Antaeus* (no. 57, Autumn), pp. 157–159. The Ecco Press, New York.

Edward O. Wilson. *In* "From Plato to Pavlov: what the well-read scientist reads." *The Sciences* 26(4) (September/October): 18.

Discovering ant language. *The Scientist,* 1(2) (November 17): 21.

The race for earthly survival: the current state of biological diversity. *Washington Book Review,* 1(11) (November): 3, 5–6.

Forum: all things great and small. *OMNI,* 9(3) (December): 18.

The organization of flood evacuation in the ant genus *Pheidole* (Hymenoptera: Formicidae). *Insectes Sociaux* 33(4): 458–469.

1987

An urgent need to map biodiversity. *The Scientist* 1(6) (February 9): 11.

The causes of ecological success: the case of the ants (The Sixth Tansley Lecture, Oxford University, 1985). *Journal of Animal Ecology* 56: 1–9.

The earliest known ants: An analysis of the Cretaceous species and an inference concerning their social organization. *Paleobiology* 13(1): 44–53.

Biology's spiritual products. *Free Inquiry* 7(2) (Spring): 13–15.

Kin recognition: an introductory synopsis. *In* D. J. C. Fletcher and C. D. Michener, eds., *Kin Recognition in Animals,* pp. 7–18. John Wiley & Sons, New York. x + 465 pp.

The evolutionary origin of mind. *The Personalist Forum* 3(1): 11–18.

Biological diversity as a scientific and ethical issue. *Papers Read at a Joint Meeting of The Royal Society and The American Philosophical Society, April 1986, Vol. 1,* pp. 29–48. The American Philosophical Society, Philadelphia.

The fossil members of the ant tribe Leptomyrmecini (Hymenoptera: Formicidae). *Psyche* 94(1–2): 1–8. (with Cesare Baroni Urbani; EOW second author)

The arboreal ant fauna of Peruvian Amazon forests: a first assessment. *Biotropica* 19(3): 245–251.

The little things that run the world (The importance and conservation of invertebrates). *Conservation Biology* 1(4): 344–346. (Also published in *Wings* [The Xerces Society], 12[(3]: 4–8.)

Religion and evolutionary theory. *In* David M. Byers, ed., *Religion, Science, and the Search for Wisdom* (Proceedings of a Conference on Religion and Science, September 1986, Bishops' Committee on Human Values National Conference of Catholic Bishops), pp. 81–102. United States Catholic Conference, Washington, DC.

Conversation between B. F. Skinner and E. O. Wilson on behavior and sociobiology (recorded November 29, 1987). (Transcript by Cambridge Transcriptions for Robert Sherman, Cambridge Center for Behavioral Studies).

The coevolution of biology and culture (based on the April 1985 Helen P. Mangelsdorf Lecture, Dept. of Biology, University of North Carolina at Chapel Hill). *Social Science* 72(2–4): 113–115.

1988

Foreword. *In* C. Scott Findlay and Charles J. Lumsden, eds., *The Creative Mind,* p. 1. *Journal of Social and Biological Structures* 11(1): 1–189.

Dense heterarchies and mass communication as the basis of organization in ant colonies. *Trends in Ecology and Evolution* 3(3): 65–68. (with Bert Hölldobler; EOW first author)

The current state of biological diversity. *In* E. O. Wilson, ed., *Biodiversity* (Proceedings of a Biodiversity Forum held in Washington, DC, September 1986), pp. 3–18. National Academy of Sciences, Washington, DC.

BioDiversity. National Academy Press, Washington, DC. 521 pp. (with Frances M. Peter; EOW first editor)

The biogeography of the West Indian ants (Hymenoptera: Formicidae). *In* James K. Liebherr, ed., *Zoogeography of Caribbean Insects,* pp. 214–230. Cornell University Press, Ithaca, NY.

The current status of ant taxonomy. *In* James C. Trager, ed., *Advances in Myrmecology,* pp. 3–10. E. J. Brill, Leiden, The Netherlands.

International conservation: The ultimate goal. *Orion* 7(3): 16–21.

Essay for "This Week's Citation Classic"® (*The Theory of Island Biogeography,* by R. H. MacArthur and E. O. Wilson, 1967, Princeton University Press, Princeton, NJ, 203 pp.). *Current Contents* 19(36) (September 5): 14.

Darwinism and ethics. *In* Erhard Geissler and Herbert Hörz, eds., *Vom Gen zum Verhalten: Der Mensch als biopsychosoziale Einheit,* pp. 243–261. Akademie-Verlag Berlin. (with Michael Ruse; EOW second author)

The diversity of life. *In* Harm J. de Blij, ed., *Earth '88: Changing Geographic Perspectives* (Proceedings of the Centennial Symposium, Washington, DC, January 1988), pp. 68–81. National Geographic Society, Washington, DC.

Foreword. *In* George D. Snell, *Search for a Rational Ethic,* pp. vii–viii. Springer-Verlag, New York. xvi + 317 pp.

1989

The coming pluralization of biology and the stewardship of systematics. *BioScience* 39(4): 242–245.

Chimaeridris, a new genus of hook-mandibled myrmicine ants from tropical Asia (Hymenoptera: Formicidae). *Insectes Sociaux* 36(1): 62–69.

Robert Helmer MacArthur (1930–1972). In *Biographical Memoirs,* vol. 58, pp. 318–327. National Academy Press, Washington, DC. (with G. Evelyn Hutchinson; EOW first author)

Threats to biodiversity. *Scientific American* 261(3) (September): 108–116.

Paths of glory: following army ants as they march through the tropics. *The Sciences* 29(6) (November/December): 18–23. (with Bert K. Hölldobler; EOW second author)

The biological basis of culture. Monographic section *Sociobiology and Sociology,* Joseph Lopreato, ed. *Revue Internationale de Sociologie (International Review of Sociology),* n.s. 3: 35–60.

1990

The Ants. Belknap Press of Harvard University Press, Cambridge, MA. 732 pp. (with Bert Hölldobler; EOW second author)

Empire of the ants. *Discover* 11(3) (March): 44–50.

Riches for humanity in the diversity of species (speaking with Alicia Moore). *Fortune* 121(7) (26 March): 66 and 70.

Deep history (acceptance speech for the Ingersoll Foundation's 1989 Richard M. Weaver Award). *Chronicles* 14(4) (April): 16–17.

Host tree selection by the giant Neotropical ant *Paraponera clavata* (Hymenoptera: Formicidae). *Biotropica* 22(2): 189–190. (with Bert Hölldobler; EOW second author)

The new environmentalism. *Chronicles* 14(8) (August): 16–18.

Foreword. *In* Xavier Bellés, *Coleoptera: Ptinidae, Gibbiinae* (Fauna Iberica, vol. 0), p. 9. Museo Nacional de Ciencias Naturales, CSIC, Madrid. 43 pp.

First word. *OMNI* 12(12) (September): 6.

Success and Dominance in Ecosystems: The Case of the Social Insects. Ecology Institute, Oldendorf/Luhe, Federal Republic of Germany. 104 pp.

Response to J. Dobrzanska and J. Dobrzanski on the raids of *Polyergus. Acta Neurobiologiae Experimentalis* 49(6): 379–380.

1991

Sociobiology and the test of time. *In* Michael H. Robinson and Lionel Tiger, eds., *Man and Beast Revisited,* pp. 77–80. Smithsonian Institution Press, Washington, DC. xxiii + 386 pp.

Biological and human determinants of the survival of species. *In* Joseph E. Earley, ed., *Individuality and Cooperative Action,* pp. 47–57. Georgetown University Press, Washington, DC. xiii + 189 pp.

Scientific humanism and religion. *Free Inquiry* 11(2): 20–23, 56.

Prefazione. *In* Joseph Lopreato, *Evoluzione e natura umana,* pp. 9–13. Rubbettino Editore, Soveria Mannelli (CZ).

Biodiversity studies: Science and policy. *Science* 253: 758–762. (with Paul R. Ehrlich; EOW second author)

Philip Jackson Darlington, Jr. (1904–1983). In *Biographical Memoirs,* vol. 60, pp. 33–44. National Academy Press, Washington, DC.

Rain forest canopy: The high frontier. *National Geographic* 180(6) (December): 78–107.

Ants. *Bulletin of the American Academy of Arts and Sciences* 45(3): 13–23.

Holism and reduction in sociobiology: lessons from the ants and human culture. *Biology and Philosophy* 6: 401–412. (with Charles J. Lumsden; EOW second author)

Biodiversity, prosperity, and value. *In* F. Herbert Bormann and Stephen R. Kellert, eds., *Ecology, Economics, Ethics: The Broken Circle,* pp. 3–10. Yale University Press, New Haven, CT. xviii + 233 pp.

1992

Sociality at the pinnacle ("This Week's Citation Classic"® *The Insect Societies,* by E. O. Wilson. 1971. Belknap Press of Harvard University Press, Cambridge, MA. 548 pp.). *Current Contents* 23(16) (April 20): 8; 25(10) (March 8): 8 (1993).

The effects of complex social life on evolution and biodiversity. *Oikos* 63: 13–18.

Social insects as dominant organisms. *In* J. Billen, ed., *Biology and Evolution of Social Insects,* pp. 1–7. Leuven University Press, Leuven, Belgium.

The Diversity of Life. Belknap Press of Harvard University Press, Cambridge, MA. 424 pp.

Communication in the primitive cryptobiotic ant *Prionopelta amabilis* (Hymenoptera: Formicidae). *Journal of Comparative Physiology* A 170: 9–16. (with Bert Hölldobler and M. Obermayer; EOW third author)

A fifty-year plan for biodiversity surveys. *Science* 258(5085): 1099–1100. (with Peter H. Raven; EOW second author)

Biodiversity: Challenge, science, opportunity. *American Zoologist* 32: 1–7.

Religion and the environment (Anniversary Symposia Offered by the Faculty). *Harvard Divinity Bulletin* 22(2): 19–20.

The return to natural philosophy (1991–92 Dudleian Lecture). *Harvard Divinity Bulletin* 21(3): 12–13,15.

Pheidole nasutoides, a new species of Costa Rican ant that apparently mimics termites. *Psyche* 99(1): 15–22. (with Bert Hölldobler; EOW second author)

Dedication: Frank Morton Carpenter. *Psyche* 99(4): 241–244.

Foreword. *In* C. J. Bibby, N. J. Collar, M. J. Crosby, M. F. Heath, Ch. Imboden, T. H. Johnson, A. J. Long, A. J. Stattersfield, and S. J. Thirgood, *Putting Biodiversity on the Map: Priority Areas for Global Conservation*, p. vi. International Council for Bird Preservation, Cambridge, England. vi + 90 pp.

1993

Foreword. *In* Laurent Keller, ed., *Queen Number and Sociality in Insects*, pp. v–vi. Oxford University Press, New York. xii + 439 pp.

Is humanity suicidal? *New York Times Sunday Magazine*, May 30, 1993, pp. 24–29. (Reprinted in *BioSystems* 31 [1993]: 235–242.)

Biophilia and the conservation ethic. *In* S. Kellert and E. O. Wilson, eds., *The Biophilia Hypothesis*, pp. 31–41. Island Press, Washington, DC. 484 pp.

The Biophilia Hypothesis, Island Press, Washington, DC. 484 pp. (with S. Kellert; EOW second editor)

Foreword: Forest ecosystems: more complex than we know. *In* Gregory H. Aplet, Nels Johnson, Jeffrey T. Olson, and V. Alaric Sample, eds. (The Wilderness Society), *Defining Sustainable Forestry*, pp. xi–xiii. Island Press, Washington, DC. xiii + 328 pp.

Analyzing the superorganism: The legacy of Whitman and Wheeler. *In* Robert B. Barlow, Jr., John E. Dowling, and Gerald Weissmann, eds., *The Biological Century: Friday Evening Talks at the Marine Biological Laboratory*, pp. 243–255. Harvard University Press, Cambridge, MA. 289 pp.

1994

Biodiversity: challenge, science, opportunity. *American Zoologist* 34(1): 4–10. (From the symposium *Science as a Way of Knowing—Biodiversity*, presented at the Annual Meeting of the American Society of Zoologists, 27–30 December 1992, at Vancouver, Canada.)

Caryl Haskins, Entomologist. *In* James D. Ebert, ed., *This Our Golden Age: Selected Annual Essays of Caryl P. Haskins, President, Carnegie Institution of Washington 1956–1971*, pp. 11–18. Carnegie Institution of Washington, Washington, DC. x + 141 pp.

Introduction. *In* Susan Middleton and David Liittschwager in association with The California Academy of Sciences, *Witness: Endangered Species of North America*, pp. 14–17. Chronicle Books, San Francisco, CA. ISBN 0-8118-0282-5 (hc).

Foreword. *In* Marjorie Harris Carr, ed., *A Naturalist in Florida: A Celebration of Eden.*, pp. ix–xi. Yale University Press, New Haven, CT. xviii + 264 pp.

Journey to the Ants. Harvard University Press, Cambridge, MA. 240 pp. (with Bert Hölldobler; EOW second author)

Naturalist. Island Press, Washington, DC. 380 pp.

1995

Science and ideology. *Academic Questions* 8: 73–91.

Foreword. *In* Scott Landis, ed., *Conservation by Design,* pp. 6–7. Woodworkers Alliance for Rainforest Protection and Museum of Art, Rhode Island School of Design.

Foreword. *In* Harriet Whelchel, ed., *American Museum of Natural History: 125 Years of Expedition and Discovery,* pp. 18–19. Harry N. Abrams, Inc., New York.

Pheidole ants and the biology of hyperdiversity. *In* P. Koomen, W. N. Ellis, and L. P. S. van dere Geest, eds., *Insekten Onderzoeken,* pp. 7–10. Nederlandse Entomologische Vereniging, Amsterdam.

Building a scientifically sound policy for protecting endangered species. *Science* 269(5228): 1231–1232. (with Thomas Eisner, Jane Lubchenco, David S. Wilcove, and Michael J. Bean; EOW third author)

Quo vadis, Homo sapiens? *GEO Extra* No. 1: 176–179.

A grassroots jungle in a vacant lot. *Wings* 18(2): 3–6.

Wildlife: legions of the doomed, *Time International* October 30, pp. 59–61.

1996

Biodiversity II: Understanding and Protecting Our Natural Resources. Joseph Henry Press, Washington, DC. (with Marjorie L. Reaka-Kudla, Don E. Wilson; EOW third editor)

Insects: The ultimately biodiverse animals. Proceedings of the XX International Congress of Entomology (Firenze, Italy), pp. xvi–xvii.

Panelist Remarks. *In* Ismail Serageldin and Alfredo Sfeir-Younis, eds., *Effective Financing of Environmentally Sustainable Development* (Proceedings of the Third Annual World Bank Conference on Environmentally Sustainable Development, Washington, DC., October 4–6, 1995), pp. 17–20. Environmentally Sustainable Development Proceedings Series No. 10, The World Bank, Washington, DC. ISBN 0-8213-3549-9 HC79.E515333 1995.

In Search of Nature. Island Press, Washington, DC.

Introduction. *In* Takuya Abe, Simon A. Levin, and Masahiko Higashi, eds. 1996. *Biodiversity: An Ecological Perspective,* pp. 1–5. Springer-Verlag, New York. xii + 294 pp.

1997

Reply to the presentation of the David S. Ingalls, Jr. Award for Excellence on May 9, 1995. *Kirtlandia* 50: 19–20 (March 1997).

Carroll Milton Williams (2 December 1916–11 October 1991), *Proceedings of the American Philosophical Society* 141(1): 115–121. (with Fotis C. Kafatos, and Daniel Branton; EOW second author)

Systematics in Asia. *TaxoNewSia* (Korean Institute for Biodiversity Research [KIBIO]) 1(1): 3.

1998

Scientists, scholars, knaves and fools. *American Scientist* 86(1): 6–7.

Consilience: The Unity of Knowledge. Knopf, New York, 367 pp.

Consilience among the great branches of learning. *Daedalus* 127(1): 131–149.

Resuming the enlightenment quest. *Wilson Quarterly* 22(1): 16–27.

Back from chaos. *The Atlantic Monthly* 281(3) (March): 41–44, 46–49, 52, 54–56.

The biological basis of morality. *The Atlantic Monthly* 281(4) (April): 53, 54, 56–59, 62, 64–68, 70.

Introduction. *In* multiple authored *Butterfly Gardening: Creating Summer Magic in Your Garden,* pp. 1–5. Xerces Society Publication with the Smithsonian Institution. Sierra Club Books, San Francisco.

Integrated science and the coming century of the environment. *Science* 279(5359): 2048–2049.

Humanity's ecological footprint. *In* Robert Livernash, William Faries, and Susan Drake Swift (eds.), *Valuing the Global Environment: Actions & Investments for a 21st Century,* pp. 84–85. Global Environment Facility, Washington, DC. x + 162 pp.

Foreword. *In* Gary Larson, *There's a Hair in My Dirt!: A Worm's Story.* HarperCollins, New York.

Chapter 1. Overview: Biodiversity: Wildlife in trouble. *In* Linda Koebner, Jane E. S. Sokolow, Francesca T. Grifo, Sharon Simpson, and Theron Cole, Jr., eds., *Scientists on Biodiversity,* pp. 1–3. American Museum of Natural History, New York.

E.T. stayed home. *Forbes ASAP Big Issue III* (November 30, 1998): 166–168.

Thoughts on biodiversity. *COSMOS 1998* 8: 155–161.

1999

Toward the ultimate synthesis. *In* Sandra Hackman et al., eds., *The Nova Reader: Science at the Turn of the Millennium,* pp. 204–206. TV Books, L.L.C., New York.

Hope and mystery. *Orion* 18(1): 28.

Foreword. *In* David S. Wilcove, *The Condor's Shadow: The Loss and Recovery of Wildlife in America,* p. xiii–xv. W. H. Freeman, New York.

Biological Diversity: The Oldest Human Heritage. New York State Museum, Albany. 58 pp.

Response to Francis Fukuyama's "Second Thoughts: The Last Man in a Bottle." *National Interest* No. 56 (Summer), pp. 35–37.

Foreword. *In* Laura Dassow Walls, ed., *Material Faith: Henry David Thoreau on Science,* p. vii–viii. The Thoreau Society, Houghton Mifflin, Boston.

The two hypotheses of human meaning. *The Humanist* September/October, 59(5): 30–31.

Hardwired for God: is our search for divinity merely a by-product of evolution? *Forbes ASAP* (Big Issue IV: The Great Convergence), October 4, pp. 132 and 134.

Responding to the reviews of Elshtain, Kaye, and Ruse. *Politics and the Life Sciences* September: 36–37.

2000

The creation of biodiversity. *In* Peter H. Raven, ed., and Tania Williams, associate ed., *Nature and Human Society (Proceedings of the 1997 Forum on Biodiversity),* pp. 22–29. National Academy Press, Washington, DC: . 625 pp.

On the future of conservation biology. *Conservation Biology* 14(1) (February): 1–3.

The age of the environment. *Foreign Policy* Summer 2000(119): 34–35.

A new chapter: saving America's living landmarks. *Nature Conservancy,* May/June 50(3): 24–25.

A global biodiversity map (editorial). *Science* 289: 2279 (29 September 2000).

Foreword: California's Living Landscapes. *In* David Wicinas, *Native Grandeur: Preserving California's Vanishing Landscapes,* p. 9. The Nature Conservancy of California, San Francisco. 136 pp.

2001

The Ecological Footprint: The Biosphere and Man (Delivered to the Kistler Prize Ceremony, Foundation for the Future, Seattle, Washington, August 15, 2000). *Vital Speeches of the Day* 67(9) (15 February): 274–277.

Biodiversity: wildlife in trouble. *In* Michael J. Novacek, ed., *The Biodiversity Crisis: Losing What Counts,* pp. 18–20. An American Museum of Natural History Book, The New Press, New York. 224 pp.

Nature matters (Commentary on Howard Frumkin, Beyond Toxicity: Human health and the natural environment, pp. 234–240). *American Journal of Preventive Medicine* 20(3): 241–242.

Hotspots: preserving pieces of a fragile biosphere. *National Geographic* 201(1) (January): 86–89.

2002

The Future of Life. Knopf, New York. 256 pp.

2003

Pheidole in the New World: A hyperdiverse ant genus. Harvard University Press, Cambridge, MA. 794 pp.

The encyclopedia of life, *Trends in Ecology and Evolution* 18(2): 77–80.

Selling out our forests. *Washington Post,* 28 August 2003, p. A.27.

Foreword. *Endangered Forests Endangered Freedoms; 10 Endangered National Forests at Risk from the Bush Administration.* National Forest Protection Agency and Greenpeace. Inside front cover.

2004

"From deep history to the century of the environment: The National Park Service as environmental leader. *The George Wright Forum* 21(1): 5–9.

Foreword to 150th Anniversary Illustrated Edition of *Walden,* by Henry David Thoreau (Houghton Mifflin, Boston).

Taxonomy as a fundamental discipline. *Philosophical Transactions of the Royal Society of London, Series B Biological Sciences* 359: 739.

Does natural selection still drive evolution? *In* J. Bindé, ed., *The Future of Values,* pp. 255–259. Berghahn Books, New York.

For the love of life. *Acta Horticulture* 642: 71–8.

Foreword. *In the Realm of Rivers: Alabama's Mobile-Tensaw Delta,* pp. 7–8, by Sue Walker. Photographs by Dennis Holt. NewSouth Books, Montgomery AL.

2005

All of man's troubles. *In* A. Ahmad and B. Forst, eds., *After Terror*, pp. 106–111, Polity Press, Malden, MA.

Ant plagues: first environmental crises of the New World. *Nature* 433: 32.

Oribatid mite predation by small ants of the genus Pheidole. *Instectes Sociaux* 52: 263–265.

Reflections on the future of life. In A. B. Brown and K. M. Poremski, eds., *Roads to Reconciliation: Conflict and Dialogue in the Twenty-first Century,* pp. 135–146. M. E. Sharpe, New York.

Sense and stability in animal names. *Trends in Ecology and Evolution* 20(8): 421–422. (with Andrew Polaszek; EOW second author)

Systematics and the future of biology. *Proceedings of the National Academy of Sciences of the United States of America* 102(suppl. 1): 6520–6521.

The Linnaean enterprise: past, present, and future, Presented at the symposium Science, Art, and Knowledge: Practicing Natural History from the Enlightenment to the Twenty-first Century. *Proceedings of the American Philosophical Society*, 149(3): 344–348.

The rise of the ants: a phylogenetic and ecological explanation. *Proceedings of the National Academy of Sciences of the United States of America* 102(21): 7411–7414. (with Bert Hölldobler; EOW second author).

Eusociality: origin and consequences. *Proceedings of the National Academy of Sciences of the United States of America*, 102(38): 13367–13371. (with Bert Hölldobler; EOW first author).

From So Simple a Beginning: The Four Great Books of Darwin, edited with introductions, W.W. Norton, New York, 1706 pp.

2006

Nature Revealed: Selected Writings, 1949–2006, Johns Hopkins University Press, Baltimore.

Ant plagues: a centuries-old mystery solved. *In Nature Revealed: Selected Writings, 1949–2006*, Johns Hopkins University Press, Baltimore. (First published here.)

Index